MODERN MASONRY

10TH EDITION

BRICK | BLOCK | STONE | CONCRETE

Timothy L. Andera | Clois E. Kicklighter

Publisher
The Goodheart-Willcox Company, Inc.
Tinley Park, IL
www.g-w.com

Library of Congress Control Number: 2024941976

ISBN 979-8-89118-817-4

2 3 4 5 6 7 8 9 – 26 – 30 29 28 27 26 25

The Goodheart-Willcox Company, Inc. Brand Disclaimer: Brand names, company names, and illustrations for products and services included in this text are provided for educational purposes only and do not represent or imply endorsement or recommendation by the author or the publisher.

The Goodheart-Willcox Company, Inc. Safety Notice: The reader is expressly advised to carefully read, understand, and apply all safety precautions and warnings described in this book or that might also be indicated in undertaking the activities and exercises described herein to minimize risk of personal injury or injury to others. Common sense and good judgment should also be exercised and applied to help avoid all potential hazards. The reader should always refer to the appropriate manufacturer's technical information, directions, and recommendations; then proceed with care to follow specific equipment operating instructions. The reader should understand these notices and cautions are not exhaustive.

The publisher makes no warranty or representation whatsoever, either expressed or implied, including but not limited to equipment, procedures, and applications described or referred to herein, their quality, performance, merchantability, or fitness for a particular purpose. The publisher assumes no responsibility for any changes, errors, or omissions in this book. The publisher specifically disclaims any liability whatsoever, including any direct, indirect, incidental, consequential, special, or exemplary damages resulting, in whole or in part, from the reader's use or reliance upon the information, instructions, procedures, warnings, cautions, applications, or other matter contained in this book. The publisher assumes no responsibility for the activities of the reader.

The Goodheart-Willcox Company, Inc. Internet Disclaimer: The Internet resources and listings in this Goodheart-Willcox Publisher product are provided solely as a convenience to you. These resources and listings were reviewed at the time of publication to provide you with accurate, safe, and appropriate information. Goodheart-Willcox Publisher has no control over the referenced websites and, due to the dynamic nature of the Internet, is not responsible or liable for the content, products, or performance of links to other websites or resources. Goodheart-Willcox Publisher makes no representation, either expressed or implied, regarding the content of these websites, and such references do not constitute an endorsement or recommendation of the information or content presented. It is your responsibility to take all protective measures to guard against inappropriate content, viruses, or other destructive elements.

Image Credits. Front cover: PC Photography/iStock/Getty Images Plus via Getty Images

Printed in Canada

Preface

Modern Masonry provides a thorough grounding in the methods of laying brick, block, stone, and concrete. In addition, the textbook presents a broad understanding of materials and their properties.

The textbook is organized into functional sections closely related to the needs of an apprenticeship training program, community college or university programs, or other related instruction. Simply and clearly written, this book covers all the important aspects of the masonry trade. Numerous step-by-step procedures for laying brick, block, stone, and concrete provide clear explanations of recommended techniques for a variety of masonry jobs. Numerous line drawings and photographs illustrate methods of bricklaying and block laying, as well as providing technical details of masonry structures.

The textbook includes the newest materials and building techniques used in the industry. Products such as insulated concrete wall forms, insulated blocks, brick veneer, and special masonry units are included. Reliance on research from the Brick Industry Association, Portland Cement Association, and the Wire Reinforcement Institute adds to the validity of the content of this textbook.

Additional features enhance the learning value of the textbook. Code Notes reference sections of the International Residential Code, building code requirements, and specifications for masonry structures. These notes familiarize students with code requirements pertaining to masonry. Additional Safety Notes address procedures and processes to improve safety during masonry or construction activities. Green Notes increase student awareness of the environmental impact and sustainability of masonry.

Finally, this textbook teaches and encourages professionalism in the field. Students will learn the factors that affect progress on the job, including work habits, safety habits, continuing education, work ethic, teamwork, and leadership. It is the goal of this book to help prepare students to embark on a successful and rewarding masonry or related career field.

Timothy L. Andera

About the Authors

Timothy L. Andera is a retired professor from South Dakota State University in Brookings, South Dakota. During his tenure, he served as the Coordinator of Career and Technical Education in the College of Education and Human Sciences, and he served as a Safety Consultant in the College of Engineering. Dr. Andera received his doctorate in Curriculum and Instruction with an emphasis in Construction from Illinois State University, Normal, Illinois. His master's degree from Bemidji State University is in Industrial Technology with an emphasis in Construction. He received a bachelor's degree in both Vocational Education and Industrial Technology in Construction from the University of South Dakota in Springfield, South Dakota.

Dr. Andera served as the Director of Education and Safety at Associated Builders and Contractors in Grand Rapids, Michigan. He taught Construction Management at Illinois State University, overseeing courses in Architecture, Construction, Scheduling, and Masonry, as well as coordinating the On-Site Construction Project. Dr. Andera oversaw the Residential Construction Program at Ellsworth College in Iowa Falls, Iowa, where students built a brick veneered residential home each year. Dr. Andera has presented nationally and internationally on construction and safety topics. He is a consultant on construction projects related to the masonry industry and construction in general.

During his career, Dr. Clois E. Kicklighter was a nationally known educator and the author of several books in the fields of architecture, drafting, and construction technology. He held a variety of industrial, teaching, and administrative positions and retired as dean and professor at the School of Technology at Indiana State University. He held the highest leadership positions in the National Association of Industrial Technology (NAIT). Dr. Kicklighter was awarded the respected Charles Keith Medal for exceptional leadership in the technology profession.

New to This Edition

The following changes have been made to the tenth edition of *Modern Masonry* to strengthen the integrated learning solution and provide up-to-date information on the latest industry and safety practices.

- Photos have been updated throughout to aid in student learning.
- Tools and equipment have been updated and added throughout to reflect the current equipment being used in the masonry field.
- Safety practices have been updated and added to reflect current code requirements and practices.
- Code references have been added to the Code Notes feature throughout to guide students to specific codes.
- A Virtual Toolbox has been added to the digital products that includes activities and assessments to help students identify key tools and equipment used in the trade.
- Understanding Measurement Videos and Activities have been added to the digital products that include instructional videos and hands-on activities to help students learn fundamental measuring skills.

Reviewers

The authors and publisher wish to thank the following industry and teaching professionals for their valuable input into the development of *Modern Masonry.*

Bill Carroll
Pikes Peak State College
Colorado Springs, CO

Charles Gallagher
Pima Community College
Tucson, AZ

Kevin Kroll
Milwaukee Area Technical College
Milwaukee, WI

Larry C. Little
North American Trade Schools
Baltimore, MD

Glenn R. Luse
Masonry Instructor
Williamsport, PA

Frank S. Morgan
Central Texas College
Killeen, TX

Oscar Ortiz
Orange Coast College
Costa Mesa, CA

Kenton Pleger
Tennessee College of Applied Technology
Nashville, TN

Stephen B. Richard
Alfred State College of Technology
Wellsville, NY

Loran Stara
Southeast Community College
Lincoln, NE

Michael Sundblad
St. Cloud State University
St. Cloud, MN

Marty Swan
Lincoln Land Community College
Springfield, IL

Bill Wilson
Eastern Westmoreland Career and Technology Center
Latrobe, PA

Acknowledgments

The authors and publisher would like to thank the following companies, organizations, and individuals for their contribution of resource material, images, or other support in the development of *Modern Masonry*.

Air Vol Block
Allen Engineering Corporation
Con-Form Industries
American Plywood Association
Anchor Wall Systems, Inc.
Anderson Corporation
Areté 3 Ltd.
Besser Company
Bomanite Corporation
Brick Industry Association
BuildBlock Insulated Concrete Forms
Cemcast Pipe & Precast, LLC, Hartford, SD
Concrete Products of New London/JME
Concrete Reinforcing Steel Institute
Cotton Angus Ranch, Volga, SD
Dayton Superior
Donley Brothers Company
Eldorado Stone Corporation
ElginButler Company
Eneco Corporation
Fireplace Professionals, Inc., Sioux Falls, SD
Fox Blocks by Airlite Plastics Co.
Gage Brothers, Sioux Falls, SD
GCC Ready Mix, Watertown, SD
General Shale Brick
Glen-Gery Corporation
Heatilator, Inc.
Indiana Limestone Institute
Interpace Corp.
ITW Ramset/Red Head
Jack Klasey
Job Corps, Denison, IA
John R. Bare & Associates
Johnson Level and Tool Manufacturing Company, Inc.
Kawneer Company, Inc.
LiteForm Technologies
Marco Manufacturing, Inc.
Marshalltown Company
MK Diamond Products, Inc.
Nancy Henke-Konopasek
National Concrete Masonry Association
National Terrazzo and Mosaic Association, Inc.
Neuvokas Corporation – Manufacturers of GatorBar®
NRG Insulated Block, Buffalo, NY
Occupational Safety & Health Administration
Oldcastle APG South, Inc. DBA Georgia Masonry Supply
Pave Tool Innovators
Portland Cement Association
Quick Headers
Rattle Stick Concrete Tools
Sarah Carroll, Boston Valley Terra Cotta
Stanley Goldblatt
Stone Products Corporation
Superior Glove
The InterDesign Group, Indianapolis, IN
The Majestic Co., Inc.
Tilt-Up Concrete Association
Trenwyth Industries, Inc.
ULTRA GREEN FLOOR SWEEP, Platte, SD
US Department of Labor
Wilkening Fireplace Company
Wire Reinforcement Institute

Credentialing Partners and Support

Goodheart-Willcox appreciates the value of industry credentials, certifications, and accreditation. We are pleased to partner with leading organizations to support students and programs in achieving credentials. Integrating industry-recognized credentialing into a career and technical education (CTE) program provides many benefits for the student and for the institution. By achieving third-party certificates, students gain confidence, have proof of a measurable level of knowledge and skills, and earn a valuable achievement to include in their résumés. For educators and administrators, industry-recognized credentials and accreditation validate learning, enhance the credibility of programs, and provide valuable data to measure student performance and help guide continuous program improvement.

Modern Masonry is correlated to the Residential Masonry & Concrete Standards credential offered by Home Builders Institute.

HBI Certifications

Home Builders Institute (HBI) is the nation's leading provider of trade skills training in residential construction. HBI provides pre-apprenticeship training, certification programs, and job placement services to secondary school students, veterans and transitioning military personnel, displaced workers, and at-risk and justice-involved youth and adults. HBI trains in carpentry, electrical, plumbing, building construction technology, HVAC, landscaping, masonry, and solar installation.

HBI is a nonprofit organization that collaborates with public, private, and nonprofit partners dedicated to workforce development. Working with its partners, HBI attracts and certifies new students through a combination of hands-on training, innovation and technology to build an emerging generation of skilled construction workers. For more information about HBI, visit www.hbi.org.

To see how *Modern Masonry* correlates to credentialing and certification standards, visit the Correlations tab at www.g-w.com/modern-masonry-2026.

HBI®

Building Careers.
Changing Lives.

Features of the Textbook

The instructional design of this textbook includes student-focused learning tools to help you succeed. This visual guide highlights these features.

Chapter Opening Materials

Each chapter opener contains a chapter outline, a list of learning outcomes, and a list of technical terms. **Learning Outcomes** clearly identify the knowledge and skills to be gained when the chapter is completed. These outcomes are labeled and correlated to related assessment elements to facilitate reporting and remediation. **Technical Terms** list the key words to be learned in the chapter. **Introduction** provides an overview and preview of the chapter content.

Additional Features

Additional features are used throughout the body of each chapter to further learning and knowledge. **Safety Notes** alert you to potentially dangerous materials and practices. **Code Notes** point out specific items from typical building codes. **Procedures** are highlighted throughout the textbook to provide clear instructions for hands-on service activities. **Trade Tips** provide advice and guidance that is especially applicable for on-the-job situations.

Illustrations

Illustrations have been designed to clearly and simply communicate the specific topic. Illustrations have been completely replaced and updated for this edition. Photographic images have been updated to show the latest equipment.

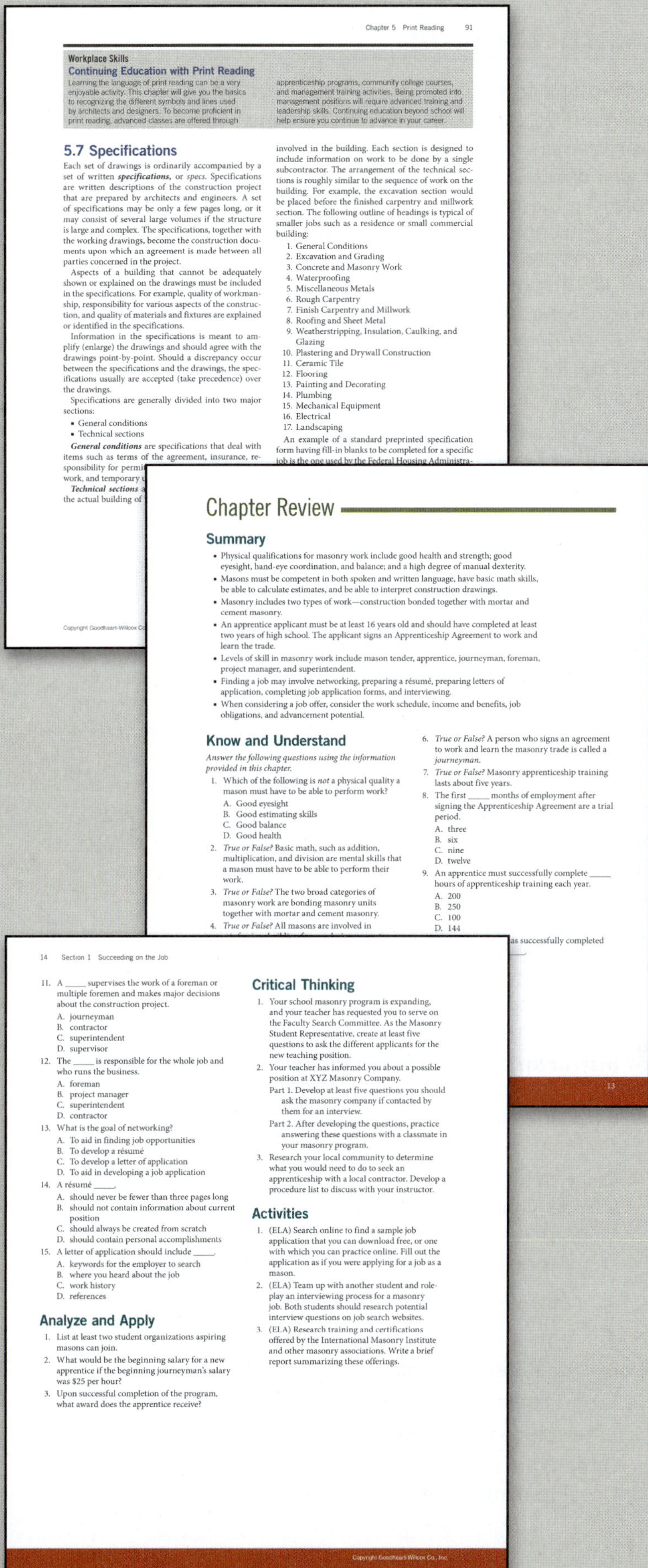

Expanding Your Learning

Workplace Skills help you understand what you can anticipate and expect in the workplace. **Green Notes** highlight key items related to sustainability, energy efficiency, and environmental issues.

End-of-Chapter Content

End-of-chapter material provides an opportunity for review and application of concepts. A concise **Summary** provides an additional review tool and reinforces key learning objectives. This helps you focus on important concepts presented in the text. **Know and Understand** questions enable you to demonstrate knowledge, identification, and comprehension of chapter material. **Apply and Analyze** questions extend learning and develop your abilities to use learned material in new situations and to break down material into its component parts. **Critical Thinking** questions develop higher-order thinking and problem-solving abilities. **Activities** extend your learning and help you apply knowledge.

TOOLS FOR STUDENT AND INSTRUCTOR SUCCESS

Student Tools

Student Text

Modern Masonry is a comprehensive text that provides a broad understanding of the properties and applications of masonry materials and practices.

Lab Workbook

- Hands-on practice includes questions and activities.
- Projects offer students opportunities to work on various masonry challenges.

G-W Digital Companion

- For digital users, e-flash cards and vocabulary exercises allow interaction with content to create opportunities to increase achievement.

Online Learning Suite

- Provides easy-to-use access and navigation
- Includes accessible resources for all learners
- Encourages exploration and discovery (or practice and repetition)
- Videos enrich learning: *Understanding Measurement Tools Videos and Activities* illustrate the use of various measurement tools.

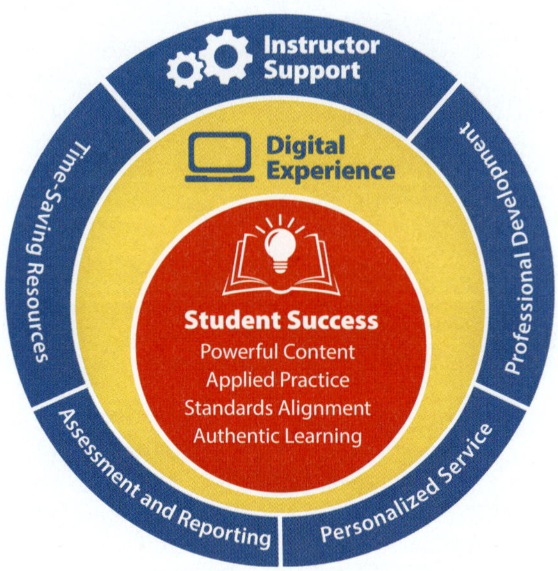

Instructor Tools

LMS Integration

Integrate Goodheart-Willcox content within your Learning Management System for a seamless user experience for both you and your students. LMS-Ready Content in Common Cartridge® format facilitates single sign-on integration and gives you control of student enrollment and data. With a Common Cartridge integration, you can access the LMS features and tools you are accustomed to using and G-W course resources in one convenient location—your LMS.

G-W Common Cartridge provides a complete learning package for you and your students. The included digital resources help your students remain engaged and learn effectively:

- **Digital Textbook**
- Online **Lab Workbook content**
- **Drill and Practice** vocabulary activities
- **Virtual Toolbox**
- **Understanding Measurement Tools Videos and Activities**

When you incorporate G-W content into your courses via Common Cartridge, you have the flexibility to customize and structure the content to meet the educational needs of your students. You may also choose to add your own content to the course.

For instructors, the Common Cartridge includes the Online Instructor Resources, which link directly to a cloud-based assessment platform that works alongside your LMS. Create, customize, edit, and print prebuilt questions and tests to meet your needs.

Online Instructor Resources

- The **Instructor Resources** provide instructors with time-saving preparation tools such as answer keys, editable lesson plans, and other teaching aids.
- **Instructor's Presentations for PowerPoint®** are fully customizable, richly illustrated slides that help you teach and visually reinforce the key concepts from each chapter.
- Easily create, customize, edit, and print assessments to meet your classroom needs using a cloud-based platform with hundreds of questions that assess student knowledge of the content in each chapter.

See www.g-w.com/modern-masonry-2026 for a list of all available formats.

Professional Development

- Expert content specialists
- Research-based pedagogy and instructional practices
- Options for virtual and in-person Professional Development

Brief Contents

Section 1—Succeeding on the Job

Section 2—Introduction to Masonry

Section 3—Materials

Section 4—Techniques

Section 5—Construction Details

Section 6—Concrete

Contents

Section 1
Succeeding on the Job

Section 2
Introduction to Masonry

Section 3
Materials

Feature Contents

Code Notes

Green Notes

Procedures

Workplace Skills

SECTION

1 Succeeding on the Job

DenGuy/iStock/Getty Images Plus via Getty Images

Irene Miller/Shutterstock.com

CHAPTER 1

Careers in the Masonry Industry

LEARNING OUTCOMES

After studying this chapter, you will be able to:

1.1-1 Describe the personal qualifications required for masonry work.
1.2-1 Describe the types of work in the masonry field.
1.3-1 Identify career preparation activities that can aid in obtaining a masonry job.
1.4-1 Recognize advancement opportunities.
1.5-1 Prepare an effective résumé and letter of application.

TECHNICAL TERMS

apprentice
contractor
foreman
journeyman
mason tender
networking
project manager
superintendent

Introduction

Masonry offers a rewarding career for people who have an interest in working with tools and materials. Masons, or bricklayers, are skilled workers who understand basic principles and practices relating to the construction industry. Masons play an important role in building homes, schools, and commercial structures. They can be proud of their skill and the fact that they can produce something that people need.

There are many benefits to being a professional mason. The work produced is functional, long-lasting, and a benefit to others. Each job is different, so the work never becomes boring. Masonry work offers the opportunity to be independent and creative while working in an outdoor environment. Masonry work offers a good salary and opportunities for advancement. See **Figure 1-1**.

LiteForm Technologies

Figure 1-1. Cement masons are placing concrete into an insulated basement wall form to create an energy-efficient home. Most of their work is performed outdoors.

1.1 Personal Qualifications

Certain physical qualifications are required to become a mason. Good health and strength are important because the work is strenuous. A mason works outdoors much of the time and must be able to lift heavy loads, bend, reach, and kneel. Masons need good hand-eye coordination and balance. Good eyesight is also essential. Masons must be able to work on scaffolding high above the ground or in small, tight places such as an elevator shaft. A mason's job is very physical and requires a high degree of manual dexterity.

Masons must also have sharp mental skills. Ability to perform basic math calculations (addition, subtraction, multiplication, and division) with whole numbers as well as decimals and fractions is required. Masons must also be able to estimate quantities, lengths, weights, and volumes. To be successful, a mason must be competent in both spoken and written language. They must be able to read construction drawings and interpret them as they pertain to the duties required. Masonry is not just manual labor; it is a skilled occupation that requires study, concentration, and continued learning.

Goodheart-Willcox Publisher

Figure 1-3. A cement mason is placing anchor bolts in a foundation.

1.2 What Masons Do

The term *masonry* has traditionally referred to the craft of building with brick. Today, it has a broader meaning and includes two types of work. The first type consists of any construction bonded together with mortar, **Figure 1-2**. The second type deals with cement masonry, **Figure 1-3**. All workers who lay brick, block, tile, or stone are called *masons*. Cement masons specialize in concrete work such as slabs, footings, and foundations.

Masons use masonry units such as brick, block, or stone. With their tools and skills, they form the units into buildings or other useful structures. The required knowledge and skills must be learned and practiced if masons are to be successful in their trade.

Masons proportion the ingredients of mortar, figure trade-related problems, mix and spread mortar, read construction prints, and handle different kinds of masonry units. Successful masons are continually willing to learn new techniques and procedures. They must cooperate with other trades.

Masons build exterior and interior walls, floors, patios, walks, columns, window and door openings, fireplaces, arches, and many other building elements. They use masonry saws and other power-driven equipment, as well as age-old tools such as trowels, brick hammers, levels, chisels, and mason's line, **Figure 1-4**.

Cement masons work with concrete. They learn the proper methods of reinforcing, building forms, placing concrete, and finishing concrete. As in any other type of masonry work, a high degree of skill is involved in quality cement mason's work.

Poured concrete construction industries employ many skilled workers as cement masons. Their job is to place concrete into forms where they will smooth and

Susan Law Cain/Shutterstock.com

Figure 1-2. A mason knows the trade and takes pride in their craftsmanship. This mason is building a brick pier.

ronfromyork/Shutterstock.com

Figure 1-4. A mason must know how to spread mortar and properly lay brick. This mason is laying brick to a line.

finish it. They will work on a wide range of construction projects, including floors, slabs, roofs, sidewalks, highways, dams, and airport runways. See **Figure 1-5**.

1.3 Education and Career Preparation

Career preparation involves developing a plan to achieve future goals. Part of this plan is taking the appropriate coursework. Classes such as drafting and blueprint reading help to develop knowledge of layout and design. Classes in industrial technology and woodworking can develop skills in hand and power tool usage and related math skills.

A

B

Wire Reinforcement Institute

Figure 1-5. Cement masons work on a wide range of construction projects. A—Cement masons placing concrete at the construction site. B—Workers placing beam shear reinforcing for a major league baseball facility.

Certain schools and colleges have internship programs where students take career classes relevant to a specific occupation. During a designated portion of the school day, students continue their learning experience at an employer's worksite. The school designates an internship coordinator to oversee the internship program. The coordinator approves the student's coursework and place of employment. They visit the student during the internship to ensure that it is a valuable learning experience.

Hiring employers evaluate previous work experience and training, but they also look at a candidate's participation in student organizations. Leadership in an organization is a particularly valuable asset because it reflects the candidate's initiative and sense of responsibility.

Two organizations with student chapters that operate in school and college settings are the Associated General Contractors (AGC) and the National Association of Home Builders (NAHB). Both groups have been chartered in a school setting to aid students who are interested in a career in construction. The student chapters usually have a very strong relationship with their local chapters, which in turn are affiliated with their national associations.

SkillsUSA is another student organization that promotes leadership and personal growth. SkillsUSA is known for conducting career competitions that revolve around specific careers, including masonry. Usually in late winter, schools participating in SkillsUSA conduct a student competition to select first, second, and third place winners in each career area. The winners from the local levels again compete at the state level in their chosen career fields. In June of each year, the first place state winners are invited to compete at the national level of SkillsUSA. At the national competition, nearly 6000 students from secondary and postsecondary schools across the country attend the competition each year in approximately 100 career and occupational areas.

The Job Corps program is controlled by the US Department of Labor. **Figure 1-6** shows skill-development projects being completed by students enrolled in such a program.

Job Corps admits individuals from 16 to 24 years old in an assortment of career offerings based on location. The masonry program is at least 18 months in length and is a performance-based curriculum where training activities are geared around individualized learning

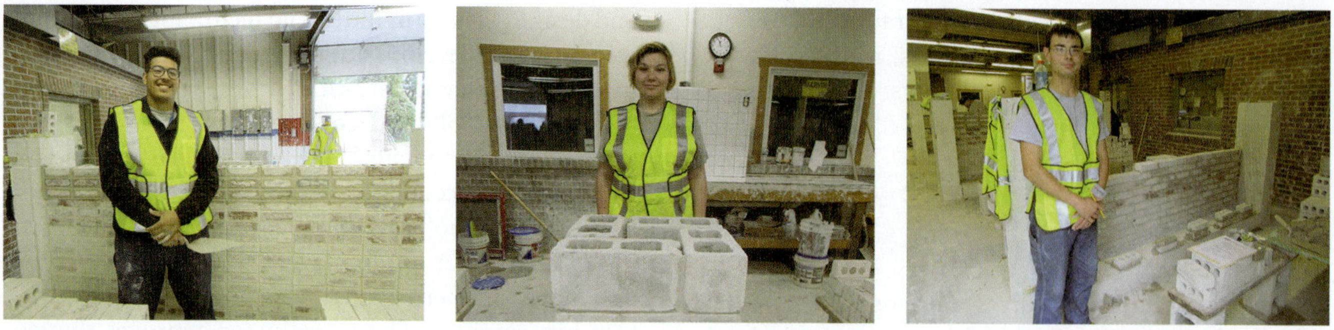

Author's images taken at Job Corps, Denison, IA

Figure 1-6. These students are in the masonry program provided by Job Corps. Shown are some of the projects students in the program must complete.

modules. As learning modules and projects are accomplished, the student will advance to the next assignment until all modules are completed.

1.4 Advancement Opportunities

The masonry trades recognize various levels of the skill. These levels are as follows:

- Mason tender
- Apprentice
- Journeyman
- Foreman
- Project manager
- Superintendent

Mason Tender

A ***mason tender*** is a person at least 16 years of age or older (depending on the type of employment) who supports professionals at higher skill levels, such as the journeyman. A tender mixes mortar, keeps materials on hand, and maintains the job area to keep it clean and safe. An individual may become a mason tender as part of an apprenticeship program, or the work activities of a mason tender can lead into a masonry apprenticeship program. This can be the first step in a lifelong career in the masonry field. A mason tender is a valuable employee who works behind the scenes performing such tasks as stocking materials on scaffolding, or by running large tools and equipment to aid the mason, increasing the overall output of the work performed during the day. Mason tenders are very valuable employees who add to the success of a masonry company.

Apprenticeship

An ***apprentice*** is a person at least 16 years old and preferably not over 24 years old who signs an Apprenticeship Agreement to work and learn the trade. An apprenticeship applicant should have completed at least two years of high school, but preferably more. Applicants younger than 18 are required to have a parent sign the Apprenticeship Agreement.

An Apprenticeship Agreement is made with a local apprenticeship and training committee or comparable group acting as the agent of the contractor. See **Figure 1-7**. The agreement may also be made directly with a contractor approved by the local joint committee or by the Department of Labor Bureau of Apprenticeship and Training.

The first six months of employment after signing the Apprenticeship Agreement is a trial period. Before the end of the period, the committee or similar apprenticeship authority will review the apprentice's ability and development. The normal term of a masonry apprenticeship varies, but can be between 4500 and 6000 hours of employment, or about three years. Apprenticeship is divided into six periods of advancement of six months each. The pay rate is a percentage of the journeyman rate and usually starts at 50% of a beginning journeyman's salary. Salary increases are based on periods of advancement throughout the apprenticeship cycle.

Each apprentice is required to take related instruction, outside of the job, for no less than 144 hours per year during each year of the apprenticeship. Subjects include print reading, math, estimating, shop practice, and safety.

A Certificate of Completion of Apprenticeship is awarded to a person who successfully completes the program. Upon successful completion of the program, the apprentice becomes a journeyman mason.

Program Registration and Apprenticeship Agreement

U.S. Department of Labor
Employment and Training Administration
Office of Apprenticeship

APPRENTICE AGREEMENT AND REGISTRATION – SECTION II OMB No. 1205-0223 Expiration Date: 06/30/2024

PART A: APPRENTICE'S INFORMATION

1. First Name Last Name

Middle Name (Optional) Suffix (Optional)

Address (No., Street, City, State, Zip Code)

Telephone Number (Optional) E-mail Address (Optional)

*Social Security Number
- -

2. Date of Birth (Mo., Day, Yr.)

3. Sex (Select One)

☐ Male ☐ Female

☐ Participant Did Not Self-Identify

Answer Both 4a. and 4b. below

4. a. Ethnicity (Select One)

☐ Hispanic or Latino
☐ Not Hispanic or Latino
☐ Participant Did Not Self-Identify

b. Race (Select One or More)

☐ American Indian or Alaska Native
☐ Asian
☐ Black or African American
☐ Native Hawaiian or other Pacific Islander
☐ White
☐ Participant Did Not Self-Identify

5. Veteran Status (Select All That Apply)

☐ Non Veteran
☐ Veteran
☐ Non Veteran, Other Eligible Individual
☐ Veteran, Eligible
☐ Participant Did Not Self-Identify

6. Education Level (Select One)

☐ Not High School graduate
☐ High School graduate
☐ (including equivalency)
☐ Some College or Associate's degree
☐ Bachelor's degree
☐ Master's degree
☐ Doctorate or professional degree

7. Employment Status of Apprentice (Select One)

☐ New Employee ☐ Current Employee

8. Did the apprentice complete a pre-apprenticeship program prior to their registration in this apprenticeship program?

☐ Yes ☐ No

If yes, please provide the Pre-Apprenticeship Program Name and Address:

PART B: PROGRAM SPONSOR'S INFORMATION

1. Program Number

Sponsor's Name and Address (No., Street, City, State, Zip Code, County)

Telephone Number Cell Phone Number (Optional)

E-mail Address

2. Occupation (The work processes listed in the standards are part of this agreement.)

a. RAPIDS Code:

b. O*NET Code:

c. Interim Credentials Offered (i.e., Career Lattice Occupation)?

☐ Yes ☐ No

Page 1 of 5 ETA 671 - Section II

US Department of Labor

Figure 1-7. The first page of an Apprenticeship Agreement between the apprentice and the local apprenticeship and training committee or designated authority.

Journeyman

A ***journeyman*** is an experienced craftsperson who has successfully completed an apprenticeship in the trade. At this point, the worker is a free agent and can work for any contractor. In most situations, the new journeyman continues working for the contractor for whom they were employed during the apprenticeship. The certificate, however, is recognized throughout the country.

Foreman

A ***foreman*** is a journeyman who has the responsibility of supervising a group of workers. This job requires not only a high degree of knowledge about the craft, but also the ability to supervise people.

Project Manager

A ***project manager*** is a professional who works in the field and is in charge of leading the project from start to finish. This person is responsible for planning and executing the construction project. The project manager works under the direction of the superintendent to keep the project on time and on budget and may also direct foremen working on the project.

Superintendent

A superintendent is generally a project manager who has been promoted because of outstanding performance. A ***superintendent*** is in charge of all the work in the field for the contractor. See **Figure 1-8**. Responsibilities include supervising the work of the subcontractors, project manager, and foreman or multiple foremen, depending on the size and scope of the job or project. A superintendent makes major decisions about the job or project under construction.

auremar/Shutterstock.com

Figure 1-8. A masonry superintendent is discussing with his lead foremen areas on the blueprint that will need their attention.

Contractor

A contractor is the person in charge of the entire job. The ***contractor*** organizes people and work, prepares bids for jobs, inspects work, and runs the business. They must be knowledgeable about all phases of the business.

Related Careers

As knowledge and experience is gained in the masonry field, opportunities expand. Masonry-related careers include masonry trainers, secondary or postsecondary level masonry instructors, building inspectors, safety specialists, and cost estimators. Masons aspiring to one of these careers may choose to pursue degrees such as engineering, building construction, construction management, or construction while continuing to work full-time or part-time.

1.5 Finding and Applying for a Job

A variety of approaches can be used to conduct a job search. Networking and résumé preparation are discussed in the following sections.

Networking

Many people find employment through networking. ***Networking*** is the exchange of information or services among individuals or groups. Your networking goal as a newcomer to the career field is to learn about possible job leads. Let people know you are looking for work and ask them to let others know.

Contacting area construction associations is a good way to network with local professional building groups in your area and research employment opportunities. Some associations may have a list of members who are looking to hire new employees. Examples of associations that pertain to construction are the Associated Builders and Contractors (ABC), Associated General Contractors (AGC), and National Association of Home Builders (NAHB). These national associations have branches at the state and local levels. Local membership is made up of building contractors representing their specific construction trades.

Contacting a local apprenticeship training facility is also a way to find employment opportunities.

Apprentices, who are usually employees of a contractor, take classes at these facilities. Apprenticeship training facilities are usually found in larger communities.

Social networking sites have become popular places to find information about companies and their available positions. Many companies network on these sites because it is an additional source of advertising for them. Users find that these sites expand their job search possibilities. In addition, these sites allow a personal exchange between users and company representatives.

Résumé

A good résumé can help convince a prospective employer to hire you. Typical information includes contact information, present job position, work history, educational background, and professional and personal accomplishments. See **Figure 1-9**.

Résumés can have content organized in order of time, starting with your current job and working through past jobs, or the content may concentrate on work skills or job functions, rather than specific jobs. Regardless of its format, your résumé should be concise (to the point), free of errors, and no more than two pages long.

Most employers receive résumés by email. You do not have to create or format your résumé from scratch. Many word processing and desktop publishing programs include résumé templates that can help you get started. There are also many examples of résumés on the internet.

Along with the résumé, develop a separate list of at least three references. A *reference* is an individual who can provide important information about you to a prospective employer. A reference can be a teacher, school official, previous employer, or any other adult outside your family who knows you well.

Always get permission from each person to use their name as a reference before actually doing so. Your list of references, along with their titles, phone numbers, and addresses, should be kept private. Share this list only with an employer who has interviewed you and asks for your references.

Letter of Application for Employment

When you send a résumé to a potential employer, include a brief letter of application. See **Figure 1-10**. This letter is often the first contact you have with a potential employer and it can make a lasting impression. The letter should be neat and follow a standard form for business letters. The paper should be clean and preferably in an ivory or white color. Be sure to check all spelling and punctuation. Ask other people to read the letter and offer advice for improving it. The letter of application should include the following items:

- Title of the job you seek
- Where you heard about the job
- Your strengths, skills, and abilities for the job
- Reasons you should be considered for the job
- Request for an interview

Job Application

A prospective employer may ask you to complete a job application form prior to an interview. The job application form highlights the information the employer needs to know about you, your education, and your prior work experience.

Fill out the form accurately, completely, and neatly. How well you accomplish the task can determine whether you get the job. When asked about salary, write *open* or *negotiable*. This means you are willing to consider offers. Do not leave any sections blank. If a section asks for information that does not apply to you, answer "not applicable," or "N/A." Otherwise, the potential employer may think the blank space was an oversight on your part.

Many employers now request electronic applications, either through the employer's website or through an independent job-search website. When filling out an online application, include keywords for which the employer may search. This will help you stand out from the many other applications the employer will receive. Examples of keywords are job titles (*brickmason, blockmason, stonemason*), job-specific skills (*precision brick cutting, foundation wall layout and construction*), personal skills (*team player, leadership ability*), foreign language skills, and industry credentials (*certifications, licenses, training*, and *degrees*).

When preparing your application, be sure to save it in the appropriate file format. If a preferred format is not given, it is best to save the application in .doc or .docx file format, or save it as a PDF file. This will enable the employer to find specific search terms in your document. Many job-search sites have sample forms on which you can practice before attempting to complete a real application.

Job Interview

If you are called for an interview, prepare for it by learning about the company you are seeking employment with. Try to learn what the company looks for when hiring new employees. Make a list of the questions you want to ask the interviewer. Research to

Robert Frank
RR 1, Box 45
Pukwana, SD 57370
(605) 555-2010
rzfrank@xyz.com

List complete contact information.

EMPLOYMENT OBJECTIVE

To obtain a position as an apprentice mason in commercial or residential construction.

Let employers know how well your needs match theirs.

RELEVANT SKILLS AND EXPERIENCE

- Apprentice supporting journeyperson in construction of three residential homes
- Masonry laborer in footing and foundation installations
- OSHA 10-Hour Construction Certificate

Include skills most appropriate to the job you are seeking.

EXPERIENCE

Apprentice employed at Edward Masonry, summer and weekends during school year, 2022 & 2023
Mason Tender employed by Edward Masonry, summer and weekends during school year, 2021
General Laborer employed by Country View Masonry, 2019–2020
(summers, weekends, and after school)

List most recent job first.

EDUCATION

Associate Degree in Applied Science in Masonry Construction
Ellsworth Community College, Iowa Falls, IA, 2023

Selected courses completed:
On-Site Construction (20 hours a week)
Masonry Theory I & II
Design & Layout Theory & Practice
Masonry Estimation & Scheduling
Blueprint Reading I & II

Chamberlain High School, Chamberlain, SD, 2021
Construction Program coursework

Highest level and most recent listed first.

ORGANIZATIONS AND ACTIVITIES

SkillsUSA Student Chapter (2021–2023)
State SkillsUSA Competition in Iowa, 2023—2nd place
Local SkillsUSA Competition at Ellsworth College, 2023—1st place

Student Chapter of National Association Home Builders of Iowa (2021–2023)

Habitat for Humanity (2021–2023)
Participated in the construction of three homes for individuals in the community.

Include all activities and memberships.

REFERENCES

Available upon request.

List references on a separate sheet of paper. Get permission from references before giving out their contact information.

Goodheart-Willcox Publisher

Figure 1-9. Example of a résumé, highlighting the main points that should be included. A résumé should show a potential employer how your skills, experience, and achievements match the requirements of the job you are seeking.

May 25, 2024

Mr. Charles Woods
President
Woods and Harper Masonry Corporation
1255 South Van Kooten Drive
Sioux Falls, SD 57105

Dear Mr. Woods:

Title of the job you seek. *Where you heard about the job.*

I am applying for the apprentice masonry position that was listed in the Sioux Falls, SD, newspaper. Please consider me a very interested candidate. I have performed a number of work activities that meet the expectations of the listed position:

- Apprentice in the construction of three residential brick homes
- Mason tender for a commercial contractor
- General laborer for a concrete contractor

During my last employment as a masonry apprentice, I was directed by journeyperson Kimberly James in a number of projects. During the construction of three brick residences, I installed batter boards and laid out footings in the preparation and placement of concrete. After footing installation, concrete block foundations were built. For each foundation, I participated in laying masonry units of different sizes. Brick veneer was placed on the exterior of the homes, and again I was involved in laying masonry units under the supervision of Ms. James.

I am a dedicated and cooperative individual who takes a great deal of pride in accomplishing goals and working as a team member. Confidence and the ability to get things done are two qualities that I always strive for as an employee.

Strengths, skills, and abilities

Please let me know if you have questions regarding my letter of application or résumé. I am available for an interview at your convenience. I can be reached at (605) 555-2010. If I am given the opportunity to work for your company, I will do my very best to do an outstanding job.

Request for an interview

Sincerely,

Robert Frank
RR1, Box 45
Pukwana, SD 57370

Goodheart-Willcox Publisher

Figure 1-10. Example of a letter of application.

find possible interview questions. Rehearse answers to questions about your skills and abilities that apply to the job requirements. Have a friend or family member interview you until you are happy with your responses.

Verify the address of the interview location by checking the site beforehand, if possible. Plan to arrive ready for the interview about 15 minutes early. This shows the interviewer you are sincere and committed to fulfilling the duties and responsibilities of the job.

Dress appropriately in clean and neat clothing, **Figure 1-11**. Do not wear jeans or work clothes to an interview unless instructed to do so by the employer.

Good preparation will make you feel more confident during the interview. Be polite, friendly, and cheerful during the process. Answer all questions carefully and as completely as you can. Be honest about your abilities. Avoid chewing gum and fidgeting. Also be aware of questions you legally do not have to answer, such as those related to age, marital status, religion, or family background.

A prospective employer may ask you to take employee tests. Some employers administer tests to job candidates to measure their knowledge or skill level under stress.

Andrey_Popov/Shutterstock.com

Figure 1-11. Wearing clothing that is clean and appropriate for an interview makes a good first impression.

Evaluating Job Offers

When considering a job offer or comparing two or more positions, you should explore the following work factors:

- **Work schedule.** Will the workdays and work hours mesh with your lifestyle? Is overtime work required?
- **Income and benefits.** Is the proposed salary fair? Will you receive benefits that are just as valuable as extra income? What is the vacation and sick leave policy? Are there medical and life insurance benefits? Is there a credit union? Will the company pay tuition for college courses or special programs related to your job?
- **Job obligations.** Will you be expected to join a union or other professional organization? If so, what are the costs? Will you be expected to attend meetings after work?
- **Advancement potential.** Is there opportunity for advancement? After demonstrating good performance, how soon can you seek a position with more responsibilities? Are training programs provided?

Chapter Review

Summary

- Physical qualifications for masonry work include good health and strength; good eyesight, hand-eye coordination, and balance; and a high degree of manual dexterity.
- Masons must be competent in both spoken and written language, have basic math skills, be able to calculate estimates, and be able to interpret construction drawings.
- Masonry includes two types of work—construction bonded together with mortar and cement masonry.
- An apprentice applicant must be at least 16 years old and should have completed at least two years of high school. The applicant signs an Apprenticeship Agreement to work and learn the trade.
- Levels of skill in masonry work include mason tender, apprentice, journeyman, foreman, project manager, and superintendent.
- Finding a job may involve networking, preparing a résumé, preparing letters of application, completing job application forms, and interviewing.
- When considering a job offer, consider the work schedule, income and benefits, job obligations, and advancement potential.

Know and Understand

Answer the following questions using the information provided in this chapter.

1. Which of the following is *not* a physical quality a mason must have to be able to perform work?
   A. Good eyesight
   B. Good estimating skills
   C. Good balance
   D. Good health
2. *True or False?* Basic math, such as addition, multiplication, and division are mental skills that a mason must have to be able to perform their work.
3. *True or False?* The two broad categories of masonry work are bonding masonry units together with mortar and cement masonry.
4. *True or False?* All masons are involved in reinforcing, building forms, placing concrete, and finishing concrete.
5. Hiring employers are interested in _____.
   A. previous work experience
   B. previous training
   C. participation in student organizations
   D. All are correct.
6. *True or False?* A person who signs an agreement to work and learn the masonry trade is called a *journeyman*.
7. *True or False?* Masonry apprenticeship training lasts about five years.
8. The first _____ months of employment after signing the Apprenticeship Agreement are a trial period.
   A. three
   B. six
   C. nine
   D. twelve
9. An apprentice must successfully complete _____ hours of apprenticeship training each year.
   A. 200
   B. 250
   C. 100
   D. 144
10. An apprentice who has successfully completed training is called a _____.
    A. foreman
    B. manager
    C. superintendent
    D. journeyman

11. A _____ supervises the work of a foreman or multiple foremen and makes major decisions about the construction project.
    A. journeyman
    B. contractor
    C. superintendent
    D. supervisor
12. The _____ is responsible for the whole job and who runs the business.
    A. foreman
    B. project manager
    C. superintendent
    D. contractor
13. What is the goal of networking?
    A. To aid in finding job opportunities
    B. To develop a résumé
    C. To develop a letter of application
    D. To aid in developing a job application
14. A résumé _____.
    A. should never be fewer than three pages long
    B. should not contain information about current position
    C. should always be created from scratch
    D. should contain personal accomplishments
15. A letter of application should include _____.
    A. keywords for the employer to search
    B. where you heard about the job
    C. work history
    D. references

Analyze and Apply

1. List at least two student organizations aspiring masons can join.
2. What would be the beginning salary for a new apprentice if the beginning journeyman's salary was $25 per hour?
3. Upon successful completion of the program, what award does the apprentice receive?

Critical Thinking

1. Your school masonry program is expanding, and your teacher has requested you to serve on the Faculty Search Committee. As the Masonry Student Representative, create at least five questions to ask the different applicants for the new teaching position.
2. Your teacher has informed you about a possible position at XYZ Masonry Company.

   Part 1. Develop at least five questions you should ask the masonry company if contacted by them for an interview.

   Part 2. After developing the questions, practice answering these questions with a classmate in your masonry program.
3. Research your local community to determine what you would need to do to seek an apprenticeship with a local contractor. Develop a procedure list to discuss with your instructor.

Activities

1. (ELA) Search online to find a sample job application that you can download free, or one with which you can practice online. Fill out the application as if you were applying for a job as a mason.
2. (ELA) Team up with another student and role-play an interviewing process for a masonry job. Both students should research potential interview questions on job search websites.
3. (ELA) Research training and certifications offered by the International Masonry Institute and other masonry associations. Write a brief report summarizing these offerings.

microgen/iStock/Getty Images Plus via Getty Images

CHAPTER 2

Job Performance, Leadership, Ethics, and Entrepreneurship

LEARNING OUTCOMES

After studying this chapter, you will be able to:

2.1-1 Identify factors that affect your job performance.
2.2-1 Evaluate your work ethic.
2.3-1 Describe the characteristics of team players and leaders.
2.4-1 Explain the importance of decision-making and problem-solving skills.
2.5-1 Identify the risks and rewards of entrepreneurship and the characteristics of a successful entrepreneur.

TECHNICAL TERMS

critical thinking skills
entrepreneur
leadership
work ethic

Introduction

There are many factors that can affect your progress on the job. These factors include job performance, ethics, teamwork, and leadership.

2.1 Job Performance

Job performance can be measured in several ways. Three primary areas that are considered when evaluating job performance are general work habits, safety record on the job, and ability to stay current in the field. This section will explore opportunities for improvement in these areas.

General Work Habits

General work habits play a large part in keeping your job as well as progressing on the job. You should always strive to improve your performance and the quality of your work. Ask yourself the following questions about your general work habits:

- Do you observe safety rules? Safety is everyone's responsibility. It involves not only your personal safety but also the safety of others. See **Figure 2-1**.
- Do you keep the work area clean and organized? A cluttered workplace encourages accidents and slows down the work.
- Do you have nonwork-related conversations that interfere with your work and the work of others?
- Do you observe company policies? Company policies are established to provide uniformity of response to foreseeable situations. They are vital to the smooth operation of the business.

Luis Santos/Shutterstock.com

Figure 2-1. Safety rules apply to everyone. It is your responsibility to work safely.

- Do you plan and execute your work efficiently? A professional comes to the job ready to work and has the necessary tools and attitude to perform the prescribed tasks.
- Do you always give your best effort to each job? One way to improve your performance is to practice good work habits. Always try to improve your skills and take pride in your work.

These are some of the questions that you should ask yourself about your general work habits. You can probably think of others. Paying attention to your work habits could result in increased pay, greater appreciation and recognition of you and your work, and job advancement. At the very least, it might preserve your job on the team.

Employers want employees who are punctual, dependable, and responsible. They want their employees to be capable of taking initiative and working independently. Other desirable employee qualities include organization, accuracy, and efficiency.

A punctual employee is always prompt and on time. This means not only when the workday starts, but also when returning from breaks and lunches. Being dependable means that people can rely on you to fulfill your word and meet your deadlines.

If you are not well, be sure to call in and let the employer know right away. If there are reasons you cannot be at work, discuss this with your employer and work out an alternative arrangement. Many people have lost jobs by not checking with their supervisors about time off.

Taking initiative means you start activities on your own without being told. When you finish one task, you do not wait to hear what to do next. Individuals who take initiative need much less supervision. They have self-motivation, or an inner urge to perform well. Generally, this motivation will drive you to set goals and accomplish them. All of these qualities together show that you are capable of working independently.

You are expected to be as accurate and error-free as possible in all that you do. Complete your work with precision and double-check it to ensure accuracy. Your coworkers depend on the careful completion of your tasks. See **Figure 2-2**.

Volt Collection/Shutterstock

Figure 2-2. It is each employee's responsibility to be as accurate as possible in their work.

Safety on the Job

Constant attention to safety guidelines is crucial to prevent injuries and safety violations. Be alert to spot possible accidents before they occur. Always tell your supervisor about unsafe conditions and safety rule violations. To develop skills in identifying potential hazards, ask to participate in jobsite walk-throughs with the foreman or supervisor.

A good safety record on the job involves using the proper safeguards for a given job. See **Figure 2-3**. Examples include protective clothing, a blade guard, goggles, a face shield, or a respirator. If you are unsure of how to operate a tool or machine, ask your supervisor for training.

Many workplace accidents occur because of carelessness. Do not leave litter underfoot where it may cause an accident. Items should not be placed on a ledge, ladder, or scaffold where they can fall and injure someone.

Poor attitudes can often cause unsafe behavior as well. Common causes of accidents include the following:

- Taking chances
- Showing off
- Forgetting safety details
- Disobeying company rules
- Daydreaming
- Losing your temper
- Falling asleep

Safety Note

If you see someone not wearing safety glasses, remind them to put on their glasses. If you see other unsafe conditions around the jobsite, say something to your supervisor. Your commitment to safety may save someone from an accident!

Susan Law Cain/Shutterstock.com

Figure 2-3. Masons are building a brick pier. They are wearing hard hats, safety glasses, and safety shoes. Bricks and mortar are stored for easy access, which can improve productivity. The site is clean with loose materials removed from area of work.

Continuing Education

When a person moves beyond the level of apprenticeship, the educational process does not stop. Workers should continue to update their skills, whether they are journeymen, foremen, project managers, superintendents, or contractors. The masonry industry is constantly evolving, which creates a demand for craftspeople and contractors to institute changes in the way they think and build. Contractors and project managers must keep their workers apprised of new products and systems and the reasons for carrying out the work in a particular way. See **Figure 2-4**.

New techniques and equipment can reduce labor costs and production time. When new equipment is purchased, it is important that workers are trained in the operation and safety procedures of the new technology. Many suppliers conduct free training for management and employees to ensure they have knowledge to operate the equipment safely and accurately.

Industry associations at the local, state, regional, and national level provide many services for those who wish to continue their education. These services include technical seminars and conferences, training sessions, university classes, technical publications, and online resources. To be successful, you must be alert, receptive to change, and have a desire to continue the educational process.

Many organizations offer educational programs and training materials for the construction industry. The following are some of these organizations:

- American Concrete Institute (ACI)
- Associated Builders and Contractors (ABC)
- Associated General Contractors (AGC)
- Brick Industry Association (BIA)

NatalieSchorr/Shutterstock.com

Figure 2-4. Reinforcing bars are protruding from the footing to be used to connect the footing to the foundation wall. One of OSHA's Standards identifies that rebar or other reinforcement needs to be capped or protected to ensure workers are not injured.

- International Masonry Institute (IMI)
- Mason Contractors Association of America (MCAA)
- Masonry Institute of America (MIA)
- Portland Cement Association (PCA)
- National Association of Home Builders (NAHB)
- National Concrete Masonry Association (NCMA)

2.2 Work Ethic

Having a good ***work ethic*** means believing in working diligently. Ethical behavior on the job means conforming to accepted standards of fairness and good conduct. Individuals and society as a whole regard ethical behavior as highly important. Integrity (firmly following your moral beliefs) and honesty are crucial aspects of ethical workplace behavior.

Work ethic pertains to a person's attitude and beliefs about work and how they handle job responsibilities. Employers are very concerned about the attitude you bring to the workplace. Work ethic includes your enthusiasm for the work, your willingness to work late to meet deadlines, and your readiness to start work on time. Work ethic is an important consideration in keeping a job or receiving a promotion.

Work ethic also includes how you relate to fellow workers and supervisors. Ask yourself the following questions to evaluate your work ethic:

- Do you talk so much that others cannot get their work done?
- Do you treat others with respect?
- Are you a gossip or a critic of other employees' work?
- Are you a constant complainer?
- Do you always have an excuse for poor work or work not finished?
- Are you willing to admit mistakes and learn from them?

These and many other related characteristics are part of your work ethic. If you expect to be happy at the workplace and progress to positions of more responsibility, then you should pay attention to your work ethic and personal behavior.

2.3 Teamwork and Leadership

A mason with strong technical knowledge and skills is of little value to a contractor if they cannot get along with others and work as part of a team. See **Figure 2-5**. Work environments are usually diverse. There will be people of different cultures, beliefs, and ages. Regardless of personal differences, show respect to your supervisor, coworkers, and any other people you interact with on the job.

Andreas G. Karelias/Shutterstock.com

Figure 2-5. Form crew members are working together to place large concrete forming panels.

Building a structure requires cooperation among many people. Much time can be wasted because three people working on a single job each think they are in charge. In actuality, different people are in charge at different times. Your performance should reflect this recognition.

Companies are always looking for people who can run a crew. Some people are skilled in running a large crew, others are capable of supervising a few other workers, and some should not be managing anyone. If you seek the opportunity to manage, observe the traits good managers have and practice these behaviors.

As you progress to a job or position that involves managing others, you will be expected to show leadership. ***Leadership*** is the ability to guide and motivate others to complete tasks or achieve goals.

The most important role of leaders is to keep the team advancing toward its goal. This role involves communicating well with others, accepting responsibility, and making decisions with confidence. A leader may need to settle disputes among members as well as encourage less productive members to participate. Leaders inspire their groups and provide the motivation to keep everyone working together. Employees with leadership skills are likely to be promoted to higher levels.

Most of the traits identified with leaders are common sense. For example, people with the ability to lead can keep track of what is being built. They can foresee potential problems and can solve these problems as they arise. They know enough to make sure that the proper materials and equipment are on hand when needed. Leaders can get along with all types of people and can communicate information clearly. They know who can handle each project and can motivate people to work hard and cooperate. They value their employees as people as well as members of the team. Most importantly, good leaders lead by example and practice the types of behavior they expect from others. See **Figure 2-6**.

Perform a self-evaluation in the areas of teamwork and leadership by asking yourself the following questions:

- How well do you receive instructions from others? If you cannot accept instructions well from others, the chances are pretty good that your instructions to others will not be accepted very well either.

ndoeljindoel/Shutterstock.com

Figure 2-6. This superintendent is wearing appropriate PPE on the construction site—hard hat, safety glasses, high visibility vest, and protective gloves. As the leader, it is their responsibility to set a good example and ensure that all employees follow safety rules.

- Are you a team player? Team players are valued more in the workplace than solo artists. Remember that most jobs or tasks involve teamwork. Reflect on what it means to be a good team player.
- Do you get into personality conflicts with other workers or the boss? You cannot completely ignore your personality traits, but you can control them. Do not see every situation as a crisis or a threat. Learn to take things in stride and place events in their total context. For example, put yourself in the other person's place and then ask yourself how you would act.
- Are you a good communicator? Many conflicts are the result of poor communication. Frequently, people take more offense at how something was said than at what was said. What is heard may be different from what is said. Think about what you wish to communicate and then put it into words that will be clearly understood.
- Do you keep your foreman or supervisor informed about potential problems on the job? A little warning about a potential problem will be appreciated and help to establish you as a valued member of the team who is interested in saving the company money and producing the best quality job possible.
- Do you have the ability to manage a crew? Others will most likely know the answer to this question even before you do. Seek their advice about improving your leadership skills. Study leadership styles or take a course in developing leadership skills. Learn from others. If you have what it takes to be a leader, it will be recognized.

Good leaders encourage teamwork, because a team that works well together is more likely to reach its goals. They listen to the opinions of others and make sure all team members are included in projects. Leaders also try to set a good example by doing a fair share of the work. In these ways, leaders cultivate a sense of harmony in the group.

Successful leaders do the following:

- Speak simply and directly
- Share the credit
- Have a vision
- Maintain a positive attitude
- Show optimism and enthusiasm
- Avoid sarcasm
- Meet problems head-on
- Check the small things
- Control their emotions
- Are always truthful
- Treat everyone fairly
- Behave in a constructive and uncritical manner
- Act as a role model

2.4 Decision Making and Problem Solving

Employers value workers who have the ability to make sound decisions and solve any problems that arise. The processes for making decisions and solving problems are similar:

1. Identify the problem or issue to be decided.
2. Brainstorm possible solutions.
3. Decide which solution to implement.
4. Implement the solution.
5. Evaluate the results.

Having the ability to solve problems on the job shows an employer that you are able to handle more responsibility. Solving problems as a group can strengthen the team and help employees feel more pride in their work.

Making decisions and solving problems both require ***critical thinking skills***. These are higher-level skills that enable you to think beyond the obvious. You learn to interpret information and make judgments. Supervisors appreciate employees who can analyze problems and think of workable solutions.

2.5 Entrepreneurship

An ***entrepreneur*** is someone who organizes, manages, and assumes the risks of a business. Some day you may want to own or manage your own business. For this reason, you may want to think about the following considerations:

- Every year many new businesses are started, but most fail. The reasons generally cited include lack of adequate financing, poor management of the enterprise, and lack of required knowledge.
- There are positive and negative aspects of owning your own business. Some advantages include being your own boss, the chance to make more money, and the satisfaction of building a successful enterprise. Disadvantages include the enormous responsibility of making the right decisions, the long hours, the risk of failure, and the responsibility for the livelihood of others.
- Business opportunities may not be opportunities at all. If it is too good to be true, it probably is. Ask questions such as:
  - Is the location good?
  - How much overhead will there be in the business (utilities, taxes, insurance)?
  - What about the competition?
  - How will you hire good employees?
  - Can you manage a workforce?
  - What business skills and knowledge do you have?
  - Can you get financing?
  - What government regulations apply to your business?
  - What type of business organization should you choose?
  - What are your goals?
  - How will you advertise your product or service?
  - Where can you get good advice?
  - What governmental agencies offer help?
- Every business should have definite, well-defined goals. Goals provide direction and help in making decisions concerning your product or service to be offered for sale. Consider these questions:
  - Will the product or service be conventional or unique?
  - How will you gain the support and respect of your employees?
  - What kind of image do you wish your company to convey to others?
  - Who will purchase your product or service?
  - How much should you charge?
  - How will you reward productive workers?
  - How will you protect your workers against injury on the job?
  - What activities will help you meet your goals?
  - How will you organize your business to maintain high efficiency and quality?
  - Do you have the strength and determination to own and manage a business?
  - What are the long-term consequences to yourself and your family?
  - Are you prepared to risk failure?

Workplace Skills

Professional Growth

You are entering a profession that can change your whole life. The opportunities are endless and the professional growth and self-satisfaction of completing a job to the best of your abilities is very rewarding. Learning and honing your basic workplace skills will help you apply for and succeed in every role in the industry, from mason tender to superintendent of a large construction company. You will learn about various workplace skills including professionalism, teamwork, and leadership.

A good idea or quality product or service is not enough to launch and sustain a successful business. A successful enterprise needs constant attention to detail and solid planning. See **Figure 2-7**. Risk of failure is always present and may be brought about by forces beyond your control. However, the rewards of running a successful business generally outweigh the liabilities.

Characteristics of Entrepreneurs

A person should possess certain characteristics to be a successful entrepreneur. The following list summarizes the major characteristics:

- **Healthy.** Long hours and much physical labor make heavy demands on a mason or masonry worker.
- **Knowledgeable.** To make a profit, an entrepreneur must know all aspects of the trade or trade specialty. In addition, they must understand the industry and the products being used in the trade.
- **Able to plan.** Running a successful business means leaving nothing to chance. An entrepreneur must be able to foresee difficulties as well as plan how to take advantage of opportunities.
- **Willing to take calculated risks.** Once a plan has been conceived that takes into account events likely to occur, an entrepreneur must have the courage to risk their money and future on making the plan work.
- **Innovative.** An entrepreneur must be successful in finding ways to improve and produce better work and thus gain the confidence of customers.
- **Responsible.** An entrepreneur must be willing to accept the consequences for decisions whether good or bad. This includes paying debts, keeping promises, and accepting the responsibility for mistakes of employees.

Szepy/iStock/Thinkstock

Figure 2-7. An entrepreneur needs strong planning ability and must attend to many details of the business.

An entrepreneur likes to set goals and works hard to achieve them. The average age of the US population is increasing, with a large number of people retiring and needing health care. The population in this group will demand many services. Housing is one of the areas predicted to be in great demand. Thus, a business designed to offer services such as specialized home construction, the rehabilitation of homes, or the building of care or retirement centers would likely be successful.

Ethics Code

Ethics can be defined as "the rules or standards governing the conduct of the members of a profession." Ethical practice has always been a concern of businesses who wish to be successful over the long term.

Many organizations have recorded their goals as a means of encouraging ethical practices. One such organization, the National Association of Home Builders (NAHB), adopted a model ethics code. The objectives of the NAHB code of ethics are as follows:

- To conduct my business affairs with professionalism and skill.
- To provide the best building value possible.
- To protect the consumer through the use of high-quality materials and building practices backed by integrity and service.
- To meet all of my financial obligations in a responsible manner.
- To comply with the spirit and letter of my business contracts, and manage all of my employees, subcontractors, and suppliers with fairness and honor.
- To keep informed regarding public policies and other essential information, which affect my business interests and those of the building industry as a whole.
- To comply with the rules and regulations prescribed by law and government agencies for the health, safety, and welfare of the community.
- To provide timely response to items covered under warranty.
- To seek to resolve any controversy which I may become involved in through a non-litigation dispute resolution mechanism.
- To not attempt to injure, with malicious intent, either directly or indirectly, the professional reputation, prospects, practice, or employment of another builder.
- To support and abide by the decisions of the association in promoting and enforcing this Code of Ethics.

Chapter Review

Summary

- Three primary areas considered when job performance is evaluated are general work habits, safety record, and ability to stay current in the field.
- A good safety record on the job involves using the proper safeguards for a given job, asking your supervisor for training if you are unsure of how to operate a tool or machine, and avoiding careless behavior.
- Workers should continue to update their skills, whether they are journeymen, foremen, superintendents, or contractors. Industry associations provide many educational and training services.
- Work ethic includes your enthusiasm for the work, your willingness to work late to meet deadlines, and your readiness to start work on time.
- Leadership is the ability to guide and motivate others to complete tasks or achieve goals. Good leaders encourage teamwork because a team that works well together is more likely to reach its goals.
- Having the ability to solve problems on the job shows an employer that you are able to handle more responsibility.
- An entrepreneur is someone who organizes, manages, and assumes the risks of a business.
- A successful enterprise needs constant attention to detail and solid planning.

Know and Understand

Answer the following questions using the information provided in this chapter.

1. Which of the following is *not* a primary area that is considered when a person's job performance is evaluated?
   A. General work habits
   B. Safety on the job
   C. Risk-taking habits
   D. Continuing education
2. *True or False?* Many workplace accidents occur due to careless behavior or poor attitudes.
3. *True or False?* The education process is completed when a mason successfully completes an apprenticeship program.
4. *True or False?* Work ethic includes an employee's enthusiasm for the work.
5. *True or False?* Complaining about management is part of a good work ethic.
6. *True or False?* Teamwork is the ability to guide and motivate others to complete tasks or achieve goals.
7. *True or False?* It is a good idea to keep your foreman informed about potential problems on the job.
8. A leader _____.
   A. can keep track of what is being built
   B. communicates information clearly
   C. sets a good example by doing a fair share of the work
   D. All are correct.
9. *True or False?* Critical thinking skills are helpful in making decisions and solving problems.
10. *True or False?* A successful entrepreneur likely takes little risk because of the possibility of failure.

Analyze and Apply

1. Why is it important to keep the work area clean and organized?
2. How do leaders keep the team advancing toward its goal?
3. List three reasons why most new small businesses fail.
4. List two advantages and two disadvantages of owning your own business.
5. Describe three characteristics of a successful entrepreneur.

Critical Thinking

1. Imagine that you are a foreman, and one of your employees is constantly late for work. What procedures would you take to handle this situation?
2. You are a supervisor on a construction site. Recently, a number of safety issues have occurred. What actions would you take with your employees?
3. Your supervisor has recognized your positive work ethic and wants to elevate you to a foreman level with a salary increase. What items should you discuss with the supervisor prior to making your decision?

Activities

1. (ELA) Visit several masonry industry association websites and report on the different types of online training and continuing education opportunities available.
2. (ELA) Use the internet to conduct a search for "Safety Toolbox Talks." Choose one of the topics you find and give a brief report on its contents in class.

SECTION

2 Introduction to Masonry

Timothy L. Andera

Igor Sokolov (breeze)/Shutterstock.com

CHAPTER **3**

Tools and Equipment

LEARNING OUTCOMES

After studying this chapter, you will be able to:

3.1-1 Identify various hand tools used in the masonry trade.
3.2-1 Describe the tools and equipment used in site preparation.
3.3-1 Identify the functions of various flatwork finishing tools.

TECHNICAL TERMS

altimeter level
block clamp
brick hammer
brick set
brick tongs
bull float
bull float extension handle
chalk box and chalk
concrete broom
concrete vibrator
curved claw hammer
darby
edger
ferrule
groover
hand float
hand tamper (aggregate or soil packer)
hand tamper (concrete)
joint raker
jointer
level case
line block
line pin
maul
masonry saw
mason's chisel
mason's level
mason's line
mason's rule
mortar bag
mortar board
mortar box
mortar hoe
mortar mixer
mortar stand
placer tool
plumb bob
plumb rule
power concrete joint saw
power screed
power trowel
pull crete tool
rattle stick
rebar bender
rebar cutter
retractable steel tape
roller tamper
rotary laser level
screed
sledgehammer
Smart Level
steel square
straight claw hammer
tamping rammer compactor
tie wire twister
tool bag
track-driven power buggy
trowel
twig
utility knife
vibratory plate compactor
wheelbarrow
ZipLevel

Introduction

Tools and equipment and their safe use are a very important part of all trades. Every tool and piece of equipment has been designed for a specific purpose. Beginning masons must learn how to use them skillfully if they are to be successful at their trade.

3.1 Masonry Tools and Equipment

This chapter describes the tools and equipment commonly used in the masonry industry and concrete work. The following sections will discuss the intended use of each tool and piece of equipment. Tool and equipment safety is covered in Chapter 4, **Safety**.

Trowels

A mason's ***trowel*** is the tool most commonly used by a mason. Trowel blades are made from flat pieces of forged steel ground to the proper balance, taper, and shape. The narrow end of the blade is called the *point* and the wide end nearest the handle is called the *heel*. The blade is connected to the handle by the shank.

Before purchasing a trowel, you will need to consider the weight, size, materials, construction, and angle of the handle to the blade. A good quality trowel should have a flexible blade of high-grade steel that will withstand long and hard use. A trowel should be light in weight and well balanced.

Trowels are available in lengths from about 3″ to 13″ and in widths from about 2″ to 7″. Mason's trowels are produced with wide, sharp heels or narrow, rounded heels. See **Figure 3-1**, which illustrates the difference in pattern and shape.

Each of the trowel patterns was developed for specific purposes. See **Figure 3-2**. Philadelphia pattern trowels allow more mortar to be placed on the blade of the trowel, so they are best suited for applying mortar to block. London pattern trowels have the standard blade type for laying brick. Wide London pattern trowels have a rounded heel, which places the mortar farther forward on the blade compared to the Philadelphia pattern trowel. A wide London pattern trowel can be used for laying brick or block.

Masons carry multiple types of trowels in order to have the appropriate trowel on hand to complete a particular project. In addition to the various types of blades, trowels vary by type of handle. Trowel handles are made of wood, plastic, or other materials. See **Figure 3-3**. Wood handles are most common, and usually have a ***ferrule*** (metal band) around the shank end. The ferrule adds strength and prevents the handle from splitting.

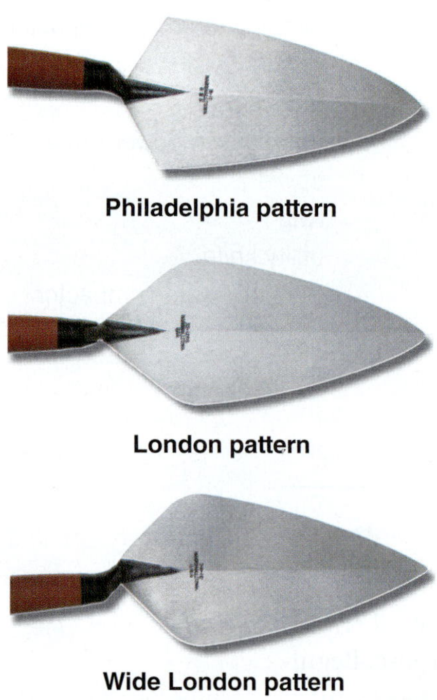

Marshalltown Company

Figure 3-1. The three major styles of trowels used by masons are the Philadelphia pattern trowel, the London pattern trowel, and the wide London pattern trowel.

Marshalltown Company

Figure 3-2. Study the different blade shapes of the three major styles of trowels.

Marshalltown Company

Figure 3-3. Trowels can be purchased in many types of patterns and handle construction.

Special-purpose trowels are designed to aid the mason in performing specialized jobs. See **Figure 3-4**.

- **Duck bill or coke.** This trowel can be used for finishing curbs or gutter faces or other rounded surfaces.
- **Gauging.** This trowel is shorter than a normal trowel and is used for gauging or mixing small amounts of mortar, thin set, or other masonry products.
- **Margin.** This trowel has a rectangular front and a V-shaped back. It is used for mixing mortar or other masonry products. It is also used to hold mortar when tuck-pointing or spreading grout or mortar.
- **Pointing.** This trowel has a distinctive pointed tip with a V-shaped back. It is used to spread mortar or similar materials in tight corners. A pointing trowel is similar to a brick trowel but considerably smaller.
- **Tuck pointer.** This trowel is used for filling or shaping mortar between brick. A margin trowel is used to hold the mortar in place, and then a tuck pointer trowel moves or pushes the mortar between the brick.

Trade Tip

A trowel must be clean to perform properly. Old mortar should be removed from the blade and shank. Form the habit of cleaning your tools at the end of each day. They will then be in good condition the next time you use them.

Jointers

Jointers are used to finish the surface of mortar joints. See **Figure 3-5**. They are also called *joint tools* or *finishing tools*. Several types and sizes of jointers are

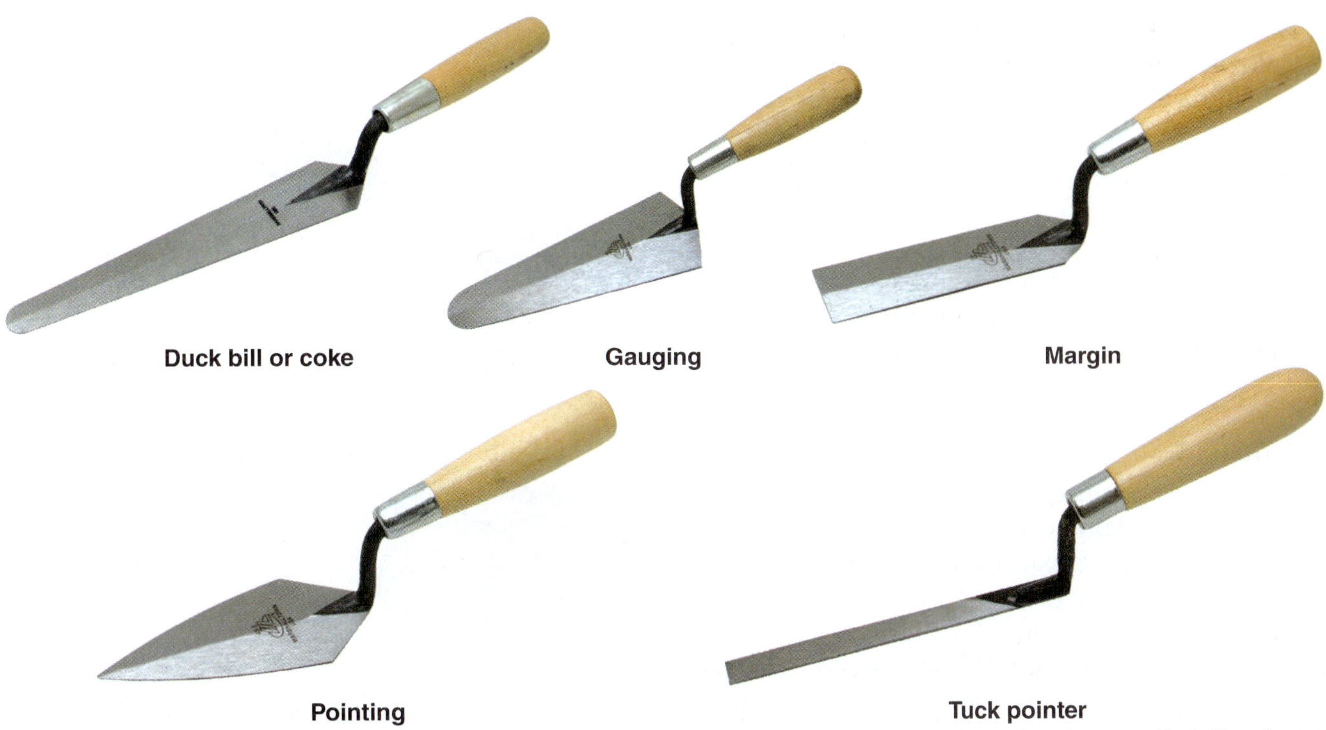

Marshalltown Company

Figure 3-4. Specialized trowels used by masons.

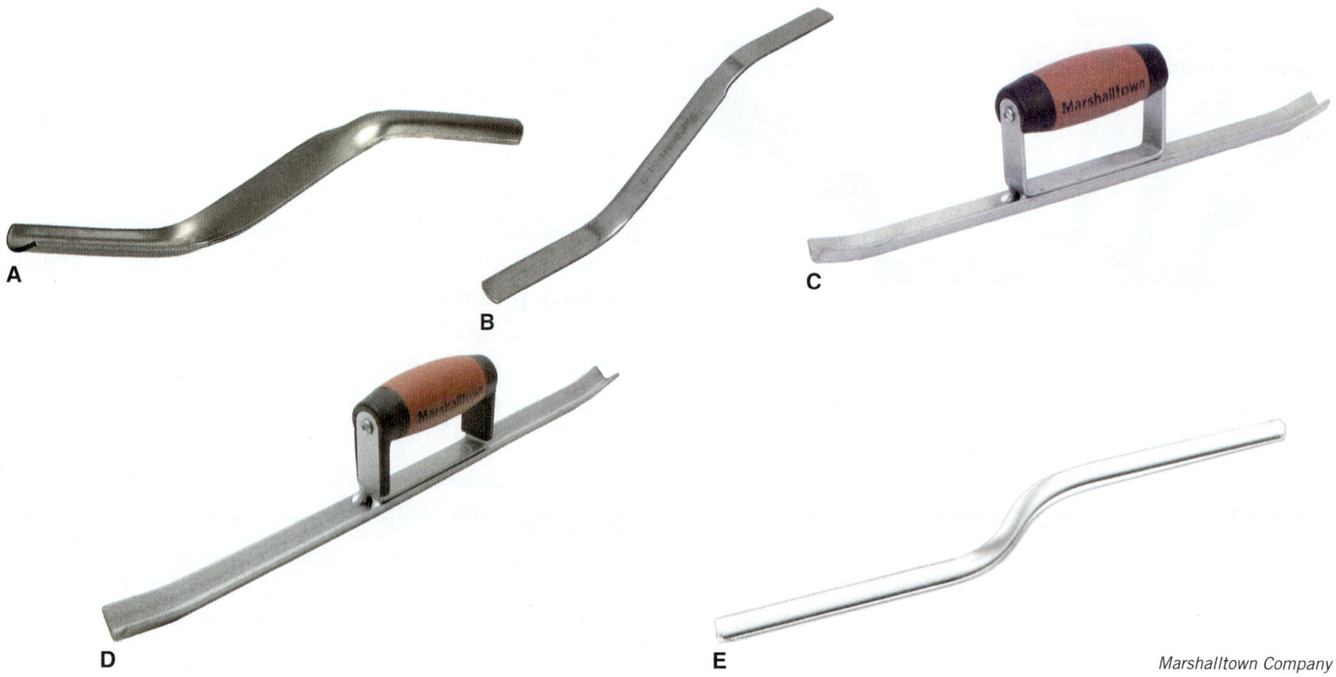

Marshalltown Company

Figure 3-5. Examples of jointers used by masons to finish mortar joints. A—Brick jointer 1/2″ × 5/8″. B—Slicker standard 1/2″ × 5/8″ solid. C—Sled runner half round 1/2″ × 15″ with a "half round" shape into the joint. D—Sled runner V groove 3/4″ × 15″ with a "V" shape into the joint. E—Acrylic 1/2″ jointer used for white mortar or glass block.

commonly used. They are usually either cast or forged rods or stamped sheet metal.

Acrylic jointers are used to finish joints for glass block or mortar that is white. This type of jointer does not leave discoloring stains that may occur when metal is used.

The cross-sectional shape of a jointer gives the mortar joint its shape. Long horizontal joints are easily finished with a sled runner (also called a *joint runner*). This tool has a handle for better control.

Joint Rakers

A ***joint raker***, **Figure 3-6**, is a tool used to remove a portion of the mortar from a joint just before the mortar hardens. Rake joints require the use of this tool.

Hammers

Hammers are classified by the style and weight of the hammer head. Hammers can be used for chipping, pounding, or ripping apart material. A ***brick hammer*** is a hammer designed to both drive nails and break or chip stone or masonry units. See **Figure 3-7**. The head is flat on one side so it may be used as a conventional hammer. The other side is drawn out to form a chisel for dressing up cuts. Most brick hammers weigh anywhere from 12 ounces to 24 ounces.

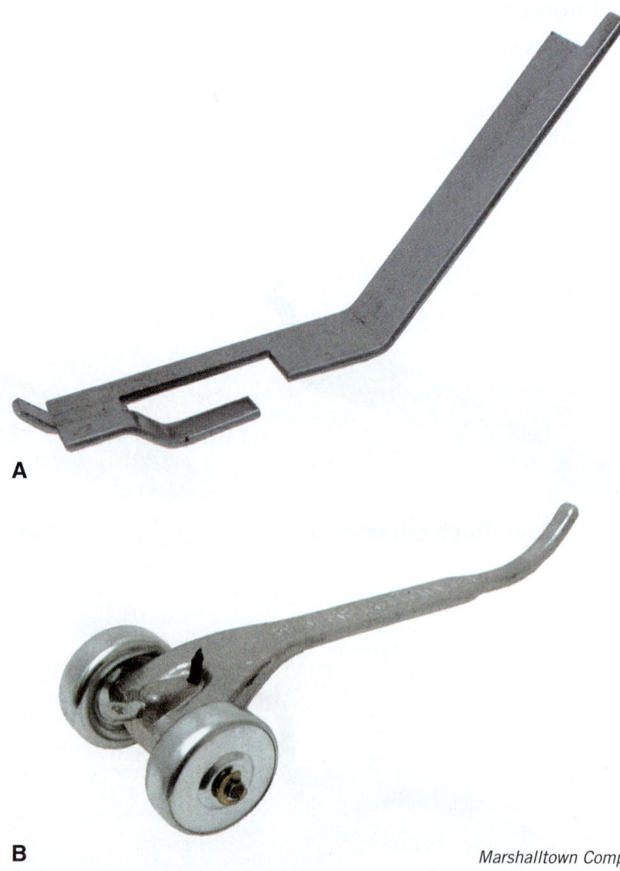

Marshalltown Company

Figure 3-6. Joint rakers produce a rake joint. A—This hand joint rake makes a 5/16″ deep cut. B—This joint raker skate wheel can be adjusted to cut different joint depths.

A

B Marshalltown Company

Figure 3-7. A—This mason's brick hammer has a soft grip handle. B—This mason's brick hammer has a wooden handle.

> **Safety Note**
> Brick hammers are not intended for striking hardened steel, such as chisels or steel stakes. Use an appropriate hammer intended for striking hardened metal. Always wear safety glasses when using hammers!

Framing hammers are used to pound nails, drive stakes, and build and dismantle formwork around driveways, sidewalks, and foundations. There are two main types of framing hammers, ***straight claw hammer*** and ***curved claw hammer***, and the heads vary in weight from 16 ounces to 28 ounces. See **Figure 3-8**. A straight claw allows the hammer to be used as a pry bar when dismantling material as well as pounding nails. A curve claw works well for pulling out and pounding nails.

A ***maul*** is a heavier hammer intended to strike wooden or steel stakes. See **Figure 3-9**. A maul has a striking area on both surfaces of the head. The head weight of the maul varies from 2 to 4 lb. Handles can be forged as part of the head or made of fiberglass or wood. Mauls are used for driving in either wood or steel stakes during the process of building forms for driveways, sidewalks, and foundations. Mauls can also be used to strike chisels when forming or shaping stone or other materials.

A B Timothy L. Andera

Figure 3-8. A—A straight claw hammer used for nailing, prying, and dismantling form material. B—A curved claw hammer used for nailing and pulling nails.

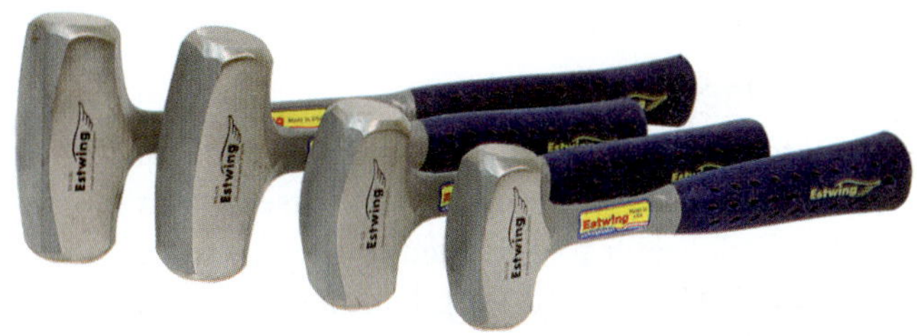

Marshalltown Company

Figure 3-9. Mauls are larger striking tools for stonework, framing, and formwork.

A ***sledgehammer*** is a larger hammer where both hands of the mason are needed to strike wooden or steel stakes. See **Figure 3-10**. The size of the sledgehammer is based on the weight of the head; the weight varies from 6 lb to 20 lb. The most common sledgehammer size is a 10 lb. Sledgehammers can be used for striking chisels, stone, or concrete. They can also be used in dismantling materials when demolishing structures.

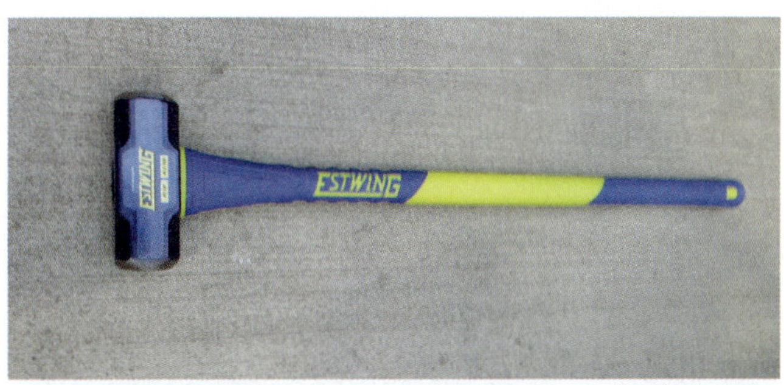

Timothy L. Andera

Figure 3-10. A sledgehammer with a fiberglass handle and a wider area next to the head to protect the handle when a hammer miss occurs. The head weight of this sledgehammer is 10 lb.

Mortar Tools

Masons, apprentices, and laborers use mortar tools as part of daily masonry activities. The larger tools, such as mortar boxes are supplied by the contractor. Smaller tools, such as a mortar board, may be owned by an employee if they have a preference when using specific tools.

Mortar Box

Masons use a ***mortar box*** to mix mortar or to hold it after it has been mixed. There are different types of mortar boxes. They can be made from metal or wood. Newer mortar boxes are made of extra high-density, UV-stabilized polyethylene, or similar materials. See **FIgure 3-11**. Contractors select this type of mortar box because it is lightweight, which aids in moving the box from one location to another on the jobsite. Another advantage is the composite mortar box is easier to clean.

Mortar Hoe

A ***mortar hoe*** is used to mix mortar in a mortar box or a mason's wheelbarrow. It is also used when mortar needs a little water to be added to the mixture.

Figure 3-12 shows a long-handled 60″ mortar hoe with straight sides. Hoes are also available in short handles and rounded sides to aid in mixing mortar in wheelbarrows or other containers that have tapered and rounded sides.

Mortar Shovel or Sand Shovel

Shovels are multi-use tools for anything from moving sand and gravel when preparing a base for a new driveway or sidewalk, to mixing mortar for a foundation. See **Figure 3-13**. In the mortar box, a mason will use a shovel to mix the materials together prior to adding water. Shovels are also used by the mason to load an electric mixer when preparing mortar for laying block or moving mortar from the wheelbarrow to the mortar board.

Timothy L. Andera

Figure 3-11. A polyethylene mortar box, which will not rust, warp, or take on moisture.

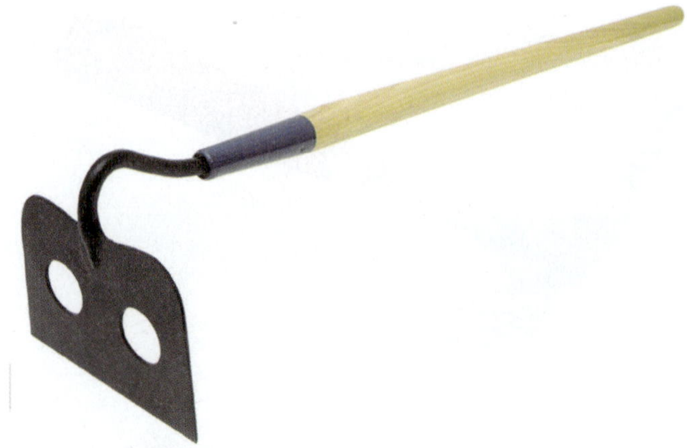

Marshalltown Company

Figure 3-12. The blade of the mortar hoe can have holes in it to allow the user to pull and push the mortar through the holes to reduce time in mixing the different components.

Timothy L. Andera

Figure 3-13. Shovels come in a variety of styles and shapes. This shovel has a wooden handle with a rubber tip on the end. Fiberglass handles are also available.

Mason's Tool Bag

Masons use ***tool bags*** to hold their hand tools and other items needed during the day at work. Mason's tool bags come in an assortment of sizes and shapes. See **Figure 3-14**. They can be made from canvas, leather, or other synthetic material such as Cordura nylon which is resistant to water and abrasions. Some bags have pockets on the inside and outside to store items such as pencils, safety glasses, and gloves.

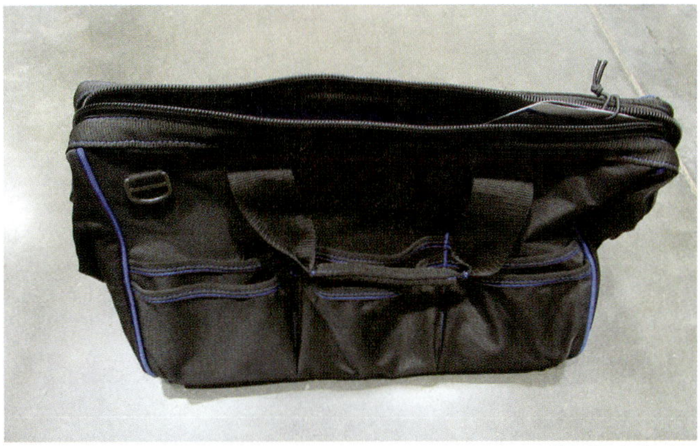

Timothy L. Andera

Figure 3-14. An example of a mason's tool bag with inside and outside pockets.

Mortar Bag

Masons use a ***mortar bag***, **Figure 3-15**, to aid in placing mortar into joints where old mortar has been removed because it was damaged or aged by weather conditions. The bag is filled with mortar and then squeezed to push mortar to the pointed front of the bag. Then the mortar bag tip is placed into the joint and mortar is forced through the tip into the mortar joint. Also, the bag can be filled with grout to fill hollow cells of masonry units during the construction process.

Mortar Stand and Mortar Boards

A ***mortar stand*** is used to support a mortar board and mortar placed on top of the board when laying brick or concrete block. **Figure 3-16** shows a mortar stand made from tubular steel. It stands 30″ tall and easily folds at the end of the day or when the job is done. The stand can be laid on its side to reduce the height when needed. It also has a chain between the legs to ensure the stand does not open past its limits. Masons can also use two concrete blocks as a support device for a mortar board and mortar.

The purpose of a ***mortar board*** is to hold mortar when a mason is laying bricks or concrete blocks. A mortar board can be made of plywood, plastic, metal, or a composite of materials. **Figure 3-16A** shows a mortar board made from impact and UV resistant polymer. The board has a slightly concaved surface to aid in keeping mortar and moisture centered on the board. This board is very rigid and water resistant.

Figure 3-16B shows a mortar board made of UV stabilized polyethylene. It is durable, and easy to clean. The board is very lightweight and dry mortar will not stick to it.

Marshalltown Company

Figure 3-15. Mortar or grout bags come in different sizes and tips. This bag has a plastic tip that can be cut to the desired diameter. More expensive models can be purchased with removable metal or plastic tips in different sizes.

Measuring and Marking Tools

The proper use and understanding of measuring and marking tools is critical for accurate results. Masons usually have two kinds of measuring tools—a folding rule and a retractable steel tape.

Mason's Rules

The 6′ folding rule or ***mason's rule*** is a standard tool used by masons. See **Figure 3-17**. On the front side is the standard US customary measurement. On the reverse side is a scale for laying brick or block. The back side has special markings representing course heights for various unit sizes and joint thicknesses. A 6′ folding rule sometimes comes with a 6″ sliding scale on the first section for inside measurements.

Measuring scales found on the back side of mason's rules include standard US customary, standard brick spacing, oversized brick spacing, and modular spacing. As an example of measurement, the numbers from 1 to 0 on the brick ruler represent the height of various brick and mortar joints when being laid to that scale.

For example, if the average measurement of bricks to be used is 2 1/16″ and the desired width of the mortar joint selected measures of 7/16″, then the total brick and joint measurement is 2 1/2″. While holding your thumbnail on the edge of the rule at 2 1/2″, turn the rule over in your hand, and the corresponding number on the back side of a standard brick rule would be the number *2* in black text. The *2* represents the scale to follow for this brick layout, which is "scale 2." The number in red on the same line tells the specific brick course.

A

B

Timothy L. Andera

Figure 3-16. A—A polymer mortar board on a steel mortar stand with a handle hole to aid in moving and storage. The size of the board is 24″ × 24″ and will hold up to 150 lb of material. B— A polyethylene board on a steel mortar stand with a lip around the perimeter to aid in keeping mortar and liquid from running over the edge.

Marshalltown Company

Figure 3-17. Rules with different measuring scales are used depending on the brick or unit size and the layout of the project. A—Standard US customary ruler. B—Oversize brick spacing ruler. C—Standard brick spacing ruler. D—Modular spacing ruler.

Since this is the first course, the red number in this instance would be *1*. To find the location on the rule for the top of the next brick course including a joint, move up the rule until "scale 2" in black is found with a corresponding red number *2*, which represents the second course. To find the top of course 3 and consecutive courses, again move up the rule to find the next occurrence of "scale 2" in black with the red course number. See **Figure 3-18**.

The type of rule used depends on the specific project being built and the layout requirements. Most masons have several types of rules for performing different masonry layouts or when working with different sizes of joints and brick. A well-constructed rule made of quality materials will have a long life and give dependable service.

Retractable Steel Tapes

Masons use a ***retractable steel tape*** to quickly measure long distances. Refer to **Figure 3-19**. Some retractable tapes have a US customary measurement on one edge and a masonry measurement on the opposite edge of the blade. Masonry tapes can be purchased in either brick, modular, or Canadian spacing. Standard retractable tapes vary in length from 10′ to 35′, but most masonry tapes are from 16′ to 25′ in length with a 1″ wide blade.

Tape Measures for Layout

Layout tapes come in different styles and lengths depending on project being constructed. Tapes are made of metal, cloth, and fiberglass. Tape lengths can be purchased from 50′ to 150′ or longer. The length will depend on the application and or the size of the project being laid out.

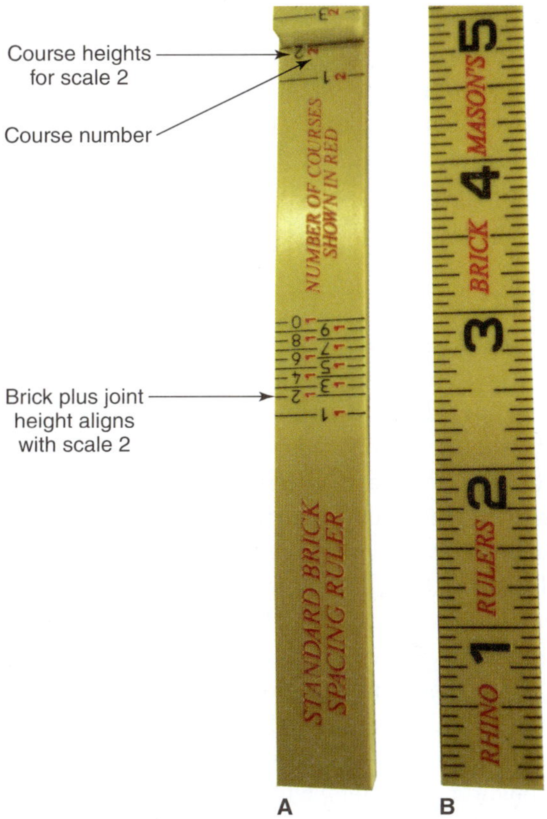

Timothy L. Andera

Figure 3-18. Mason's rule. A—Mason's standard brick rule measuring joint and brick that shows the mason's scale as number 2. B—Mason's rule measuring a masonry joint and one brick totaling 2 1/2″.

Marshalltown Company

Figure 3-19. Retractable steel tapes are used by masons to measure longer distances such as when laying out concrete forms.

Chalk Box and Chalk

Figure 3-20 shows a ***chalk box and chalk***. A chalk box is used to mark lines on a concrete footing or similar areas to identify where the foundation walls or other structures are to be located. A chalk container is used to refill the chalk box once the chalk level gets low in the storage area of the box. Chalk can be purchased in a number of colors. Blue is a nonpermanent color, where black and red are considered permanent.

Plumb Bob and Line

A ***plumb bob*** is used in combination with a mason's line to transfer locations from batter board lines to footings and foundations. See **Figure 3-21**. Batter

Timothy L. Andera

Figure 3-20. This chalk box has a 3 to 1 gear ratio to aid in quick retrieval of the line. Normal boxes have a 1 to 1 gear ratio. The chalk shown is a nonpermanent blue type.

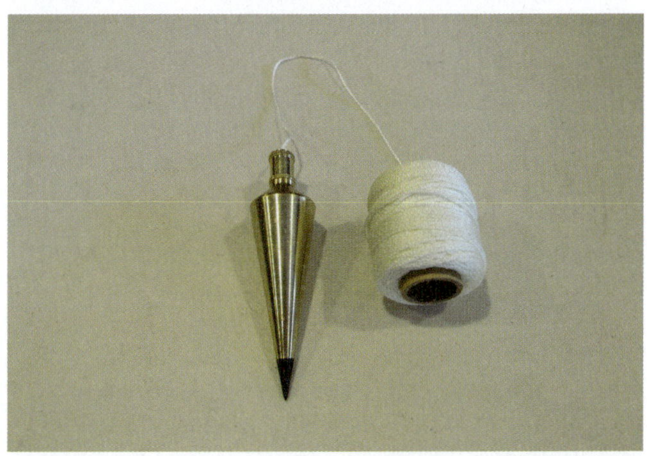

Timothy L. Andera

Figure 3-21. Plumb bob and mason's line are used to transfer building lines from the elevation at the batter boards to footings and foundation walls. This plumb bob is made of brass with a steel tip for extra strength.

boards are laid out to identify corners of a building. Batter boards are further discussed in Chapter 12, **Laying Brick**. Where mason's lines intersect identifies the outside corners of the building. At the intersection of the two lines, a mason's line and plumb bob are attached and lowered down into the excavated area to transfer the outside corners of the building to locate the footing and foundation walls. See **Figure 3-22**.

Steel Squares

A ***steel square*** is made in the form of a right angle. See **Figure 3-23**. The square has two arms (the body and the tongue) that form a 90° angle or *right angle*. The body of the square is 2″ wide and 24″ long. The tongue is 1 1/2″ wide and 16″ long. The outer and inner edges of the square are divided in inches and fractions of an inch for measuring purposes. A steel square has a multitude of uses, including laying out a 90° corner, drawing straight lines, cutting a straight line, or measuring distances less than 24″.

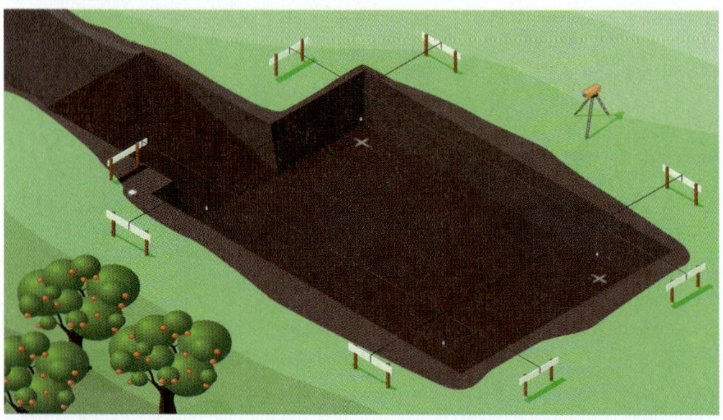

Viniarskyi/Shutterstock.com

Figure 3-22. A graphical representation of an excavation where batter boards identify future wall locations. A plumb bob and mason line are being used to locate the footings and foundation walls.

Timothy L. Andera

Figure 3-23. A steel square is used by masons for a number of applications. Measuring can be performed on the inside and outside edges of the square.

Levels and Cases

A ***mason's level***, or ***plumb rule***, is a level that is used to check a wall (or other structures) to ensure it is built vertical and level. It is the most delicate piece of equipment that a mason uses. See **Figure 3-24**. Several spirit bubble vials are built into the level to permit plumb (vertical) and level (horizontal) use. The vials are adjusted precisely and are generally embedded in a plaster of paris setting.

Masonry levels are usually 24″ to 48″ in length. A good level is lightweight and absolutely straight. It may be made from wood, metal, or a combination of the two. **Figure 3-25** shows a 2′ level, which is used in locations where a shorter level is needed. This level is made of bamboo wood with metal bands to protect the edges, and it also has a rubber bumper on each end to aid in protection.

Stanley Goldblatt

Figure 3-24. Common type of 4′ wood level used in masonry work.

Johnson Level and Tool Manufacturing Company, Inc.

Figure 3-25. A 2′ wood level with three sets of vials used in masonry work.

Level cases are used to protect delicate levels when not in use or being transported from one jobsite to another. Level cases come in different style and sizes. See **Figure 3-26**.

Electronic Levels

Electronic levels can be purchased in 2′ and 4′ lengths. Other lengths may be available depending on the manufacturer. An electronic level usually has a reading port in the middle of the level. The measurement of angle that is displayed in the reading port can be adjusted to set in different readings—degrees, percent, inches/feet/fractions, inches/feet/decimal, or metrics. Electronic levels are extremely accurate. Certain models have a lighted display for use when conditions warrant.

Timothy L. Andera

Figure 3-26. The black case for a 2′ level has a snap lid. The yellow case is for a 4′ level and has a twist top that locks in place. Both are made of a plastic composite that is sturdy and resistant to damage.

Rotary Laser Levels

Rotary laser levels are extremely accurate and timesaving construction tools. See **Figure 3-27**. Applications of laser levels in the masonry field include site preparation for grading, construction of batter board and stake layout, and concrete leveling. Rotary laser levels can also be used to shoot lines when constructing forms for footings and walls.

A rotary laser produces a 360° laser beam that rotates around the perimeter of the unit. Models produce a laser beam in either red or green light. Some models have features of self-leveling and a remote control to turn the unit on and off. The units operate on single or multiple batteries. Some models use a removable rechargeable power pack that can vary from 12 to 18 volts.

Safety Note

Always wear eye protection when using a laser level. Most laser units include eye protection in the kit. Always read the manufacturer's instructions before operating a laser. Refer to the manufacturer's instruction manual to determine necessary eye protection for others in the area.

Johnson Level and Tool Manufacturing Company, Inc.

Figure 3-27. This rotary laser level was set up on a construction site to aid in positioning footing forms for a residential home. The laser level was used to ensure the top of the footings are level and at a specific grade elevation.

Altimeter Levels

Altimeter levels are precision instruments that measure differences in elevation heights. These types of levels can be adjusted based on feet, inches, and fractions or decimals of a foot or can be set to metric measurements. Altimeter levels do not need line of sight as required by laser levels. An altimeter level has a base station where a coiled hose is stored and then dispensed as needed to record elevation variations. The hose can be from 50′ to 100′ depending on the manufacturer and unit selected. Attached to the end of the hose is a readout unit that registers changes in elevation. Once the base station is positioned in a central location where elevations are to be taken, the station is then calibrated to that precise spot. After calibration, the readout unit is removed from the station and the appropriate length of hose is reeled out to perform recordings. These recordings will identify elevation changes from the base station to the readout unit.

One type of altimeter level is a ***Smart Level***. It can be used for 3D terrain mapping and image mapping, then the data can be uploaded to a computer-aided drafting (CAD) program. A lighted display can be adjusted for brightness depending on the situation. See **Figure 3-28**.

ZipLevel is another type of altimeter level that measures change in elevation between the base unit and readout unit. See **Figure 3-29**. ZipLevel offers different units depending on the need of the masonry contractor. Units come in either 75′ or 100′ of hose to offer the client price options when selecting a specific model. The advantage of the longer hose length allows the mason to operate around the base a distance of 100′. This means with correct placement of the base unit, a mason can cover an area of 200′. The benefit of using an altimeter level is that a mason can work alone to operate and record elevation data when designing a masonry project.

Pave Tool Innovators

Figure 3-28. Smart Level is an altimeter device that records different elevations to aid the mason in determining grades when laying out driveways and sidewalks.

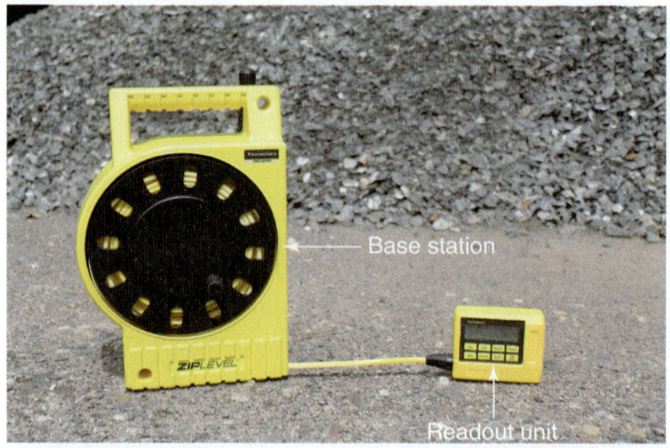

Pave Tool Innovators

Figure 3-29. ZipLevel is an altimeter level that allows a mason to work alone and record elevations accurately when designing a masonry project.

Chisels and Utility Knife

Masons use various chisels for different kinds of materials. A ***brick set*** is used to cut brick to exact dimensions. See **Figure 3-30A**. ***Mason's chisels*** are designed to cut brick, block, and stone. A mason's chisel is available in a number of sizes, shapes, and weights. See **Figure 3-30B**. There are a variety of mason's chisel types designed for scoring or splitting stone. See **Figure 3-30C**.

All chisels must be kept sharp and straight if they are to work properly. Burrs must be removed from the top end to prevent injury during use. Quality chisels are made from tempered tool steel. They are very hard and must be sharpened by grinding.

A ***utility knife*** is a type of knife with a blade that can be opened to reveal the blade and closed to conceal it inside the handle. See **Figure 3-31**. Utility knives have removable blades and some models have extra blades in the handle. Some allow a new blade to be installed without dismantling the knife. Read the manufacturer's instructions and follow procedures to remove and install a new blade. The mason uses the utility knife for several projects, such as cutting mason's line or polyethylene film when installing formwork.

> **Safety Note**
>
> Wear safety glasses when you are using a chisel. When sharpening chisels or removing burrs, always wear safety glasses and a face shield. When striking chisels, use an appropriate hammer; do not use a brick hammer.
>
> When working with a utility knife be extra careful, the blade is razor sharp and will cut skin just as easily as it will cut paper or plastic.

A *Marshalltown Company*

B *Marshalltown Company*

C *Stanley Goldblatt*

Figure 3-30. Types of chisels. A—A brick set is a type of chisel used to cut brick to specified length. The chisel is 4″ wide. B—A mason's chisel is used for cutting stone and concrete block. This chisel is 1 3/4″ wide. C— Stone mason's chisels are used to split stone.

Line and Holders

A ***mason's line*** is a strong nylon cord that is used to keep each course (row of bricks or concrete block) level and the wall "true" and "out-of-wind" (no bulges or hollows). See **Figure 3-32A**. When performing long layouts, it is advantageous to have a mason's line with a special winder that allows for feeding out the line without getting tangled and then to be rewound when the line is not needed. See **Figure 3-32B**.

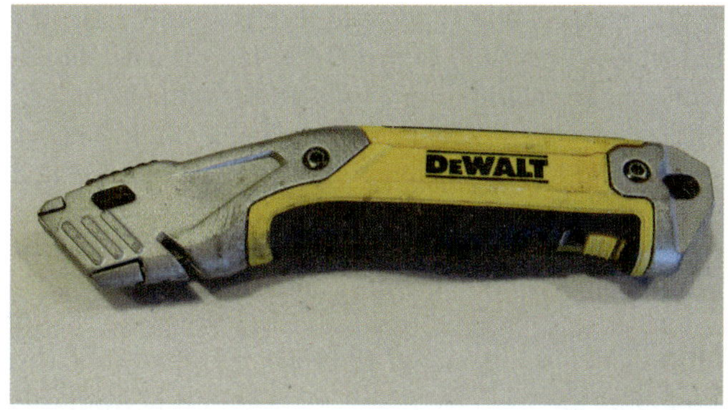

Timothy L. Andera

Figure 3-31. This utility knife has an adjustable blade with extra blades inside the handle. A mason can remove an old blade and insert a new one with the touch of the front button. It also has a slot that can be used to cut mason line by just slipping the line into the underside slot.

A *Stanley Goldblatt*

B *Timothy L. Andera*

Figure 3-32. A—Nylon mason's line on plastic winders in three highly visible colors. B— Mason's line and holder allows for quick dispensing and retrieval. When the line becomes damaged, the holder allows for a new line spool to be installed.

The mason's line is secured at either end with line holders or line pins, **Figure 3-33**. Line is available in strengths from 100 lb to 350 lb and lengths from 100′ to 1000′. Various colors are available, including white, yellow, and green. Mason's line comes in braided and unbraided line. The unbraided line is less expensive but has a tendency to unravel when one of the strands is cut or frayed.

Line Pins

Line pins are used to secure a mason's line from one outside masonry corner to the other corner of a building. See **Figure 3-34**. Line pins are hammered into a mortar joint or driven into the ground for a location to secure the mason's line. This is usually performed at the masonry corner leads so a mason's line can be strung between corners to aid in laying brick or concrete block between the outside corners or corner leads.

Marshalltown Company

Figure 3-34. Line pins are made of hardened steel. Notches are cut into the pins to aid in attaching mason's line.

Line Blocks and Twigs

Line blocks are used in combination with a mason's line and line pins. After the line is attached and stretched between the pins, the line is fed into the slot of each line block. The blocks are then hooked on the outside of each corner to create a line stretched from corner to corner. The line blocks are hooked on each corner to aid the mason in laying bricks or concrete blocks at the correct height. See **Figure 3-35**.

Twigs are used to hold the mason's line at the appropriate location and height when laying masonry units between the corners. The line is fed between the legs of the twigs (metal twig) or snapped into the connection on the twigs (plastic twig). Two twigs are used in combination, one at each corner lead. Each twig is set on top of the masonry course to string a line between corners to give the mason a guideline to follow when laying masonry units between the two corner leads. A brick or similar weight is set on the top of the twig to ensure the twig and line do not move from the correct location on the lead.

Stanley Goldblatt

Figure 3-33. A pair of adjustable line holders and braided nylon line. The line is braided to last longer.

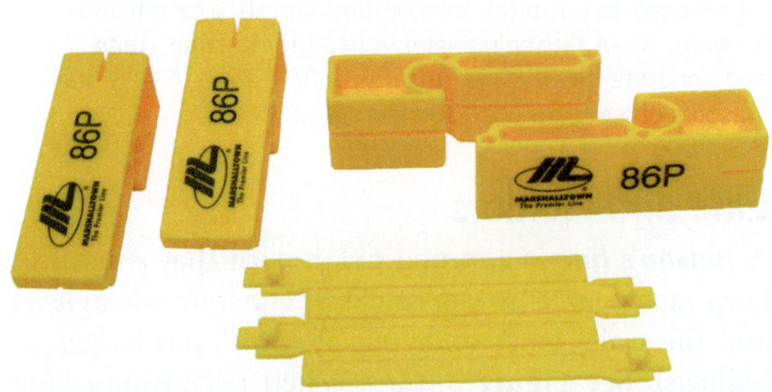

Marshalltown Company

Figure 3-35. Line blocks are made of plastic with grooves to receive the mason line. Twigs are made of plastic and can be purchased as part of a package of line blocks and twigs. Other types are made of metal and can be purchased separately.

Brushes and Sponges

Brushes and sponges are produced in a variety of shapes, sizes, and textures, **Figure 3-36**. They are used for the following purposes:

- To remove mortar from the masonry units after the wall has been constructed
- To wash and scrub brick surfaces with muriatic acid
- To add water to surfaces for cleaning
- For general cleaning of brick, stone, and concrete surfaces
- To place a finish texture on the surface of fresh masonry and concrete surfaces

Brick Tongs and Block Clamps

Brick tongs are used to carry brick. They are designed in such a way that they do not chip the brick. Brick tongs are adjustable for various size units. See **Figure 3-37**.

Block clamps are used to carry block and hardscape block. Block clamps are adjustable and rated based upon the level they will support. **Figure 3-38** shows BL-100 clamps being used by a mason. A mason can adjust the BL-100 clamp to fit different sizes of block being used on the construction site. **Figure 3-39A** shows a BL-180 clamp is being used by a mason to lift a block. **Figure 3-39B** is a machine carrying the block to a location directly above the wall. A mini-crane, small backhoe, or forklift can be used to lift and move blocks while the mason positions them in the wall.

Safety Note

In construction, lifting is a daily requirement. For example, a 8″ × 16″ hollow two-cell concrete block can weigh 38 lb to 40 lb. The same size block but solid, can weigh up to 54 lb. Hardscape blocks, which vary in length and height, can weigh 25 lb to 80 lb per unit. It is important to lift heavy objects correctly as to not injure your back or other parts of your body. The Occupational Safety and Health Administration (OSHA) administers rules and regulations regarding safety and health for construction and general industry activities. For safe lifting, OSHA recommends:

- Standing as close to the load as possible
- Planting your feet shoulder-width apart with one foot slightly ahead of the other
- Bending at the hips and knees only until you're deep in a squatting position
- Keeping your head up and straight with your shoulders back, as to keep your back straight
- Holding the load close to your body at the waist height
- Engaging your core muscles as you push against the ground and straighten your legs

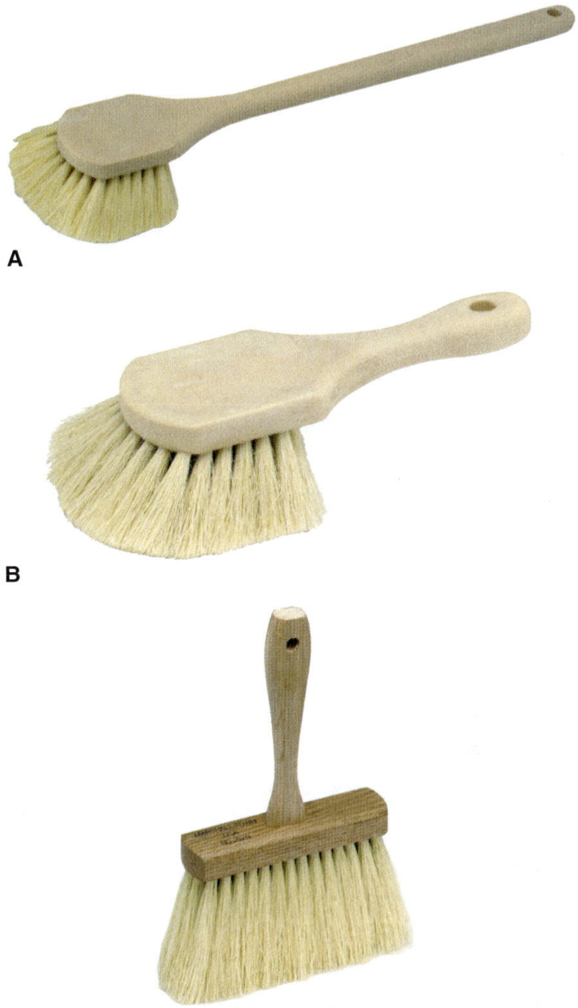

A

B

C

Marshalltown Company

Figure 3-36. Masonry brushes have bristles with various degrees of stiffness or softness, depending on their purpose. The brushes can be used for cleaning brick or wetting down masonry materials. A—Long-handled acid brush for general cleaning of brick and other surfaces. This brush works well with a plastic bucket containing acid water solution for cleaning walls. B—Short-handled acid brush for cleaning brick and other products. C—Masonry brush for general cleaning, wetting surfaces, or other uses.

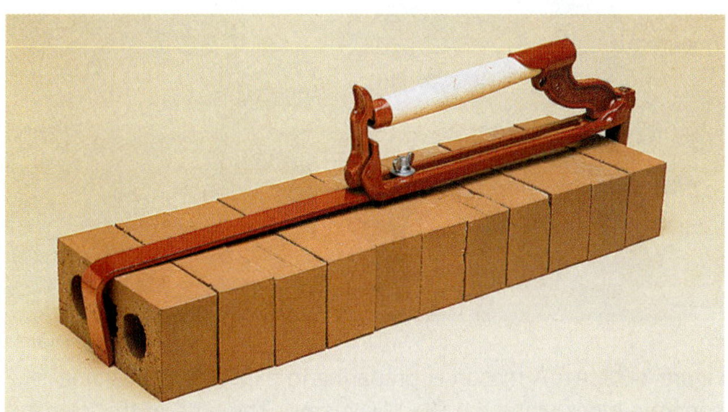

Stanley Goldblatt

Figure 3-37. Brick tongs make carrying brick easier.

Pave Tool Innovators

Figure 3-38. A mason carrying two BL-100 clamps. The advantage of the clamp is it allows the worker to carry multiple blocks around uneven surfaces with minimal effort. The limit of this clamp is 100 lb.

A

B

Pave Tool Innovators

Figure 3-39. A—A mason is preparing to pick up a hardscape concrete block using the BL-180 clamp. The limit of this clamp is 180 lb. B—A mason is setting a hardscape block in a retaining wall using the BL-180 clamp and mini-crane to aid in lifting.

Scaffolding

Few masonry projects are at a height where they can be reached from the ground. For this reason, much of a mason's work is done on scaffolding. Scaffolding makes masons' work easier by providing a place for materials and tools at a convenient height.

Several types of scaffolding are used by masons—tubular, tower, and swing stage scaffolding. **Figure 3-40** shows the tubular type.

Safety Note

Wear appropriate safety gear when working on or near scaffolding. This includes a hard hat, safety glasses, and safety footwear.

A scaffold supports workers and masonry materials. See **Figure 3-41**. When erecting scaffolding, a competent person should oversee the construction process so a level structure is maintained (kept plumb) and workers are safe. As the project is constructed, the scaffold is assembled to meet the needs of the masons.

During the construction of the scaffold, ladders are used so masons can access the work area. Certain scaffolds have built-in ladders when they are manufactured. As the height of the building increases, so too will the scaffolding structure.

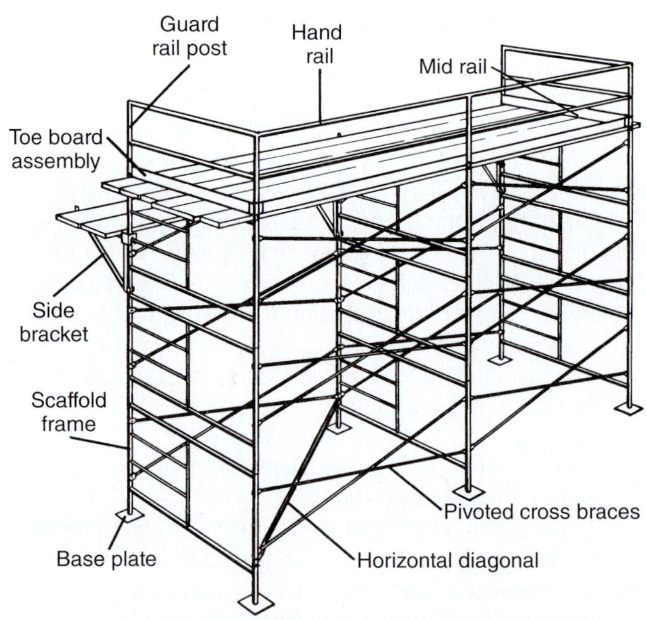

Goodheart-Willcox Publisher

Figure 3-40. Tubular scaffolding is frequently used by masons to reach high places. Side brackets hold planking to provide a deck for workers. Materials are never placed here but on the upper deck.

Daniel Kourey/iStock/Thinkstock

Figure 3-41. Scaffolding is erected to aid masons in constructing masonry walls.

Safety Note

Training is required prior to working on scaffolding to a level that meets Federal Safety Standards. OSHA **Standard 1926.451(g)(2)** states:

"The employer shall have a competent person determine the feasibility and safety of providing fall protection for employees erecting or dismantling supported scaffolds." For more detailed information on construction in general and specifically on scaffolding, see OSHA's 1926 Construction Code of Federal Regulations (CFR) Standards manual.

The scaffold is positioned approximately 22″ to 26″ away from the finished exterior wall surface. This distance depends on the size of the side brackets attached to the scaffold. Once the side brackets are in position, planking is placed on them. Side brackets and planking create a surface for masons to stand on when performing their work. As blocks or bricks are laid on the building's wall structure, the side brackets and planks are moved up the side of the scaffold. In this way, the masons' work area is kept directly in front of them to improve production.

Safety Note

When placing materials on the scaffold, position only what will be used that day. Severe accidents can occur if scaffolding is overloaded with materials.

Mortar Mixers

Mortar is usually prepared in a mechanical mixer, **Figure 3-42**. ***Mortar mixers*** may be electric or gasoline powered, depending on contractor preference and availability of electricity at the construction site. Mixers are produced in several sizes, but a typical size mixes about 4 cubic feet (cu ft) of mortar at a time. For best results, the dry materials are mixed first, and then the water is added.

Do not allow mortar to harden in the mixer. Hardening prevents proper mixing of ingredients because the metal surfaces become clogged with mortar.

Marshalltown Company

Figure 3-42. A mechanical mortar mixer can be driven by either a gasoline engine or an electric motor.

Masonry Saws

Cutting masonry units with a masonry saw is faster and more accurate than using a mason's chisel and hammer. A typical ***masonry saw*** is shown in **Figure 3-43**. Masonry saw blades vary from 6″ to 24″ in diameter and are about 1/8″ thick. Blades are made of very hard material, such as silicon carbide or industrial diamond.

Most masonry saws can saw materials dry or wet. Water is used to cool the blade and eliminate dust. Water is pumped from the tray below the movable table where brick and block are positioned prior to cutting. The water flows over the top of the blade and returns back into the water tray to be pumped again.

MK Diamond Products, Inc.

Figure 3-43. Masonry saw used for cutting brick and concrete block.

Safety Note

A power masonry saw is a dangerous piece of equipment to operate. Safety glasses or goggles are required. A face shield and protective gloves should also be worn to protect from flying chips. Ear protection is mandatory because of the amount of noise created by sawing brick or block. When sawing materials dry, wear a respirator to protect from silica dust created by sawing.

3.2 Site Preparation Tools and Equipment

Compaction of the subgrade is a very important step that needs to occur prior to concrete being ordered for delivery. If the subgrade is not compacted sufficiently, the overlying concrete driveway, sidewalk, or other projects will crack or settle over time because the compaction process was not performed correctly. The following are examples of tools and equipment to aid masonry workers in performing an important step in site preparation.

Tampers, Packers, and Compactors

Soil ***hand tampers*** are used to compact aggregate, sand, soil, or other materials during the time of backfilling around a foundation or structure. Hand tampers are used also in preparation for concrete to be placed in forms for sidewalks, patios, concrete floors, or a number of other applications. See **Figure 3-44**.

Figure 3-45 shows a ***tamping rammer compactor***. It is used to compact materials, similar to a hand tamper, but will perform the job more quickly. The device has a hammering effect when compacting the material. The compactor is smaller than other packing devices, which will allow it to get into tighter spaces like backfilling around a foundation.

In a ***vibratory plate compactor***, **Figure 3-46**, the bottom plate vibrates to compact the material with the weight of the machine aiding in the compaction process. The primary use of this compactor is to prep subbases for footings, driveways, sidewalks, and basement floors.

Material Handling Equipment

Wheelbarrows come in many shapes, sizes, and wheel configurations. **Figure 3-47** shows a wheelbarrow that has wooden handles with a metal frame. It has one wheel attached to the frame and arms. This device is

Marshalltown Company

Figure 3-44. Hand tamper has a cast iron base to add weight and strength to the tool. The handle is made of wood that can be replaced when needed. The base of the hand tamper is 8″ × 8″ but other sizes can be purchased.

Allen Engineering Corporation

Figure 3-45. Tampering rammer compactor is powered by a 3 hp gasoline engine. The rammer hammers blows to the surface being compacted. It operates at 650 to 700 blows per minute (BPM).

Allen Engineering Corporation

Figure 3-46. Vibratory plate compactor operates on a 3.5 hp. gasoline engine. This model vibrates at 5540 vibrations per minute (VPM) with a plate size of approximately 20″ × 14″.

Timothy L. Andera

Figure 3-47. This wheelbarrow has a 7 cu ft metal tray (bin) capacity. The wheels are air-filled. The tray has been powder coated, similar to paint but will withstand construction practices better than normal paint.

used to move sand, gravel, soil, mortar, concrete or a number of other products around the jobsite. Wheelbarrows can also have two wheels to make it easier to maneuver around an unlevel jobsite.

Figure 3-48 shows a ***track-driven power buggy*** that has a capacity of carrying 14 cu ft of material. Similar to a wheelbarrow, this device is powered by a gasoline engine that can carry two to three times the amount a wheelbarrow can. Buggies are available with wheels or tracks. Buggies have the ability to lift and dump materials; more advanced machines can lift and turn the bucket to dump the load more precisely. The operator stands on a platform that is attached to the back of the buggy. It is easier for a track buggy to operate on surfaces that are wet or when carrying heavy loads up steep hills. Buggies have made it easier for construction workers to transport materials around the jobsite.

Allen Engineering Corporation

Figure 3-48. Model AT 14S track buggy has a steel bucket that can be lifted and turned to dump the contents. This buggy has a 20 hp gasoline engine. The work speed of the buggy is 4 mph and 6 mph travel speed. The maximum payload of the AT 14S is 1540 lb.

3.3 Concrete Masonry Tools and Equipment

There are numerous concrete masonry tools. Some are used for prep work prior to concrete being placed in forms. Other tools are designed for use in finishing horizontal concrete surfaces. These tools are known as *flatwork finishing tools.*

Screeds

A ***screed*** is a straightedge or strike-off rod made of any straight piece of wood or metal that has sufficient rigidity (stiffness). It is the first finishing tool used by a cement mason after the concrete is placed. A screed is used to strike off the concrete surface to the proper level. Screeds come in many sizes and materials based on the application and size of the job. Most common metal screeds are made of either aluminum alloy or magnesium alloy. They vary in size from 1″ × 4″ to 2″ × 5″ with lengths varying from 6′ to 24′, depending on the width of the concrete project being constructed. Aluminum and magnesium metal screeds will not rust or rot, and are warp-resistant. Metal screeds are lighter because they are hollow. **Figure 3-49** shows screeds with endcaps to keep concrete from entering the screed.

Safety Note

When using metal equipment, such as a screed or bull float handle, it is important to keep the tool away from overhead electrical wires. Severe harm or death may occur if the tool comes in contact with overhead energized wires.

A ***power screed***, shown in **Figure 3-50**, is used for most concrete work because it saves time. Through the vibration that it creates, the screed drives the surface rock into the mixture and brings a layer of liquid concrete, sometimes called *cream*, to the surface of the concrete mixture.

Marshalltown Company

Figure 3-49. Magnesium screeds come in different lengths, widths, and thicknesses depending on the application and the size of the project.

Marshalltown Company

Figure 3-50. A power screed is used to remove excess concrete and to vibrate the surface of freshly placed material. When screeding, the blade rides on top of the forms to level the concrete.

Rebar Benders, Cutters, and Wire Tie Devices

A ***rebar bender*** is used to bend reinforcement bars to fit concrete forms, flatwork, and vertical walls. See **Figure 3-51**. A quality bender will allow the mason to bend rebar quickly and easily. It is important to select an appropriate rebar bender based on the maximum bar size it is intended to bend.

A ***rebar cutter*** is used to cut reinforcement to the correct lengths when forming concrete work. See **Figure 3-52**. This tool makes quick work of cutting rebar and other material. Selection and type are based on the size of the rebar being cut for the project.

Marshalltown Company

Figure 3-51. A rebar bender is used to form rebar when installing reinforcement in driveways, floors, and forms.

Marshalltown Company

Figure 3-52. A rebar cutter is used to cut bars when preparing to install reinforcement in concrete driveways, floors, and forms where rebar is used as reinforcement.

A ***tie wire twister*** is used to twist or tie form wire around reinforcement. This operation occurs when reinforcement is used in preparing driveways, sidewalks, concrete floors, and wall forms prior to concrete being placed. See **Figure 3-53**.

Concrete Placement Tools or Crete Tools

Concrete placement tools or ***placer tools*** are used to position concrete either by pulling or pushing the mix once it has been placed in the form. It is important to use placer tools instead of a common garden rake with teeth. Rakes will separate larger rocks from moist concrete material and small aggregate in the concrete mix. Placer tools are also called ***pull crete tools*** and are available in different widths and sizes with handles available in wood and fiberglass. Crete tools have been designed to aid the mason in keeping the concrete mix together when moving it in the forms. There are two main types of pull crete tools, ones with a hook directly above the handle and ones void of the hook. See **Figure 3-54**.

Tampers

Concrete ***hand tampers*** are used to compact concrete into a dense mass. These tools are used on flatwork construction with low-slump concrete, which is

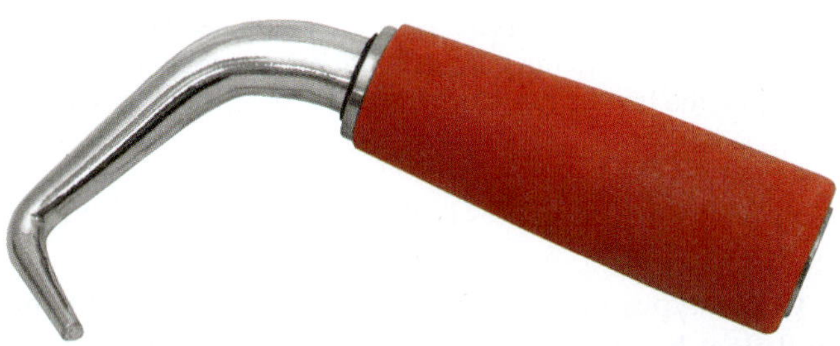

Marshalltown Company

Figure 3-53. Tie wire twisters secure wire around rebar or other reinforcement prior to placing concrete in framework.

A

B

Marshalltown Company

Figure 3-54. A—A pull crete tool with a hook used to aid the mason to ensure reinforcement does not move to the bottom of the concrete. B—A pull crete tool without a hook used to move concrete without segregating the mix in the form.

usually stiff and hard to work. A tamper also pushes large aggregate below the surface so it is not exposed, resulting in a smooth finish. See **Figure 3-55**.

Roller tampers serve the same purpose as hand tampers but cover larger surface areas. A roller tamper uses the same handle that attaches to a bull float. See **Figure 3-56**.

Darbies

A ***darby*** is a long, flat, rectangular piece of wood, aluminum, or magnesium used to remove any high or low spots left by the screed. Darbies are usually 30″ to 80″ long and from 3″ to 4″ wide with a handle on top. See **Figure 3-57**. A darby is used to float the surface of the concrete slab immediately after it has been screeded. It also helps to embed the course aggregate for later floating and troweling.

Bull Floats

A ***bull float*** is a large, flat, rectangular piece of wood, aluminum, or magnesium that does essentially the same job as the darby but enables the mason to float a much larger area. A bull float is usually 8″ wide and 42″ to 60″ long with a long handle.

Figure 3-58 shows a magnesium float 8″ × 48″ with a bracket that allows different types and lengths of

Marshalltown Company

Figure 3-55. A hand tamper aids in driving aggregate down into the concrete to bring the liquid concrete to the surface.

handles to be screwed into the float very easily. Another type of bracket can be purchased where the handle snaps into the float bracket rather than being threaded.

A bull float is particularly suited for outdoor use since the long handle (up to 16′) is difficult to use inside. **Figure 3-59A** shows a ***bull float extension handle*** made of fiberglass that screws into the bull float bracket. This type of handle is lighter than the normal metal handle because it is made of fiberglass. The handle is not

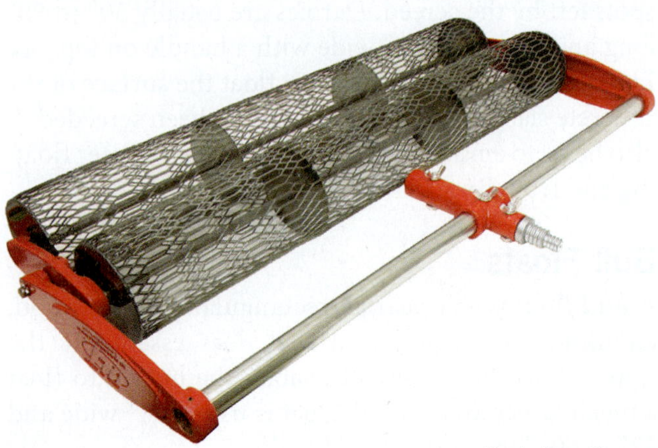

Marshalltown Company

Figure 3-56. A roller tamper serves the same purpose as a hand tamper but covers larger surface areas.

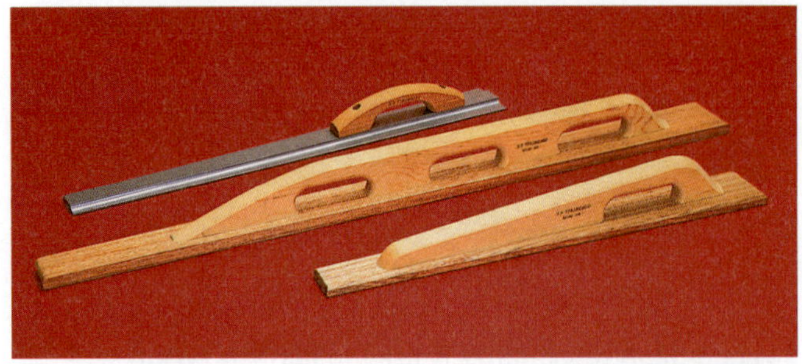

Stanley Goldblatt

Figure 3-57. Darbies are used to float the surface of concrete immediately after it has been screeded.

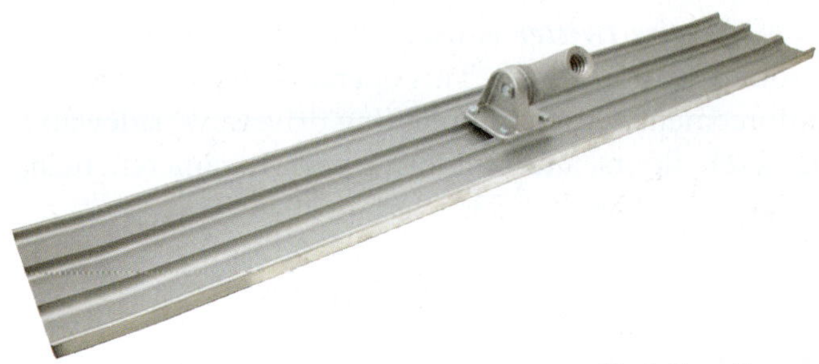

Marshalltown Company

Figure 3-58. A lightweight magnesium bull float. The ends of the float are turned up slightly to reduce lines when bull floating the fresh concrete.

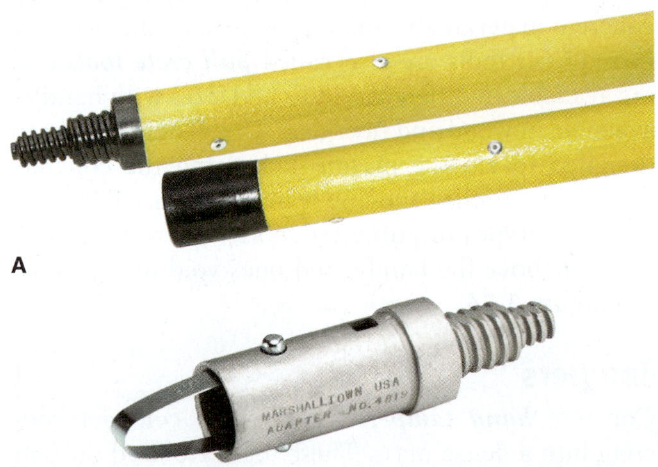

Marshalltown Company

Figure 3-59. Bull floats. A—Bull float extension handle is in two parts, one end screws into the bull float and the second part can be added to extend the handle when extra length is needed to float large projects. B—A bull float handle adapter. One end screws into the bull float bracket and the other end slips into the handle to allow for quick attachment and release.

conductive, which makes it safer near overhead electrical power sources.

Other types of handles instead of screwing into the bull float bracket will have a quick button attachment. All that is needed to release the handle from the bull float bracket is to push on the button. See **Figure 3-59B**.

Tampers, darbies, and bull floats all have a similar purpose, which is to aid in pushing exposed aggregate below the surface of the concrete and to bring a thin layer of liquid concrete (cream) to the top. This layer is troweled later in the process to create a smooth, hard finish.

Figure 3-60 shows a ***rattle stick***, a device that is attached between the bull float bracket and the handle. The rattle stick vibrates when turned on to aid in driving the aggregates below the surface in the mixture. The rattle stick also assists in bringing the concrete cream to the top to create a smooth surface on the concrete with less time spent.

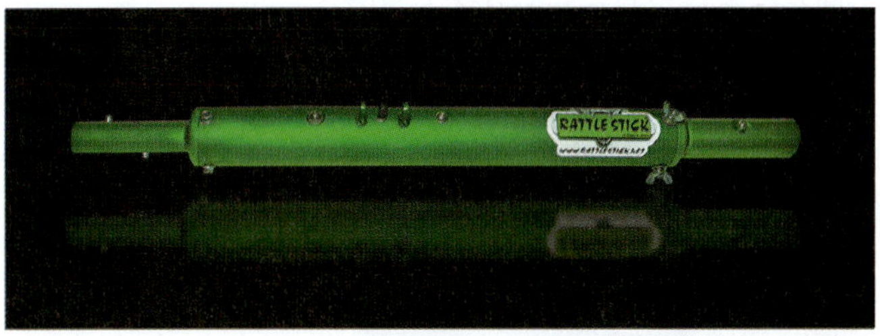

Rattle Stick ConcreteTools

Figure 3-60. The rattle stick uses a quick attachment to connect to the bull float and handle.

Edgers

Edgers are tools used to produce a radius on the edge of a slab. Several sizes and styles of edgers are available. See **Figure 3-61**. A typical length is 6″, and widths vary from 1 1/2″ to 4″. Edgers have radii from 1/8″ to 1 1/2″. The radius improves the appearance of the slab and reduces the risk of damage to the edge. The curved-end edger is very popular.

Groovers

Groovers are used to cut a joint partly through fresh concrete to control the location of any possible cracks. See **Figure 3-62**. The purpose of the joint is to weaken the concrete so cracks will occur at the bottom of the cut joint rather than appearing on the surface of the finished concrete. Groovers can be handheld or attached to a handle for larger surfaces. Groovers are

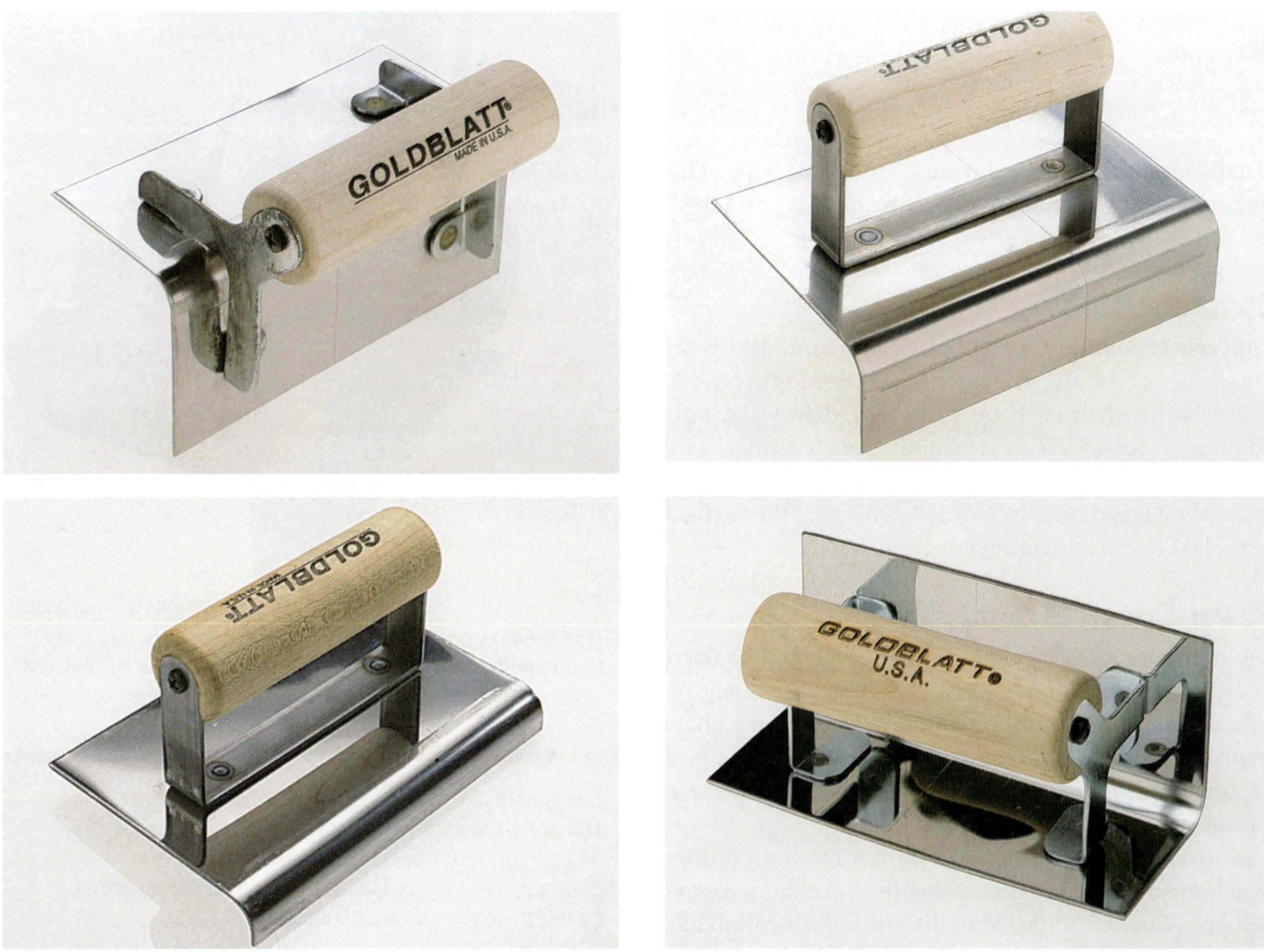

Stanley Goldblatt

Figure 3-61. An assortment of edgers used to produce a radius on edges of concrete slabs.

A

B

Marshalltown Company

Figure 3-62. Two popular styles of groovers. A—Groover size is 3″ × 6″ with a cutting depth of 1/2″. B—Groover size is 4 3/8″ × 8″ with a cutting depth of 1″. This groover is made to be attached to a handle to allow for grooving areas such as driveways.

usually about 6″ long and from 2″ to 4 1/2″ wide. The cutting edge or bits are available in depths from 3/8″ to 1 1/2″.

Concrete Brooms

Concrete brooms are used to place a textured finish on the surface of the still pliable concrete. See **Figure 3-63**. A broom finish is used on sidewalks, driveways, and other surfaces to create a slightly rough finish. The rough finish reduces the chance of slipping on the concrete when it is damp, wet, or icy. *Brooming* the surface is the last step in finishing concrete.

Power Concrete Joint Saw

A ***power concrete joint saw***, **Figure 3-64**, can be used to cut joints in concrete slabs. The machines may be either electric or gasoline powered. They have a shatterproof abrasive or diamond blade that produces a narrow joint. Joints are usually cut when the concrete has hardened four to twelve hours.

Most concrete saws have an option for wet cutting, which greatly reduces the sawing dust problem. A garden hose can be connected to the saw if the location has a water source. Using water also aids in cooling the saw blade, which lengthens its life.

Marshalltown Company

Figure 3-63. A concrete broom has soft bristles so it does not cut into freshly placed concrete. This broom is 24″ wide, but brooms range in size from 18″ to 48″.

Marshalltown Company

Figure 3-64. A power concrete joint saw used for large jobs. This model has an 11 hp motor and cuts to a depth of 4 5/8″.

Safety Note

The concrete joint saw must be handled carefully to avoid injury from flying particles. When operating a concrete saw, safety glasses and ear protection must be worn. When sawing concrete dry, a respirator or approved dust mask must be worn because the dust that is created contains silica, which is a health hazard.

Hand and Power Floats

Hand floats are used to prepare concrete surfaces for troweling. The floats are made from wood, aluminum, magnesium, cork, or molded rubber. They range in size from 10″ to 18″ long and 3 1/8″ to 4 1/2″ wide. See **Figure 3-65**.

Some floats are powered by electricity or gasoline engines. They have a rotating disk about 2′ in diameter that performs the same task as the hand float.

Vibrating Tools for Concrete

Concrete vibrators are tools that masonry contractors use to aid in removing air pockets or voids in concrete that has been placed in walls, floors, footings, or other types of concrete forms. The use of a concrete vibrator will aid in reducing voids or air pockets that can occur in or next to the concrete form. These pockets or voids are visible when the forms are removed from the concrete structure. A concrete vibrator moves or oscillates in such a way that it consolidates the concrete into a uniform mass.

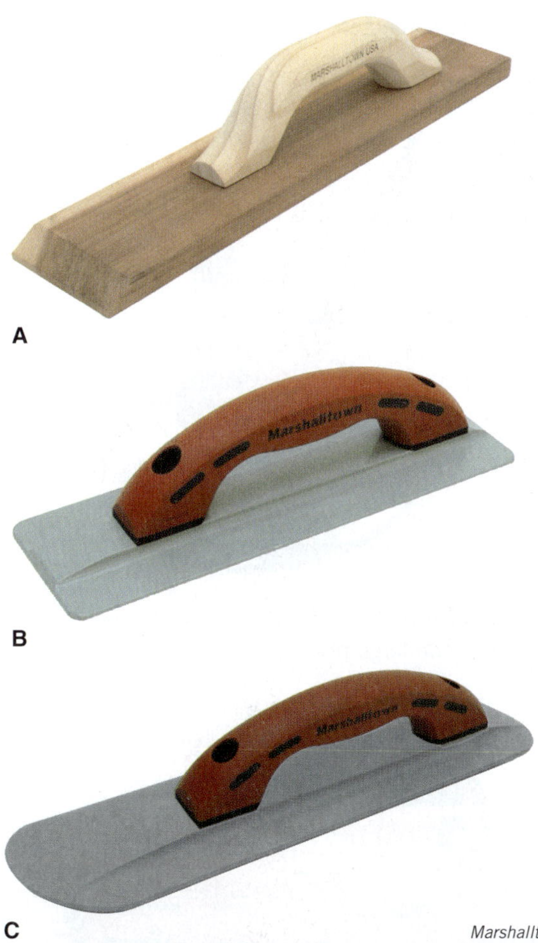

Marshalltown Company

Figure 3-65. Hand floats are made from various materials, including wood, aluminum, and cork. A—A 3 1/2″ × 16″ redwood float with a wood handle. B—A 3 3/4″ × 12″ magnesium float with a soft grip handle. C—A 3 1/8″ × 16″ magnesium float with a soft grip handle. This float has rounded edges, which help keep the trowel from gouging into the concrete that has been floated.

Trade Tip

Never over-vibrate the concrete, as it will segregate sand, gravel, and crushed stone in the mixture. Over-vibrating will also separate water from the concrete mix.

Operating a concrete vibrator is simple. Once the device is turned on, the head of the shaft is fed into the fresh concrete and allowed to move down through the concrete. After reaching the appropriate depth, the shaft is raised and moved to the next location to perform the operation again as needed until the concrete in the structure has adequately been vibrated.

Safety Note

When operating a concrete vibrator, it is important to wear ear protection because of the vibration noise the device emits. Average decibels may be 85 or higher depending on the device, the manufacturer, and the material being vibrated. Wear waterproof hand protection because the chemicals in the concrete can cause burns to the skin. Safety glasses or goggles and hand protection are necessary because the cement slurry can cause burns to the eyes and hands.

Handheld Concrete Vibrators

Figure 3-66A shows an electric flex shaft handheld concrete vibrator. Handheld concrete vibrators operate on 115 V of electrical power. The vibrating head in this model has a red rubber coating; the head can also be purchased in a steel version. Flexible shafts can be purchased in lengths from 2′ to 21′, depending on the need or application of the concrete project.

Figure 3-66B shows a high-frequency concrete vibrator. This model is a compact vibrator controlled by an all-electronic circuitry in the control box. The head diameter is 2.2″ and approximately 15″ in length. The shaft is approximately 16′ in length.

Backpack Concrete Vibrators

Figure 3-67 shows a backpack concrete vibrator. This unit is beneficial when the construction site has limited electrical power or access to electrical power is not available at the construction site. The concrete vibrator's power source is a 2.5 hp gasoline engine. Shafts can be purchased in lengths from 2′ to 21′. Heads are available in rubber or steel.

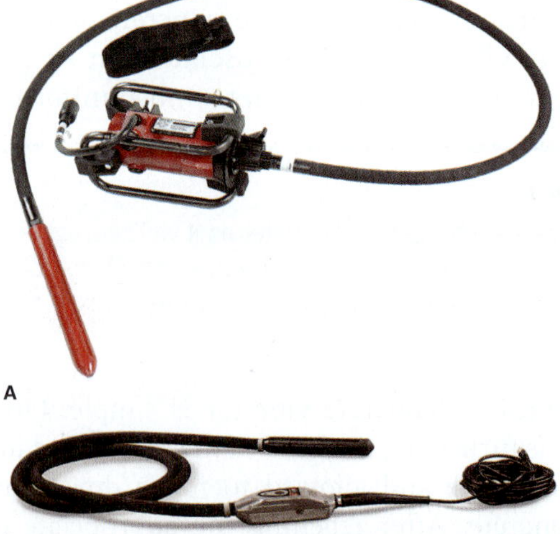

A

B

Allen Engineering Corporation

Figure 3-66. A—This device has built-in handles to make it easier to carry and operate when moving the head into the concrete and around the jobsite. B—This high-frequency concrete vibrator has an extra-long power cord of 49′ to reach most locations on a concrete project.

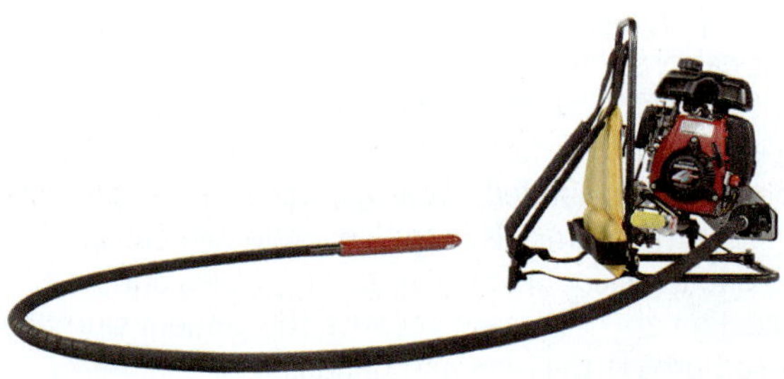

Allen Engineering Corporation

Figure 3-67. The backpack concrete vibrator oscillates from 11,000 to 12,000 vibrations per minute.

Hand and Power Trowels

A concrete mason's hand finishing trowel is the last tool used in the finishing process of a concrete slab. These trowels range in size from 10″ to 20″ long and 3″ to 4 3/4″ wide. See **Figure 3-68**. The first troweling of a slab is generally performed with a wide trowel that is 16″ to 20″ long. The last few troweling operations are usually done with a "fanning" trowel, which is 14″ to 16″ long and 3″ to 4″ wide.

Power trowels have three or four rotating steel trowel blades. Sizes vary from 36″ to 48″ in diameter. See **Figure 3-69**. They may be powered by electricity or a gasoline engine. The purpose of troweling is to give the surface a smooth, dense finish.

When large concrete projects are to be troweled, a riding trowel can be used. **Figure 3-70** displays a riding towel that has two powered rotating blade systems with a diameter of 46″ each. The panning path width of the machine unit is 101.5″, or 8′-5 1/2″. The trowel is controlled by a hydraulic system that runs on a diesel engine.

Safety Note

A power masonry trowel can be dangerous. Fuel-powered engines give off carbon monoxide fumes during the running process. The hazardous fumes can cause serious health conditions or death. Carbon monoxide sickness can occur when gas-powered equipment is operated inside buildings or in confined spaces where fresh air is limited.

Stanley Goldblatt

Figure 3-68. Mason's concrete finishing trowel.

Marshalltown Company

Figure 3-69. Power trowels have rotating steel blades. These machines are used on large projects.

Allen Engineering Corporation

Figure 3-70. This riding trowel is operated by hydraulic steering, which aids the operator in maneuvering the machine across the concrete surface.

Chapter Review

Summary

- When using masonry equipment, make sure it is being used for the correct purpose and you are properly trained.
- Before purchasing equipment such as a trowel, jointer, or hammer, consider the weight, size, materials, construction, and type of job.
- A brick hammer is used to drive nails and break or chip masonry materials, not for striking hardened steel, such as chisels or steel stakes.
- Larger mortar tools such as mortar boxes are supplied by the contractor. While smaller tools, such as a mortar board, may be owned by the employee if they have a specific preference.
- A mason's level, or plumb rule, is used to check a wall or other structures to ensure they are vertical and level.
- Tamping, packing, and compacting are all necessary to prepare a site for masonry work.
- A screed is used after concrete is placed to strike off or screed the concrete surface to the proper level.
- A bull float does essentially the same job as a darby but allows the mason to float a much larger area.
- Edgers are tools used to produce a radius on the edge of a slab and to push stones below the surface.
- Groovers are used to cut a joint partly through fresh concrete to control the location of any possible cracks.
- Hand floats are used to prepare a concrete surface for troweling.
- Concrete vibrators are used to aid in removing air pockets or voids in concrete that has already been placed.

Know and Understand

Answer the following questions using the information provided in this chapter.

1. A _____ is used by masons to finish the surface of mortar joints.
   A. margin trowel
   B. jointer
   C. gauging trowel
   D. joint raker
2. What is a type of tool used to remove mortar from a joint?
   A. Tuck pointer
   B. Brick jointer
   C. Sled runner
   D. Joint raker
3. The measuring tool used most frequently by the mason is the _____.
   A. tape measure for layout
   B. retractable steel tape
   C. folding rule
   D. steel square
4. What tool is used to mark long lines on footings or concrete for identifying locations of walls or other structures?
   A. Retractable steel tape
   B. Folding rule
   C. Sled runner
   D. Chalk box
5. A mason uses a _____ to check if a wall is plumb or level.
   A. brick jointer
   B. mason's level
   C. sled runner
   D. folding rule
6. *True or False?* A rotary laser level can be used to set elevations of footings.
7. The _____ is a small device that has a hammering effect when compacting material.
   A. hand tamper
   B. tamping rammer compactor
   C. vibratory plate compactor
   D. packer

8. *True or False?* A standard track-driven power buggy will carry 22 cu ft of material.
9. *True or False?* Most tools used in concrete masonry are designed for use in finishing horizontal concrete surfaces.
10. _____ are used to prepare concrete surfaces for troweling.
    A. Hand tampers
    B. Bull floats
    C. Concrete brooms
    D. Hand floats

Apply and Analyze

1. When using mortar to fill voids in joints between brick, one tool is used to hold mortar and the other is used for pushing mortar into the void in the joint. What are the two tools used to perform this task?
2. What is the difference between a high-frequency concrete vibrator and a backpack concrete vibrator?

Critical Thinking

1. The power trowel is wearing out, and the owner of the company has asked you to research different manufacturers and models to aid in purchasing a new power trowel. What evaluation criteria should be considered when researching a new machine?
2. If you are a foreman of a layout crew, what items do you need to discuss and train new employees on before allowing them to operate the rotary laser level?

Activities

1. (MATH) Find several concrete masonry tool kits online or in a hardware store. Prepare a chart that compares the prices of each tool, the number of tools, and then the overall price of each kit.
2. (ELA) Research safety precautions to be taken when using a masonry saw. Prepare an oral presentation to give in class.

Timothy L. Andera

CHAPTER 4
Safety

LEARNING OUTCOMES

After studying this chapter, you will be able to:

4.1-1 Describe the various methods used by employers to ensure safety on a jobsite.
4.2-1 Differentiate between safe and unsafe work practices.
4.3-1 Recall basic first aid rules.

TECHNICAL TERMS

anchorage point
competent person
cardiopulmonary resuscitation (CPR)
crystalline silica
face shield
fall protection harness
first aid
ground-fault circuit interrupter (GFCI)
high-visibility vest
job hazard analysis (JHA)
lockout
Occupational Safety and Health Administration (OSHA)
personal protective equipment (PPE)
safety data sheets (SDS)
safety glasses
safety goggles
safety hard hat
safety helmet
safety meetings
safety programs
scaffold
tagout
toolbox talks

Introduction

Most occupations have some hazards. It is important to recognize occupational hazards and protect against them. Accidents frequently occur in the construction industry. These accidents often result in lost time on the job, partial or total disability, or even loss of life. Accidents can be reduced if each person works safely and takes the required precautions for the work being done. To advance a safety culture in the company, employers discuss safety procedures, demonstrate safe work practices, develop an employee handbook, develop a set of company safety rules, and enforce safe work practices.

4.1 Jobsite Safety

Construction is a dangerous business! In fact, construction is cited by the ***Occupational Safety and Health Administration (OSHA)*** as one of the most dangerous occupations in the United States. OSHA is part of the US Department of Labor and develops and enforces safety and health standards to ensure safe and healthy working conditions for workers.

Among all occupations, an employee sustains an on-the-job injury every 11 seconds, and approximately 12 workers are killed every day in the United States. Job safety must be a primary concern for every employer and every employee involved in construction.

Most accidents are caused by either unsafe conditions or unsafe actions. Unsafe workers find a way to injure themselves regardless of safety gear, safety rails, red warning signs, safety nets, and guards. The solution is to train workers to work safely and to teach them to identify hazards in the workplace. Working safely involves a thorough understanding of tools, equipment, and materials used on the jobsite. It also involves *thinking* safety.

Job Hazard Analysis

Employers are responsible for training employees about jobsite hazards. A technique used by safety-conscious employers is to conduct a site walk-through to identify and discuss possible hazards that currently exist or ones that may appear on the job. This procedure is sometimes called ***job hazard analysis (JHA)***. JHA is used to determine existing hazards and unsafe conditions that may be present on a jobsite or construction project.

In the first step of the JHA process, a foreman or supervisor performs the walk-through, or JHA, with selected employees and then discusses the actual or potential hazards that were found. Supervisors should discuss with *all* employees the hazards discovered and possible procedures to reduce or eliminate those hazards. For training purposes, the selection of employees to participate in the walk-through should change each time a JHA is conducted. In this way, all employees participate in a selected number of walk-throughs and learn to identify hazards and the procedures required to correct them.

The next step in a JHA process is to identify ways to engineer hazards out of the building or procedural process and ways to eliminate them from reoccurring in the future. Because the construction jobsite is constantly changing and the size of jobs or projects differ, multiple JHAs may be performed during the day, usually one in the morning and another directly after lunch. On large jobs, such as arenas and multistory buildings, there may be 100 to 250 employees on a specific site. In these cases, superintendents and representatives from all subcontractors and selected employees should perform the JHA walk-through. They then return to the job trailer or selected location to discuss hazards found, procedures to correct hazards, and mechanisms to stop them from occurring in the future.

Safety Training

Contractors who are concerned about safety and reducing injuries conduct safety training when new employees are hired. Yearly training, as prescribed by OSHA, is also conducted. Training may include hazard identification, use of hand and power tools, equipment operation, working from scaffolding, and other safety topics.

To ensure that employees are adequately trained, written and hands-on or dexterity testing may be administered after completion of training. If an employee fails the testing, retraining or follow-up training occurs to ensure that all employees have the basic knowledge and skills to perform work and operate equipment and tools safely.

To keep safety uppermost in the minds of all workers, short safety talks called ***toolbox talks*** are given by a foreman or supervisor. Toolbox talks, which generally last from 5 to 15 minutes, are usually conducted at the jobsite or work area at the beginning of the day. The talks relate to the proper use of tools, machines, and equipment. Other topics may include unsafe conditions at the site and hazards related to the changing site environment.

Safety Meetings

On large jobs, as well as smaller projects, superintendents should schedule regular ***safety meetings***. Safety meetings are scheduled meetings that are usually attended by superintendents, subcontractors, foremen, supervisors, and other supervisory staff. The meetings may be scheduled on a weekly or daily basis, depending on the size of the job and overall scope of the project. The meetings should begin prior to the start of construction at the site and continue until the project is finished.

Safety meeting activities may include scheduling events, discussing safety concerns and previous injuries, identifying hazards, and making corrections. Topics of discussion may include the changing size of the construction project, traffic occurring on the project, training of new workers who come to the site, and changes occurring in the work environment.

Safety Programs

In addition, companies implement and continually revise ***safety programs*** to reduce accidents and hazards. OSHA requires companies to provide certain safety programs. Examples of safety programs that are required by contractors are fall protection, emergency action plan, and personal protective equipment (PPE).

A safety program is not simply a written and distributed document; it is an ongoing effort among a contractor, crew leaders, and the workers on the site. As part of a safety program, emergency procedures should be discussed and practiced with workers in the event of a jobsite accident or emergency. The list of procedures is customized to reflect individual company needs. A few of the items that should be addressed in a safety program are as follows:

- Where is the closest emergency room?
- What procedures should be followed when an accident occurs?
- What is the phone number for the fire department, hospital, or ambulance?
- What is the address of the construction site to share with emergency staff?
- Where is the closest tornado shelter?

If an accident occurs, everyone should participate in a discussion about the particulars—how it happened and what could have prevented it. Safety-conscious employers conduct accident drills to prepare employees for how to respond to an accident if one should occur.

4.2 Safe Work Practices

Everyone on the job must know where accidents are likely to occur and how to prevent them. Above all, there must be a keen sense of responsibility for other workers. Learning safe work practices is just as much a part of learning a trade as using the tools.

Always follow these safety guidelines:

- If you are unsure how to perform a task or operate a tool or machine, contact your supervisor for training. You should receive training prior to performing any new operation or using a new tool or piece of equipment.
- Ask to participate in jobsite walk-throughs with the foreman or supervisor in order to develop skills in identifying potential hazards.
- Be alert and spot possible accidents before they occur.
- Do not take shortcuts or expose yourself to danger unnecessarily.
- Ensure that all safety guards on tools and equipment are working properly prior to use. If a safety guard is not working properly, take it out of service and contact your supervisor.
- Use the proper safeguards for a given job. Protective clothing, a blade guard, or goggles could save a hand, an eye, or a foot.
- Be sure to unplug tools or equipment prior to changing blades or bits or performing adjustments.
- Keep the work area clean. Many accidents are caused by litter underfoot.
- Never engage in horseplay on the job.
- Pick up your tools and store them properly. This should become a habit.
- Lift heavy objects with your legs rather than your back. Keep the arms and back as straight as possible when lifting. If the object is too heavy, get help.
- Never place articles on a ledge ladder or scaffold where they may fall and injure someone. Check scaffolds and ladders for loose articles before moving them.
- Get first aid for all injuries immediately. Do not chance getting an infection or aggravating a wound.
- Tell your supervisor about unsafe conditions and safety rule violations.
- If something *looks* unsafe, it most likely *is* unsafe. Don't do it!

The concern for job safety prompted the creation of OSHA, but OSHA sets only minimum standards. Some state requirements are stricter and should be reviewed regularly for the benefit of everyone on the job.

Proper Dress and Personal Protective Equipment (PPE)

Personal protective equipment (PPE) is the protective clothing, footwear, eye protection, head protection, and other safety equipment that an employee or worker needs to wear during the time spent at the jobsite or shop. The level of protection may vary depending on the job or activities being performed. For example, an employee may need to wear ear, eye, and face protection when sawing brick or concrete block. If sawing dry (without water) an approved respirator is also required. OSHA identifies that it is the employer's responsibility to purchase and supply workers with the appropriate PPE to perform the assigned work. Wearing the appropriate clothing and/or equipment is an important part of safety on the job. Safety shoes with a protective toe are a must. See **Figure 4-1**. Never wear shoes with a loose sole. They can hang on a scaffold board and cause a fall. Trouser legs should not be too long because they may get caught on a projecting edge and cause an accident.

OSHA requires certain PPE to perform selected jobs or tasks associated with construction. The following **OSHA Standard 1926.28(a)** pertains to PPE as it relates to construction.

"The employer is responsible for requiring the wearing of appropriate personal protective equipment in all operations where there is an exposure to hazardous conditions or where this part indicates the need for using such equipment to reduce the hazards to the employees."

Arina P Habich/Shutterstock.com

Figure 4-1. Workers finishing a concrete city street are wearing appropriate PPE—hardhats, safety glasses, protective gloves, rubber boots or work boots, and hi-vis vests. The safety cones will be placed in locations to direct traffic to protect workers on the site.

For example, when working on a jobsite where employees may be struck in the head with material or equipment, it is a requirement to wear hard hats. Where employees may receive particles in the eyes, they are required to wear safety glasses. In both situations, OSHA requires the employer to ensure that employees are wearing hard hats and safety glasses during the time spent working on the construction site.

Certain construction sites or jobs require workers to wear high-visibility apparel. Refer to **Figure 4-1**. A ***high-visibility vest*** or *hi-vis vest* is worn by workers on building and highway construction sites so the wearer can be more readily seen. Increasingly, workers on construction jobs are being required to wear hi-vis vests 100% of the time to protect them from injury.

The following are general guidelines for selecting and wearing proper dress:

- Keep clothing in good repair.
- Roll up or button shirtsleeves.
- Tuck shirttails in the pants.
- Turn pant cuffs down or wear pants without cuffs.
- Do not wear short pants.
- Protect your feet with approved safety shoes. Thick-soled shoes are a must.
- Wear a hard hat or safety helmet (required on all construction sites).
- Wear eye protection—either goggles or safety glasses. Also, wear a face shield when hazardous materials are present, such as when grinding or cutting concrete or other materials.
- Wear gloves to protect the hands from chemicals, abrasive materials, and cold weather. On certain construction jobsites, protective gloves may be required to be worn 100% of the time.
- Wear an approved respirator for the job being performed. See **Figure 4-2**. For example, an approved respirator protects a masonry saw operator against hazardous silica dust.
- Long, loose hair is a hazard. Cut it or tie it back and tuck it under your hard hat.

Head Protection Safety

A ***safety hard hat*** is a type of PPE worn to protect a worker's head from a direct impact from above. See **Figure 4-3**. A safety hard hat could save your life if a brick were to fall from above and strike your head. A safety hard hat has an adjustable headband to accommodate different head sizes. The hard hat is available in different styles and multiple colors. Some construction companies use different colors of hard hats to identify different levels of workers, such as a superintendent, foreman, journeyman, and apprentices.

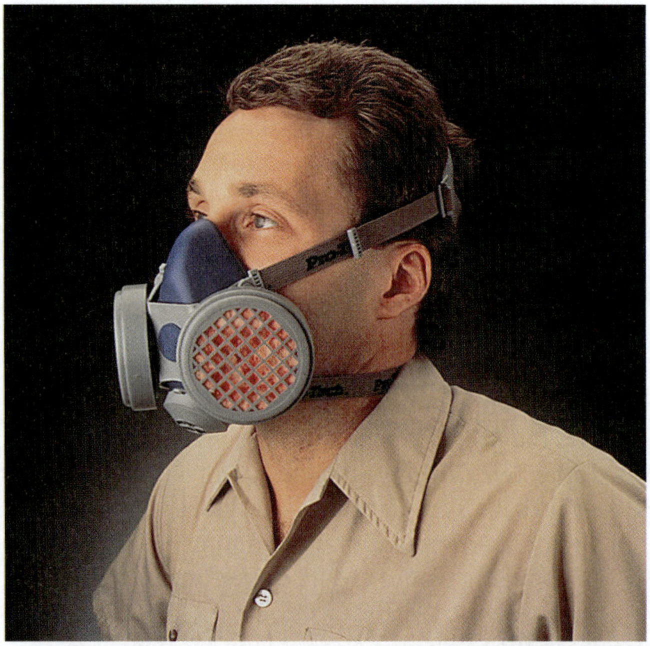

Susan Law Cain/Shutterstock.com

Figure 4-2. The proper respirator should always be worn when hazardous materials are airborne.

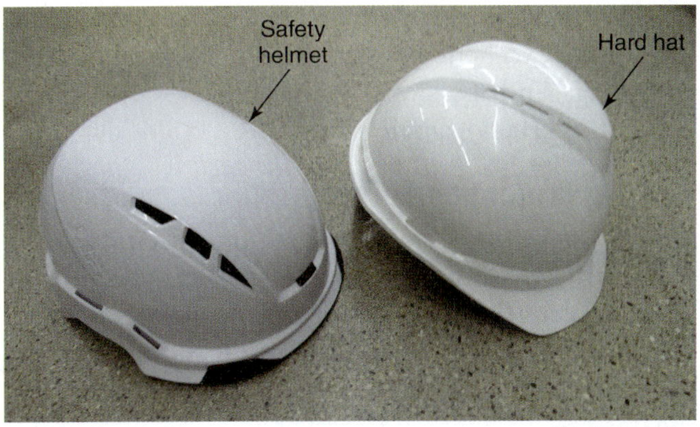

Timothy L. Andera

Figure 4-3. Two types of head protection, safety helmet and hard hat.

A new type of head protection is a ***safety helmet***. See **Figure 4-3**. The safety helmet was created to protect the worker's head from above impacts but also to aid in protecting the employee from side impacts. Helmets come with a chin strap to secure the helmet to the head. Multiple variations of helmets are available that accommodate accessories, such as eye protection, ear protection with and without communication capabilities and ventilation ports.

Eye and Face Protection Safety

In the masonry industry almost every task a mason performs requires some type of eye and face protection.

OSHA Construction Standard 1926.102(a)(1) requires workers to wear eye and face protection when injury may occur. Manufacturers of ***safety glasses*** have created an assortment of eye protection, allowing some safety glasses to fit over prescription eyewear. See **Figure 4-4**. ***Safety goggles*** are another type of eye protection that comes in different types and costs. Safety goggles offer a higher level of protection from dust and particulates but can fog up if they are not vented properly. They are usually made to fit over prescription eyewear and are available in clear or tinted lenses. See **Figure 4-5**. Certain safety glasses and goggles have tinted lenses for added protection against ultraviolet light.

Face protection is another concern to protect workers from flying particles because of grinding, sawing, or other tasks. A ***face shield*** is adjustable for size and can be rotated up when not needed. It can be purchased in clear or tinted versions. See **Figure 4-6**. It is important to work with your employer to determine which type of eye and face protection is appropriate for the type of work being performed. In certain situations, safety glasses are worn under a face shield to

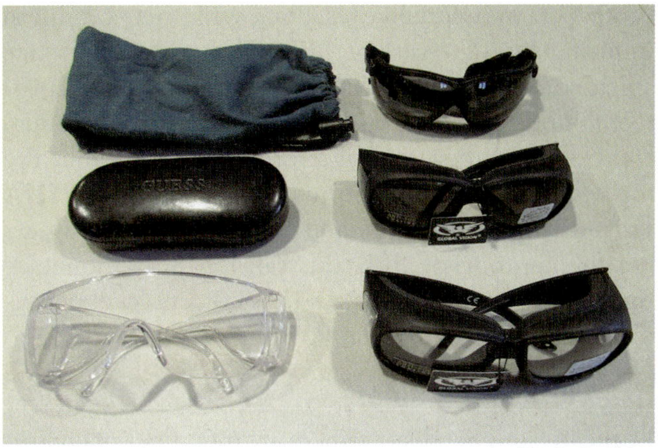

Timothy L. Andera

Figure 4-4. Different types of safety glasses. The top two on the left are examples of cases. The clear lenses (bottom left) can fit over prescription eyewear. The examples on the right are tinted and have a foam gasket. The safety glasses on the bottom right are able to go over prescription eyewear.

Timothy L. Andera

Figure 4-5. Two types of safety goggles. The tinted pair on the left have a foam gasket, while the pair on the right are clear.

Timothy L. Andera

Figure 4-6. A clear face shield.

protect an individual's eyes along with their face. It is important to take care of your safety glasses and face shield. They scratch easily, so it is important to protect them when not in use. In the construction industry, OSHA lists eye and face protection as one of the top 10 citations that continue to occur. When OSHA visits a construction site and finds workers not wearing the appropriate PPE, this can lead to a monetary fine against the employer. OSHA's current level per violation is $16,131.

Hand Protection Safety

In the masonry industry, there are a number of chemicals and products that a mason can come in contact with during their daily activity. That is why a number of construction companies are requiring everyone on the jobsite to wear hand protection 100% of the time while they are at work. **OSHA Standard 1926.95(a)** identifies that protection is required when employees are exposed to chemical hazards or irritants, which can cause an injury or impairment to the skin.

> **Safety Note**
> Masonry cement and Portland cement have potentially hazardous components; calcium carbonate and trace amounts of hexavalent chromium. Exposure to wet cement can cause serious skin injuries from caustic burns to allergic dermatitis.

Figure 4-7 shows a type of glove that can be used to protect the hands from the caustic chemicals in mortar and concrete. This glove is a fully nitrile-coated glove made with HPPE, which is a high-performance polyethylene. Nitrile is a synthetic rubber coating. Nitrile offers excellent physical resistance to punctures, cuts, snags, and abrasions, along with resistance to oils, fuels, and certain organic solvents.

Superior Glove

Figure 4-7. One type of glove that can be worn by masons to protect their hands against chemicals found in mortar and concrete. This glove also can protect hands from cuts, punctures, and abrasions.

Hearing Protection Safety

Almost every tool or piece of equipment emits noise. If it is difficult to understand normal conversation at an arm's length, then hearing protection is needed. If that noise is not reduced by some type of hearing protection, hearing loss can occur over time. Hearing loss is gradual and irreversible, meaning once your hearing is affected, it cannot be brought back. Selecting the correct type and size of hearing protection is important. Earplugs can be purchased in different sizes to accommodate the user. Ear canal sizes vary from one individual to another, so it is important to find a set of earplugs that fit properly. To install earplugs correctly, follow the manufacturer's instructions found on the package, or contact your supervisor. **Figure 4-8** shows different types of hearing protection.

Tool and Equipment Safety

Keeping tools and equipment in good repair and using them properly is the best way to avoid accidents with them. Also, when working near others, remember that a trowel or chisel is sharp. Take care not to injure anyone.

Electrical and mechanical equipment is widely used in the masonry trades. This equipment presents a special hazard because of moving parts. Safety is important

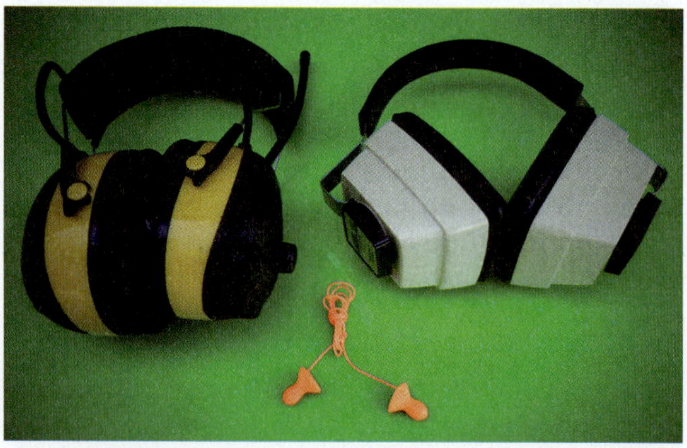

Timothy L. Andera

Figure 4-8. Three different styles of hearing protection. The left and right are earmuffs. The earmuffs on the left can play radio stations. They also have a jack that can be attached to an external audio device. The center earplugs are reusable and made of a rubber-type material. An attached cord allows the earplugs to hang around the neck when not in use.

when using any power equipment, including mixers, masonry saws, sandblasters, and motorized wheelbarrows. See **Figure 4-9**.

Follow these rules when using a power mixer:

- Wear eye protection.
- Wear a respirator when mixing dry materials, sand, or concrete products.
- Take precautions when working with cement, concrete, or stone products. The amount of silica in these products may create a severe health hazard.
- Wear ear protection, earplugs, or earmuffs.
- Never allow your shovel to extend past the mixer grate when adding ingredients.
- Never reach inside the mixer for any reason when it is running.
- Do not exceed the capacity of the machine.
- Be sure the machine is in a stable position before operating.
- Check the machine daily to ensure that all guards are in place to protect the user from belts, pulleys, and sprockets.
- Always clean the machine properly when you are finished.

Using a power masonry saw decreases work time and produces a nice finish, but it can be a dangerous machine if it is not operated properly. Observe the following rules:

- Wear eye protection that is approved by American National Standards Institute (ANSI). Approved safety glasses show "ANSI Z87" on the safety glasses along with the manufacturer's name.
- Wear a respirator when sawing concrete or stone products. Concrete or stone products may contain large amounts of silica and are a severe health hazard.
- Wear ear protection, earplugs, or earmuffs.
- Properly secure the masonry unit before cutting.
- Maintain the blade and do not force the cut.
- Never place your hands in the path of the blade.
- Never operate the machine when someone is standing in the path of an object that might be thrown from the machine.
- Follow the manufacturer's instructions for safe operation.

Arina P Habich/Shutterstock.com

Figure 4-9. Concrete workers are building a new street. One worker is driving a motorized wheelbarrow, and another is directing him to the proper location to unload the concrete. Note the personal protective equipment being worn: hard hats, hi-vis vests, safety boots, and protective work gloves.

Sandblasting is sometimes used to clean new walls, even though other methods are preferred. When using a sandblasting machine, observe the following rules:

- Wear eye protection.
- Wear a face shield to protect the face and gloves to protect your hands.
- Wear protective clothing to protect exposed skin.
- Wear a respirator when sandblasting. Concrete or stone products may contain large amounts of silica, which can create a severe health hazard.
- Wear ear protection, earplugs, or earmuffs.
- Never point the stream at someone else.
- Be sure the pressure is properly adjusted.
- Do not use a higher pressure than necessary to get the job done.
- Use the proper abrasive for the particular job.
- Be sure hoses are clear of walkways and work areas.
- Properly secure the machine when you are finished with it.

Electrical Safety

Electricity is the power source for much of the mechanical equipment used on construction jobs. Electricity can be deadly if used improperly. Be careful not to touch any bare wire!

If an electrical accident occurs, never touch someone who may have live current flowing through them. You may be severely injured or killed. If possible, shut off the power. If this cannot be done, use a dry piece of wood to break the contact.

At the beginning of each day and before use, examine extension cords to ensure that the ground plug is attached. Also, check the cord for cracks or damaged exterior insulation that may cause electrical shocks. If a damaged cord is found, take it out of service and contact your supervisor. Before using any portable equipment, be sure to check all guards to determine if they are operating correctly. Check for worn or defective insulation, loose or broken connections, and a bad ground wire connection.

OSHA requires lockout/tagout prior to repair of a machine or piece of equipment. Ensure that the equipment is unplugged and the end of the cord is in your control. ***Lockout*** is the process of locking the machine or locking the electrical power out so the equipment will not operate. ***Tagout*** involves placing a tag within the lock to explain why the machine is out of service.

For example, if a masonry saw needs to be taken out of service for repair, a lockout container is used to secure the plug so the masonry saw cannot be plugged into a power source. A tag is then placed inside the lock to identify the problem with the saw. The tag includes the name of the employee who locked out the machine. See **Figure 4-10**.

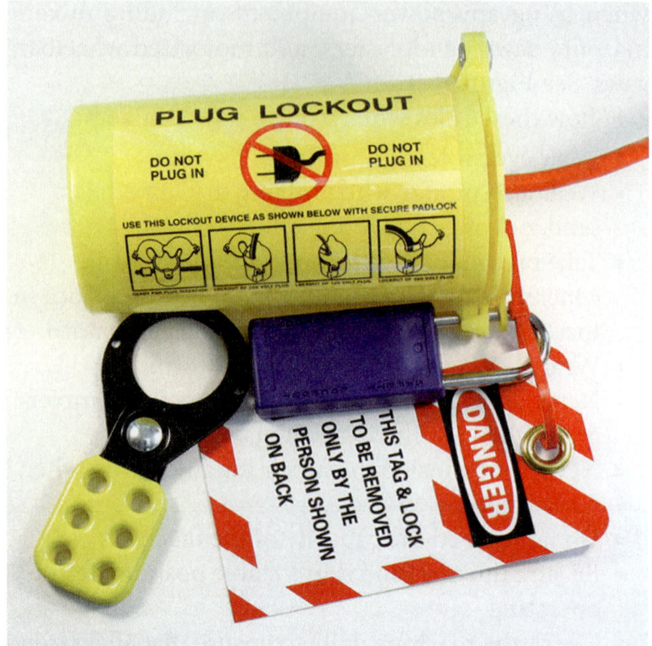

dcwcreations/Shutterstock.com

Figure 4-10. A lockout container used to secure an electrical plug. A tag is installed within the lock prior to closing.

Lockout/tagout is also performed on power supply devices, such as an electrical panel. A breaker switch in an electrical panel that controls a machine or piece of equipment is shut off or tripped to the off position. The power that supplies the electricity to the circuit powering the tool or machine is cut. See **Figure 4-11**.

digitalreflections/Shutterstock.com

Figure 4-11. Lockout/tagout being installed on an electrical panel to secure and control power delivered to equipment.

OSHA requires all temporary power on a construction site to incorporate a ***ground-fault circuit interrupter (GFCI)*** as part of the electrical system. A GFCI protects workers from receiving a severe electrical shock or being electrocuted by activating and tripping to interrupt the electrical current. A GFCI breaker can be installed into the electrical panel or a GFCI electrical cord can be used. A GFCI extension cord has a built-in protection device as part of the unit. See **Figure 4-12**.

Safety Note
Be sure that all electrical equipment is grounded appropriately. Keep electrical wires off the ground. Never operate electrical equipment in wet locations without proper grounding and safeguards.

Handling and Storing of Materials

Your own safety, and the safety of others, depends on proper handling and storage of materials. See **Figure 4-13**. Observe the following rules for handling and storing materials:

- When lifting heavy loads, lift with your legs, not your back. The proper procedure is to squat down and pick up the load by straightening your legs. Keep the back straight. If the load is too heavy, get help.

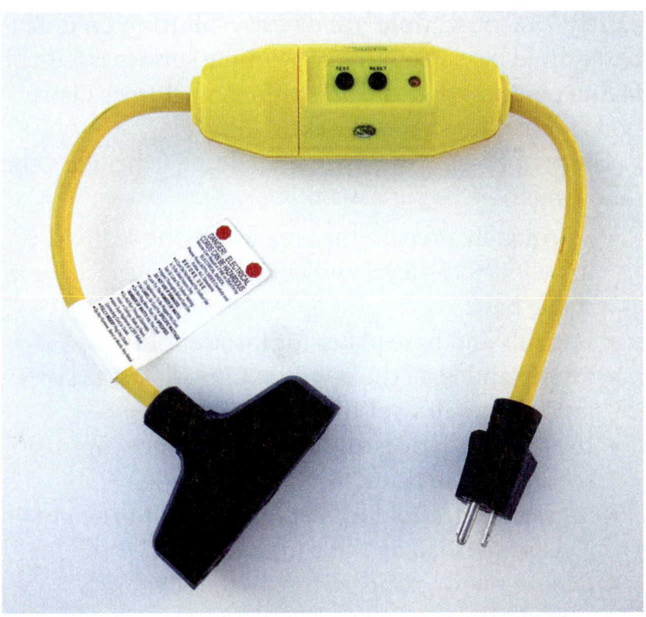

Timothy L. Andera

Figure 4-12. A GFCI protection cord has a built-in test button to determine if the unit is operating correctly.

- Store materials such as brick, concrete blocks, and bags of cement or mortar on a paved surface, on boards, on pallets, or on shipping paper. Cement and mortar will take on moisture from a concrete floor or the ground. This is why they need to be separated or elevated.
- Be careful not to overload a wheelbarrow or power buggy. Place the load over the wheel, not toward the handles. Be sure the wheelbarrow or buggy is in proper working condition. Do not use equipment that has a cracked, loose, or broken handle or an improperly inflated tire.
- Do not exceed the maximum height of 7′ for open, unsupported stacks of brick. A setback of 1″ for every 1′ of height above 4′ is the recommended practice.
- Concrete block and brick should always be stacked in tiers that rest on a solid foundation. Tall stacks over 4′ high should be stepped back, braced, and propped.
- Store cut stone for sills and trim on level boards. Proper support of stone is necessary to prevent breakage. A cracked piece of stone could break off and become a hazard.
- Stacked bags of cement or mortar should be positioned with tops facing inward. Lay every other layer crosswise and do not exceed 10′ high. Be sure the stack is stable at all times and covered from the elements.
- Keep materials clear of walkways, doors, and hoists. Also, keep materials at least 10′ away from the edge when they are above the first floor of a building.
- Protect masonry units from rain or snow by covering with plastic or tarp.

Scaffold Safety

Much of a mason's work is performed on some type of ***scaffold***. A scaffold is a temporary structure that holds workers and materials while they are erecting, repairing, or doing other work on a building. See **Figure 4-14**.

The three main sources of injury associated with scaffolding are falling from the scaffold, being struck by tools or materials falling from the scaffold, and faulty scaffolding.

Scaffolding must be erected properly and designed to support the load it is expected to carry. **OSHA Standard 1926.451(f)(7)** states the following:

"Scaffolds shall be erected, moved, dismantled, or altered only under the supervision and direction of

Lev Kropotov/Shutterstock.com

Figure 4-13. A construction site showing bricks stored close to the location where they will be needed. The site is kept clean of clutter and refuse that can cause accidents.

a competent person qualified in scaffold erection, moving, dismantling or alteration. Such activities shall be performed only by experienced and trained employees selected for such work by the competent person."

A ***competent person*** is one who is capable of identifying existing and predictable hazards in the surroundings or working conditions that are unsanitary, hazardous, or dangerous to employees as defined by **OSHA Standard 1926.450(b).** A competent person is one who has been authorized to take prompt corrective measures to eliminate unsafe conditions.

The scaffolding process has many inherent risks associated with layout, erection, stocking of masonry products, and dismantling of scaffolding. Constant vigilance is required through each scaffolding process to ensure safety. **OSHA Standard 1926.451(f)(3)** states, "Scaffolds and scaffold components shall be inspected for visible defects by a competent person before each work shift, and after any occurrence which could affect a scaffold's structural integrity."

OSHA construction standards state that approximately 75% of scaffold accidents result from an unstable scaffold base, planking or working surface issues, or falling objects. Always read the manufacturer's instructions for erection procedures.

Safety procedures for scaffolding include the following:

- Adequately prepare the area where the scaffold is to be located—level and compact the surface for a stable base.
- Use sills and base plates for the scaffold.
- Plumb and level the scaffold. Use adjustable screw jacks for leveling of frames.
- Install cross braces and ensure the braces are secured to the frames.
- Secure the scaffold to the structure at the height level prescribed by the manufacturer.
- Ensure the working surface of the scaffold is completely planked or decked.
- Install and use scaffold ladders to access different levels. Never climb cross-braces.

marilyn barbone/Shutterstock.com

Figure 4-14. This multiple level scaffold includes parts that create a safe work platform—handrails, midrails, toeboards, and cross-bracing. Scaffold stairs aid workers in moving from one level to another.

- Do not use a stepladder on scaffolds.
- When installing scaffold side brackets, follow the manufacturer's specifications.
- When standing on the planking and side brackets, always place a gate at the end of the area to prevent falls.
- Use handrails, midrails, and toeboards.
- Keep the scaffold away from power lines.
- Do not overload a scaffold. Follow the manufacturer's recommendations.

Fall Protection

Falls are the leading cause of deaths in the construction industry. OSHA requires that construction workers be protected when working at levels 6′ and above for most types of construction activities. From OSHA's data, there is a total of nearly 1000 deaths each year in the construction industry. Approximately 400 of those deaths, or 40%, are caused by falls. The most cited violation in the construction industry according to OSHA is lack of fall protection to protect employees from injury. It is the employer's responsibility to supply safety equipment to their employees and train them how to use it. This includes fall protection equipment.

Figure 4-15 shows an employee wearing a fall protection harness. A ***fall protection harness*** helps distribute fall arrest to the whole body should an individual fall. Numerous manufacturers make fall protection harnesses and related equipment. Select equipment from a manufacturer who has a track record of producing quality equipment. Be sure to select a harness that fits properly and is matched to the activity the employee will be performing. Never mix components from one manufacturer to another when selecting fall protection equipment. This is because one manufacturer's equipment may not work correctly with a different producer's fall protection gear. The employer will need to select one fall protection manufacturer and purchase the appropriate equipment to create a system that meets the needs of the company and its employees.

When working at heights using fall protection equipment, select an anchorage point that can withstand the shock of someone falling. An ***anchorage point*** is a secure location of attachment for lanyards or other devices to connect to when wearing fall protection equipment. **Figure 4-16** shows an employee wearing fall protection equipment that has been designed to protect the construction worker when working at heights.

Safety Nets

A safety net is a device designed to catch a worker who falls from a high place. Nets made from synthetic fibers are commonly used. They are placed in elevator

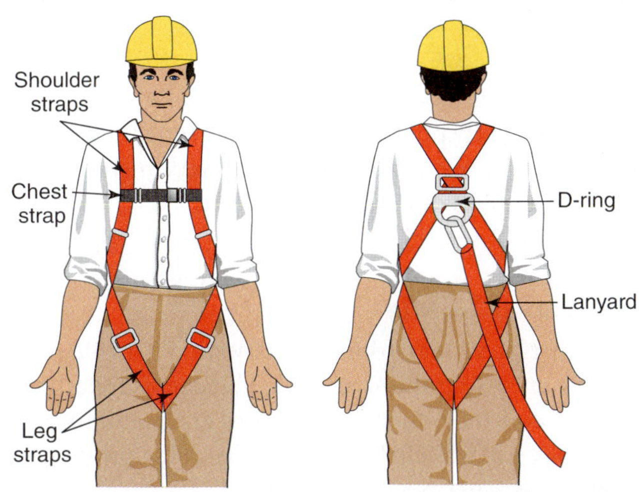

Emre Terim/Shutterstock.com

Figure 4-15. A fall protection harness should be adjusted to fit over the shoulders and between the legs to distribute the forces across the body in an event of a fall.

Richard Thornton/Shutterstock.com

Figure 4-16. A masonry worker is installing reinforcing bars while wearing a fall protection body harness. A D-ring is attached to the harness located between workers shoulders. A lanyard is attached to the harness D-ring and to the metal framework being constructed. The lanyard is an approved safety line to protect the worker in the event of a fall.

shafts, between floors, or under rafters. Be sure safety nets are properly positioned before working high above the ground. Always follow manufacturer's instructions regarding installation and inspections.

Enclosure Safety

In many places, year-round construction is made possible by enclosing the structure. The enclosure allows work to continue during adverse weather conditions. A tubular scaffold enclosure provides some protection from the weather. See **Figure 4-17**.

Enclosures are generally heated with temporary heaters. Heaters used for enclosures are sometimes called *salamander heaters* or *torpedo heaters* because of their shape. See **Figure 4-18**. Most temporary heaters use propane, natural gas, or kerosene. Some heating devices are electric. Make sure an electric heating device is properly grounded.

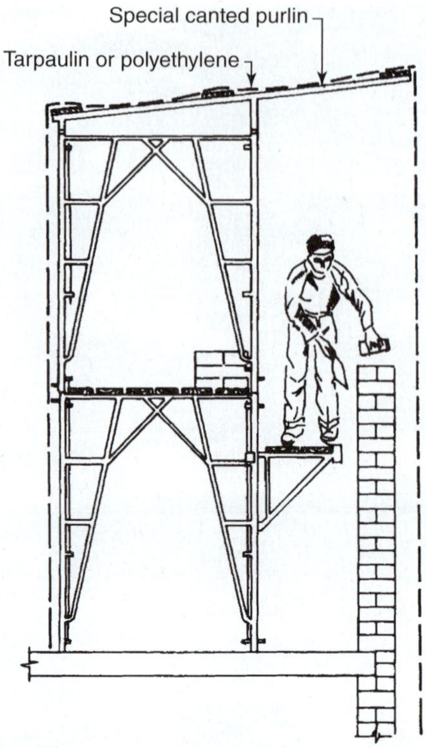

A

Brick Industry Association

B

Brian Guest/iStock/Thinkstock

Figure 4-17. Scaffold enclosures. A—This type of tubular scaffold enclosure may be used to provide some protection during cold or rainy weather. B—Scaffold covers or enclosures can be made of polyethylene film. Some types have nylon line or thread embedded into the film for extra strength.

Timothy L. Andera

Figure 4-18. Torpedo or salamander heaters are used for temporary heating of enclosures.

A potentially dangerous condition for workers is created if an enclosure is not properly vented. Before using a temporary heater, always read the instruction manual to be sure the device is properly installed and vented.

Safety Note

During the burning process, heaters emit deadly carbon monoxide gas that can accumulate in the heated structure. Carbon monoxide is odorless, colorless, and can be a lethal gas. It is a sound safety practice to install carbon monoxide detectors in situations where enclosures are used.

When heaters are not in use, disconnect the power, shut off the gas, and then disconnect the valve from the tank. To keep connections clean, replace the plug cover on the tank and place the gas valve in a plastic bag.

Ladder Safety

Masons use ladders frequently in their work. Accidents are greatly reduced when the following ladder safety rules are observed:

- Be sure the ladder is in good repair. If a ladder has cracked or broken rungs or damaged legs, remove it from service and contact your supervisor.
- Keep ladders out of walkways and traffic lanes.
- Place the ladder on a firm and level surface.
- Avoid placing a ladder on slippery surfaces.
- Be sure the ladder extends at least 3′ above the point where you plan to step off it, such as a roof.
- The ladder should be placed against a strong support at a sufficient angle to prevent tipping.
- The area around the bottom of the ladder must be clear of clutter.
- Fiberglass ladders are an excellent selection to be used on the construction site. Avoid aluminum ladders if possible because they conduct electricity.
- When climbing a ladder, transport tools in an over-the-shoulder bag.
- Never stand on the highest rung of a ladder. A ladder is required to extend 3′ above a surface or roof when accessing that level.
- It is good practice to attach the ladder at the top so it does not slip. If there is a chance of the ladder slipping at the bottom, secure it or have someone hold the bottom of the ladder.
- Stepladders are intended to be used when folded open and spreader arms are locked in position. Never use stepladders that are leaning against a wall or other structure.
- Never stand or sit on the top or next-to-the-top step on a stepladder. It is an unsafe practice that can cause the ladder to tip over.

Silica and Respirable Crystalline Silica Dust Exposure

OSHA has implemented a new safety standard that pertains to crystalline silica. ***Crystalline silica*** is a common mineral found in the earth's crust. Silica is a very hard, glassy mineral that can be found in a number of forms, including quartz, opal, and sand.

Respirable crystalline silica is a very small particle 100 times smaller than ordinary beach sand. It is created by cutting, grinding, sawing, drilling or crushing products containing silica. Because of the small size of the particles, workers who inhale this dust are at an increased risk of developing serious silica-related diseases. These include lung cancer, silicosis (an incurable lung disease), and kidney disease. All of these diseases can be fatal.

A respirable crystalline silica dust cloud is created by many activities, including cutting, grinding, and drilling. The cloud is a health hazard if the worker is not protected with appropriate PPE. An approved respirator is required to protect the worker. Refer to **Figure 4-2** for an appropriate respirator to use when exposure exists to silica dust. **Figure 4-19** shows dust created by a masonry saw cutting concrete.

Housekeeping as it relates to silica can also be problematic for the employer and their employees. OSHA standards identify housekeeping as a concern when silica can be present in the area being cleaned. Two OSHA standards that discuss housekeeping rules are **Construction 1926.1153(f)** and **General Industry 1910.1053(h)**. To meet OSHA's housekeeping rules, a

A RAW4/Shutterstock.com

B mikeledray/Shutterstock.com

Figure 4-19. A masonry saw is being used to saw a street made of concrete. A—An employee is dry-sawing concrete. B—A wet saw is being used to saw a concrete slab. At the same time, a wet vacuum is being used to pick up wastewater that contains silica.

wet or moist sweeping compound will be needed when cleaning floors. **Figure 4-20** shows one type of sweeping compound.

> **Safety Note**
> When hand-sweeping floors, remember to keep the dust on the floor and not allow it to become airborne. Use a pushing motion to keep bristles of the broom in contact with the floor. This is especially important when sweeping materials that may contain silica, as OSHA standards specify sweeping requires the application of a compound to eliminate airborne dust.

A HEPA vacuum is required to vacuum in areas that may contain silica. A HEPA vacuum should be used to remove concrete dust from an employee's clothes. There are many different types and styles of HEPA-rated vacuums. **Figure 4-21** shows a HEPA canister vacuum.

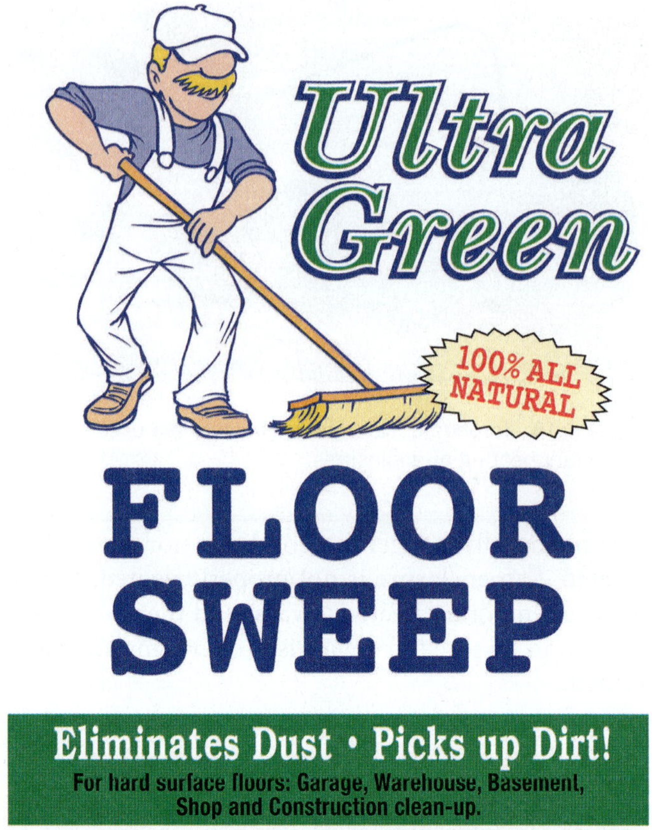

ULTRA GREEN FLOOR SWEEP, Platte, SD

Figure 4-20. Ultra Green Floor Sweep is an all-natural product made from soybeans. The floor sweeping compound comes in different size containers depending on the project or application.

Floor sweepers come in different models and styles. Single- or twin-brush varieties use brushes to direct dust into the sweeper. Indoor sweepers are usually powered by rechargeable batteries. Larger models can be purchased in gasoline, liquefied petroleum gas, or diesel engines. **Figure 4-22** shows a single-brush model powered by rechargeable batteries.

Safe Use of Chemicals and Materials

Before using any chemical or material that contains chemicals, read and follow the manufacturer's directions. More chemicals are being used by masons, and some can cause serious burns or loss of sight. If you are not familiar with the use of a product, ask your supervisor or foreman for training and proper handling instructions.

Wear the proper protective clothing when using chemicals. Rubber gloves and safety glasses are standard items. There may be other PPE that should be worn, such as a face shield and long-sleeve shirt. If the situation is appropriate, have a bucket of water or

Timothy L. Andera

Figure 4-21. A HEPA canister vacuum is on wheels to aid in moving the device around the facility or construction site.

Author's image taken at Cemcast Pipe & Precast, LLC. Hartford, SD

Figure 4-22. The plant manager is demonstrating how a HEPA floor sweeper is used to vacuum concrete dust off the plant manufacturing floor. Because it is a HEPA sweeper, employees are not required to wear a respirator. The company requires everyone to wear appropriate PPE: hard hat, hi-vis vest, protective gloves, safety glasses, and safety shoes.

garden hose available to wash off any chemicals that come in contact with the skin. Know what to expect from any chemical you use. Know the proper safety and clean-up procedures in case of a spill or other accident.

Safety Data Sheet

It is the contractor's responsibility to accumulate product documentation of the materials or chemicals being used on a construction site. These documents are called ***safety data sheets (SDS)***. An SDS is a written document that outlines the product information and procedures to be used when working with or handling the product or chemical. The SDS also includes emergency procedures and manufacturer contact information.

Figure 4-23 is an OSHA example of an SDS. This SDS example is a total of 10 pages with the first page being shown. Section 1 of the SDS identifies the substance or mixture and its uses. This area also lists the address and email of the manufacturing company, emergency contact information, and SDS version.

Symbols (pictograms) indicate the type of risk pertaining to a product. Pictograms, which are found on safety data sheets and on product labels, serve to convey health, physical, and environmental hazards. A black hazard symbol inside a red diamond-shaped frame depicts the health or risk hazard. A signal word, either "Danger" or "Warning," is located under the pictogram frame. "Danger" is used for more severe hazards, and "Warning" is used for less severe hazards. Only one signal word is listed in an SDS document or on a label. This word relates to the most severe hazard, even if there are also less hazardous chemicals contained in the product.

In the previous OSHA example, Section 2 identifies and classifies the hazards associated with this SDS product. **Figure 4-24** shows three pictographs OSHA uses to identify the hazards. In this example, the first pictogram is flame, meaning the product is flammable, emits a flammable gas, and can be self-heating or igniting.

CHYMEIA

According to EC-Regulation 1907/2006 (REACH)

SAFETY DATA SHEET

SECTION 1: Identification of the substance/mixture and of the company/undertaking

1.1. Product identifier

Trade name
MyProduct - SDS example, not valid for use
Product no.
Produkt 123456
REACH registration number
Not applicable

1.2. Relevant identified uses of the substance or mixture and uses advised against

Relevant identified uses of the substance or mixture
Laboratory Chemicals (PC 21)
Use as laboratory reagent (PROC 15)
Scientific research and development (SU 24)
Wide dispersive indoor use of processing aids in open systems (ERC8a)
Uses advised against
-
The full text of any mentioned and identified use categories are given in section 16

1.3. Details of the supplier of the safety data sheet

Company and address
CHYMEIA ApS
Universitetsparken 2
DK-4000 Roskilde
tlf: +45 7240 1622
Contact person
Salgschef, Lars Bugge
E-mail
info@chymeia.com
SDS date
2016-02-23
SDS Version
1.0

1.4. Emergency telephone number

Use your national or local emergency number
See section 4 "First aid measures"

SECTION 2: Hazards identification

2.1. Classification of the substance or mixture

Flam. Liq. 2; H225
Asp. Tox. 1; H304
Skin Irrit. 2; H315
Eye Irrit. 2; H319
STOT SE 3; H336
Repr. 2; H361
STOT RE 1; H372
See full text of H-phrases in section 2.2.

Occupational Safety & Health Administration

Figure 4-23. The first page of an OSHA SDS.

Occupational Safety & Health Administration

Figure 4-24. Examples of hazard pictographs used in the OSHA SDS from Figure 4-23.

The second pictogram is the exclamation mark which identifies the product as an irritant to skin and eyes. It is also toxic and an irritant to the respiratory tract. The third pictogram is a health hazard.

Hazard statements are also included to describe the nature and degree of the hazard(s) related to the product. An example of a hazard statement is "Causes damage to kidneys through prolonged or repeated exposure when absorbed through skin."

OSHA requires manufacturers to create SDS forms for all the products and chemicals they manufacture or sell. All safety data sheets are comprised of 16 different sections, but each individual SDS varies in length depending on the product and the hazards associated with it.

4.3 First Aid

First aid is the assistance provided to someone who has been injured or become ill. The goal of first aid is to preserve life, prevent the deterioration of the individual injured, and promote recovery. ***Cardiopulmonary resuscitation (CPR)*** is an emergency technique used on individuals when the heart has stopped beating.

Individuals attend training courses that teach the proper techniques to administer first aid and CPR. Certificate cards are given to those who complete the training. You can obtain training individually or with a group. Some companies will pay for their employees to be trained in first aid and CPR.

OSHA has addressed who needs to be trained in first aid and CPR as well as the level of training to ensure if an accident occurs that adequate measures are in place to protect employees. OSHA acknowledges the difference between working in an office and on a construction site and specifies that time is of the essence when applied to construction because of the many accidents and hazards that could occur. If a construction activity is longer than three to four minutes, then it is the responsibility of the company to ensure there is a qualified employee trained in first aid and CPR present.

Before construction work begins, either at the shop, construction site, or work vehicle, it is required by OSHA standards to have a first aid kit accessible. Some of the more common items found in a first aid kit include gauze pads, adhesive bandages, tweezers, adhesive tape, latex gloves, antiseptic treatment, cold pack, and eyewash.

When at the site or in the shop, report and have all injuries treated immediately. If chemicals such as cement, lime, or cleaning agents get into your eyes, flush immediately with lots of clean water. Then contact your supervisor and see a doctor immediately. Refer to a first aid manual for all first aid procedures.

Contact with wet (plastic) concrete, cement, mortar, grout, or cement mixtures can cause skin irritation, severe chemical burns, or serious eye damage. Wear waterproof gloves, a long-sleeved shirt, full-length trousers, and appropriate eye protection when working with these materials. If you must stand in wet concrete, wear high-top waterproof boots. Wash wet concrete, mortar, grout, cement, or cement mixtures from your skin immediately. If any of the wet cement mixture soaks through your clothes, remove them and wash the skin immediately. Seek medical attention if you experience a reaction when coming in contact with these materials.

Always keep in mind these three basic first aid rules:

1. Seek first aid immediately after an injury occurs.
2. See a doctor for a serious injury or any eye injury.
3. Report all accidents to your supervisor.

Chapter Review

Summary

- Accidents are caused by unsafe actions as well as unsafe conditions.
- Employers are responsible for training employees about jobsite hazards.
- Job hazard analysis (JHA) involves a walk-through of the jobsite by a foreman or supervisor and selected employees to look for and document hazards and unsafe conditions. Afterwards, actual and potential hazards are discussed, as well as procedures to reduce or eliminate those hazards.
- Company safety programs are an ongoing effort between a contractor, crew leaders, and the workers.
- Training is required before the operation of an unfamiliar tool or piece of equipment.
- OSHA mandates that appropriate personal protective equipment (PPE) be worn in operations where there is exposure to hazardous conditions or where such equipment is needed to protect an employee from injury or health hazards.
- Electrical safety includes checking cords and guards prior to using them.
- The three main sources of injury associated with scaffolds are falling from the scaffold, being struck by tools or materials falling off the scaffold, and faulty scaffolding.
- Following all ladder safety rules can greatly reduce accidents. Report a damaged ladder and do not use it.
- Before using any chemical or material, read and follow the manufacturer's product safety data sheet (SDS) for usage and safety procedures. If unsure about proper usage of the chemical or material, contact your supervisor for training.
- First aid should be sought immediately after an injury or accident. A doctor should be seen for serious injuries and eye injuries. All accidents should be reported to a supervisor.

Know and Understand

Answer the following questions using information provided in this chapter.

1. _____ is the organization that enforces safe work practices.
   A. OSHA
   B. MHS
   C. JHA
   D. PPE
2. *True or False?* Job hazard analysis involves conducting an inventory of materials on the job site.
3. *True or False?* A safety program is a written and distributed document.
4. Heavy objects should be lifted _____.
   A. with both the arms and back bent
   B. with the back as straight as possible
   C. without using the leg muscles
   D. with the legs straight and the back bent
5. *True or False?* When potential exposure exists, a hard hat is required on construction sites.
6. Which of the following is *not* a safe practice when working with a power mixer?
   A. Check that the machine is in a stable position before operating it.
   B. Allow your shovel to extend past the grate when adding ingredients.
   C. Wear a respirator when mixing dry ingredients.
   D. Wear ear protection.
7. Lockout/tagout is performed on _____.
   A. power supply devices
   B. chemicals
   C. tools
   D. crystalline silica
8. *True or False?* Bags of cement or mortar can be stored on concrete floors.
9. *True or False?* According to OSHA standards, a foreman must direct the erection of a scaffold.
10. According to OSHA, the most cited violation in the construction industry is lack of _____.
    A. scaffolding
    B. fall protection
    C. hearing protection
    D. lockout/tagout materials

11. *True or False?* Gas-fired heaters give off deadly carbon monoxide.
12. *True or False?* To use a ladder safely, be sure that it extends at least 2′ above the point where you plan to step off it.
13. *True or False?* Documents that outline the product information and procedures to be used when working with chemicals or materials are called MHS documents.
14. According to OSHA, if a construction site is at a greater time distance than _____, the company must ensure a qualified employee trained in first aid and CPR is present.
    A. 1-2 minutes
    B. 3-4 minutes
    C. 10 minutes
    D. 30 minutes
15. *True or False?* Contact with cement can cause skin irritation.

Apply and Analyze

1. Why should a respirator be worn when you are sawing concrete or stone products?
2. Where should the load in a wheelbarrow be placed?
3. What is the maximum height allowed for an open, unsupported stack of bricks?
4. What is the purpose of a safety net?
5. Why should the use of aluminum ladders be avoided?
6. Explain what can happen when exposed to wet concrete or mortar.

Critical Thinking

1. Research in greater detail the cause of cement burns and present your findings to the class. Also identify safety procedures pertaining to cement exposure that can be implemented in the lab or construction site to improve safety conditions for your classmates or other workers.
2. Currently, your masonry program does not have fall protection equipment. Research some of the different fall protection manufacturers to determine which one would be the best company to purchase equipment as it relates to your program. Use your research to justify the company being selected. Develop a proposal that can be presented to your class on the research finding and recommendations.
3. You are a masonry foreman in a small block and brick laying company. You have noticed that a number of employees are not wearing hearing protection when operating masonry saws. What steps do you take as the foreman to ensure they will wear appropriate PPE?
4. What first aid measures should be immediately taken if you get a chemical such as lime, cement, or a cleaning agent in your eyes?

Activities

1. (SCIENCE) Use the internet to find an example of a safety data sheet pertaining to muriatic acid used to clean brick or another chemical commonly used in the masonry field. Using the information in the SDS, prepare a report summarizing the health hazards of the acid, the personal protection equipment needed to handle the acid, and safe storage methods.
2. (ELA) Observe the work activities taking place at a construction site. Make an oral report to the class explaining the safety precautions and equipment you observed being used. Also include in your report any unsafe conditions you may have observed.

The InterDesign Group, Indianapolis, Indiana

CHAPTER 5

Print Reading

LEARNING OUTCOMES

After studying this chapter, you will be able to:

5.1-1 Define *construction drawings.*
5.2-1 Identify the different types of lines used on construction drawings.
5.3-1 Recognize the symbols and abbreviations used on construction drawings.
5.4-1 Interpret the scales and read dimensions used on construction drawings.
5.5-1 Explain how the metric system could be used in construction drawings.
5.6-1 Differentiate among the various types of working drawings.
5.7-1 Explain the purpose of specifications.

TECHNICAL TERMS

Alphabet of Lines
architectural rendering
border lines
center lines
climate control plan
construction drawings
crosshatch lines
cutting plane lines
detail drawings
dimension lines
dimensions
drawing notes
electrical plan
elevation drawings
extension lines
floor plan
foundation plan
framing plan
general conditions
hidden lines
landscaping plan
long break lines
mechanical plan
object lines
perspective drawings
preliminary drawings
presentation drawings
roof plan
section lines
shop drawings
short break lines
site plan
specifications
technical sections
working drawings

Introduction

In order to understand drawings, you need to be able to visualize what the architect has drawn. You must know the meaning of the lines, symbols, abbreviations, and notes shown. Together, these items make up the language of the design and construction industry. Learning to read drawings is essentially learning a new language.

5.1 Construction Drawings

Construction drawings describe the size, shape, location, and specifications of the elements of a structure. These drawings may be called *working drawings, prints, blueprints,* or *plans.* A skilled worker must be able to read construction drawings and understand the information they contain in order to build the structure as the designer intended. The drawings use standard symbols and notes that are recognized and understood by all the professionals associated with the construction field.

5.2 Line Symbols

Lines make up a large part of the symbols used on construction drawings. The different types of lines are called the ***Alphabet of Lines***. See **Figure 5-1**. A specific line symbol is used to communicate more precisely than just a plain line. It is important to learn the use of each line.

Border Lines

Border lines are very heavy lines that are drawn to form a boundary for the drawing. They assure the reader that no part of the drawing has been removed. They also give the drawing a finished appearance.

Object Lines

Object lines are heavy lines that show the outline of the visible features of an object. They should be easily seen since they represent important elements. Walls, windows, patios, and roof lines are represented by object lines.

Hidden Lines

Hidden lines are medium-weight lines that represent an edge or intersection of two surfaces that are not visible in a given view. For example, the foundation wall and footings are represented by hidden lines on an elevation because they are below grade and, therefore, not visible. Also, hidden lines may be used to indicate features above the cutting plane, such as wall cabinets in a kitchen or an archway. Hidden lines are usually not as thick as object lines.

Center Lines

Center lines are thin lines that indicate the center of symmetrical objects. For example, a window or door may have a center line through it on the floor plan. Center lines simplify dimensioning and are used for location of features.

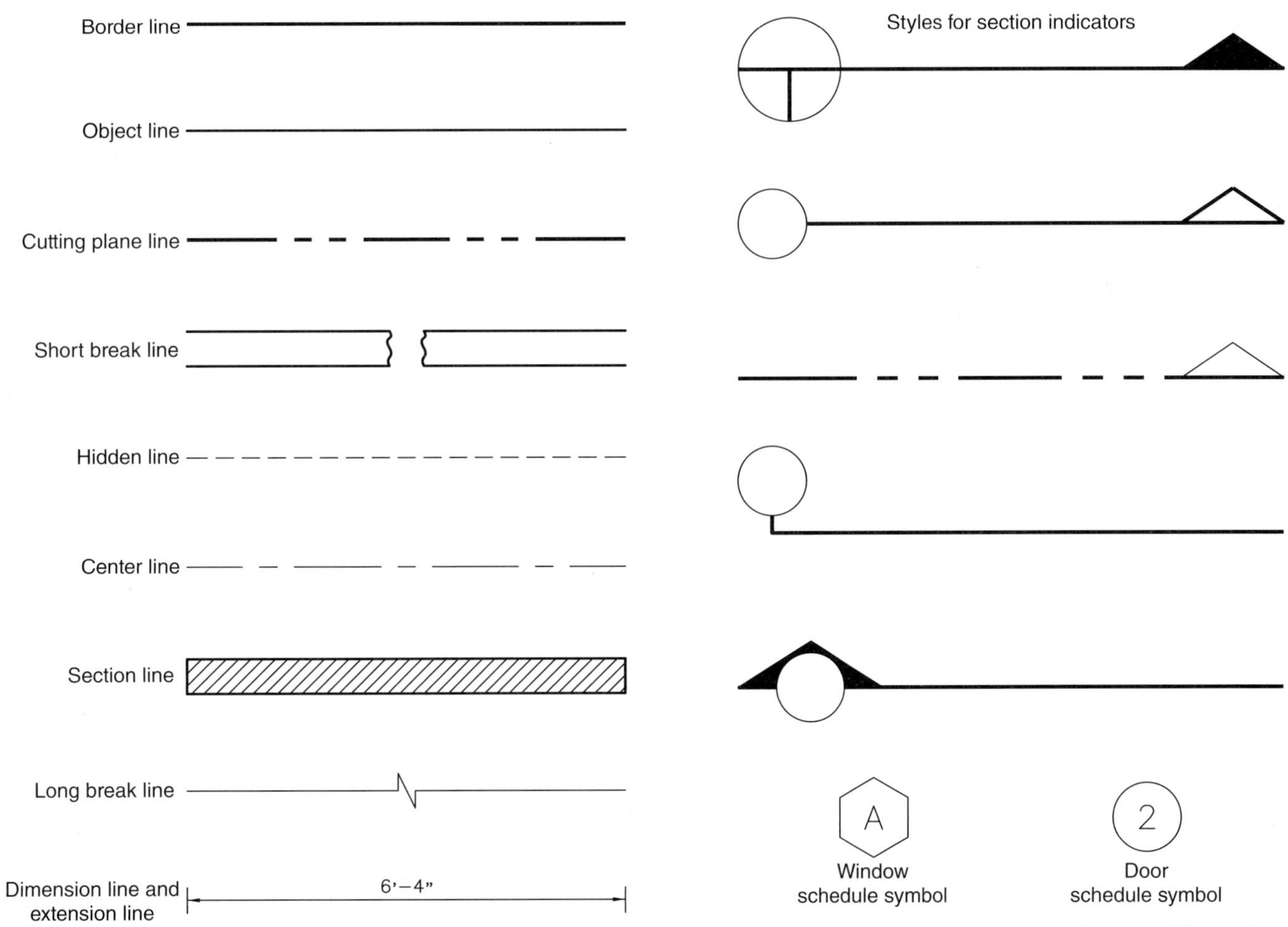

Goodheart-Willcox Publisher

Figure 5-1. The Alphabet of Lines is used on construction drawings.

Extension Lines

Extension lines are thin lines used to show where a dimension line ends. They extend from a portion of the object past the dimension line about 1/16″. They extend the object for dimensioning purposes.

Dimension Lines

Dimension lines are thin lines used to show size or location of a feature of the structure. They may be placed outside or, if there is sufficient space, inside the object. All dimension lines have a dimension figure (number or letter) about halfway between the ends. Each end has some type of termination (ending) symbol. See **Figure 5-2**.

Long Break Lines

Long break lines are thin lines used to show that all of the part is not shown. They extend past the object about 1/16″ on either side and have a zigzag symbol in the center.

Short Break Lines

Short break lines are heavy lines used when part of the object is shown broken away to reveal a hidden feature. They normally have a curvy, irregular shape.

Cutting Plane Lines

Cutting plane lines are heavy lines used to indicate where the object has been sectioned to show internal features. These lines are generally attached to a section symbol that identifies the detail drawing where the section view is shown.

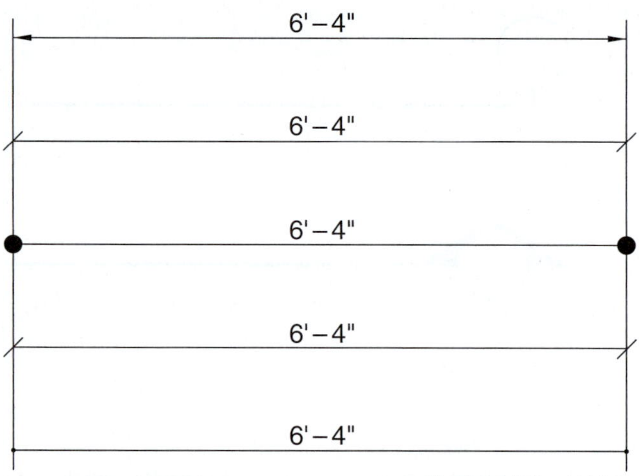

Goodheart-Willcox Publisher

Figure 5-2. Various methods of terminating a dimension line.

Section Lines

Section lines, or ***crosshatch lines,*** are very thin lines used to show that the feature has been sectioned. Section lines may represent a specific material or may be a general symbol. General section lines are usually drawn at 45° and about 1/16″ to 1/8″ apart.

The various line symbols are identified on a simple floor plan in **Figure 5-3**. Study these lines. They are the basic symbols used on all construction drawings.

5.3 Symbols and Abbreviations

The purpose of symbols and abbreviations is to communicate efficiently and to conserve space. Some symbols look much like the actual feature they represent, while many others do not. Many symbols are found on each set of construction drawings, including the following types:

- Building material symbols
- Topographical symbols
- Plumbing symbols
- Climate control/mechanical symbols
- Electrical symbols

Figure 5-4 shows examples of the types of symbols listed above.

Symbols may represent an elevation (front, side, or rear) or a section view of the material or feature. Elevation symbols are easiest to recognize because they are picture-like. Section symbols are not pictorial representations. They must be memorized or otherwise identified.

Trade Tip

Some symbols represent several different materials or conditions. Be careful when reading them. For example, the face brick section symbol in **Figure 5-4** is the same as the general material symbol. The sand symbol (a series of dots) is the same in elevation and section. The sand symbol is also exactly like the section of cut stone, plaster, and the elevation of cast concrete. Study the location of the symbol and type of drawing before deciding what the symbol represents.

Abbreviations save space and are widely used on construction drawings. Tradespeople must be able to read and interpret all abbreviations that have anything to do with their specific function or trade. The *Reference Section* of this book lists many frequently used abbreviations.

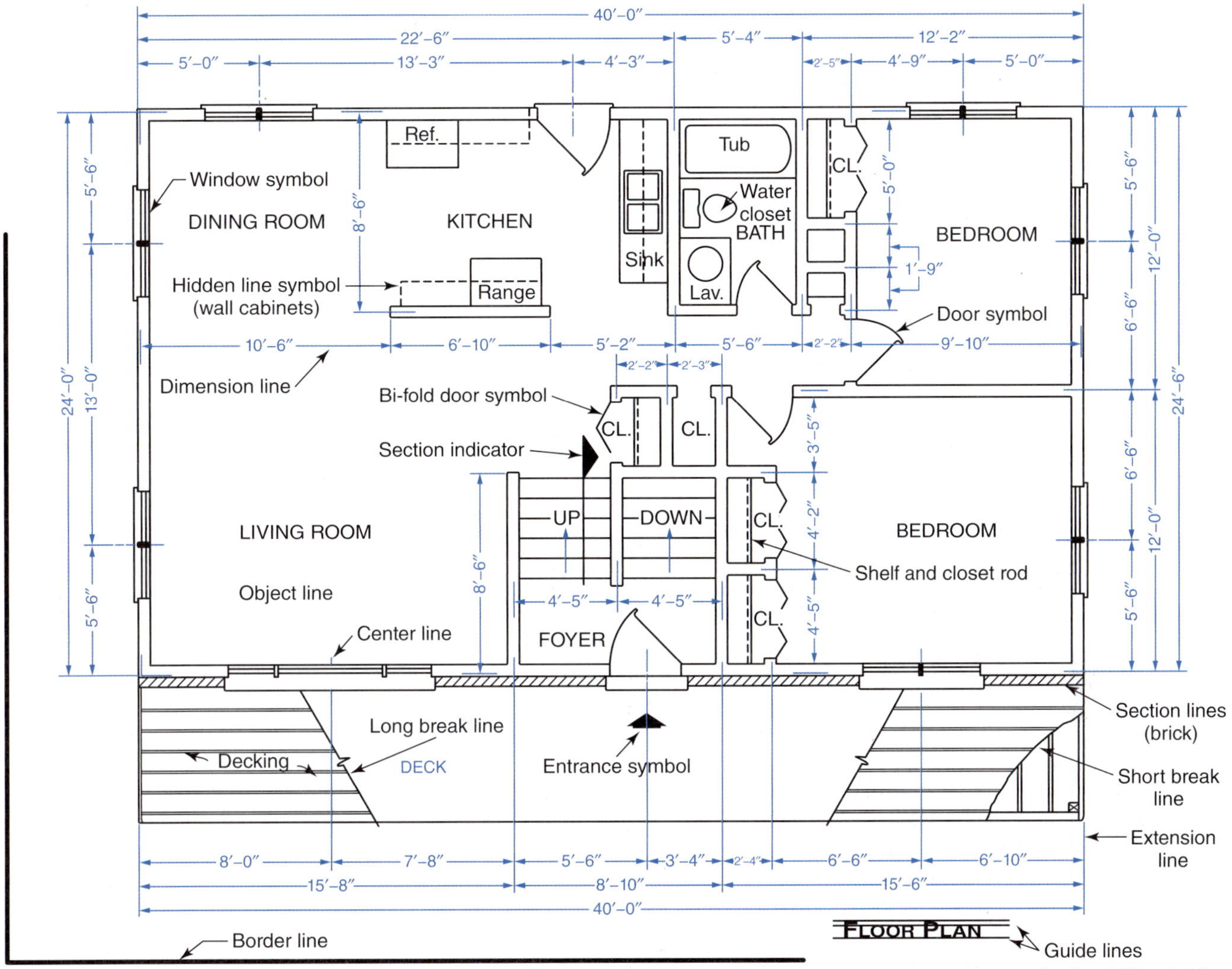

Goodheart-Willcox Publisher

Figure 5-3. Simple floor plan with various line symbols identified.

5.4 Scale and Dimensions

Most drawings are drawn at a particular scale. Residential floor plans, elevations, and foundation plans are generally drawn at 1/4″ = 1′ scale. This means that every 1/4″ on the drawing is equal to 1′ on the house. Details are almost always drawn at a larger scale, such as 1/2″ = 1′ or 1″ = 1′. Do not confuse the terms *size* and *scale*. Size means inches to inches (1/4″ to 1″ or quarter size) and scale means inches to feet (1/4″ to 1′ or quarter scale).

For example, if the side of a building were 40′ long and drawn at 1/4 scale, the side of the building would be 10″ on the drawing. However, if it were drawn at 1/4 size, it would be 10′ long on the drawing.

Drawings of large commercial buildings are often drawn at scales other than 1/4″ = 1′. A scale of 1/8″ = 1′ is common. Be sure to look for the scale on the drawing so you can visualize the relative size of the building elements.

Trade Tip

As a rule, never measure distance on a drawing to determine a length. If a needed dimension is not listed on a drawing, use the dimensions that are provided to calculate the needed dimension. If the drawing does not include enough information to determine the distance, ask the architect or builder to provide the missing information.

Dimensions, which indicate size and location of building elements, must be followed very accurately. Dimensions are usually shown in feet and inches. However, they may be shown in inches if the length is less than 1′ or is a standard distance that is more easily recognized in inches.

Interior wood frame walls may be dimensioned several ways, **Figure 5-5**. They may be dimensioned to the center of the wall, to the outside of the studs, or to the outside of the finished wall.

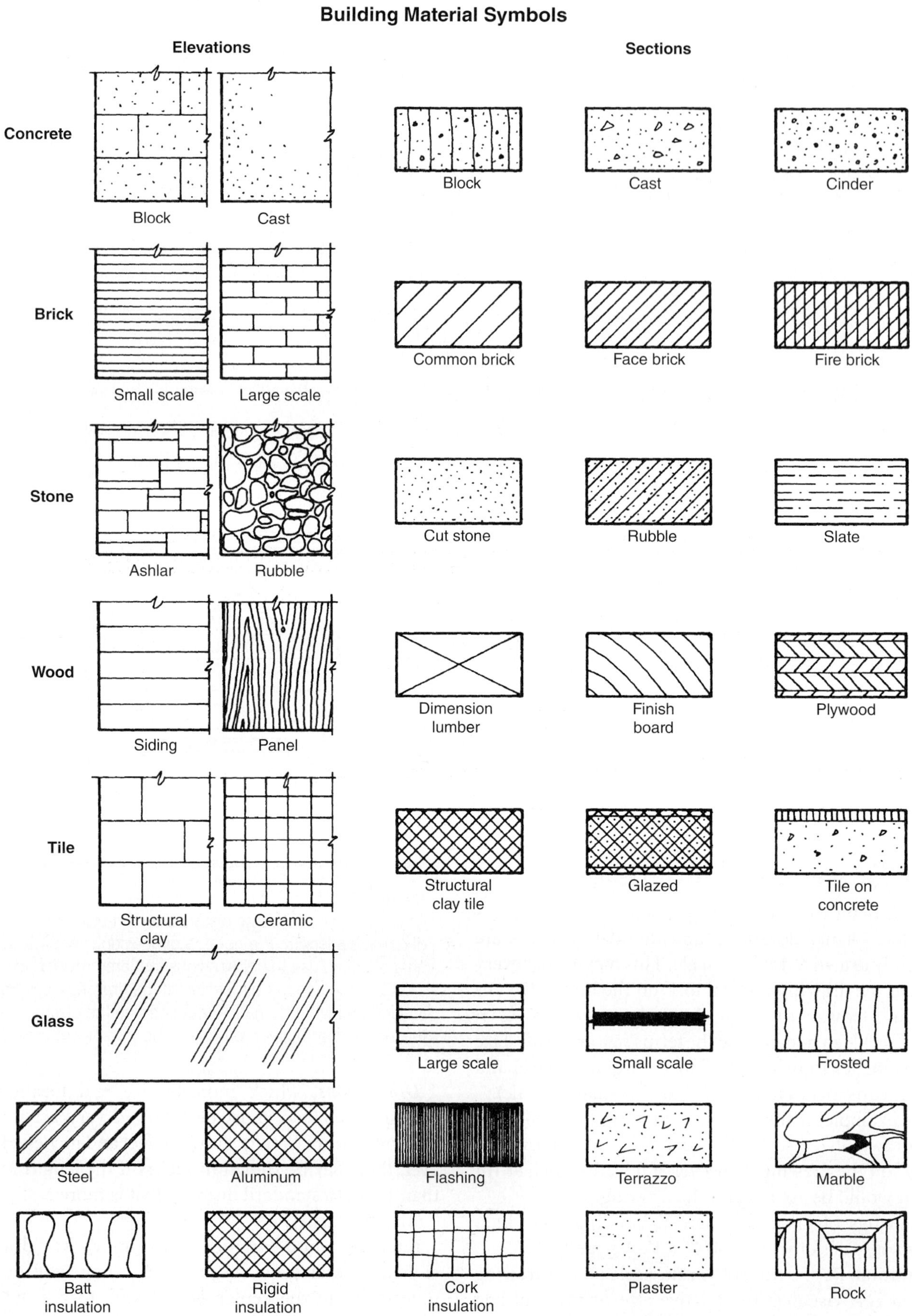

Goodheart-Willcox Publisher

Figure 5-4. Symbols commonly found on construction drawings. Some are shown in elevation and section drawings.

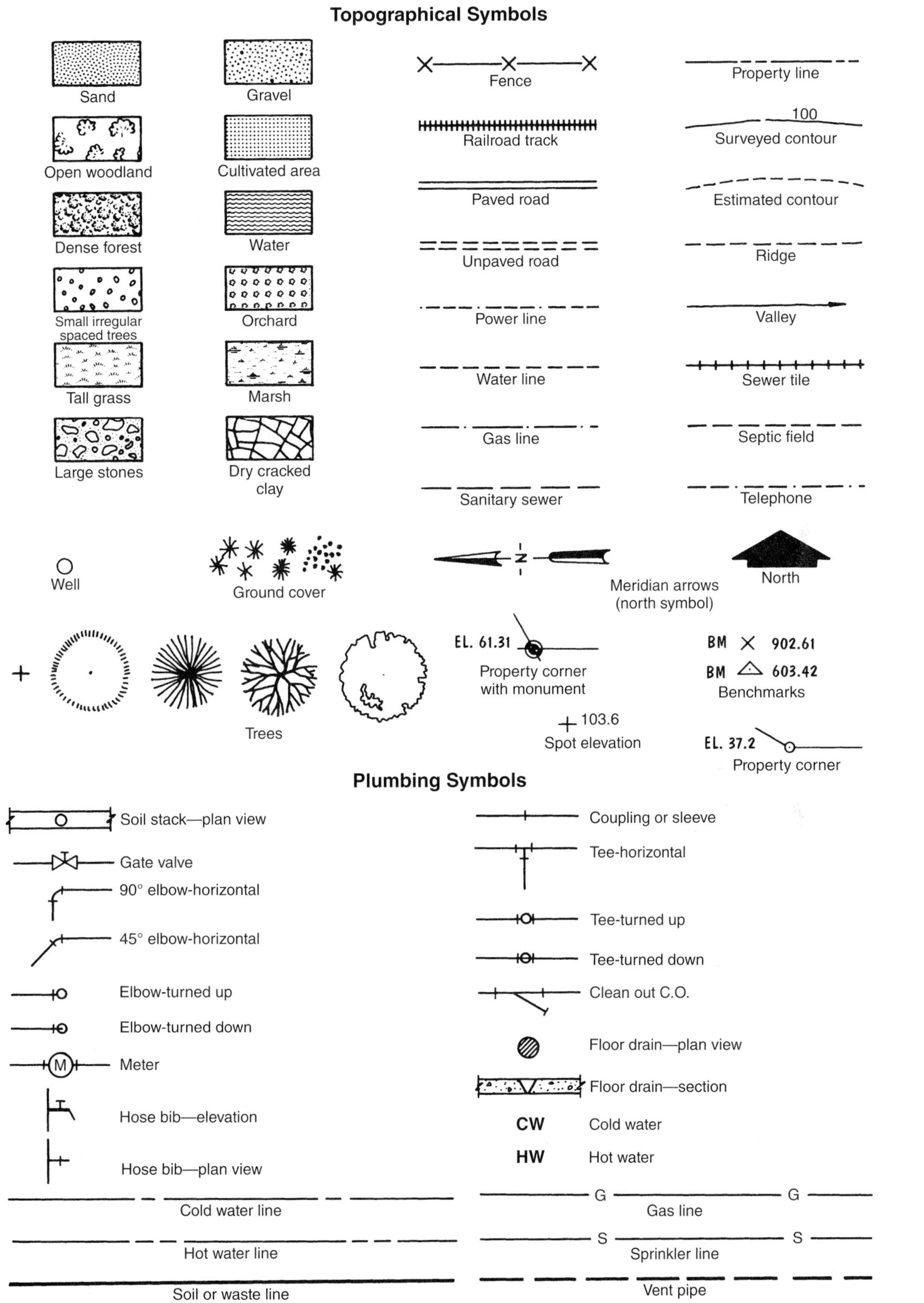

Goodheart-Willcox Publisher

Figure 5-4 (Continued).

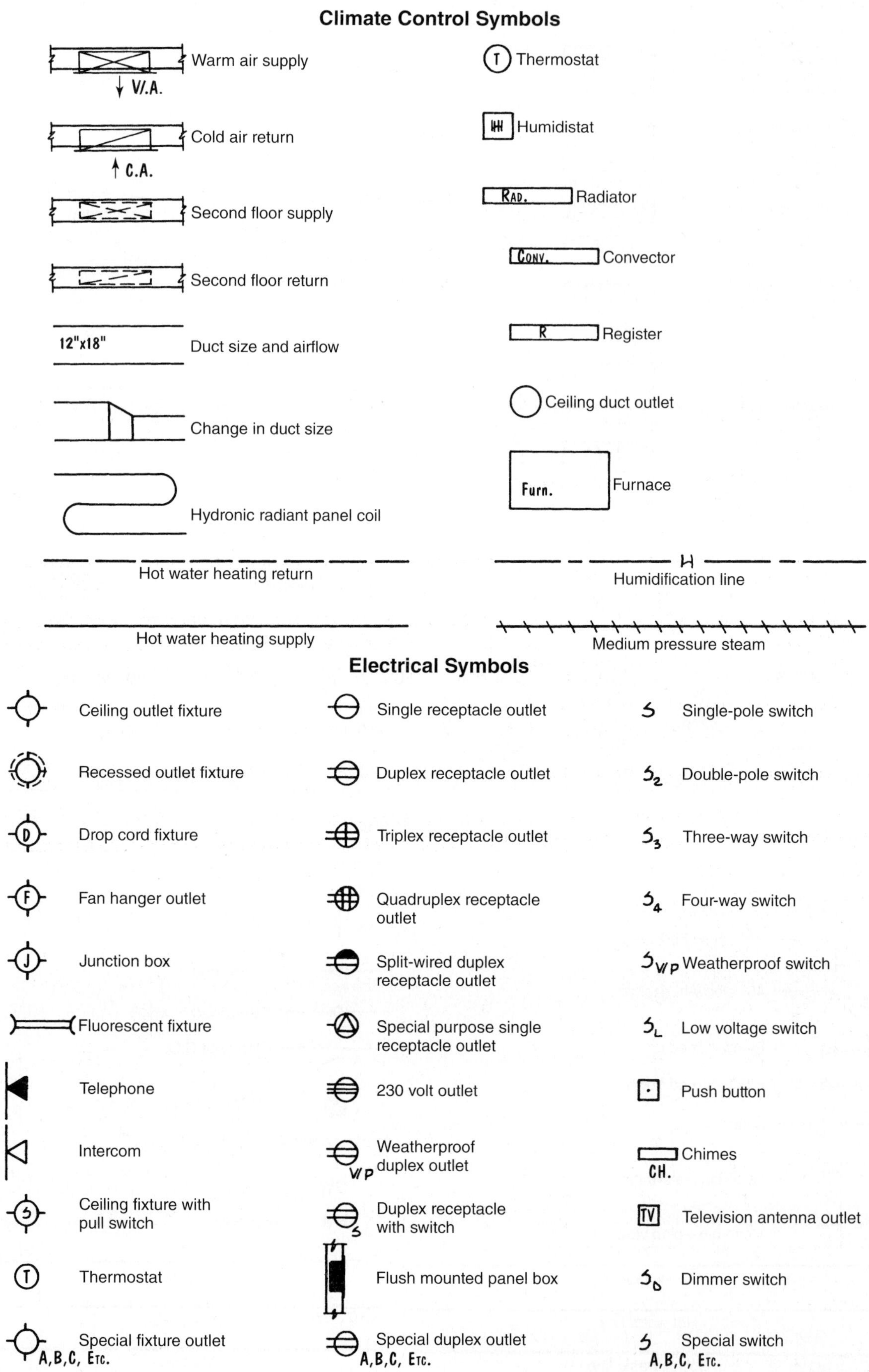

Goodheart-Willcox Publisher

Figure 5-4 (Continued).

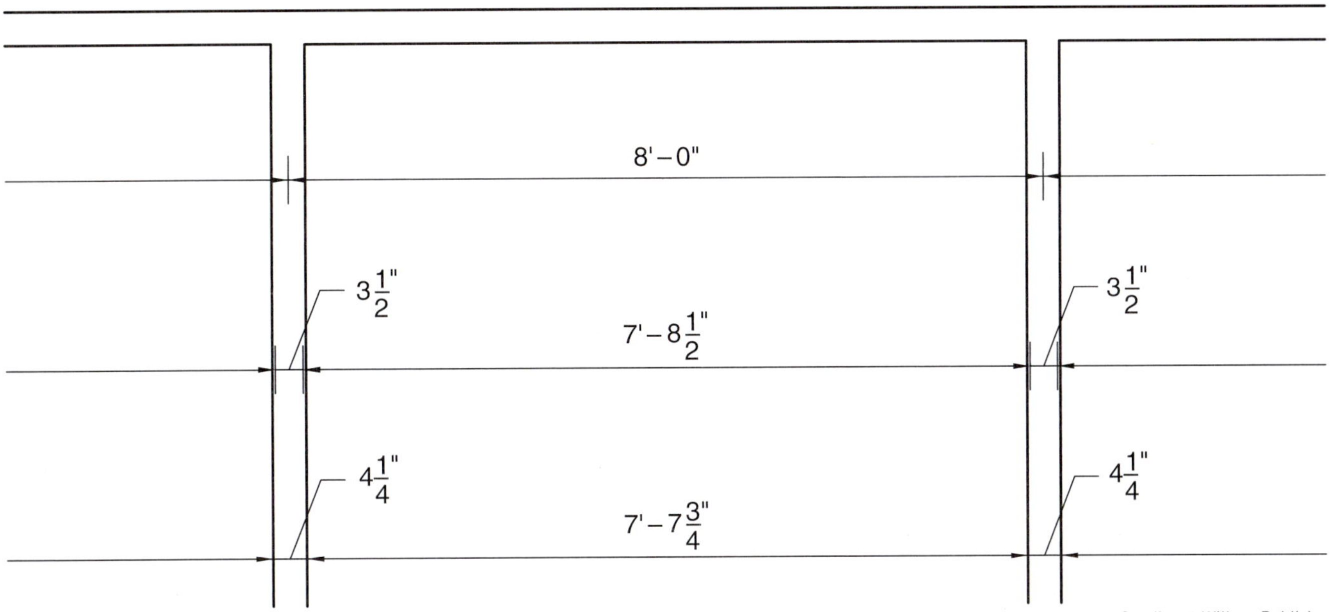

Goodheart-Willcox Publisher

Figure 5-5. Accepted methods of dimensioning interior frame walls.

The technique used may be easily determined by remembering that a 2 × 4 stud is actually 1 1/2″ × 3 1/2″. The thickness of finished wall material can vary from 1/4″ to 1″ or more. Dimensioning to the center of the wall is the most commonly used method.

Interior masonry or concrete walls are dimensioned to the outside of the walls. See **Figure 5-6**. They are never dimensioned to the center.

Exterior wood frame walls are dimensioned to the outside of the stud wall. This dimension usually includes the sheathing but not the siding. See **Figure 5-7**. Again, masonry walls are dimensioned to the outside of the wall. Brick veneer is dimensioned to the outside of the stud wall.

Drawing notes are frequently required in order to present information that cannot be represented by a conventional dimension or symbol. Be sure to read all drawing notes. They contain information that is needed to build the structure.

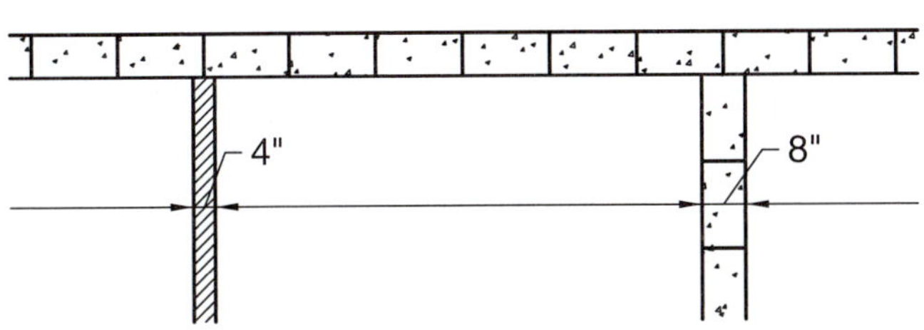

Goodheart-Willcox Publisher

Figure 5-6. Interior masonry and concrete walls are dimensioned to the outside of the walls. Wall thickness is also shown.

Trade Tip

Study the entire set of plans and specifications before starting work on a building. Any questions or problems should be cleared up before work begins.

5.5 Metric System of Dimensioning

The metric system of measurement is used in most countries. A simple metric floor plan is shown in **Figure 5-8**.

In construction, the basic metric unit of measure is the meter. Units for linear measurement in the construction industry should be restricted to the meter (m) and the millimeter (mm). On drawings, whole number dimensions normally indicate millimeters, and decimal numbers (to three places of decimals) normally indicate meters.

Before universal adoption of the metric system can be made by the building industry, new lumber sizes and standards must be approved. Clay products, concrete masonry products, and reinforcing steel can be specified in metric sizes. These nominal metric sizes are examples of a soft conversion, which means the actual sizes of the products have not changed, but approximate metric sizes are used.

5.6 Working Drawings

Working drawings are the drawings used to construct a building. They are sometimes called a *set of working drawings*. ***Preliminary drawings*** are drawings that are

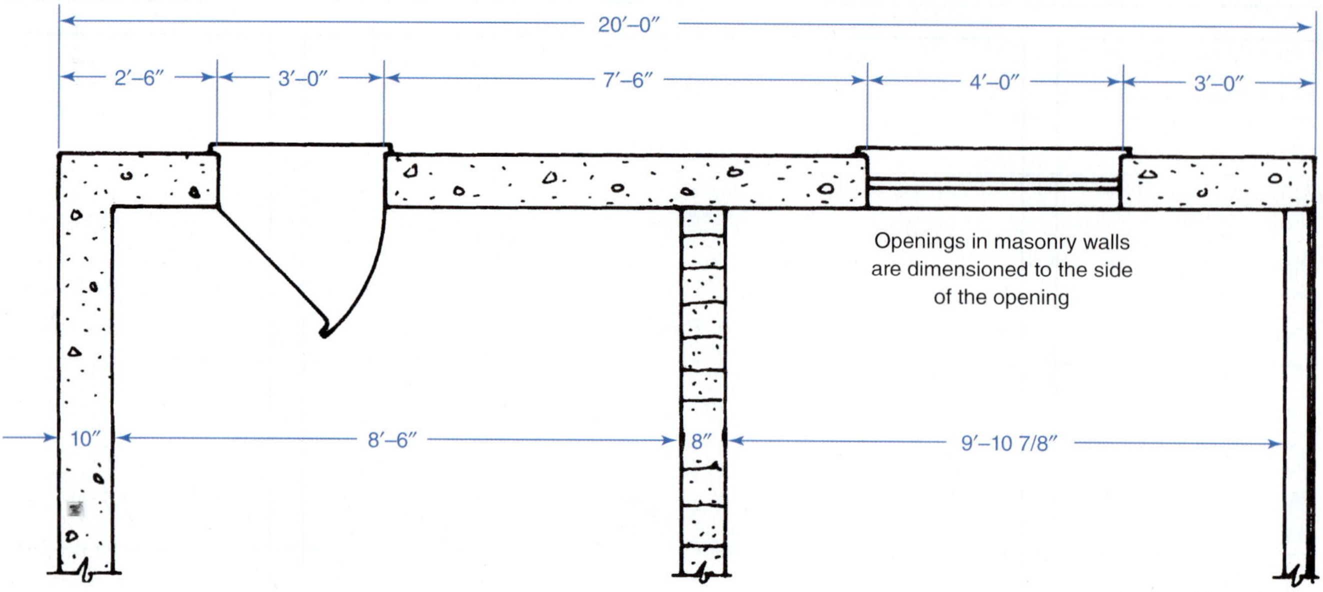

Goodheart-Willcox Publisher

Figure 5-7. Exterior frame walls are dimensioned to outside of the stud wall. Sheathing is usually included but siding is not. Exterior masonry or concrete walls are dimensioned to the outside with thickness shown.

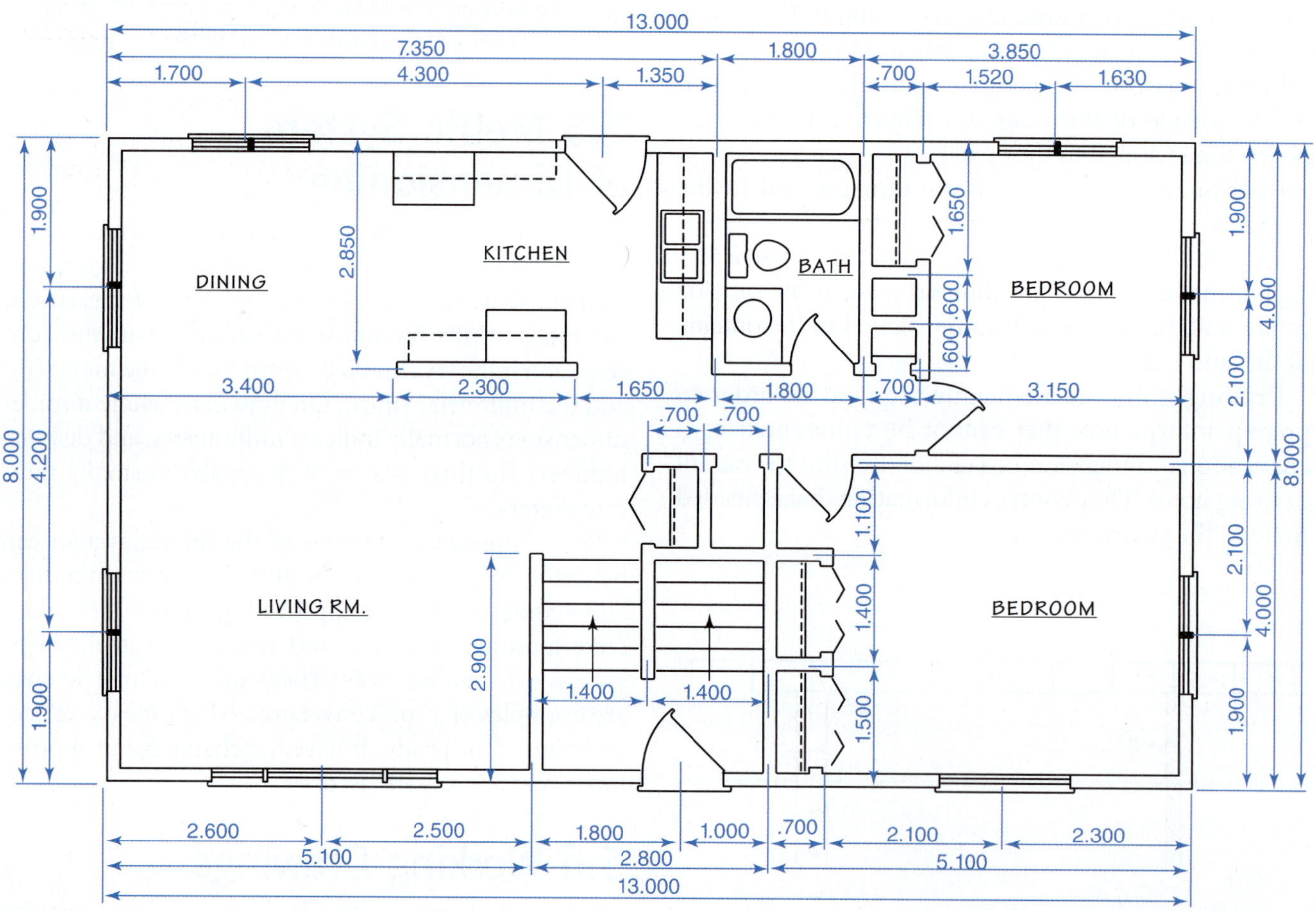

Goodheart-Willcox Publisher

Figure 5-8. Units for linear measurement in the construction industry should be restricted to the meter (m) and the millimeter (mm).

often prepared during the promotional stage of a building's design. Corrections and revisions are recorded on the preliminary drawings. These drawings form the basis for the working drawings.

Presentation drawings are generally pictorial (picture-like) drawings. A pictorial drawing shows several faces of the structure to be built, giving a better idea of how the structure will actually look.

There are several types of pictorial drawings. The one used most often by architects is the ***perspective drawing***, sometimes called an ***architectural rendering***. A perspective is more natural in appearance, like a photograph. In a perspective drawing, parallel lines move closer together as they move away from the foreground. You can observe the same phenomenon when studying a photograph of a building.

Many other types of working drawings are necessary to construct a building. Shop drawings are usually prepared by the trades participating in the project in order to complete their part of the work.

Working drawings are generally prepared by the architect or designer. These drawings show the size, shape, location, quantity, and relationship of the parts of the building in graphic form. They are produced in sufficient quantity so each subcontractor or individual trade area has access to a set of prints.

A set of working drawings for a typical structure might include the following:

- Site plan
- Foundation plan
- Floor plan
- Elevations
- Electrical plan
- Mechanical plan
- Construction details

The plans for a small concession building illustrate the type of information commonly shown on the various drawings. See **Figures 5-9** through **5-19. Figure 5-9** shows a pictorial drawing of the concession building. The following paragraphs briefly describe each working drawing.

The ***site plan***, **Figure 5-10**, shows the location of the building on the site. It may also show utilities, topographical features, site dimensions, other buildings on the property, landscaping, streets, walks, drives, retaining walls, and property lines.

The ***foundation plan***, **Figure 5-11**, shows the foundation walls and their sizes and materials. It may also give information about the excavation, reinforcing materials, special wall construction procedures, waterproofing, and supporting structures such as footings and piles. The accuracy of this drawing is very important. Upper parts of the building depend on the layout and construction of the foundation.

The ***floor plan***, **Figure 5-12**, shows all exterior and interior walls, doors, windows, patios, walks, decks, fireplaces, built-in cabinets, and appliances. A separate plan view is drawn for each floor. A floor plan is a section drawing usually cut about 4′ above the floor. Small buildings are usually drawn at 1/4″ = 1′ scale. Larger buildings may be drawn at 1/8″ = 1′ or some other scale.

The InterDesign Group, Indianapolis, Indiana

Figure 5-9. A pictorial view of a school concession building.

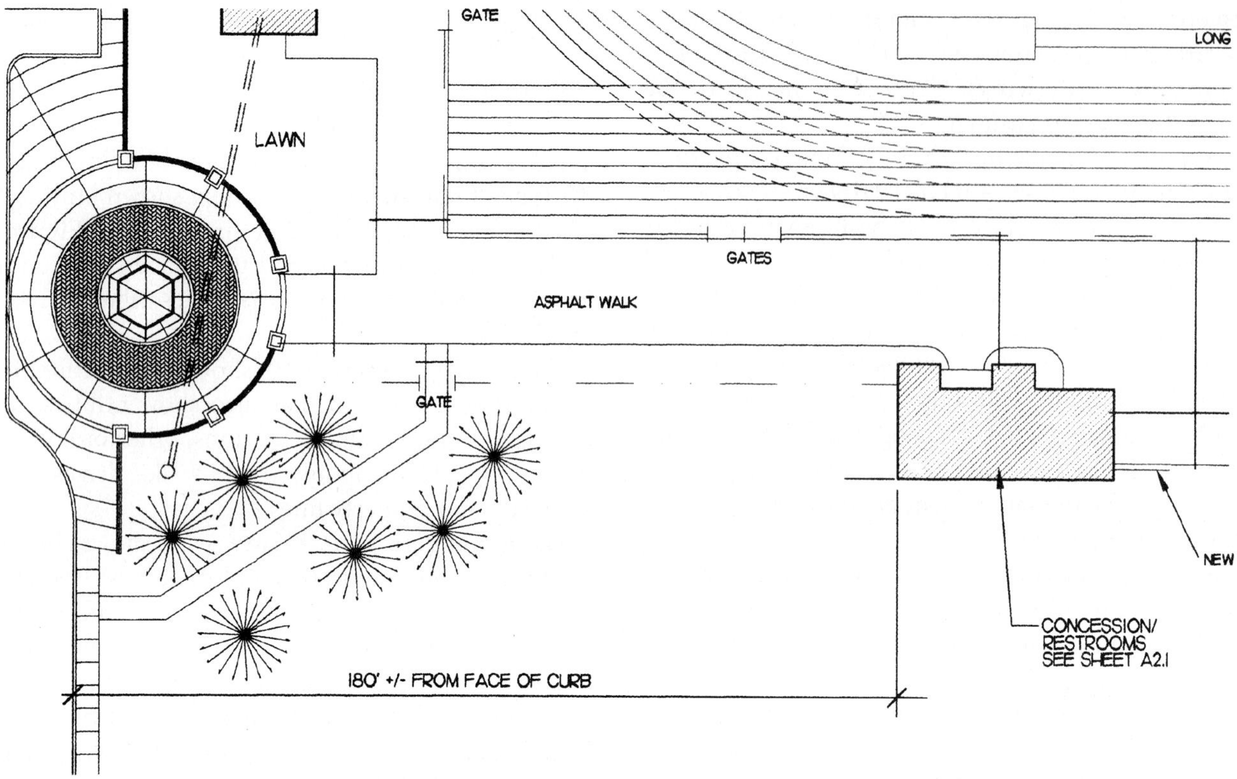

The InterDesign Group, Indianapolis, Indiana

Figure 5-10. Site plan for the concession building.

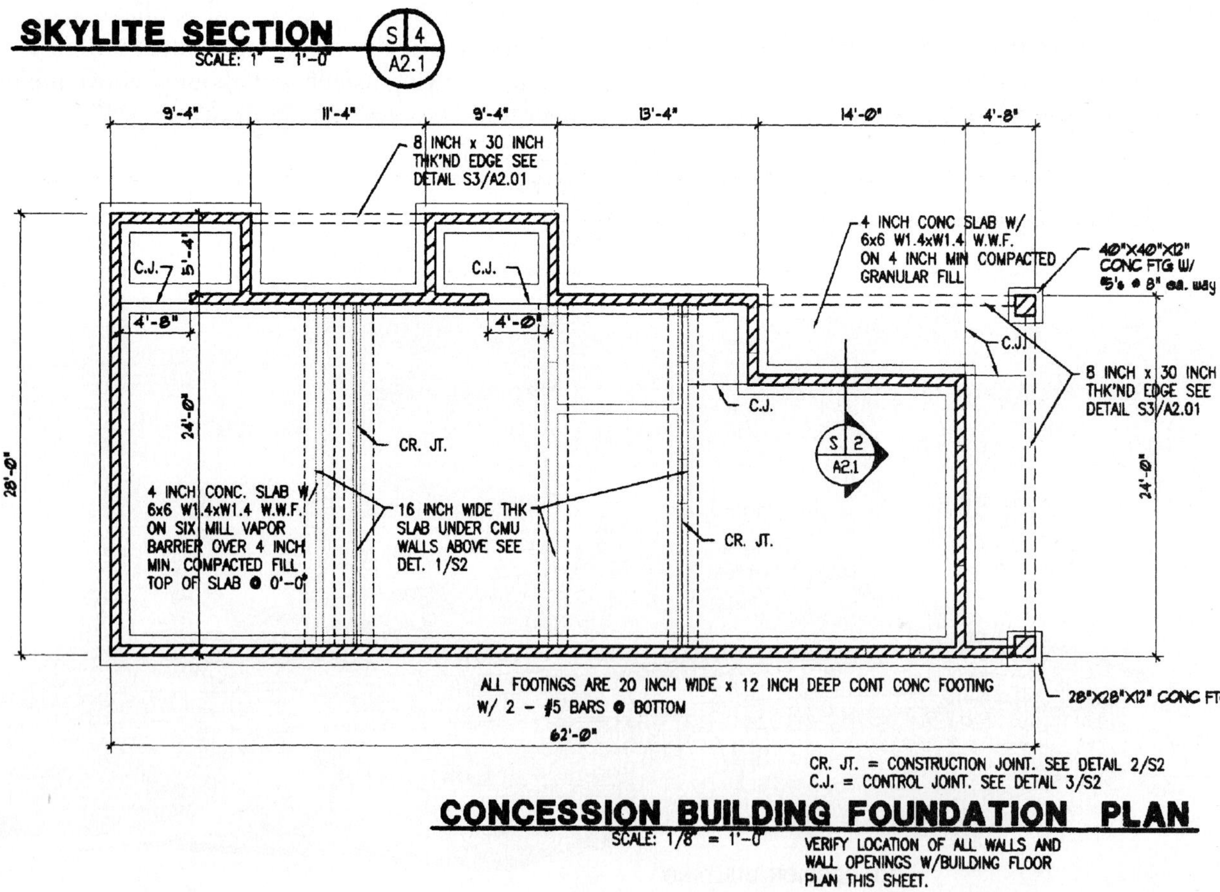

The InterDesign Group, Indianapolis, Indiana

Figure 5-11. Foundation plan for the concession building.

Elevation drawings, **Figure 5-13**, are drawn for each side of the structure. These drawings show outside features such as placement and height of windows, doors, chimney, and roof lines. Exterior materials are indicated, as well as important vertical dimensions. Section lines can also be shown on elevation drawings, as well as on the foundation plan, floor plan, and other plan views. Section views and detail drawings will be discussed later in this chapter.

The ***electrical plan***, **Figure 5-14**, is drawn from the floor plan. It locates switches, convenience outlets, ceiling outlet fixtures, TV jacks, service entrance

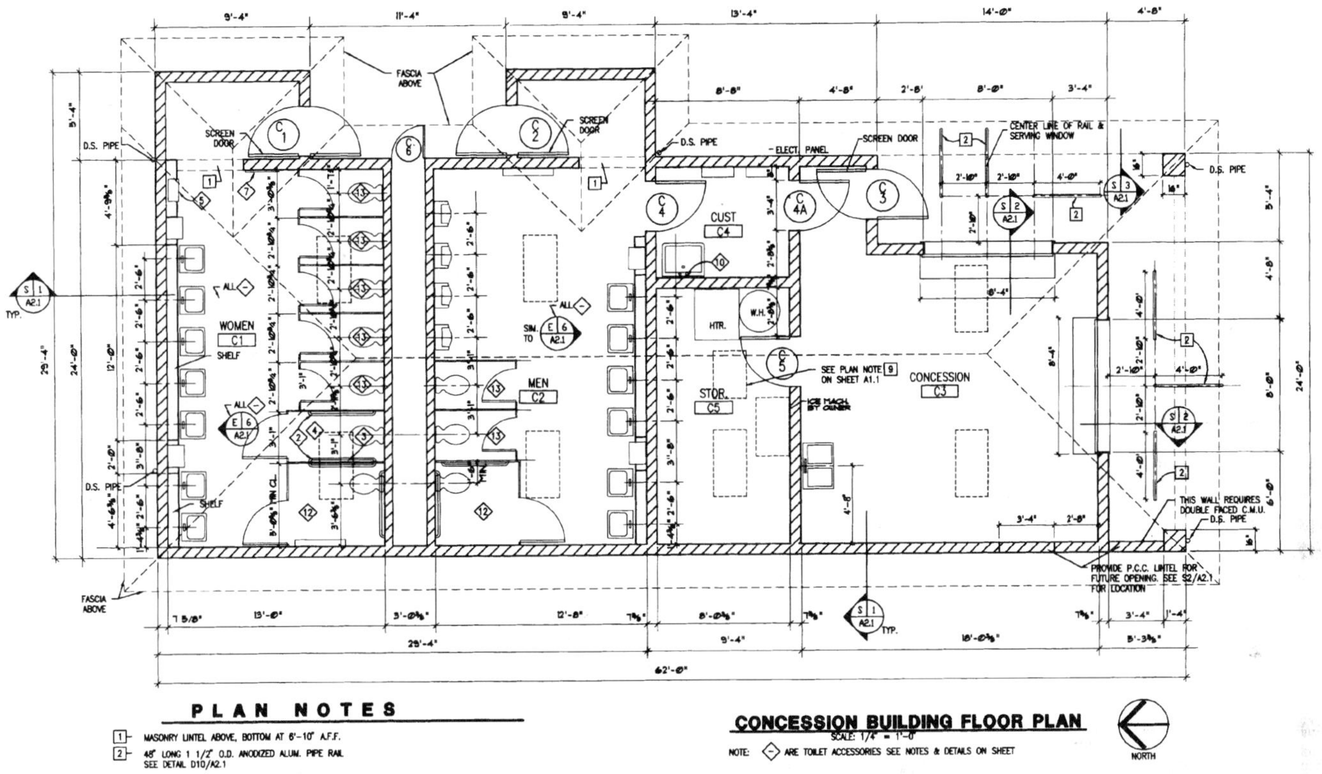

The InterDesign Group, Indianapolis, Indiana

Figure 5-12. Floor plan for the concession building.

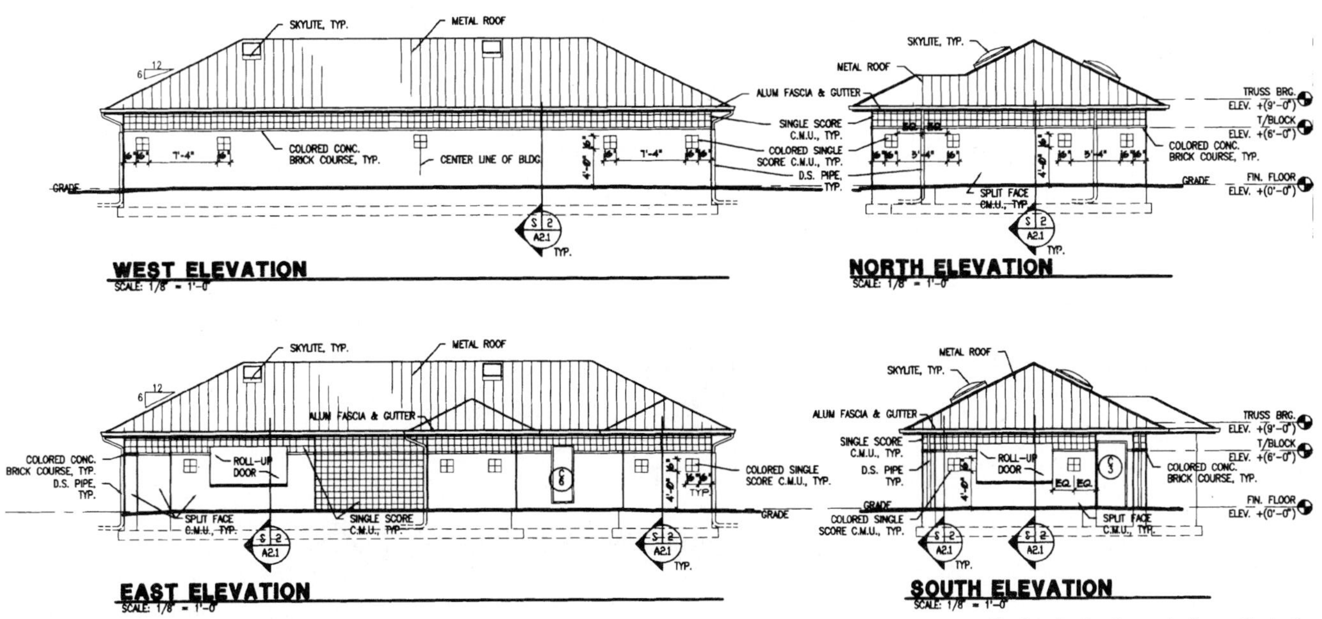

The InterDesign Group, Indianapolis, Indiana

Figure 5-13. Elevation drawings plans for the concession building.

location, and panel box. It also gives general information concerning circuits and special installations.

The ***mechanical plan***, **Figure 5-15**, shows the plumbing, heating, and cooling systems. Each of these systems may be shown on separate plans designated as the plumbing plan, heating plan, and cooling plan. A plan showing heating, ventilation, and air-conditioning is sometimes called an *HVAC plan*.

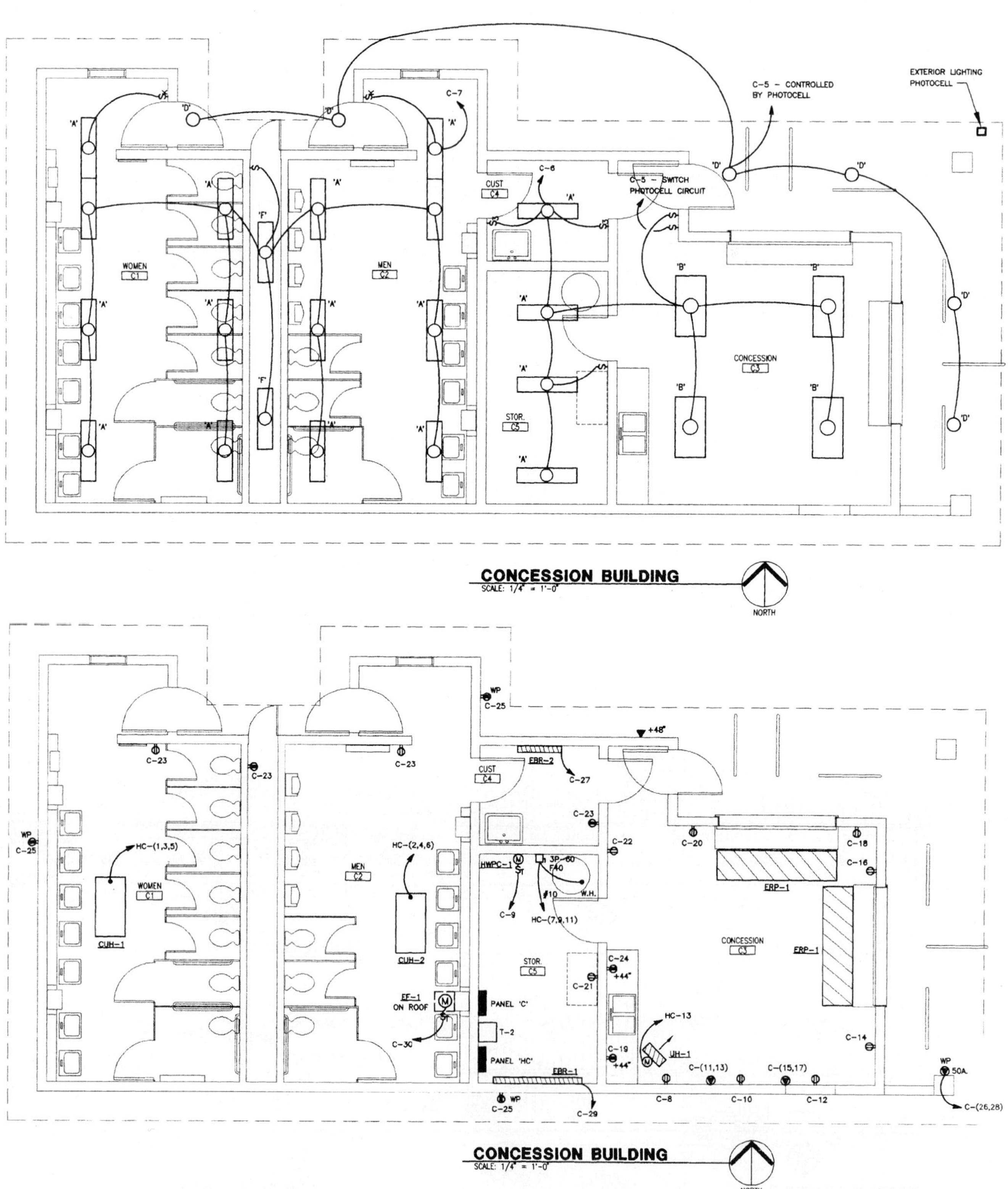

The InterDesign Group, Indianapolis, Indiana

Figure 5-14. Electrical and lighting plans for the concession building.

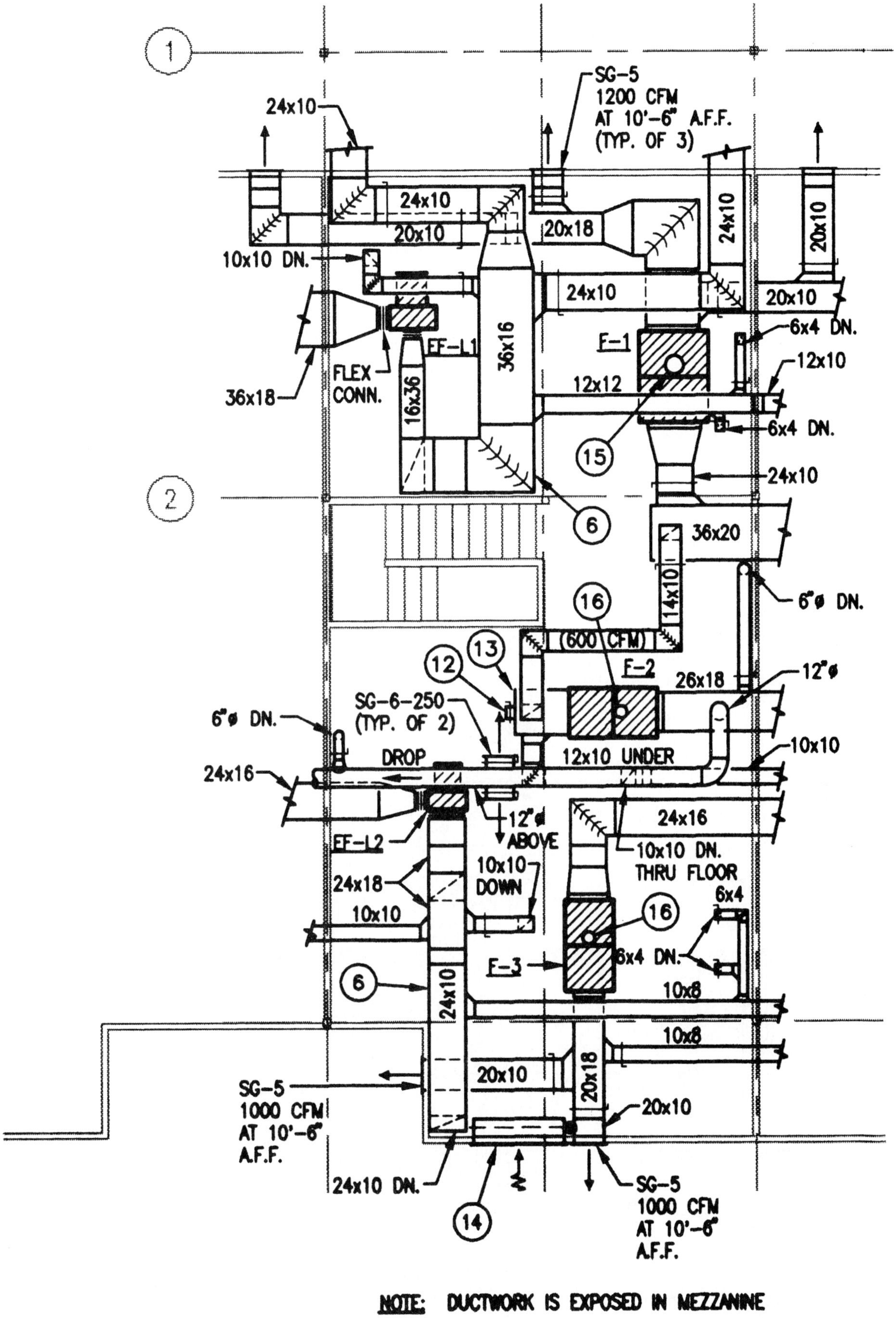

The InterDesign Group, Indianapolis, Indiana

Figure 5-15. Mechanical plans for the concession building.

Other processes that may be included in an HVAC plan include humidification, dehumidification, and air-cleaning equipment. Another name for a mechanical plan is a ***climate control plan***.

Detail drawings, **Figure 5-16**, provide construction details of specific areas of a building. Detail drawings document the architect's vision of the internal and external images in print form, which then are viewed by the builder to accurately recreate the "plan" through the construction process. These drawings show the builder how different building materials are assembled to form a specific area of a structure.

A section drawing reveals the internal makeup of the wall or other area that needs to be shown in greater detail. Section views are generally drawn of areas for which more information is needed to construct the specific area. Typical detail drawings include section views that are drawn through the entire structure, wall sections, stair sections, fireplace sections, and fireplace details. Other special masonry features

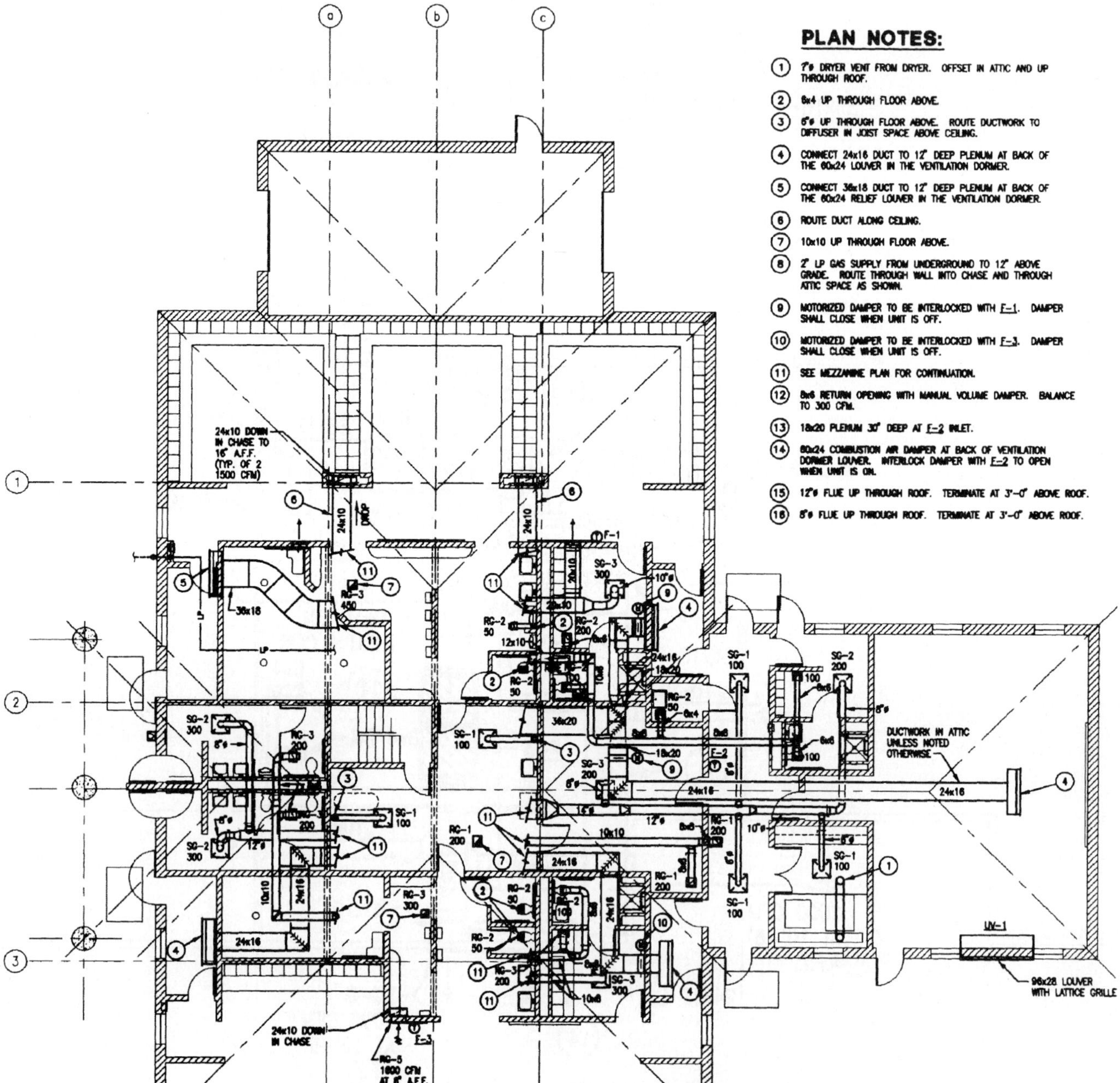

The InterDesign Group, Indianapolis, Indiana

Figure 5-15 (Continued).

or unique construction processes can be shown in detail drawings.

To fully understand the relationship between a plan drawing and its detail drawings, refer back to **Figure 5-12**, which shows the floor plan of the concessions building. Find the section indicator *S2/A2.1*. This symbol is a circle with the *S* and the *2* in the top half of the circle and the *A2.1* in the bottom half. The *S2/A2.1* symbol refers

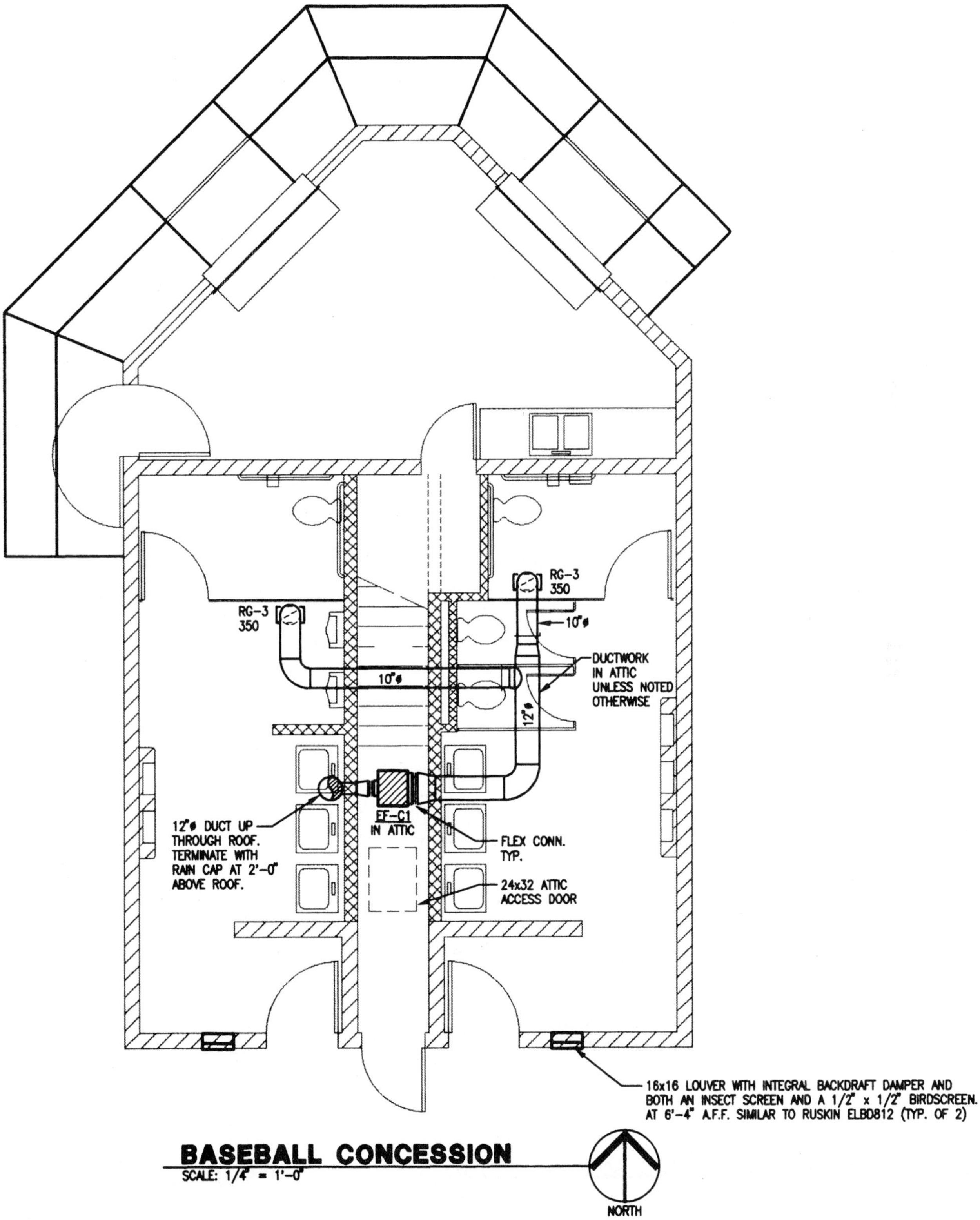

The InterDesign Group, Indianapolis, Indiana

Figure 5-15 (Continued).

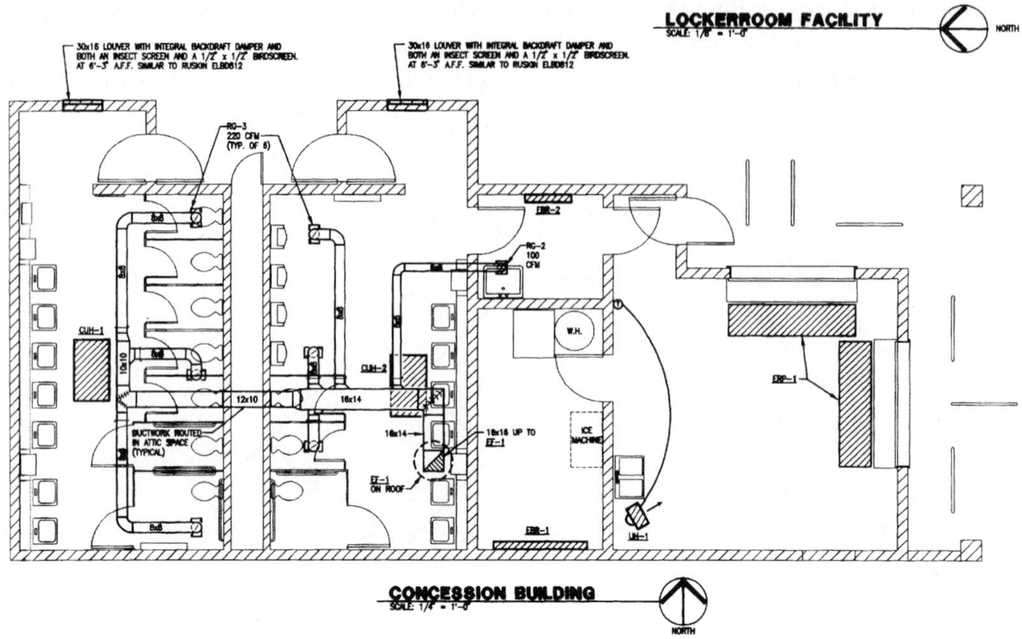

The InterDesign Group, Indianapolis, Indiana

Figure 5-15 (Continued).

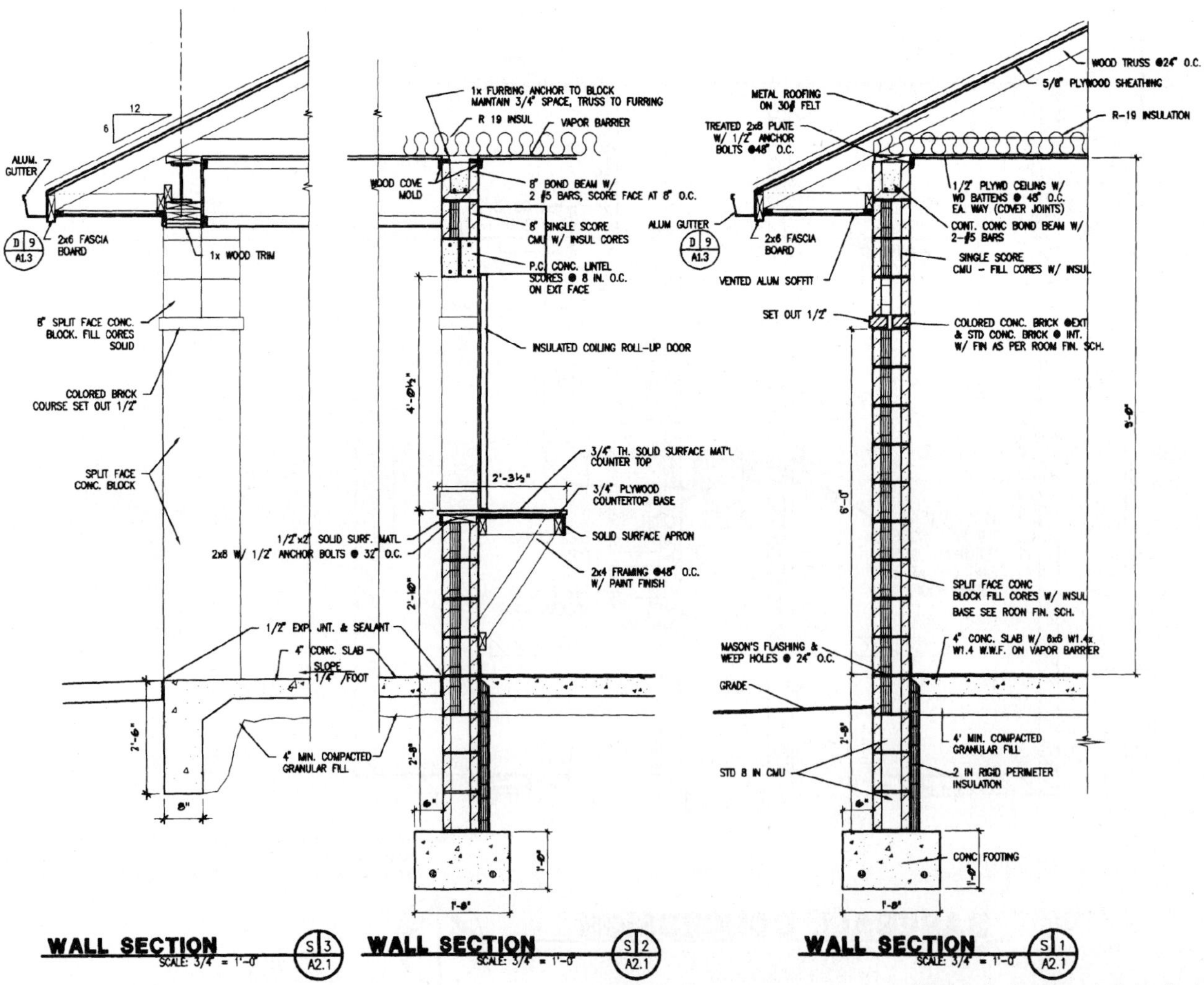

The InterDesign Group, Indianapolis, Indiana

Figure 5-16. Construction details of wall sections. Each of the three illustrations is a representation of a cutting plane line and is shown as a wall section drawing.

to one of the detail drawings, specifically the detail *S2* on drawing number *A2.1*. The line attached to the circle of S2/A2.1 cuts through the wall at that point in the floor plan drawing. This line is a ***cutting plane line*** that allows the builder to view the internal parts of the wall at that specific point in a detail drawing. This detail is shown in **Figure 5-16**.

Refer to **Figure 5-11**, the foundation plan. Symbol S2/21 is displayed there also. Note that the symbol *S2/A2.1* is in the same place on both the floor plan and the foundation plan. This indicates that the section drawing shows both the foundation plan and the floor plan information, giving the tradesperson a graphical view of how the two floors are assembled.

In addition to the drawings described previously, a roof plan, framing plan, landscaping plan, presentation plan, and shop drawings may be needed. A complex structure usually requires more drawings.

A ***roof plan***, **Figure 5-17**, is included if the roof is complex or not clearly shown on other drawings. A roof plan shows ridges, valleys, skylights, and other features of the roof. It also displays the sheathing and roofing material, such as shingles or metal standing seam roof. Sometimes the roof features may be shown as part of the site plan.

A ***framing plan***, **Figure 5-18**, is designed to show in detail the framing required for a roof, floor, or other framed area.

A ***landscaping plan***, sometimes combined with the site plan, locates and identifies trees, plants, hardscape, and other elements included in landscaping. (The term *hardscape* refers to materials made from clay or concrete such as pavers, specialized concrete brick and block, and other units.)

A ***presentation drawing***, or architectural rendering, shows how the finished structure will appear. A two-point perspective is commonly used. Refer to **Figure 5-9**.

Shop drawings provide additional information about such things as reinforcing steel, complex cabinetwork, and electronic systems. This information may be furnished by suppliers or by the various subcontractors involved in the construction or installation process.

Drawing notes are included in a set of construction drawings to clarify details that might otherwise cause confusion. **Figure 5-19** shows the drawing notes associated with the concession building.

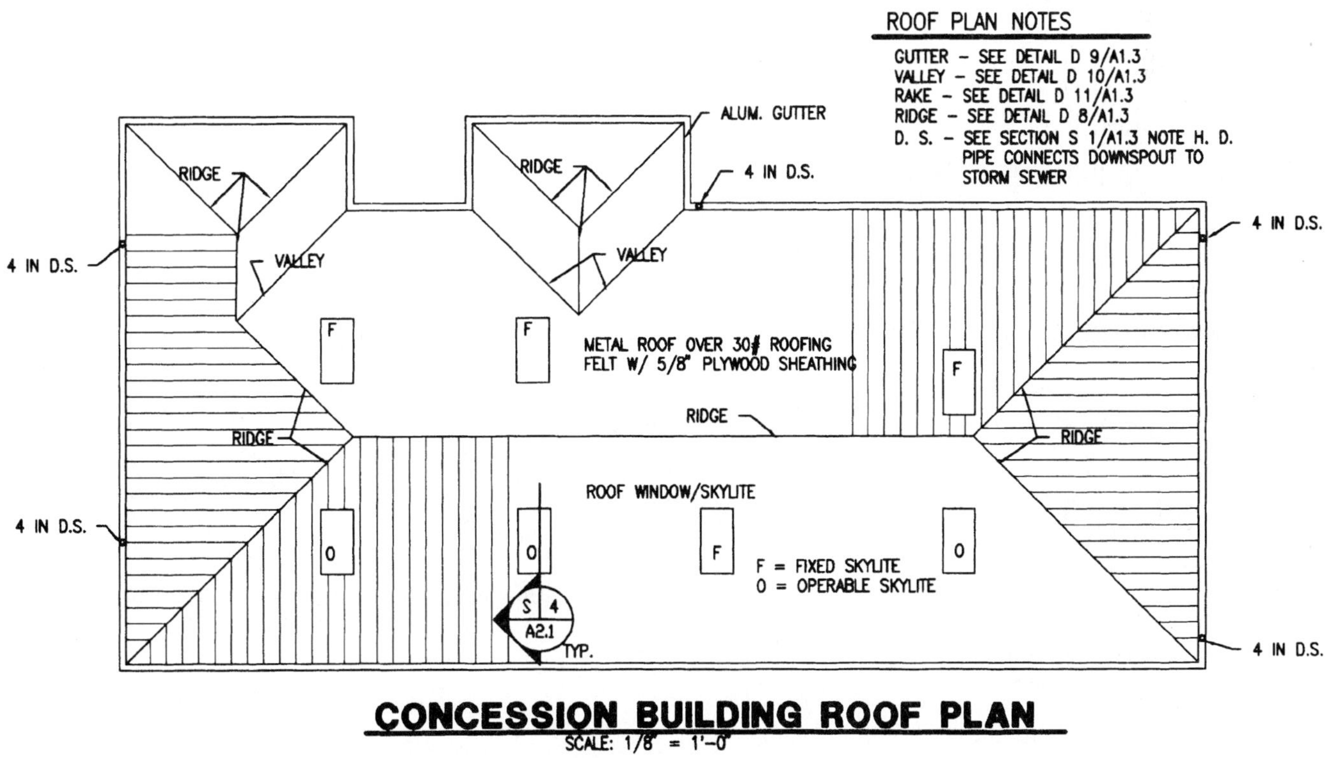

The InterDesign Group, Indianapolis, Indiana

Figure 5-17. Roof plan showing ridges, valleys, and skylights.

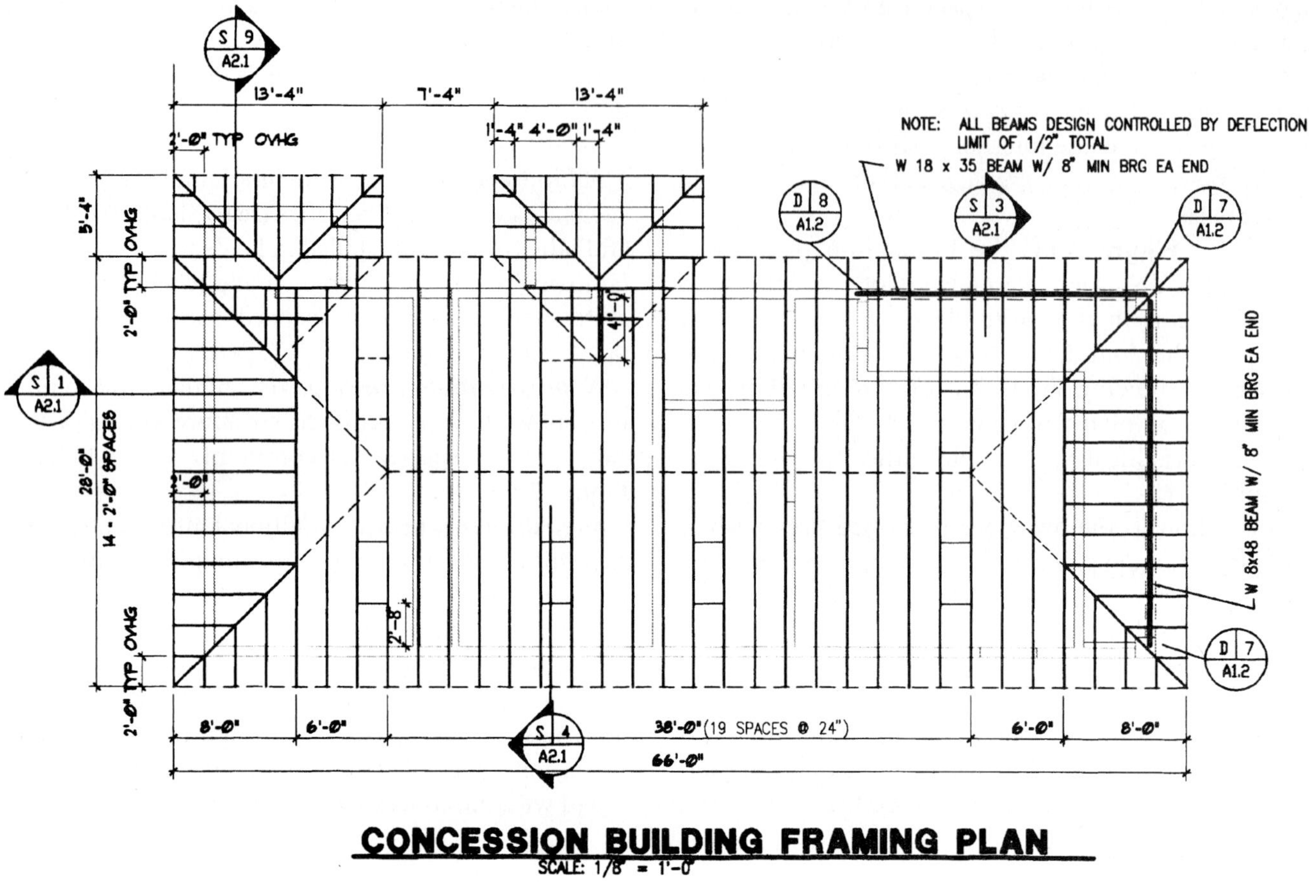

The InterDesign Group, Indianapolis, Indiana

Figure 5-18. Framing plan including beam sizes and locations, trusses, section locations, notes, and scale.

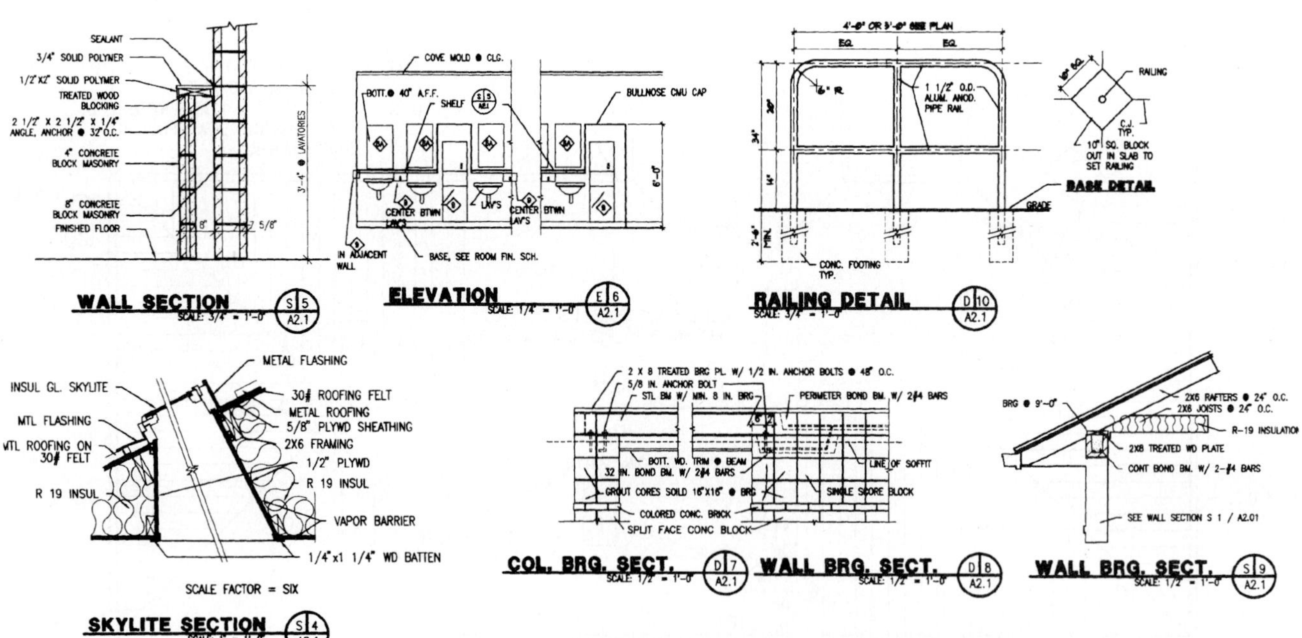

The InterDesign Group, Indianapolis, Indiana

Figure 5-19. Drawing notes for the concession building.

Workplace Skills

Continuing Education with Print Reading

Learning the language of print reading can be a very enjoyable activity. This chapter will give you the basics to recognizing the different symbols and lines used by architects and designers. To become proficient in print reading, advanced classes are offered through apprenticeship programs, community college courses, and management training activities. Being promoted into management positions will require advanced training and leadership skills. Continuing education beyond school will help ensure you continue to advance in your career.

5.7 Specifications

Each set of drawings is ordinarily accompanied by a set of written ***specifications,*** or *specs*. Specifications are written descriptions of the construction project that are prepared by architects and engineers. A set of specifications may be only a few pages long, or it may consist of several large volumes if the structure is large and complex. The specifications, together with the working drawings, become the construction documents upon which an agreement is made between all parties concerned in the project.

Aspects of a building that cannot be adequately shown or explained on the drawings must be included in the specifications. For example, quality of workmanship, responsibility for various aspects of the construction, and quality of materials and fixtures are explained or identified in the specifications.

Information in the specifications is meant to amplify (enlarge) the drawings and should agree with the drawings point-by-point. Should a discrepancy occur between the specifications and the drawings, the specifications usually are accepted (take precedence) over the drawings.

Specifications are generally divided into two major sections:

- General conditions
- Technical sections

General conditions are specifications that deal with items such as terms of the agreement, insurance, responsibility for permits, supervision, payment for the work, and temporary utilities.

Technical sections are specifications that deal with the actual building of the structure and the tradework involved in the building. Each section is designed to include information on work to be done by a single subcontractor. The arrangement of the technical sections is roughly similar to the sequence of work on the building. For example, the excavation section would be placed before the finished carpentry and millwork section. The following outline of headings is typical of smaller jobs such as a residence or small commercial building:

1. General Conditions
2. Excavation and Grading
3. Concrete and Masonry Work
4. Waterproofing
5. Miscellaneous Metals
6. Rough Carpentry
7. Finish Carpentry and Millwork
8. Roofing and Sheet Metal
9. Weatherstripping, Insulation, Caulking, and Glazing
10. Plastering and Drywall Construction
11. Ceramic Tile
12. Flooring
13. Painting and Decorating
14. Plumbing
15. Mechanical Equipment
16. Electrical
17. Landscaping

An example of a standard preprinted specification form having fill-in blanks to be completed for a specific job is the one used by the Federal Housing Administration (FHA) and the Veterans Administration (VA). See **Figure 5-20**.

OMB Control No. 2900-0788
Respondent Burden: 30 Minutes
Expiration Date: 10/31/2021

Department of Veterans Affairs

DESCRIPTION OF MATERIALS

PRIVACY ACT NOTICE: VA will not disclose information collected on this form to any source other than what has been authorized under the Privacy Act of 1974 or Title 38, CFR 1.576 for routine uses (for example: Authorizing release of information to Congress when requested for statistical purposes) as identified in the VA system of records, 55VA26, Loan Guaranty Home, Condominium and Manufactured Home Loan Applicant Records, Specially Adapted Housing Applicant Records, and Vendee Loan Applicant Records - VA, 17VA26, Loan Guaranty Fee Personnel and Program Participant Records - VA, and published in the Federal Register. Your obligation to respond is required to obtain or retain benefits.

RESPONDENT BURDEN: We need this information to establish the value and or cost of adaptations or new construction before work begins. Title 38 U. S. C. authorizes collections of this information. We estimate that you will need an average of 30 minutes to review the instructions, find the information and complete this form. VA cannot conduct or sponsor a collection of information unless a valid OMB control number is displayed. You are not required to respond to a collection of information if this number is not displayed. Valid OMB control numbers can be located on the OMB Internet Page at www.reginfo.gov/public/do/PRAMain. If desired, you can call 1-800-827-1000 to get information or where to send comments or suggestions about this form.

☐ PROPOSED CONSTRUCTION ☐ UNDER CONSTRUCTION CASE NO. ____________

PROPERTY ADDRESS *(Include City and State)*

NAME AND ADDRESS OF LENDER OR SPONSOR

NAME AND ADDRESS OF CONTRACTOR OR BUILDER

INSTRUCTIONS

1. For additional information on how this form is to be submitted, number of copies, etc., see the instructions in the VA Lender's Handbook.
2. Describe all materials and equipment to be used, whether or not shown on the drawings, by marking an X in each appropriate check-box and entering the information called for each space. If space is inadequate, enter "See misc." and describe under item 27 or on an attached sheet. **The use of paint containing more than the percentage of lead by weight permitted by law is prohibited.**
3. Work not specifically described or shown will not be considered unless required, then the minimum acceptable will be assumed. Work exceeding minimum requirements cannot be considered unless specifically described.
4. Include no alternates, "or equal" phrases, or contradictory items. (Consideration of a request for acceptance of substitute materials or equipment is not thereby precluded.)
5. Include signatures required at the end of this form.
6. The construction shall be completed in compliance with the related drawings and specifications, as amended during processing. The specifications include this Description of Materials and the applicable Minimum Property Requirements.

1. EXCAVATION

Bearing soil, type ____________

2. FOUNDATIONS

Footings concrete mix ______ strength psi ______ Reinforcing ______
Foundation wall material ______ Reinforcing ______
Interior foundation wall material ______ Party foundation wall ______
Columns material and sizes ______ Piers material and reinforcing ______
Girders material and sizes ______ Sills material ______
Basement entrance areaway ______ Window areaways ______
Waterproofing ______ Footing drains ______
Termite protection ______
Basement space ground cover ______ insulation ______ foundation vents ______
Special foundations ______
Additional Information ______

3. CHIMNEYS

Material ______ Prefabricated *(make and size)* ______
Flue lining material ______ Heater flue size ______ Fireplace flue size ______
Vents *(material and size)* gas or oil heater ______ water heater ______
Additional Information ______

4. FIREPLACES

Type ☐ solid fuel ☐ gas-burning ☐ circulator *(make and size)* ______ Ash dump and clean-out ______
Fireplace facing ______ lining ______ hearth ______ mantel ______
Additional Information

VA FORM OCT 2018 **26-1852** SUPERSEDES VA FORM 26-1852, SEP 2018, WHICH WILL NOT BE USED. Retain this record for three years Page 1 of 6

Goodheart-Willcox Publisher

Figure 5-20. The first page of a specification form used by the Federal Housing Administration and Veterans Administration. The form includes most of the areas required for small buildings.

Chapter Review

Summary

- Construction drawings describe the size, shape, location, and specifications of the elements of a structure.
- Lines make up a large part of the symbols used on construction drawings. The different types of lines are called the Alphabet of Lines.
- Symbols found on construction drawings include building material symbols, topographical symbols, plumbing symbols, climate control/mechanical symbols, electrical symbols, and others. Symbols may represent an elevation or a section view of a material or feature.
- Residential floor plans, elevations, and foundation plans are generally drawn at 1/4″ = 1′ scale.
- Dimensions, which indicate size and location of building elements, are usually shown in feet and inches.
- When the metric system is used in construction, the basic metric unit of measure is the meter.
- A set of working drawings for a typical structure might include a site plan, foundation plan, floor plan, elevations, electrical plan, mechanical plan, and construction details. In addition, a roof plan, framing plan, landscaping plan, presentation plan, and shop drawings may be needed.
- Drawing notes are included in a set of construction drawings to clarify details that might otherwise cause confusion.
- Specifications are written descriptions of the construction project prepared by architects and engineers. They are generally divided into two major sections—general conditions and technical sections.

Know and Understand

Answer the following questions using the information provided in this chapter.

1. Construction drawings include each of the following *except* _____.
   A. size
   B. schedule
   C. shape
   D. location
2. *True or False?* A specific line symbol communicates more precisely than a plain line.
3. _____ lines show the outline of an object's visible features.
   A. Border
   B. Object
   C. Hidden
   D. Center
4. Foundations and footings are represented by _____ lines.
   A. border
   B. outline
   C. hidden
   D. center
5. _____ lines show size or location of a feature.
   A. Extension
   B. Elevation
   C. Border
   D. Dimension
6. Use _____ lines to show a feature has been sectioned.
   A. crosshatch
   B. long break
   C. short break
   D. dimension
7. *True or False?* Section symbols are pictorial representations.
8. Residential floor plans are generally drawn at _____ = 1′.
   A. 1/16″
   B. 1/2″
   C. 1/3″
   D. 1/4″
9. An exterior brick veneer wall should be dimensioned to the _____.
   A. center of the wall
   B. outside of the brick
   C. outside of the stud wall
   D. inside of the stud wall

10. *True or False?* When the metric system is used in the construction industry, the basic unit of measurement is the meter.
11. *True or False?* The floor plan shows the location of the structure on the site.
12. *True or False?* The foundation plan shows the footings, wall size, and structural materials.
13. The _____ plan shows the panel box.
    A. electrical
    B. mechanical
    C. detail
    D. service
14. Written descriptions of construction projects are _____.
    A. specifications
    B. drawing notes
    C. engineering notes
    D. details
15. General conditions do *not* include _____.
    A. insurance
    B. payment
    C. tradework
    D. supervision

Apply and Analyze

1. What are four other names for construction drawings?
2. List five line symbols that appear in the Alphabet of Lines.
3. What kind of line is used to represent the visible features of a building?
4. What kind of line is used to show the length of a wall?
5. What is the purpose of a cutting plane line?
6. Symbols may be shown in elevation or section. Which type is usually easiest to recognize?
7. If a plan is drawn at 1/4″ = 1′ scale, how long on the drawing would a 40′ wall be?
8. How long would a 40′ wall be if it were drawn at 1/4 size?
9. List the seven drawings that usually compose a set of working drawings for a building.
10. Which drawing shows the location of switches, convenience outlets, and TV jacks?

Critical Thinking

1. Explain in detail why construction details are necessary.
2. You are a foreman reviewing prints, and you determine that there is an error in the foundation print. What should you do before the construction crew begins their work?
3. You are the superintendent of a new building to be constructed. Prior to construction, you are reviewing the drawing and specifications, where a number of discrepancies have been found between the two documents. What should you do?

Activities

1. (MATH) Determine the size of the following as shown on a drawing:
    A. A 40′ building drawn at 1/8 scale.
    B. A 24′ wall drawn at 1/4 scale.
    C. A 5′ wall drawn at 1/4 size.
2. (ENGINEERING) Draw the following symbols:
    A. Batt insulation
    B. Water line
    C. Power line
    D. Block (elevation symbol)
    E. Block (section symbol)
    F. Fence
    G. Gas line (topographical symbol)
    H. Gas line (plumbing symbol)
    I. Junction box

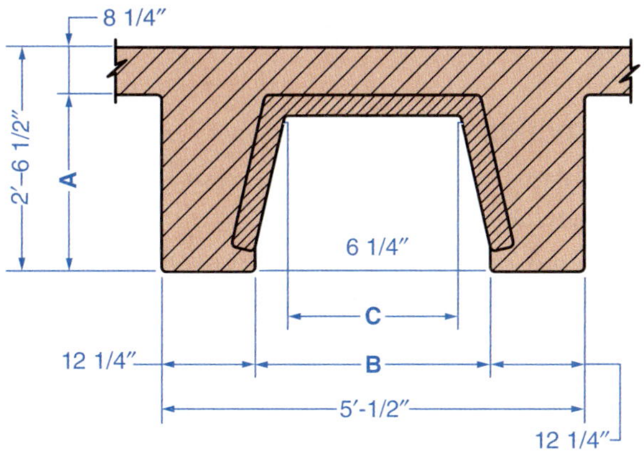

Goodheart-Willcox Publisher

CHAPTER **6**

Math for Masonry Careers

LEARNING OUTCOMES

After studying this chapter, you will be able to:

6.1-1 Calculate addition, subtraction, multiplication, and division of whole numbers.
6.2-1 Calculate addition, subtraction, multiplication, and division of fractions.
6.3-1 Calculate addition, subtraction, multiplication, and division of decimals.
6.4-1 Calculate percentage and interest.
6.5-1 Define applied geometry terms commonly used in masonry.
6.6-1 Calculate areas and volumes.
6.7-1 Differentiate between metric and US customary units of measure.
6.8-1 Explain the importance of weights and measures in masonry applications.
6.9-1 Apply calculations to estimate brick masonry.

TECHNICAL TERMS

3-4-5 method
acute angle
addition
angle
arc
circle
circumference
decimal
decimal mixed number
denominator
diameter
difference
dividend
division
divisor
fraction
hard metric conversion
hypotenuse
interest
International System of Units (SI)
line
mixed number
multiplication
net wall area
numerator
obtuse angle
percentage
perpendicular
point
principal
product
Pythagorean Theorem
quotient
radius
rate of interest
remainder
right angle
right triangle
sector
segment
semicircle
soft metric conversion
straight angle
subtraction
sum
tangent line
triangle
whole number
wythe

Introduction

A skilled mason or apprentice must be able to take measurements and perform various calculations related to the trade. These calculations require an understanding of certain basic mathematical concepts. The purpose of this chapter is to review some of those basic concepts and give examples of their application using trade terms and figures.

6.1 Whole Numbers

A ***whole number*** is any one of the natural numbers, such as 1, 2, and 5. Numbers represent quantities of anything and can be added, subtracted, multiplied, or divided.

Addition

Addition is the process of combining two or more quantities (numbers) to find a total. The total in addition is the ***sum***. Addition is indicated by the plus (+) sign and may be written as 2 + 2. The sum may be indicated by using the equal (=) sign. For example, 2 + 2 = 4. Another way of writing this problem showing the sum of 4 is as follows:

$$\begin{array}{r} 2 \\ +\ 2 \\ \hline 4 \end{array} \quad \textbf{\textit{sum}}$$

Problems—Addition

The following problems are included to refresh your memory of basic addition in trade terms.

1. Three bricklayers working together on a job each laid the following number of bricks in one day. The first bricklayer laid 887 bricks, the second bricklayer laid 1123, and the third bricklayer laid 1053. How many bricks did all three lay that day?

   Answer: 887 + 1123 + 1053 = 3063 bricks
2. A mason was paid the following sum of money for each of five days of work: $96, $105, $87, $97, and $85. How much money was he paid for the five days' work?

   Answer: $96 + $105 + $87 + $97 + $85 = $470
3. Weight slips on four loads of gravel delivered to one job were 6272 lb, 6098 lb, 6138 lb, and 6193 lb. What was the total weight of the gravel delivered to the job?

   Answer: 6272 + 6098 + 6138 + 6193 = 24,701 lb

Subtraction

Subtraction is the process of taking one number away from another number. The portion that remains after the subtraction is the ***difference***. The sign that indicates that one quantity (number) is to be subtracted from another is the minus (–) sign. Example: 6 – 4. In this example, 4 is being subtracted from 6. The difference is 2, or 6 – 4 = 2. Another way of writing the same thing is as follows:

$$\begin{array}{r} 6 \\ -\ 4 \\ \hline 2 \end{array} \quad \textbf{\textit{difference}}$$

Problems—Subtraction

The following problems are included to refresh your memory of basic subtraction in trade terms.

1. A mason ordered 75 bags of cement and used 68 bags on the job. How many bags of cement are left?

   Answer: 75 – 68 = 7 bags
2. A 36″ wide chimney has two flues that are 8″ and 16″ wide. The thickness of the chimney walls is 4″. How much space is between the flues?

   Answer: 36 – (8 + 16 + 4 + 4) = 36 – 32 = 4″
3. A mason contracts a job for $700. Labor cost is $316 and material cost $203. Other expenses total $78. How much profit is made on the job?

   Answer: 700 – (316 + 203 + 78) = 700 – 597 = $103

Multiplication

Multiplication is the process of repeated addition using the same numbers. The answer to multiplication is the ***product***. For example, if 2 + 2 + 2 + 2 + 2 were to be added, the shortest method would be to multiply 5 times 2 to get the product of 10. The sign used to indicate multiplication is the times (×) sign. For example, 5 × 2 = 10. This may also be written as follows:

$$\begin{array}{r} 5 \\ \times\ 2 \\ \hline 10 \end{array} \quad \textbf{\textit{product}}$$

Problems—Multiplication

The following problems are included to refresh your memory of basic multiplication in trade terms.

1. If a bricklayer can lay 170 bricks an hour, how many bricks can he lay in four hours?

   Answer: 170 × 4 = 680 bricks
2. One type of brick costs $9 per hundred. If 14,000 bricks were ordered, how much would they cost?

   Answer: $9 × 140 = $1260. Note: The bricks were $0.09 each, $9 per hundred, or $90 per thousand. Therefore, the answer could have been determined by multiplying $0.09 × 14,000, or $90 × 14, or $9 × 140.
3. If five bricks covered 1 sq ft of wall space, how many bricks would be required to cover a wall 8 ft high by 10 ft long?

   Answer: First, find the wall area by multiplying 8 × 10. The wall area is 80 sq ft. The number of bricks required is 5 × 80 = 400 bricks.

Division

Division is the process of finding how many times one number (***divisor***) is contained within another number (***dividend***). The division symbol is (÷). For example, when we wish to find how many times 3 is contained in 9, we say 9 divided by 3 equals 3 or 9 ÷ 3 = 3. The answer to a division problem is the ***quotient***. If a number is not contained in another number an equal number of times, the amount left over is the ***remainder***. The problem is generally written as follows:

```
              2
divisor     4|9      dividend
              8
              1      remainder
    or 2 1/4  quotient
```

Notice that the remainder is placed over the divisor to form a fraction. This fraction is part of the quotient as shown in the division example above.

Problems—Division

The following problems are included to refresh your memory of basic division in trade terms.

1. If a set of steps has 8 risers and the total height of all the steps (total rise) is 56″, what is the height of each step?

```
  7
8|56
  56
   0
```

Answer: 7″

2. Four bricklayers laid 4160 bricks on a job one day. What was the average number of bricks laid by each bricklayer that day?

```
  1040
4|4160
  4
  01
   0
   16
   16
   00
    0
    0
```

Answer: 1040 bricks per bricklayer

3. If a brick veneer wall requires 5 bricks to lay up 1 sq ft, how many square feet would 587 bricks cover?

```
  117
5|587
  5
  08
   5
   37
   35
    2
```

Answer: 117 2/5 sq ft of wall

6.2 Fractions

A ***fraction*** is one or more parts of a whole. See **Figure 6-1**. Fractions are written with one number over the other, such as 1/4, 1/2, or 3/4. The top number of a fraction is the ***numerator***. The numerator indicates the number of parts of the whole that are of concern. The bottom number of a fraction is called the ***denominator***. The denominator identifies the number of parts into which the whole is divided.

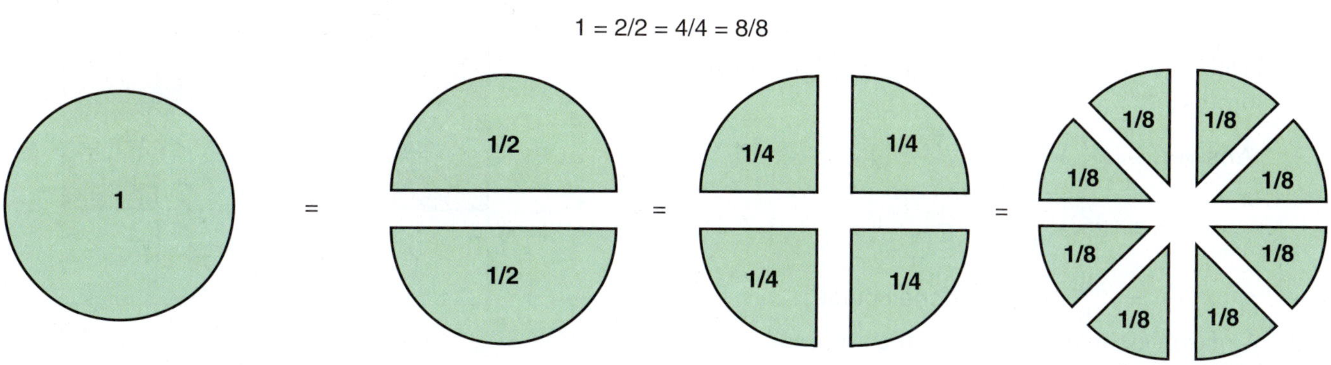

Goodheart-Willcox Publisher

Figure 6-1. Fractions are parts of a whole.

In reading a fraction, the top number is always read first. For example, 1/2 would be read "one-half," 3/4 would be read "three-fourths," and 3/8 would be read "three-eighths."

A fraction should always be reduced to its lowest terms. For example, the fraction 8/12 can be reduced by dividing the numerator and the denominator by the largest number that can be divided into both exactly. In this case, that number is 4, because 8 divided by 4 is 2, and 12 divided by 4 is 3. The reduced fraction is, therefore, 2/3.

Another example is the fraction 3/2, which is not in the lowest form, or reduced to its lowest terms. It should be 1 1/2 because 2/2 = 1 and 1 + 1/2 = 1 1/2. The 1 1/2 is a ***mixed number*** (a whole number followed by a fraction). When the numerator and denominator are the same number, such as 1/1, 2/2, or 3/3, the fraction is equal to 1.

Adding Fractions

The easiest fractions to add are those that have the same denominators, such as 1/8 and 3/8. Simply add the numerators together and keep the same denominator. For example, 1/8 + 3/8 = 4/8, which reduces to 1/2. Reducing the fraction to its lowest terms is preferred. Another example of reducing to the lowest terms is 8/24 = 1/3, because 24 can be divided by 8 three times.

When fractions to be added have different denominators, multiply both the numerator and denominator of each fraction by a number that makes the denominators equal. For example, in 1/3 + 3/5, multiply the numerator and denominator of 1/3 by 5, and multiply the numerator and denominator of 3/5 by 3. The result is 5/15 + 9/15. Observation indicated that 15 was the smallest number that could be divided evenly by both denominators. To complete the example, 5/15 + 9/15 = 14/15. Therefore, the sum of 1/3 and 3/5 is 14/15.

Problems—Addition of Fractions

1. What is the height of one stretcher course of brick if the bricks are 2 1/4″ high and the mortar joint is 3/8″?

Answer:
$$2\frac{1}{4} + \frac{3}{8}$$
$$= 2\frac{2}{8} + \frac{3}{8}$$
$$= 2\frac{5}{8}'' \text{ height for one course}$$

2. A mason estimated the following amounts of mortar required for a job—5 1/2 cu yd, 11 1/3 cu yd, and 6 1/4 cu yd. What is the total amount of mortar required for the job?

Answer:
$$5\frac{1}{2} + 11\frac{1}{3} + 6\frac{1}{4}$$
$$= 5\frac{6}{12} + 11\frac{4}{12} + 6\frac{3}{12}$$
$$= 22\frac{13}{12}$$
$$= 23\frac{1}{12} \text{ cu yd of mortar}$$

3. A brick wall on a residence is 8 1/16″ thick, the furring is 3/4″, plaster board is 3/8″, and the plaster is 1/2″. What is the total thickness of the wall?

Answer:
$$8\frac{1}{16} + \frac{3}{4} + \frac{3}{8} + \frac{1}{2}$$
$$= 8\frac{1}{16} + \frac{12}{16} + \frac{6}{16} + \frac{8}{16}$$
$$= 8\frac{27}{16}$$
$$= 9\frac{11}{16}''$$

Subtracting Fractions

When subtracting fractions, change all fractions to the same common denominator as was done for adding fractions. When the denominators are the same, subtract the numerators.

Problems—Subtraction of Fractions

1. What is the dimension of "A" in **Figure 6-2**?

Answer:
$$2'\text{-}6\frac{1}{2}'' - 8\frac{1}{4}''$$
$$= 2'\text{-}6\frac{1}{2}'' = 30\frac{1}{2}''$$
$$\text{Therefore, } 30\frac{1}{2}'' - 8\frac{1}{4}''$$
$$= 30\frac{2}{4}'' - 8\frac{1}{4}''$$
$$= 22\frac{1}{4}'' \text{ (or } 1'\text{-}10\frac{1}{4}''\text{)}$$

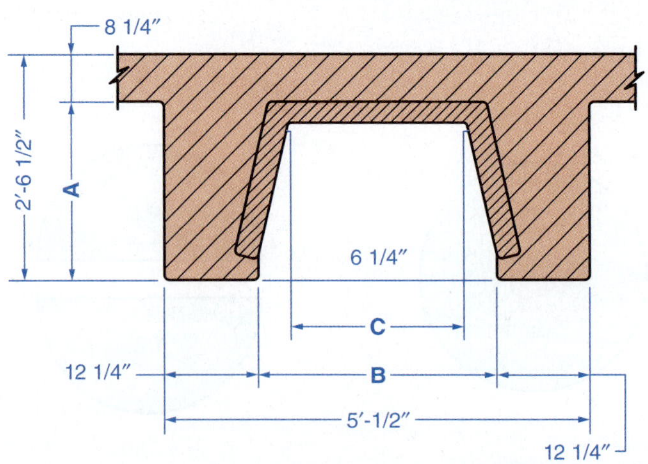

Goodheart-Willcox Publisher

Figure 6-2. Plan view of a fireplace with dimensions.

2. What is the dimension of "B" in **Figure 6-2**?

Answer: $5'\text{-}\frac{1}{2}'' - (12\frac{1}{4}'' + 12\frac{1}{4}'')$

$= 5'\text{-}\frac{1}{2}'' = 60\frac{1}{2}''$

Therefore, $60\frac{1}{2}'' - (12\frac{1}{4}'' + 12\frac{1}{4}'')$

$= 60\frac{1}{2}'' - 24\frac{1}{2}''$

$= 36''$ (or $3'$)

3. What is the dimension of "C" in **Figure 6-2**?

Answer: $36''$ (from Question 2) $- (6\frac{1}{4}'' + 6\frac{1}{4}'')$

$= 36'' - 12\frac{1}{2}''$

$= 23\frac{1}{2}''$ (or $1'\text{-}11\frac{1}{2}''$)

Multiplying Fractions

The procedure for multiplying fractions is to multiply the numerators together to find the numerator for the answer. Then, multiply the denominators together to find the denominator for the answer. If the product is a fraction, it should be reduced to its lowest form. For example, in 1/2 × 3/4, multiply the 1 of 1/2 by the 3 in 3/4 to get 3. Then, multiply the 2 of 1/2 by the 4 of 3/4 to get 8. The product of this problem is 3/8.

Whole numbers must be changed to fractions by placing the whole number over a 1. For example, in 4 × 5/8, the 4 is changed to 4/1. Then, 4/1 × 5/8 equals 20/8. The fraction 20/8 reduces to 2 4/8, which reduces to 2 1/2.

Problems—Multiplication of Fractions

1. If 2 1/4″ thick standard bricks are used to lay a wall with 3/8″ mortar joints, what is the height of the wall after nine courses?

Answer: First, add the thickness of one mortar joint to the thickness of one brick. Then multiply that sum times 9 to find the height.

$2\frac{1}{4}'' + \frac{3}{8}''$

$= 2\frac{2}{8} + \frac{3}{8}$

$= 2\frac{5}{8}$

$= 2\frac{5}{8} \times 9$

$= \frac{21}{8} \times \frac{9}{1}$

$= \frac{189}{8}$

$= 23\frac{5}{8}''$

2. If a set of steps is five risers high, and each riser is 7 1/4″, what is the total rise of the steps?

Answer: $7\frac{1}{4}'' \times 5$

$= \frac{29}{4} \times \frac{5}{1}$

$= \frac{145}{4}$

$= 36\frac{1}{4}''$

3. What is the length of a wall consisting of 28 stretchers if each stretcher is 7 1/2″ and the mortar joint is 1/2″?

Answer: $7\frac{1}{2}'' + \frac{1}{2}'' = 8''$

$= 8'' \times 28 = 224''$

$224'' \div 12 = 18\frac{2}{3}'$

convert $\frac{2}{3}$ to inches

$= \frac{2}{3} \times \frac{12}{1} = \frac{24}{3} = 8''$

$= 18'\text{-}8''$

Dividing Fractions

The process of dividing fractions is accomplished by inverting (turning upside down) the divisor and then multiplying. For example, for 3/8 ÷ 3/4, change the 3/4 to 4/3, and then multiply 3/8 by 4/3. Therefore, 3/8 ÷ 3/4 = 3/8 × 4/3 = 12/24 = 1/2.

Problems—Division of Fractions

1. How many risers 7 1/2″ high are required to construct a flight of concrete steps 3′-1 1/2″ high?

Answer: Change $3'\text{-}1\frac{1}{2}''$ to inches.

Divide that number by $7\frac{1}{2}''$.

$3'\text{-}1\frac{1}{2}'' = 37\frac{1}{2}''$

$37\frac{1}{2}'' \div 7\frac{1}{2}''$

$= \frac{75}{2} \div \frac{15}{2}$

$= \frac{75}{2} \times \frac{2}{15}$

$= \frac{150}{30}$

= 5 risers

2. If a brick mantel is corbeled out 4 1/2″ in six courses, how much does each course project past the previous course?

Answer: $4\frac{1}{2} \div 6$

$= \frac{9}{2} \div \frac{6}{1}$

$= \frac{9}{2} \times \frac{1}{6}$

$= \frac{9}{12}$

$= \frac{3}{4}''$

3. If a story pole is 8′-11 1/4″ long and divided into 39 equal spaces, what is the length of each space?

Answer: $8'\text{-}11\frac{1}{4}'' \div 39$

$= 107\frac{1}{4}'' \div 39$

$= \frac{429}{4} \div \frac{39}{1}$

$= \frac{429}{4} \times \frac{1}{39}$

$= \frac{429}{156}$

$= 2\frac{117}{156}$

$= 2\frac{3}{4}''$

6.3 Decimals

Decimals are types of fractions that only have numerators. The denominators are implied by the place of the last digit to the right of the decimal point. For example, 0.1 is equal to 1/10, 0.01 is equal to 1/100, and 0.001 is equal to 1/1000.

The zeros that may follow the significant digits in a decimal number do not increase the value of the fraction. Thus, 0.250 and 0.2500 both equal 0.25. The added zeros indicate an increased accuracy, but not an increase in size. However, zeros to the left of the number and before the decimal are very significant. For example, 0.25 is equal to 25/100, 0.025 is equal to 25/1000, 0.0025 is equal to 25/10,000, and 0.00025 is equal to 25/100,000.

When there is one number to the right of the decimal, it is read as so many tenths (0.5 is five-tenths). Two numbers to the right of the decimal are read as hundredths (0.05 is five-hundredths). Three numbers to the right of the decimal indicate thousandths, four indicate ten thousandths, etc.

A ***decimal mixed number*** has numbers on both sides of the decimal. The word *and* is generally used to represent the decimal when reading a decimal mixed number. For example, 6.125 is read as "six *and* one hundred twenty-five thousandths." The word *point* is also used to denote the decimal. In this case, 6.125 would be read as "six *point* one two five."

Problems—Addition of Decimals

1. If the cost of materials for a masonry job is $35.50 for sand, $162.38 for cement, $9.63 for one steel lintel, and $338.19 for brick, what is the total cost of the materials for the job?

Answer:

$$\begin{array}{r} 35.50 \\ 162.38 \\ 9.63 \\ +\ 338.19 \\ \hline \$545.70 \end{array}$$

2. A mason needs to determine if the total cross-sectional area of three pieces of reinforcing steel meet the specifications for a reinforced masonry lintel. The cross-sectional areas of the three pieces are 0.11 sq in, 0.3 sq in, and 1.000 sq in. What is the total cross-sectional area of the three pieces of reinforcing steel?

Answer:

$$\begin{array}{l} 0.11 \\ 0.3 \\ +\ 1.000 \\ \hline 1.410 \text{ sq in} \end{array}$$

Problems—Subtraction of Decimals

1. A mason placed a bid on a job for $1247.38. He allowed $413.75 for materials and $500.00 for labor. How much did he allow for profit and miscellaneous costs?

Answer: $1247.38 – ($413.75 + $500.00)
= $1247.38 – $913.75
= $333.63

2. A cement contractor was to stake out a foundation on a site that was 125.00′ wide and 200.00′ deep. The plot plan showed that the front of the house (placed along the width of the lot) faced south and was 50.00′ from the lot line. The east side of the house was 28.35′ from the lot line and the west side of the house was 36.65′ from the lot line. How wide is the front of the house?

Answer: 125.00′ – (28.35′ + 36.65′)
= 125.00′ – 65.00′
= 60.00′ (or 60′ as it is written on a construction drawing)

Problems—Multiplication of Decimals

1. What is the cost of 20 bags of cement if each bag costs $3.00?

   Answer: Add up the total number of spaces to the right of the decimal point in both parts of the problem. Count over the same number in the answer and place the decimal there.

   $$\begin{array}{r} \$3.00 \\ \times\ \ 20 \\ \hline \$60.00 \end{array}$$

2. How much should a mason charge for laying 400 bricks if each brick costs $0.09 each and the mason gets $0.18 for laying each brick? Include the cost of the bricks in the problem.

   Answer: (400 bricks × $0.09)
   + (400 bricks × $0.18)

   $$\begin{array}{r} 400 \\ \times\ 0.09 \\ \hline \$36.00 \end{array} \qquad \begin{array}{r} 400 \\ \times\ 0.18 \\ \hline 3200 \\ 400 \\ \hline \$72.00 \end{array} \qquad \begin{array}{r} 36.00 \\ +\ 72.00 \\ \hline \$108.00 \end{array}$$

Problems—Division of Decimals

1. If a mason charges $2100.00 for laying 14,000 bricks, how much is received per brick?

   Answer: $2100.00 ÷ 14,000
   = 0.15 (0.15$ each)

   $$\begin{array}{r} 0.15 \\ 14000\overline{)2100.00} \\ \underline{14000} \\ 70000 \\ \underline{70000} \\ 0 \end{array}$$

2. Clay flue lining is sold in 2′ lengths. What is the cost of one 2′ length if 56′ of lining costs $98.00?

   Answer: 56 ÷ 2 = 28 pieces of lining
   $98.00 ÷ 28 = $3.50 each

   $$\begin{array}{r} 3.50 \\ 28\overline{)98.00} \\ \underline{84} \\ 140 \\ \underline{140} \\ 00 \\ \underline{0} \\ 0 \end{array}$$

Converting Fractions to Decimals

Common fractions can be changed to decimals by dividing the numerator by the denominator. See **Figure 6-3**. For example, the fraction 3/4 can be changed to a decimal by dividing 3 by 4. The answer is 0.75. Therefore, 0.75 is equal to 3/4.

If the division does not come out even (with no remainder), it is carried out as far as desired. For example, the fraction 1/3 is changed to a decimal by dividing 1 by 3. The answer is 0.333 repeated.

When a mixed number such as 3 1/2 must be changed to a decimal form, the mixed number is first changed to a fraction (3 1/2 = 7/2), and then the division is performed:

$$\begin{array}{r} 3.50 \\ 2\overline{)7.00} \\ \underline{6} \\ 10 \\ \underline{10} \\ 00 \\ \underline{0} \\ 0 \end{array}$$

Fraction-Decimal Equivalents

Fraction	Decimal
1/16	0.0625
1/8	0.1250
1/6	0.1600
3/16	0.1875
1/4	0.2500
5/16	0.3125
1/3	0.3333
3/8	0.3750
7/16	0.4375
1/2	0.5000
9/16	0.5625
5/8	0.6250
2/3	0.6667
11/16	0.6875
3/4	0.7500
13/16	0.8125
5/6	0.8300
7/8	0.8750
15/16	0.9375

Goodheart-Willcox Publisher

Figure 6-3. Decimal equivalents of fractions.

Problems—Converting Fractions to Decimals

1. What is the labor cost of laying 12 3/4 cu yd of rubble stone if the cost to lay rubble stone is $9.50 per cu yd?

   Answer: 12 3/4 = 12.75
   12.75 × $9.50 = $121.13

2. What is the cost for 24 1/4 cu yd of fill if it sells for $2.25 per cu yd?

   Answer: 24 1/4 = 24.25
   24.25 × $2.25 = $54.56

Converting Decimals to Fractions

A decimal may be changed to a fraction by dropping the decimal point and using the number as the numerator of the fraction. The denominator is written with a "1" and as many zeros after it as there were decimal places in the original decimal number. Thus, 0.125 may be changed to a fraction by writing 125 over 1000, or 125/1000. When reduced, 125/1000 equals 1/8.

Problems—Converting Decimals to Fractions

1. A mason reads on a drawing that the reinforcing steel required for a certain situation was 0.375″ in diameter. What is the fractional size of this steel?

   Answer: 375/1000 = 3/8″ in diameter

2. What is the width of a piece of steel angle that is specified as 3.875″? Answer should be a fraction rather than a decimal.

   Answer: 3.875 = 3 875/1000,
   which reduces to 3 7/8″

Another method to convert a decimal of an inch to a fraction of an inch can be accomplished by multiplying the decimal by the denominator of which the mason would like to work in as part of the calculation. For example, if the mason would like to have the decimal converted to the closest 1/16 of an inch, then the decimal would be multiplied by 16. As a fraction, there are sixteen-sixteenths in a whole inch. Written as a fraction, it is shown as 16/16″. When the fraction is reduced to its lowest terms, it becomes 1″.

For example, if you want to find the fraction of an inch from the decimal 0.58″, multiply that number by 16:

$$\begin{array}{r} 0.58 \\ \times\ 16 \\ \hline 9.28 \end{array}$$

The answer is 9.28. Because the mason wants the fraction to be the closest to 1/16 of an inch, the 0.28 will be rounded to the nearest whole number. Because 0.28 is less than 0.50, the 28 will be dropped off, so the answer is 9, or 9/16″.

As another example, if you want to find the fraction from the decimal 0.75″, multiply it by 16.

$$\begin{array}{r} 0.75 \\ \times\ \ 16 \\ \hline 12.00 \end{array}$$

The answer is 12.00, or 12. Place the 12 into a fraction form, 12/16″. The 12/16 needs to be reduced to its lowest form, or 3/4″.

6.4 Percentage and Interest

Percentage is a number used to denote parts of one hundred. Interest is a type of percentage that represents the cost of borrowing money.

Percentage

Percentage is indicated by the % sign. For example, 3% of a number is three hundredths of the number. It may be indicated as 3/100, 0.03, or 3%.

Finding percentage is done by multiplying. The number should be written in the decimal form (0.03) when computing the answer. To determine 4% of 1452, you would multiply 1452 by 0.04. The answer, or product, is 58.08.

Study the following conversions before trying the problems:

0.25 = 25%
0.75 = 75%
0.2 = 20%
0.7 = 70%
0.364 = 36.4%
0.5748 = 57.48%

1/4 = 0.25 = 25%
3/8 = 0.375 = 37.5%

Problems—Percentage

1. A mason calculated that 5 cu yd of sand was needed for a bricklaying job. Because of waste, the mason added 5%. How much sand should be ordered?

   Answer: 0.05 × 5 cu yd = 0.25 cu yd
   5 + 0.25 = 5.25 cu yd should be ordered

2. It was estimated that 12,000 bricks would be required for a job. An additional 2% were added because of breakage, bats, and salmon brick. How many bricks should be ordered?

   Answer: 0.02 × 12,000 = 240
   12,000 + 240 = 12,240 bricks

Interest

Interest is usually considered to be the cost of borrowing money. The amount borrowed is the ***principal***. The percentage charged per year is the ***rate of interest***. Interest may be figured by the year, the month, or the day.

To illustrate, if \$500 was borrowed for one year (12 months) at 8% interest, the interest to be paid would be \$500.00 × 0.08 = \$40.00. The same amount of money and the same rate of interest would cost \$6.66 for two months. (\$500.00 × 0.08) ÷ 6. The total is divided by 6 because two months is 1/6 of a year.

Problems—Interest

1. If a mason deposited \$1000 in a savings account at 5% interest per year, how much interest would he earn at the end of 12 months?

   Answer: \$1000 × 0.05 = \$50.00

2. A contractor purchased a new piece of equipment for \$2500 and agreed to pay 8% interest per year. How much interest would be owed at the end of the year?

   Answer: \$2500 × 0.08 = \$200.00 in interest for 12 months

3. How much interest for 6 months?

   Answer: \$200.00 ÷ 2 = \$100.00 in interest for 6 months

6.5 Applied Geometry

The properties of points, lines, and planes are a definite concern of the mason. See **Figure 6-4**. This section is designed to clarify terms commonly used in geometry that have a direct application to the masonry careers.

Points

Points are basic elements of lines that have no width, length, or height, but merely indicate a location. **Figure 6-4** shows recommended symbols for a point. Points may be identified with a name, number, letter, or other symbol.

Lines

A ***line*** has only length and is the distance between two or more points. A line may be either straight or curved. A horizontal line is a level line. A vertical line is a plumb line and is perpendicular to a horizontal line. ***Perpendicular*** means that the line forms a 90 degree (°) angle, or right angle, with respect to another line. The symbol for a perpendicular line is shown in **Figure 6-4**.

A diagonal line is one joining two opposite angles. Parallel lines remain a constant distance apart and never cross. Curved lines may be either arcs with a single center or irregular curves with many centers. A circle is an example of a curve with a single center. It is a 360° arc. An irregular or free curve is represented in **Figure 6-4**.

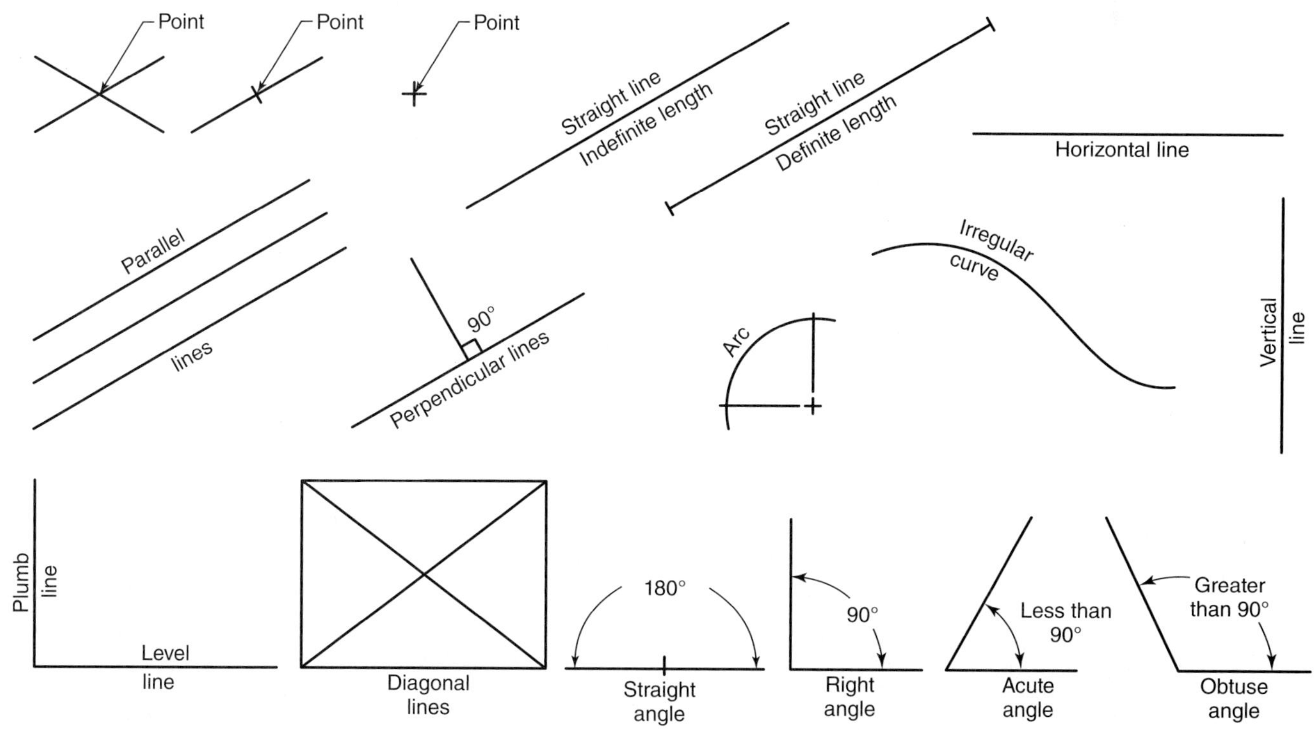

Figure 6-4. Lines, points, and planes. These elements combine to form edges, corners, and surfaces of all objects.

Angles

An ***angle*** is formed by two intersecting lines. An ***obtuse angle*** is an angle greater than 90°. A ***straight angle*** is an angle of 180°. A ***right angle*** is an angle with exactly 90°. An ***acute angle*** is an angle less than 90°.

Angles are measured in degrees (°), minutes (′), and seconds (″). Each degree is divided into 60 minutes, and each minute is divided into 60 seconds. For example, an angle measurement of 125 degrees, 35 minutes, 40 seconds is written as 125°35′40″.

Circles

A ***circle*** is a continuous curve maintaining an equal distance from its center at all points. **Figure 6-5** identifies the parts of a circle. The ***circumference*** of a circle is the distance around it. A ***semicircle*** is equal to half the circumference. An ***arc*** is any portion of the circumference. The ***diameter*** of a circle is the distance across a circle through its center. The ***radius*** is the distance from the center to any point on the circumference or one-half the diameter. A ***sector*** is the portion of a circle between two radii (more than one radius). A ***segment*** is the portion of a circle contained by a straight line and the circumference that it cuts off. A ***tangent line*** is a line that touches a circle, but does not cut it and is at right angles to a straight line from the center.

The circumference of a circle has a direct relationship to the circle's diameter. The circumference is equal to a constant of 3.1416 multiplied by the diameter. This constant is called "pi" and is represented by the symbol (π). Thus, the formula is $C = \pi D$, where C is circumference, π is pi or 3.1416, and D is diameter.

Triangular Relationships

A ***triangle*** is a geometric shape that has three angles that total 180°. When two angles are known, the third may be found by subtracting the sum of the two angles from 180°.

A ***right triangle*** has one 90° angle. Therefore, the sum of the other two angles is 90°. The longest side of a right triangle is called the ***hypotenuse***. The hypotenuse is always opposite of the 90° angle of the right triangle. See **Figure 6-6**.

The ***Pythagorean Theorem*** is a mathematical formula that aids masons in laying out a building. The Pythagorean Theorem is displayed as $a^2 + b^2 = c^2$. The square of the hypotenuse of a right triangle is equal to the sum of the squares of the other two sides. For example, if a right triangle had two sides equal to 3′ and 4′, then the other side (hypotenuse) would equal 5′. ($\text{Hypotenuse}^2 = \text{Side a}^2 + \text{Side b}^2$). Therefore, $\text{Hyp}^2 = 3^2 + 4^2$ equals $25 = 9 + 16$. Five multiplied by itself is 25. Thus, the hypotenuse is 5′. This is also called a 3-4-5 triangle or the ***3-4-5 method*** of layout and is very

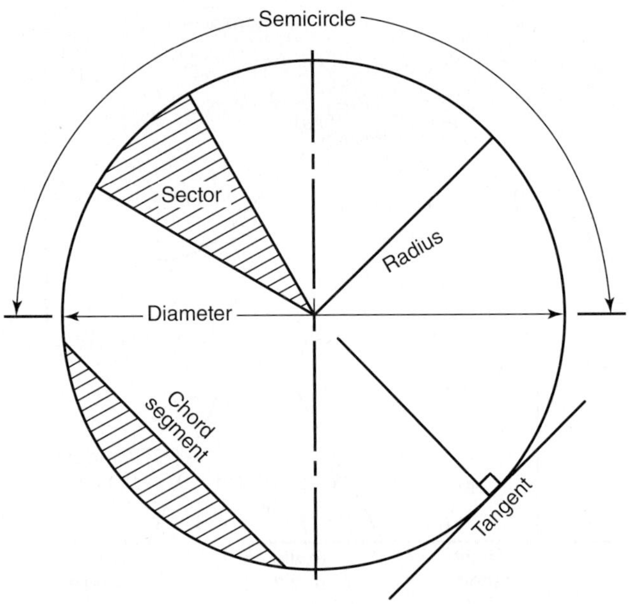

Goodheart-Willcox Publisher

Figure 6-5. The circle and its parts.

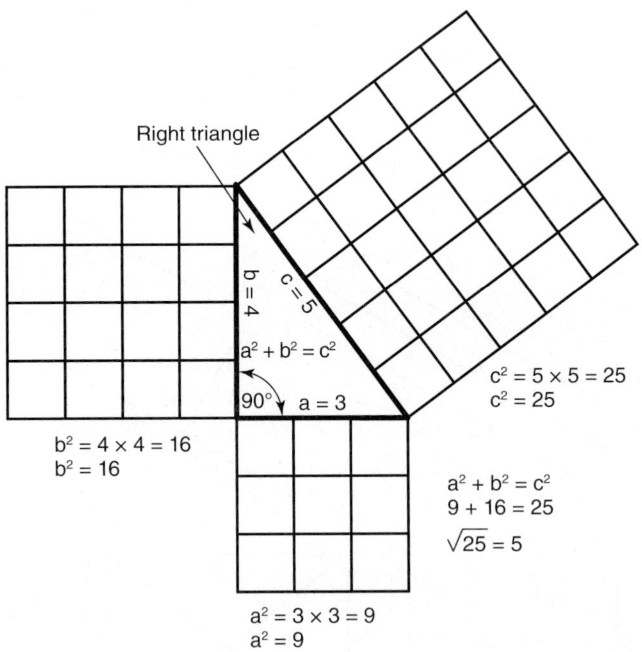

Goodheart-Willcox Publisher

Figure 6-6. The blocks outside the legs and the hypotenuse represent the squaring of the lengths of the sides. Leg a = 3, and 3^2 is shown as three blocks long and 3 blocks wide, for a total of 9 blocks. Leg $b^2 = 4^2 = 16$ blocks. Using the formula $a^2 + b^2 = c^2$ to find the hypotenuse squared, 9 + 16 = 25. The hypotenuse is the square root of 25, or 5.

useful for laying out square corners for foundations. For greater accuracy on large projects, use multiples of the 3-4-5 method to position corners of the building in the layout process. The use of the 3-4-5 method is discussed in Chapter 12, **Laying Brick**.

6.6 Areas and Volumes

Much of the math presented thus far is used to find areas and volumes related to a job. Areas are two-dimensional (length × width) and are shown in square measure such as 1296 sq in, 9 sq ft, or 1 sq yd. Volumes are three-dimensional quantities (length × width × height) and are shown in cubic measure (46,656 cu in, 27 cu ft, or 1 cu yd). **Figure 6-7** shows several plane and solid figures with the formulas for calculating their areas or volumes.

Problems—Calculating Areas

1. A large window opening is 72 3/8″ × 60 3/8″. What is the area of the opening?

   Answer: $A = L \times H$

   $A = 72\ 3/8'' \times 60\ 3/8''$

   $A = 72.375'' \times 60.375''$

   $A = 4369.64$ sq in. Divide by 144 to find sq ft

   $4369.64 \div 144 = 30.34$ sq ft

2. What is the area of a circular slab with a radius of 6′?

   Answer: $A = \pi R^2$ (π is a constant 3.1416 or 22/7)

   $A = 3.1416 \times 6^2$ (6^2 means 6×6)

   $A = 3.1416 \times (6 \times 6)$

   $A = 3.1416 \times 36$

   $A = 113.10$ sq ft of slab

Problems—Calculating Volumes

1. A common brick is 8″ × 2 1/4″ × 3 3/4″. What is its volume in cubic inches?

   Answer: Volume = $L \times W \times H$

   $V = 8'' \times 2\ 1/4'' \times 3\ 3/4''$

   $V = 8'' \times 2.25'' \times 3.75''$

   $V = 67.5$ cu in

2. How many cubic yards of concrete would be required to fill a cylindrical form 12′ long and 18′ in diameter?

   Answer: Volume = $\pi R^2 \times H$

   (radius is 1/2 the diameter)

   $V = 3.1416 \times 9^2 \times 144''$ (must change feet into inches)

   $V = 3.1416 \times 81$ sq in $\times 144''$

   $V = 36{,}644.788$ cu in or 21.21 cu ft or 0.79 cu yd

   Divide cubic inches by 1728 to determine cubic feet and divide by 27 cu ft to determine cubic yards.

   1 cu yd = 27 cu ft = 46,656 cu in

6.7 Metric Measurement and US Conversions

The ***International System of Units (SI)*** has been adopted by all industrialized countries except the United States. It is, therefore, important to become familiar with metric units and their relationship to US customary units.

Metrics in Construction

The Metric Conversion Act was passed in 1975. It called for voluntary compliance, and volunteers were few. The Omnibus Trade and Competitiveness Act of 1988 declared that the metric system was to be the preferred system for US trade and commerce. This Act was also mostly ignored by the public. However, the Omnibus Act ordered all US government agencies to "go metric," and one-by-one they complied. The CIA, NASA, the military, and the State Department have long been using metric. However, the conversion was hardly noticed in the private sector until the General Services Administration (GSA), which oversees all federal building construction, began rejecting bids that did not have metric specifications.

Because of the problems in adopting the 1975 law and difficulties that were occurring in the masonry industry and construction in general, a new law was passed. The Savings in Construction Act of 1996, P.L. 104-298, placed limitations on federal agencies that were requiring hard metric conversions and implementation of the metric system by the masonry industry. The new law was intended to prevent masonry and construction companies from being forced out of business due to the financial burden of retooling equipment and products to comply with the Metric Conversion Act. Adherence to the 1975 law would have forced the construction industry to manufacture and stockpile two sizes of materials and products—US customary and metric.

Areas of Plane Figures

Parallelogram

$A = B \times H$

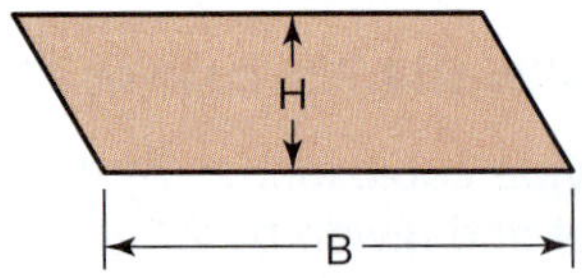

Trapezoid

$A = \frac{B + C}{2} \times H$

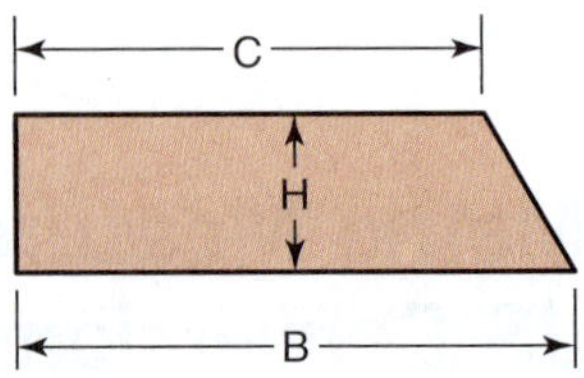

Triangle

$A = \frac{B \times H}{2}$

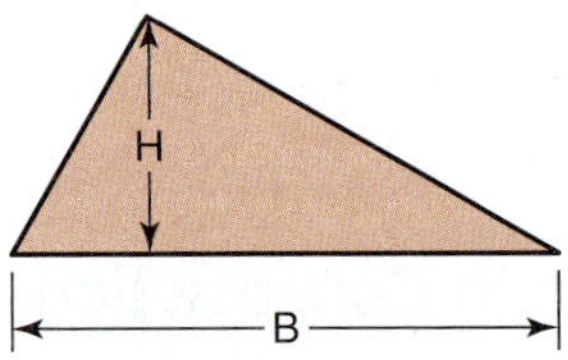

Regular polygon

$A = \frac{\text{Sum of sides (S)}}{2} \times R$

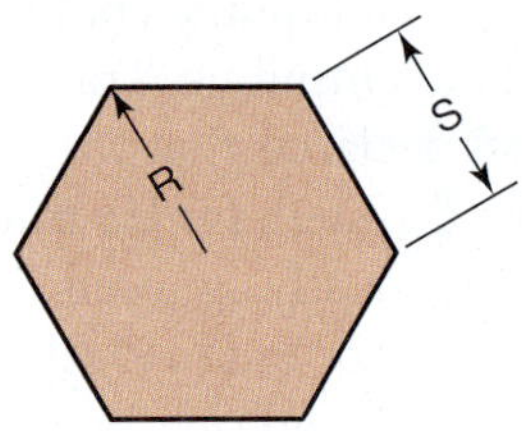

Circle

$A = pR^2$

$A = 0.7854 \times D^2$

$A = 0.0796 \times C^2$

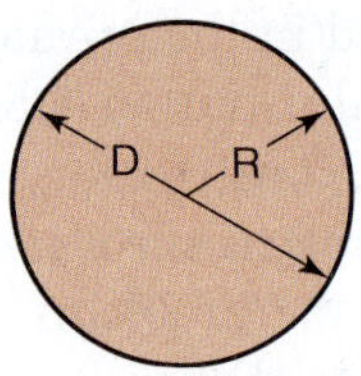

Ellipse

$A = M \times m \times 0.7854$

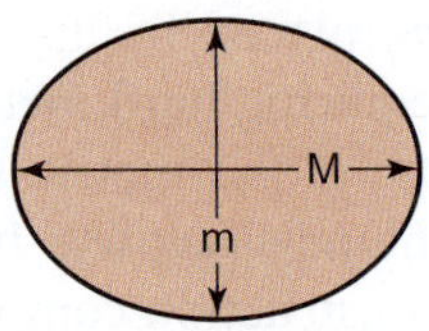

Volumes of Solid Figures

Rectangular solid

$V = L \times W \times H$

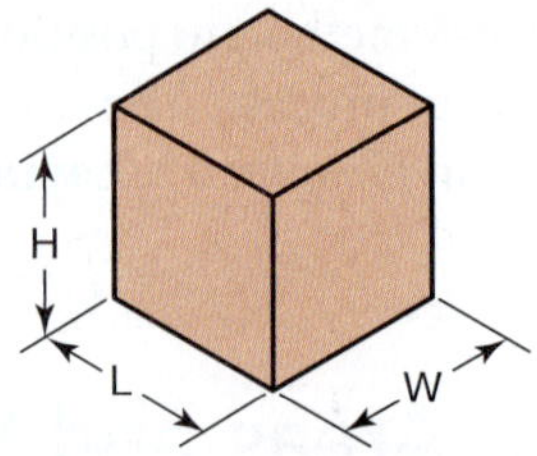

Prism

$V = \text{Area of end} \times H$

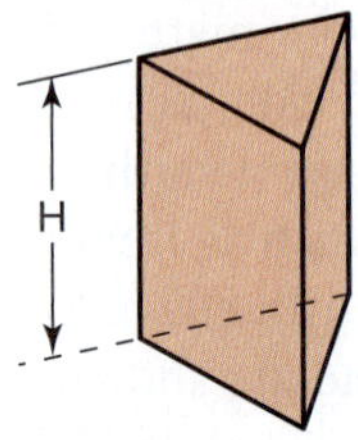

Cylinder

$V = \pi R^2 \times H$

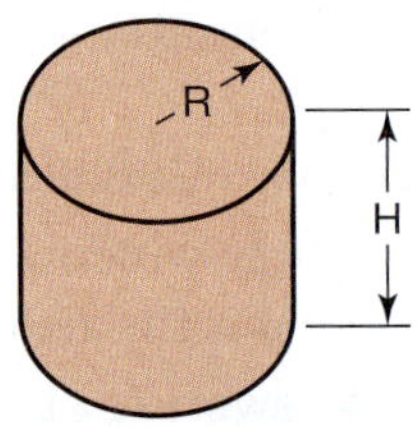

Cone

$V = \frac{\pi R^2 \times H}{3}$

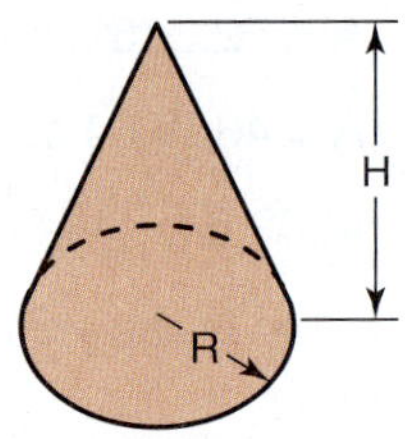

Pyramid

$V = \text{area of base} \times \frac{H}{3}$

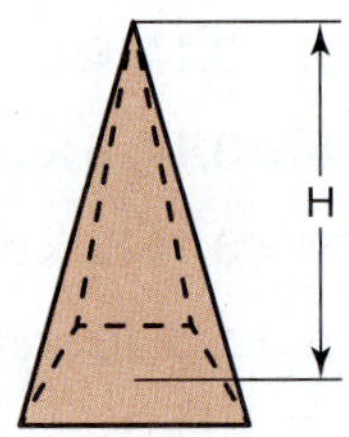

Sphere

$V = \frac{1}{6} \times \pi D^2$

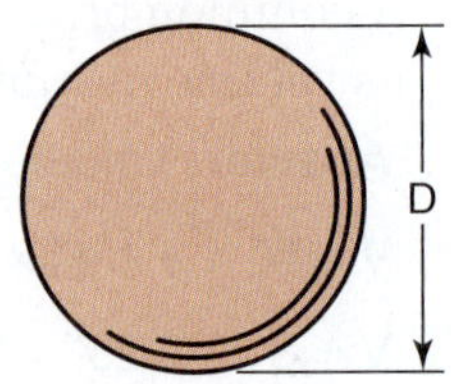

Goodheart-Willcox Publisher

Figure 6-7. Formulas for calculating areas or volumes of typical geometric shapes.

Implementation of the metric system in the masonry and construction industries continues to progress. As brick and block manufacturers convert to the metric system, they gain the opportunity to sell products to other countries. Architects are incorporating the metric system into building design when securing materials with metric dimensions is feasible.

There are two methods of implementing the metric system into the US construction industry:

- ***Soft metric conversion***. Products are relabeled from the US customary to the metric system without physically changing the product dimension.
- ***Hard metric conversion***. Products are resized to new dimensions based on metric measurement.

SI Units of Measure

SI metric is a convenient system that uses a base of 10 to form the various units of measure. The units of measure include the following:

- Base units
- Supplemental units
- Derived units that have been identified by name
- Non-SI units

Base Units

There are seven base units to which all other units are related. These units, with the exception of the kilogram, are based on natural phenomena that can be duplicated in the laboratory and require no international prototypes that must be stored by national standards organizations. The seven base units are shown in **Figure 6-8**.

Derived Units

Derived units are formed by multiplication or division of two or more SI base units. Some of the SI derived units are as shown in **Figure 6-9**.

Quantity	SI Unit	SI Symbol
Length	meter	m
Mass (weight)	kilogram	kg
Time	second	s
Temperature	degree Kelvin	K
Electric current	ampere	A
Luminous intensity	candela	cd
Amount of substance	mole	mol

Goodheart-Willcox Publisher

Figure 6-8. Base units of the SI units of measure.

Quantity	SI Unit	SI Symbol
Electrical charge	coulumb	C
Electrical potential (voltage)	volt	V
Electrical resistance	ohm	Ω
Electrical conductance	siemens	S
Electrical capacitance	farad	F
Electrical inductance	henry	H
Energy (heat or work)	joule	J
Force	newton	N
Frequency	hertz	Hz
Illumination	lux	lx
Luminous flux	lumen	lm
Magnetic flux	weber	Wb
Magnetic flux density	tesla	T
Power	watt	W
Pressure	pascal	Pa

Goodheart-Willcox Publisher

Figure 6-9. Derived units can be found by multiplying or dividing multiple base units of SI measure.

SI Prefixes

SI prefixes are formed by multiplying or dividing base units by powers of 10. This is an advantage because it eliminates insignificant digits and decimals. The accepted SI prefixes are shown in **Figure 6-10**. Note that multiples of 1000 of the base unit are to be used in preference to others.

Non-SI Units

Certain units of measure that are not part of the SI system are so widely used and accepted that they have been recognized. These units include those shown in **Figure 6-11**.

Multiplication Factors	Prefix	Symbol
1 000 000 000 000 = 10^{12}	tera	T
1 000 000 000 = 10^{9}	giga	G
1 000 000 = 10^{6}	mega	M
1 000 = 10^{3}	kilo	k
100 = 10^{2}	hector	h
10 = 10^{1}	deka	da
BASE UNITS 1 = 10^{0}		
0.1 = 10^{-1}	deci	d
0.01 = 10^{-2}	centi	c
0.001 = 10^{-3}	milli	m
0.000 001 = 10^{-6}	micro	μ
0.000 000 001 = 10^{-9}	nano	n
0.000 000 000 001 = 10^{-12}	pico	p
0.000 000 000 000 001 = 10^{-15}	femto	f
0.000 000 000 000 000 001 = 10^{-18}	atto	a

Goodheart-Willcox Publisher

Figure 6-10. SI prefixes are based on powers of 10. Note that the multiples of 1000 of the base unit are preferred over others.

Quantity	Name	Unit	Base Unit
Liquid volume	liter	L	meter
Area	square meter	m^2	meter
Solid volume	cubic meter	m^3	meter
Temperature	celsius	C	kelvin
Angle	degree	°	radian
Angle	minute	′	radian
Angle	second	″	radian
Time	minute	min	second
Time	hour	h	second
Time	day	d	second
Mass	tonne	t	kilogram

Goodheart-Willcox Publisher

Figure 6-11. Non-SI units of measure.

Converting US Customary and Metric Units

Until the United States converts to the SI system of measurement, the need to convert US customary to metric and metric to US customary will exist. **Figure 6-12** shows factors for converting from US customary to metric units.

To convert a quantity from US customary to metric units, use the following procedure:

1. Change US customary measurements in fractional form to decimal form.
2. Multiply the quantity by the factor shown in **Figure 6-12**.
3. Round off the result to the degree of accuracy required.

The meter is the standard of length for the metric system and all divisions or multiples related to it. Study the following prefixes, which are used for every weight and measure in the metric system.

1 meter = 1000 *milli*meters (thousandths)
1 meter = 100 *centi*meters (hundredths)
1 meter = 10 *deci*meters (tenths)
10 meters = 1 *deka*meter (tens)
100 meters = 1 *hecto*meter (hundreds)
1000 meters = 1 *kilo*meter (thousands)

Figure 6-13 shows factors for converting from metric to US customary units. Use the following procedure to convert a quantity from metric to US customary units:

1. Multiply the quantity by the factor shown in **Figure 6-13**.

US to Metric Conversions

Lengths
1 inch (in) = 2.540 centimeters
1 foot (ft) = 30.48 centimeters
1 yard (yd) = 91.44 centimeters or 0.9144 meters
1 mile (mi) = 1.609 kilometers

Areas
1 square inch (in^2) = 6.452 square centimeters
1 square foot (ft^2) = 920.0 square centimeters or 0.0929 square meters
1 square yard (yd^2) = 0.8361 square meter

Volumes
1 cubic in (in^3) = 16.39 cubic centimeters
1 cubic ft (in^3) = 0.02832 cubic meter
1 cubic yard (yd^3) = 0.7646 cubic meter

Weights
1 ounce (oz) = 28.35 grams
1 pound (lb) = 435.6 grams or 0.4536 kilograms
1 (short) ton = 907.2 kilograms

Liquid Measurements
1 (fluid) ounce = 0.0295 liter or 28.35 grams
1 pint = 473.2 cubic centimeters
1 quart = 0.9263 liter
1 (US) gallon = 3785 cubic centimeters or 3.785 liters

Power Measurements
1 horsepower (hp) = 0.7457 kilowatt (kW)

Temperature Measurements
To convert degrees Fahrenheit to degrees Celsius, use the following formula: C = 5/9 × (F – 32)

Goodheart-Willcox Publisher

Figure 6-12. US customary to metric conversions.

2. Round off the result to the degree of accuracy required.
3. Convert the decimal part of the number to the nearest common fraction for most trade applications.

Problems—US Customary and Metric Conversions

1. A residential structure is 40′ × 60′. What are its dimensions in meters?

   Answer: 40′ × 30.48 cm = 1219.2 cm

   1 m = 100 cm; therefore, divide by 100

   1219.2 cm ÷ 100 = 12.192 m

   40′ = 12.192 m

   60′ × 30.48 cm = 1828.8 cm ÷ 100

   = 18.288 m

   60′ = 18.288 m

   40′ × 60′ = 12.192 m × 18.288 m

Metric to US Conversions		
Lengths		
1 millimeter (mm)	=	0.03937 inch
1 centimeter (cm)	=	0.3937 inch
1 meter (m)	=	3.281 feet or 1.0937 yard
1 kilometer (km)	=	0.6214 miles
Areas		
1 square millimeter	=	0.00155 square inch
1 square centimeter	=	0.155 square inch
1 square meter	=	10.76 square feet or 1.196 square yard
Volumes		
1 cubic centimeter	=	0.06102 cubic inch
1 cubic meter	=	35.31 cubic feet or 1.308 cubic yard
Weights		
1 gram (G)	=	0.03527 ounce
1 kilogram (kg)	=	2.205 pound
1 metric ton	=	2205 pound
Liquid Measurements		
1 cubic centimeter (cm^3)	=	0.06102 cubic inch
1 liter (1000 cm^3)	=	1.057 quarts, 2.113 pints, or 61.02 cubic inches
Power Measurements		
1 kilowatt (kw)	=	1.341 horsepower (hp)
Temperature Measurements		
To convert degrees Celsius to degrees Fahrenheit, use the following formula: F = (9/5 × C) + 32		

Goodheart-Willcox Publisher

Figure 6-13. Metric to US customary conversions.

2. A metric drawing showed a recess to be 10.16 cm deep by 20.32 cm wide. What are its dimensions in inches?

   Answer: 2.54 cm = 1 in; therefore, divide each dimension by 2.54 cm

   10.16 cm ÷ 2.54 = 4″

   20.32 cm ÷ 2.54 = 8″

   The recess is 4″ deep by 8″ wide

6.8 Weight and Measures

Various weights and measures are used in estimating, proportioning, and ordering materials. Masons must be familiar with common weights and measures so estimates are accurate. Some of the more frequently used lengths, quantities, and their equivalents are shown in **Figure 6-14**.

144 sq in = 1 sq ft
9 sq ft = 1 sq yd
1296 sq in = 1 sq yd
1728 cu in = 1 cu ft
27 cu ft = 1 cu yd
1 cu ft of water weighs 62.5 lb
1 cu ft of damp loose sand equals 80 lb of dry sand
1 cu ft of cement weighs approximately 100 lb
1 cu ft of cement equals 1 bag; 4 bags equal 1 barrel
1 cu ft of hydrated lime weighs approximately 100 lb
1 1/4 cu ft of hydrated lime equals one 50 lb sack
1 cu ft of brickwork equals 120 lb
1 cu ft of concrete weighs approximately 150 lb
1 cu ft of packed earth weighs approximately 100 lb
1 cu ft of gravel weighs approximately 110 lb

Goodheart-Willcox Publisher

Figure 6-14. Common weights and measures.

Problems—Weights and Measures

1. What part of a square foot does the side of a clay tile represent? The side is 8″ × 12″.

   Answer: 8″ × 12″ = 96 sq in

   (1 sq ft = 144 sq in)

   96 ÷ 144 = 0.66 sq ft = 2/3 sq ft

2. If brickwork weighs 112 lb per cu ft, how many tons would a 16″ × 16″ × 12′ pier weigh?

   Answer: 16″ × 16″ = 256 sq in

   256 sq in × 144″ = 36,864 cu in

   (1 cu ft = 1728 cu in)

   36,864 cu in ÷ 1728 cu in

   = 21.33 cu ft

   (1 ton = 2000 lb)

   112 lb × 21.33 cu ft = 2389.33 lb

   2389.33 lb ÷ 2000 = 1.19 tons

6.9 Estimating Brick Masonry

According to the Brick Industry Association, almost all bricks produced in the United States fit the 4″ modular grid system, except for nonmodular "standard bricks" (3 3/4″ × 2 1/4″ × 8″) and some oversize bricks (3 3/4″ × 2 3/4″ × 8″). Modular size masonry units greatly simplify the job of estimating the number of units required for a given wall area.

The most widely used procedure for estimating brick masonry is the "wall-area" method. It consists of multiplying the quantity of materials needed per square foot by the net wall area. ***Net wall area*** is the gross area less (minus) all wall openings.

The three standard modular joint thicknesses are 1/4″, 3/8″, and 1/2″. For a given nominal size masonry unit, the number of modular masonry units per square foot of wall is the same regardless of the mortar joint thickness. This is true if the units are laid with the joint thickness for which they were designed. However, the number of nonmodular standard bricks required per square foot of wall varies with the thickness of the mortar joint.

When estimating, determine the net quantities of all materials before adding any allowances for waste and breakage. As a general rule, 5% is added for bricks and 10% to 25% for mortar. These factors may vary considerably with job conditions.

Figure 6-15 shows the net quantities of brick and mortar necessary to construct walls one wythe in thickness with various modular and nonmodular brick sizes using either 1/2″ or 3/8″ mortar joints. A ***wythe*** is each continuous vertical section of masonry one unit in thickness.

Quantities shown in **Figure 6-15** are for running or stack bond that contain no headers. Bonds that require full headers must use the correction factor shown in **Figure 6-16**. Multiwythe walls also require a correction factor for the collar joint between the wythes. See **Figure 6-17**.

Mortar Yield

Accurate mortar yield calculations are based on absolute volume. These calculations are complex and require extensive data about the materials being used. Most masons do not have the data, nor do most jobs

Modular Brick					
Nominal Size of Bricks (in) T H L	Number of Bricks per 100 Sq Ft	Cubic Feet of Mortar per 100 Sq Ft (3/8″ Joints)	Cubic Feet of Mortar per 100 Sq Ft (1/2″ Joints)	Cubic Feet of Mortar per 1000 Bricks (3/8″ Joints)	Cubic Feet of Mortar per 1000 Bricks (1/2″ Joints)
4 × 2 2/3 × 8	675	5.5	7	8.1	10.3
4 × 3 1/5 × 8	563	4.8	6.1	8.6	10.9
4 x 4 x 8	450	4.2	5.3	9.2	11.7
4 × 5 1/3 × 8	338	3.5	4.4	10.2	12.9
4 × 2 × 12	600	6.5	8.2	10.8	13.7
4 × 2 2/3 × 12	450	5.1	6.5	11.3	14.4
4 × 3 1/5 × 12	375	4.4	5.6	11.7	14.9
4 × 4 × 12	300	3.7	4.8	12.3	15.7
4 × 5 1/3 × 12	225	3	3.9	13.4	17.1
6 × 2 2/3 × 12	450	7.9	10.2	17.5	22.6
6 × 3 1/5 × 12	375	6.8	8.8	18.1	23.4
6 × 4 × 12	300	5.6	7.4	19.1	24.7
Nonmodular Brick					
Nominal Size of Bricks (in) T H L	**Number of Bricks per 100 Sq Ft**	**Cubic Feet of Mortar per 100 Sq Ft (3/8″ Joints)**	**Cubic Feet of Mortar per 100 Sq Ft (1/2″ Joints)**	**Cubic Feet of Mortar per 1000 Bricks (3/8″ Joints)**	**Cubic Feet of Mortar per 1000 Bricks (1/2″ Joints)**
3 3/4 × 2 1/4 × 8	655	5.8	8.8	—	—
3 3/4 × 2 1/4 × 8	616	—	7.2	—	11.7
3 3/4 × 2 3/4 × 8	551	5.0	—	9.1	—
3 3/4 × 2 3/4 × 8	522	—	6.4	—	12.2

Goodheart-Willcox Publisher

Figure 6-15. Modular and nonmodular bricks and mortar required for single-wythe walls in running or stack bond. These figures do not allow for waste and breakage.

Bond	Correction Factor
Full headers every fifth course only	1/5
Full headers every sixth course only	1/6
Full headers every seventh course only	1/7
English bond (full headers every second course)	1/2
Flemish bond (alternate full headers and stretchers every course)	1/3
Flemish headers every sixth course	1/18
Flemish cross bond (Flemish headers every second course)	1/6
Double-stretcher, garden wall bond	1/5
Triple-stretcher, garden wall bond	1/7

Goodheart-Willcox Publisher

Figure 6-16. Correction factors for the walls with full headers. Add to the facing and subtract from the backing. Correction factors are applicable only to those bricks that are twice as long as they are wide.

Cubic Feet of Mortar per 100 Square Feet of Wall		
1/4″ Joint	3/8″ Joint	1/2″ Joint
2.08	3.13	4.17

Goodheart-Willcox Publisher

Figure 6-17. Correction factors for collar joints. Quantities are given in cu ft of mortar for collar joints. Note that cu ft/1000 units = 10 × 11 cu ft/100 sq ft of wall ÷ the number of units/sq ft of wall.

require critical mortar yield calculations. Therefore, a rule-of-thumb may generally be used. For each 1 cu ft of damp loose sand, the mortar yield is 1 cu ft.

Problems—Estimating Brick Masonry

1. How many bricks (nominal size of 4″ × 2 2/3″ × 8″) are required to veneer a wall 24′ long and 4′ high? A 3/8″ mortar joint is to be used. The wall has no openings.

   Answer: First, find the wall area: 24′ × 4′ = 96 sq ft

   Second, find the number of bricks per sq ft in **Figure 6-15**

   The table shows 675 bricks/100 sq ft or 6.75 bricks/sq ft

   Next, multiply the number of bricks/sq ft by the wall area

   6.75 × 96 = 648 bricks

   Note that 5.28 cu ft of mortar is required to lay these bricks (96% of 5.5)

2. How many bricks and how much mortar are needed to lay a wall 40′ long and 8′ high using 6″ × 4″ × 12″ SCR bricks with 1/2″ mortar joints? The wall has three window openings 3′ wide by 4′ high.

   Answer: Gross wall area = 40′ × 8′ = 320 sq ft

   Net wall area = 320 sq ft – (3′ × 4′ × 3)

   = 320 – 36 = 284 sq ft

   Brick/sq ft (**Figure 6-15**) = 300 per 100 sq ft or 3.0/sq ft

   Therefore, 284 × 3 = 852 brick

   Mortar required = 7.4 per 100 sq ft or 0.074/sq ft

   0.074 × 284 = 21.02 cu ft of mortar

Chapter Review

Summary

- Whole numbers are natural numbers and represent quantities of anything that can be added, subtracted, multiplied, or divided.
- A fraction is one or more parts of a whole and should always be reduced to its lowest terms.
- Before adding or subtracting fractions with different denominators, change the fractions so they have the same (common) denominator.
- Multiply fractions by first multiplying the numerators to find the numerator for the answer, then multiply the denominators together to find the denominator for the answer.
- To divide fractions, invert the divisor and then multiply the fractions.
- The word *and* is generally used to represent the decimal when reading a decimal mixed number (numbers on both sides of the decimal).
- Percentage is a number used to denote parts of one hundred.
- A triangle is a geometric shape that has three angles that total 180°. When two angles are known, the third can be found by subtracting the sum of the two angles from 180°.
- The Pythagorean Theorem is a mathematical formula that aids masons in laying out a building. The formula is $a^2 + b^2 = c^2$.
- Calculating areas and volumes is a crucial skill in the masonry industry.
- Metric measurement is used by most countries. SI metric units of measure include base units, supplemental units, derived units, and non-SI units.
- Brick masonry is estimated by multiplying the quantity of materials needed per square foot by the net wall area. Net wall area is the gross area minus all wall openings.

Know and Understand

Answer the following questions using the information provided in this chapter.

1. If four masons each average laying 358 bricks a day, how many bricks would they average laying in five days?
   A. 7250 bricks
   B. 7155 bricks
   C. 7160 bricks
   D. 7120 bricks
2. If a mason ordered 5 1/2 cu yd of sand and used 1/4 cu yd, how much is left?
   A. 2 1/4 cu yd
   B. 2 1/2 cu yd
   C. 3 1/2 cu yd
   D. 1 3/4 cu yd
3. What is the height of one stretcher course of concrete block if the block is 7 5/8″ high and the mortar joint is 3/8″?
   A. 8 1/8″
   B. 7 7/8″
   C. 8 1/4″
   D. 8″
4. If one isolated footing requires 1/8 yd of concrete, how much concrete is needed for a footing 3/4 as large?
   A. 5/32 yd
   B. 1/3 yd
   C. 1/4 yd
   D. 3/32 yd
5. Four pieces of stone trim weigh 10.5 lb, 17.4 lb, 23.8 lb, and 5.6 lb. What is the total weight of the four pieces of trim?
   A. 55.5 lb
   B. 60.2 lb
   C. 54.3 lb
   D. 57.3 lb
6. Which of the following quantities is the largest?
   A. 0.25
   B. 0.025
   C. 0.0025
   D. 2.2500
7. _____ is 12% of 692.
   A. 69
   B. 34.6
   C. 173
   D. 83.4

8. The circumference of a circle that has a diameter of 6.54″ is _____.
   A. 134.4″
   B. 20.55″
   C. 18.7″
   D. 2.05″
9. If the hypotenuse of a right triangle is 15′ and one side is 12′, what is the length of the other side?
   A. 19.2′
   B. 81′
   C. 9′
   D. 12′
10. The volume of a cylindrical pipe 6′ long with an inside diameter of 4″ is _____.
    A. 904.8 cu ft
    B. 75.4 cu ft
    C. 75.4 cu in
    D. 904.8 cu in
11. A wall area of 23 sq ft required 115 bricks to face it. How many bricks were needed for each square foot of area?
    A. 5 bricks
    B. 4 bricks
    C. 4.5 bricks
    D. 6 bricks
12. One meter is equal to _____.
    A. 1000 millimeters
    B. 100 centimeters
    C. 10 decimeters
    D. All are correct.
13. One inch is equal to _____.
    A. 2.54 cm
    B. 36 m
    C. 1/38 m
    D. 100 mm
14. What part of a square foot does the side of a tile represent? The side is 4″ × 12″.
    A. 1/2 sq ft
    B. 1/3 sq ft
    C. 1/4 sq ft
    D. 2/3 sq ft
15. *True or False?* Accurate mortar yield calculations are based on area.

Apply and Analyze

1. If 1 1/2 yd of mortar is divided among three masons, how much mortar does each get?
2. What is the sum of the following quantities: 0.12, 2.03, 12.004, 9.0?
3. A masonry job that lasted four days required 13.5 cu yd of mortar. Records show that 6.8 cu yd were used the first two days. How many yards were used the last two days of the job?
4. A mason purchased 5.5 cu yd of washed masonry sand for $7.50 per yd. How much was paid for the sand?
5. If a mason charged $0.25 per brick and the total bill was $3500.00, how many bricks were laid?
6. What is the decimal equivalent of 3 11/16″?
7. What is the fractional size of a piece of steel that is 2.375″?
8. If 3% were allowed for breakage on 10,000 bricks, how many bricks could be used?
9. How much money would be earned on a $1000 savings certificate in one year at 3% interest per year?

Critical Thinking

1. Explain the process of laying out a square corner using the 3-4-5 method. One leg of the corner is 3′ and the other leg is 4′. Create a step-by-step procedure that could be used to follow as if you were laying out a building that is 3′ × 4′. To aid in understanding the layout process, cut three pieces of string, one 3′, another 4′, and finally a string 5′ in length.

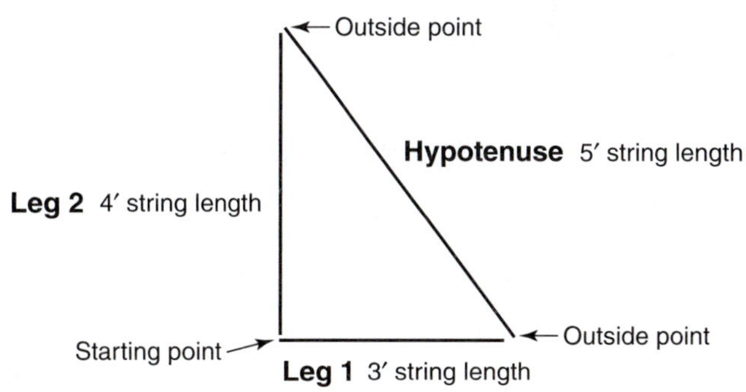

Activities

1. (MATH) Calculate the following. Show your work.
   A. Convert 3/5 to a decimal number.
   B. Convert 2 1/8 to a decimal number.
   C. Calculate 5% of 120.
   D. Calculate the metric equivalent of 3 inches.
   E. Calculate the metric equivalent of 4 cubic feet.
2. (MATH) Use a rule or tape measure to determine the length and width of a room in your home in US customary units. Calculate the area of the room, and then convert the result to SI metric units.

SECTION

3 Materials

frantic00/Shutterstock.com

Lex20/iStock via Getty Images

CHAPTER 7

Clay Masonry Materials

LEARNING OUTCOMES

After studying this chapter, you will be able to:

7.1-1 Recall the standards that apply to masonry construction.
7.2-1 Discuss the three classifications of structural clay products.
7.3-1 Explain the three processes used to manufacture brick.
7.4-1 Describe the process of drying and firing bricks.
7.5-1 Recall the classifications of brick.
7.6-1 Differentiate between modular and nonmodular bricks.
7.7-1 Describe solid masonry units and their properties.
7.8-1 Interpret the meaning of the term bond as it refers to masonry.
7.9-1 Explain the purpose of special shapes of brick.
7.10-1 Describe types of hollow masonry units.
7.11-1 Recall the classifications of architectural terra cotta.

TECHNICAL TERMS

adhesion ceramic veneer
American bond
anchored ceramic veneer
bat
bearing surfaces
block bond
brick panel
ceramic glazed bricks
chemical-resistant masonry units
common bond
compressive strength
concave joints
course
double stretcher garden wall bond
Dutch bond
efflorescence
end construction
English bond
English cross
extruded joint
facing bricks
feldspar
Flemish bond
firebox brick
flush joint
garden wall bond
glazed brick
glazed clay facing tile
Grade MW brick
Grade NW brick
Grade SW brick
green bricks
grout
hollow bricks
hollow masonry unit
industrial floor brick
kiln
king closure
modular unit
mortar bond
ornamental terra cotta
pattern bond
quarter closure
queen closure
raked joint
rough cut joint
running bond
SCR acoustile
side construction
snap headers
split
stack bond
struck joint
structural bond
structural clay products
terra cotta
thin brick veneer
three-quarter closure
tooled joint
troweled joint
unglazed facing tile
V-shaped joints
weathered joint

Introduction

The variety of masonry materials available for use in construction has greatly increased. There are many new shapes, textures, colors, and applications. This is due, in part, to the variety, versatility, durability, and beauty of masonry materials. See **Figure 7-1**. For instance, it is estimated that brick alone is made in at least 10,000 different combinations of colors, textures, and shapes.

Masonry construction provides permanence. Few, if any, materials are as maintenance-free and long-lasting as brick, block, stone, and concrete.

Masonry units and materials may be grouped under several basic categories or products:

- Structural clay products (brick, clay tile, and terra cotta)
- Concrete masonry units
- Sand-lime brick
- Glass block
- Stone
- Mortar
- Masonry anchors, ties, and joint reinforcement

The classification, characteristics, sizes, and applications of each of these basic materials will be discussed later in this chapter.

7.1 Building Codes for Masonry Structures

The American Concrete Institute (ACI) and the American Society of Civil Engineers (ASCE) have adopted a consensus standard for the design of masonry structures that also includes a standard construction specification. These documents, known as ACI 530-08/ASCE 5-08 *Building Code Requirements for Masonry Structures*, and ACI 530.1-08/ASCE 6-08 *Specification for Masonry Structures*, were developed to replace existing standards for the design and construction of masonry structures.

History and Development

The development of a sole masonry standard for the design and construction industry began in 1977. The several design standards for masonry in use at that time

General Shale; Glen-Gery Corporation; Robinson Brick Company

Figure 7-1. These clay masonry applications show only a few of the many textures, colors, and shapes possible in masonry construction.

did not have consistent requirements. Concerned individuals representing masonry materials and the design profession saw the need for a single, consensus document for the design and construction of all types of masonry.

ACI and ASCE agreed to jointly develop a consensus standard for masonry design with the support of the masonry industry. Public review began in 1988 with the final approval of the code and specifications in 1989. The Standard Building Code Congress International (SBCCI) adopted the Standard Building Code in 1988. Then the Building Officials and Code Administrators (BOCA), who published the National Building Code, adopted the Standard Building Code in 1989. In order to avoid duplication BOCA, ICBO (International Conference of Building Officials), and SBCCI agreed in 1994 to join to form the International Code Council (ICC). BOCA, ICBO, and SBCCI agreed to adopt the new codes being created by the International Code Council that were to be named the International Building Code (IBC). At that time, state and local governments began to adopt the newly developed IBC codes.

The International Codes, or I-Codes, are widely used building codes. The following are examples of I-Codes:

- International Building Code (IBC)
- International Residential Code (IRC)
- International Plumbing Code (IPC)
- International Mechanical Code (IMC)
- International Energy Conservation Code (IECC)
- International Fire Code (IFC)

The IBC codes replaced existing masonry design standards. For clay brick masonry, the previous standards were ANSI A41.1 *Empirical Design of Reinforced Masonry*; ANSI A41.2 *Design for Reinforced Masonry*; and BIA-69 *Building Code Requirements for Engineered Brick Masonry*. Other masonry standards affected by the adoption of the IBC Codes were NCMA TR-75B, ACI 531 and ACI 531.1, which address only concrete masonry design and construction. The standards listed in the codes contribute to consistent design of masonry and a higher degree of quality in construction. These documents incorporate the latest research on material properties and performance. They reduce design procedures that are arbitrary and limiting to masonry construction.

Masonry Standards Joint Committee (MSJC)

The Masonry Standards Joint Committee (MSJC) is a joint effort between The Masonry Society (TMS), the American Concrete Institute (ACI), and the Structural Engineering Institute of the American Society of Civil Engineers (SIE/ASCE). The MSJC Code covers the design and construction of masonry structures. The specifications provide minimum construction requirements of masonry in structures being built. Building codes for masonry structures are covered in (TMS 402-ACI 530/ASCE 5) and the specifications for masonry structures are found in (TMS 602-ACI 530.1/ASCE 6). The year of the code follows the specific code number. For example, in the code (TMS 402-16/ACI 530-16/ASCE 5-16), the number *16* indicates 2016.

7.2 Structural Clay Products

There are three classifications of structural clay products:

- Solid masonry units (brick)
- Hollow masonry units (tile)
- Architectural terra cotta

Structural clay products are made from clay or shale. See **Figure 7-2**. These materials are found throughout

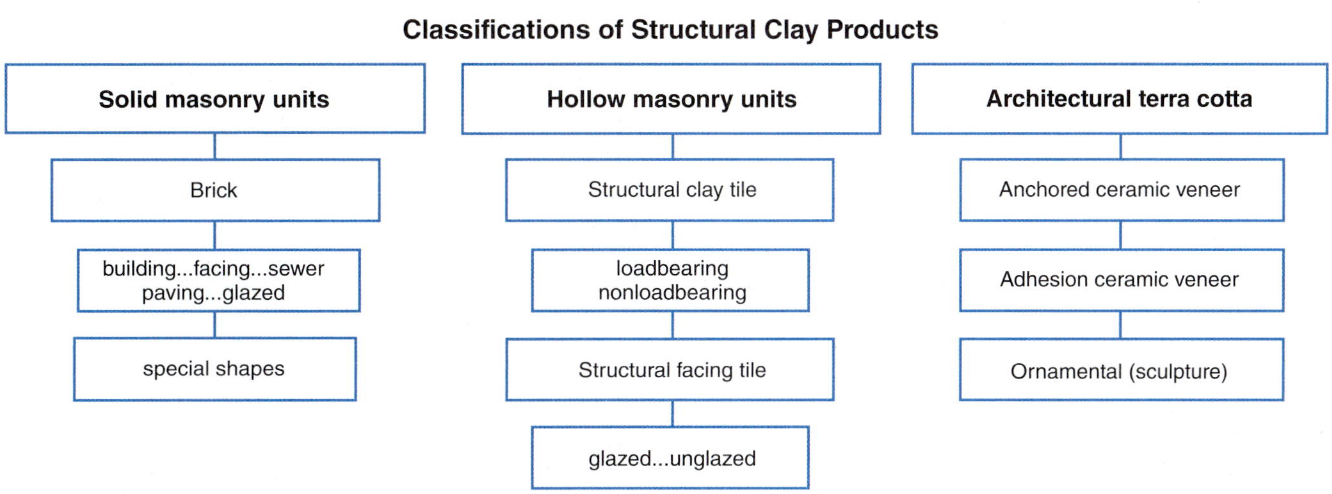

Goodheart-Willcox Publisher

Figure 7-2. Structural clay products can be grouped under three basic types.

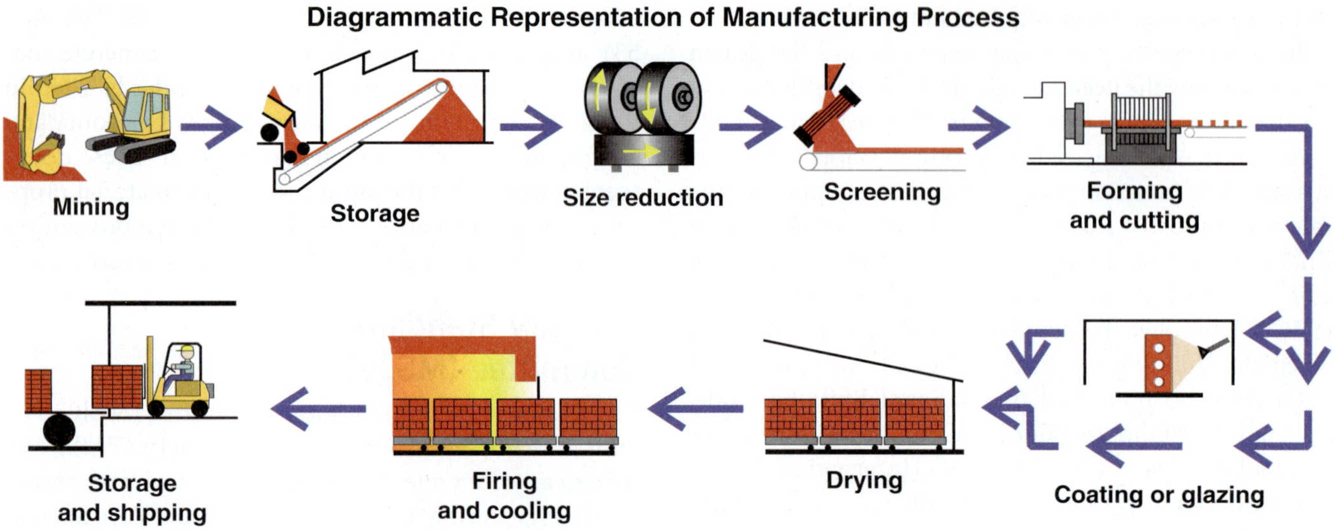

Brick Industry Association

Figure 7-3. Diagrammatic representation of the brick manufacturing process.

the United States. Clays and shales suitable for the manufacture of brick fall into three groups:

- Surface clays
- Shales
- Fire clays

Clays are produced naturally as rocks weather. Pure clay is mainly silica and alumina. It is made up of tiny crystals of certain rocks, chiefly feldspar. ***Feldspar*** is a crystalline mineral made up of aluminum silicates with either potassium, sodium, calcium, or barium.

Shales are compressed clay. They are denser than clay. Fire clays have higher percentages of aluminum silicates, flint, and feldspar. The different characteristics of these materials make a great variety of clay products possible.

7.3 Manufacture of Brick

As shown in **Figure 7-3**, the brick manufacturing process includes the following general phases:

- Mining of raw materials
- Crushing and storage of raw materials
- Preparation of raw materials
- Forming and cutting units
- Coating or glazing
- Drying
- Firing and cooling
- Storage and shipping

Bricks are commonly manufactured by three methods:

- Soft mud process
- Stiff mud process
- Dry-press process

Soft Mud Process

The two variations of the soft mud process are the water-struck method and sand-struck method. Both methods prevent the clay from sticking to the mold. The clay is mixed wet and forced into a mold. The thickness of the brick is formed by scraping off the top of the mold.

In the water-struck process, the brick face is usually rather smooth, **Figure 7-4**. A sand-struck brick has a sandpaper-like surface. See **Figure 7-5**. Particles of sand stick to the brick material during removal from the mold and create this roughness.

Stiff Mud Process

In the stiff mud process, the clay is just wet enough to stick together. It is forced through a die and extruded

Goodheart-Willcox Publisher

Figure 7-4. The surface of this water-struck brick is smooth.

Goodheart-Willcox Publisher

Figure 7-5. This sand-struck brick has a sandpaper-like surface.

Shi Yali/Shutterstock.com

Figure 7-7. Bricks in the green state are arranged in pallets. The bricks can be either air-dried or dried in a kiln prior to being fired.

(shaped) in the form of a ribbon or column. The thickness and width of the ribbon is the same as the dimensions of the brick being produced. Wires are used to cut the ribbon to lengths desired. **Figure 7-6** shows the brick produced by this method.

Dry-Press Process

In the dry-press process, the clay is dry and loose. The clay is forced under high pressure into molds and forms the brick. These bricks are very dense.

7.4 Drying and Burning or Firing

Green bricks are bricks that have not been fired. **Figure 7-7** shows green brick either being air-dried or waiting to be dried in a ***kiln***. A kiln is an oven or furnace that heats the brick to a predetermined temperature, depending on if the furnace is being used for drying or firing. Brick contains from 5% to 30% moisture as it comes from the molding or extruding machines. Most of this moisture must be evaporated before the brick is burned or fired. Therefore, the first step is to dry the brick in a kiln for 24 hours to 48 hours, depending on the original moisture content and the size of the brick. Brick may also be allowed to air-dry, but this takes longer. If the brick is fired too soon, it cracks because the moisture is leaving the brick too rapidly. When the brick reaches the appropriate moisture content during the drying process, it can then be moved to the firing kiln. Even distribution of heat is crucial for uniform results.

The most popular types of kilns used for drying brick are periodic and tunnel kilns. The primary difference between the two is that the brick remains stationary in the periodic kiln and the temperature is changed for each stage of the burning. In the tunnel kiln, the bricks are placed on moving cars that pass through different temperature zones.

Burning time may be as little as 50 hours in a tunnel kiln and as long as 150 hours in a periodic kiln. Cooling may require over 50 hours. Proper cooling time prevents checking (small cracks) and maintains the right color.

Goodheart-Willcox Publisher

Figure 7-6. Wire-cut brick has a rough surface.

Green Note

Environmental Impact of Brick Manufacture

The heat for most kilns in brick manufacturing plants is generated by burning natural gas, which generates significant carbon emissions. Many manufacturing plants are now using bio-based fuel resources, such as methane or sawdust. Burning methane produces fewer emissions than burning natural gas. Adding sawdust to the raw materials used to manufacture brick also reduces the use of fossil fuels. The sawdust burns in the firing process.

7.5 Classification of Brick

Brick is usually placed into the following categories:

- Building brick (often referred to as *common brick*) (ASTM C62)
- Facing brick (ASTM C216)
- Hollow brick (C652)
- Pedestrian and light traffic paving brick (C902)
- Heavy vehicular paving brick (C1272)
- Ceramic glazed brick (C126)
- Glazed brick, single fired (C1405)
- Thin brick veneer units (C1088)
- Firebox brick, residential fireplaces (C1261)
- Chemical-resistant masonry units (C279)
- Sewer and manhole brick (C32)
- Industrial floor brick (C410)

Building Brick

Building, or common, bricks are used chiefly as structural material where durability and strength are of more importance than appearance. Three grades of building bricks are available. ***Grade SW brick*** is brick that is highly resistant to frost action. It is used where bricks are exposed to freezing weather. An example is brick used for foundation courses or retaining walls. ***Grade MW brick*** is brick that is used where there may be exposure to temperatures below freezing, but where the brick is not likely to be permeated with water. An example is brick used in the face of a wall aboveground. ***Grade NW brick*** is brick designed to be used as a backup on interior masonry.

Facing Brick

Facing bricks are used on exposed surfaces where appearance is an important consideration. Such brick must also be strong and durable. Appearance is covered in three grades. Type FBX brick is used for general use in exposed exterior and interior walls and partitions where a high degree of perfection is required. This brick must be uniform in color and size. Type FBS brick is for general use in exposed exterior and interior masonry walls and partitions where wider color ranges and greater variation in sizes are permitted. Type FBA brick is manufactured and selected to produce architectural effects resulting from nonuniformity in size, color, and texture.

Hollow Brick

Hollow bricks are identical to facing bricks but have a larger core area. Most hollow bricks are used in the same applications as facing bricks. Hollow bricks with very large cores are used in walls that are reinforced with steel and grouted solid. Larger cores or cells in hollow bricks allow reinforcing steel and grout to be placed in these units, which would be difficult to do with building bricks, facing bricks, or some hollow bricks.

Pedestrian and Light Traffic Paving Brick

Pedestrian and light traffic paving bricks are intended for use as a paving material to support pedestrian and light vehicular traffic. They must meet additional requirements for abrasion resistance or alternate performance requirements. Paving bricks are classified into different types according to their resistance to abrasion and ability to withstand traffic. Type I pavers can be exposed to extensive abrasion, such as on driveways or public entries. Type II pavers can be exposed to high levels of pedestrian traffic, such as in stores, restaurant floors, or exterior walkways. Type III pavers can be exposed to pedestrian traffic, such as floors or patios.

Heavy Vehicular Paving Brick

Heavy vehicular paving bricks are intended for heavy traffic and equipment applications. Type R pavers are intended to be set in mortar or asphalt with an adequate subbase that is compacted and well drained. Type R pavers must be at least 2 1/4″ in thickness to support the heavy traffic that will be placed on the brick. Type F pavers are intended to be set in a base of sand with sand joints. Type F pavers must be at least 2 5/8″ in thickness. This is because a sand base does not have the supportive strength of an asphalt or mortar base. The resistance to abrasion for Type F and Type R pavers is the same.

Ceramic Glazed Brick and Glazed Brick, Single Fired

Ceramic glazed bricks and ***glazed bricks*** are units with a glaze fused to the body. These units are used as facing brick. Glazed brick may be used on walls and the underside of large openings on buildings, **Figure 7-8**. The glaze or ceramic finish is available in standard colors as well as a variety of custom colors. The body of the brick may be either facing brick or other solid masonry units.

Ceramic glazed brick and glazed brick, single fired are two similar brick classifications, but they have different characteristics and strengths. The Brick Industry Association's Technical Notes on Brick Construction, *9A Specifications for Classifications of Brick* discusses applications and strengths of brick. Refer to this document when selecting ceramic glazed brick or glazed brick, single fired.

Elgin-Butler Company

Figure 7-8. Ceramic glazed brick on an elementary school.

Ceramic or glazed finish has the following properties:

- Imperviousness
- Opacity
- Resistance to fading
- Resistance to crazing (fine hairline cracks in the ceramic surface after firing of the unit)
- Fire resistance
- Heat resistance (withstands high temperatures without distorting, melting, or releasing toxic fumes)
- Smoke density (smoke does not enter the hard surface of the glazing)
- Hardness
- Resistance to abrasion

Thin Brick

Thin brick veneer units are fired clay units with normal face dimensions but a reduced thickness. They are used in adhered veneer applications where a thinner than normal brick width is desired. Thin brick can be used either in interior or exterior applications. Refer to the manufacturer's specifications for the appropriate installation procedure based on the application, since procedures may differ depending on whether the thin brick is installed in an interior or exterior setting. Most thin bricks have an irregular rear surface to aid in securing them to the backing or substrate surface. See **Figure 7-9**.

Gage Brothers in Sioux Falls, SD

Figure 7-9. Thin brick structure showing face and edge profiles. The end view shows the irregular rear surface that will aid in securing the brick to the substrate.

Brick Panels

Brick panels are a combination of thin brick, reinforcement, insulation, and concrete. The panels are created in a factory setting where thin brick is placed in molds that position the brick in a specialized form for creating wall panels. Once the thin brick is in position, reinforcement is installed. Concrete is then placed in the form to create the completed panel. Insulation may be included in the internal part of the panel, depending on the application and the future owner's preference.

Once the panels have gained the required strength, they are transported from the factory to the jobsite. The wall sections are removed from the trailer using a crane. The panels may be stored on the jobsite or erected at the time of delivery. The panels are lifted and set in position, then attached to other panels to form the structure of the building. See **Figure 7-10**.

Gage Brothers in Sioux Falls, SD

Figure 7-10. Brick panel being removed from the transport trailer to be set in place by a crane.

Firebox Brick

Firebox brick is intended for use in the lining of fireboxes in residential fireplaces. These bricks are manufactured from clay, fireclay, shale, or other similar materials. The materials are subjected to a heat treatment to develop the strength and durability required to meet standards for firebrick. To qualify as firebrick, the brick must pass a number of assessments, including a rupture test, size test, and warpage test, which will ensure that the brick can withstand the heat generated in the firebox of a fireplace.

Chemical-Resistant Masonry Units

Chemical-resistant masonry units are brick and tile made specifically for locations where chemicals and acids are present and may be at a significant level. The units are made in forms of brick or tile with a composition of clay, shale, or a suitable combination of the two. Chemical-resistant masonry units are classified into three types:

- **Type I.** For use where low absorption and high acid resistance are not major factors.
- **Type II.** For use where lower absorption and higher acid resistance are required.
- **Type III.** For use where minimum absorption and maximum acid resistance are required.

Chemical-resistant masonry units are used in many locations, including the following:

- Acid processing areas
- Bakeries
- Battery plants
- Breweries
- Chemical plants
- Meat packing plants
- Petroleum refineries

Sewer and Manhole Brick

Sewer and manhole bricks are intended for use in drainage structures for the conveyance of sewage, industrial wastes, and storm water. They are also used in related structures such as manholes and catch basins. The laying of manhole brick and sewer brick usually occurs when repairs are needed on older sewer systems, **Figure 7-11**. In most modern sewer systems, manufactured concrete products are used because they reduce the time of installation. See **Figure 7-12**.

appalacian trail/Shutterstock.com

Figure 7-11. Manhole cover being positioned over freshly laid manhole brick.

A Foto76/Shutterstock.com

B Yuangeng Zhang/Shutterstock.com

Figure 7-12. Manufactured concrete products are used in most modern sewer systems. A—Concrete riser and metal sleeve ready to be used for a sewer system. B—Manhole cover positioned in a metal sleeve with a concrete outer ring.

Industrial Floor Brick

Industrial floor brick is intended for the surface of industrial floors. This brick is made of clay, shale, or a combination of the two. Industrial floor brick fall into four categories—T, H, M, and L. The properties of the various categories are as follows:

- **Type T.** Brick with a high resistance to thermal and mechanical shock.
- **Type H.** Brick with a resistance to chemicals and thermal shock.
- **Type M.** Brick with a low absorption rate, limited mechanical shock resistance, and high resistance to abrasion.
- **Type L.** Brick with a minimal absorption rate, a high degree of chemical resistance, very limited thermal resistance, limited mechanical shock resistance, and high resistance to abrasion.

7.6 Brick Sizes and Nomenclature

For many years, bricks were manufactured in only a few sizes—Standard, Roman, and Norman. Many more sizes are now available. In the past, a given brick size may have been known by several names due to regional variations. A joint committee of the Brick Industry Association and the National Association of Brick Distributors developed standard nomenclature for bricks. This nomenclature represents roughly 90% of all brick sizes currently manufactured. Brick sizes can be separated into two groups—modular and nonmodular.

A ***modular unit*** is based on a measurement of 4″. It could include a measurement of 2″ (one-half of the module) or 8″ (twice the module). However, in bricks, this size is only nominal. The dimension is not exactly 4″. Some allowance has to be made for the thickness of the mortar joints. Thus, a nominal 4″ may be actually 3 7/8″ or some other dimension.

Bricks are available in sizes ranging in thickness from a nominal 3″ to 12″. Heights vary from a nominal 2″ to 8″, and lengths up to 16″ are common.

Figure 7-13 shows the most typical sizes of nonmodular brick. Sizes of modular brick are shown in **Figure 7-14**. Few manufacturers produce all the sizes, while other sizes are produced by some manufacturers. As design requirements change, new sizes may be added and less popular sizes dropped. For these reasons, it is advisable to check the sizes available in a given region before purchasing.

With the exception of the nonmodular standard, oversize, and 3″ units, most bricks are produced in modular sizes. The nominal dimensions of modular bricks are equal to the manufactured dimensions plus the thickness of the appropriate mortar joint. See **Figure 7-15**. The mortar joint is usually 3/8″ or 1/2″ thick. The standard joint thickness (3/8″ or 1/2″) is subtracted from the length, height, and thickness of the masonry unit to obtain the standard or specified dimensions of the unit. However, the actual dimensions may vary from the standard, usually by the following amounts:

- Plus or minus 1/16″ in thickness
- Plus or minus 1/8″ in width
- Plus or minus 1/4″ in length

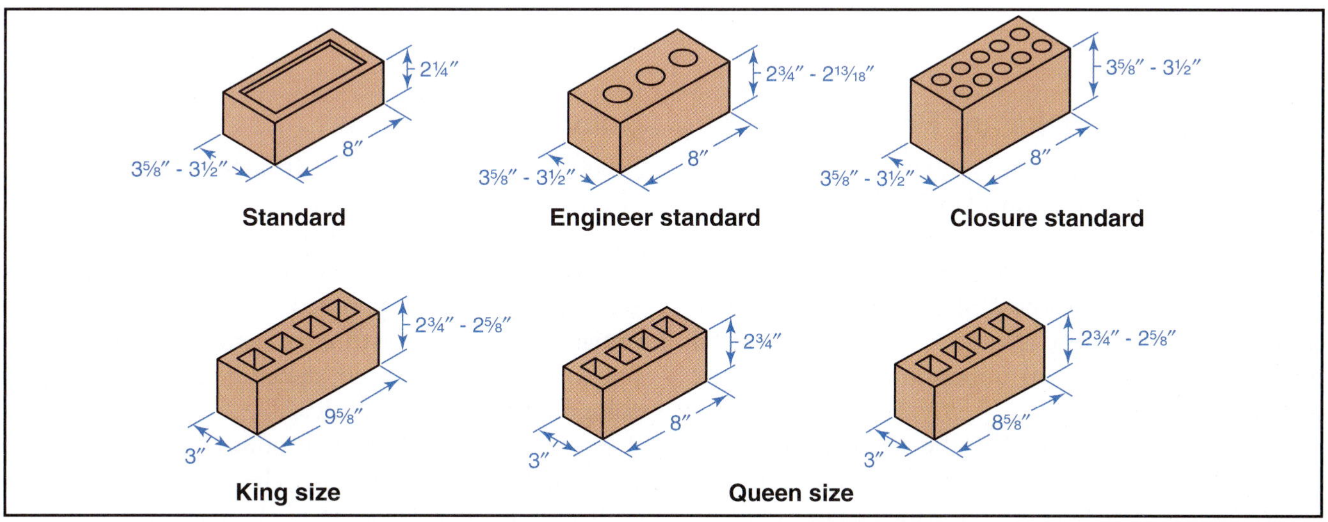

Figure 7-13. Nonmodular bricks with actual dimensions shown.

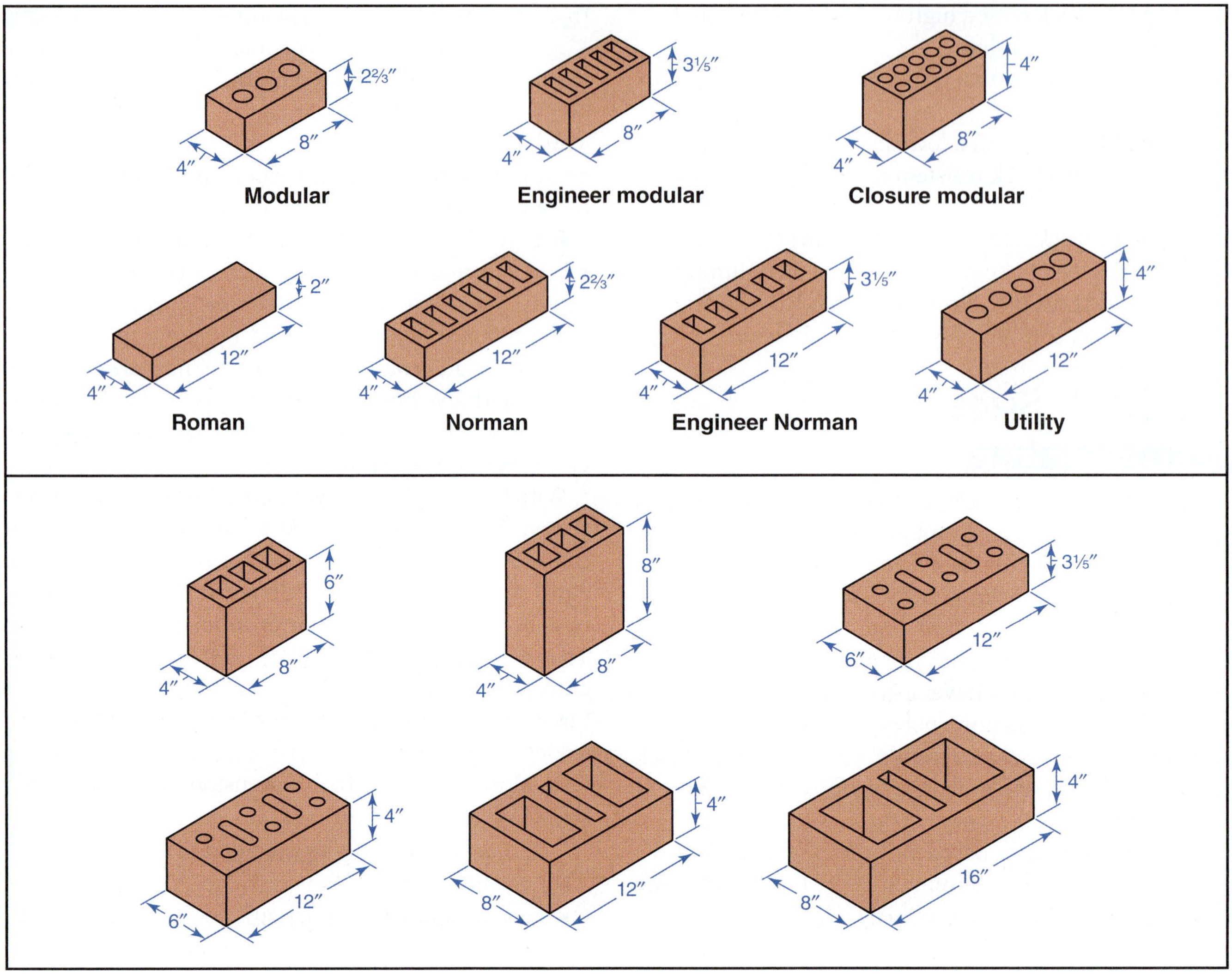

Brick Industry Association

Figure 7-14. Modular brick sizes shown in nominal dimensions.

Why Modular Bricks are Popular

Modular brick sizes became popular because they provide a quick and simple way of estimating the number of bricks needed. Since the nominal dimensions used in modular brick consist of the actual dimensions of the brick plus the thickness of one mortar joint, the estimator can measure the area to be bricked and divide it by the nominal dimensions of the specific bricks to be used.

For example, suppose that the standard modular brick is 4″ × 2 2/3″ × 8″. If a mortar joint of 1/2″ is used, the actual manufactured dimensions would be 3 1/2″ × 2 1/4″ × 7 1/2″. These dimensions should be used when ordering brick. Common brick sizes are shown in **Figure 7-16.** The standard Brick Industry Association nomenclature for brick sizes is shown in **Figure 7-17.**

Colors and Finishes

Facing bricks are produced in many colors, textures, and finishes. In fact, no two bricks are exactly alike. For this reason, ASTM specifications for bricks and structural clay facing tile state that texture and color shall conform to an approved sample showing the full range of color and texture. In general, from three to five samples are required, depending on the range of color and texture in the bricks.

About 99% of all brick produced is in the range of reds, buffs, and creams. The actual color varies with the chemical makeup of the clay, which usually depends on the source. Color is also affected by the method of manufacture and the length and temperature of the burn.

Addition of chemicals and the method of molding and cutting the bricks determine texture. Most

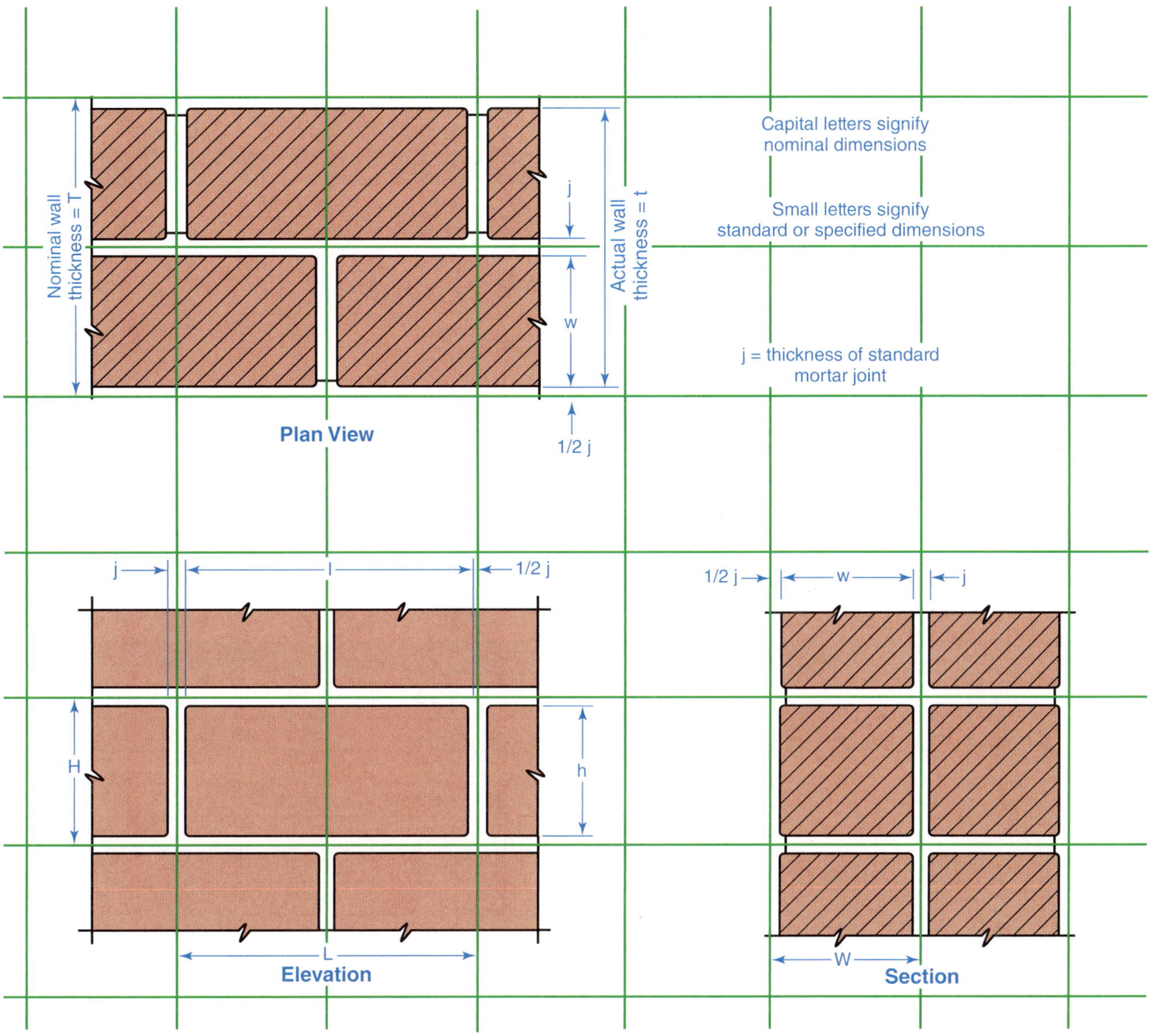

Goodheart-Willcox Publisher

Figure 7-15. The nominal dimensions of modular brick are equal to the actual dimensions plus the thickness of the mortar joint.

manufactured brick has a smooth or sand finish texture. However, many other textures are produced on bricks. **Figure 7-18** shows several brick textures created during the production process. A large sample of popular bricks is included at the end of this chapter to illustrate the variety of colors and textures available.

7.7 Properties of Brick

The term ***structural clay products*** generally refers to burned clay units that are used primarily in building construction. These units may help support the structure or may serve only as a decorative finish. Technically, when assembled in a structure, burned clay units should support their own weight. Structural clay products may be loadbearing or nonloadbearing.

Because brick masonry is bonded into an integral mass by mortar and grout, it is considered to be a homogeneous construction. It is the behavior of the combination of materials that determines the performance of the masonry as a structural element. However, the performance of a structural masonry element depends on the properties of the constituent materials and the interaction of the materials as an assemblage.

Standard Nomenclature for Brick Sizes[1]								
Unit Designation	**Width in inches (nominal)**	**Height in inches (nominal)**	**Length in inches (nominal)**	**Joint Thickness[2] (in)**	**Width in inches (specified)[3]**	**Height in inches (specified)[3]**	**Length in inches (specified)[3]**	**Vertical Coursing**
Modular Brick Sizes								
Modular	4	2 2/3	8	3/8	3 5/8	2 1/4	7 5/8	3C = 8 in
				1/2	3 1/2	2 1/4	7 1/2	
Engineer Modular	4	3 1/5	8	3/8	3 5/8	2 3/4	7 5/8	5C = 16 in
				1/2	3 1/2	2 13/16	7 1/2	
Closure Modular	4	4	8	3/8	3 5/8	3 5/8	7 5/8	1C = 4 in
				1/2	3 1/2	3 1/2	7 1/2	
Roman	4	2	12	3/8	3 5/8	1 5/8	11 5/8	2C = 4 in
				1/2	3 1/2	1 1/2	11 1/2	
Norman	4	2 2/3	12	3/8	3 5/8	2 1/4	11 5/8	3C = 8 in
				1/2	3 1/2	2 1/4	11 1/2	
Engineer Norman	4	3 1/5	12	3/8	3 5/8	2 3/4	11 5/8	5C = 16 in
				1/2	3 1/2	2 13/16	11 1/2	
Utility	4	4	12	3/8	3 5/8	3 5/8	11 5/8	1C = 4 in
				1/2	3 1/2	3 1/2	11 1/2	
Nonmodular Brick Sizes								
Standard	—	—	—	3/8	3 5/8	2 1/4	8	3C = 8 in
				1/2	3 1/2	2 3/4	8	
Engineer Standard	—	—	—	3/8	3 5/8	2 1/4	8	5C = 16 in
				1/2	3 1/2	2 13/16	8	
Closure Standard	—	—	—	3/8	3 5/8	3 5/8	8	1C = 4 in
				1/2	3 1/2	3 1/2	8	
King	—	—	—	3/8	3	2 3/4	9 5/8	5C = 16 in
					3	2 3/4	9 5/8	
Queen	—	—	—	3/8	3	2 3/4	8	5C = 16 in

1 1 in = 25.4 mm; 1 ft = 0.3 m

2 Common joint sizes used with length and width dimensions. Joint thicknesses of bed joints vary based on vertical coursing and specified unit height.

3 Specified dimensions may vary within this range from manufacturer to manufacturer.

Brick Industry Association

Figure 7-16. Nominal and manufactured dimensions for most common modular and nonmodular bricks. Metric common bricks have a nominal size of 215 mm × 102.5 mm × 65 mm. Mortar joints are 10 mm.

Therefore, it is important to first consider the properties of the constituent materials—clay and shale units, mortar, grout, and steel reinforcement.

> **Code Note — Secondhand Units**
>
> The International Building Code (IBC) **2103.1.1 Secondhand units**, specifies that masonry units are not to be reused unless they conform to the requirements of new units. The units should be free from cracks or defects that would interfere with proper laying or use. Old mortar should be cleaned from the units before reusing.

Structural clay products must meet the following standards established by the Federal Trade Commission:

- The composition is primarily of clay or shale or mixtures.
- The ingredients have been fused together as a result of the application of heat.

When products do not meet these two requirements, manufacturers must clearly show what the product is made of. For example, concrete brick, coral brick, plaster brick, sand-lime brick, and concrete structural tile are not ceramic (made of clay and hardened by heat) products.

BIA Standard Nomenclature for Brick Sizes			
Unit Name	**Width in inches (mm)**	**Height in inches (mm)**	**Length in inches (mm)**
Modular	3 1/2–3 5/8 (89–92)	2 1/4 (57)	7 1/2–7 5/8 (190–194)
Standard	3 1/2–3 5/8 (89–92)	2 1/4 (57)	8 (203)
Engineer Modular	3 1/2–3 5/8 (89–92)	2 3/4–2 13/16 (70–71)	7 1/2–7 5/8 (190–194)
Engineer Standard	3 1/2–3 5/8 (89–92)	2 3/4–2 13/16 (70–71)	8 (203)
Closure Modular	3 1/2–3 5/8 (89–92)	3 1/2–3 5/8 (89–92)	7 1/2–7 5/8 (190–194)
Closure Standard	3 1/2–3 5/8 (89–92)	3 1/2–3 5/8 (89–92)	8 (203)
Roman	3 1/2–3 5/8 (89–92)	1 5/8 (41)	11 1/2–11 5/8 (292–295)
Norman	3 1/2–3 5/8 (89–92)	2 1/4 (57)	11 1/2–11 5/8 (292–295)
Engineer Norman	3 1/2–3 5/8 (89–92)	2 3/4–2 13/16 (70–71)	11 1/2–11 5/8 (292–295)
Utility	3 1/2–3 5/8 (89–92)	3 1/2–3 5/8 (89–92)	11 1/2–11 5/8 (292–295)
King Size	3 (76)	2 5/8–2 3/4 (67–70)	9 1/2–9 5/8 (241–244)
Queen Size	3 (76)	2 3/4 (70)	7 5/8–8 (194–203)

Brick Industry Association

Figure 7-17. Brick Industry Association nomenclature for brick sizes in US customary and metric.

Structural clay products may be classified as solid masonry units and hollow masonry units. The classification depends on how much solid material is in the units.

Solid Masonry Units

ASTM International (formerly the American Society for Testing and Materials) has set the standard for solid masonry units. A solid masonry unit is "one whose cross-sectional area in every plane parallel to the bearing surface is 75% or more of its gross cross-sectional area measured in the same plane."

The ***bearing surfaces*** are the tops and bottoms of the bricks or blocks. To visualize a parallel plane of the bearing surface, imagine that you have placed a brick on its side and have sliced it into a number of thin pieces. The sides of each slice would be parallel planes. If you measured the area of the slices, then subtracted the area of the holes in the brick, you could determine what percentage is solid material. See **Figure 7-19**.

Many bricks are solid, but many facing bricks have holes running through them. These cored bricks are usually classified as solid masonry units because 75% of their area is solid material. The cored holes reduce weight, permit easier handling, and make the mortar joint stronger.

Hollow Masonry Units

A ***hollow masonry unit*** is a brick with less than 75% solid material in its net cross-sectional area in any plane. Structural clay tiles are included in this classification.

Weight of Brick

The weight of a brick unit depends on materials used, method of manufacture, burning, and the size of the brick. Since there are so many variables involved, be sure to check information secured from the manufacturer. Building bricks weigh about 4 1/2 lb to 5 lb each.

Goodheart-Willcox Publisher

Figure 7-18. These brick textures are made during the production process.

Durability

Several conditions affect the weathering (durability) of a brick. These include heat, cold, wetting, drying, and the action of soluble salts.

Efflorescence

Efflorescence is a white powder or salt-like deposit on masonry walls, which is caused by water-soluble salts. See **Figure 7-20**. These salts collect on the wall's surface as water evaporates. The salts may have been mixed into the masonry units, mortar, or plaster.

Several things must happen for efflorescence to develop:

- There must be soluble salts in the masonry wall.
- Moisture must pass through the wall and carry the salts to the surface.
- There must be a path for the salts to migrate to the surface.
- The moisture must evaporate to deposit the salts.

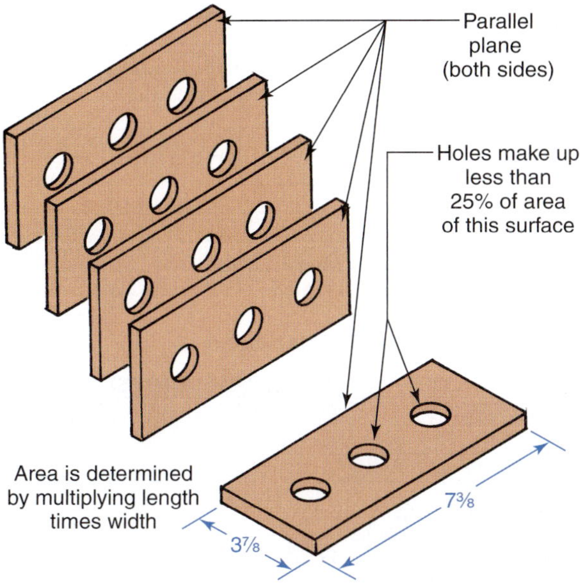

Goodheart-Willcox Publisher

Figure 7-19. Bricks with holes in them are still considered solid brick if 75% of the area in any parallel plane is solid material. Imagine that the brick is sliced like the drawing above. Each cut surface is a parallel plane. To find the area, multiply the length by the width. The area of the holes is found by multiplying 3.14 by the square of the radius.

Masonry structures must be designed in such a way as to keep water or moisture from entering the masonry cavity or wall structure. This can be achieved by flashing, sealing, or capping the masonry structure with metal or a waterproof material.

Compressive Strength

Compressive strength is the ability of a masonry unit to stand up under heavy weight without crumbling. The material properties of brick and structural clay tile that have the most significant effect on structural performance of masonry are compressive strength and those properties affecting the bond between the unit and mortar. These properties include rate of water absorption and surface texture. Compressive strength is affected by the following factors:

- Strength of the units (bricks)
- Strength of the mortar
- Quality of workmanship
- Proportion of gross area that is given a bearing surface at bed joints

High compressive strength is developed by using high-strength brick and mortars and making well-filled joints. In most construction today, compressive strengths do not exceed 100 psi (pounds per square inch). Most bricks have compressive strengths of over 4500 psi. Typical mortars have strengths from 750 psi to 2500 psi. Therefore, most masonry walls have ample compressive strength for current construction practices.

Goodheart-Willcox Publisher

Figure 7-20. Efflorescence on masonry walls is caused by soluble salts in wet brick being carried to the surface as water evaporates.

7.8 Bonds and Patterns

The strength, durability, and appearance of a brick wall depend to a great extent on the bond used to lay up the wall. The word *bond*, when it refers to masonry, has three meanings:

- ***Structural bond*** is the method by which individual units are interlocked or tied together to form a wall. See **Figure 7-21**.
- ***Pattern bond*** is the pattern formed by the masonry units and the mortar joints on the face of a wall. The pattern may be the result of the structural bond used or may be purely for decorative purposes.
- ***Mortar bond*** (joint) is the adhesion of mortar to the masonry units or reinforcing.

Brick Industry Association

Figure 7-21. These three typical brick and tile walls illustrate various structural bonding methods using headers.

Structural Bonds

Structural bonding in brick masonry may be done in a number of ways:

- The brick may be arranged in an overlapping fashion.
- Ties may be embedded (buried) in the mortar joints. See **Figure 7-22**.
- Grout may be applied to adjacent wythes of masonry. A wythe is a single vertical tier or stack of masonry.

The overlapping method is best illustrated by two traditional bonds known as English bond and Flemish bond. See **Figure 7-23**. The English bond consists of alternating courses of headers and stretchers. Alternating is defined as switching back and forth between two things. A ***course*** of bricks is a continuous level row of bricks. The Flemish bond consists of alternating headers and stretchers in every course, so the headers and stretchers in every other course appear in vertical lines.

Stretchers are the bricks laid with the length of the wall. They develop longitudinal (lengthwise) bonding strength. Headers, laid across the width of the wall, bond the wall across its thickness. **Figure 7-24** identifies the name of six brick positions. **Figure 7-25** identifies the various brick positions in a wall.

Names have been given to various sizes of cut or broken bricks, **Figure 7-26**. For example, a half brick is called a ***bat*** or *half*. Three-fourths of a brick is called a ***three-quarter closure***, and one-fourth of a brick is called a ***quarter closure***. When one corner is cut off, the remaining brick is called a ***king closure***. A brick cut lengthwise across the end is a ***queen closure***. Finally, a brick cut lengthwise parallel to the bed is called a ***split***.

Modern building codes specify that masonry-bonded brick walls must have no less than 4% headers in the wall surface. The distance between nearest headers should not exceed 24″ vertically or horizontally.

Structural bonding of masonry walls with metal ties is acceptable for solid wall and cavity wall construction.

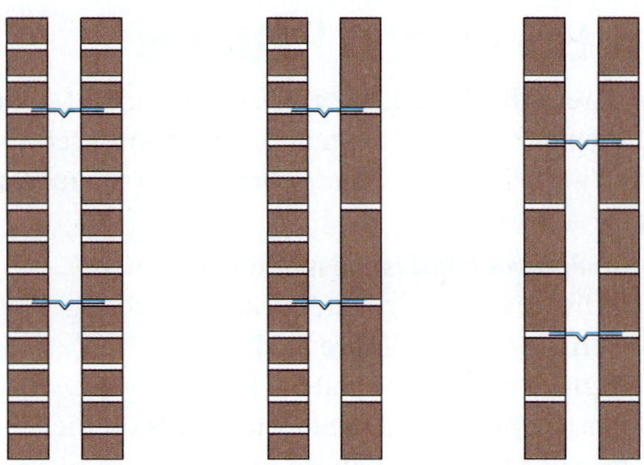

Brick Industry Association

Figure 7-22. Structural bonding of cavity walls using metal ties.

Brick Industry Association

Figure 7-23. These two traditional bonds are used in brick masonry.

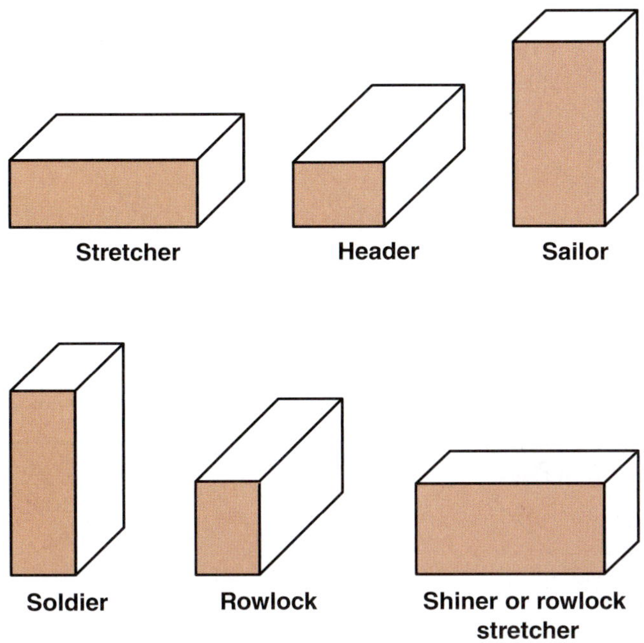

Goodheart-Willcox Publisher

Figure 7-24. Each brick position has a specific name.

See **Figure 7-27**. No less than one 3/16″ diameter metal tie for each 4 1/2 sq ft of wall surface should be used.

Ties in alternate courses should be staggered with the distance between adjacent ties no more than 24″ vertically or 36″ horizontally. Additional bonding ties should be placed around the perimeter of openings. These ties should be spaced no more than 3″ apart and within 12″ of the opening. If smaller diameter ties are used, the spacing between them should be reduced.

Solid and reinforced brick masonry walls may be structurally bonded by pouring grout into the cavity between wythes of masonry. ***Grout*** is a liquid cement mixture used to fill voids.

The method of bonding depends on the wall type, use requirements, and other factors. Metal ties are usually recommended for exterior walls because they make construction easier and provide greater resistance to rain penetration. Metal ties also allow for slight movements of facing and backing, which may prevent cracking.

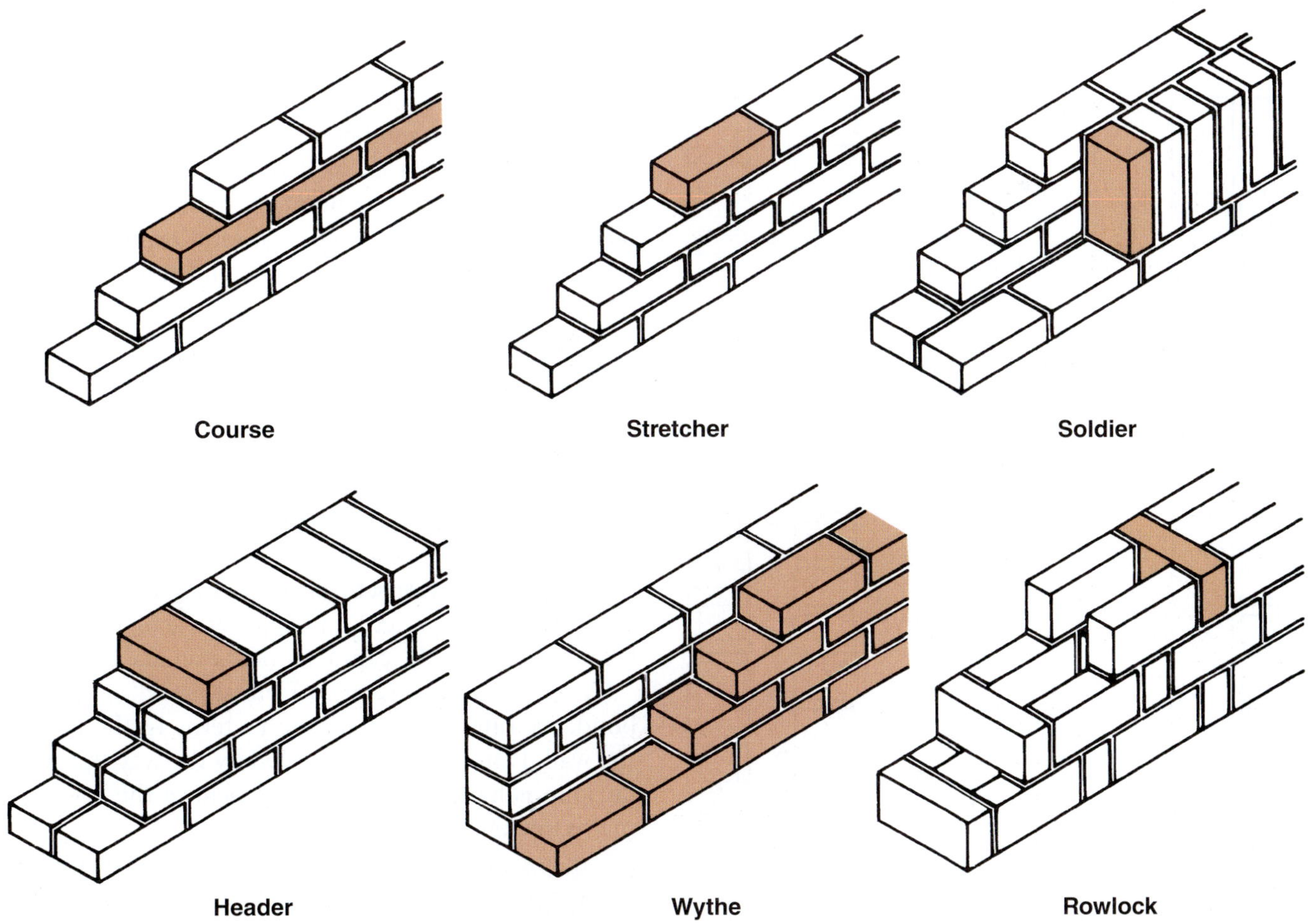

Goodheart-Willcox Publisher

Figure 7-25. Brick positions in the wall and terms used in masonry wall construction.

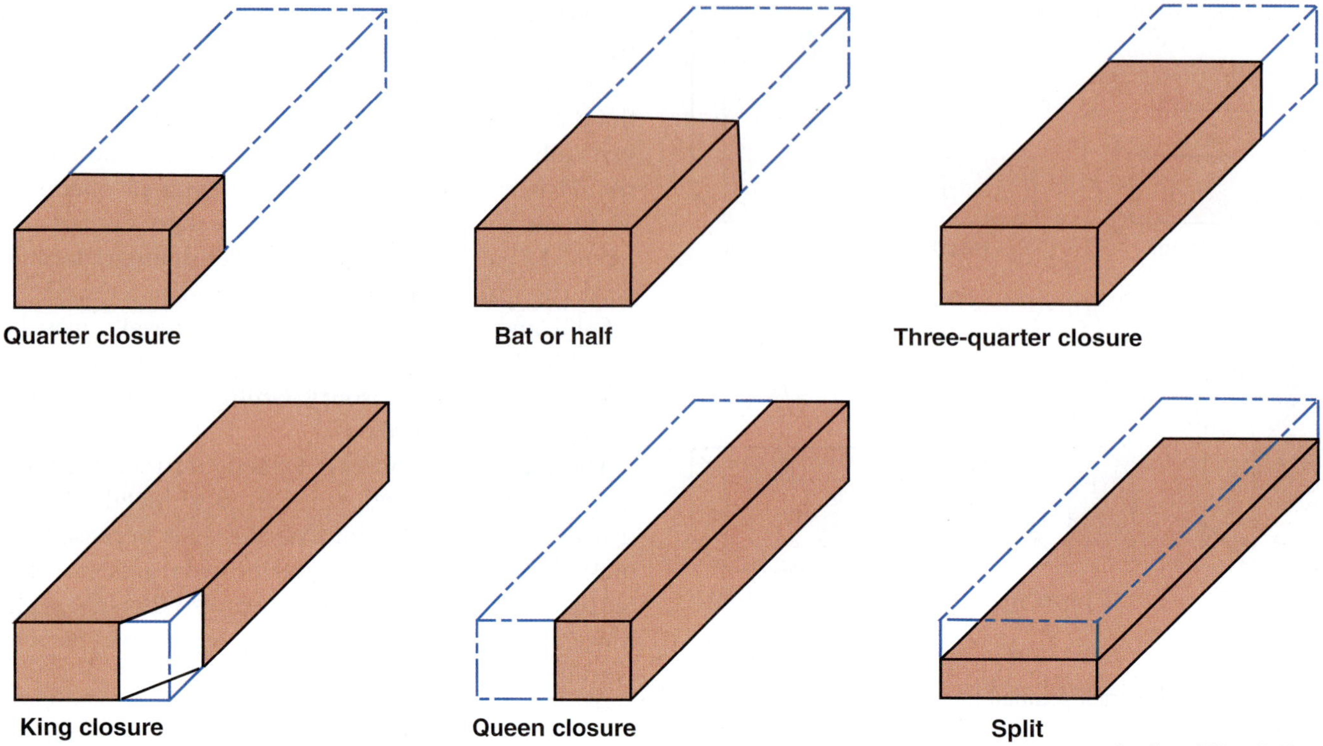

Goodheart-Willcox Publisher

Figure 7-26. Names used to describe cut or broken brick pieces.

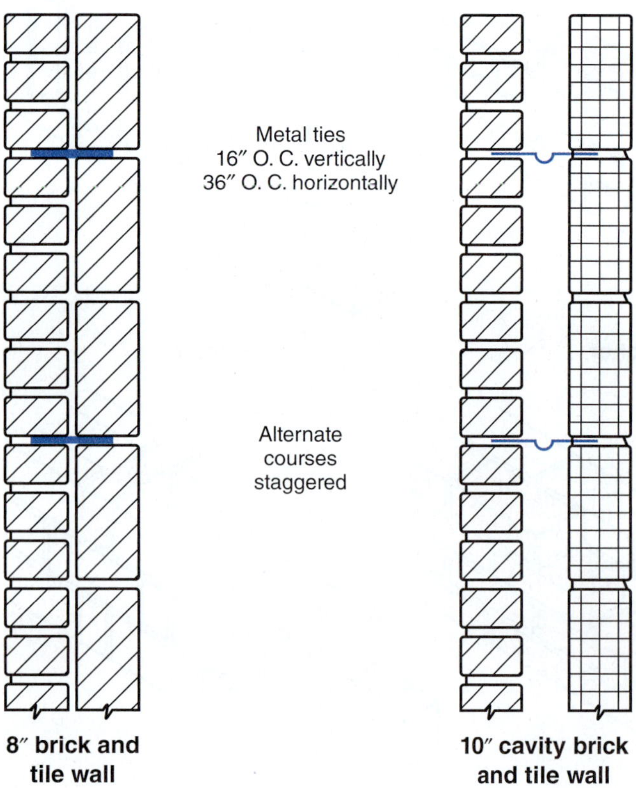

Goodheart-Willcox Publisher

Figure 7-27. A solid masonry wall and cavity wall bonded with metal ties.

Unit texture and absorption affect the bond strength of the masonry assembly. In general, mortar adheres better to rough surfaces, such as wire-cut surfaces, than to smooth surfaces. The bond strength of sanded surfaces depends on the amount of sand on the surface, the sand's adherence to the unit, and the absorption rate of the unit at the time of laying.

Pattern Bonds

The term *pattern bonds* refers to the process of using brick and joint selection to create different appearances of the finished masonry wall. "Pattern" refers to the changes in or varied arrangement of brick texture, color, or joints in the face of the wall being created. By selecting different structural bonds, masons can drastically change the pattern and appearance of a wall surface, and thereby the character of the finished product.

Structural bonds, such as Flemish bond or English bond, are often used to create patterns in the face of the wall. Patterns may be created by the way the mortar joint is handled. The arrangement of the brick remains unchanged.

It may also be desirable to produce a pattern bond by allowing certain brick to stick out beyond or sink below the plane (surface) of the wall. Pattern bonds can also be created by varying the texture of brick units.

Five Basic Bonds

Five structural bonds create different patterns in masonry walls. These bonds include the running bond, the common or American bond, the Flemish bond, the English bond, and the block or stack bond.

Color and texture in both brick and joints create endless variety. **Figure 7-28** shows many different patterns. **Figure 7-29** illustrates the basic structural bonds.

Running bond consists of all stretchers and is used largely in cavity wall and veneer wall construction. It is the simplest of the basic pattern bonds. See **Figure 7-30**.

Common bond or ***American bond*** is a variation of the running bond. It has a course of full-length headers at regular intervals to provide structural bonding as well as pattern. Header courses generally are used every fifth to seventh course. See **Figure 7-31**.

Goodheart-Willcox Publisher

Figure 7-28. This wall is decorated with a collection of pattern bonds.

Goodheart-Willcox Publisher

Figure 7-30. This reclaimed brick is laid in running bond.

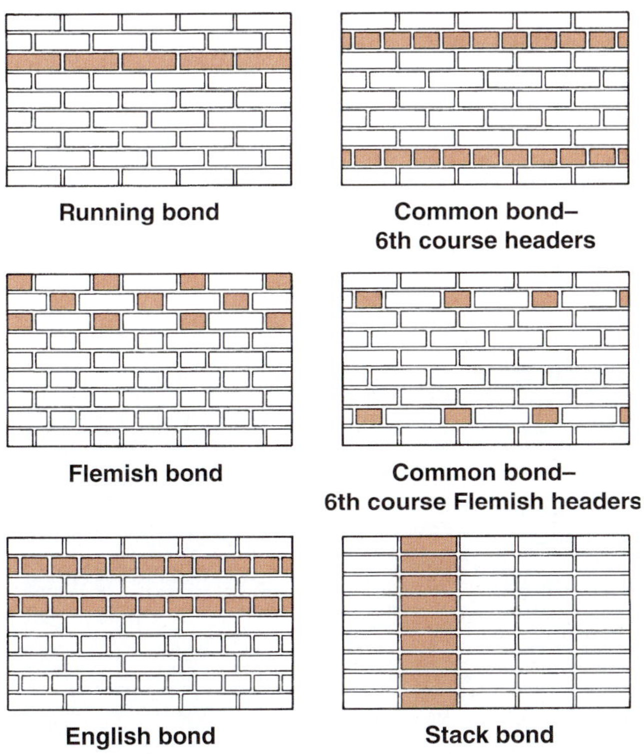

Goodheart-Willcox Publisher

Figure 7-29. The basic structural bonds.

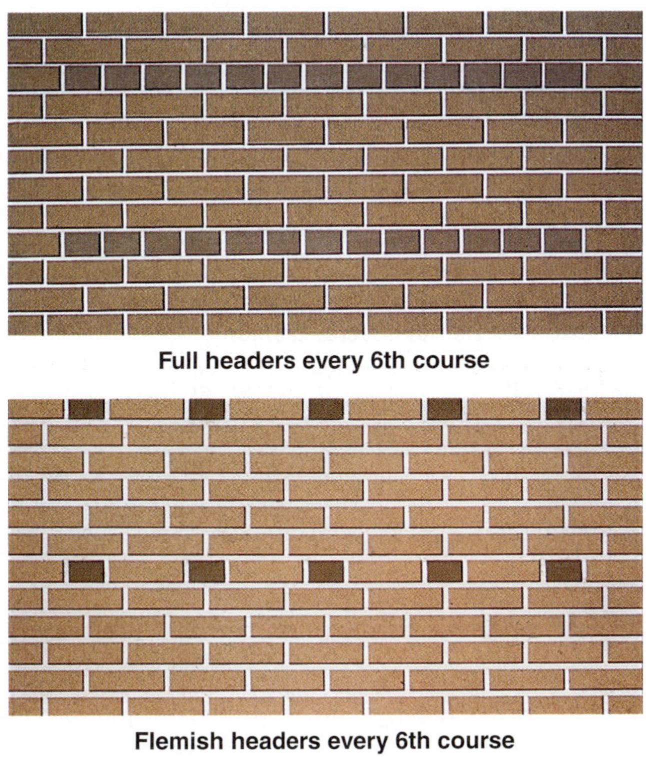

Brick Industry Association

Figure 7-31. The common or American bond is a variation of the running bond. Two variants are shown.

Flemish bond alternates courses of stretchers and headers. The headers in alternate courses are centered over the stretchers in the course between them.

If the headers are not needed for structural bonding, as in veneer wall construction, half bricks may be used. These half bricks are called clipped or ***snap headers***.

The Flemish bond may be varied by increasing the number of stretchers between headers in each course. Three stretchers alternating with a header is known as a ***garden wall bond***. Two stretches between headers is called a ***double stretcher garden wall bond***. Four or five stretchers between headers may be laid to form other garden wall bonds. See **Figure 7-32**.

English bond is made up of alternate courses of headers and stretchers. The headers are centered on the stretchers and joints between stretchers in all courses are aligned vertically. Snap headers may be used when structural bonding is not necessary.

English cross or ***Dutch bond*** is a variation of English bond. The only difference is that vertical joints between the stretchers in alternate courses do not align vertically. These joints center on the stretchers themselves in the courses above and below.

Block or ***stack bond*** is purely a pattern bond. All vertical joints are aligned. There is no overlapping of units. See **Figure 7-33**. Because the bond lacks the strength that other bonds provide, it is usually bonded to the backing with rigid steel ties. In loadbearing construction or large wall areas, it is advisable to reinforce the wall with steel. This is placed in the horizontal mortar joints.

Figure 7-34 shows a pattern that was made through variations in color of the brick used. This is a common way to develop patterns.

Wall Texture

Many new and interesting patterns and effects are created by projecting and recessing units, **Figure 7-35**, and omitting units to form perforated walls or screens, **Figure 7-36**. These techniques greatly extend the traditional patterns and add a new dimension to brick masonry. **Figure 7-37** shows a brick sculptured wall that incorporates a variety of colors and textures.

Mortar Joints

Mortar joints use mortar between brick units to perform one or more of the following functions:

- Bond units together and seal spaces between them
- Bond to reinforcing steel and cause it to be part of the wall
- Compensate for variations in the size of brick units
- Provide a decorative effect

Mortar joint finishes are either troweled or tooled joints. In a ***troweled joint***, the excess mortar is struck (cut off) with a trowel and finished with the same tool. The ***tooled joint*** is made with a special tool to compress and shape the mortar in the joint. Several types

Garden wall bond with units in dovetail fashion

Double stretcher garden wall bond with units in diagonal lines

Goodheart-Willcox Publisher

Figure 7-32. Two variations of the garden wall bond.

Goodheart-Willcox Publisher

Figure 7-33. In a stack bond, all vertical joints line up.

of mortar joints are typically used in brick masonry, **Figure 7-38**.

Concave joints are created by first striking off the excess mortar on a joint, and then running a steel jointing tool with a concave surface to compact the mortar, creating a concave appearance to the joint. ***V-shaped joints*** are formed by striking a joint to remove excess mortar, then running a V-shaped steel joint tool

Goodheart-Willcox Publisher

Figure 7-34. Glazed tiles of various colors were used to create this pattern.

Goodheart-Willcox Publisher

Figure 7-36. Brick screens are highly decorative. They also help reduce noise and control light.

Goodheart-Willcox Publisher

Figure 7-35. Pattern bonds may be produced by projecting or recessing certain bricks from the plane surface of the wall.

General Shale Brick

Figure 7-37. A panoramic view created with sculptured brick.

to compact the mortar. Because of the shape of the tool, a recessed "V" joint is formed. Because these two joints are compacted, they effectively prevent rain from penetrating into the joint. Their use is recommended where heavy rains and high winds are likely to occur. The concave joint is the most popular mortar joint.

A ***weathered joint*** requires care since it must be worked from below. It is a compacted joint that readily sheds water.

A ***struck joint*** is common for ordinary brickwork. The joint is struck with the trowel and some compacting occurs, but it does not shed water as well as other joints.

A ***rough cut joint***, or ***flush joint***, is the simplest joint to make. It is made by holding the trowel flat against the brick and cutting in any direction. The joint is not compacted and leaves a small hairline crack where the mortar is pulled away from the brick by the trowel. This joint may not be watertight.

The ***raked joint*** is made by removing some of the mortar while it is still soft. This joint is difficult to make watertight even though slightly compacted. It is not recommended for areas with heavy rains, high winds, and freezing temperatures because it creates a ledge for moisture. Sealing the raked joint can be improved by striking the joints once they have been raked. The striker packs the joint and also seals it to reduce water absorption. Raking joints creates a warm shadowing effect on regular brick, and this joint is effective when laying irregular brick as it tends to enhance its character.

The ***extruded joint***, sometimes called a *weeping joint*, is made by squeezing out the mortar as the unit is laid. It is not trimmed off but is left to harden in its extruded form. This joint is best suited for dry climates. See **Figure 7-39**.

Colored mortars may be used to accent brick masonry patterns. See **Figure 7-40**. It may be applied in two ways:

- The entire mortar joint may be colored. This method increases the cost of the mortar but takes less labor than the tuckpointing method.
- A portion of the joint is colored; this is called the *tuckpointing method*. In this method, the entire wall is finished with a 1″ deep raked joint and the color mortar is filled in later.

When using either method, it is important to take special care to keep the colored mortar off the brick.

7.9 Special Shapes

One of the primary reasons for the popularity of brick with architects, designers, and builders is its ability to be adapted to a wide variety of design criteria. See **Figure 7-41**. Brick can be shaped and used in traditional and contemporary detailing—from arches to sills, water table, copings, columns, and treads. See **Figure 7-42** and **Figure 7-43**. Special shapes add a unique and distinctive touch, projecting individual styling and taste.

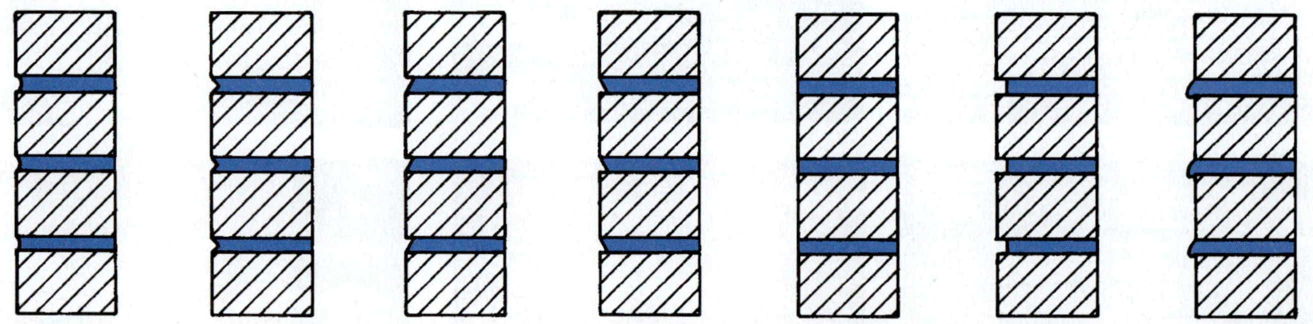

Goodheart-Willcox Publisher

Figure 7-38. Typical mortar joints used in masonry work.

Darryl Brooks/Shutterstock.com

Figure 7-39. The extruded mortar joint or weeping joint is best suited for dry climates because its ledges do not shed water well.

Goodheart-Willcox Publisher

Figure 7-41. Handmade bricks have been used in a variety of shapes to produce this example of classic architecture.

General Shale Brick

Figure 7-42. The water table on this traditional home was formed with specially shaped bricks.

Brick Industry Association

Figure 7-40. The color of the brick is the same, but the color of the mortar is changed to accent the brick pattern.

General Shale Brick

Figure 7-43. These traditional steps with rounded nosing were made possible with specially shaped bricks.

Many shapes are manufactured by large brick companies, but other shapes may be desired. See **Figure 7-44**. Custom shapes that conform to the designer's specification can be produced in extruded, molded, or handmade brick. Consult with the manufacturer before specifying unique shapes to determine the most practical and cost-effective method of achieving the desired shape. **Figure 7-45** shows a residential structure that incorporates special shapes in the jack arches, treads, circular arch, and water table.

7.10 Hollow Masonry Units (Tile)

Hollow masonry units are a machine-made product that is extruded (squeezed) through a die and cut to the desired height or length. The raw materials may be surface clay, shale, fire clay, or combinations of these. The materials are pulverized (crushed), mixed with water, and then extruded. After the moisture content is reduced to a satisfactory level, they are fired in a kiln at temperatures from 1750°F (954°C) to 2500°F (1093°C).

Glen-Gery Corporation

Figure 7-44. Specially shaped bricks, such as these pool coping units, are available for specific applications.

General Shale Brick

Figure 7-45. This traditional home incorporates several examples of specially shaped and custom brick units.

Unlike brick, tile is hollow. Hollow in this instance means that each masonry unit has a net cross-sectional area in any plane parallel to the bearing surface less than 75% of its gross cross-sectional area measured in the same plane. Hollow clay tile is classified as either structural clay tile or structural clay facing tile.

Structural Clay Tile

Structural clay tile is produced as loadbearing and nonloadbearing types. Both loadbearing and nonloadbearing tile is produced with vertical or horizontal cells, which are the open spaces through the tile.

When the tile is to be laid with the cells in a horizontal plane, the unit is called ***side construction*** or *horizontal cell tile*. When the tile is to be laid with the cells in a vertical plane, the unit is called ***end construction*** or *vertical cell tile*. The size, number, shape, and thickness of cells vary from one manufacturer to another.

Loadbearing Tile

Structural clay loadbearing tile is suitable for use in walls as backup or in partitions carrying superimposed loads (weight of other parts). Loadbearing wall tiles are produced in two grades—LB and LBX. Grade LB tile is designed for areas not exposed to frost action. Grade LBX tile is suitable for use in areas exposed to frost action.

Nonloadbearing Tile

Structural clay nonloadbearing tile is suitable for use as fireproofing, furring, or the construction of partitions that do not support superimposed loads. Nonloadbearing tile is available in one grade: NB.

Sizes and Shapes

Structural clay tile is most often used in actual thickness of 3″ and nominal thicknesses of 4″, 6″, and 8″. However, nominal 2″, 10″, and 12″ thick units are available. Heights for clay tile units are 5 1/3″, 8″, and 12″ nominal.

Tile unit length is generally 12″ nominal. However, 8″ and 16″ units are available in some areas.

Structural clay tile is available in a variety of sizes and shapes, **Figure 7-46** and **Figure 7-47**. Other shapes include header units, jamb units, closures, and kerfed units for use in reinforced bond beams or lintels.

All structural clay tile is designed to be used with 1/2″ mortar joints. All are modular sizes.

Various Sizes and Shapes of Structural Clay Tile

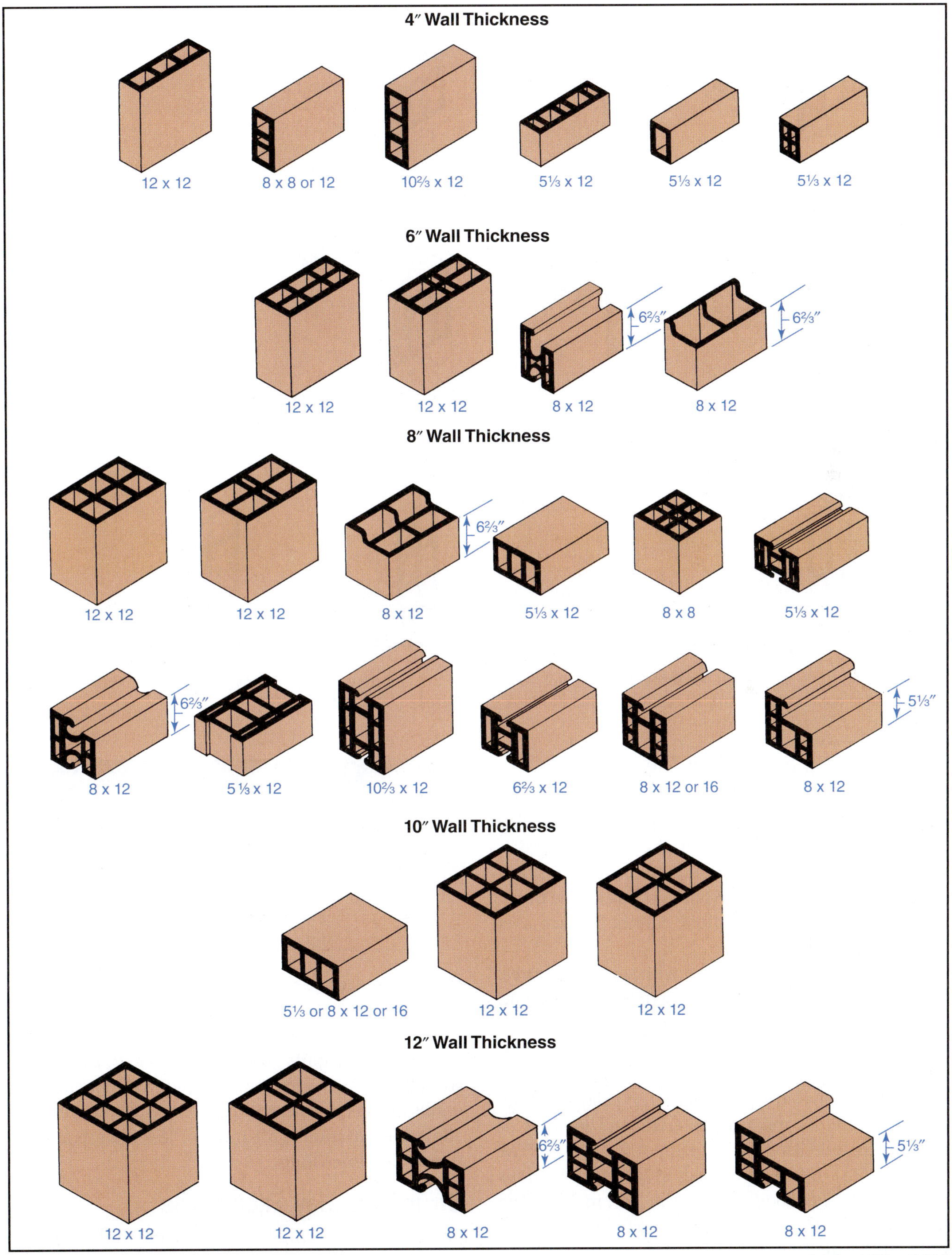

Goodheart-Willcox Publisher

Figure 7-46. All tile dimensions are nominal. Face dimensions are given height by length.

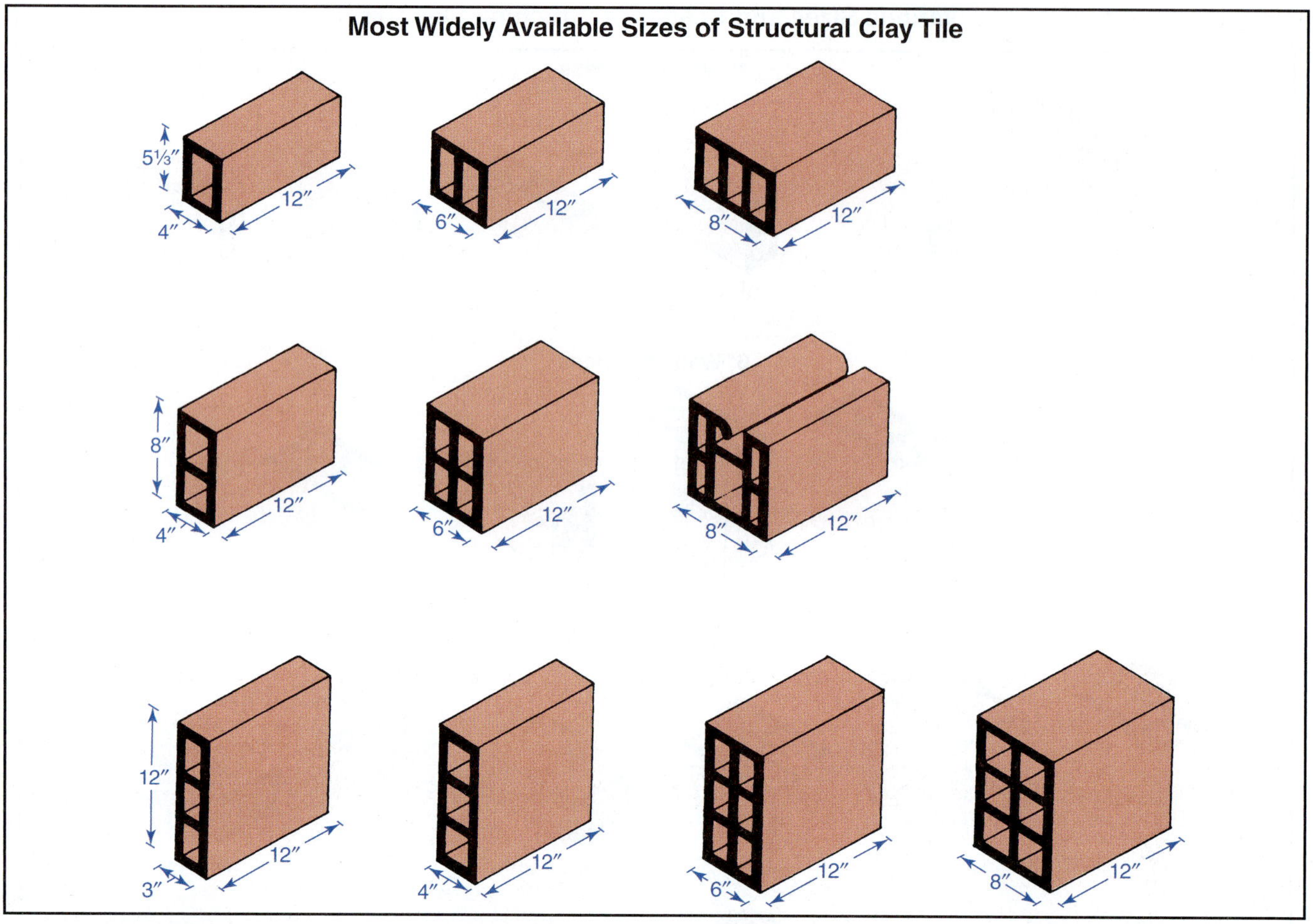

Goodheart-Willcox Publisher

Figure 7-47. All units shown are horizontal cell tile, but vertical cell tile is also available. Dimensions are nominal. This basic group provides compatibility with all facing tile and modular brick sizes.

Properties of Structural Clay Tile

Properties of structural clay tile include color, texture, strength, and variations in dimensions. Structural clay tile is produced in a wide range of colors. The greatest influence on the color is the chemical composition of the raw materials and the temperature of the burn. Due to these factors, color alone cannot be used as a measure of quality.

Structural clay tile is produced with surface textures. The two most generally seen are scored and wire cut. Wire cut is known as *universal finish* in some regions. Both surfaces readily receive plaster, and the wire cut may be painted or left exposed.

The compressive strength of structural clay tile is affected not only by the raw materials and method of manufacture, but also by the design of the unit. ASTM standard C34—62 (Reapproved 1970), *Specifications for Structural Clay Loadbearing Wall Tile*, includes the requirement of the average compressive strength, as shown in **Figure 7-48.**

Average Minimum Compressive Strength		
Grade	End Construction Tile (in psi)	Side Construction Tile (in psi)
LBX	1400	750
LB	1000	750

Goodheart-Willcox Publisher

Figure 7-48. Average compressive strength requirements for clay tile.

Most clay tile units produced today have compressive strengths of two to four times these requirements. Variations in sizes are also controlled by ASTM. The biggest allowable variation is 3% more or less than required by the actual dimensions.

Structural Clay Facing Tile

Structural clay facing tile consists of two broad types: glazed and unglazed. See **Figure 7-49**.

Goodheart-Willcox Publisher

Figure 7-49. Structural facing tile is produced with either a glazed or unglazed face.

Glazed clay facing tile is produced from high-grade, light-burning fire clay suitable for the application of ceramic glaze. The tile may also be left unglazed. There are two grades of glazed facing tile—S and G.

Grade S (select) is intended for use with somewhat narrow mortar joints. Grade G (ground edge) is used where face dimensions must be exact. An example of this would be a stacked bond, where units are placed in the wall so that both the horizontal and vertical joints are continuous. Both grades may be purchased as single-faced units or two-faced units. A single-faced unit has one face exposed. In a two-faced unit, two opposite finished faces are exposed.

Unglazed Facing Tile

Unglazed facing tile is made from either light-burning fire clay or shale and other darker-burning clays. These tiles are not glazed and have either a smooth or a rough-textured finish. They can be purchased as standard or heavy duty tile. The difference is in the thickness of the face shells.

Another classification for unglazed facing tile is based on factors that control appearance of the finished wall. This classification includes FTX and FTS types.

FTX unglazed facing tile is suitable for general use in exposed exterior and interior masonry walls and partitions. It is easily cleaned, resists staining, and is low in absorption. This type is also best where the following characteristics are important:

- A high degree of mechanical perfection
- A narrow color range
- A minimum variation in face dimensions

FTS unglazed facing tile is smooth or rough textured. This tile is suitable for the following applications:

- In exposed exterior or interior masonry walls and partitions
- In special conditions where tile of moderate absorption are required
- Where moderate variation in face dimensions and medium color range is acceptable
- Where minor defects in surface finish or small handling chips are acceptable

SCR acoustile is designed to absorb sound. Distinguishing characteristics are the perforations (holes) through the faces and the fibrous glass pads behind the faces, **Figure 7-50**. The perforations can be circular or slotted, equal or unequal in size, and in a regular or random pattern.

All types of structural clay facing tile, glazed or unglazed, can have an acoustile. These units may be used in wall or ceiling applications.

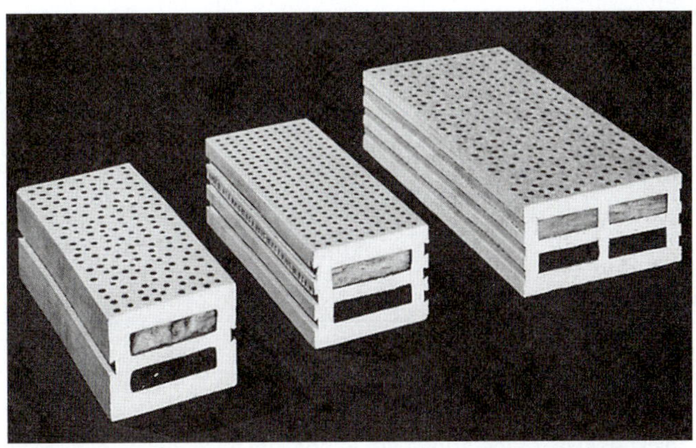

Brick Industry Association

Figure 7-50. These are typical SCR acoustile units designed to absorb sound. Note the use of an insulating blanket behind the face.

Sizes and Shapes

In addition to the typical sizes, **Figure 7-51**, some makers also produce structural clay facing tile with a nominal 4″ × 12″ face size. These units are designed to be laid with a 1/4″ mortar joint. The specified unit dimensions are 1/4″ less than the nominal dimensions.

Nominal modular sizes of structural clay facing tile are shown in the bottom part of **Figure 7-51**. Units shown in nominal modular sizes are usually designed to be laid with 1/4″, 3/8″, or 1/2″ joints.

The basic shapes of structural clay facing tile, some of which are shown in **Figure 7-52**, are as follows:

- Stretchers
- Corners and closures
- Starters and miters

A wide variety of supplementary shapes are also available for complex wall layouts. These shapes include the following:

- Sills
- Caps and lintels
- Cove base stretchers and fittings
- Coved internal corners
- Octagons
- Radials

All structural clay facing tile is modular, which means that all units are multiples of 4″. Because the units fit together with each other and with all other modular building materials, cutting is reduced. However, some units have a dimension of 5 1/3″, which is 1/3 of 16″, rather than a multiple of 4″. Clay tile is also produced in various shapes for use as decorative screens. See **Figure 7-53**.

7.11 Architectural Terra Cotta

Architectural ***terra cotta*** is fired clay, usually brownish-red in color. Terra cotta is used for architectural and structural units. The products or units can be made by machine, by hand, or by combination of the two, depending on the design. Terra cotta can be unglazed or glazed, depending on the application and desired finished appearance.

Architectural terra cotta is a custom product. It is produced in sizes and shapes to meet specific requirements. An unlimited color range of high-fired, ceramic vitreous (glass-like) glazes are available. See **Figure 7-54**. The configurations (arrangement of parts), forms, and radius shapes in which architectural terra cotta can be made are almost unlimited.

Structural Clay Facing Tile Available Sizes		
Series	Nominal face dimensions in inches	Nominal thickness in inches
6T	5 1/3 by 12	2, 4, 6, 8
4D	5 1/3 by 8	2, 4, 6, 8
4S	2 2/3 by 8	2, 4
8W	8 by 16	2, 4

Nominal Modular Sizes*	
Face Dimensions	
Height by length in inches	Height by length in inches
4 by 8	8 by 8
4 by 12	8 by 12
5 1/3 by 8	8 by 16
5 1/3 by 12	12 by 12

Thickness: All of the above are in nominal thicknesses of 4, 6, and 8 inches
* Nominal sizes include the thickness of the standard mortar joint for all dimensions.

Elgin-Butler Company

Figure 7-51. The typical sizes of structural clay facing tile. Joint thickness should be deducted from the dimensions above in determining actual size of the units.

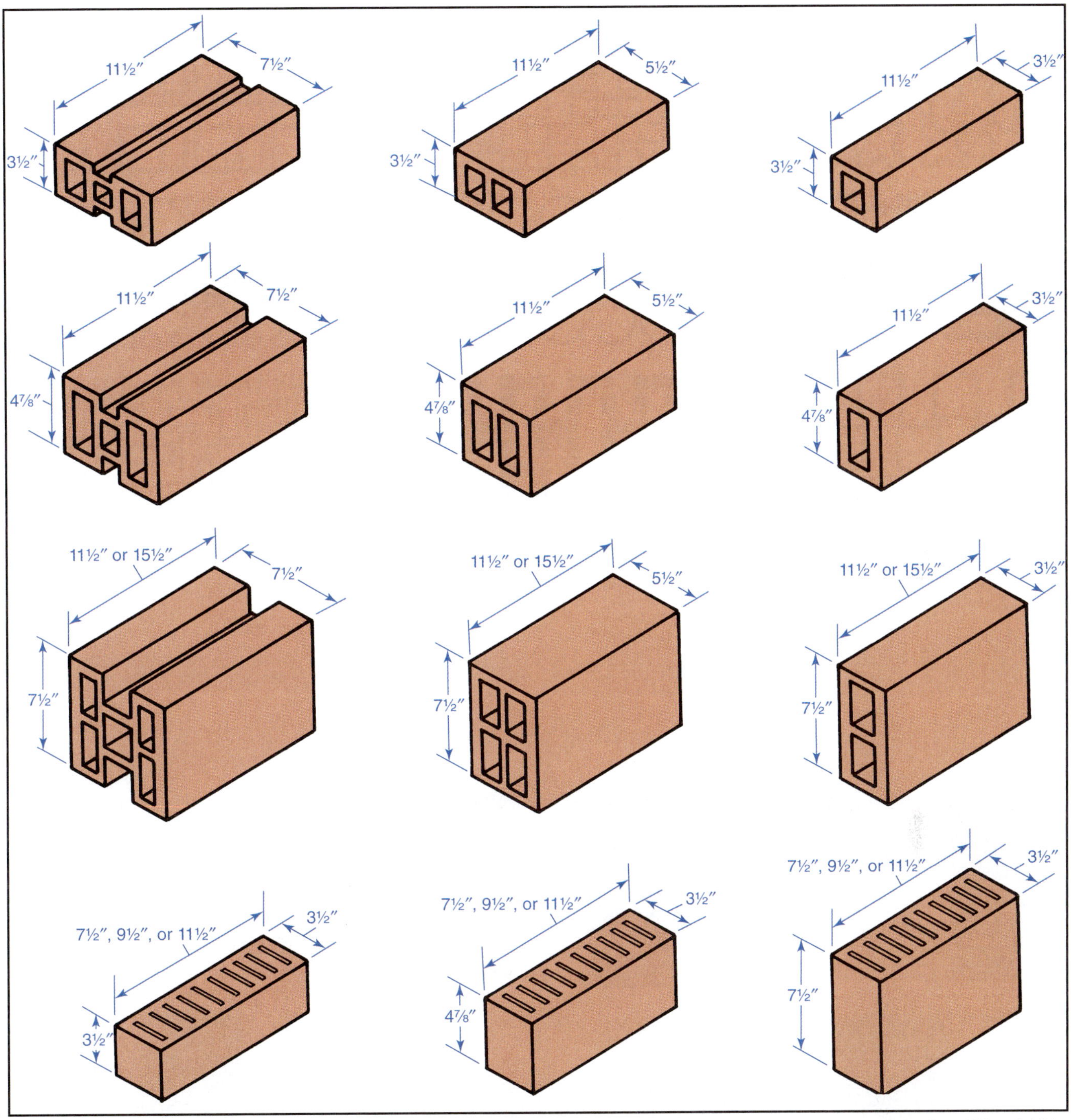

Goodheart-Willcox Publisher

Figure 7-52. Typical shapes of structural clay facing tile. Included are stretchers, jamb units, and corner units.

> **Code Note — Terra Cotta**
> International Building Code (IBC) **1404.9**, specifies that anchored terra cotta or ceramic units not less than 1 5/8″ (41 mm) thick should be anchored directly to masonry, concrete, or stud construction. The IBC lists specific requirements for anchoring and for tied terra cotta or ceramic veneer units.

Classification of Architectural Terra Cotta

There are several classes of architectural terra cotta products. These classes include anchored ceramic veneer, adhesion ceramic veneer, and ornamental or sculptured terra cotta.

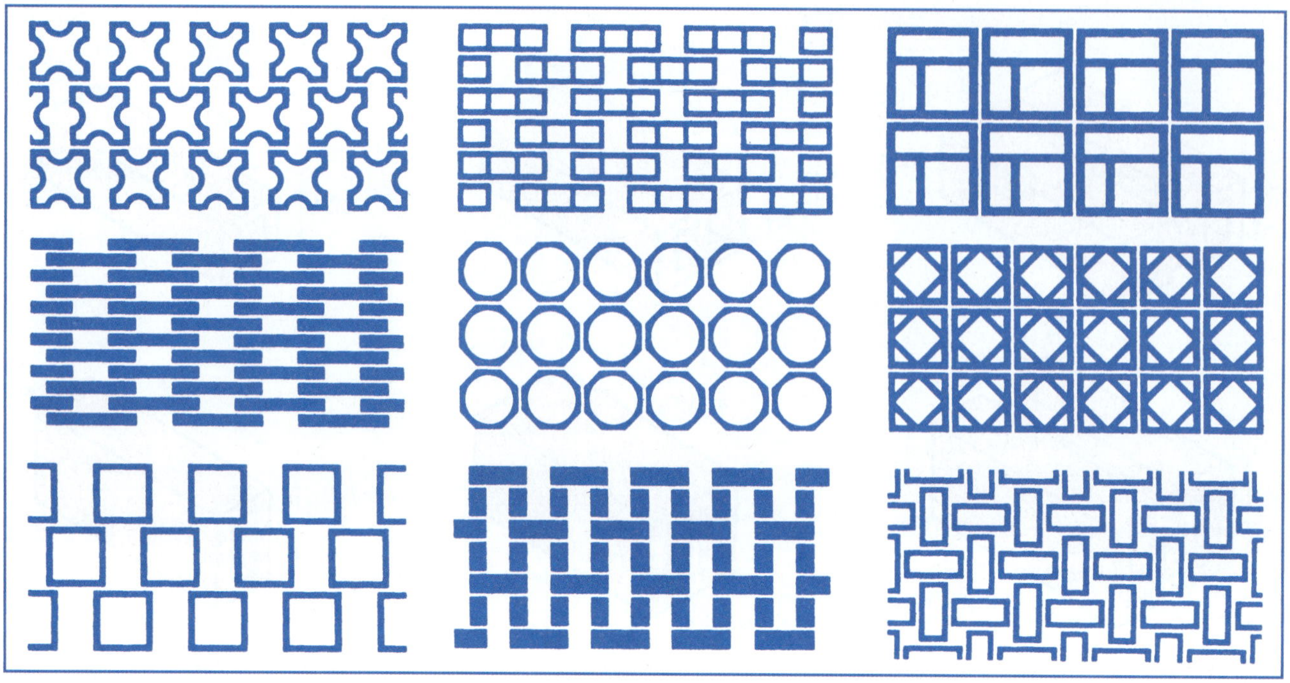

Goodheart-Willcox Publisher

Figure 7-53. Clay tile in shapes used for decorative screens.

Interpace Corp.

Figure 7-54. Examples of the broad color range in which architectural terra cotta is produced.

Anchored Ceramic Veneer

Anchored ceramic veneer is usually produced in thick sections held in place by grout and wire anchors or ties connected to its backing. It is least expensive to use the standard tie sizes. Anchored ceramic veneer may be used as facing for exterior and interior walls, column facings, soffits, copings, and sills.

Adhesion Ceramic Veneer

Adhesion ceramic veneer is usually produced in thin sections. It is held in place without metal anchors by a mortar backing. Thickness is ordinarily 1″. Adhesion ceramic veneer is produced in a variety of maximum finish sizes.

Adhesion ceramic veneer can be installed over concrete masonry, metal, or wood. It is used primarily as facing for exterior and interior walls and coping units. See **Figure 7-55**.

Ornamental or Sculptured Terra Cotta

Ornamental terra cotta is produced in sculptured patterns or freestanding sculpture. It is used frequently as cornices and column capitals on large buildings.

Ornamental terra cotta can be either anchored or adhesion veneer, depending on the application. Molded pieces may also be made in larger dimensions than facing veneer. **Figure 7-56** shows examples of ornamental cornices being used on two different architectural styles.

Interpace Corp.

Figure 7-55. Adhesion ceramic veneer is available in a variety of colors for use as facing and coping units.

A Francois Roux/Shutterstock.com

B lazyllama/Shutterstock.com

Figure 7-56. A—An example of terra cotta used as architectural facade located in New York. B—An example of terra cotta used to trim a Spanish-style church built in 1887, in St. Augustine, FL.

The sculptures shown in **Figure 7-57**, **Figure 7-58**, and **Figure 7-59** are manufactured at Boston Valley Terra Cotta. The terra cotta produced is made from a special clay mixture. Architectural drawings, photo documentation, laser scanning, and site surveys are used to create 3D CAD (computer-aided drawing) and CAM (computer-aided manufacturing) designs. The company also replicates designs using damaged façade fixtures or figurines. Because of past designs and fabrications, the company knows exactly how much clay shrinkage will occur during the drying cycle of the raw clay and also when the fixture is fired (placed in an oven similar to the process of manufacturing brick). During the design and manufacturing process, some of the larger fixtures have images on the front exposed surfaces, and the back resembles a concrete block with a hollow core. This reduces the amount of clay, which decreases the overall weight of the finished architectural façade or sculpture. During firing of the terra cotta, the clay products are heated to 2200°F (1204°C) in kilns for three to five days.

A

B

Sarah Carroll, Boston Valley Terra Cotta

Figure 7-57. A—A sculptor is using an original terra cotta fixture as a reference to create a mold for replacement pieces. The new fixture is larger than the original because of the shrinkage that will occur in the firing process. B—The original terra cotta piece was removed from the building. The sculptor has placed it next to a mold being recreated for replacement pieces. Eventually, the replacement pieces will be installed on the building.

Sarah Carroll, Boston Valley Terra Cotta

Figure 7-58. A terra cotta sculpture is laid out for quality control checks.

A B C

Sarah Carroll, Boston Valley Terra Cotta

Figure 7-59. A—The cloth on the Lion is used to keep the clay from drying out prior to the figure being completed. B— The Lion sculpture is completely finished and ready to be installed at the renovation building site. C—Close-up of the completed Lion sculpture shows how intricate the clay terra cotta can be as a building product.

Chapter Review

Summary

- The American Concrete Institute (ACI) and the American Society of Civil Engineers (ASCE) have adopted a consensus standard for the design of masonry structures. A standard construction specification is included.
- Structural clay products are classified as solid masonry units (brick), hollow masonry units (tile), and architectural terra cotta. These products are made from clay or shale.
- Three common brick manufacturing methods are the soft mud process, the stiff mud process, and the dry-press process.
- Green bricks are bricks that have not been fired.
- Bricks are dried before firing in periodic kilns and tunnel kilns. Bricks can also be air-dried, but this takes considerably more time than kiln-drying.
- Bricks are commonly classified as building (common) brick; facing brick; hollow brick; pedestrian and light traffic paving brick; heavy vehicular paving brick; ceramic glazed brick; glazed brick, single fired; thin brick veneer units; firebox brick, residential fireplaces; chemical-resistant masonry units; sewer and manhole brick and industrial floor brick.
- A modular unit is based on a measurement of 4″.
- Structural clay products may be solid masonry units or hollow masonry units.
- Efflorescence is a white powder or salt-like deposit on masonry walls that is caused by water-soluble salts.
- Compressive strength is the ability of masonry units to stand up under heavy weight without crumbling. High compressive strength is achieved by using high-strength brick and mortars and making well-filled joints.
- The word "bond" in masonry has three meanings: structural bonds, pattern bonds, and mortar bonds (joints).
- Structural bonding of masonry walls with metal ties is acceptable for solid wall and cavity wall construction.
- The five basic structural bonds are the running bond, the common (American) bond, the Flemish bond, the English bond, and the block or stack bond.
- Mortar joint finishes are either troweled or tooled joints. Mortar joints include concave and V-shaped joints, weathered joints, struck joints, rough cut (flush) joints, raked joints, and extruded joints.
- Brick can be shaped and used in traditional and contemporary detailing. Many shapes are manufactured by large brick companies, but other shapes may be customized.
- Hollow clay tile is classified as either structural clay tile or structural clay facing tile. Structural clay tile may be loadbearing or nonloadbearing.
- Structural clay facing tile may be glazed or unglazed.
- SCR acoustile is designed to absorb sound.
- Architectural terra cotta is a custom product that includes anchored ceramic veneer, adhesion ceramic veneer, and ornamental or sculptured terra cotta.

Know and Understand

Answer the following questions using the information provided in this chapter.

1. The _____ organization published the National Building Code.
   A. SBCCI
   B. IBCQ
   C. BOCCA
   D. ACI
2. The three main classifications of structural clay products are solid masonry, hollow masonry, and _____.
   A. surface clays
   B. shales
   C. fire clays
   D. architectural terra cotta

3. _____ are produced naturally as rocks weather.
   A. Tiles
   B. Clays
   C. Shales
   D. Bricks
4. *True or False?* The three methods of manufacturing brick are the soft mud process, the stiff mud process, and the wet-press process.
5. The _____ process forces clay under high pressure into molds to form the brick.
   A. dry-press
   B. soft mud
   C. stiff mud
   D. wet-press
6. _____ bricks have not been fired.
   A. Peach
   B. Beige
   C. Green
   D. Blue
7. Burning time in a tunnel kiln may be as little as _____ hours.
   A. 2
   B. 10
   C. 25
   D. 50
8. When appearance is important as well as durability and strength, _____ bricks are used.
   A. building
   B. facing
   C. hollow
   D. solid
9. _____ bricks are classified according to their resistance to abrasion and ability to withstand traffic.
   A. Paving
   B. Building
   C. Solid
   D. Firebox
10. Modular units are based on a measurement of _____.
    A. 1″
    B. 2″
    C. 4″
    D. 6″
11. The ability of a masonry unit to stand up under a heavy weight without crumbling is _____.
    A. efflorescence
    B. durability
    C. structural strength
    D. compressive strength
12. A continuous level row of brick is called a _____.
    A. wythe
    B. common bond
    C. running bond
    D. course
13. *True or False?* Custom shapes that conform to the designer's specification can be produced in extruded brick.
14. _____ is suitable for use in fireproofing and furring.
    A. Solid masonry
    B. Facing tile
    C. Loadbearing tile
    D. Nonloadbearing tile
15. *True or False?* Ornamental terra cotta can be either anchored or adhesion veneer.

Apply and Analyze

1. What is the difference between hollow brick and facing brick?
2. Identify three factors that determine brick color.
3. What is the weight of a typical standard building brick?
4. Identify the three methods of producing a structural bond.
5. Name four types of mortar joints.
6. Structural clay facing tile may be divided into what two broad types?
7. What is the function of SCR acoustile?
8. Identify the three classifications of architectural terra cotta.

Critical Thinking

1. A residential home with brick veneer was completed in the past year. The homeowner called the president of XYZ Masonry Company, where you are employed as a foreman. The homeowner is complaining of a white substance that has appeared on the surface of the brick. The president has directed you to schedule a visit with the homeowner.
   A. Research in this chapter to determine the name for the staining and what causes the stain. List the different causes for the stain to appear.
   B. Identify and list the discussion you should have with the homeowner.
   C. Search the internet to determine methods of cleaning the wall and identify those methods.
   D. Develop procedures that can be used to reduce the chances of staining from reoccurring.

Activities

1. (SCIENCE) The color of fired clay depends upon its chemical composition, the firing temperatures, and the method of firing. Write a brief paper describing how these factors affect the color of the brick.
2. (TECH) Research the manufacture of brick to determine what technological advancements have improved the efficiency of brick plants and the quality of their products. Form small groups in class and share your findings.

Basic white sand mold brick

Sand mold brick in zinc-flashed pinks and cross sets

Smooth reds

Smooth finished in ebony with iron spots

Sand finished with soft pinks and dusty rose

Smooth gray with manganese specks

Goodheart-Willcox Publisher

Matt textured fireclay brick

Textured charcoals

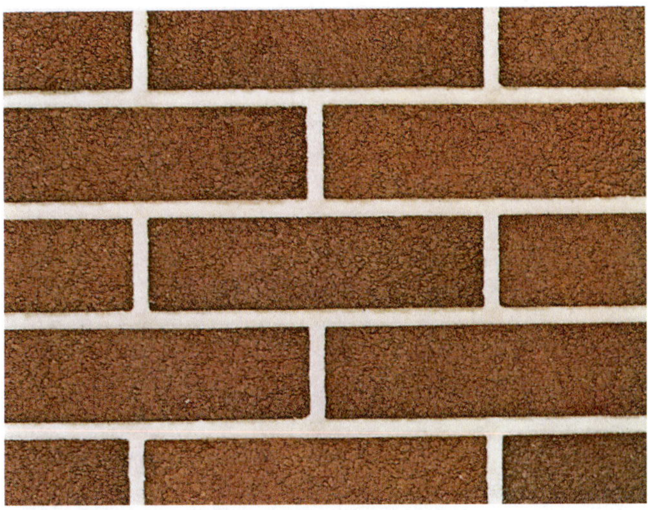

Matt textured burgundy

White velour texture

Tan flashed shades and cross sets

Water struck brick in fine sand-finished texture

Goodheart-Willcox Publisher

Textured brick in tan and dark brown

Semismooth texture in grays and flashed shades

Matt-textured winter white

Standard brick, smooth finish

Matt texture in buff

Semismooth in reds, cross sets, hearts, browns, and tans

Goodheart-Willcox Publisher

Matt textured multicolors

Matt texture in pinks and tans

Sand finished in buff pink, tan, and flashed shades

Matt texture in gray shades

Rustic sand finish

Antique finish

Goodheart-Willcox Publisher

Textured snow white

Sand finish in pinks, reds, and flashed shades

Sand texture in brown

Double-fired, glazed, smooth, and textured

Sand mold brick in reds, tans, cross sets, and earth tones

Velour texture in pink

Goodheart-Willcox Publisher

Molded, textures in brown

Textured in grays, green, and brown

Sand molded in grays and tans

Sand molded in earth tones

Colonial russet brown

Standard striated in brown

Goodheart-Willcox Publisher

Michal Bellan/Shutterstock.com

CHAPTER 8
Concrete Masonry Units

LEARNING OUTCOMES

After studying this chapter, you will be able to:

8.1-1 Explain how concrete masonry units are made.
8.2-1 Describe the properties and uses of the various types of concrete masonry units.
8.3-1 Differentiate between soft and hard metric conversion.

TECHNICAL TERMS

admixtures
aggregate
architectural concrete masonry units (ACMU)
cells
compaction
concrete masonry unit (CMU)
cores
cross web
ears
face shell
Grade MW
Grade N
Grade NW
Grade S
Grade SW
ground face units
hard conversion
hardscape
lightweight aggregate
normal weight aggregate
prefaced concrete masonry units
retaining wall block
scoria
soft conversion
split face masonry units
stretchers
voids

Introduction

The term *concrete masonry* includes all the sizes and types of hollow or solid block, brick, and concrete building tile, as long as they are made from concrete and are laid by masons.

There is no doubting the usefulness of concrete masonry. Block manufacturing plants produce hundreds of different types of units in a wide variety of sizes and shapes. See **Figure 8-1**. Concrete masonry is popular because it is inexpensive to make, can be bought almost anywhere, and is easy to work with. It is a good insulator, has excellent sound reduction properties, is fire resistant, and stands up well in all kinds of weather.

8.1 How Concrete Masonry Units Are Made

Concrete masonry units (CMU) are hollow or solid units made from a relatively dry mixture of Portland cement, aggregates, water, and often admixtures. Sometimes other cement-like materials are used in place of Portland cement. ***Aggregate*** refers to sand and gravel or a suitable substitute. Units are often referred to by the type of aggregates used in their manufacture: cinder block, lightweight block, etc. ***Admixtures*** include coloring agents, air-entraining materials, accelerators, retarders, and water repellents. These materials are mixed, and the mixture then is molded into desired shapes through ***compaction*** (squeezing action to make a material denser) and vibration. See **Figure 8-2**. The

Trenwyth Industries, Inc.

Figure 8-1. Concrete block materials include cracked block, screen block, typical concrete blocks, and prefaced blocks.

units are then cured under controlled moisture and temperature conditions. After a period of aging, they are ready for use.

Besser Company

Figure 8-2. In this machine, a concrete mixture is molded into desired shapes through compaction and vibration.

Green Note
Fly Ash

Fly ash is produced when coal is burned to produce electricity. This by-product can be used in concrete as a replacement for part of the Portland cement. The environmental benefits of this replacement include conservation of resources and energy, reduction of greenhouse gas emissions, and reduced costs. In addition, fly ash improves the workability of the fresh concrete; requires less water, which reduces cracking; and makes the final product more durable because permeability is reduced.

Aggregates

Aggregates normally make up about 90% of concrete masonry units by weight. Desirable properties of aggregates include:

- Uniform amounts of fine and coarse sizes
- Toughness, hardness, and strength to resist impact abrasion and loading
- Ability to resist the forces exerted by freezing, thawing, expansion, and contraction
- Cleanliness and lack of foreign materials that would reduce strength or cause surface roughness

The two classes of aggregates by weight are normal weight (dense) and lightweight. ***Normal weight aggregate*** includes sand, gravel, crushed limestone, and air-cooled slag. ***Lightweight aggregate*** is expanded shale or clay, expanded slag, coal cinders, pumice, and scoria. ***Scoria*** is the scrap left after the melting of metals. Different aggregate types differ in weight, strength, and other characteristics. See **Figure 8-3**.

Code Note **Combined Units**

International Residential Code **R606.9.1** specifies that when walls or other structural members are made up of different kinds or grades of units, materials, or mortars, the maximum stress should not exceed the allowable stress for the weakest unit, material, or mortar.

8.2 Classification of Concrete Masonry Units

Concrete masonry units are classified into six main groups based on their intended use, size, and appearance:

- Concrete Brick, ASTM C55-11
- Loadbearing Concrete Block, ASTM C90-14
- Nonloadbearing Concrete Block, ASTM C129-11
- Calcium Silicate Face Brick, ASTM C73-10
- Prefaced Concrete Units, ASTM C744-11
- Units for Catch Basins and Manholes, ASTM C139-11

These six groups of masonry units are covered by six ASTM standards that apply to units intended primarily for construction of concrete masonry walls, beams, columns, or specialty applications. The first part of an ASTM standard number is the fixed designation for that standard. For example, ASTM 90 is the fixed designation for loadbearing concrete masonry units. The number immediately following the fixed designation indicates the year of the last revision. Units for catch basins and manholes are not covered in this text. The classification of concrete masonry units shown in **Figure 8-4** presents another useful categorization.

Concrete Brick

Concrete bricks are solid brick-size units, solid concrete veneer, and facing units larger than brick size. They are similar in function to clay brick. Concrete bricks are completely solid or have a shallow depression called a *frog*. See **Figure 8-5**. The frog is designed to reduce weight and provide for a better bond when the bricks are laid in mortar.

There are two grades of concrete bricks included in ASTM C55—Grade N and Grade S. ***Grade N*** is concrete brick used for architectural veneer and facing units in exterior walls. It is also used where high strength and resistance to moisture penetration and severe frost action are desired. ***Grade S*** is concrete brick used where moderate strength and resistance to frost action and moisture penetration are required.

Concrete bricks are meant to be laid with a 3/8″ mortar joint. The most popular nominal modular dimension is 4″ × 8″. The thickness of horizontal mortar

Properties of Concrete Block Made of Different Aggregates					
Aggregate (Graded: 3/8″ to 0)					
Type	Density (air-dry) lb/cu ft	Weight lb/cu ft of concrete	Weight 8″ × 8″ × 16″ unit	Compressive strength (gross area) psi	Water absorption lb/cu ft of concrete
Sand and gravel	130–145	135	40	1200–1800	7–10
Limestone	120–140	135	40	1100–1800	8–12
Air-cooled slag	100–125	120	35	1100–1500	9–13
Expanded shale	75–90	85	25	1000–1500	12–15
Expanded slag	80–105	95	28	700–1200	12–16
Cinders	80–105	95	28	700–1000	12–18
Pumice	60–85	75	22	700–900	13–19
Scoria	75–100	95	28	700–1200	12–16

Goodheart-Willcox Publisher

Figure 8-3. This chart shows the effect of aggregate type on weight, strength, and other characteristics of concrete block.

Classification of Concrete Masonry Units		
Concrete Brick	**Concrete Block**	**Special Units**
Solid brick-size units Solid concrete veneer Facing units larger than brick size	Solid loadbearing Hollow loadbearing Solid nonloadbearing Hollow nonloadbearing	Ground face units Split face block Prefaced concrete block Decorative block

Goodheart-Willcox Publisher

Figure 8-4. A chart of 11 types of concrete masonry products.

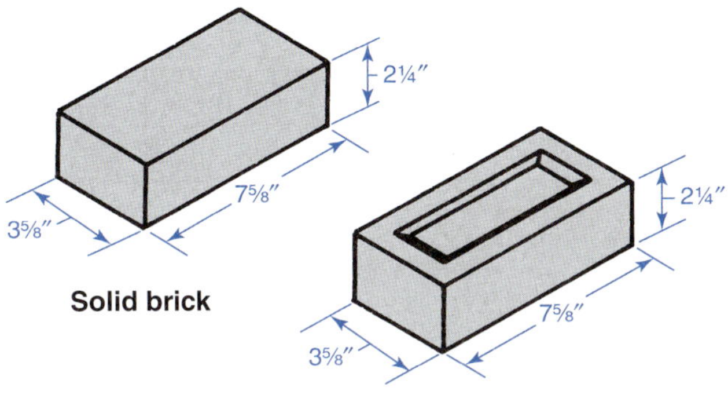

Goodheart-Willcox Publisher

Figure 8-5. In size, concrete building bricks are much like clay brick. The bricks may be solid or frogged.

joints is such that three courses (three brick and three bed joints) equal a height of 8″. Some manufacturers produce oversized jumbo and double brick units for special applications.

Slump Brick

Slump brick or slump block, **Figure 8-6**, has a rounded face surface to mimic old adobe brick or block. Slump brick or block is produced from a mixture that is wet enough to cause the units to sag or slump when removed from the molds. In modern manufacturing, the slump is produced by an additional compression placed on the unit immediately after it is ejected from the mold to create a more uniform brick or block height. Resulting faces are irregular, and the surface texture and appearance may vary greatly depending on the type and style of the unit. Slump bricks and blocks produce special effects and resemble stone when laid in a wall or structure. See **Figure 8-7**.

Concrete Block

Concrete block units are usually produced in four classes:

- Solid loadbearing
- Hollow loadbearing
- Solid nonloadbearing
- Hollow nonloadbearing

Each of these classes may be purchased with either normal weight or lightweight aggregates. Strength and absorption requirements vary for concrete masonry units using lightweight concrete and normal weight concrete, **Figure 8-8**.

A solid concrete block is one in which the hollow parts in a cross section are not more than 25% of the

Air Vol Block

Figure 8-6. This concrete slump brick has the appearance of stone.

Air Vol Block

Figure 8-7. Slump blocks are used to produce an attractive masonry planter and wall.

Strength and Absorption Requirements for Concrete Masonry Units								
Type of Unit	ASTM Designation	Grade of Unit	Average of Three Units[1]	Individual Unit[1]	Lightweight Concrete[2], Less than 85 pcf	Lightweight Concrete[2], Less than 105 pcf	Medium-Weight Concrete[2], 105 to 125 pcf	Normal-Weight Concrete[2], 125 pcf
Concrete Brick	C55	N	3500*	3000*	—	15	13	10
Concrete Brick	C55	S	3500*	2000*	—	18	15	13
Solid Loadbearing Block	C145	N	1800	1500	—	18	15	13
Solid Loadbearing Block	C145	S	1200	1000	20	—	—	—
Hollow Loadbearing Block	C90	N	1000	800	—	18	15	13
Hollow Loadbearing Block	C90	S	700	600	20	—	—	—
Hollow and Solid Nonloadbearing Block	C129	—	600**	500**	—	—	—	—

* Concrete brick tested flatwise.
** Minimum compressive strength, psi, on average net area.
[1] Minimum compressive strength, psi, on average gross area.
[2] Maximum water absorption, pcf, based on oven-dry unit weight.

Portland Cement Association

Figure 8-8. Compressive strength is an important property of concrete masonry and its related applications.

total cross-sectional area. A hollow concrete block is one in which the core or hollow area is greater than 25% of its total cross-sectional area. The core area in hollow units is generally from 40% to 50% of the gross area.

The ASTM C90 standard covers specifications required for concrete masonry units and the load bearing of those units. Under ASTM C90, concrete units are listed in three classes based on the overall weight of the concrete being used to create the different units. See **Figure 8-9**.

These units are suitable for both loadbearing and nonloadbearing applications. Compressive strength of CMU is based on gross area of the units, with a minimum average of 2000 pounds per square inch requirement for a test sample of three blocks. The 2000 psi strength is required for all three classes of concrete masonry units.

ASTM C90 Concrete Classifications		
Density Classification	Oven-Dry Density of Concrete (lb/ft^3)	Oven-Dry Density of Concrete (kg/m^3)
Normal weight	125 or more	2000
Medium weight	105 to 124	1680 to 2000
Lightweight	Less than 105	Less than 1680

Goodheart-Willcox Publisher

Figure 8-9. Concrete units are listed in classes based on the overall weight of the concrete.

Uses of Concrete Block

Solid loadbearing blocks are most often used to support great loads. Hollow loadbearing blocks have many uses. They combine high compressive strength with light weight and flexible design, size, and shape. A large percentage of the block produced is hollow loadbearing block.

Hollow nonloadbearing block is thin-shelled and lightweight. It is intended to be used mostly in nonloadbearing partitions. If used on nonloadbearing exterior walls, the block should be protected from the weather.

Concrete masonry units are available in such a large number of shapes, sizes, textures, colors, and profiles that they are used in almost all aspects of masonry construction. The great variety of patterns and designs that are possible allow a wide range of wall surfaces and architectural treatments to be produced. The following list of applications illustrates the broad use of concrete masonry units:

- Exterior loadbearing walls (**Figure 8-10**)
- Interior loadbearing walls

Goodheart-Willcox Publisher

Figure 8-10. Concrete blocks are used in this residential structure as exterior loadbearing walls and columns.

- Curtain walls, party walls, and fire walls
- Solar screens, panel walls, and partitions
- Veneer or nonstructural facing for concrete, masonry, wood, or steel
- Backing for brick, stone, stucco, and exterior insulation finish systems
- Fire protection of structural steel members
- Pilasters, columns, and piers (**Figure 8-10**)
- Bond beams, lintels, and sills
- Firesafe enclosures of stairwells, elevator shafts, and storage vaults
- Chimneys and fireplaces
- Retaining walls and slope protection
- Paving and turf block
- Catch basins

Finish and Appearance

Provisions found in ASTM C90-14 relating to finish and appearance prohibit defects that impair the strength or permanence of the construction. Blocks with defects such as a crack that continues through the web or large chipped corners or edges are not allowed to be used. However, provisions permit minor cracks incidental to the usual manufacturing methods of the units.

For units in exposed wall construction, the presence of objectionable imperfections is based on viewing the face or faces from a distance of 20′ under diffused lighting. The specification requires that color and texture be specified by the purchaser. An approved sample consisting of not less than four units, representing the range of color and texture permitted, is used to determine conformance.

Sizes and Shapes

Concrete building units are made in many sizes and shapes to fit different construction needs. It is common practice to specify the width first, the height second, and the length of the stretcher block last.

Sizes are usually given in their nominal dimensions. A unit measuring 7 5/8″ wide, 7 5/8″ high, and 15 5/8″ long is known as an 8″ × 8″ × 16″ unit. The mortar joint is intended to be 3/8″ to fit the modular design based on a 4″ module. Concrete blocks are commonly available in widths of 2″, 4″, 6″, 8″, 10″, and 12″. Standard block heights are 4″ and 8″, but 12″ units are also available. Lengths include 8″, 12″, 16″, 20″, and 24″. Popular shapes and sizes are produced in full and half-length units. See **Figure 8-11**.

Some block units have a two-core design rather than three. The two-core design has the following advantages:

- The shell is thicker at the center web. This increases the strength of the unit and reduces the tendency to crack from shrinkage.
- Heat conduction is reduced by 3% to 4%.
- Blocks are about 4 lb lighter.
- Hollow portions provide more space for placing conduit or other utilities.

The following are several popular types of concrete block (refer to **Figure 8-11**):

- **Stretcher block** is the most common and used on exterior surfaces.
- **Corner block** has one flush end that may be used for corners, simple windows, and door openings.
- **Double corner** or **pier block** is designed for piers or pilasters or any other place where both ends are exposed.
- **Bullnose block** is used the same way as the corner block but has a rounded corner.
- **Jamb** or **sash block** is used to make installation of windows or other openings easier.
- **Solid top block** has a solid top for use as a bearing surface in the finishing course of a foundation wall.
- **Header block** has a recess to receive the header unit in a masonry bonded (brick and block) wall.
- **Metal sash block** is used for window openings in which a metal sash is to be used. The slot anchors the jamb.
- **Partition block** (4″ or 6″) is used in constructing nonloadbearing partition walls.
- **Lintel block** is used to construct horizontal lintels or beams. It is generally U- or W-shaped.
- **Bond beam block** is used to construct bond beams to add strength to the wall. A portion

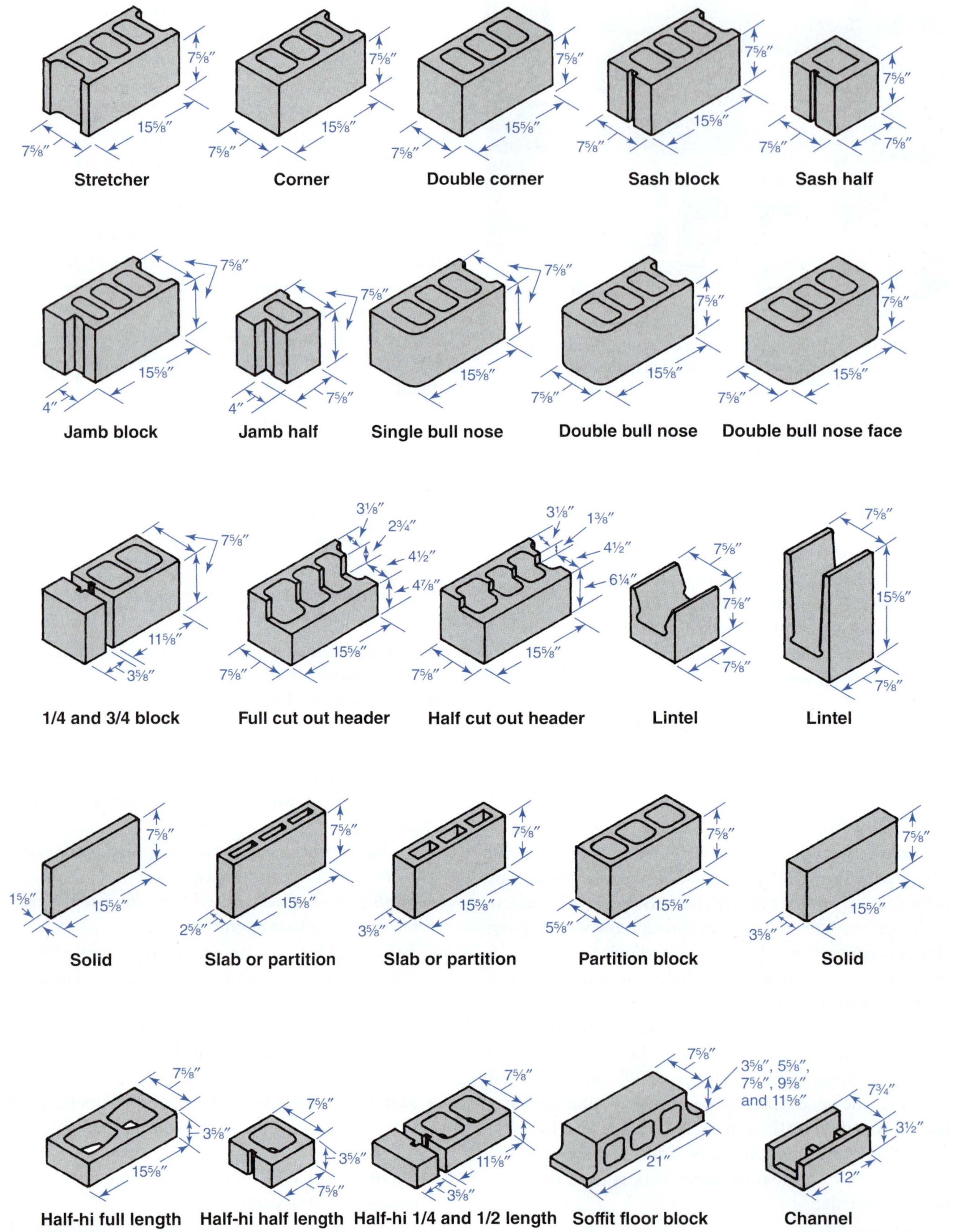

Goodheart-Willcox Publisher

Figure 8-11. A large selection of the most popular shapes and sizes of concrete block.

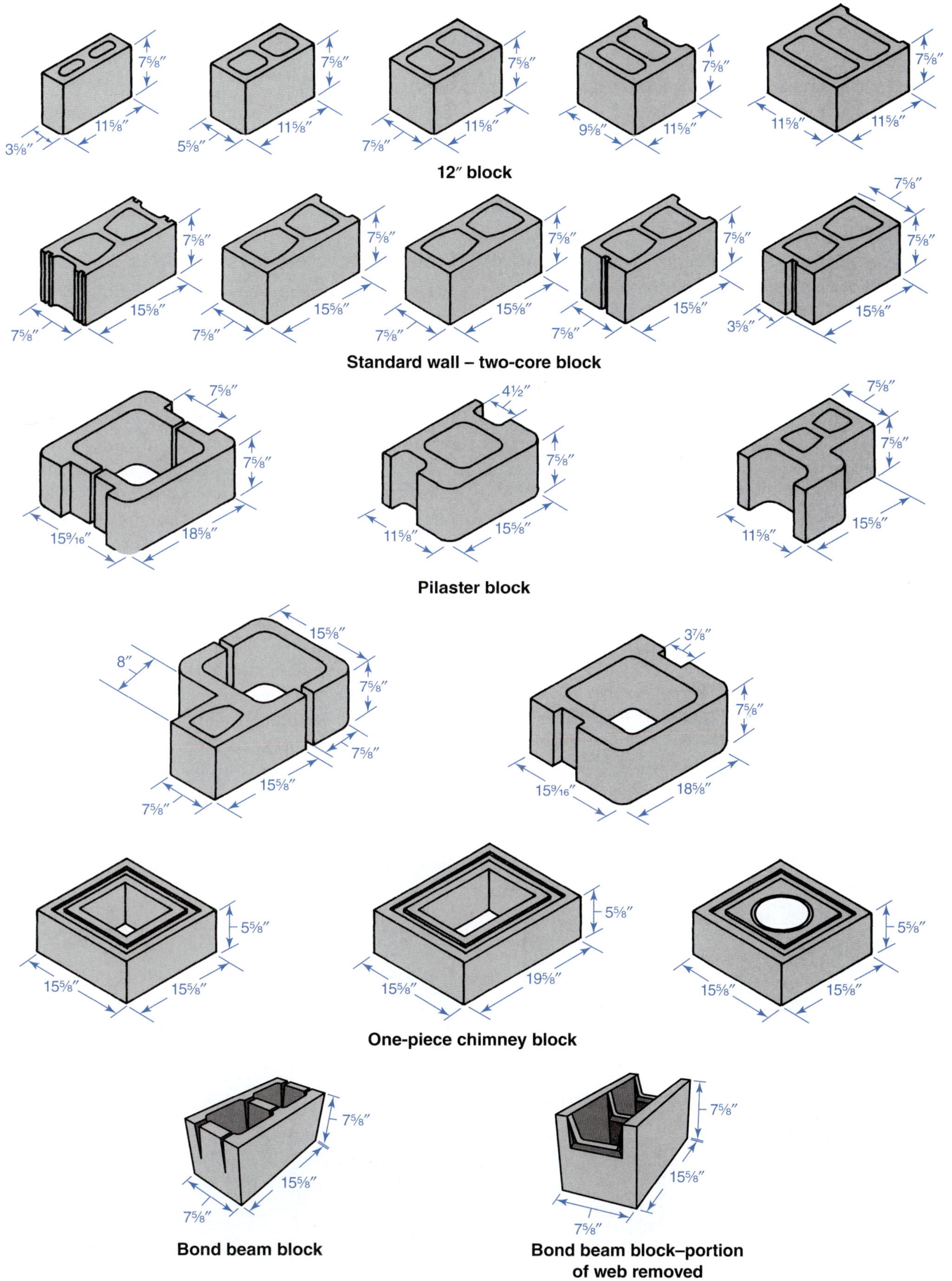

Goodheart-Willcox Publisher

Figure 8-11 (Continued).

of the web is removed to allow for horizontal reinforcement. Vertical reinforcement is also used to strengthen the wall vertically. Both horizontal and vertical rebar are tied together prior to grout, mortar, or concrete being used to fill the cores of the block.

- **Control joint block** is used to construct vertical shear-type control joints.

Block Terminology

Several terms are used to describe parts of standard concrete blocks, **Figure 8-12**. Blocks may be plain or flat on the end or concave in shape. Plain end blocks are used at corners, piers, or other locations where a plain end is necessary or desirable. Blocks with concave ends have two ***ears***. Blocks are produced with ears on one or both ends. Blocks with ears on both ends are called ***stretchers***. Stretchers are used along the course between the corners.

The openings in blocks are called ***cells***, ***cores***, or ***voids***. The wall between the cells is called a ***cross web***. The outside face of a block is called the ***face shell***. Standard blocks are produced with two and three cells. Half blocks have a 4″ thickness and may be solid, two-cell, or three-cell blocks.

Surface Texture

Surface texture is important in blocks because it affects sound absorption, appearance, and painting characteristics. Coarse textures look nicer and absorb sound better than smooth textures. Fine textures may absorb less sound but also take less paint than coarse textures. Some concrete masonry units have face surfaces ground smooth to produce interesting aggregate colors.

Color

The typical color of concrete masonry units varies from light to dark gray to tints of buff, red, or brown, depending on the color of aggregate, cement, and other ingredients used. The curing method can also affect the color. Variations in color are common.

Prefaced Concrete Masonry Units

Prefaced concrete masonry units have a specially prepared surface material applied to one or more surfaces to provide color, pattern, or texture. See **Figure 8-13**. Shapes such as stretchers, cove bases, bond beam units, caps or sills, and lintel blocks are frequently selected for prefacing. See **Figure 8-14**. ASTM C744 is the applicable standard for prefaced concrete masonry units.

The most common binders used to face masonry units are resin, resin and fine sand, and Portland cement and inert fillers. Glazed masonry units that have ceramic or mineral glazes are also produced. Facing material can add as much as 1/4″ to the block dimension. As a result, mortar joints are reduced to 1/4″ to maintain modular dimensions.

Since facings are resistant to water penetration, cleaning detergents, and abrasion, prefaced units are frequently used in locations where cleanliness, decoration, or low maintenance are desirable. Examples of applications include school corridors, locker rooms, hospitals, and food-processing areas.

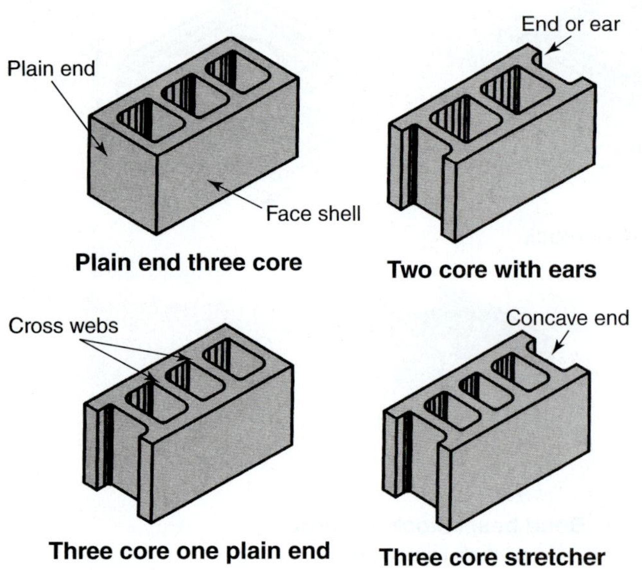

Goodheart-Willcox Publisher

Figure 8-12. Concrete block terminology.

Trenwyth Industries, Inc.

Figure 8-13. A sampling of prefaced concrete masonry units showing a variety of colors, textures, and shapes.

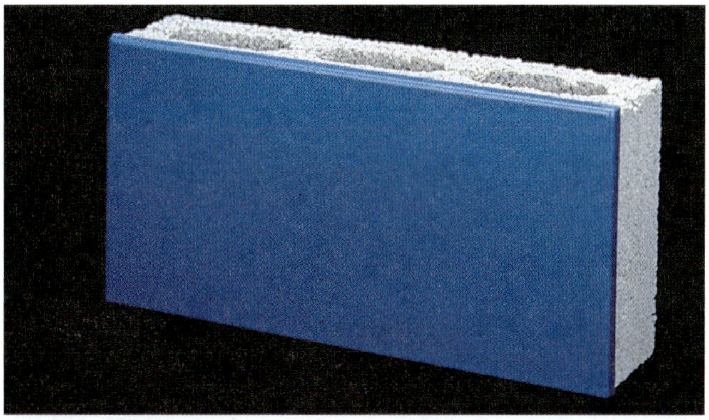
A

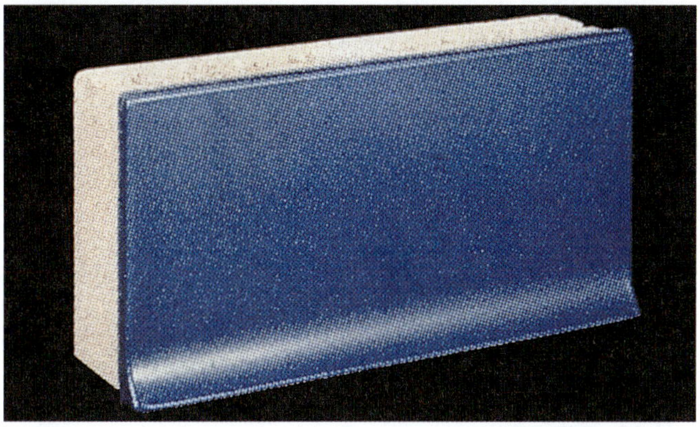
B

Trenwyth Industries, Inc.

Figure 8-14. Prefaced units include stretchers and cove bases. A—Concrete block stretcher prefaced unit. B—Concrete block prefaced cove based unit.

Ground Face Units

Ground face units, **Figure 8-15**, include oversized and specialty units that may be used to create walls, columns, and arches of monumental proportions. Lengths of 11 5/8″, 15 5/8″, 17 5/8″, and 23 5/8″; widths of 3 9/16″ and 7 9/16″; and heights of 7 5/8″, 11 5/8″, and 15 5/8″ are available in hollow and solid shapes.

Goodheart-Willcox Publisher

Figure 8-15. This close-up view of a ground face concrete block unit shows the aggregate that gives this product its character.

Special shapes of ground face units are also produced. These shapes include keystone arch units, water table shapes, cornice units, sculpted sill shapes, column wraps, chamfered sill units, slant units, scored units, chamfer and quoining units, radial units, corbel and medallion units, cap units, and fluted units.

Decorative Blocks

Architectural concrete masonry units (ACMU) are masonry units that have been designed for aesthetic appeal. ACMU are used in interior and exterior applications for walls, partitions, terrace walls, and other enclosures. Some units are designed to have the surface treatment on both sides to serve as an interior and exterior finished surface. This process may reduce the overall cost of the project because both surfaces are finished. A number of concrete masonry units are classified as architectural concrete masonry units. Examples include split face units, scored units, ribbed units, ground face units, sandblasted units, glazed units, and slump block.

Fluted and scored or ribbed units are concrete masonry units that have a grooved exterior surface created during the manufacturing process. The striations applied to the exterior of the block can be used to create many kinds of patterns for finished wall surfaces. See **Figure 8-16** and **Figure 8-17**. Striations are a series of parallel ridges or grooves. The accuracy that is achieved in machine production makes it possible to

Goodheart-Willcox Publisher

Figure 8-16. Fluted and ribbed concrete blocks are used to achieve unique textured surfaces.

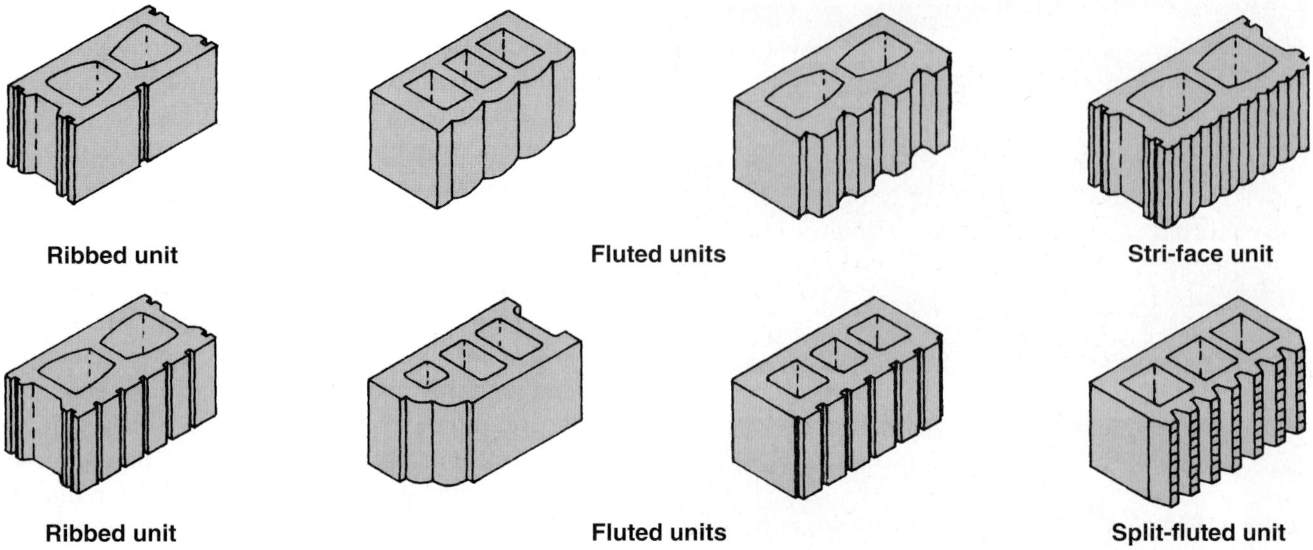

Goodheart-Willcox Publisher

Figure 8-17. Split-fluted concrete block units form attractive wall patterns.

produce the effect of long, vertical, straight lines. These blocks can be quite decorative.

Blocks with recessed corners and raised patterns add endless effects to concrete masonry wall construction. See **Figure 8-18** and **Figure 8-19**.

Split face masonry units look like rough quarried stone. They are produced by splitting solid concrete block lengthwise, **Figure 8-20**. Several sizes and color selections are available. Architects and designers select split face masonry units to enhance the exterior and interior appearance of exposed finished surfaces. Split face blocks are used in locations where special hollow block as well as ribbed units can be split to produce unusual effects.

Goodheart-Willcox Publisher

Figure 8-18. Even though this sculptured face concrete block wall is painted a uniform color, it develops the effect of several shades.

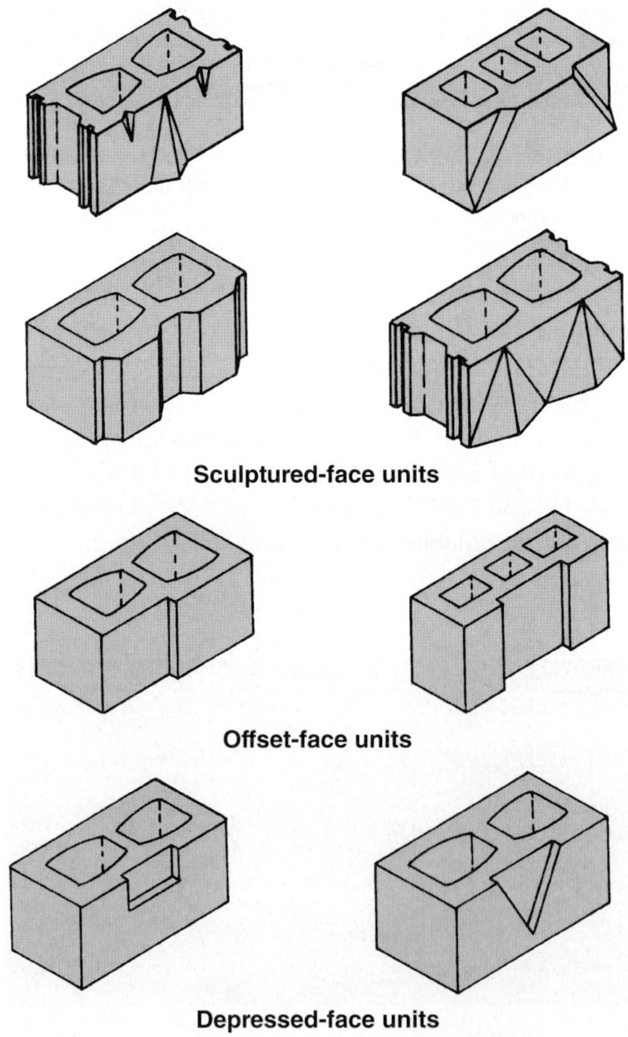

Goodheart-Willcox Publisher

Figure 8-19. Concrete block with offset, depressed, or sculptured faces.

Goodheart-Willcox Publisher

Figure 8-20. Split concrete block produces the effect of quarried stone.

Screen Block

Screen blocks are highly decorative, **Figure 8-21**. They can also fill several useful functions, such as building facades, ornamental room dividers and partitions, garden fences, and patio screens. Screen blocks, also called *grille units*, can provide only limited, directional vision or produce a balance between privacy and openness from within or without. They also screen out some light and produce a degree of shade.

Screen blocks range in size from 4″ to 16″ square and 8″ × 16″ rectangles. Even though a wide variety of screen block patterns exists, the patterns available in any given area may be limited. Check with local suppliers before planning to use a particular design. See **Figure 8-22**.

Some screen blocks can be used in several positions in a wall to produce different patterns. These units can be combined in many different ways to create interesting designs. They can also be mixed with common block shapes to add variety. The use of screen block is not as predominant as it has been in the past.

Hardscape Blocks and Pavers

Hardscape refers to hard landscape materials used to create structures that are part of an outside landscape. Hardscape can include freestanding walls, retaining walls, stairs, walkways, driveways, and patio areas. Hardscape derived its name from *hard-wearing products*. Dimensions of hardscape products can vary from one manufacturer to another based on the company's product designs and manufacturing processes. See **Figure 8-23**.

Hardscape materials are made similar to concrete block but are usually made of solid materials because of the necessity to survive severe weather. If hardscape blocks had

Timothy L. Andera

Figure 8-21. Screen concrete block has been added to the top of the concrete block wall for decoration.

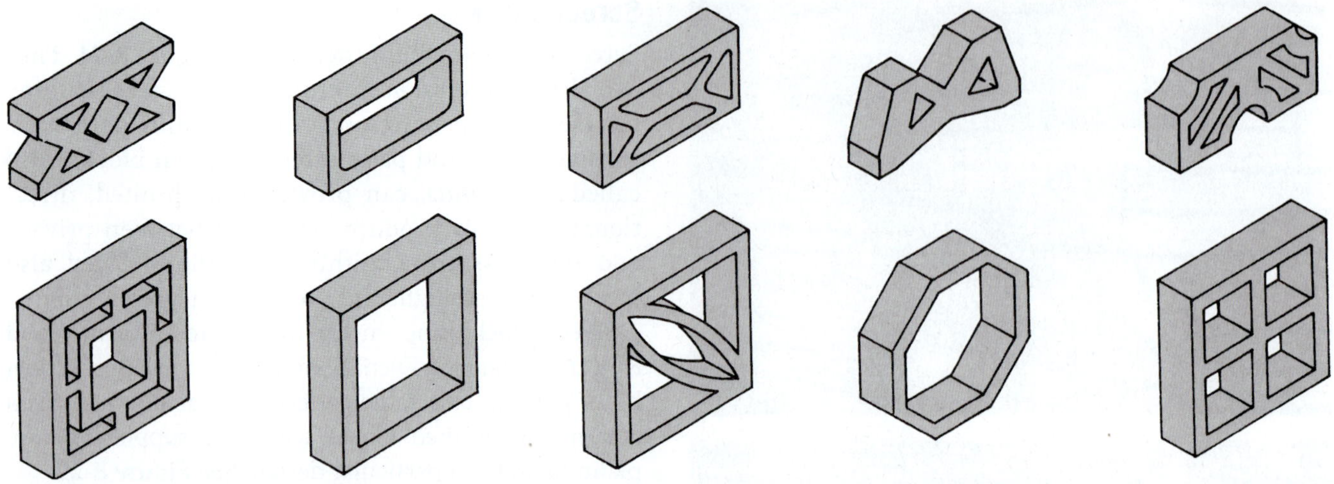

Goodheart-Willcox Publisher

Figure 8-22. Typical screen block patterns.

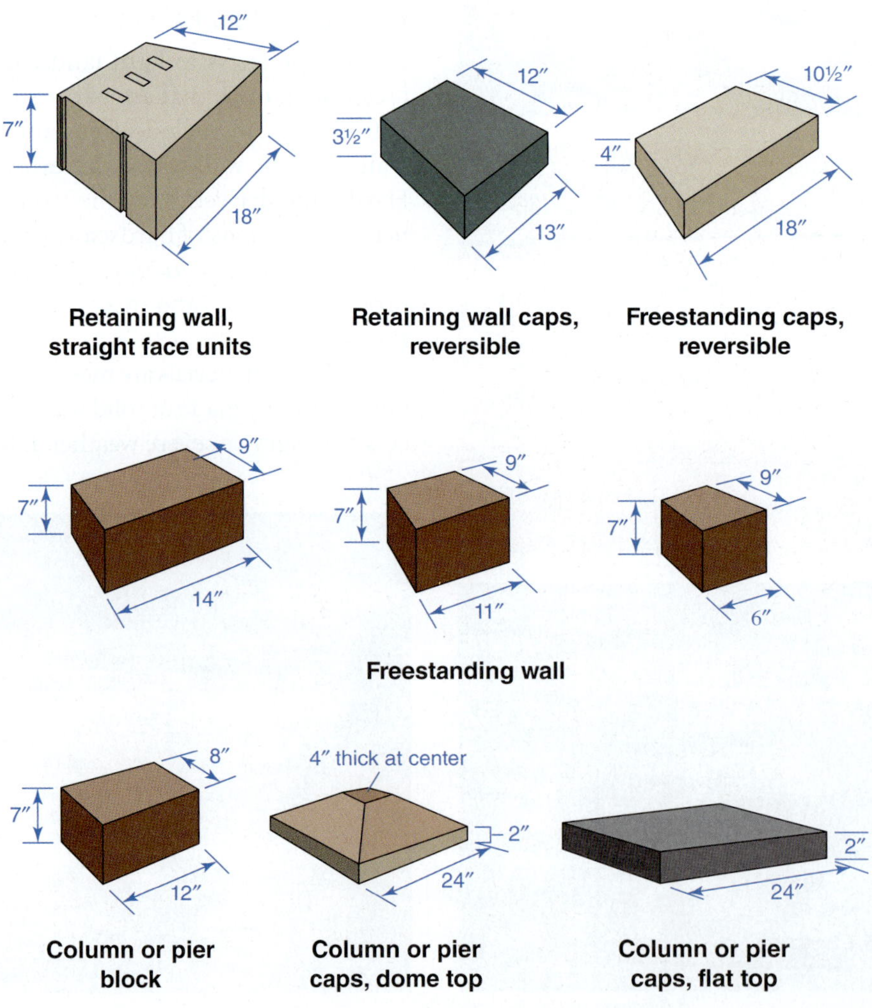

Goodheart-Willcox Publisher

Figure 8-23. Block sizes and shapes will vary depending on the block manufacturer and design type. Notice that the middle column or pier cap has a 5″ × 5″ flat top, for a light fixture to be mounted.

Workplace Skills
Interviewing

The interview is an essential part of the hiring process, for both the employee and the employer. This step generally involves the first face-to-face meeting, so making a strong impression is key. The interview can often be a stressful experience for some, but with the right preparation, you will feel confident from beginning to end.

To prepare for a job interview, learn everything you can about the position and the company. Researching online or referring to those in your professional network may help you find information about the position or company. Additionally, prepare answers for common interview questions that you expect your interviewer to ask. Think about your short- and long-term goals, your strengths, your weaknesses, why you are interested in the job and working for this company, etc. Practice answering these questions by conducting a mock interview with a friend or colleague. Also, think of some questions that you would like to ask your interviewer. For instance, ask about the duties of the position you are applying for, company policies, or any on-the-job training required. Employers will appreciate your interest in the company, so be sure you have questions ready to ask.

Another essential step in preparing for an interview is wearing professional attire. Employers expect interviewees to present their best qualities in an interview, including their appearance. Remember, a good rule of thumb is to dress more conservatively than needed as opposed to dressing too casually. As the saying goes, "Dress for the job you want!"

full cores as regular blocks, then moisture could enter the voids to precipitate cracking from freezing and thawing temperatures. ***Retaining wall blocks*** allow changes to be made in grade elevation when the terrain has a large slope or a hill. See **Figure 8-24**. Retaining wall blocks are designed with extra strength, weight, and size since they will be holding back earth, gravel, and soil.

Certain hardscape blocks have voids on the underneath side. This serves to reduce the weight of the block, reduce the concrete material being used to form the block, and reduce the cost of the individual hardscape block. See **Figure 8-25**.

Mortar is not normally used as part of the hardscape process. Instead, specially manufactured mastic is placed on the top smooth surface of the block. This mastic adheres the individual hardscape components into a single structural mass. During the manufacturing process of the block, coloring agents can be added to the hardscape mixture to enrich the appearance of the final product. Some hardscape retaining wall blocks have a rough exposed face. This is achieved by cracking a larger block into two equal pieces to form two separate blocks. Once cracked, the two will have a mirror-like surface. The rough surface of the retaining wall block will be laid toward the exposed side of the wall to display the texture which will be viewed by the homeowner and others. See **Figure 8-26**.

Pier or column blocks are specially made to create a perfect square structure. Sizes of pier blocks vary depending on the manufacture but normally are around 7″ high, 8″ wide and 12″ long. In **Figure 8-27**, you can

Timothy L. Andera

Figure 8-25. This retaining wall block is upside down to show the two voids in the block. A pronounced lip was created during the manufacturing process. The lip is used to catch the backside of the previous course laid. This ensures that the block does not move forward after constructing the wall.

Elmar Langle/Shutterstock.com

Figure 8-24. A two-tiered retaining wall to aid in making grade changes for this property. Retaining wall blocks and wall cap blocks are used in this design.

Timothy L. Andera

Figure 8-26. Retaining wall blocks with the rough face to the front. The left block is upside down displaying the two voids and the lip. The right block shows the smooth surface to receive mastic and the next course of block.

Author's image of manufacturer, Concrete Products of New London/JME

Figure 8-27. The pier has five exposed courses and two buried courses. A domed cap of 2″ at the edge and 4″ in the center. A light fixture is attached to the cap. A hole has been drilled in the domed cap to allow an exterior electrical cable to be fed through the center of the pier to energize the fixture.

see two blocks that create the width of the pier. One block is laid lengthways and measures 12″ long and the other sideways being 8″. Adding the two blocks together creates a pier that is approximately 20″ wide. Laying the next two blocks in the course creates a pier that is 20″ × 20″. The pier cap is usually approximately 24″ square, creating a 2″ lip over the pier. Four pier blocks make up each course. The space in the center of the pier can be used to run electrical wiring. See **Figure 8-28**.

Freestanding wall blocks vary in size from one manufacturer to another. Short block unit lengths average 6″, medium length block 11″, and long block can measure 17″. See **Figure 8-29**. The width of the block varies, but 9″ is a common size. Heights of the units are usually around 7″. The weight of the units varies from 35 lb to 75 lb.

Freestanding walls can be combined with piers to create a unique design that acts as a divider between one landscape area and another. See **Figure 8-30**. Piers can be used as corners where two walls come together from different directions. Piers can also be added into long freestanding walls to enhance the look and also add strength to the wall structure.

Paving bricks come in many different sizes and colors depending on the application and usage. See **Figure 8-31**. Some pavers are manufactured for pedestrian traffic and others specifically for vehicle traffic. Paving brick should be laid flat on a subbase of aggregate that has been compacted to withstand the weight of the intended traffic.

Paving can be used for many applications, including enhancing a patio. Other applications include

Timothy L. Andera

Figure 8-28. The domed cap has been removed from a pier and it is resting on the freestanding wall. In this instance, it is low-voltage wiring attached to a cap light. Once the hardscape cap is replaced, the light will rest between the column and cap to light the walkway below.

A Author's image of manufacturer, Concrete Products of New London/JME

B Timothy L. Andera

Figure 8-29. A—Hardscape freestanding wall blocks in three different lengths, manufactured as solid units with tapered ends to accommodate laying and spacing. Each block has a split rough face on the front and back because both sides of the wall are exposed to viewing. B— Three freestanding blocks positioned to demonstrate how they would be placed in a hardscape wall.

driveways, using multiple brick colors or by creating designs through the use of different types of brick. See **Figure 8-32**.

Sound Block

A sound block is made to absorb sound. Molded or sawed openings conduct the sound into the cores and absorb it. See **Figure 8-33**. Sound blocks are often used in schools and universities, gymnasiums, plants, subways, bowling alleys, or any place where noise is a problem.

Author's image of manufacturer, Concrete Products of New London/JME

Figure 8-30. The freestanding wall has three different lengths of blocks. The pier is constructed of one block size laid at right angles to each other to create individual courses. A 4″ pier cap is installed on the pier.

Author's image of manufacturer, Concrete Products of New London/JME

Figure 8-31. The width of the pavers are 6 1/2″, thickness is 2 3/8″, and lengths are 6 1/2″ and 9 1/2″.

Vladimir Trynkalo/Shutterstock.com

Figure 8-32. Paving brick used to create a driveway and sidewalk. Different colors of brick are used to accent the surfaces. The laying pattern of the brick changes from the narrow part of the driveway to where it widens.

Trenwyth Industries, Inc.

Figure 8-33. Sound block absorbs sound and is useful where noise is a problem.

photo: courtesy NRG Insulated Block, Buffalo, NY

Figure 8-34. Insulated blocks have high thermal performance because no concrete webs extend through the block to conduct heat.

Insulated Block

Concrete blocks are available with plastic foam inserts that increase their resistance to heat gain and heat loss. Some types have inserts that can be placed in the cores as the blocks are laid. Others have foam inserts cast into the block as it is made. See **Figure 8-34**. The foam insert in these blocks isolates one side of the block from the other, thus creating a better insulator. Typical webs are not present to transmit heat or cold from one side to the other. The exterior appearance of a building built with insulated blocks looks similar to regular masonry construction.

Portland Cement Association

Figure 8-35. Projecting concrete blocks produce a dramatic and decorative effect.

Pattern Bonds

In addition to decorative effects that center on characteristics of the individual block shape, texture, size, or form of the block, still other decorative effects can be achieved through pattern bonds. See **Figure 8-35** and **Figure 8-36**. Pattern bonds are classified as follows:

- Running bond
- Stacked bond
- Coursed or patterned ashlar
- Diagonal bond
- Basket weave
- Diagonal basket weave

Each classification includes several variations. The number of variations may be further increased through the use of different size blocks.

Sand-Lime Brick

Calcium silicate face brick (sand-lime brick), as the name indicates, is manufactured from sand and lime. A small amount of cement is sometimes added.

Among the advantages of sand-lime brick are uniform size, shape, and weight. These bricks are not burned. Shrinkage, warping, and color variation is not as great a problem as in clay and shale brick. To produce various colors, the manufacturer adds coloring to the mixture.

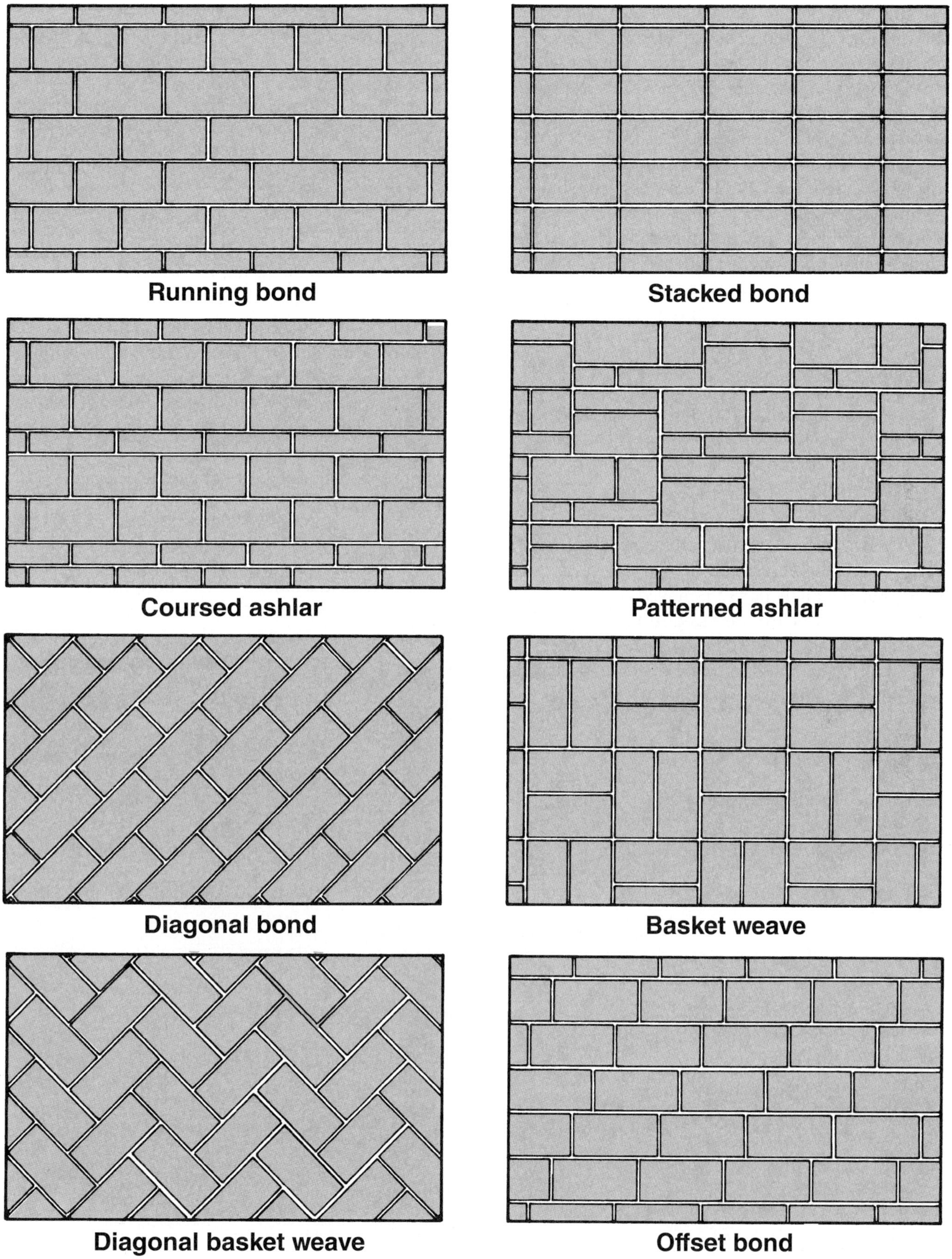

Goodheart-Willcox Publisher

Figure 8-36. Common concrete block bond patterns.

Sand-lime bricks are mainly used as a backup masonry product. The term *backup* refers to the masonry material behind the finished surface. There are three grades of sand-lime bricks—SW, MW, and NW. ***Grade SW*** includes brick that can be exposed to freezing temperatures and water saturation. This grade is used in foundations, retaining walls, and exposed piers. Grade SW bricks can be used almost anywhere that burned-clay bricks are used.

Grade MW bricks may be exposed to freezing conditions, but not to water saturation. Applications of this grade include wall faces in areas where the frost conditions are not too severe and where the climate is rather dry.

Grade NW includes bricks that may be exposed to light frost action with very little moisture present. This grade can be used in exterior wall construction in warm, dry climates.

Glass Block

Glass blocks are hollow, partially evacuated units of clear pressed glass, **Figure 8-37**. "Partially evacuated" means that the air is partly removed from inside the block. Glass blocks are produced in two sections and sealed to provide a dead airspace that acts as insulation. Square, rectangular, radial, and other special shapes are available.

Grynoid/Shutterstock.com

Figure 8-37. Glass blocks have rough edges to form a good bond with the mortar.

Code Note — Hollow Glass Units

The International Residential Code **R607.2** specifies that hollow glass units be partially evacuated and have a minimum average glass face thickness of 3/16" (5 mm). A polyvinyl butyral coating or latex-based paint is used to treat the surface of units in contact with mortar. The use of reclaimed units is not allowed.

The construction of glass block allows designs to be imprinted on both sides. These designs distribute and control the direction of light rays. This is one of the most important properties of glass block. Other advantages are good insulation value, low maintenance cost, and elimination of dust that comes through windows.

Mortar-bearing surfaces of glass block are corrugated (rippled) with a gritty, alkali- and moisture-resistant substance. This coating acts as a bond between the block and the mortar.

There are three categories of glass block:

- Functional
- General purpose
- Decorative

Functional glass block controls both the distribution and diffusion (scattering) of light. See **Figure 8-38**.

Goodheart-Willcox Publisher

Figure 8-38. Glass blocks are functional as well as decorative.

This type of glass block is often used in schools, hospitals, and basement walls.

General purpose glass block is installed where direct light transmission is required. Bathrooms, windows in homes such as entrance side panels, and interior partitions would require this type of glass block.

Decorative glass block is best when architectural design is important. This type of glass block may be used in applications such as the front entrance of an office building, decorative screens, corridor partitions, and stairwell walls.

Glass blocks are manufactured in three popular sizes—5 3/4 inches square, 7 3/4 inches square, and 11 3/4 inches square. See **Figure 8-39**. All of these sizes are 3 7/8″ thick. A 1/4″ mortar joint is required. The maximum panel area, using chase construction (reinforced masonry frame around a glass block panel), is 144 sq ft (maximum height is 20′; maximum width is 25′). The maximum panel area using wall anchor construction is 100 sq ft (maximum height is 10′; maximum width is 10′). Panel and curtain wall sections may be erected up to 250 sq ft if properly braced to limit movement and settlement.

8.3 Metric Concrete Masonry Units

Metric conversion for some materials, such as concrete masonry units, means inch-pound dimensions are stated in metric units, but the material sizes currently produced are not required to change. This is called a ***soft conversion***. A ***hard conversion*** requires a physical change in the size of the product.

Because concrete masonry uses soft metric conversion, the metric equivalents of concrete masonry unit dimensions are usually the exact metric conversions of the inch-pound unit dimensions. For example, an 8″ × 8″ × 16″ concrete block is considered to be 200 mm × 200 mm × 400 mm. These are the nominal dimensions specified by the Brick Industry Association. However, the exact metric equivalents are 203 mm × 203 mm × 406 mm for this nominal size block. The nominal size of a block includes the mortar joint.

Glass Block Layout Table

No. of Blocks	6″, 5 3/4 × 5 3/4 × 3 7/8	8″, 7 3/4 × 7 3/4 × 3 7/8	12″, 11 3/4 × 11 3/4 × 3 7/8
1	0′–6″	0′–8″	1′–0″
2	1′–0″	1′–4″	2′–0″
3	1′–6″	2′–0″	3′–0″
4	2′–0″	2′–8″	4′–0″
5	2′–6″	3′–4″	5′–0″
6	3′–0″	4′–0″	6′–0″
7	3′–6″	4′–8″	7′–0″
8	4′–0″	5′–4″	8′–0″
9	4′–6″	6′–0″	9′–0″
10	5′–0″	6′–8″	10′–0″
11	5′–6″	7′–4″	11′–0″
12	6′–0″	8′–0″	12′–0″
13	6′–6″	8′–8″	13′–0″
14	7′–0″	9′–4″	14′–0″
15	7′–6″	10′–0″	15′–0″
16	8′–0″	10′–8″	16′–0″
17	8′–6″	11′–4″	17′–0″
18	9′–0″	12′–0″	18′–0″
19	9′–6″	12′–8″	19′–0″
20	10′–0″	13′–4″	20′–0″
21	10′–6″	14′–0″	21′–0″
22	11′–0″	14′–8″	22′–0″
23	11′–6″	15′–4″	23′–0″
24	12′–0″	16′–0″	24′–0″
25	12′–6″	16′–8″	25′–0″

Goodheart-Willcox Publisher

Figure 8-39. This chart shows the course height of glass block in 6″, 8″, and 12″ sizes. For example, 12 courses of 8″ glass block with a 1/4″ mortar joint produces an 8′–0″ high wall. An 8′–0″ long wall also requires 12 blocks.

Chapter Review

Summary

- Concrete masonry units are made from a relatively dry mixture of Portland cement, aggregates, water, and other admixtures.
- Concrete bricks are solid brick-size units, solid concrete veneer, and facing units larger than brick size.
- Slump brick or block has a rounded face surface to simulate old adobe brick or block.
- Concrete blocks may be solid loadbearing, hollow loadbearing, solid nonloadbearing, or hollow nonloadbearing.
- Sand-lime brick is made from sand and lime to which a small amount of cement may be added.
- Glass blocks are hollow units of clear, pressed glass that is partially evacuated to provide insulating dead airspace.
- Soft conversion means that inch-pound dimensions are stated in metric units, but the material sizes currently produced are not changed.

Know and Understand

Answer the following questions using the information provided in this chapter.

1. One type of aggregate is _____.
   A. water repellent
   B. coloring agent
   C. Portland cement
   D. gravel
2. *True or False?* Aggregates normally make up about 50% of CMUs by weight.
3. _____ is the scrap left after the melting of metals.
   A. Fly ash
   B. Slag
   C. Scoria
   D. Gravel
4. The two grades of concrete bricks included in ASTM C55 are _____.
   A. M and N
   B. N and S
   C. C and S
   D. C and M
5. A hollow concrete block is one in which the core is greater than _____ of its total cross-sectional area.
   A. 10%
   B. 25%
   C. 40%
   D. 50%
6. *True or False?* A concrete block whose nominal dimensions are 8″ × 8″ × 16″ has an actual size of 7 5/8″ × 7 5/8″ × 15 5/8″.
7. Grade _____ bricks are used in areas where the frost conditions are not too severe and where the climate is rather dry.
   A. MW
   B. MN
   C. NW
   D. SW
8. _____ glass block is best when architectural design is important.
   A. Functional
   B. Decorative
   C. General purpose
   D. Architectural
9. *True or False?* Concrete blocks that have plastic foam inserts to increase their resistance to heat gain and heat loss are called insulating blocks.
10. *True or False?* The nominal dimensions of an 8″ × 8″ × 16″ concrete block in metric units is 200 mm × 200 mm × 500mm.

Apply and Analyze

1. Identify four reasons why concrete masonry is popular.
2. What are the three primary ingredients in concrete?
3. List the six main classifications of concrete masonry units.
4. What concrete masonry unit is similar in function to clay brick?
5. List the three classifications of concrete block units based on the overall weight of the unit.

6. Standard concrete blocks are produced in two-core and three-core designs. Identify four advantages of two-core blocks.
7. What are stretchers?
8. What are prefaced concrete masonry units?
9. Identify five typical pattern bonds used with concrete block.
10. Which grade of sand-lime brick may be exposed to freezing conditions, but not to water saturation?
11. What is glass block?
12. What are the three categories of glass blocks?
13. What is the difference between soft conversion and hard conversion when referring to metrics?

Critical Thinking

1. You are the owner of a small foundation company. A customer has contacted you to discuss constructing a foundation for a new home. The home will have a walkout basement. The area has severe winters where the temperature can reach 40 below zero or greater.
   A. Identify the type or types of concrete block you would recommend.
   B. List justifications you will discuss with the homeowner as to your recommendations.
2. A local high school has asked Edward Masonry Company to present to the freshman Career Class about the process of making concrete blocks. The president of Edward Masonry Company has assigned you to represent the firm. Use this chapter as a basis for developing your presentation. The internet also has valuable resources for you to use when discussing procedures, processes, and images of manufacturing concrete block units.

Activities

1. (SCIENCE) Research admixtures used in concrete. Write a report about the various types and functions of admixtures.
2. (TECH) Write a report describing the different types of lightweight aggregates and the applications of lightweight aggregate concrete.

Artazum/Shutterstock.com

CHAPTER 9
Stone

LEARNING OUTCOMES

After studying this chapter, you will be able to:

9.1-1 Identify stone based on the process of formation.
9.2-1 Recall the four grades of Indiana limestone.
9.3-1 Describe the various surface finishes possible with stone.
9.4-1 Distinguish between the three groups of stone wall patterns.
9.5-1 Identify the various applications of stone.
9.6-1 Apply manufactured stone to masonry surfaces.

TECHNICAL TERMS

acidic
ashlar
basic
coursed stone
dressed
gneiss
granite
Indiana limestone
limestone
magma
manufactured stone
marble
pointed up
quartzite
quoins
range
rubble
sandstone
schist
slate
traprock
uncoursed stone

Introduction

Stone is one of the oldest building materials. The use of stone as the sole building material for a structure has declined over the years. Today, stone is almost always used as a nonstructural material. Generally, it is applied as a facing, as a veneer, or for decorative purposes. See **Figure 9-1**. Other uses, such as sandwich and panel systems, are increasing.

Green Note
Stone as a Green Building Product

Because stone is an all-natural product, it is an environmentally friendly building material. Once quarried, stone can be used as a building product for structural and veneer applications. In the process of quarrying stone, the larger pieces can be used for construction, and the smaller materials crushed for paving or road applications. Stone can be reused very easily. For example, when an old building is being demolished, the stone can be salvaged and then used again, adding to its value as a green building product.

Eldorado Stone Corporation; Stone Products Corporation

Figure 9-1. Four different types of stone in traditional and modern applications.

9.1 Classification of Stone

Stone is divided into three categories based on the process of formation. See **Figure 9-2**. The categories are as follows:

- Igneous (volcanic)
- Sedimentary
- Metamorphic

Classification of Stone		
Igneous	**Sedimentary**	**Metamorphic**
Granite	Sandstone	Marble
Traprock	Limestone	Slate
—	—	Schist
—	—	Gneiss
—	—	Quartzite

Goodheart-Willcox Publisher

Figure 9-2. Stone is classified into three main types based on the process of formation.

The three categories are in many situations related. A given sample of rock may have resulted from more than one of the formation processes.

Stone is made up of minerals in various mixtures. Individual samples vary greatly in composition. These variations greatly affect the strength, color, texture, and durability of stone. This is especially true of marble, limestone, and sandstone. See **Figure 9-3**.

Igneous Stone

Igneous stone is classified by its texture, mineral content, and origin. All igneous stone comes from magmas generally found deep in the earth. ***Magma*** is a molten mixture of minerals. Igneous stone usually contains ferro-magnesian minerals and feldspar or quartz. When composed primarily of light minerals (quartz and potash feldspar), igneous stone is called ***acidic***. Acidic igneous stone is light in color and weight.

Igneous stone with a greater degree of ferro-magnesian minerals is called ***basic***. It is darker and heavier than acidic stone. In texture, igneous stone ranges from those with large crystals to glassy stone with no crystals at all. Two popular types of igneous stone are granite and traprock.

Building Stone Characteristics						
Characteristic	Igneous Rock, Granite	Igneous Rock, Traprock	Sedimentary Rock, Sandstone	Sedimentary Rock, Limestone	Metamorphic Rock, Marble	Metamorphic Rock, Slate
Color	Almost white to pink and white or gray and white gray, olive green to black	Dark gray to black; also dark olive green	Very light buff to light chocolate brown or brick red; may tarnish to brown	White, light gray to light buff	Highly varied; snow white to black; also blue-gray and light to dark olive green	Grayish-green, brick red or dark brown, usually gray; may be banded
Texture	Fine or coarse crystals, crystals may be varicolored	Usually coarsely crystalline but may be fine grained	Granular, showing sand grains, cemented together	Fine to crystalline; may have fossils	Finely granular to very coarsely crystalline showing flat-sided crystals	Finely crystalline; flat crystals give slately fracture
Weight	156 to 170 pcf	180 to 185 pcf	135 to 155 pcf	147 to 170 pcf	165 to 178 pcf	170 to 180 pcf
Hardness	Harder than limestone and marble; keeps a good edge	Similar to granite; keeps a good edge	Fairly hard if well cemented	Fairly soft and easily scratched by steel	Slightly harder than limestone	Softer than granite or quartzite; scratches easily
Strength	15,000–30,000 psi	20,000 psi	3000–20,000 psi	4000–20,000 psi	10,000–23,000 psi	10,000–15,000 psi
Chief Uses	Building stone and panels	Crushed stone	All building uses	All building uses	Building stone and panels	Roofing and paving

Goodheart-Willcox Publisher

Figure 9-3. Summary of the characteristics of popular building stones.

Granite

Granite is an igneous stone that is usually light colored. It is formed mainly of potash feldspar, quartz, and mica. Granite is hard and takes a good polish. It is the best known of the deeper igneous stone.

Fine granite has a salt-and-pepper pattern, but feldspar may redden it. Fine granite may be purchased in a broad range of colors, including white, gray, pink, green, and various combinations. See **Figure 9-4**.

Goodheart-Willcox Publisher

Figure 9-4. This fine granite sample has the typical grain pattern of most granites.

Traprock

Traprock is an igneous stone that is hard and durable. Traprock is the quarryman's term for diabase, basalt, or gabbro. This stone is used in construction on a limited basis because it contains iron minerals. The iron minerals create rusty stains as the stone weathers. Traprock is excellent as a crushed rock.

Sedimentary Stone

Sedimentary stone is formed in layers. Deposits of shell, disintegrated stone, or sand become cemented together under pressure to form the stone. Sedimentary stone differs widely in texture, color, and composition. See **Figure 9-5**. Two popular types are sandstone and limestone.

Sandstone

Sandstone is sedimentary stone composed mainly of grains of quartz cemented by silica, lime, or iron oxide. Silica cement can produce a hard, durable sandstone, but other cements are not as resistant. Porous sandstone may not weather well, and special treatment may be required in cold regions. Color and texture are available in such varieties as bluestone, brownstone, silica sandstone, lime sandstone, and many mixed varieties. See **Figure 9-6**.

Stone Products Corporation

Figure 9-5. Sedimentary stone varies in texture and composition.

hdere/istock/Getty Images Plus via Getty Images

Figure 9-6. Thin sandstones form an attractive facing for a fireplace.

The South Amherst quarries in Lorain County, Ohio are the largest sandstone quarries in the United States and are known for the quality of the stone produced. Stone from Amherst is generally light drab in color, homogeneous in texture, and composed of nearly pure silica. It is resistant to fire and weathering and is generally considered a fine building stone.

Limestone

Limestone is sedimentary stone that varies widely in color and texture. It consists mainly of the calcium carbonate mineral *calcite* and is usually marine in origin. About 70% of the calcite comes from the skeletal material that forms the framework grains, and about 20% comes from the cementing material that binds the grains together. The remainder of the rock, about 5%, is composed of porosity and small amounts of non-calcareous material. Limestone weathers rapidly in humid climates but very slowly where it is dry.

Several types of limestone are used in construction, including dolomitic, oolitic, crystalline, and travertine. A high-quality, widely used building stone is limestone from south-central Indiana, called ***Indiana limestone*** or *Bedford limestone*. See **Figure 9-7**.

Characteristics of Indiana limestone include uniformity of composition and structure. Indiana limestone is white, even-textured, and sometimes packed with small fossils. A high internal elasticity helps the stone to weather extreme temperature changes without damage.

Indiana limestone is easily machinable. It is a freestone, which means it does not have pronounced cleavage planes and can be easily cut in any direction. It does, however, have a grain running horizontally in the deposit. Stone is said to be set on its *natural bed* when it is set in a building with its grain running horizontally. Stone set with the grain running vertically is on *edge*.

Metamorphic Stone

Metamorphic stone is formed through reconstitution due to great heat and pressure. Changes may be barely visible or may be so great that it is impossible

Goodheart-Willcox Publisher

Figure 9-7. Bedford limestone from central Indiana.

to determine what the original stone once was. Metamorphic stone includes marble, slate, schist, gneiss, and quartzite.

Marble

Marble is recrystallized limestone that can be white, yellow, brown, green, or black. See **Figure 9-8**. Marble does not often develop the parallel bonding and mineral arrangement seen in slate and schist. Marble is a classic stone used for the finest work. See **Figure 9-9**. However, it is softer and less resistant to weathering than granite.

Slate

Slate is frequently a blue-gray color, but may be green, red, or brown. See **Figure 9-10**. Slate may be split into sheets used for roofing or flagstones. Most United States slate comes from Vermont, Maine, and Pennsylvania.

Schist

Schist is a rather coarse-grained stone that contains large amounts of mica. Several kinds of schist are classified according to the most characteristic mineral present. Examples of schist include mica schist, hornblende schist, chlorite schist, and quartz schist.

Gneiss

Gneiss, pronounced "nice," is difficult to define or describe because it is so varied. In general, it is a coarse-textured stone with minerals in parallel streaks or bands. It is relatively rich in feldspar and usually contains mica.

Goodheart-Willcox Publisher

Figure 9-8. Marble is available in a variety of colors and patterns.

Shelia Say/Shutterstock.com

Figure 9-9. The marble face of this fireplace displays a beautiful pattern.

pics721/Shutterstock.com

Figure 9-10. Two different examples of slate being used in homes. It is a favorite flooring material because it is easy to clean, resists soil and stains, and is extremely durable.

Quartzite

Quartzite is sandstone that has been recrystallized. The grain structure in quartzite is not nearly as clear as in sandstone. It is very hard and durable.

9.2 Grades of Stone

Stone is generally graded into four groups for construction purposes. These groups, in descending order of quality, are statuary, select, standard, and rustic. Fineness of grain or texture is the basis for this classification.

Classification of Indiana Limestone

The Indiana Limestone Institute of America (ILI) classifies Indiana limestone into two colors and four grades based on granular texture and other natural characteristics. See **Figure 9-11**. When specifying Indiana limestone, it is necessary to identify the color and grade required as well as the surface finish to be applied to the stone.

Color Descriptions

There are two color descriptions of Indiana limestone—buff and gray. Buff varies from a light creamy shade to a brownish buff. Gray varies from a light silvery gray to shades of bluish gray.

Grade Descriptions

ILI classifications are based on the degree of fineness of the grain particles and other natural characteristics of the stone. The structural soundness of each of the grades is essentially identical.

Indiana limestone naturally contains some fossils or shell formations, calcite streaks or spots, pit holes, reed formations, open texture streaks, honeycomb formations, iron spots, and grain-formation changes. The following are four grades recognized by the Indiana Limestone Institute of America based on these characteristics:

- **Select.** Fine to average grained stone having a controlled minimum of the above characteristics.
- **Standard.** Fine to moderately large grained stone permitting an average amount of the above characteristics.

Grade and Color Classifications of Indiana Limestone	
Buff	**Gray**
Select	Select
Standard	Standard
Rustic	Rustic
Variegated	Variegated

Indiana Limestone Institute of America

Figure 9-11. Grade and color classifications of Indiana limestone.

- **Rustic.** Fine to very coarse grained stone permitting an above average amount of the above characteristics.
- **Variegated.** An unselected mixture of grades Select, Standard, and Rustic permitting both the buff and gray colors.

Trade Tip
For the best possible color control, all stone for each project should be furnished from a single quarry.

9.3 Stone Surface Finishes

Many surface finishes are possible with stone, **Figure 9-12**. The following are popular stone finishes:

- **Gang sawed.** A moderately smooth surface with visible saw marks. It is inexpensive to produce because it requires no further finishing after leaving the saw. This finish may be used on all types of stone.
- **Shot sawed.** A variable or slightly rough finish according to the amount of shot used and the texture of the stone. This finish should be used with softer stone.
- **Machined (planer).** Smooth finish with some texture from tool marks. This finish may be used with most types of stone.
- **Machine tooled.** Finish that has two to ten grooves per inch. Grooves are parallel and concave in shape. This finish may be used on all types of stone but is expensive.
- **Plucked.** Surface obtained by rough planing, then textured by "plucking" out small particles of the stone. This finish is used mainly for limestone.

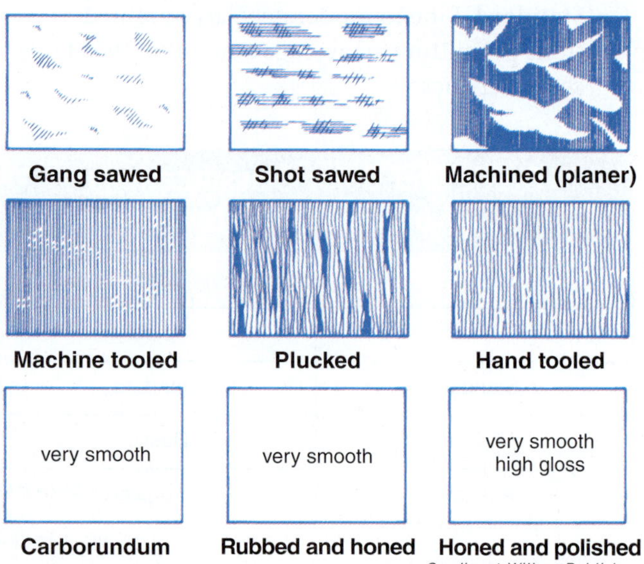

Goodheart-Willcox Publisher

Figure 9-12. Graphic symbols for typical stone finishes.

- **Hand tooled.** Finish that may be applied in a regular or random pattern. This finish is expensive and should be used only in instances that require a special accent.
- **Carborundum.** Smooth finish attained by using a carborundum machine rather than a planer. This finish is used primarily for limestone.
- **Rubbed and honed.** Smooth finish—smoother than the carborundum finish. This finish is most often used for interior marble and granite.
- **Honed and polished.** The smoothest of all finishes. This finish has a "high gloss" sheen and is used on marble and granite.

9.4 Stone Wall Patterns

Stone wall patterns fall roughly into three groups—rubble, roughly squared stone, and dimensioned, or ashlar. See **Figure 9-13** for examples of the patterns discussed in the following sections. When stones are laid, there are two common methods of placement. ***Coursed stone*** is stone laid in regular levels or courses to form a specific pattern or style. ***Uncoursed stone*** is stone laid in an irregular pattern where stones are placed randomly in the wall.

Rubble

Rubble is stone that consists of broken fragments of irregular shapes and textures. With appropriate planning and layout, rubble can be used to construct walls or other structures. Uncoursed field stone or random rubble (common rubble) consists of stones of many sizes and shapes as they are found in fields and streams. Even though there is no coursing, the stones are usually laid in a horizontal position.

Uncoursed cobweb or polygonal rubble are stones dressed with relatively straight line edges. The stones are selected to fit a place in the wall that has a particular shape. Mortar joints are approximately the same width.

Code Note **Rubble Stone Wall Thickness**
IRC **R606.4.2 Rubble Stone Masonry Walls** specifies the minimum thickness specification of rough, random, or coursed rubble stone walls as 16″ (406 mm).

Roughly Squared Stone

Roughly squared stone may be coursed or uncoursed. In an uncoursed pattern, both large and small stones may be used. The stones in a coursed pattern vary in size but are generally rectangular in shape, and horizontal coursing begins to appear.

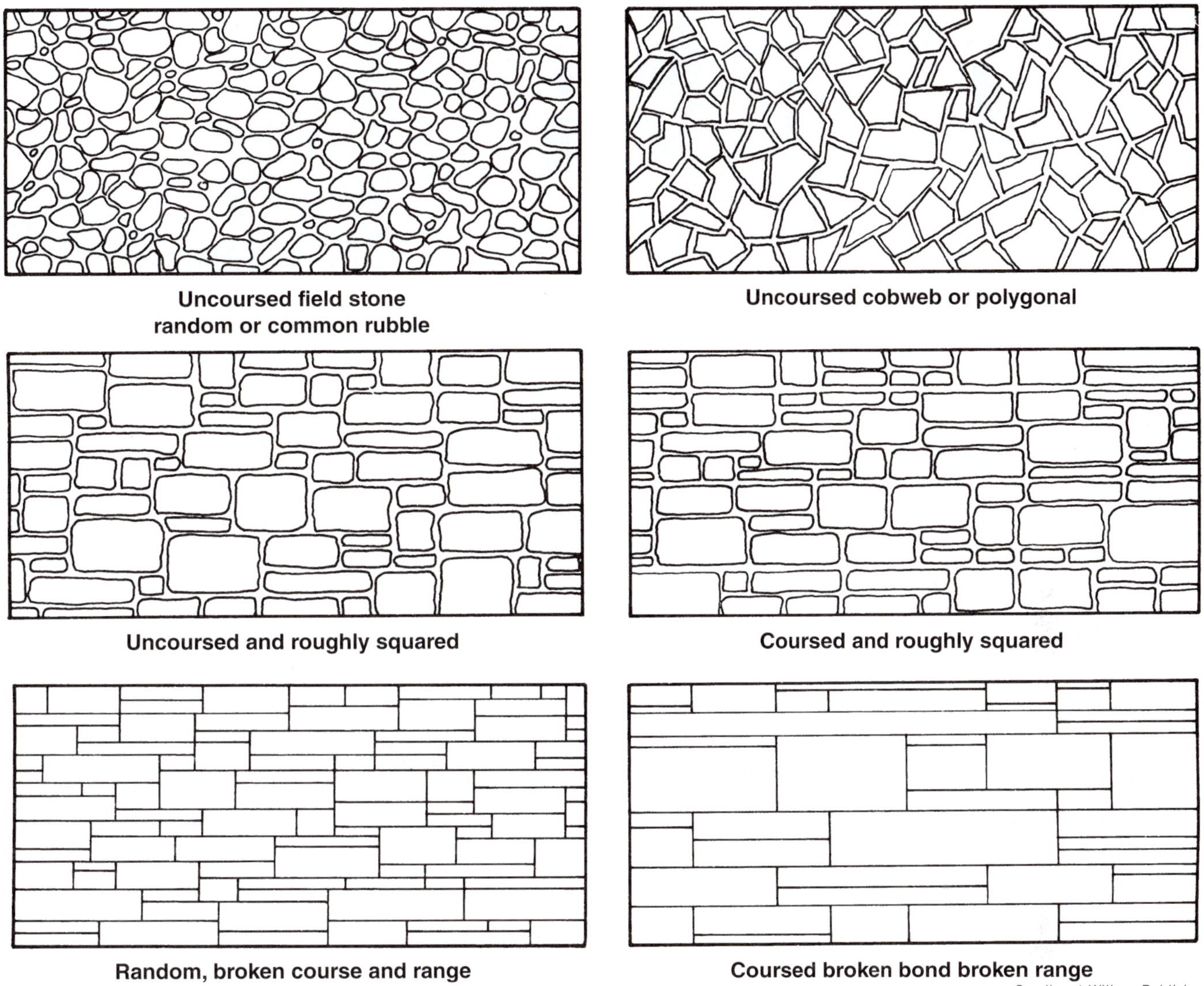

Goodheart-Willcox Publisher

Figure 9-13. Representative patterns used in stone masonry.

Ashlar

Dimensioned stone, or ***ashlar***, is cut stone of specific dimensions. It is cut, dressed, and finished to precise job requirements at the mill, and then transported to the site as a finished product. Stone that is ***dressed*** is shaped to the desired dimension and size with the face of the stone being smooth and ready for installation.

An ashlar pattern may be coursed or random. An example of a random pattern is the random, broken course and range pattern. In this pattern, no attempt is made to maintain coursing or range. ***Range*** is squared stones laid in horizontal courses of even height.

An example of a coursed ashlar pattern is the coursed broken bond, broken range pattern. This pattern keeps the basic coursing without maintaining it fully. Again, the range is broken.

9.5 Stone Applications

Stone was once widely used for foundations and exterior walls. However, stone masonry is slow and difficult due to the different sizes of stone, and in some styles, odd shapes. Therefore, stone has become mostly limited to veneer, trim, floor, and paving applications.

Limestone is an excellent sill material that can be used on exterior masonry walls. See **Figure 9-14**. The fact that it is one piece ensures better protection against water leakage.

Continuous courses of stone just above a foundation, or at other levels, can add to the appearance as well as the function of a masonry wall. A grade course of stone prevents moisture from the earth from entering the brickwork. The stone may function as a water table to divert water away from the foundation wall.

Goodheart-Willcox Publisher

Figure 9-14. A—Limestone slip sill in brick masonry. B—Limestone lug sill. C—Close-up detail of a limestone slip sill in a brick veneer wall.

Stone is still used extensively for copings (coverings or tops) on masonry walls. It prevents moisture from entering through the top. See **Figure 9-15**. Stone is also popular in the form of ***quoins*** (large squared stones set at the corners of buildings) for buildings constructed with certain architectural styles.

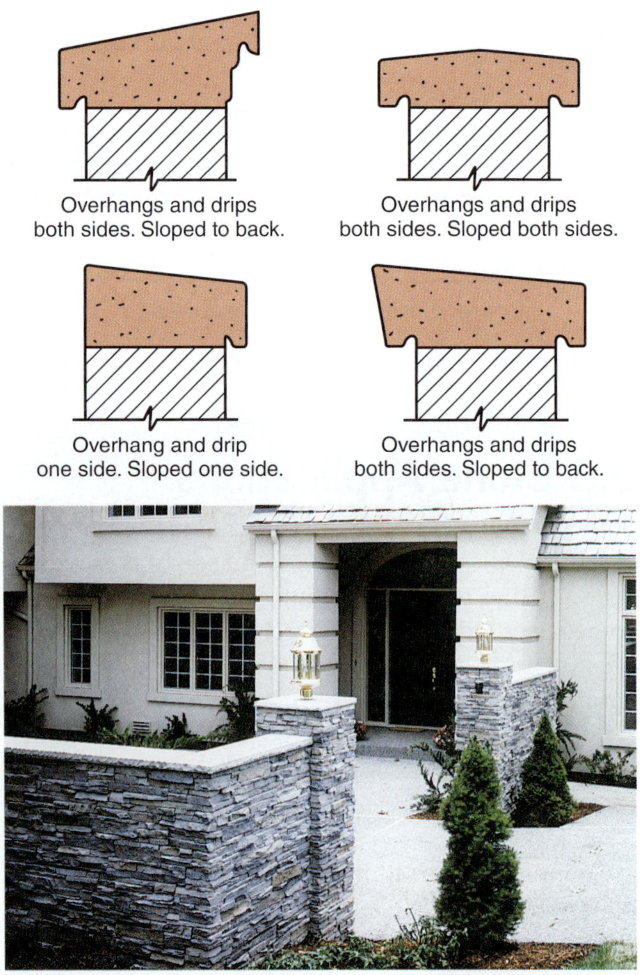

Eldorado Stone Corporation

Figure 9-15. Stone copings on top of a masonry wall prevent moisture from entering the wall.

Stone paving of patios, courts, entryways, and walks is also popular. See **Figure 9-16**. Slate and marble are favorites for paving because they provide a durable, hard surface that requires minimum maintenance.

9.6 Manufactured Stone

Manufactured stone is a simulated stone veneer made from lightweight concrete. It is colorfast, is weatherproof, and has the look and feel of natural stone. Produced in a variety of colors and textures, manufactured stone simulates various types of natural stone. See **Figure 9-17**.

Manufactured stone may be applied directly to a base coat of stucco, concrete block, brick, concrete, or any masonry surface that has not been treated or sealed and is rough enough to provide a good mechanical bond. See **Figure 9-18**. If the structure has outside corners, specially manufactured corner stones are employed to cover the corners. In this situation, the corner stone is the first unit laid on the wall and then other manufactured stones are trimmed and fitted around the corner stone. **Figure 9-19** shows the front and back side of a corner stone and regular stone. As the mason's work progresses, new corner stones are laid and regular stones fitted next to them.

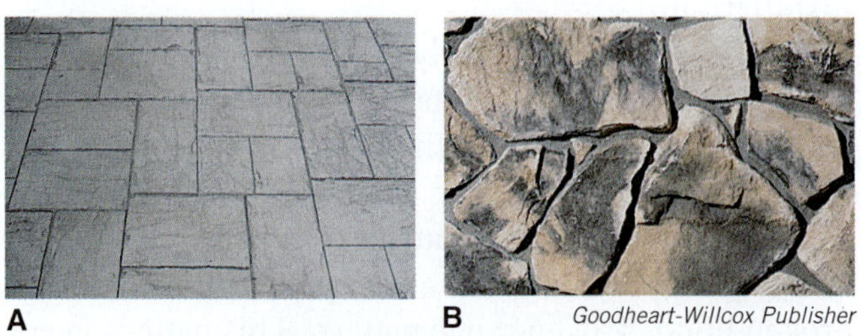

Goodheart-Willcox Publisher

Figure 9-16. A—Roughly squared stones create an attractive formal paving. B—Polygonal stone paving presents a more casual appearance.

Stone Products Corporation

Figure 9-17. Manufactured stone is produced in several colors and textures to simulate natural stone.

Stone Products Corporation

Figure 9-18. Manufactured stone being applied to a concrete block wall.

Wallboard, plaster, or wood surfaces must be covered with wire mesh or metal lath before manufactured stone is applied. If the structure is outside or exposed to moisture, prior to installing wire mesh or metal lath, two layers of water-resistant barrier (WRB) materials need to be applied to protect the structure from moisture. Painted or sealed stucco and masonry surfaces require the attachment of wire mesh, sand blasting, or a coating of suitable masonry adhesive.

Mortar used for manufactured stone should be a mixture of one part cement to three parts clean sand. To this, from 1/4 to 1/2 part lime, mortar cream, or fire clay is added. Packaged dry mix mortar that requires the addition of water also works well with manufactured stone. In any case, the mortar should be mixed to

Workplace Skills

Attention to Detail in Stonework Careers

This chapter can help you prepare for a possible career in this division of masonry. Becoming a stonemason will take talent, a strong back, and a good eye. As with many careers in the masonry industry, attention to detail is a crucial skill. Attention to detail includes observational abilities as well as patience. Surveying stones laying on the ground to select the correct stone can be an overwhelming task. The technique of laying stone is similar to completing jigsaw puzzles, selecting the right stone, and then trimming it to fit precisely into wall, one stone at a time. Stonemasons and apprentices take great pride in building their stone walls. With desire and talent, you can also become a stonemason.

Timothy L. Andera

Figure 9-19. A—Different color variations on the front side of a corner stone and regular stone. B—Back side of the stones have rough surfaces to aid in adhering mortar once it is applied.

a consistency similar to that of brick mortar. It may be applied either to the wall surface, the back of the stone, or both. Applying mortar to both ensures there is an adequate bond between the stone and the wall surface.

Code Note

Installation of Manufactured Stone

The Masonry Veneer Manufacturers Association (MVMA), Installation Guide for Adhered Concrete Masonry Veneer, identifies procedures and requirements for installation of manufactured stone and other related veneering materials. Refer to their publication about the installation and detailing options for manufactured stone veneer.

Manufactured stone should be laid out in a convenient location to the work area to ensure even distribution of sizes and colors. Laying out the stones will also allow the mason to sort and rearrange them as if a jigsaw puzzle was being built. Performing this process will lessen the trimming and cutting on the stones to reduce time in laying the manufactured stone wall. The stone can be cut with a brick trowel, hatchet, or similar tool to achieve sizes and shapes required for fitting stones and keeping mortar joints to a minimum.

While the mortar is still soft and pliable, the stone should be pressed into place with enough pressure so the mortar is squeezed out around the edges of the stone. This ensures a good bond between the stone and wall surface.

When the mortar joints become firm, the joints should be ***pointed up*** (worked with a wood or metal striking tool). The excess mortar should be raked out and the stones sealed around the edges to give the finished job the appearance of a natural stone wall.

Trade Tip

Mortar joints become firm after one to three hours, depending on climate and suction of the base.

At the end of each day, the finished wall should be broomed or wire brushed to remove loose mortar and to clean the face of the stone. A wet brush should not be used to treat the mortar joints for at least 24 hours. Using a wet brush too soon is likely to cause staining and discoloration that is difficult to remove.

Chapter Review

Summary

- The variations of minerals in stone affect its strength, color, texture, and durability.
- Based on the process of formation, stone is categorized as igneous (volcanic), sedimentary, or metamorphic.
- Two popular types of igneous stone are granite and traprock.
- Two popular types of sedimentary stone are sandstone and limestone.
- Metamorphic stone includes marble, slate, schist, gneiss, and quartzite.
- The four grades of Indiana limestone are Select, Standard, Rustic, and Variegated.
- Many surface finishes are possible with stone.
- Stone wall patterns include rubble, roughly squared stone, and ashlar (dimensioned).
- The use of stone has become limited to veneer, trim, floor, and paving applications.
- Manufactured stone may be applied directly to any masonry surface that has not been treated or sealed and is rough enough to provide a good mechanical bond.

Know and Understand

Answer the following questions using the information provided in this chapter.

1. The three categories of stone are igneous, sedimentary, and _____.
   A. granite
   B. limestone
   C. metamorphic
   D. slate
2. The _____ grade of Indiana limestone is characterized by fine to very coarse grained stone.
   A. Variegated
   B. Rustic
   C. Standard
   D. Select
3. The _____ surface finish is obtained by rough planing and is used mainly for limestone.
   A. gang sawed
   B. shot sawed
   C. plucked
   D. rubbed and toned
4. A moderately smooth surface with visible saw marks is a _____ finish.
   A. gang sawed
   B. shot sawed
   C. plucked
   D. rubbed and toned
5. The _____ finish is expensive and should only be used in instances that require a special accent.
   A. machine tooled
   B. hand tooled
   C. carborundum
   D. honed and polished
6. _____ is stone laid in regular levels to form a specific pattern.
   A. Coursed stone
   B. Uncoursed stone
   C. Rubble
   D. Field stone
7. Stone that is _____ is shaped to the desired size with the face of the stone being smooth and ready for installation.
   A. dressed
   B. coursed
   C. uncoursed
   D. dimensioned
8. _____ is an excellent sill material that can be used on exterior masonry walls.
   A. Ashlar
   B. Gneiss
   C. Traprock
   D. Limestone
9. *True or False?* Manufactured stone is not recommended for use because it is susceptible to all types of weather.

10. Mortar used for manufactured stone should be a mixture of _____.
    A. one part lime to three parts cement
    B. 1/4 part fire clay to one part cement
    C. 1/2 part lime to three parts clean sand
    D. one part cement to three parts clean sand

Apply and Analyze

1. Identify five metamorphic stones that are used in construction.
2. What metamorphic stone is frequently a blue-gray color and is used for roofing or flagstone?
3. Stone is generally classified according to what four grades?
4. List the four grades of limestone recognized by the Indiana Limestone Institute of America.
5. What are the two colors of Indiana limestone?
6. What is manufactured stone made of?

Critical Thinking

1. Explain why it is important to use coping for masonry walls.
2. As a foreman of a masonry crew, you have been directed to prepare a wood frame chimney prior to applying manufactured stone to the structure. What procedures do you need to take to prepare the plywood frame structure to receive manufactured stone?

Activities

1. (ELA) Research two famous buildings or structures constructed of Indiana Limestone. Also, try to find one building in your local community that has limestone as part of its design. Report on how the limestone was used in those structures.
2. (SCIENCE) List several exterior and interior applications of granite. What properties of granite make it desirable for these applications? What problems may occur due to the natural characteristics of granite and conditions of exposure?

Timothy L. Andera

CHAPTER 10 Mortar and Grout

LEARNING OUTCOMES

After studying this chapter, you will be able to:

10.1-1 Identify the basic ingredients of mortar.
10.2-1 Recall the proper methods of measuring mortar ingredients.
10.3-1 Mix mortar using appropriate procedures.
10.4-1 Distinguish between plastic (wet) mortar properties and hardened mortar properties.
10.5-1 Identify the properties and uses of the five types of mortars.
10.6-1 Explain how to color mortar.
10.7-1 Use appropriate methods to work with mortar in extreme weather.
10.8-1 Estimate the amount of mortar needed for a job.
10.9-1 Recall the three main purposes of grout.

TECHNICAL TERMS

bleeding
bond
curing
dryout
fireclay
grout
high-lift grouting
low-lift grouting
mortar
pigment
plasticity
repointing
retempering
slaked
tuckpointing
Type K mortar
Type M mortar
Type N mortar
Type O mortar
Type S mortar

Introduction

The bonding agent that ties masonry units into a strong, well-knit, weathertight structure is ***mortar***. It secures each of the units into a wall or other building element. Mortar must be strong, durable, and capable of keeping the wall intact. It must also help to create a water-resistant barrier. These requirements are influenced by the composition, proportions, and properties of mortar.

10.1 Mortar Materials

Historically, mortars have been made from a variety of materials. Burned gypsum and sand were used to make mortar in ancient Egypt, while lime and sand were used extensively in this country before the 1900s. Currently, basic mortar ingredients include Portland cement, hydrated lime, and sand. Masonry cement is sometimes used to replace Portland cement and hydrated lime or combined with Portland cement to make mortar. Most of these materials make a definite contribution to mortar performance.

Mortar serves four functions:

- It bonds the units together, sealing the spaces between them.
- It makes up for differences in sizes of the units.
- It provides bonding for metal ties or other types of reinforcement.
- It provides aesthetic (attractive) qualities by creating pleasing lines and color effects.

The material properties of mortar that influence the structural performance of masonry are compressive strength, bond strength, and elasticity. Because the compressive strength of masonry mortar is less important than bond strength, workability, and water retentivity, the latter properties should be given principal consideration in mortar selection.

ASTM International provides standards for materials commonly used in mortars. See **Figure 10-1**. Mortar is generally made up of cementitious (cement-like) materials together with sand and water.

Cementitious Materials

Cementitious materials include Portland cement (Type I, II, or III), masonry cements, and hydrated lime (Type S).

Portland Cement

Mortar is mostly Portland cement. Portland cement is a hydraulic material, meaning that it hardens under water. ASTM C150, *Standard Specification for Portland Cement*, covers eight types. Only the following three types are recommended for use in mortar:

- **Type I.** For general use when the special properties of Types II and III are not needed.
- **Type II.** For use when moderate sulfate resistance or moderate heat of hydration (the chemical reaction between water and cement) is desired.
- **Type III.** For use when high early strength is desired.

ASTM Standards for Mortar Materials	
Material	**ASTM Designation**
Portland Cement (Types I, II, III)	C150
Air-Entraining Portland Cement (Types IA, IIA, IIIA)	C175
Blended Cement (Types IS, ISA, IP, IPA, S, SA)	C595
Masonry Cement	C91
Quicklime	C5
Hydrated Lime	C207
Aggregate	C144

Goodheart-Willcox Publisher

Figure 10-1. ASTM standards for materials used in mortars.

Types IA, IIA, and IIIA are not recommended for masonry mortar because air entrainment reduces the bond between the mortar and masonry units or reinforcement. Types IS, ISA, IP, IPA, S, and SA are not recommended for masonry mortar unless strengths of masonry constructed with mortar containing such cements are first established by appropriate tests.

Masonry Cements and Premixed Mortars

Masonry cements usually contain Portland cement, ground limestone as a filler, and additives that provide workability, water retentivity, and air entrainment. The ingredients of masonry cements can vary widely among different brands. Therefore, brands of masonry cement should not be changed during the construction of a specific project. Each brand may have a slightly different composition and vary in color and performance. The mixing process for masonry cements or Portland cement–lime mortars must be consistent from batch to batch to ensure that the mortar being used is uniform throughout the building process.

> **Trade Tip**
> Masonry constructed with Portland cement–lime mortar exhibits greater flexural bond strength than masonry constructed with masonry cement mortar or air-entrained Portland cement–lime mortar of the same type. Masonry cement mortars may not be used in seismic zones 3 and 4.

Hydrated Lime

Hydrated lime is quicklime that has been slaked before packaging. ***Slaked*** means the quicklime has been formed into putty by combining it with water. This converts the calcium oxide to calcium hydroxide. Hydrated lime can be mixed and used immediately. For this reason, it is more convenient to use than quicklime. Quicklime, which is essentially calcium oxide, must be mixed with water and stored for as long as two weeks before using.

Hydrated lime is available in two types—S and N. Only Type S is recommended for masonry mortar because unhydrated oxides and ***plasticity*** (ease of working with and applying mortar) are controlled, which improves the mortar bond to the masonry units. Unhydrated oxides and plasticity are not controlled in Type N.

The use of lime in mortar is a primary contributor to the workability of the mortar. Lime provides the lubrication in the mortar and at the same time the stickiness needed to keep the mortar on the trowel until it is applied to a masonry unit. Lime also aids in the adherence of the mortar to the masonry unit once it has been applied.

Sand (Aggregate)

Sand is the primary aggregate used in masonry mortar. Either natural or manufactured sand may be used. There are recommended sand gradation limits for both natural and manufactured sand, **Figure 10-2**. If sand grains passing through a sieve fall within the percentage range given, the ASTM gradation standards are met. Well-graded aggregate retards (holds back) separation of material in a plastic mortar mix, resulting in reduced bleeding and improved workability. Sand that has too few fine particles generally produces harsh (difficult to spread) mortars. Sand with too many fine particles results in weak mortar.

Trade Tip

Bleeding is the separation of water during the time that the mortar is in the plastic state prior to setting. When preparing mortar, ensure that the correct amount of water is measured for the mixture. Excessive water in the mix can cause bleeding of the mortar.

Manufactured sand is made by crushing stone, gravel, or air-cooled blast furnace slag. Natural sand has rounder, smoother particles than manufactured sand. Manufactured sand, with its sharp and angular particle shapes, can produce mortars with workability properties different from mortars made with natural sand.

Trade Tip

Cementitious materials and aggregates should be stored in such a manner as to prevent deterioration or contamination by foreign materials. Cement or mortar should be placed on pallets or other raised surfaces to keep them dry. They should also be covered with a waterproof tarp or other waterproof material.

Recommended Sand Gradation Limits

Sieve Size	Percent Passing	
	Natural Sand	Manufactured Sand
No. 4	100	100
No. 8	95–100	95–100
No. 16	60–100	60–100
No. 30	35–70	35–70
No. 50	15–35	20–40
No. 100	2–15	10–25
No. 200	–	0–10

Goodheart-Willcox Publisher

Figure 10-2. Recommended sand gradation limits for both natural and manufactured sand.

Water

Water for masonry mortar is required by ASTM C270 to be clean and free of harmful amounts of acids, alkalis, or organic materials. Whether or not the water is potable (safe for drinking) is of no concern. However, water from city mains or private wells is generally suitable for mixing mortar.

10.2 Measuring Mortar Materials

When mixing mortar, consistency from mix to mix with respect to yield, workability, and color is important. This consistency can be achieved by using proper methods to measure the ingredients.

Even though proportions of aggregates are usually expressed in terms of loose volume, sand can vary greatly in volume due to moisture bulking. For example, fine sands can increase in volume by almost 40% when the percentage of moisture is only 5% by weight. Coarse sands are not affected as much, but may experience an 18% increase in volume under the same moisture conditions. ASTM C270 states that 1 cu ft of loose, damp sand contains 80 lb of dry sand. Most sand on the construction site contains from 4% to 8% moisture content.

The common practice of measuring sand by the shovel can result in oversanding or inconsistent batches of mortar. Measurement of sand by shovel should not be permitted. Cubic foot measuring boxes are recommended for the jobsite to provide accurate mixes.

Ingredients such as Portland cement, blended cement, masonry cement, and hydrated lime are sold in bags that are labeled by weight. This presents a problem because mortar is proportioned by volume. Add cement or lime by the bag.

Water can be measured accurately by using a pail. When the desired consistency is reached, the amount can be recorded and used each time.

10.3 Mixing Mortar

The ingredients of mortar must be thoroughly mixed to obtain good workability and performance of plastic (wet) and hardened mortar. Mortar on the jobsite can be mixed by machine or by hand.

Safety Note

When mixing mortar, it is important to wear safety glasses to protect your eyes. Wear protective gloves to shield your hands from chemicals and equipment. Use a respirator to protect against airborne silica and dust.

Hand Mixing

Hand mixing is usually done when the job is small and a mechanical mixer is not needed at the jobsite or when the mason determines it is easier to do the job by hand. Mortar boxes are made of steel or polyethylene and vary in sizes from 3 cu ft to 10 cu ft. The advantages of a high-density polyethylene mortar box are it does not rust and dry mortar is less likely to stick to the box. Hand mixing can also be done in a wheelbarrow. The following are the procedures for hand mixing mortar:

- Make sure the mortar box or wheelbarrow is secure.
- Measure the dry ingredients and place in the mortar box.
- Mix the dry ingredients with a mason's hoe. Work the material from one end of the box or wheelbarrow to the other several times.
- Add two-thirds to three-fourths of the water and mix until it is uniformly wet. Add additional water until the proper consistency is reached. See **Figure 10-3**.

Trade Tip
Add additional water to the mortar carefully. Remember that a little water can make a big difference in the mortar consistency.

- Allow the batch to stand for about five minutes.
- Mix again.

Author's image taken at Job Corps, Denison IA

Figure 10-3. An apprentice mixing mortar to the correct consistency using a mortar hoe and a wheelbarrow. Take note that the apprentice is wearing protective gloves, hi-vis vest, safety shoes, and safety glasses.

Trade Tip
When mixing mortar, a good technique is to place the mortar hoe or shovel in a bucket of water to stop it from drying out when not in use.

Machine Mixing

On larger jobs mixing mortar with the use of a machine can save time and increase productivity of the masonry company and its workers. Mortar mixers can be purchased in either gasoline or electric versions, depending on the masonry contractor's preference. See **Figure 10-4**. Most mortar mixers have a capacity from 4 cu ft to 7 cu ft and have a rotating spiral or paddle-blade design with a tilting drum. Ingredients should be mixed for three to five minutes. Too little mixing may result in poor workability, nonuniformity, low water retention, and too little air content. Too much mixing can reduce the strength of the mortar.

There is no one correct way to load the mixer. Some masons prefer to add about three-fourths of the water, one-half of the sand, and all of the cementitious materials, and then mix briefly before adding the remaining sand and water. Other procedures are also used. However, mixing is most effective when the mixer is charged to its capacity. Do not exceed the design specifications. Also, never begin a new batch of mortar when some mortar remains in the mixer from the previous batch.

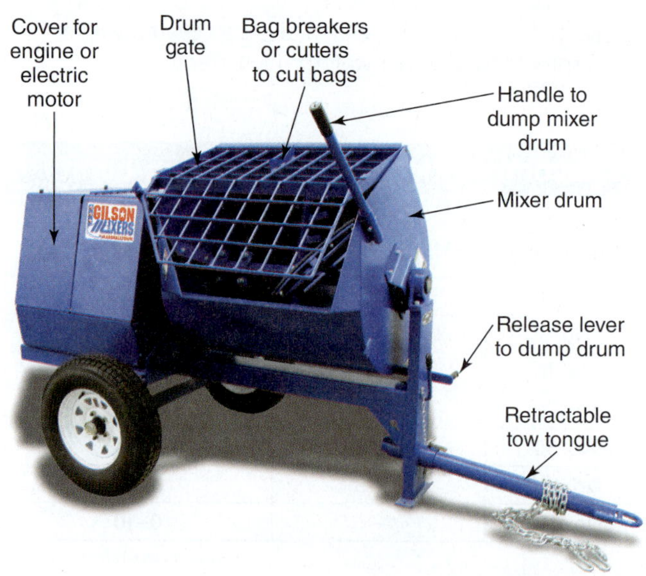

Marshalltown Company

Figure 10-4. Mortar mixer for large masonry projects.

Safety Note

When using a mortar mixer, always ensure that the safety grill is closed during the mixing process. The safety grill is made of steel bars and attached to the tub. The safety grill protects the person who is mixing the mortar from the turning paddles in the tub of the mortar mixer.

Safety and Operation Procedures for Power Mixers

Always read the manufacturer's operation manual and safety manual prior to running the machine or servicing the machine. While the following are basic rules found in all manuals, it is not a comprehensive list.

- When selecting a location for the mixer, find a level area to ensure the mixer does not tip over when emptying the mixture or operating the machine.
- Clean the area where the mixer is to be located to ensure tripping hazards are removed.
- Always wear safety glasses and ear protection when operating the mixer.
- Never add material while the drum of the mixer is stopped.
- Never reach into the drum while the mixer is operating. First, shut off the engine, or if electric, unplug the motor.
- Keep unauthorized people away from the machine.
- Never run the machine unattended.
- Never move the mixer while the motor is operating.
- If electric, never pull on the cord to disconnect. Use the plug-ins to separate cords.

Retempering Mortar

Mortar should be prepared at the rate it is being used to maintain best workability. Mortar that has stood for a while before being used may dry out and stiffen due to loss of water. Workability can be restored by adding the lost water and mixing briefly (***retempering***). Mortars that have stiffened because of evaporation of water from the mortar should be retempered by adding water as frequently as needed to restore the required consistency. Mortar that has begun to harden due to hydration should be discarded. Mortar should be used and placed in final position within 2 1/2 hours after mixing to avoid hardening due to hydration.

Code Note **Mortar and Glass Masonry**

The International Residential Code (IRC) **R607.8** specifies that glass masonry is to be laid with Type S or Type N mortar. The mortar should not be retempered after initial set. Mortar that is not used within 1 1/2 hours after remixing should be discarded.

10.4 Mortar Properties

Mortars have two distinct sets of properties—plastic (wet) mortar properties and hardened mortar properties. Plastic mortar properties are workability and water retentivity. Hardened mortar properties are compressive strength, bond strength, and durability.

Properties of Plastic Mortar

Plastic mortar is mortar that has just been mixed. It must be workable. Plastic mortar is uniform, cohesive, and of a usable consistency (easy to spread). It clings to vertical faces of masonry units, readily extrudes from the mortar joint, but does not drop or smear. The mortar also supports the weight of the masonry unit. See **Figure 10-5**.

Trade Tip

Water content has a great deal to do with the workability of masonry mortar. To determine if the mortar is of proper consistency, scoop up a medium load of mortar on a trowel, and then lightly "snap" the towel to set the mortar on the trowel. Then, slowly tilt the trowel until it is vertical. The mortar should stick to the trowel without slipping off.

Water retention is a mortar's ability to prevent rapid loss of mixing water to an absorptive masonry unit or to the air on a hot, dry day. Water retention aids in the reduction of bleeding mortar. Retaining moisture longer in the mortar improves the hydration process, which increases the bonding of the mortar to the masonry unit.

Portland Cement Association

Figure 10-5. A good workable mortar is soft but has good body. It adheres readily to masonry units and will not drop away while handling.

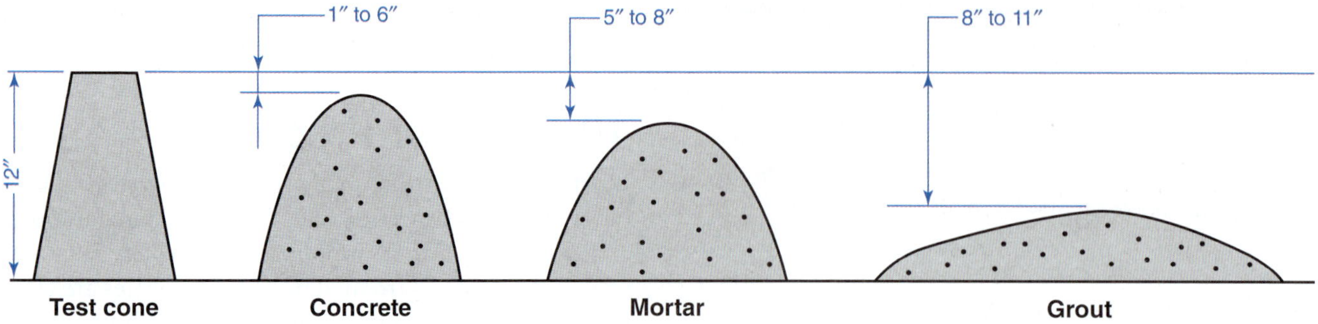

Figure 10-6. Slump test comparisons of concrete, mortar, and grout.

Water retentivity improves with higher lime content and the addition of fine sand. However, the mixture must be within allowable gradation limits.

Water content is a misunderstood aspect of masonry mortar, most likely due to the similarity between mortar and concrete materials. Mortar specification can be mistakenly based on the assumption that mortar requirements are similar to concrete requirements, especially with regard to the water-cement ratio. **Figure 10-6** shows a comparison of the desired slump for concrete, mortar, and grout.

Many specifications incorrectly require mortar to be mixed with the minimum amount of water consistent with workability. Often, retempering of mortar is also prohibited. These provisions result in mortars that have higher compressive strengths but lower bond strengths. Mixing mortar with the maximum amount of water consistent with workability provides maximum bond strength within the capacity of the mortar. Retempering is permitted, but only to replace water lost by evaporation. Water loss due to evaporation can usually be controlled satisfactorily by requiring that all mortar be used within 2 1/2 hours after mixing.

Properties of Hardened Mortar

The compressive strength of mortar increases as the cement content is increased and decreases as lime is increased. Air entrainment also reduces compressive strength. Structural failures due to compressive loading are rare. Bond strength is generally more critical.

Bond is the property of hardened mortar that holds masonry units together. Bond strength is the most significant property of hardened mortar, but it is also the most difficult to predict. Bond failure results from the masonry structure being subjected to these conditions:

- Eccentric (not exactly vertical) gravity loads
- Loads from earth pressure, winds, and other lateral (sideways) loads

Close contact between mortar and the masonry unit surface is essential for a good bond. To ensure a good bond, use of mortar with good workability is essential. Mortar should be placed in final position within 2 1/2 hours after the original mixing as required by ASTM C270.

Movement of a masonry unit after mortar has begun to harden destroys the bond. The mortar is not plastic enough to reestablish the bond.

Durability of mortar is measured primarily by its ability to resist repeated cycles of freezing and thawing under natural weather conditions. Mortars of high compressive strength have been found to have good durability.

Air-entrained mortar is durable and highly resistant to freezing. The tiny air bubbles absorb the expansive forces of freezing. Laboratory tests have shown that mortars with adequate air entrainment can withstand hundreds of freeze-thaw cycles, while other mortars soon fail.

10.5 Mortar Proportions and Uses

Choosing the right mortar type depends on the type of masonry and its location in the structure. No mortar produces the highest rating in all properties. Proportion adjustments in the mix to improve one property usually impairs other properties. **Figure 10-7** shows proper proportioning of materials.

The five mortar types recommended for various classes of masonry construction are designated by the letters M, S, N, O, and K. See **Figure 10-8**. The mortar designations are derived from every other letter in the word *masonwork*.

Type M Mortar

Type M mortar has high compressive strength and somewhat greater durability than other mortar types. It is useful for unreinforced masonry below grade and in contact with earth. Such structures include foundations, retaining walls, walks, sewers, and manholes.

Mortar Proportions by Volume			
Mortar Type	**Portland Cement* or Portland Blast Furnace Slag Cement** (parts by volume)**	**Masonry Cement (parts by volume)**	**Hydrated Lime or Lime Putty (parts by volume)**
M	1	1 (Type II)	–
	1	–	1/4
S	1/2	1 (Type II)	–
	1	–	Over 1/4 to 1/2
N	–	1 (Type II)	–
	1	–	Over 1/2 to 1 1/4
O	–	1 (Type I or II)	–
	1	–	Over 1 1/4 to 1 1/2
K	1	–	Over 2 1/2 to 4

* Types I, II, III, IA, IIA, IIIA
** Types IS, ISA
Data from ASTM C270

Goodheart-Willcox Publisher

Figure 10-7. Mortar types with the proper proportions of materials by volume. Note that for aggregates of any mortar type, measured in a damp loose condition, they should be not less than 2 1/2 and not more than 3 times the sum of the volumes of the cement and lime used.

Type S Mortar

Type S mortar is a medium-high-strength mortar. It is intended for use where Type M is recommended but where bond and lateral strength are more important than high compressive strength. Tests indicate that the tensile bond strength between brick and Type S mortar is near the best obtainable with cement-lime mortars. Type S is recommended for the following applications:

- Reinforced masonry
- Unreinforced masonry where maximum flexural (bending) strength is required
- Where mortar adhesion is the sole bonding agent between facing and backing (for example, ceramic veneers)

Mortar Types for Classes of Construction	
ASTM Mortar Type Designation	**Construction Suitability**
M	Masonry subjected to high compressive loads, severe frost action, or high lateral loads from earth pressure, hurricane winds, or earthquakes. Structures below grade, manholes, and catch basins.
S	Structures requiring high flexural bond strength, but subject only to normal compressive loads.
N	General use in above grade masonry. Residential basement construction, interior walls and partitions. Concrete masonry veneers applied to frame construction.
O	Nonloadbearing walls and partitions. Solid loadbearing masonry of allowable compressive strength not exceeding 100 psi.
K	Interior nonloadbearing partitions where low compressive and bond strengths are permitted by building codes.

Goodheart-Willcox Publisher

Figure 10-8. The type of mortar required is shown for a variety of classes of construction.

Type N mortar

Type N mortar is a medium-strength mortar suitable for general use in exposed masonry above grade. It is best suited for parapet walls, chimneys, and exterior walls that are subjected to severe exposure. Its principal property is weather resistance. Type N mortar may not be suitable in seismic zones 3 and 4.

Type O Mortar

Type O mortar is a low-strength mortar. It is suitable for general interior use in nonloadbearing masonry. Type O mortar can be used for loadbearing walls of solid masonry where compressive stresses do not exceed 100 psi (pounds per square inch). In such cases, however, exposures must not be severe. In general, do not use Type O mortar where it is subjected to freezing.

Code Note **Type O Mortar**

Type O mortar is not recommended by the Masonry Standards Joint Committee (MSJC) Code.

Type K mortar

Type K mortar is a very low-strength mortar. It is best used in interior nonloadbearing partition walls where high strength is not needed.

Code Note **Type K Mortar**

Type K mortar is not recommended by the Masonry Standards Joint Committee (MSJC) Code.

Fireclay Mortars

Certain applications require mortars that can withstand high temperatures. ***Fireclay*** is a type of clay that resists the effects heat can have on regular mortar. Fireclay can withstand temperatures up to 3000°F. High temperatures occur when combustible materials are burned in a fireplace or masonry stove. Fireclay can be used to create brick as well as fireclay mortars. These types of mortars are utilized in constructing fireplaces, chimneys, and stoves, **Figure 10-9**. For more information on fireclay mortars refer to ASTM C1655-06 *Standard Classification of Fireclay and High-Alumina Mortars.*

Nagy-Bagoly Arpad/Shutterstock.com

Figure 10-9. A mason is building a fireplace using fireclay and firebrick, both can withstand the temperatures associated with a fireplace.

Educational Mortar

Educational or training mortar is used when laying brick, block, and stone. The advantage of this type of mortar is that it can be reused again and again. Students complete individualized projects with guidance from their instructor. Once the student completes a specific project, the instructor evaluates the work to determine advancement to the next project. After successfully completing the project, the student dismantles it by removing the brick and block and scraping the educational mortar from the masonry units.

Admixtures

No air-entraining admixtures or cementitious materials containing air-entraining admixtures or agents shall be used in the mortar. Antifreeze compounds, accelerators, retarders, water-repellent agents, or other admixtures should not be added to mortar. Calcium chloride or admixtures containing calcium chloride should not be added to mortar. Mortar colors may be added to the mortar as specified.

Specific Mortar Uses

The following specific uses are recommended based on the properties of the various mortars.

Cavity Walls

For cavity walls in areas where wind velocities may exceed 80 mph, use Type S mortar. For locations where lesser winds are expected, use Type S or Type N mortars.

Facing Tile

For facing tile, any mortar type except Type K is acceptable. Where 1/4″ joints are specified, all aggregate (sand) should pass through a No. 16 sieve. For white joints, use white Portland cement and white sand in construction mortar. An alternate method is to rake all joints to a depth of 1/2″ and point or grout with mortar of the desired color.

Tuckpointing and Repointing Mortar

Tuckpointing is the process of filling a mortar joint with fresh mortar. Over time, masonry joints may be affected

Susan Law Cain/Shutterstock.com

Figure 10-10. A mason is using a tuckpointing trowel along with a margin trowel to hold the mortar while filling in the joint with new mortar.

by the age of the mortar and weather conditions. The life expectancy of mortar joints varies depending on environmental conditions and the quality of the original mortar. Mortar can last 30 years or longer, while brick may last over 100 years. Tuckpointing is performed to extend the life of masonry. Tuckpointing is one of the last steps in the process of repairing damage to masonry joints. See **Figure 10-10.**

Repointing is the complete process of repairing a defective masonry joint. Repointing starts with cutting out or removing defective mortar from a joint. The removed mortar is then replaced with fresh mortar. It is important to repoint mortar joints once they are determined to be defective, because damaged, cracked, or deteriorated joints allow moisture to enter the brick face. Moisture entering the brick wall over time will start to destroy the integrity of the masonry structure. The depth of the mortar removed should be equal to or exceed two times the thickness of the original joint. After removing the old mortar, it is necessary to clean the area prior to tuckpointing.

Safety Note

Repointing involves the use of joint chisels and electrical masonry grinders to remove old mortar joints. Flying debris is a safety hazard. When repointing, always wear eye protection, a face shield, and ear protection. It is also important to wear a respirator because old mortar may contain silica, which is a severe health hazard if inhaled.

The following are typical portions by volume for mixing mortar to be used for tuckpointing:

- Type N—1 part Portland cement, 1 part hydrated lime, and 6 parts sand
- Type O—1 part Portland cement, 2 parts hydrated lime, and 9 parts sand
- Type K—1 part Portland cement, 4 parts hydrated lime, and 15 parts sand

Trade Tip

Two documents that address mortar are ASTM C270 *Standard Specifications of Mortar for Unit Masonry*, and The Brick Industry Association's Technical Note 8A.

Coloring the mortar may be required to match the existing mortar on the original structure. ***Pigment*** is a coloring agent that can be added to replacement mortar to make it match the original mortar.

Prior to tuckpointing, perform a color match test to determine if the new mortar matches the existing mortar. Prepare a sample of replacement mortar with pigment added to be used as a test material. Wet a section of the existing wall as a test area, and then allow the test area and mortar sample to dry. Compare the mortar sample with the existing mortar to determine if the color blended sample is acceptable. If not, perform an adjustment to the mixture and repeat the test until an acceptable mixture is created.

Dirt-Resistant Mortar

When a dirt-resistant or stain-resistant mortar is required, add aluminum tristearate, calcium stearate, or ammonium stearate to the mortar. The amount added should be 3% of the weight of the Portland cement.

Grout

Grout is used in brick masonry to fill cells of hollow units or spaces between wythes of solid unit masonry. Grout increases the compressive, shear, and flexural strength of the masonry element and bonds steel reinforcement and masonry together. Grout that is used in brick or structural clay tile masonry should conform to the requirements of ASTM C476, *Specification for Grout for Masonry*.

Code Note **Grout**

The International Residential Code (IRC) **R606.2.12** specifies that Type M or Type S mortar can be used as grout when a sufficient amount of water has been added to produce a pouring consistency.

The amount of mixing water and its migration from the grout to the brick or structural clay tile determines the compressive strength of the grout and the amount of grout shrinkage. Grouts with high initial water content exhibit more shrinkage than grouts with low initial water content. Consequently, a nonshrink grout admixture is recommended to minimize the number of flaws and shrinkage cracks in grout while still producing a grout slump of 8″ to 11″, unless otherwise specified.

MSJC specifications require grout compressive strength to be at least equal to the specified compressive strength of masonry, but not less than 2000 psi (13.8 MPa) as determined by ASTM C1019, *Test Method of Sampling and Testing Grout.* In general, the compressive strength of ASTM C476 grout by proportions is greater than 2000 psi (13.8 MPa).

10.6 Coloring Mortar

Mortar can be colored with aggregates or pigments. Colored aggregates are preferred. White sand, ground granite, marble, or other stone usually have permanent color and do not weaken the bond or mortar. For white joints, use white sand, ground limestone, or ground marble with white Portland cement and lime.

Mortar pigments must be capable of resulting in the desired color when used in permissible quantities. The pigments must also be sufficiently fine to disperse throughout the mix and be nonreactive with other ingredients. These requirements are usually met by metallic oxides. Iron oxides, manganese oxides, chromium oxides, carbon black, and ultramarine blue have been used successfully as mortar colors.

Trade Tip

Do not use organic colors, particularly those containing Prussian blue, cadmium lithopone, and zinc and lead chromates. Paint pigments may not be suitable for mortars.

Use the minimum quantity of pigments necessary to produce the desired results. Too much can impair strength and reduce durability. Carbon black, a popular pigment, impairs mortar strength when used in amounts greater than 2% or 3% of the cement weight.

Trade Tip

For best results, dry mix color with Portland cement in large controlled quantities. This ensures a more uniform color blend than when mixing smaller quantities at the jobsite.

10.7 Climactic Conditions

The ideal temperature of mortar is between 60°F and 80°F (15.6°C and 26.7°C). Masonry work performed at temperatures outside the ideal conditions requires special procedures to ensure that the materials perform to manufacturer's specifications. Because of the special procedures necessary to protect masonry when working in cold and hot conditions, check local or state building standards. Special precautions should be implemented to protect masonry from extreme weather conditions.

Code Note Cold and Hot Weather Masonry

The International Building Code (IBC) refers to provisions found in MSJC specifications of TMS-602/ACI 530.1/ASCE 6 pertaining to precautions needed to protect masonry when the ambient temperature falls below 40°F (4.4°C) or rises above 100°F (37.8°C).

Cold Weather

Cold weather conditions may require heating of materials to ensure appropriate bonding and curing. If temperatures are at or below freezing, enclosures may be necessary to protect newly laid masonry units from freezing.

Heating the mixing water and heating the aggregate are two of the easiest methods of raising mortar temperature. However, the water or sand should not be heated above 140°F (60°C) to prevent the danger of "flash set," which occurs when the heated water or sand comes in contact with the cement that is part of the mix.

Workplace Skills

Networking

Networking is a great way to learn about different career opportunities in a field and can often help one find employment. *Networking* is the exchange of information or services among individuals or groups. It is often daunting in any professional field for newcomers who are looking for work. Networking can help narrow down your job search and learn about possible job opportunities. A great resource to begin your search is on social networking sites. Many companies use these sites to promote their services, job openings, and communicate with other professionals in their field. Job fairs are another great resource for networking. Many companies attend these events for the purpose of bringing on new talent.

Trade Tip

Flash set is undesirable rapid hardening of Portland cement paste, mortar, or concrete. Plasticity cannot be restored by further mixing without adding water. To avoid flash set, keep materials covered in cold weather and do not heat materials above recommended temperatures.

In freezing weather, moisture in the sand turns to ice. The sand must be thawed before using. Never use sand that has been scorched from overheating.

In the past, calcium chloride has been widely used to speed up the setting time of mortar and grout. This method is not recommended because it causes corrosion of metals used in masonry such as ties, anchors, and reinforcement.

Antifreeze materials should not be used to lower the freezing point of mortar. Such large quantities would be required that the mortar strength and other properties would be seriously impaired.

The following cold and hot weather construction information has been referenced from MSJC *Building Code Requirements and Specifications for Masonry Structures* (TMS 602-13/ACI 530.1-13/ASCE 6-13) Specifications 1.8 – Project Conditions.

Cold Weather Construction Procedures

When the air temperature falls below 40°F (4.4°C), follow these procedures to protect masonry:

- Do not heat water or aggregates that will be used for mortar or grout above 140°F (60°C).
- Do not lay glass units when temperatures are below 40°F (4.4°C).
- Remove ice and snow from existing masonry to prepare for new work. Heat the surfaces where new masonry will be laid to a temperature above freezing. Use methods that will not damage existing masonry.
- Do not lay masonry units that have a temperature below 20°F (–6.7°C) or that contain ice or snow.

When the temperature is from 40°F to 32°F (4.4°C to 0°C), follow these procedures:

- For mortar, heat sand and water to a temperature to produce a mix not above 120°F (49°C).
- Do not heat grout materials unless they are below 32°F (0°C).

When the temperature is from 32°F to 25°F (0°C to –3.9°C), follow these procedures:

- For mortar, heat sand and water to a temperature to produce a mix not above 120°F (49°C).
- Maintain mortar temperature above freezing until it is used in masonry.
- For grout, heat aggregates and water to a temperature not above 120°F (49°C) at the time of mixing.
- Maintain grout temperature above 70°F (21.1°C) until the time of placement.

When the temperature is from 25°F to 20°F (–4°C to –6.7°C), comply with the requirements for temperatures from 32°F to 25°F (0°C to –3.9°C), as well as the following:

- Heat masonry surfaces where new units are to be placed to a temperature level above 40°F (4.4°C).
- Before grouting masonry units, heat the masonry to a minimum of 40°F.
- Use a windbreak or enclosure when winds exceed 15 mph.

When the temperature is 20°F (–6.7°C) and below, comply with the requirements for temperatures 25°F to 20°F (–4°C to –6.7°C), as well as the following:

- Build an enclosure to protect masonry and use auxiliary heat to maintain the temperature in the area at a level above 32°F (0°C) for at least a 24-hour period of time.

Masonry must also be protected after it is erected. In cold weather, follow these procedures:

- Keep newly laid glass units above 40°F (4.4°C) for 48 hours.
- 40°F to 25°F (4.4°C to –3.9°C)—Keep newly laid masonry covered with weather-protective materials for 24 hours.
- 25°F to 20°F (–3.9°C to –6.7°C)—Keep newly laid masonry covered with weather-protective insulating blankets for 24 hours. For grouted masonry, extend the period of time of protection to 48 hours.
- 20°F (–6.7°C) and below—Maintain masonry above 32°F (0°C) for at least 24 hours by constructing enclosures and heaters or other acceptable means. For grouted masonry, extend the time of protection to 48 hours.

Hot Weather

Extreme temperatures and wind can affect the bonding and curing of newly laid masonry units. Masonry units that are hot from being stored in the sun or extreme temperatures absorb more water from mortar.

Trade Tip

In hot weather, keep units and mixing materials in shaded areas or covered. Use cool or cold water to reduce the temperature of the sand when mixing a batch. Do not spread mortar beds more than 4′ ahead of masonry units. Lay units within one minute after spreading mortar.

Dryout is the process of mortar or grout losing moisture to extremely hot and dry masonry units or brick. Dryout adversely affects the strength of the bond when the masonry is curing.

Hot Weather Construction Procedures

When the ambient air temperature exceeds 100°F (37.8°C), or 90°F (32.2°C) with a wind velocity greater than 8 mph, follow these procedures prior to performing masonry work:

- Keep sand damp and in loose condition, covered or located in a shady area.
- Implement procedures to produce mortar having a temperature below 120°F (49°C).

When the ambient air temperature exceeds 100°F (37.8°C), or 90°F (32.2°C) with a wind velocity greater than 8 mph, follow these procedures while performing masonry work:

- Keep mortar and grout at temperatures below 120°F (49°C).
- Rinse mixer, wheelbarrow, and mortar boards with cool water before mixing or placing mortar or grout.
- Retemper mortar to original consistency with cool water when needed.
- Mix mortar in small batches so it is used within two hours of mixing.

When the air temperature exceeds 115°F (46°C), or 105°F (40.6°C) with a wind velocity greater than 8 mph, follow these procedures:

- Use cool water for mixing mortar and grout.
- Add ice to the mixing water to reduce temperature. The ice must be melted prior to the mortar being used.

Protect newly laid masonry in hot weather. When the average daily temperature exceeds 100°F (37.8°C), or 90°F (32.2°C) with a wind velocity greater than 8 mph, mist newly laid masonry until units are damp. Repeat the misting three times a day. Continue misting until the masonry is three days old.

10.8 Estimating Mortar Quantities

Estimating the amount of mortar needed for a specific job is difficult for a novice. Most estimates are too low. Job delays can result from poor estimates, and waste occurs when too much material is ordered. The objective is to order the proper amount of all ingredients. Many students are surprised to learn that 0.33 cu ft of cementitious material and 0.99 cu ft of sand makes 1 cu ft of mortar.

Estimating Mortar for a Concrete Block Wall

The amount of mortar needed for 100 sq ft of wall for various sizes of concrete masonry units is shown in **Figure 10-11**. There are practical reasons to order a greater amount of mortar than calculations would indicate. Extra mortar may be needed for the following reasons:

- Shell thickness may be 50% greater at the top than at the bottom, thus requiring more mortar.
- Some mortar falls to the ground as droppings.
- Some masons just use more mortar.
- Some mortar sticks to the mixer, wheelbarrow, and mortar board.
- Some sand cannot be used because it was dumped on the ground.
- Some cement, lime, or mortar is spilled, lost, or spoiled.
- Mortar left at the end of the day, or at lunchtime, must be discarded.
- Weather turns bad before mortar can be used.
- Human error causes waste.
- The mortar is rejected—the mix is incorrect.
- Variation in masonry units may require more mortar.
- Cutouts for plumbing or electrical are closed with mortar.
- The first course requires a full mortar bed.
- Solid cores around anchor bolts require more mortar.
- Door and window jambs are filled with mortar.
- Change orders may require more materials.

Mortar Needed for Concrete Masonry Units

Nominal Height and Length of Units in Inches	Cubic Feet of Mortar per 100 Sq Ft
8 × 16	6.0
8 × 12	7.0
5 × 12	8.5
4 × 16	9.5
2 1/4 × 8	14.0
4 × 8	12.0
5 × 8	11.0
2 × 12	15.0
2 × 16	15.0

National Concrete Masonry Association

Figure 10-11. Mortar requirements for 100 sq ft of single wythe wall area. Note that this chart shows considerably more mortar needed than does the calculation method.

Estimating Mortar for a Single Wythe Brick Wall

Estimating the mortar required for a typical single wythe brick wall may be more confusing than estimating for a concrete block wall because more sizes exist and mortar coverage is different. Charts can be used for estimating mortar. See **Figure 10-12** and **Figure 10-13**. Rule-of-thumb estimates for masonry units, mortar, and sand are shown in **Figure 10-14**.

10.9 Grout

Grout is a primary element in reinforced brick and block masonry construction. It is generally used in reinforced loadbearing masonry walls or in spaces that contain steel reinforcement. The grout bonds the steel and masonry units together so they function as a single unit to resist imposed loads. Grout may also be used in some or all of the cores in nonreinforced loadbearing masonry walls to provide added strength.

Grouting brick and block walls accomplishes three main purposes:

- Bonds the wythes of masonry together
- Transfers stress from the masonry to the reinforcing steel when lateral forces are present
- Increases the cross-sectional area of the wall to aid in the support of vertical loads and lateral shear loads

Nonmodular Brick				
Size of Brick in Inches t × h × l	3/8″ Mortar Joints, Number of Brick per 100 Sq Ft	3/8″ Mortar Joints, Cubic Feet Mortar per 100 Sq Ft	1/2″ Mortar Joints, Number of Brick per 100 Sq Ft	1/2″ Mortar Joints, Cubic Feet Mortar per 100 Sq Ft
2 3/4 × 2 3/4 × 9 1/4	455	3.2	432	4.5
2 5/8 × 2 3/4 × 8 3/4	504	3.4	470	4.1
3 3/4 × 2 1/4 × 8	655	5.8	616	7.2
3 3/4 × 2 3/4 × 8	551	5.0	522	6.4

Brick Industry Association

Figure 10-12. Nonmodular brick and mortar required for single wythe walls in running bond with no allowances for waste or breakage.

Modular Brick			
Nominal Size of Brick in Inches t × h × l	Number of Brick per 100 Sq Ft	3/8″ Joints, Cubic Feet of Mortar per 100 Sq Ft	1/2″ Joints, Cubic Feet of Mortar per 100 Sq Ft
4 × 2 2/3 × 8	675	5.5	7.0
4 × 3 1/5 × 8	563	4.8	6.1
4 × 4 × 8	450	4.2	5.3
4 × 5 1/3 × 8	338	3.5	4.4
4 × 2 × 12	600	6.5	8.2
4 × 2 2/3 × 12	450	5.1	6.5
4 × 3 1/5 × 12	375	4.4	5.6
4 × 4 × 12	300	3.7	4.8
4 × 5 1/3 × 12	225	3.0	3.9
6 × 2 2/3 × 12	450	7.9	10.2
6 × 3 1/5 × 12	375	6.8	8.8
6 × 4 × 12	300	5.6	7.4

Brick Industry Association

Figure 10-13. Modular brick and mortar required for single wythe walls and running bond with no allowance for waste or breakage.

Masonry Estimating Guide	
Brick and Block Type	**Units**
Standard block	1.125 block per sq ft of wall area
Face brick modular	7 brick per sq ft of wall area
Oversized brick	6 brick per sq ft of wall area
Utility brick	3 brick per sq ft of wall area
Mortar Type	**Quantity**
Block	3 bags per 100 blocks
Face brick	7 bags per 1000 bricks
Oversized brick	8 bags per 1000 bricks
Utility brick	10 bags per 1000 bricks
Sand Type	**Quantity**
Sand (1 yd = 1.25 tons)	1 cu yd per 7 bags of mortar
–	1 yd per 1000 bricks
–	1 yd per 200 blocks
Volume of Grout to Fill Cores in Block	**Quantity**
8″ × 8″ × 16″	2 core, 0.25 cu ft/block
10″ × 8″ × 16″	2 core, 0.33 cu ft/block
12″ × 8″ × 16″	2 core, 0.39 cu ft/block

Oldcastle APG South, Inc. DBA Georgia Masonry Supply

Figure 10-14. A guide to estimating quantities of brick and block units, mortar, and sand.

Codes, Specifications, and Requirements

Grout used in masonry wall construction should comply with the requirements of ASTM C476; *Building Code Requirements for Masonry Structures*, ACI 530/ASCE; and *Specifications for Masonry Structures*, ACI 530.1/ASCE 6.

The ingredients of grout should meet the requirements of ASTM C150 for cement, ASTM C5 for lime putty, ASTM C207 for hydrated lime, and ASTM C404 for aggregates. All of the materials covered in ASTM C476 may be used in grout. Most grout used in large quantities is purchased from ready mixed concrete suppliers.

Admixtures

When concrete masonry units are highly absorbent, use of a grouting-aid admixture can reduce early water loss to the masonry units, promote bonding, and produce a slight expansion. Chlorine admixtures should not be used in grout because they can cause corrosion of reinforcement, metal ties, or anchors.

Fluid Consistency

Grout should be fluid enough to be pumped or poured, but not so fluid as to cause segregation. The grout should flow around reinforcing steel and into all cavities and joints without leaving voids. Further, plastic grout should not cause bridging or honeycombing.

Grout consistency is measured using a slump test (ASTM C143) and should be based on the rate of absorption of the masonry units, ambient temperature, and humidity conditions. The desired slump is between 8″ for units with low absorption and 10″ for units with high absorption. As with concrete and mortar, the cone height is 12″.

Strength

Grout material proportions are covered in ASTM C476. These proportions control the strength, but grout can be proportioned to produce a compressive strength equal to or greater than the recommended strength of masonry as specified in ASTM C1019. Grout strength should not be less than 2000 psi.

The amount of water greatly affects the strength of grout. Water that is absorbed by the masonry units before the grout has set reduces the water-cement ratio and increases the compressive strength.

Mixing

ASTM C94 covers ready-mixed concrete. It is recommended that grout should be batched, mixed, and delivered in accordance with this standard. Ready-mixed grout should be continuously agitated after mixing and until placement to prevent segregation. Its high slump causes grout to be susceptible to segregation. On-site mixing of grout is generally not recommended. Grout should be placed within 1 1/2 hours after water is first added.

Placement

It is good practice to consolidate grout by vibration or rodding to ensure that it fills all voids and adheres completely to reinforcing steel. Pours up to 12″ high can be consolidated by rodding or vibration. Pours higher than 12″ should be vibrated and then reconsolidated after settlement and initial water loss occurs.

The volume of grout is slightly reduced after placement due to water absorbed by the masonry units. To remedy this condition, shrinkage-compensating admixtures or expansive cement is sometimes recommended for high-lift grouting situations. The maximum height of grout lifts is usually 5′ but depends on building codes and construction practices.

Trade Tip

The term ***low-lift grouting*** refers to grouting in multiple lifts (segments) that can vary from two to three courses up to 4′ in height of the wall structure. Grouting is performed as the walls are constructed to add strength.

The term ***high-lift grouting*** refers to grouting a structure in one pour, usually after the walls have been constructed. A pump truck lifts (pumps) the mortar up to the top of the wall. The grout is then poured into the cavities to add strength to the wall structure.

Curing

Curing is the process of controlling the moisture loss of cementitious materials (mortar, grout, and concrete) over time to attain design standards of strength and bond adherence. Adequate moisture for curing grout is usually provided by the water absorbed by masonry units due to the high water content of masonry grout. However, in dry areas with high winds, some moist curing through fogging or plastic sheet protection may be required.

Grout is especially vulnerable to freezing because of its high water content. Therefore, care should be taken in cold weather to prevent freezing. Sand and water used in the grout mixture can be heated, or heated enclosures can be provided to protect the masonry during these adverse conditions.

Sampling and Testing

ASTM C1019 should be used for quality control and as a guide in selecting grout proportions. Specimens of grout for compressive tests should be prepared as specified in ASTM C1019 rather than cast in cylinder molds used for concrete samples. For grout slump and strength tests, a minimum of 1/2 cu ft of grout should be sampled from materials being placed in the wall. One grout sample should consist of three 3 1/2″ × 3 1/2″ × 7″ specimens. A sample should be taken whenever there is a change in mix proportions, materials used, or the method of mixing is changed. Also, a sample must be taken for each 5000 sq ft of masonry.

Safety Note

Contact with wet (plastic) concrete, cement, mortar, grout, or cement mixtures can cause skin irritation, severe chemical burns, or serious eye damage. Wear waterproof gloves, a long-sleeved shirt, full-length trousers, and eye protection when working with these materials. If you must stand in wet concrete, wear high-top waterproof boots. Wash wet concrete, mortar, grout, cement, or cement mixtures from your skin immediately. Flush eyes with clear water immediately upon contact. Seek medical attention if you experience a reaction to contact with these materials.

Chapter Review

Summary

- Mortar bonds units together, compensates for differences in unit sizes, provides bonding for metal ties or other reinforcement, and creates pleasing lines and color effects.
- Basic mortar ingredients include Portland cement, hydrated lime, and sand.
- The compressive strength, bond strength, and elasticity of mortar affect the structural performance of masonry.
- During the construction of a specific project, consistency from mix to mix with respect to yield, workability, and color is important. This consistency can be achieved by using proper methods to measure the ingredients.
- Mortar should be used within 2 1/2 hours after mixing to avoid hardening due to hydration.
- Mixing mortar with the maximum amount of water consistent with workability provides maximum bond strength within the capacity of the mortar.
- Read the manufacturer's operation manual and safety manual prior to running or servicing the mortar mixer.
- Plastic (wet) mortar properties are workability and water retentivity. Hardened mortar properties are compressive strength, bond strength, and durability.
- The five types of mortar recommended for masonry construction are Types M, S, N, O, and K. The type of mortar selected is based on the type of masonry and its location in the structure.
- Grout is used in brick masonry to fill cells of hollow units or spaces between wythes of solid masonry.
- Mortar can be colored using aggregates or pigments. Colored aggregates are preferred.
- Mortar temperature should be between 60°F and 80°F (15.6°C and 26.7°C) when the mortar is used. Masonry work performed at temperatures outside the ideal conditions requires special procedures to ensure that the materials perform to manufacturer's specifications.
- The objective of making estimates is to order the proper amount of all ingredients. Job delays can result from poor estimates, and waste occurs when too much material is ordered.
- Grout is generally used in reinforced loadbearing masonry walls or in those spaces that contain steel reinforcement.
- The fluid consistency of grout should be great enough to be pumped or poured, but not so fluid as to cause segregation.
- It is good practice to consolidate grout by vibration or rodding to ensure that it fills all voids and adheres completely to reinforcing steel.

Know and Understand

Answer the following questions using the information provided in this chapter.

1. *True or False?* Mortar is generally composed of cementitious materials, gravel, and water.
2. Type _____ Portland cement should be used when moderate sulfate resistance is desired.
   A. I
   B. II
   C. III
   D. IV
3. The primary aggregate used in masonry mortar is _____.
   A. sand
   B. gravel
   C. limestone
   D. slag
4. ASTM C270 states that 1 cu ft of loose, damp sand contains _____ lb of dry sand.
   A. 20
   B. 40
   C. 50
   D. 80

5. *True or False?* Overmixing mortar can increase the strength of mortar.
6. Workability can be restored to mortar by _____.
   A. hand mixing
   B. power mixing
   C. retempering
   D. removing excess water
7. _____ is a property of wet mortar.
   A. Bond strength
   B. Durability
   C. Compressive strength
   D. Workability
8. The most significant property of hardened mortar is _____.
   A. bond strength
   B. water retentivity
   C. compressive strength
   D. workability
9. Which mortar type is suitable for general interior use in nonloadbearing masonry?
   A. M
   B. N
   C. O
   D. S
10. Which mortar type has greater durability than other types?
    A. M
    B. N
    C. O
    D. S
11. *True or False?* When coloring mortar, use the greatest quantity of pigment available.
12. *True or False?* The temperature of mortar should be between 60°F (15.6°C) and 80°F (26.7°C) when used.
13. Enclosures are necessary _____.
    A. in extreme cold
    B. in extreme heat
    C. in both hot and cold weather
    D. always
14. To make 1 cu ft of mortar, you need _____.
    A. 0.5 cu ft of cementitious material and 0.5 cu ft of sand
    B. 0.3 cu ft of cementitious material and 0.7 cu ft of sand
    C. 0.33 cu ft of cementitious material and 0.99 cu ft of sand
    D. 0.33 cu ft of cementitious material and 0.67 cu ft of sand
15. Grout consistency is measured using a _____ test.
    A. flash set
    B. dryout
    C. cure
    D. slump

Apply and Analyze

1. What is the function of mortar?
2. What is meant by the term *hydraulic material*?
3. What three types of Portland cement are recommended for use in mortar?
4. What is hydrated lime?
5. What might be the result of measuring sand by the shovel?
6. When should mortar be discarded?
7. List the three properties of hardened mortar.
8. How is the durability of mortar measured?
9. What are the five types of mortars recommended for various classes of masonry construction?
10. What type of mortar has high-compressive strength and somewhat greater durability than other mortar types?
11. Why is antifreeze *not* recommended for use in mortar mixtures?

Critical Thinking

1. Describe an easy method of raising the temperature of mortar in cold weather.
2. Identify the three main purposes for grouting brick and block walls.
3. Why should chlorine admixtures not be used in grout?
4. Describe the proper fluid consistency of grout.

Activities

1. (ELA) Conduct research on slump tests. How is the test performed? What apparatus is needed? How are the results interpreted? Prepare a class presentation that answers these questions.
2. (MATH) Determine the amount of mortar in cubic feet needed to construct a block wall 12′ tall and 30′ long. Concrete blocks being used for the wall are 8″ × 16″ units. Use data from **Figure 10-11**.

Pamela Au/Shutterstock.com

CHAPTER 11

Anchors, Ties, and Reinforcement

LEARNING OUTCOMES

After studying this chapter, you will be able to:

11.1-1 Recognize different types of metal ties and their applications.
11.2-1 Identify types of reinforcing rods and bars and their applications.
11.3-1 Identify types of anchors and their applications.
11.4-1 Recall the shapes, sizes, and applications of joint reinforcement.

TECHNICAL TERMS

adjustable ties
anchor
anchor bolt
concrete spacers
fiber reinforced polymer (FRP)
hardware cloth
mesh wall ties (MWT)
rebar positioners
reinforcing bars (rebar)
reinforcing chairs or stools
reinforcing rods
wall ties

Introduction

Many types of masonry anchors, ties, and joint reinforcement are used in masonry construction, **Figure 11-1**. These materials are placed inside a masonry wall to give it greater strength or to hold it in place. The following sections describe metal ties, reinforcing rods or bars, anchors, and joint reinforcement.

11.1 Metal Ties or Wires

Wall ties are strips of metal or metal wire used to tie masonry wythes together or to tie masonry veneer to a concrete frame or wood frame wall. Wall ties normally perform three functions—transfer lateral loads, provide a connection, and permit or restrain movement between different materials in a wall. In addition, metal wall ties can serve as horizontal structural reinforcement. However, these functions can be achieved only if the tie system is securely attached and embedded, has the required stiffness to transfer lateral loads, has little mechanical play, is corrosion-resistant, and is easily installed.

Code Note — **Wall Tie Installation**

The International Residential Code **R606.13.2.1** specifies that ends of wall ties are to be embedded in mortar joints. There should be a minimum of 5/8″ (15.9 mm) mortar coverage from the exposed face. Wall ties should not be bent after being embedded in grout or mortar. Wall ties for solid masonry units, solid grouted hollow units, or hollow units in anchored masonry veneer are to be embedded at least 1 1/2″ (38 mm) in the mortar bed. Wall ties for hollow masonry units in other than anchored masonry veneer should engage outer face shells by at least 1/2″ (13 mm).

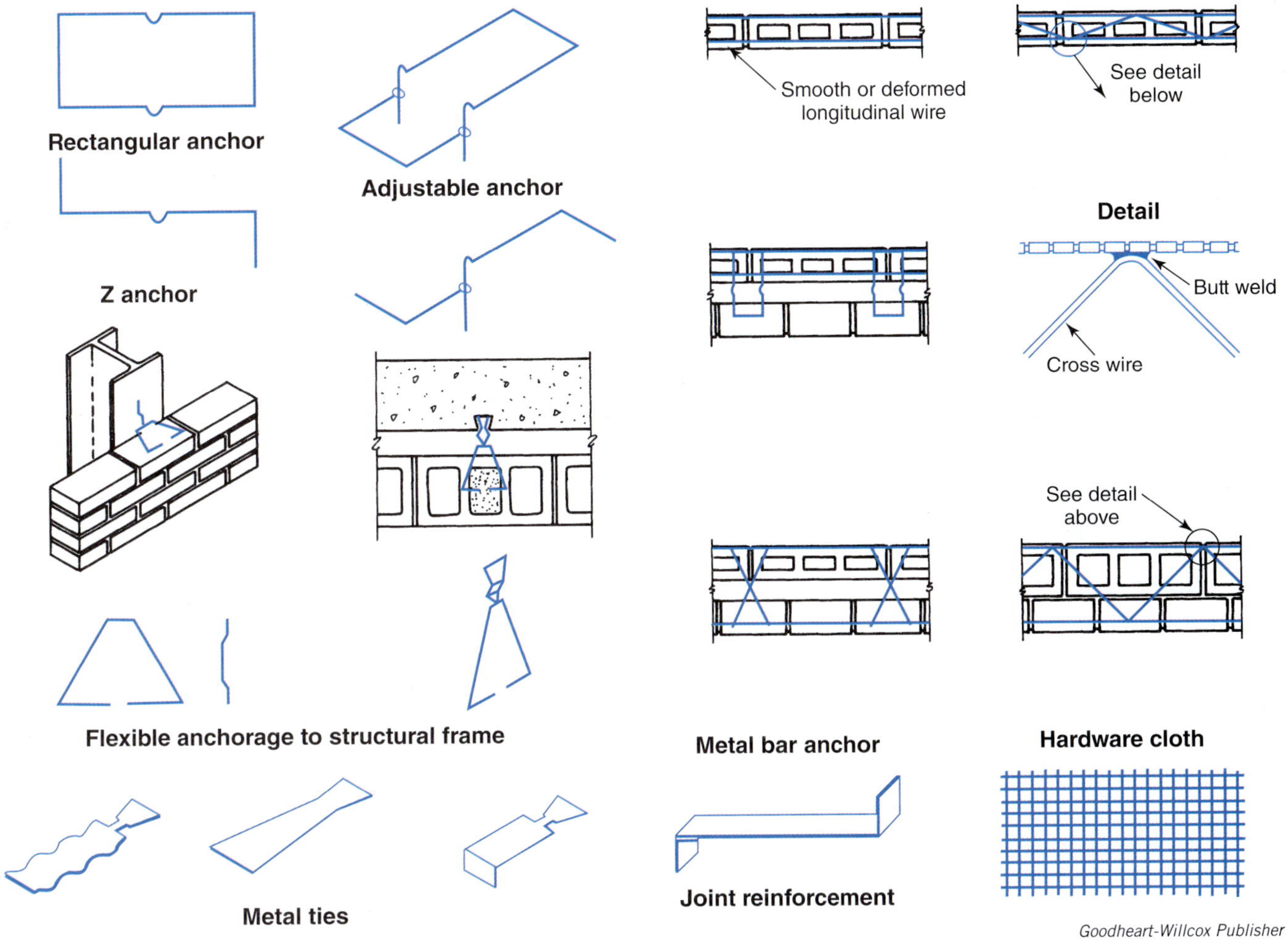

Goodheart-Willcox Publisher

Figure 11-1. Common masonry anchors, ties, and joint reinforcement used in masonry wall construction.

Structural bonding with metal ties or wires is widely used in three types of construction:

- Masonry veneer to frame
- Solid masonry
- Cavity wall (**Figure 11-2**)

Portland Cement Association

Figure 11-2. Rectangular anchors used in cavity wall construction.

Types of Ties

The basic types of metal wall ties include unit ties, continuous horizontal joint reinforcement, adjustable ties, and re-anchoring systems.

Unit Ties

Three types of unit ties are used in masonry wall construction. They are rectangular ties, Z ties, and corrugated ties. See **Figure 11-3, Figure 11-4,** and **Figure 11-5**. Rectangular and Z ties are generally made from cold-drawn steel wire or copper-clad wire that meets ASTM standard A82 (cold-drawn steel) or ASTM B227 (copper-clad). Stainless steel rectangular and Z ties are also available for highly corrosive applications. Corrugated ties are usually made from sheet steel, grade D, but are also available in copper and stainless steel.

Horizontal Joint Reinforcement

Continuous horizontal joint reinforcement is ordinarily made from #8, #9, #10, or #11 gage wire or 3/16″ diameter wire. See **Figure 11-6**. Lengths available

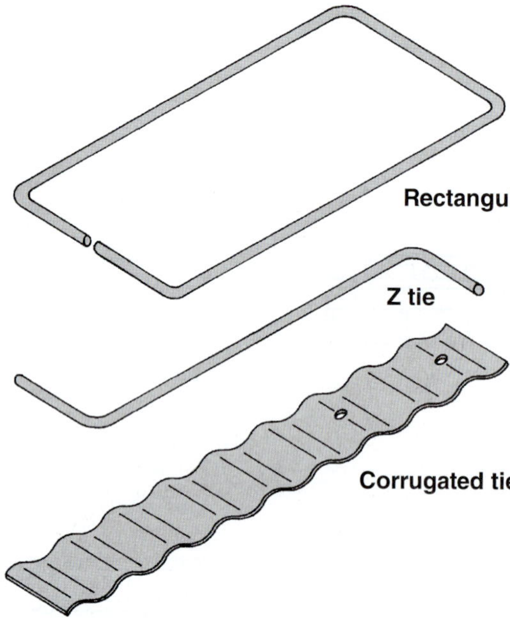

Brick Industry Association

Figure 11-3. Typical unit ties used in masonry construction.

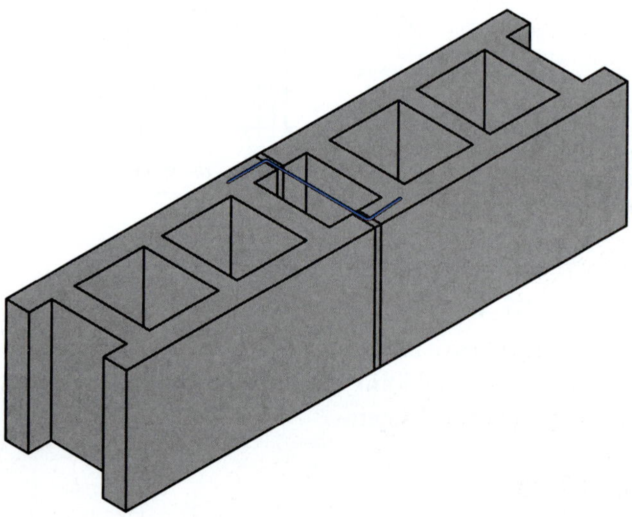

Goodheart-Willcox Publisher

Figure 11-4. Z ties may be used in concrete block wall construction.

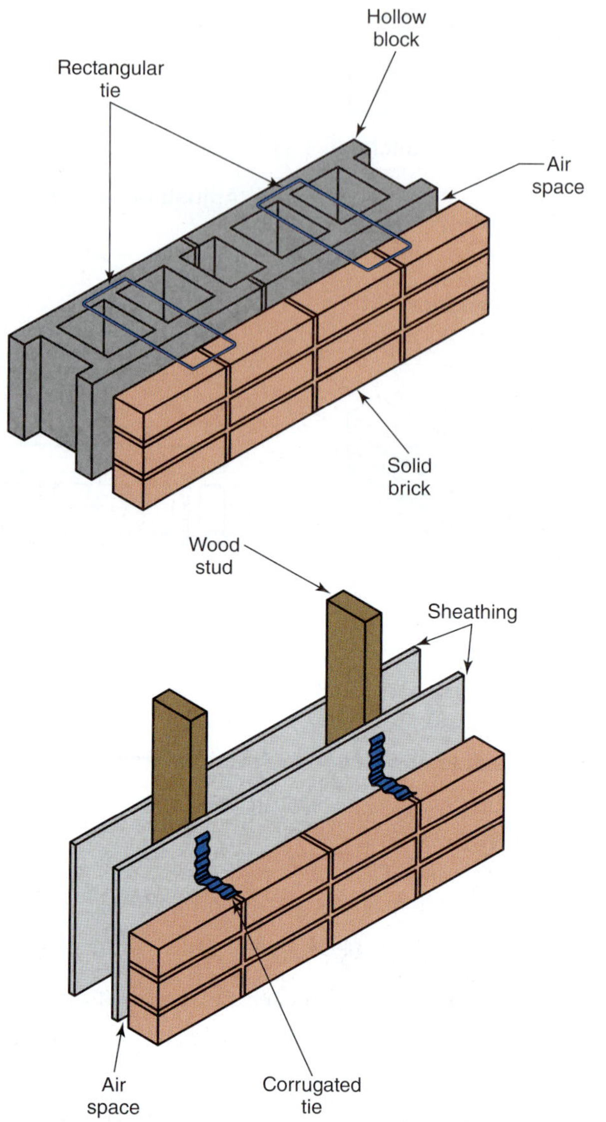

Goodheart-Willcox Publisher

Figure 11-5. Typical applications of unit ties.

Steel Reinforcement Wire Sizes		
Designation	Diameter in (mm)	Area sq in (sq mm)
W1.1 (11 gage)	0.121 (3.1)	0.011 (7)
W1.7 (9 gage)	0.148 (3.8)	0.017 (11)
W2.1 (8 gage)	0.162 (4.1)	0.021 (14)
W2.8 (3/16 in)	0.188 (4.8)	0.028 (18)
W4.9 (1/4 in)	0.250 (6.4)	0.049 (32)

Goodheart-Willcox Publisher

Figure 11-6. Steel wire reinforcement sizes used in masonry construction.

are 10′ to 12′. Styles of horizontal joint reinforcement include ladder, truss, and tab types. See **Figure 11-7** and **Figure 11-8**.

When positioning the continuous joint reinforcement on the top of the block course, it is important to install the metal so that the edges are spaced equally from the sides of the concrete block. Refer to **Figure 11-9**. If the reinforcement is not correctly positioned, the metal can be exposed in the masonry joint, affecting the strength and quality of the masonry job.

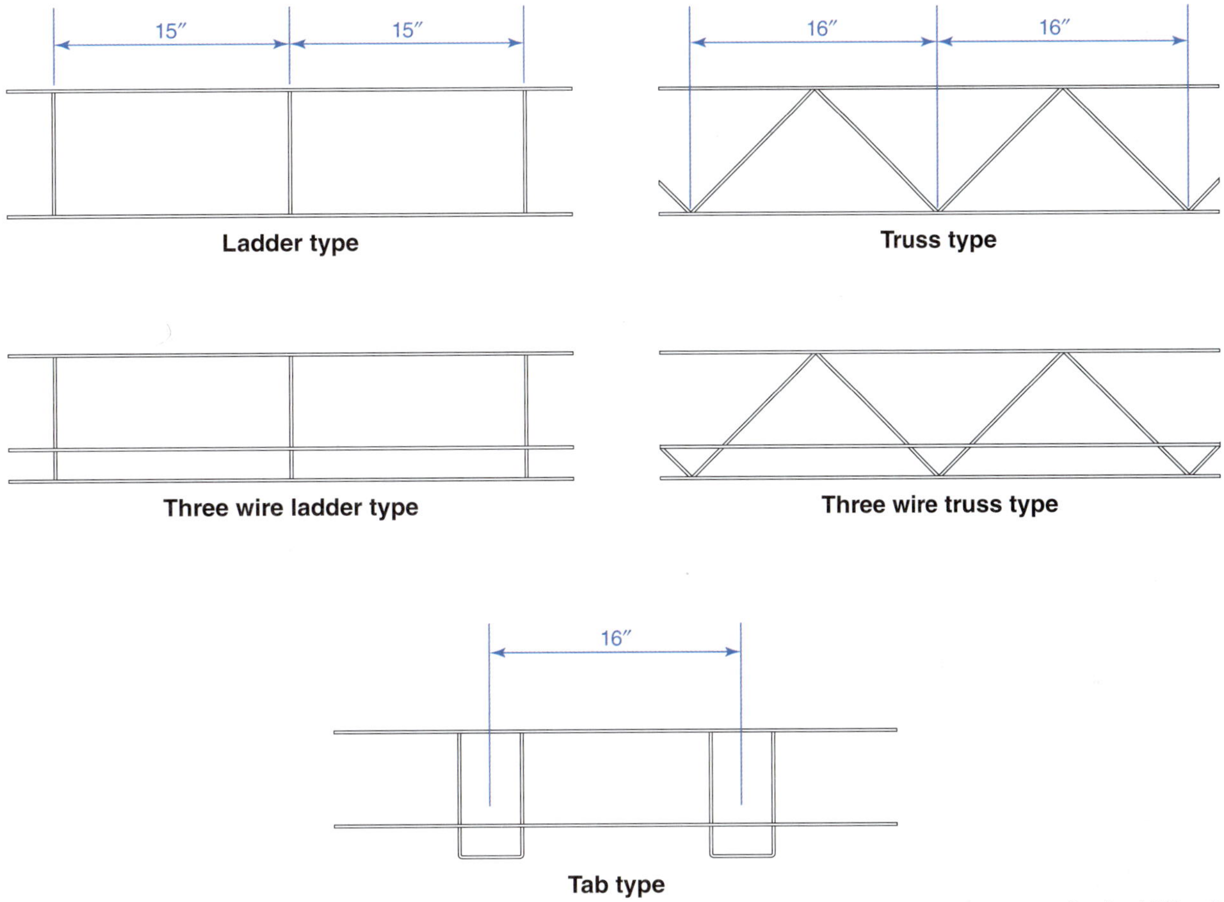

Goodheart-Willcox Publisher

Figure 11-7. Types of continuous joint reinforcement.

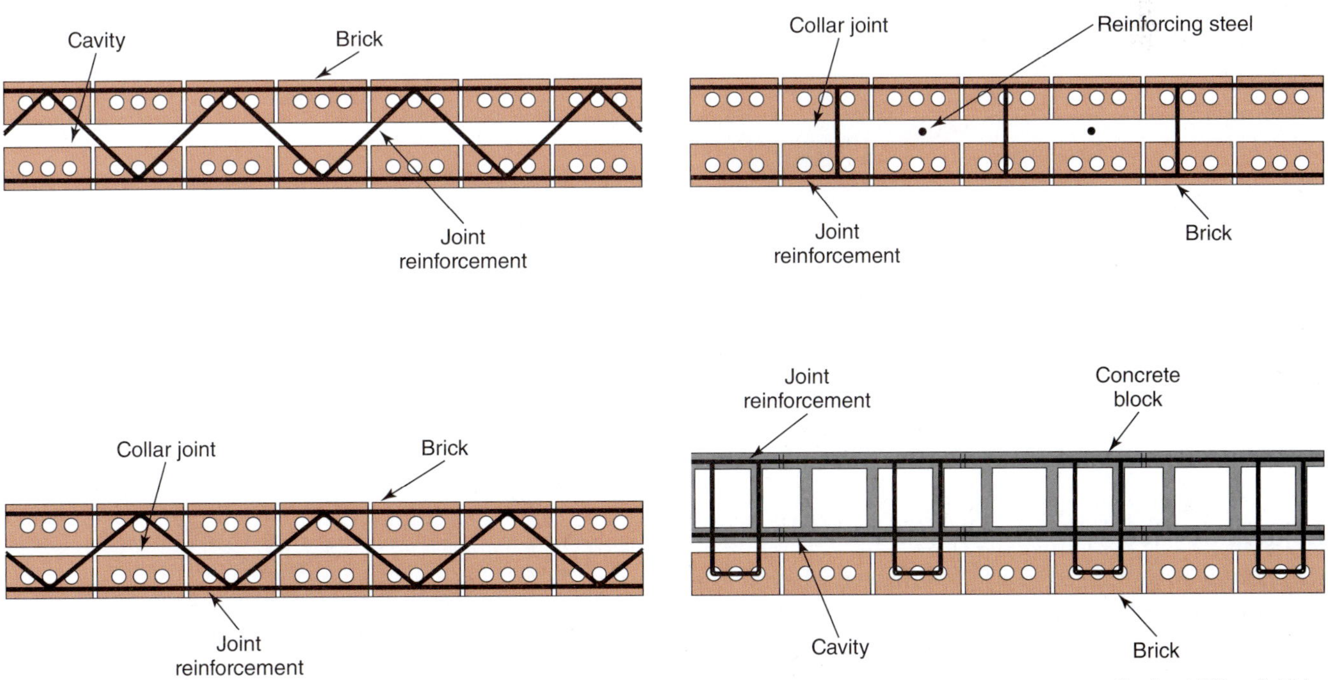

Goodheart-Willcox Publisher

Figure 11-8. Joint reinforcement shown for typical applications.

Code Note **Horizontal Continuous Joint Reinforcement**

When continuous joint reinforcement is required, the National Concrete Masonry Association (NCMA) requires continuous reinforcement wires to be lapped at least 6″ at the splice locations. It is also important to check with local or state codes because the reinforcement overlapping amount may be greater than 6″. When placing reinforcement in different courses, splices should be staggered to ensure the splices are not located directly above each in the wall to ensure maximum strength in the masonry structure. If the block cores or cells are to receive reinforcing bars and grout, it is recommended to use ladder-type reinforcement rather than truss-type reinforcement because the diagonal wires may interfere with the placement of the bars in the cores and filling them with grout. For more information, refer to *NCMA TEK 12-2B Joint Reinforcement for Concrete Masonry.*

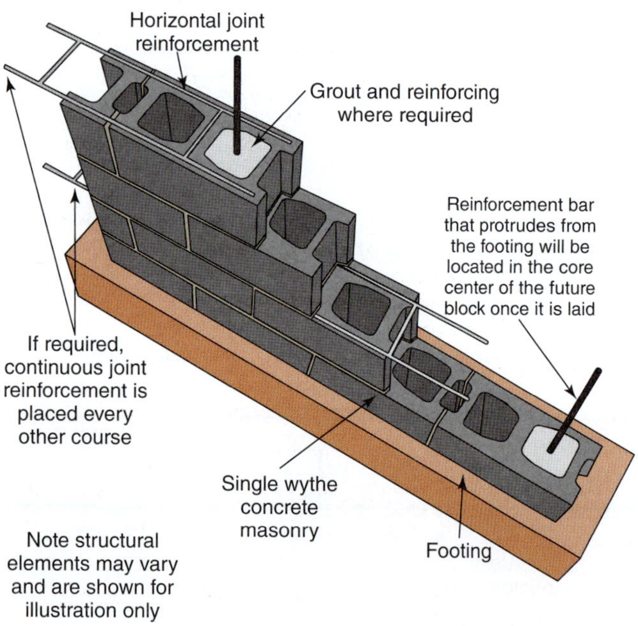

Goodheart-Willcox Publisher

Figure 11-9. Continuous joint reinforcement is placed horizontally every two courses and steel reinforcement bars vertically with grout to increase strength of this concrete block wall.

Adjustable Ties

Adjustable ties are used to attach a masonry facing to a backup system of masonry or typical stud construction. The use of adjustable ties speeds construction because the ties consist of two pieces. One piece is installed as the backup is constructed, and the other piece is installed as the facing is constructed. One disadvantage of adjustable ties is that they generally provide less strength and stiffness than unit ties or joint reinforcement.

Adjustable ties are available as unit ties and adjustable assemblies. See **Figure 11-10**. Adjustable unit ties are intended for use with masonry backup, concrete backup, and steel studs or frames. Adjustable assemblies are available in ladder and truss styles. These are intended for use in masonry-backed cavity, veneer, and grouted walls. See **Figure 11-11**.

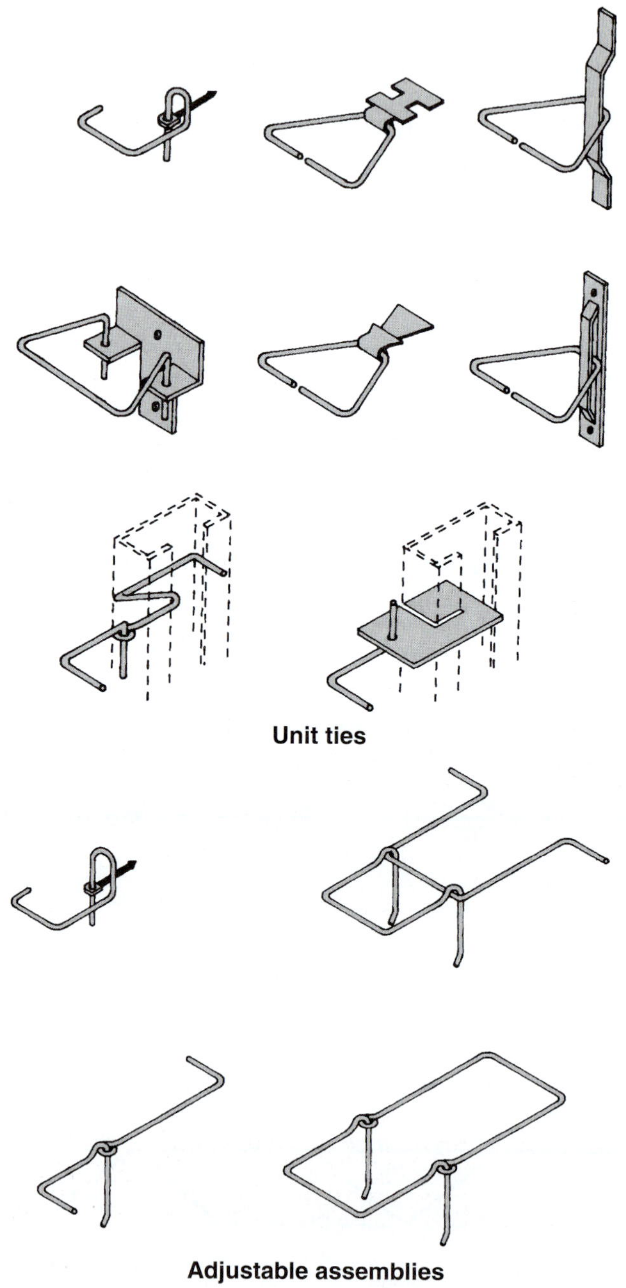

Brick Industry Association

Figure 11-10. Top—Adjustable unit ties for steel, concrete, and stud wall backup. Bottom—Typical adjustable unit ties for masonry backup.

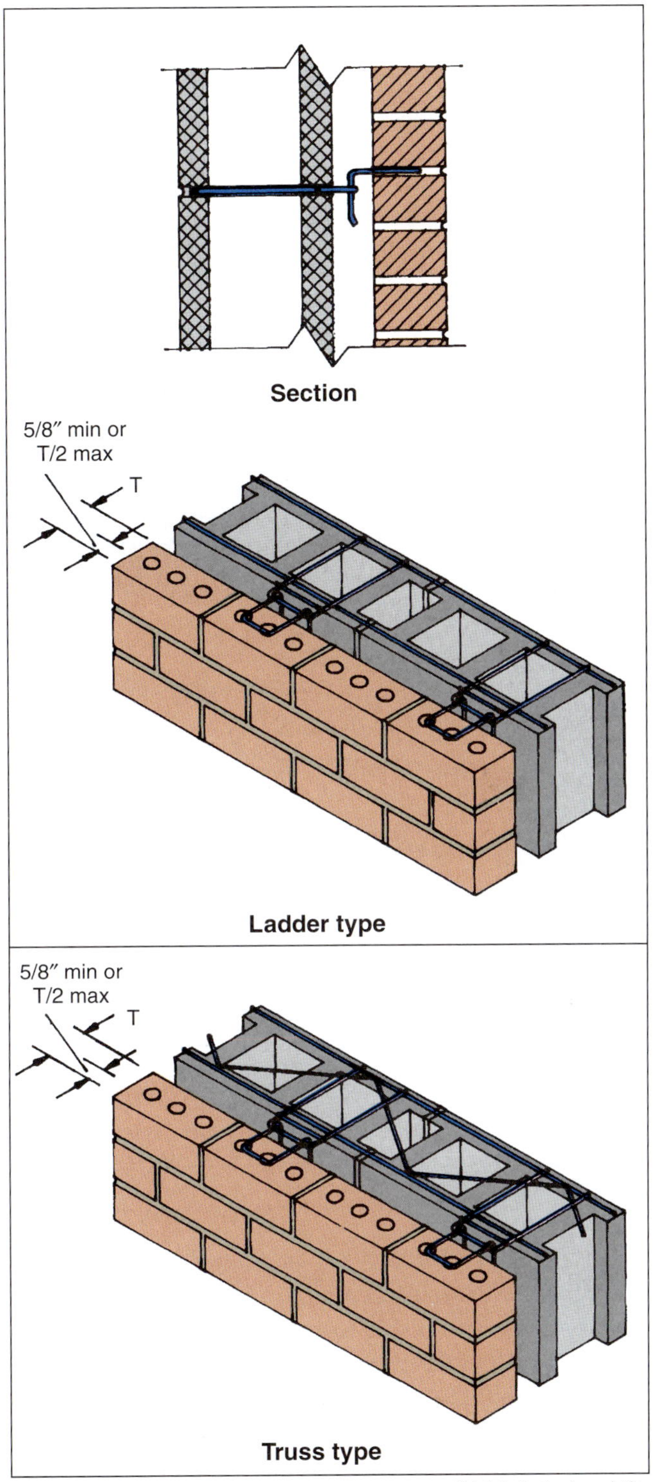

Brick Industry Association

Figure 11-11. Adjustable assemblies in typical applications.

Masonry Re-Anchoring Systems

Typical re-anchoring approaches include mechanical expansion systems, screw systems, and epoxy adhesive systems. Masonry re-anchoring systems are used where ties were not installed during the original construction or to replace a failed system.

11.2 Reinforcing Rods or Bars and Chairs

Concrete has limited tensile strength. To improve the qualities of concrete, reinforcement is added; either steel or fiber reinforced polymer (FRP). Reinforcement products come in many sizes and lengths depending on the specific application.

Reinforcing Rods or Bars

Reinforcing rods (re-rods) or ***reinforcing bars (rebars)*** are reinforcement used to increase the tensile strength of concrete. When reinforcing is used in a structure, the steel or FRP used act together with the masonry to resist forces being exerted on the wall or surface, such as wind, load, and backfill pressure.

Reinforcing rods are manufactured in various diameters and lengths. They are used in masonry construction elements such as footings, walls, collars, pilasters, and beams. Reinforcing rods and bars are placed in concrete, grouted cavities, pockets, cores, cells, or bond beams of masonry. See **Figure 11-12**.

Architects and engineers specify the reinforcing bars needed to add strength and stability to masonry elements, such as foundation footings and masonry walls. Reinforcing rods can be added to footings or concrete slabs to increase strength, where rebars cross over each other the rods are tied together using wire ties. See **Figure 11-13**. The same procedure is performed when rebars are placed vertically in the footing to connect to wall reinforcement. The vertical part of the rebar that stays in the footing is usually bent at a 90° angle and is attached to the horizontal reinforcement in the footing. This ensures that horizontal and vertical reinforcement aids in securing the concrete wall to the footing.

Safety Note

When placing and installing reinforcing rods, the exposed ends can be dangerous. OSHA construction regulation **1926.701(b)** requires all protruding reinforcing steel, onto and into which employees could fall, be guarded to prevent the risk of impalement. Protective safety caps are required to be placed on the exposed ends to protect workers from injury. When selecting safety caps, ensure that they meet OSHA standards. This OSHA Standard also applies to the newer fiber reinforced polymer (FRP) rebars.

When installing horizontal rebar in footings, floors, and driveways, it is important that the reinforcement stays at the appropriate elevation or height in the

Igorsky/Shutterstock.com

Figure 11-12. Reinforcing bars are placed in a footing to tie the masonry wall to the footing. These exposed rebars have safety caps placed on the ends to protect workers from injury. In conditions where footings are installed in cold weather, straw can be used to insulate the footing during the period of concrete curing.

Vadim Ratnikov/Shutterstock.com

Figure 11-13. A concrete mason installing metal wire ties that connect horizontal rebars in footing and floors. The mason is using a tie wire twister to ensure the ends of the reinforcement are securely attached to each other. The irregular surface of the rebars increases the connection between the reinforcement and concrete placed in the forms.

concrete form. To achieve this, reinforcement chairs or stools are used. Chairs and stools will be discussed later in this chapter.

Once the concrete footings are poured and set, the vertical rods protruding from the footings are used to attach the wall reinforcement. This occurs in poured concrete walls as well as concrete block walls to create a bond between footings and walls. See **Figure 11-14**. Common practice is to place vertically reinforcing bars every 4′ in footings to tie the footing to the concrete block wall or concrete wall. If reinforcement bars are to continue up through the block cores, it is customary to position them in the same locations where the reinforcement protrudes from the footing into the first course of block. Reinforcement bars will continue up through the individual courses and the cores containing the vertical reinforcing bars will be filled with grout at a later time.

A type of reinforcing bar is ***fiber reinforced polymer (FRP)***. FRP reinforcing bars are made by combining fiberglass and resins to create manufactured rebar, **Figure 11-15**. FRP rebars are lighter weight, higher tensile strength, and more durable than its steel counterpart. FRP rebars are non-corrosive, meaning they do not rust like reinforcement made from steel. This type of rebar is non-conductive, meaning they do not conduct electricity as steel does, making them less of an electrocution hazard. One company who manufactures FRP is GatorBar, located in Ahmeek, Michigan. GatorBar products are 100% made in the USA.

Reinforcing bars can be placed in masonry walls vertically, horizontally, or both, depending on the requirements specified in the construction plans or building specifications. When bars are laid horizontally in a

Lorraine Cline/Shutterstock.com

Figure 11-14. Reinforcing bars are used to strengthen a concrete wall. Bars originate in the footing and run up through the block cores to bond the footing and wall together.

Photo Courtesy of Neuvokas Corporation – Manufactures of GatorBar©

Figure 11-15. A concrete mason is adjusting FRP reinforcement bars with a plastic lift chair. The chair was tilted during the installation of the FRP reinforcement. It is important to observe the height of the rebar over the base.

Safety Note

It is important when handling reinforcing bars made of steel that they do not come in contact with overhead powerlines. Steel rebars come in lengths 20′ and greater. So when moving them from one place to another, it is important never to carry them vertically. Wear protective gloves when handling steel or FRP rebars because the outer surface can be abrasive and the ends can have steel or fiberglass fragments from cutting.

pryzmat/Shutterstock.com

Figure 11-16. Stretcher blocks in this wall have depressions to receive reinforcing bars.

masonry block wall, portions of the cross web may need to be cut and removed. The cutting of the cross web depends on the type of block being laid. Certain block types have the top of the web depressed to receive reinforcing material. See **Figure 11-16**. Standard block requires a portion of the top of the web to be removed using a masonry saw.

Portions of blocks may also need to be removed at intersections of walls or outside corners of walls. See **Figure 11-17**. Once cuts are made in the block, areas are removed to create a trench. See **Figure 11-18**. When the trench in the block is complete, reinforcing bars are positioned around the perimeter of the top of the foundation wall.

Standard lengths of reinforcing bars can be purchased in 20′, 30′, 40′, and 60′. The reason for the longer lengths is to reduce splices when building large structures. Shorter lengths can be ordered and cut to length to meet the request of the purchaser. The diameters of reinforcing bars permitted by the MSJC are indicated by numbers. See **Figure 11-19**. Bar sizes #3 through #8 are the number of eighths of an inch of the nominal diameter of the bars. For example, #5 is 5/8″ in diameter. Sizes #9 through #11 are 1 1/8″, 1 1/4″, and 1 3/8″ in diameter, respectively. Bar sizes continue up to #18, but these larger sizes are not generally used in masonry construction.

Coated bars or FRP rebars are used when corrosion protection is needed. Coatings are generally epoxy or zinc (galvanized).

Reinforcing bars have deformations on the surface that aid in bonding. See **Figure 11-20**. Reinforcing bars

pryzmat/Shutterstock.com

Figure 11-17. A masonry saw is being used to remove a portion of the block where reinforcing rods will be placed. For protection, the mason is wearing gloves, safety glasses, and a respirator.

pryzmat/Shutterstock.com

Figure 11-18. A mason is lightly tapping on the side of the block to remove material where reinforcement bars will be placed.

Steel Reinforcement Bar Sizes

Designation	Diameter in (mm)	Area sq in (sq mm)
No. 3	0.375 (10)	0.110 (71)
No. 4	0.500 (13)	0.196 (126)
No. 5	0.625 (16)	0.307 (198)
No. 6	0.750 (19)	0.442 (285)
No. 7	0.875 (22)	0.601 (388)
No. 8	1.000 (25)	0.785 (506)
No. 9	1.128 (29)	1.000 (645)
No. 10	1.270 (32)	1.267 (817)
No. 11	1.410 (36)	1.561 (1007)

Goodheart-Willcox Publisher

Figure 11-19. Steel bar reinforcement sizes used in masonry construction.

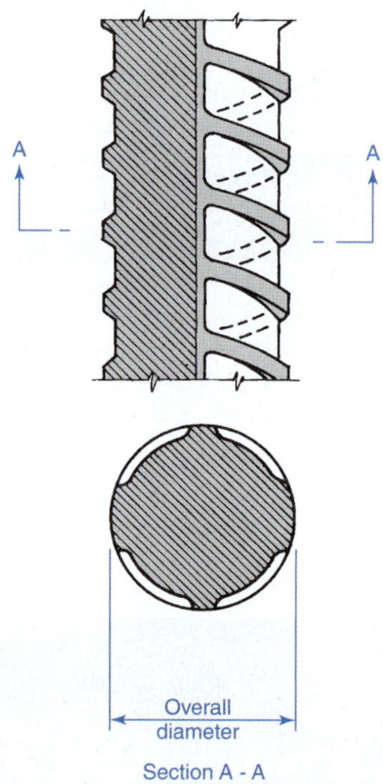

Concrete Reinforcing Steel Institute

Figure 11-20. Bar diameters are nominal, with the actual diameter outside of deformations being somewhat greater. This deformation style is one typical type.

are formed from rolled steel and are similar to those used in reinforced concrete construction.

Identification marks on ASTM standard bars are rolled into the surface of one side of the bar. The marks denote the producer's mill designation, the bar size, and the type of steel. Yield strength is indicated by grade marks, which are used for indicating grade 60 and grade 75 bars. Grade 40 and grade 50 bars show the identification marks of the producing mill (usually an initial), bar size number (#3 through #18), and type. See **Figure 11-21**. Grade marks are not used for grade 40 and grade 50 bars.

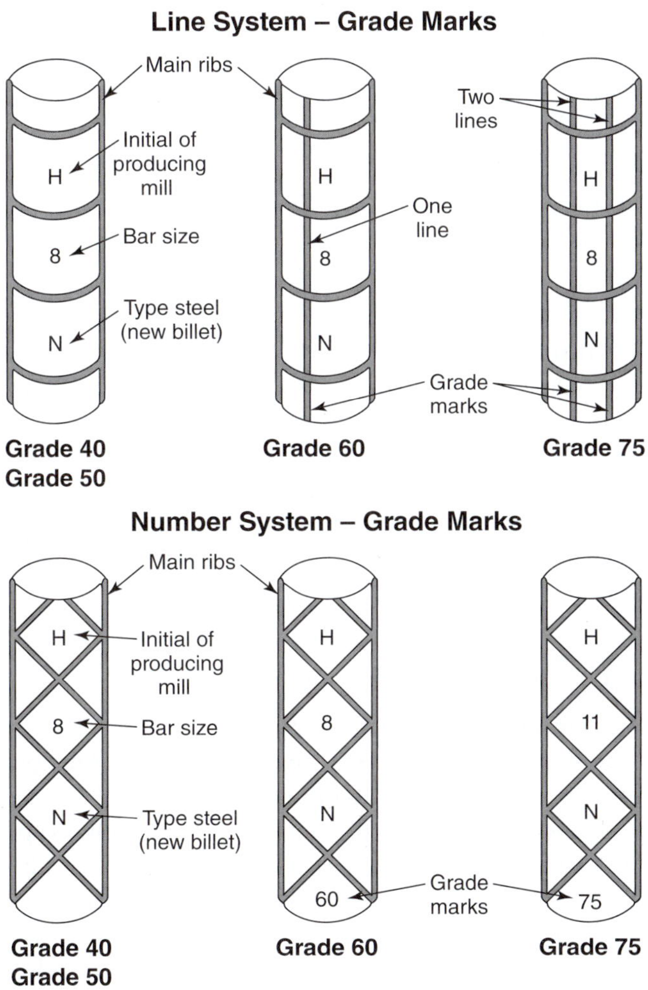

VARIATIONS: Bar identification marks may also be oriented to read horizontally (at 90 to those illustrated above.)
Grade mark lines must be continued at least 5 deformation spaces.
Grade mark numbers may be placed within separate consecutive deformation spaces to read vertically or horizontally.

Concrete Reinforcing Steel Institute

Figure 11-21. Two systems of identification marks are used for ASTM standard bars—line system and number system.

Reinforcing Chairs or Stools

Reinforcing chairs or stools are used to ensure rebars do not sink below the recommended or required height above the base material. Concrete chairs are used to support and elevate the reinforcement, so it stays at the appropriate location when placing concrete in the forms. See **Figure 11-22**. They are called chairs or stools because they resemble a stool setting next to a counter in a kitchen.

Chairs or stools become a permanent part of the concrete since they are embedded in the structural pour. The base material under the concrete can be gravel, sand, or other products. If plastic is used between the base and concrete, chairs will sit on top of the plastic. Plastic film is used to create a moisture barrier between the gravel base and the underside of the concrete. Concrete is a porous material, and the plastic is used to reduce moisture from wicking into the concrete. Plastic is rolled out over the base material and chairs are placed on-top of the plastic. Rebars are then placed into the chairs.

Chairs are used to support reinforcement in various situations such as footings, driveways, sidewalks, and concrete floors. Chairs come in many different sizes, shapes, and materials. Metal wire, plastic, concrete, or other materials are used when manufacturing chairs or stools. Metal chairs or stools usually have 4 legs to support the reinforcement. The height of the chairs will depend on the application. See **Figure 11-23**. Common metal bar chair heights vary from 1 1/2″ to 2″, taller ones from 2 1/4″ to 12″, or larger can be purchased.

Photo Courtesy of Neuvokas Corporation – Manufactures of GatorBar©

Figure 11-22. An example of plastic chairs being used to lift the reinforcement and support rebar at the correct height prior to placing concrete. Notice the plastic film placed on top of the base prior to installing rebars and chairs. FRP rebars are tied together using twisted wire.

laurello/Shutterstock.com

Figure 11-23. A metal chair being used to support reinforcement bars prior to placing concrete in formwork.

theLIMEs/Shutterstock.com

Figure 11-24. Concrete circular spacers made of plastic are used to position reinforcement in concrete walls being fabricated. Slots in the spaces allow the mason to work the rebar into the center of the spacer.

Concrete Spacers and Rebar Positioners

Rebars placed in vertical concrete walls need to be securely positioned so the surface of the rebar is not exposed once the forms are removed. ***Concrete spacers*** are used by masons to hold rebars in their appropriate locations in the concrete formwork. See **Figure 11-24.** Spacers can be made of plastic, metal wire, or concrete. Spacers come in different sizes based on wall thickness and reinforcement requirements. Electrical conduit or other pipes may be located in the formwork, and they too need to be secured by spacers to ensure their position stays the same when concrete is placed in the form. When building masonry block walls, the application of similar devices is used to secure rebar placed in the cores of the block wall. ***Rebar positioners*** are devices used to hold reinforcement in a specific location in the core or cell of the block. See **Figure-11-25.**

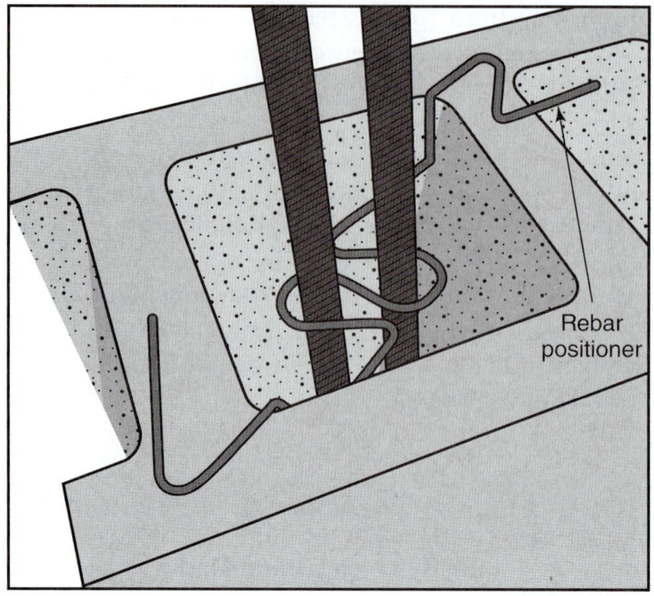

Goodheart-Willcox

Figure 11-25. A rebar positioner is securing bars prior to filling the core with grout, mortar, or concrete. This positioner is made of heavy wire where part of the unit sits on top the block and the remaining wire is lowered into the core to lock it into position.

11.3 Anchors

Anchors are devices, usually made of steel, that connect masonry to a structural support or frame. Anchors usually connect two dissimilar materials. They are produced in a variety of shapes and sizes. One type of anchor is corrugated with a dovetail on one end for veneer construction. These corrugated anchors are made of galvanized steel. The dovetail fits into a slot formed in the concrete beam or column, and the other end is attached to the veneer.

Another type of anchor is an ***anchor bolt.*** Anchor bolts are placed in fresh concrete and are used to attach sill plates to masonry work. See **Figure 11-26.** They are available in various diameters or gages and lengths. Some are made of band steel and have a bent

Chris Kryzanek/iStock/Thinkstock

Figure 11-26. Anchor bolts were installed in this foundation directly after concrete was placed. Later, a wood sill will be installed on top of the concrete wall and bolted down.

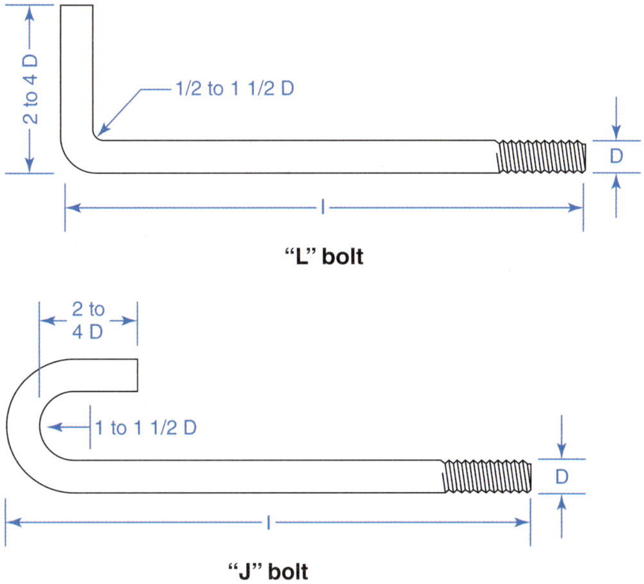

Goodheart-Willcox Publisher

Figure 11-27. These are "L" and "J" bent bar anchors.

Goodheart-Willcox Publisher

Figure 11-28. A strap anchor being put into place connecting a partition wall to a bearing wall.

end for better holding power. Others are round and may have a head or bend on the embedded end. See **Figure 11-27**.

Bearing walls that intersect are frequently connected with a strap anchor. See **Figure 11-28**. Strap anchors are usually 1 1/4″ × 1/4″ × 30″. Three-inch, right-angle bends are made at each end for embedding in mortar-filled cores of masonry block.

Concrete anchor devices intended for installation after the concrete has hardened generally involve some type of expansion device. **Figure 11-29** shows common types of these anchors.

11.4 Joint Reinforcement

Joint reinforcement is produced in many shapes and sizes for various applications. It is especially designed for placement in the horizontal joints of masonry walls to provide greater lateral strength. One popular style of joint reinforcement is made in the shape of a truss or a ladder. Refer back to **Figure 11-7**.

Smooth or deformed wire or steel rod is used in the fabrication of truss and ladder style reinforcements. Joint reinforcement must be corrosion-resistant. Cross wires are welded perpendicularly or diagonally to the longitudinal wires, generally at a 16″ spacing.

Joint reinforcement is available in most widths and in lengths of 10′ to 12′. The longitudinal and cross

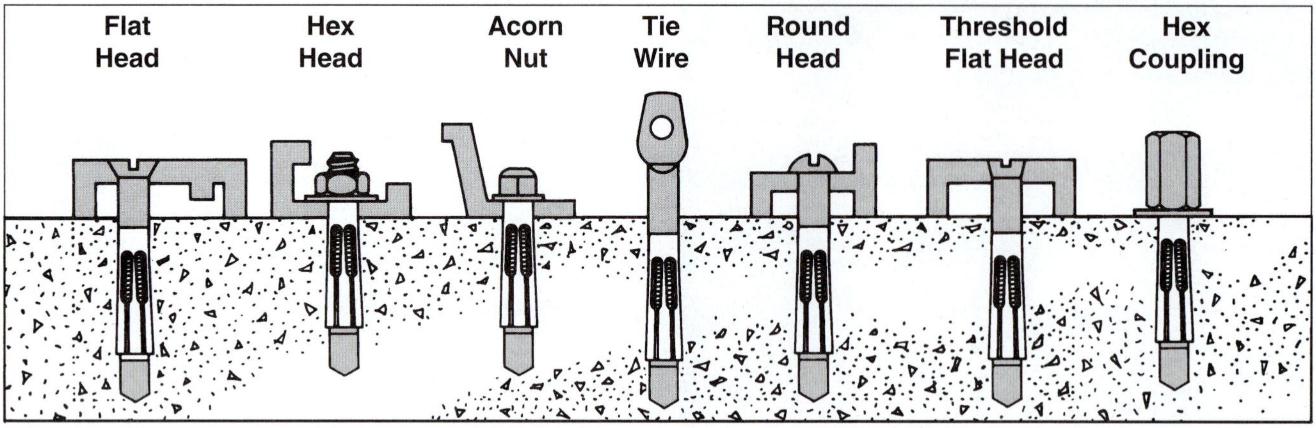

ITW Ramset/Red Head

Figure 11-29. Available head styles of common anchor devices.

wires are generally #8 or #9 gage or 3/16″ diameter. The maximum size of the wire is one-half the mortar joint thickness.

Hardware cloth is metal fabric that adds extra bonding strength, usually where one block wall butts up against another wall. Hardware cloth is also called ***mesh wall tie (MWT)***. This material is used to reinforce masonry joints. See **Figure 11-30**. MWT is frequently used with a metal tie bar to tie nonload-bearing walls to intersecting walls. MWT is usually galvanized and may be purchased in various widths and lengths.

Many specialized anchors and ties are available for unique construction applications. See **Figure 11-31**.

Portland Cement Association

Figure 11-30. Hardware cloth can be used to tie intersecting walls together.

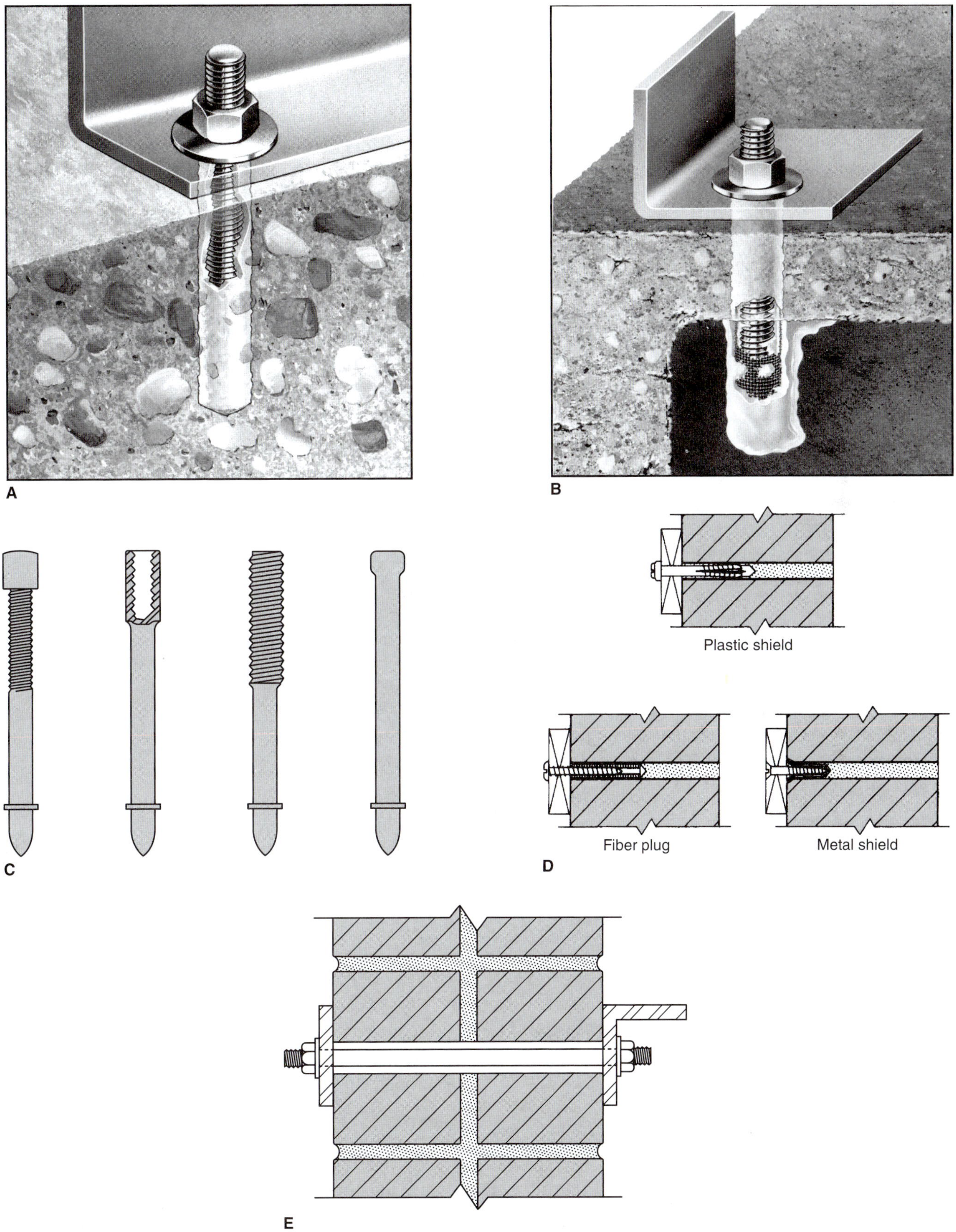

ITW Ramset/Red Head; Brick Industry Association

Figure 11-31. A—Threaded stud embedded in a concrete slab using epoxy adhesive. B—Threaded stud attached to a hollow concrete masonry unit using epoxy adhesive. C—Power-driven pins. D—Screw shields and plugs. E—Through bolt.

Chapter Review

Summary

- Wall ties are strips of metal or metal wire used to tie masonry wythes together or to tie masonry veneer to a concrete frame or wood frame wall.
- Wall ties transfer lateral loads, provide a connection, and permit or restrain movement between different materials in a wall.
- Metal ties or wires are widely used for structural bonding in masonry veneer to frame, solid masonry, and cavity wall construction.
- Metal wall ties include unit ties, continuous horizontal joint reinforcement, adjustable ties, and re-anchoring systems.
- Reinforcing rods (re-rods) or reinforcing bars (rebars) are steel used to increase the tensile strength of concrete. The rods or bars are used in masonry elements such as footings, walls, collars, pilasters, and beams.
- Reinforcing bars can be placed in masonry walls vertically, horizontally, or both.
- Anchors are devices, usually made of steel, that connect masonry to a structural support or frame.
- Anchor bolts are used to attach roof plates and sills to masonry work.
- Strap anchors are often used to connect bearing walls that intersect.
- Joint reinforcement is designed for placement in the horizontal joints of masonry walls to provide greater lateral strength.
- Hardware cloth, or mesh wall tie (MWT), is metal fabric that adds extra bonding strength, usually where one block wall butts up against another wall.

Know and Understand

Answer the following questions using the information provided in this chapter.

1. The IRC specifies that the ends of wall ties should be _____ mortar joints.
   A. embedded in
   B. tied to
   C. bent into
   D. left off of
2. _____ are made from #8, #9, #10, or #11 gage wire.
   A. Adjustable ties
   B. Re-anchoring systems
   C. Horizontal joint reinforcements
   D. Unit ties
3. Rectangular ties, Z ties, and corrugated ties are types of ____.
   A. adjustable ties
   B. re-anchoring systems
   C. horizontal joint reinforcements
   D. unit ties
4. Vertically reinforcing bars should be placed every _____ in footings to tie the footing to the concrete block wall.
   A. 1″
   B. 6″
   C. 1′
   D. 4′
5. A standard length of reinforcing bar includes _____.
   A. 20′
   B. 25′
   C. 50′
   D. 75′
6. *True or False?* Fiber-reinforced plastic rebars or coated bars are used when corrosion protection is required.
7. A(n) _____ is placed in fresh concrete used to attach sill plates to masonry work.
   A. corrugated anchor
   B. strap anchor
   C. bearing anchor
   D. anchor bolt

8. Bearing walls that intersect are frequently connected with a(n) _____.
   A. corrugated anchor
   B. strap anchor
   C. bearing anchor
   D. anchor bolt
9. *True or False?* Joint reinforcement is especially designed for placement in the vertical joints of masonry walls.
10. *True or False?* A #8 sized reinforcing bar measures 3/4″ in diameter.

Apply and Analyze

1. List four functions of wall ties.
2. Metal ties or wires are widely used for structural bonding in what three types of construction?
3. Identify the four basic types of metal wall ties.
4. List three applications of adjustable tie assemblies.
5. Identify the devices used to protect employees from getting injured by exposed reinforcement bars protruding from concrete footings.
6. What three identification marks are shown on grade 40 and grade 50 bars?
7. What type of anchor is usually used to attach roof plates and sills to masonry work?

Critical Thinking

1. Imagine you are a foreman, and the superintendent of the project explains the following:
   - Every other course (16″ in height) a continuous joint reinforcement wire will be placed horizontally on top of that course.
   - Continuous joint reinforcement will be lapped according to code specifications.
   - Currently, reinforcement bars are protruding every 4′ from the top of the footing. When blocks are laid in the locations of the footing reinforcement, those concrete blocks will receive reinforcement that will continue up through the courses to be filled with grout at a later time.

   Research this chapter to create instructions to be presented to the masons for the following:
   A. Procedures to install continuous joint reinforcement
   B. Procedures to meet building code specifications where splices are to occur in the joints
   C. Correct joint reinforcement when rebars are used in specified block cores that are then filled with grout

Activities

1. (ELA) Conduct research on one type of wall reinforcement. Determine the benefits and disadvantages of that type of reinforcement. Are there any new designs or techniques related to this type reinforcement? Prepare a class presentation that answers these questions.
2. (MATH) Determine the amount of rebar needed for a building footing that is 30′ wide and 76′ long. Two rebars will be used to reinforce the footing. An extra 5% will be required for lapping of rebar at splices and for waste.
   A. Calculate the total footage length of rebar needed to reinforce the footing. (Remember, there are two rebars running the entire footing.)
   B. Add 5% for splices and waste to the footage length of the rebar. What is the total amount with 5% added?
   C. Rebar is ordered in 20′ lengths. Determine the amount of rebar to be ordered to complete the job. Round up your answer if it includes a portion of a complete rebar.

SECTION

4 Techniques

Alison Hancock/Shutterstock.com

Lex20/iStock via Getty images

CHAPTER 12

Laying Brick

LEARNING OUTCOMES

After studying this chapter, you will be able to:

12.1-1 Explain how to lay brick, spread mortar, form a head joint, cut brick, and use mason's line.
12.2-1 Summarize how to locate a building line and erect batter boards.
12.3-1 Discuss the methods of laying several common brick walls.
12.4-1 Describe advanced techniques, including corbeling a 12″ wall and laying a hollow brick pier.
12.5-1 Recall appropriate methods of cleaning new masonry.

TECHNICAL TERMS

batter boards
buttering
corbeling
corner pole
cross joint
drop line
gauge stick
header high
lead
mortar fin
plaster back
racking
spalling
spotting the brick
weep hole

Introduction

Laying brick masonry units involves knowledge of planning and materials, and the ability to use tools properly. Laying masonry is an art that requires a great deal of practice. However, the sense of satisfaction gained from doing a job well is worth the effort. This chapter is designed to help you master the proper techniques used in laying several common types of brick walls.

12.1 Basic Operations

A number of basic operations must be mastered before a masonry structure can be built. These basic operations include planning and preparing the site, spreading mortar, holding a brick, using both hands at the same time, forming a head joint, cutting brick, leveling and plumbing brick, and using a mason's line. Much practice is needed to master these operations.

Planning and Site Preparation

Prior to any masonry work being performed, the construction company plans for the materials to be used during the construction project. An estimator determines quantities and then orders required materials for the project. The supervisor has the materials delivered to the construction site when preparation has concluded, and the site is ready for work to begin. As for the brick, keep in mind there are many different types, sizes, and color selections that can be chosen. Some may take a few weeks to be delivered, others may take a month or longer depending on the type, uniqueness, and quantity selected.

One of the first brick operations to occur is organizing the site for brick laying. This can be for practicing, such as in an educational setting where bricks are laid in a masonry shop or at an outside location. The other type of preparation is to organize the construction site in advance of the work beginning. The foreman or supervisor clears areas near the project so mortar stands and mortar boards can be erected close to the locations where bricks are to be laid. Cubes of bricks are positioned or *spotted* at different locations around the site to reduce repositioning in the future and decrease time to stock or restock the masons with bricks. See **Figure 12-1**. A mortar mixer is positioned in an area that is easily accessible and centrally located. There may be other devices or equipment located at the jobsite, such as masonry saws and a construction trailer.

Usually, large quantities of bricks are delivered to a construction site on a tractor-trailer. Bricks are delivered stacked as a cube. The number of bricks in a cube varies depending on the size and shape selected. A brick cube usually has plastic bands that hold the individual bricks together in a uniform cube. Depending on the manufacturer, some brick bands may be made of metal. Brick cubes are usually unloaded from delivery trucks using a construction forklift. This same forklift can be used to spot brick cubes at appropriate locations for easy access around the construction site. See **Figure 12-2**.

Author's image taken at Job Corps, Denison, IA

Figure 12-1. This brick cube is attached to a wooden pallet to aid in moving the cube and to keep the brick clean when the jobsite is muddy. The cube is wrapped in a plastic covering to protect the brick and keep them dry from inclement weather.

Miljan Zivkovic/Shutterstock.com

Figure 12-2. A construction forklift is used to move brick cubes and other materials around the construction site.

Spreading Mortar

Spreading mortar is part of all masonry construction. With practice, you will be able to determine just how much mortar to use for a given joint. Placing the unit is easier if the right amount of mortar has been laid down.

Trade Tip

A mason should know the different parts of the tools they are working with, especially the trowel, which is the most used tool. **Figure 12-3** identifies parts of a trowel.

Select a trowel that is best for the job being performed and that "feels good" to you. If a trowel is too small, effort is wasted in needless movements. If it is too large, you will tire easily. You may want to begin with a standard mason's trowel, about 10″ long.

Marshalltown Company

Figure 12-3. The different parts of a masonry trowel are identified.

Grasp the trowel in your dominant hand. Fingers should be under the handle and the thumb on top of the ferrule as shown in **Figure 12-4**.

Timothy L. Andera

Figure 12-4. This London pattern trowel is being held correctly, with the fingers under the handle and thumb on top of the ferrule.

Loading a Trowel

Mortar may be loaded on a trowel in several ways:

- From the side of the board
- From the middle
- From the top of the pile

Figure 12-5 shows the steps for one method of loading a trowel. Develop the method that works best for you and perfect it. It is important to wear appropriate gloves when working with mortar or similar products because of the caustic chemicals contained in the mixture.

Spread the mortar with a quick turn of the wrist toward the body and a backward movement of the arm. When the trowel is nearly empty, tip the trowel blade even more to help the remaining mortar slide off. See **Figure 12-6**. The kind of mortar and its consistency most often determines the speed of the stroke.

When solid masonry units are used or the first course of brick or concrete block is being laid, the mortar is

Step 1: Work the mortar into a pile in the center of the mortar board.

Step 2: Smooth off a place with a backhand stroke.

Step 3: Cut a small amount from the larger pile with a pulling action.

Step 4: Scoop up the small pile with a quick movement of the trowel.

Author's images taken at Job Corps, Denison, IA

Figure 12-5. Steps in loading a trowel.

A

B

Author's images taken at Job Corps, Denison, IA

Figure 12-6. Steps in emptying a trowel. A—The trowel shows a portion of the mortar already unloaded off the edge. B—The mason has completely emptied his trowel to create a mortar bed very close to the snapped chalk line on the concrete.

spread and furrowed with the point or toe of the masonry trowel, **Figure 12-7**.

Furrowing helps to form a uniform bed to lay the masonry on. When mortar is laid down, it should be slightly heavier on the outside (face) edge than on the inside in order to force out enough mortar to the edge and up into the next cross joint when the following brick is laid. A ***cross joint*** is another name for the brick head joint in courses being laid. Mortar is placed on the leading end of the brick, and then set on top of the bed of mortar. After being positioned, the brick is pushed against the previously laid brick to create the head joint, or cross joint. At this point, the mason looks at the spacing of the head joint and bed joint to ensure that they are correct in width. Adjustments are made accordingly. Mortar hanging over the sides should be cut off using the front half of the trowel and returned to the mortar board or applied to the next brick to be laid.

Author's image taken at Job Corps, Denison, IA

Figure 12-7. Furrowing the first course to form a uniform bed.

Mortar Consistency

Mortar on the board should be kept well-tempered (sprinkled lightly) with water until it is used. Avoid constant working of the mortar, which causes it to dry out and become stiff. Stiff mortar does not spread well. Use the mortar before it begins to set up. Mortar that is not used within 2 1/2 hours should be discarded.

Safety Note

Contact with wet (plastic) concrete, cement, mortar, grout, or cement mixtures can cause skin irritation, severe chemical burns, or serious eye damage. Wear waterproof gloves, a long-sleeved shirt, full-length trousers, and eye protection when working with these materials. If you must stand in wet concrete, wear high-top waterproof boots. Wash wet concrete, mortar, grout, cement, or cement mixtures from your skin immediately. Flush eyes with clear water immediately upon contact. Seek medical attention if you experience a reaction when coming in contact with these materials.

Holding a Brick

Bricks should be held correctly when applying mortar and laying the brick on the mortar bed. The thumb should curl down over the top edge of the brick and be slightly away from the brick face. The fingers should be extended outward. It is important not to touch the line when laying units to a mason's line. The hand position described allows the line to fit between the thumb and the top edge of the brick. When laying a veneer facing, this position allows the line to fit between the fingers and top edge of the brick. See **Figure 12-8**. This hand position also does not interfere with cutting off excess mortar.

Author's image taken at Job Corps, Denison, IA

Figure 12-8. The proper way to hold a brick when laying to the line.

Author's image taken at Job Corps, Denison, IA

Figure 12-9. Applying mortar to a brick to form a head joint.

Using Both Hands at Once

Learn to use both hands at the same time when laying brick. For example, if you are right-handed, pick up the brick with your left hand and load the trowel with the right when laying a veneer facing. Both hands should be used to pick up bricks and stack them on the finished back wythe of a solid 8″ wall. If the bricks are stacked before any mortar is spread, less bending and reaching is necessary. Using both hands at the same time increases efficiency and reduces fatigue.

Forming a Head Joint

Mortar for a head joint should be applied to the brick before it is placed on the mortar bed. The first brick is placed on the mortar bed with no mortar on it. Mortar for the head joint is applied to each brick thereafter using a swiping or throwing action across the end of the brick. This motion attaches the mortar to the brick and forms the mortar into a wedge shape. See **Figure 12-9**. The brick is then pressed into place, squeezing out some of the mortar. Getting just the right amount of mortar on the brick takes practice.

Trade Tip

The application of mortar on a masonry unit with a trowel is referred to as ***buttering***.

Cutting Brick

Most bricklaying jobs require some brick to be cut to size for a particular application. Various names are used to describe cut or broken brick pieces, **Figure 12-10**. Brick can be cut using a brick hammer, brick set chisel, trowel, or masonry saw.

Safety Note

When performing any cutting procedure, be sure to wear safety glasses or safety goggles. When using a masonry saw, always wear ear protection and a face shield. If the saw is set up for wet sawing, wear rubber gloves. When dry sawing, always use a respirator because brick may contain silica, which is a severe health hazard.

Cutting Brick with a Brick Hammer

A brick hammer is the tool used most often by masons to cut brick. The recommended procedure is to hold the brick in one hand, keeping the fingers away from the side where cutting takes place. With the other hand, strike the brick with light blows. The blows are struck with the blade edge of the brick hammer along the line where the cut is planned. Turn the brick over to the adjacent edge and continue the cut along that side. Repeat the process until all sides have been scored. Then strike the face of the brick with a sharp blow with the hammer. The brick should break along the scored lines. This procedure produces a rough cut but is acceptable for applications where the cut edge is hidden by mortar.

Cutting Brick with a Brick Set Chisel

When a more accurate, straight cut is needed, a brick set chisel is used. First, mark the brick with a pencil or other marking device where the cut is to be made. Place the brick on a soft surface such as soil or a board. Hold the chisel of the brick set vertically with the flat side of the blade facing the direction of the finished cut. Strike the brick set sharply with the brick hammer. The resulting cut should be relatively smooth and should not require additional chipping before use in the wall.

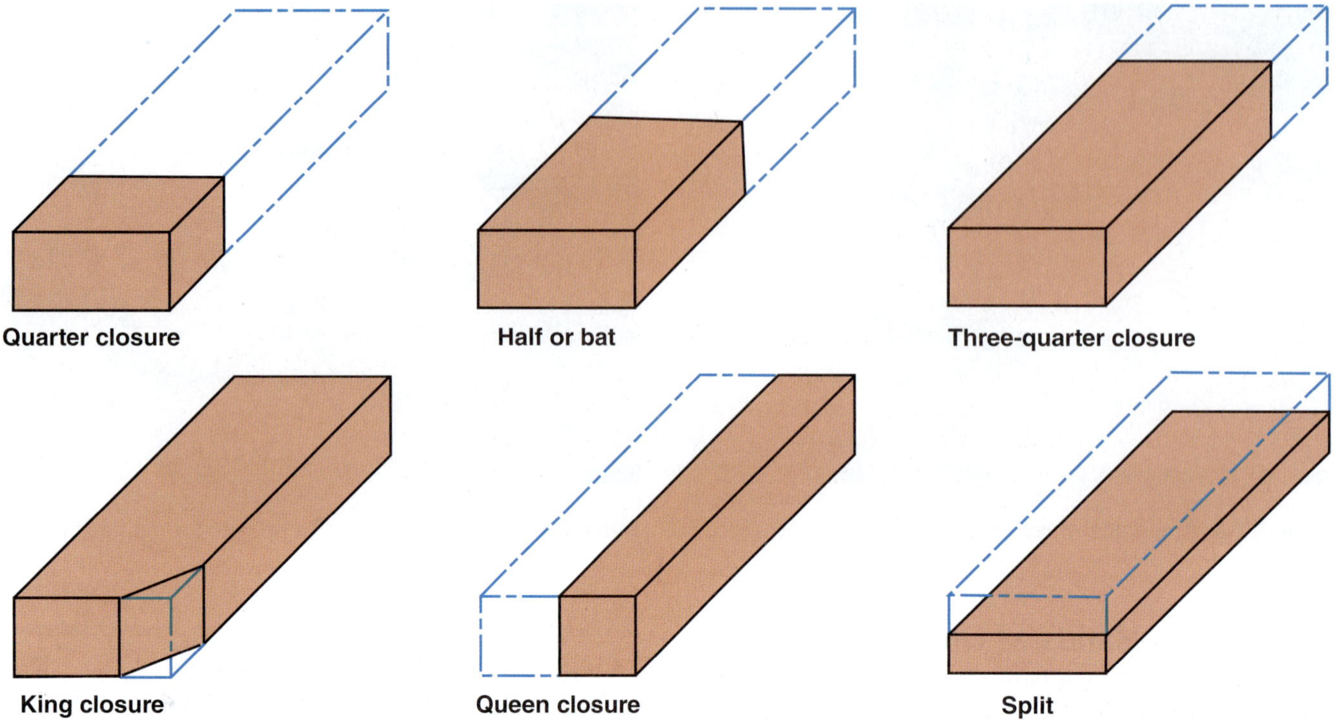

Goodheart-Willcox Publisher

Figure 12-10. Names used to describe cut or broken brick pieces.

Cutting Brick with a Trowel

A brick trowel can be used to cut softer brick, but is not recommended for most cutting, especially hard brick. Even though it is a common practice to cut brick with a trowel when a brick set chisel and hammer are not within reach, a trowel is not intended as a brick cutting tool.

When cutting soft brick with a trowel, hold the brick in one hand, keeping the fingers away from the side where the cutting is to be done. Hold the brick down and away from the face. Strike the brick with the edge of the trowel using a quick, sharp blow at the spot where the cut (break) is intended. See **Figure 12-11**.

Cutting Brick with a Masonry Saw

When an exact, smooth cut is required, a power masonry saw is used. This process is slower but produces the highest quality cut. Be sure you have had proper instruction on the use of the masonry saw. Move the saw slowly through the cut. Move the saw back to

A B

Author's images taken at Job Corps, Denison, IA

Figure 12-11. The mason is holding a brick securely just prior to it being struck. Then, the brick has been struck and is cut into two separate pieces.

its original position before removing the pieces. Some masonry saws have a foot pedal to move the saw and blade up and down during the cutting process and a table that slides back and forth. Practice using the saw table and foot pedal prior to actually cutting brick. See **Figure 12-12**.

Leveling and Plumbing Brick

To level brick, place a level horizontally on the top surface of the brick or bricks. Then look at the glass vial in the level to determine if the internal bubble is exactly in the middle between the two indicator lines. If the bubble is exactly in the middle, the surface being measured for angle is level. If the bubble is not centered, lightly tap on the high end of the brick or bricks to move them down into the mortar. Continue to tap the bricks until the bubble is exactly in the middle of the two lines.

Turn the level 90° to check if the brick is level side to side. If the brick is not level, tap the brick (not the level). Once the brick is level side to side, check the bricks along their length to ensure they have remained level.

Plumbing is the process of determining if the side and end of a brick or bricks are vertically plumb. If the top surface of the brick is level end to end and level side to side, the bricks should be very close to being plumb. Set the level vertically and place it on the side of the brick. Check the vertical vial to see if the bubble is in the middle. Tap lightly on the brick to adjust the side. Check the end of the brick the same way and make adjustments. Lay the level horizontally across the surface of the brick or bricks again to recheck them to ensure they have stayed level.

Author's image taken at Job Corps, Denison, IA

Figure 12-12. Masonry student has marked the brick to the correct length prior to starting the electric masonry saw. The student is wearing appropriate PPE: rubber gloves, safety glasses, and a hi-vis vest.

Multiple Brick Leveling and Plumbing

When leveling multiple bricks, ensure that all are flush across the top and flush with the level when it is laid on top of the brick. If one or more bricks are not flush with others, make adjustments until all are flush, the level rests on all brick uniformly, and the vial indicates levelness. When plumbing multiple courses, ensure that all bricks are again flush with the level and that no one brick sticks out beyond another. Make adjustments to correct bricks that are protruding and to ensure that the brick as a unit is plumb.

Using Mason's Line

Any straight wall longer than the plumb rule (4′) is usually laid to a line. A line is a guide that helps bricklayers build a straight, level, and plumb wall. Mason's line is a light, strong cord that can be stretched taut with little or no sag. Pull the line to the same degree of tautness for each course. This is necessary so all bed joints are uniform and parallel.

A ***lead*** is usually built first at each corner to establish proper height and provide a place to attach a mason's line. See **Figure 12-13**. The height of each course is determined by a mason's rule or a corner pole with height marks on it. On long walls, it is impossible to eliminate sagging in the line. The proper course height should be checked with a corner pole to ensure accuracy.

Avoid laying brick too close to the line. Stay a line width away so you do not risk moving the line, thereby destroying its usefulness as a reference. After laying the brick, ensure that it does not touch the line. If the brick is touching the line, the straightness of the wall will be affected.

> **Trade Tip**
> Laying brick too close to the line is known as *crowding the line*.

12.2 Related Tasks

The tasks discussed in this section may or may not be performed by masons. These tasks include locating the building line, erecting batter boards, and using a corner pole and gauge stick.

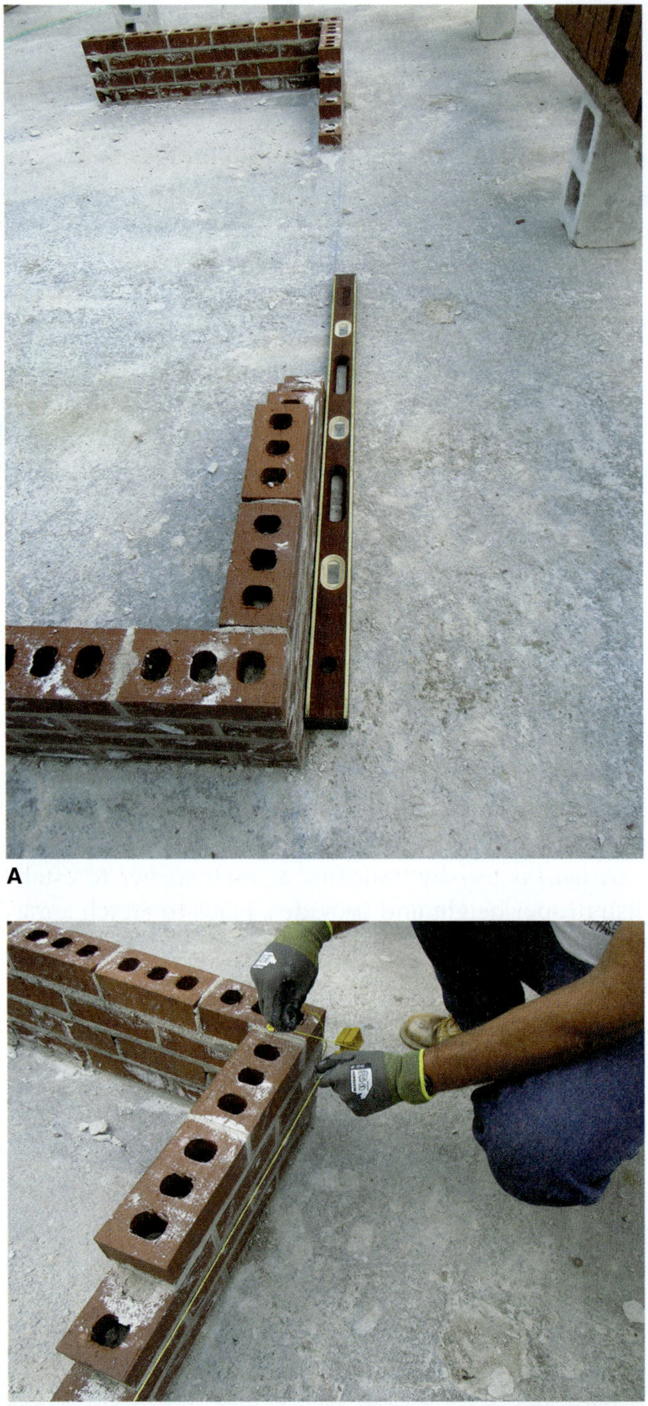

A

B

Author's image taken at Job Corps, Denison, IA

Figure 12-13. A—Leads are the built-up sections of the wall at each corner. Leads establish proper height for each course and provide a place to attach a mason's line. The blue chalk line will also be used to aid in determining the position of the first course of brick. B—Mason is attaching a corner block to the outside of the corner lead to establish height for the first course of brick to be laid between the corner leads.

Locating the Building Line

The building line is generally the face or side of the building being constructed. It is an important line because all crafts involved in the construction of the building refer to it. The line is frequently referenced on drawings. In residential construction, the contractor usually establishes the building line for the foundation walls. In commercial construction, a surveying crew will most likely locate the building line as defined on the plot plan.

In some towns and cities, the zoning commission has already determined the distance the building line or setback is to be from the property line, center of the street, or some other reference point. The building line makes the front of all buildings on the street conform to a straight line. In most situations, the actual building line can be at a greater distance than the minimum required setback, but not closer than the specified distance listed in the code. Therefore, building lines must be established exactly to prevent a code violation or other complications.

Trade Tip

It is good practice to check with the local building official to determine setback locations and to ensure accuracy. In certain communities, the building official may need to approve the location of the building lines prior to excavation.

Erecting Batter Boards

Batter boards are boards and stakes that locate and preserve the building lines during excavation and construction. Right angle boards are formed by nailing two boards to three stakes to form the angle. See **Figure 12-14**. Straight batter boards are made with one board and two stakes. Material used for batter boards is usually 1 × 4 or 1 × 6 boards, and the stakes are 2 × 4s.

Batter boards are located in each corner of the structure to position corners of the building. The front edge and one side are measured from property lines or other points on the property. The 3-4-5 method is used to ensure that the layout of the first corner is square. The 3-4-5 method of checking for squareness is explained in Chapter 6, **Math for Masonry Careers**.

Stakes are driven into the ground outside the area where the excavation will be performed and later the foundation walls will be built. Stakes must be placed well beyond the excavation area, (minimum 6′ to 8′) to ensure that the boards are not disturbed during excavation. If the boards are disturbed, the squareness of the building will be affected. The excavation process can remove from 3′ to 4′ outside the building lines. The extra space is needed to position the footing, erect forms, and build the foundation walls.

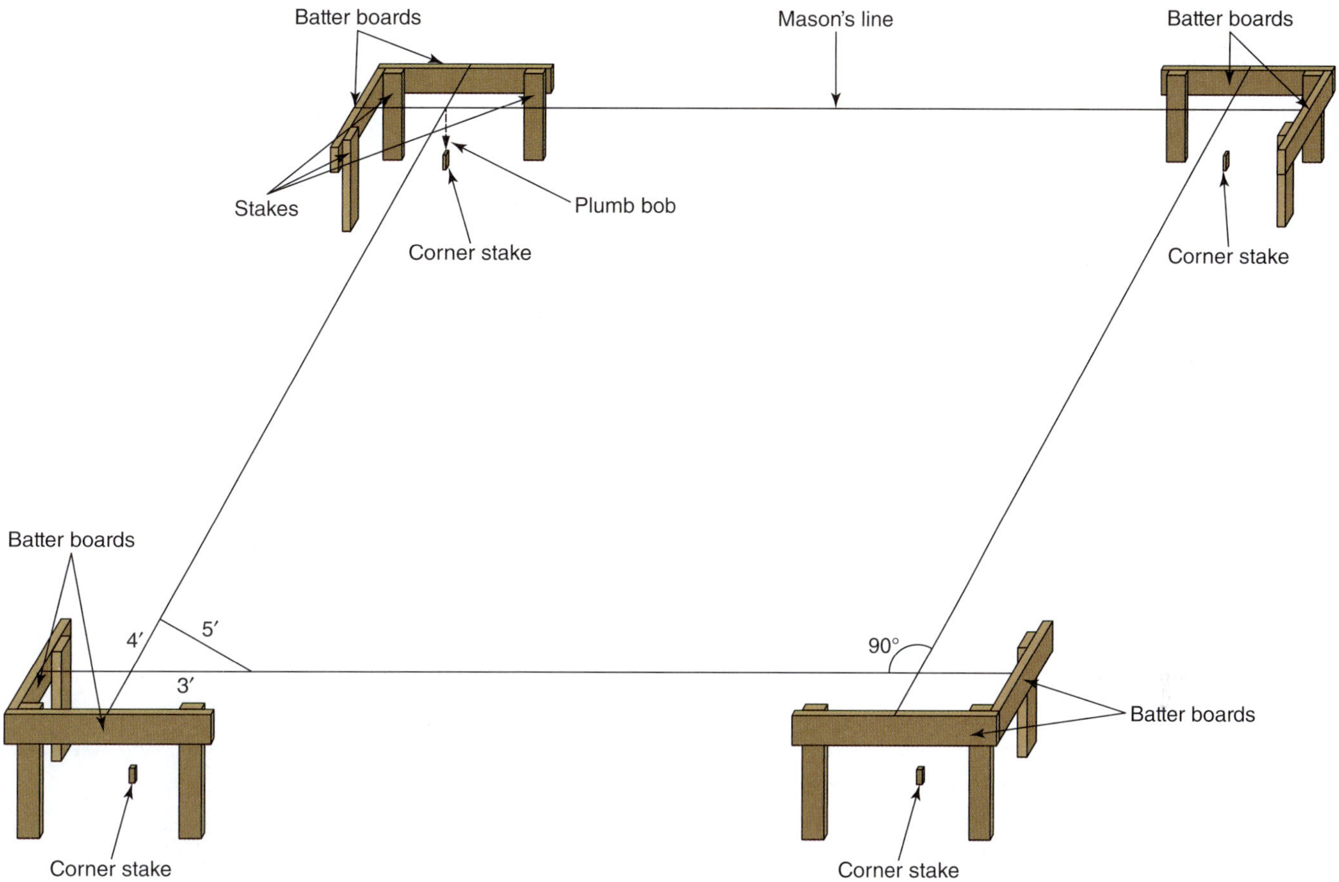

Goodheart-Willcox Publisher

Figure 12-14. Batter boards are located in each corner of the structure to position corners of the building. The 3-4-5 method is used to ensure that the layout of the first corner is square.

A builder's level, transit, or laser level is used to set the elevation height of the top edge of batter boards. The batter boards are usually installed at the future final height of the top of the foundation walls.

Mason's line is normally used when installing batter boards. The locations where lines intersect or cross over each other represent future wall corners in the structure. The locations of batter boards in the layout process are determined by the walls and corners shown on a foundation blueprint.

The following is the general procedure for installing batter boards.

PROCEDURE

Erecting Batter Boards

1. String mason's lines between batter boards to create a representation of the structure found on the blueprint.
2. Adjust location of lines and lengths of lines on the batter boards for accuracy.
3. Compare wall lines formed by the mason's lines to the blueprint to determine if they are accurate.
4. Once the line measurements correspond to the blueprint, measure diagonally from corner to corner between the lines and compare dimensions to ensure that the building layout is square.

Continued

PROCEDURE *(continued)*

Trade Tip

The building layout must be square and not a parallelogram. A parallelogram is a shape similar to a square or rectangle but the corners are not at 90° to each other.

5. If the dimension readings are the same, the building is square. If not, adjust the back corner lines on the batter boards, moving the lines the same amount in the same direction. Then recheck lengths of mason's lines that represent walls to ensure they are accurate compared to the blueprint.
6. Remeasure diagonally and compare dimensions to determine if they are the same. Continue these procedures until the walls are appropriate lengths and diagonal dimensions are the same.
7. Once squareness is achieved, mark locations with nails or saw kerfs to permanently register them on the batter boards.

After the excavation is performed, a drop line and plumb bob will be attached to locations where intersecting lines create corners. ***Drop line*** is a mason's line attached to a plumb bob to transfer line intersections down into the excavation to locate footings and walls. The drop line and plumb bob aid in determining the locations of footing and walls when preparing to install forms. Masons and other workers can refer to the lines created by the batter boards whenever they need to locate the different corners or face of the building.

Laying Out Wall Lines by Using 3-4-5 Method

The following information identifies procedures to follow when laying out building lines using the 3-4-5 method. For more information on the 3-4-5 method, refer to Chapter 6, **Math for Masonry Careers**.

Prepare the site where brick leads will be built. Determine and mark the first wall corner location on the surface. Measure the length and mark the second outside corner position. Hold the end of a chalk box line at one corner location, **Figure 12-15**. Extend the chalk box past the second corner positon and hold the line against the concrete. Raise the chalk line and snap it to mark the future brick wall location. See **Figure 12-16**. Recheck the length of the first wall and remark if needed. Use the 3-4-5 method to identify the first leg of the triangle. Measure 3′ from the corner and mark the location on the chalk line. To determine the second leg of the triangle, lay two tape measures on the chalk line. See **Figure 12-17**. All measurements will be performed

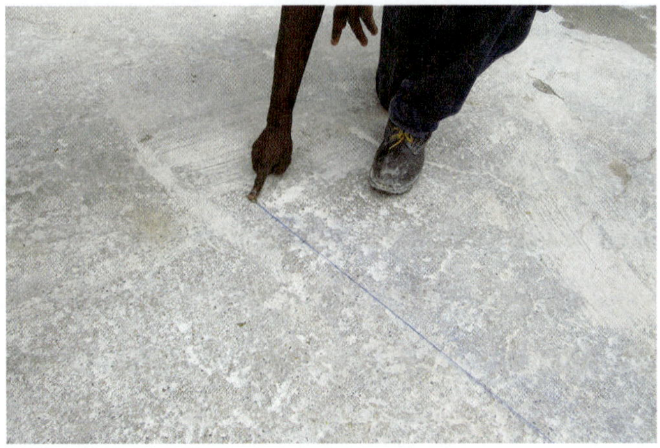

Author's image taken at Job Corps, Denison, IA

Figure 12-15. Hold the end of a chalk box line at one corner location.

Author's image taken at Job Corps, Denison, IA

Figure 12-16. Extend the chalk box past the second corner position and hold the line against the concrete. Raise the chalk line and snap it to mark the future brick wall location.

Author's image taken at Job Corps, Denison, IA

Figure 12-17. To determine the second leg of the triangle, lay two tape measures on the chalk line. All measurements will be performed from the inside edges of the tape measures.

from the inside edges of the tape measures. Hold the first tape at the mark representing the outside corner of the triangle. Hold the second tape at the 3′ mark on the chalk line on the concrete. Position tape measures to create a triangle so the two tapes intersect at 4′ and 5′. Mark the concrete at the intersecting inside edges of the two tapes at 4′ and 5′. This will identify the top part of the triangle. Position one end of the chalk box over the mark on concrete, representing the outside corner of the brick wall. The other end of the chalk box should be run past the mark where the intersection of 4′ and 5′ occurred. Snap the line. This line will be the second line of the 3-4-5 triangle. **Figure 12-18** shows that the three tape measures have been laid out to demonstrate the 3-4-5 triangle. Using the 3-4-5 method creates a perfect right angle as represented by the framing square next to the tape measures. See **Figure 12-19**. If larger distances are needed, multiples of 3′, 4′, and 5′ can be used.

Author's image taken at Job Corps, Denison, IA

Figure 12-18. The three tape measures have been laid out to demonstrate the 3-4-5 triangle.

Author's image taken at Job Corps, Denison, IA

Figure 12-19. Using the 3-4-5 method creates a perfect right angle as represented by the framing square next to the tape measures.

Using a Corner Pole and Gauge Stick

A ***corner pole***, sometimes referred to as a *story pole*, is a narrow board that is generally 1″ × 2″ and 8′–10′ long (about one story in height). A corner pole is used to measure the mortar joint heights of each course of masonry. See **Figure 12-20**. Each joint is marked on three sides of the board (pole), as well as window heights, door heights, sill heights, or any other feature that needs reference. A corner pole saves time when building corner leads by determining individual course heights.

Some jobs require more than one corner pole because different materials are used or because features such as windows and doors are at different elevations. A bench mark (nail or other permanent object) is generally used to ensure proper placement of the corner pole with respect to the building elevation. It is always positioned at that point for measurements.

A ***gauge stick*** is similar to a corner pole, but has the advantage of being shorter and less bulky, measuring about 4′ long and 3/4″ square. A gauge stick helps masons to lay all brick courses in uniform bed joints. Gauges are generally marked in terms of four brick courses. For example, an 11″ gauge would require 1/2″ bed joints for 2 1/4″ high brick (4 bricks × 2 1/4″ + 4 mortar joints × 1/2″ each = 11″). Masons also use a 6′ mason's folding rule to maintain proper course spacing.

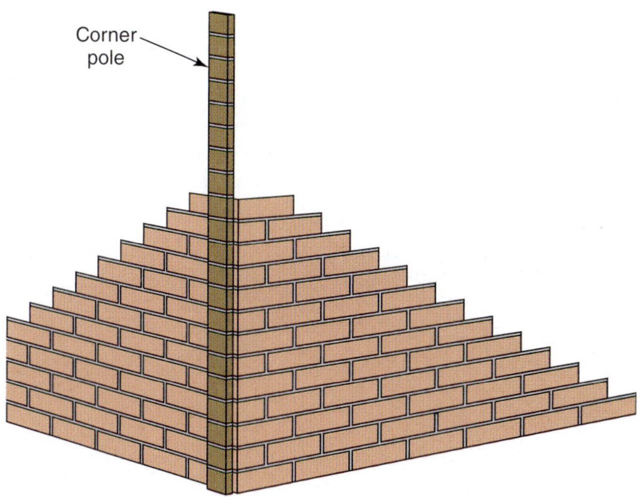

Goodheart-Willcox Publisher

Figure 12-20. Corner pole being used to set the height for the first brick in each course of the corner lead.

12.3 Laying Common Brick Walls

The procedures for laying several common brick walls are presented in the following sections. The types of walls covered in this chapter include the following:

- 4″ common bond wall with leads
- 8″ common bond, double wythe wall with leads
- 12″ common bond solid wall with leads
- Corner layout in Flemish bond
- Corner layout in English bond
- Cavity wall
- Single wythe brick bearing wall

Laying a Four Course, Single Wythe, Running Bond Lead

Corners (leads) must be carefully constructed because they are generally used as a guide for the entire wall. Laying the corners is called *laying the leads*. Two leads are used, one at each end of the wall. The first course contains three to five bricks laid in each direction from the corner. After the first course in the lead is completed, a second one is begun, starting again at the corner. The first brick in the second course will be laid in the opposite direction of the first brick in the first course. This procedure will create an overlapping process from one course to another. The last brick in the second course ends back from the first course. This technique is called ***racking***, racking back the lead, or rack back lead.

Figure 12-21 shows a rack-back lead, demonstrating how each course is stepped back, or rack-backed, from the previous course. A mason's line is stretched between the leads to serve as a guide for laying the rest of the brick between the corners. Sometimes an intermediate lead or single brick is positioned or laid with the appropriate masonry joint between the corner leads to reduce line sagging. This technique is sometimes called ***spotting the brick***. Placing a midpoint between corner leads reduces the chance of the mason's line sagging when corner leads are at a long distance apart.

Author's image taken at Job Corps, Denison, IA

Figure 12-21. The lead has been racked-back to create a stair-step effect. Eventually, the leads will identify height and position of each course to be laid between the corners.

PROCEDURE

Laying Up a Four Course, Single Wythe, Running Bond Lead

1. Snap a chalk line where the face of the masonry wall is to be located along adjacent sides. See **Figure 12-22**, Step 1. Mark the actual head joints along the chalk line for proper placement of the masonry units.
2. Start the first course by laying the first unit from the corner. Note that the placement marks for the head joints are visible even when the unit is in place. See **Figure 12-22**, Step 2.
3. Complete the first course along both legs of the lead. Each leg should be about the same length. Leads are generally about seven or nine courses high, but a shorter lead may be used for practice, as in this case. See **Figure 12-22**, Step 3.
4. Lay up the second and third courses, alternating the pattern of brick placement. Lay up the third course similar to the first course. See **Figure 12-22**, Step 4.
5. Lay up the fourth course to complete the lead. Two bricks on the highest course of the lead provide enough resistance to maintain a tight line. See **Figure 12-22**, Step 5.

Laying a 4″ Running Bond Wall with Leads

Place materials and arrange your workspace for efficient work. The mortarboard or boards should be located in the center of the workspace about 24″ from the wall. Stack brick on both sides of the mortar board or boards as shown in **Figure 12-23**.

PROCEDURE

Laying a 4″ Running Bond Wall with Leads

1. Establish a wall line using a chalk line. See **Figure 12-24**. The line is coated with colored chalk. Hold the line taut, then snap it to produce a chalked line on the surface you wish to mark. Check the slab or footing to ensure that it is square before snapping the wall line. The diagonal distance from corner to

Continued

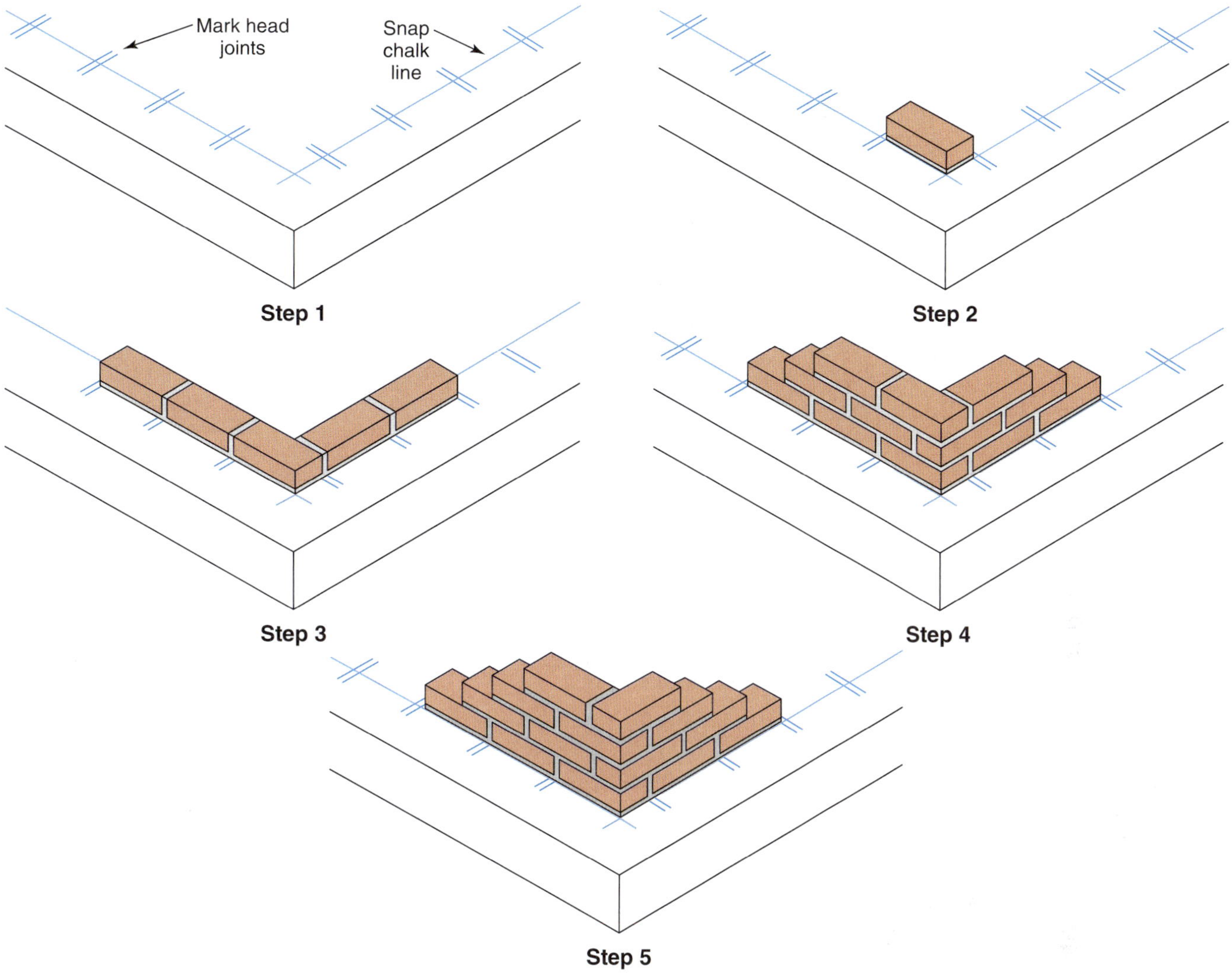

Goodheart-Willcox Publisher

Figure 12-22. Steps in building a four-course corner lead.

Author's image taken at Job Corps, Denison, IA

Figure 12-23. Materials have been arranged for efficient work. Multiple locations of brick, mortar boards, and stands have been positioned to reduce time when the mason is laying brick at the different locations. Planks are used to elevate brick to make it easier when selecting brick to lay.

PROCEDURE *(continued)*

corner should be the same. The squareness of a building is based on a right triangle. The 3-4-5 method can also be used to check for squareness.

Trade Tip

For larger layouts, multiples of the 3-4-5 method can be used to ensure corners are at 90°. For example, doubling the method would be 6-8-10, and tripling it would be 9-12-15.

2. After the wall line has been drawn or chalked, lay out a dry course of brick from corner to corner as shown in **Figure 12-25**. Arrange the bricks to eliminate as much cutting of units as possible. Allow for a mortar joint between each brick. Mark the joints and then move the bricks aside. See **Figure 12-26**.

Continued

PROCEDURE *(continued)*

A

B

Author's images taken at Job Corps, Denison, IA

Figure 12-24. A—Establishing the first wall line by using a chalk box and line. The line should be taut, then raised and snapped carefully to create one continuous line. B—Both brick wall lines have been located and are being marked with a chalk line.

3. Spread the mortar on one side of the corner and furrow it as shown in **Figure 12-27**.
4. Lay the corner brick exactly on the point where the corner is located. See **Figure 12-28A**. The corner brick must be set level and square with the wall line. Cut excess mortar and return it to the mortar board. See **Figure 12-28B**.
5. Lay the remaining four or five bricks of the lead corner. This is called *tailing out* the lead of the corner. After the bricks have been laid, level them with a plumb rule. After leveling, remove excess mortar and return it to the mortar board. This will help keep the level clean. See **Figure 12-29**.

Author's image taken at Job Corps, Denison, IA

Figure 12-25. The mason is using chalk lines as a guide to dry lay the brick. Joint spacings are added between each brick. Another reason for dry laying the brick is to reduce cutting as much as possible.

Author's image taken at Job Corps, Denison, IA

Figure 12-26. Mark the location of each brick, front and back, to identify the space for the mortar joint.

Author's image taken at Job Corps, Denison, IA

Figure 12-27. The mortar has been placed and is being furrowed. Be careful not to cover the chalk line as it is a guide for placing brick.

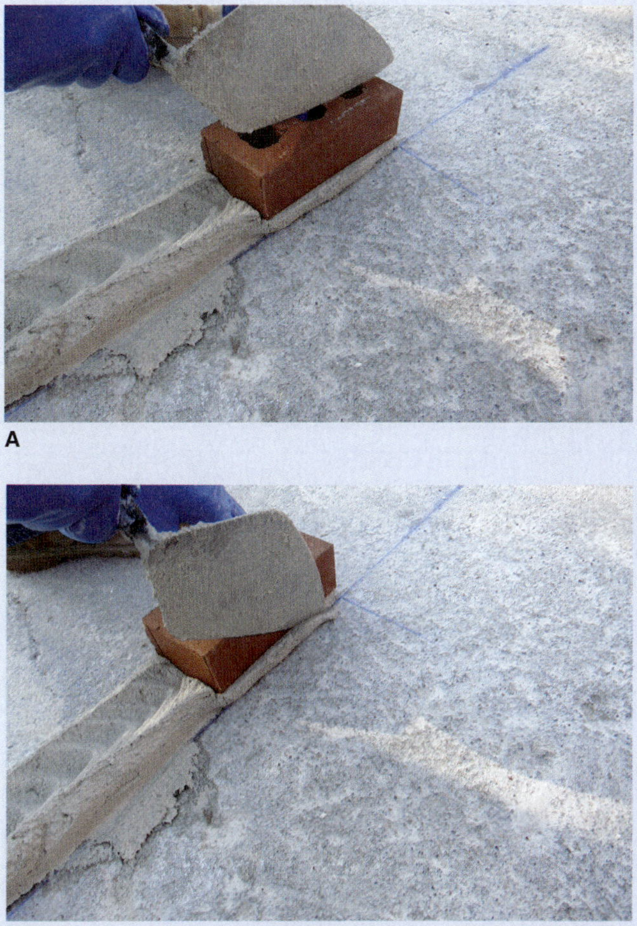

A

B

Author's images taken at Job Corps, Denison, IA

Figure 12-28. A—The corner brick is laid at the exact corner of the wall. Tap the top of the brick with the trowel to gently position it in the mortar. B—After positioning the brick, cut mortar and remove it to the mortar board. Cutting mortar and then picking it up reduces the chances of smearing the clean brick with mortar.

6. Plumb the corner brick. See **Figure 12-30**. Then plumb the tail end. See **Figure 12-31**. Leveling is done on the outside and top edge of the brick.
7. Line up the bricks between the two plumb points. See **Figure 12-32**. Here the level is used as a straightedge and the bubbles are disregarded. When building a corner, always follow the same sequence—level the unit, then plumb, and then line up.
8. After one side of the corner has been laid and trued, start the other side of the corner as shown in **Figure 12-33**. Again, spread the mortar and furrow it. Lay three or four bricks. Level, plumb, and line them up. Do not tap the level with a trowel or hammer; use your hand. This time the corner brick does not require plumbing because it has already been plumbed. However, if you think you might have moved it, recheck it. Also check to ensure that the two bricks that form the corner are flush across the top. This can

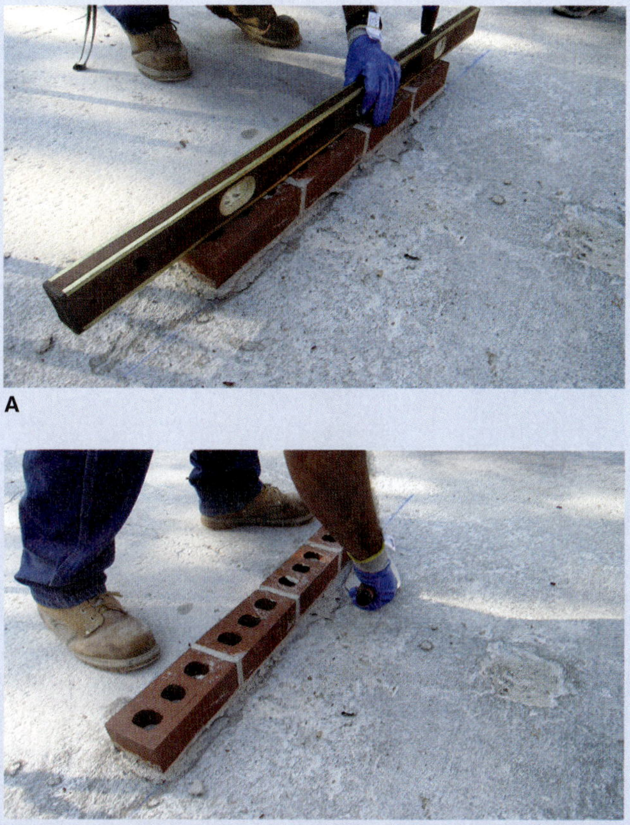

A

B

Author's images taken at Job Corps, Denison, IA

Figure 12-29. A—Level the first four brick with a mason's level. B—Remove excess mortar and return to mortar board. Performing this will aid in keeping the masonry level clean when plumbing and straight-edging the brick.

Author's image taken at Job Corps, Denison, IA

Figure 12-30. After leveling, plumb the corner brick to make its edge exactly vertical. The mason is lightly tapping on the outer edge of the brick to bring it into plumb.

Continued

PROCEDURE *(continued)*

Author's image taken at Job Corps, Denison, IA

Figure 12-31. Next, plumb the tail end brick. Note how the mason braces their arm against their knee to steady the mason's level. Brick is tapped lightly with the trowel to adjust it to proper height and position.

Author's image taken at Job Corps, Denison, IA

Figure 12-32. The first four brick are straight-edged with the mason's 4′ level. They lightly tap on the brick to adjust them into alignment.

be done by using the trowel lying flat and running it across the top of the two bricks. The trowel should glide easily from one to another; if not they are not flush across the top.

9. Lay the second course following the same sequence used for the first course. Check for proper height just after the course has been laid, but before the brick has been leveled. See **Figure 12-34**. If the bricks

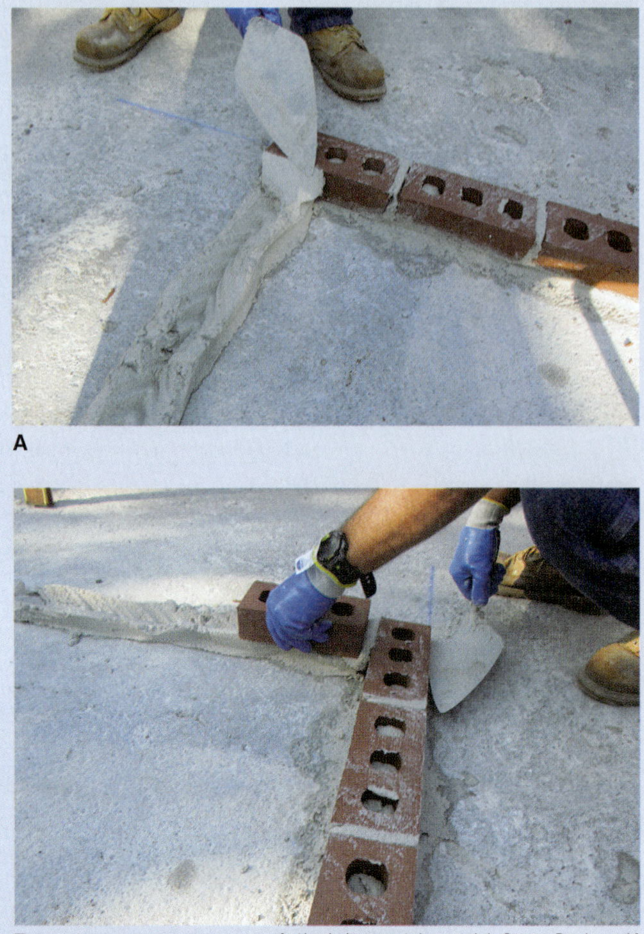

A

B

Author's images taken at Job Corps, Denison, IA

Figure 12-33. A—The other side of the corner lead is begun by spreading mortar and furrowing. Then, place mortar on the side of the first brick as shown. B—Place the first brick. Note how the mason supports the corner with the trowel.

Author's image taken at Job Corps, Denison, IA

Figure 12-34. Checking for the proper height with a rule. Hold the rule straight up and down for an accurate measurement.

are too high, tap them down as they are leveled. If they are too low, remove them and add more mortar before leveling.

10. Repeat the sequence until the corner is built. See **Figure 12-35**.
11. Straight-edge the rack of the lead (the brick alignment at the tail end of the courses on the lead), as shown in **Figure 12-36**, to eliminate any wind, belly, or cave-in in the wall. The rack refers to the "steps" formed by the last brick in each course. If the rule rests on each of them, the spacing is proper.

Trade Tip

Wind refers to twisting of brick, belly is the bowing out of brick, and cave-in is the bowing in of brick.

12. Lay the second corner of the wall following the same procedure used for the first. See **Figure 12-37**.
13. Stretch the mason's line between the corners at the top of the first course. See **Figure 12-38**. Secure the line with line pins or corner blocks. Pull the line taut enough that it is straight and level. Be sure to use the same amount of tension each time.

Author's image taken at Job Corps, Denison, IA

Figure 12-35. The completed right corner lead. The two corner leads will be used together as guides for the rest of the wall.

Author's image taken at Job Corps, Denison, IA

Figure 12-36. The rack of the corner lead is straight-edged to check for proper spacing of the brick, as well as wind, belly, or cave-in in the wall.

Trade Tip

When laying the first brick, leave a small space between the brick and the line. If the brick touches the line at any location between the two leads, the straightness of the wall will be affected. Maintain a uniform amount of space between the brick and line to lay the brick consistently.

A

B

C

Author's images taken at Job Corps, Denison, IA

Figure 12-37. Begin the second corner lead by using the same procedure performed in first corner lead.

Continued

PROCEDURE *(continued)*

A

B *Author's images taken at Job Corps, Denison, IA*

Figure 12-38. Laying the first course to the line. A—Install the mason's line into the twigs to guide the line so it is flush with the top of the corner lead course. The twigs will be held in place with a half brick on top of each twig. B—Lay bricks between leads, then install closure brick to finish first course.

14. Continue laying the wall from the lead toward the center, one course at a time, **Figure 12-39**. Each brick on the second course should be centered over the cross joint of the first course. The cross joints must be of uniform width or the closer brick (the last brick laid in the course) will not fit correctly. The course should be laid from both leads toward the center. Level, plumb, and line up this course as you lay it.

A

B *Author's images taken at Job Corps, Denison, IA*

Figure 12-39. Laying the second course from both ends toward the middle. A—The mason's line is in position. The twigs will be installed next. B—Bricks are being laid to the line. Note how the mason's fingers are holding the brick. Also, the twig and line are being held in place using a half-brick on end.

15. Tool or strike the joints when the mortar can be indented with a thumbprint without too much pressure ("thumbprint hard"). See **Figure 12-40**. If the mortar is too wet, the brick will be smeared. If too dry, the mortar will turn black from metal worn off the jointer. Use the short jointer for vertical or head joints. These joints are done first. See **Figure 12-41**. Then strike the bed joints using the long jointer. See **Figure 12-42**. With the edge of the trowel, cut off the tags of mortar that were forced out by the jointers. See **Figure 12-43**. To remove the tags, move the trowel horizontally along the joints.

Trade Tip

Be careful not to smear the tags or other mortar into the brick. Once adhered to the brick, mortar is difficult to remove.

16. Brush the wall when the mortar is stiff enough, **Figure 12-44**. Brushing may be done at the end of the day or more frequently, if desired. Brushing reduces the amount of cleaning required later.
17. Lay successive courses in the same manner until the wall is the required height. See **Figure 12-45**.

Code Note **Tooling Mortar Joints**

The National Concrete Masonry Association (NCMA) identifies the process of striking mortar joints. They stipulate that head joints should be struck first and bed joints second. NCMA identifies striking joints too early will cause them to shrink away from the masonry units being laid. Tooling too late will cause the joints to darken because of the pressure required from the tool. They recommend using the "thumb" test to determine the appropriate time to strike joints. Mortar joints should be struck when a thumb is pressed into the surface of the joint without leaving cement paste on the thumb. For more information on mortar joints, refer to TEK 05-16 publication from NCMA.

Author's image taken at Job Corps, Denison, IA

Figure 12-40. A masonry student is demonstrating the procedure of using a thumb print to determine if the mortar joint is ready to strike.

Author's image taken at Job Corps, Denison, IA

Figure 12-41. When the mortar joints are ready, they are struck with an appropriate jointing tool. A short jointer is used to strike the head joints. Head joints are struck first.

Author's image taken at Job Corps, Denison, IA

Figure 12-42. Bed joints are struck last with the long jointer, sometimes called a *sled-runner*.

Author's image taken at Job Corps, Denison, IA

Figure 12-43. Remove the mortar tags by moving the blade of the trowel forward along the head and bed joints.

Continued

PROCEDURE *(continued)*

Author's image taken at Job Corps, Denison, IA

Figure 12-44. Brush the wall when the mortar is stiff enough. Make sure that the brush selected has soft bristles; it is important not to damage the joints that have been struck.

Author's image taken at Job Corps, Denison, IA

Figure 12-45. The completed project is shown with the instructor and masonry student.

Laying an 8″ Common Bond, Double Wythe Wall with Leads

Common (American) bond is strong and can be laid up quickly. It consists of a course of headers between every five to seven courses of stretchers. The heading course serves as a bond between the inside and outside 4″ tiers. The mortar joint is 3/8″ or 1/2″ thick.

PROCEDURE

Constructing an 8″ Common Bond Wall with Leads

1. Lay out the wall location as was done with the 4″ wall. In addition, locate a second line (about 8″) inside the first wall line. This distance is equal to the average length of the brick being used.
2. Lay out the bond to eliminate excessive cutting for windows and doors. This is an important step!
3. Lay the corner bricks as shown in **Figure 12-46**. The header course is starting at the second course, but it can start wherever specified.
4. Lay the first course of stretchers for the corner. Four or five stretchers should be sufficient. Lay the header course as shown in **Figure 12-46**.
5. Lay successive courses of stretchers until you reach the next header course. Level, plumb, and line up each course as it is laid.
6. Lay the header course, breaking the bond in the usual manner.
7. Finish laying the first corner and complete the second in the same manner.
8. Construct the first stretcher course of the wall by laying the outside first and then the inside. The stretcher courses should be laid from the leads to the center. Level, plumb, and line up each course before beginning the header course. See **Figure 12-47**. A line should be used on all backing courses (the inside tier) to ensure a good face.
9. Lay the header course from each lead toward the center. Level, plumb, and line up the header course.
10. Lay the outside tier up to the next header course keeping it straight, level, and plumb. This is called ***header high***.

Trade Tip

The term *header high* refers to the distance up a masonry wall to the top of the brick course directly under a header course.

11. Lay the inside tier up to the same height. Be careful to keep it level with the outside tier as it is laid up. This is important because a level surface is required for the header course.
12. Continue as before until the desired height is reached. Strike the joints and brush the wall at the proper time.

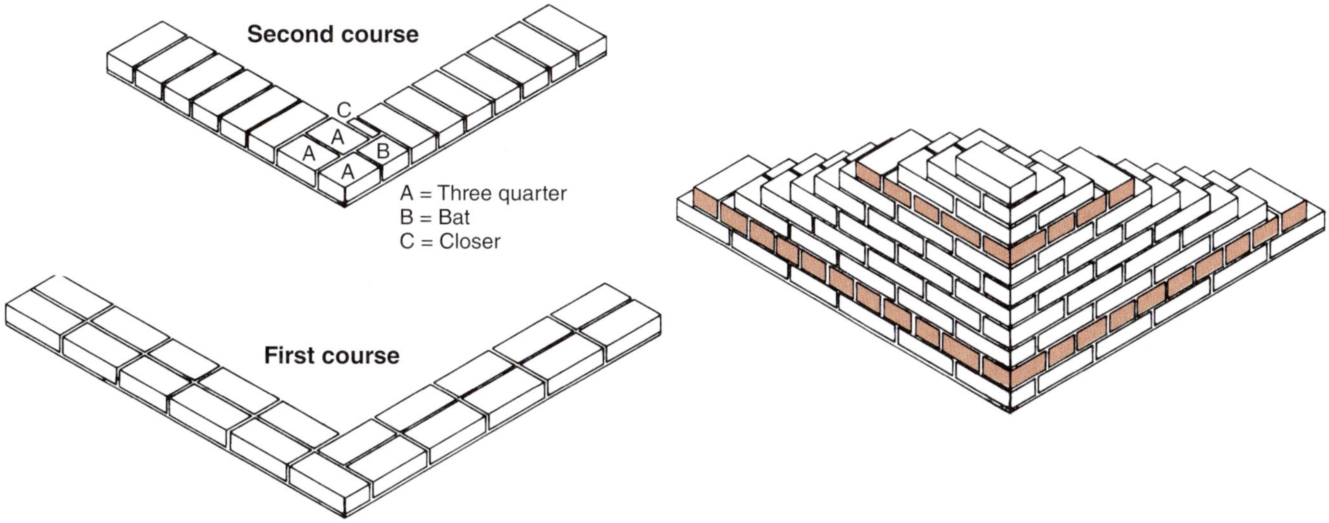

Goodheart-Willcox Publisher

Figure 12-46. An 8″ common bond wall lead with first and second courses shown in detail.

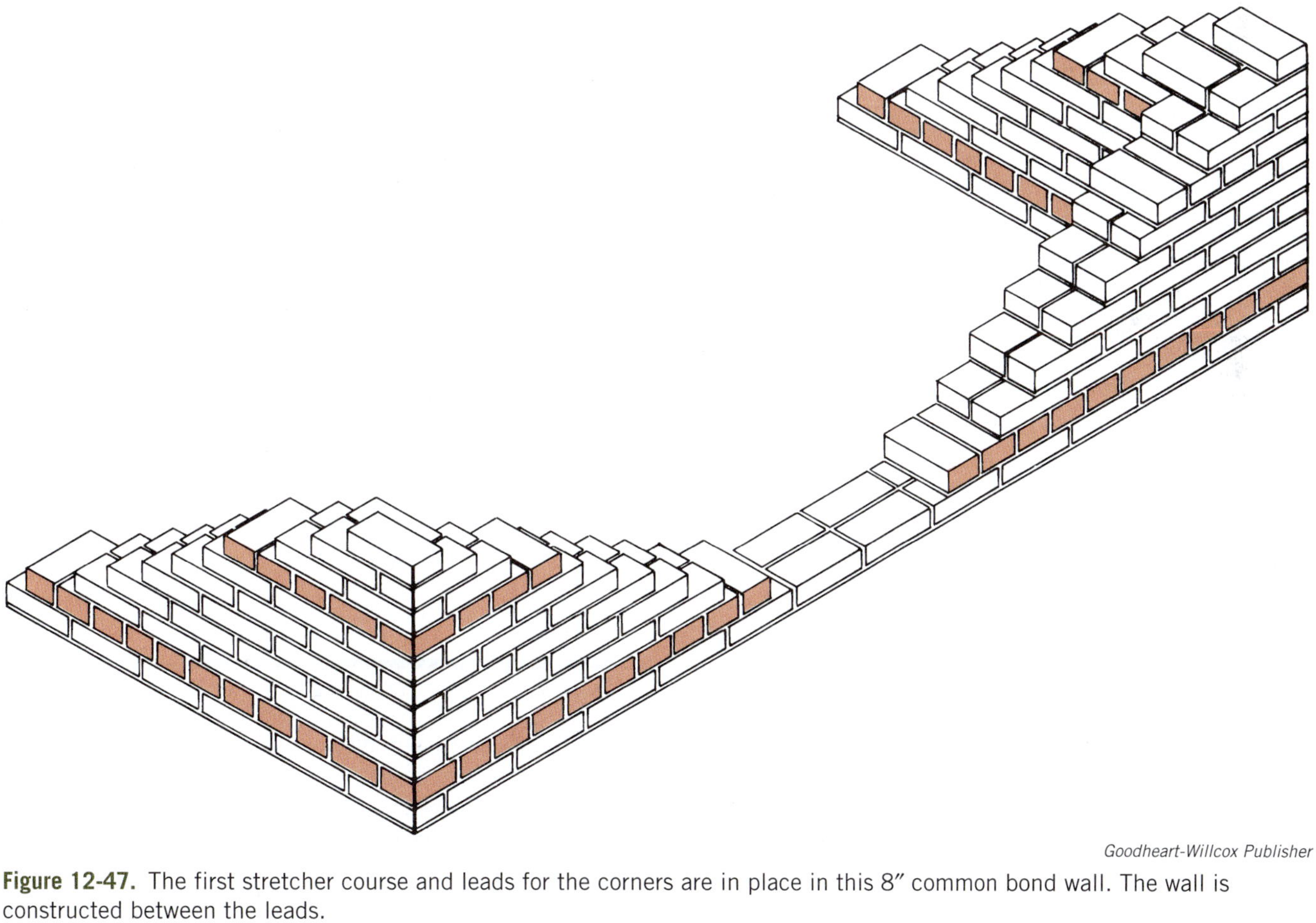

Goodheart-Willcox Publisher

Figure 12-47. The first stretcher course and leads for the corners are in place in this 8″ common bond wall. The wall is constructed between the leads.

Laying an 8″, Two Wythe Intersecting Brick Wall

Intersecting brick walls are often necessary in the construction of brick structures. The walls are laid out so the courses are tied together to form an integrated unit. Individual units are placed so they interlock the wall segments together. Reinforcement ties (Z ties) are generally used.

The following procedure can be used to construct an 8″, two wythe intersecting brick wall in standard running bond with joints broken so they do not line up vertically. Refer to the numbered bricks in **Figure 12-48**.

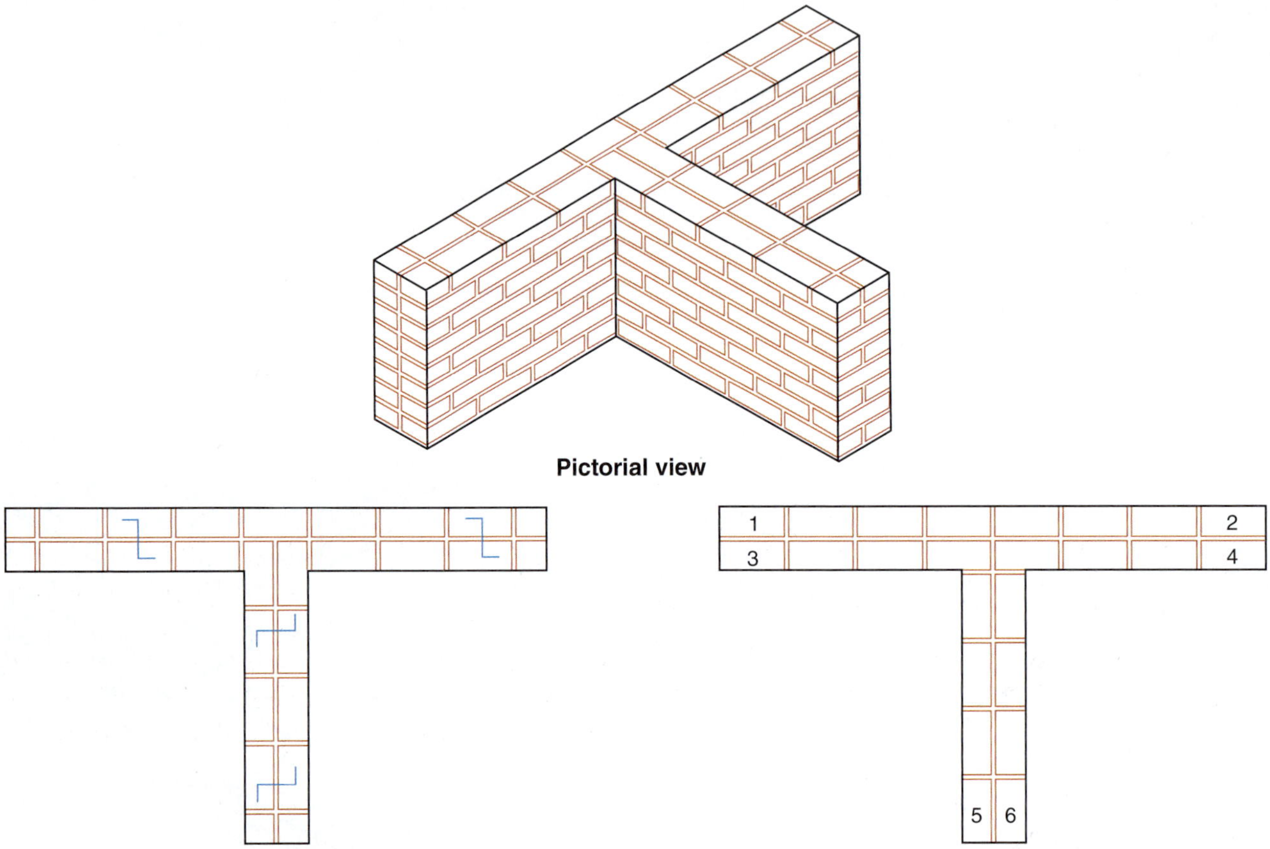

Goodheart-Willcox Publisher

Figure 12-48. The procedure for laying an 8″, two-wythe intersecting brick wall.

PROCEDURE

Laying an 8″, Two Wythe Intersecting Brick Wall

1. Locate the face of the walls on the floor and snap a chalk line to preserve the location.
2. Lay out the first course using no mortar (dry bond) to maintain proper spacing and identify any difficulties or problems.
3. Spread the mortar bed and lay out the face wythe beginning with brick #1 through brick #2 in **Figure 12-48**. Level, align, and plumb the course of brick.
4. Lay the backing wythe beginning with brick #3 and continuing with brick #4. Do not forget to butter the back side of each brick where the backing wythe sits against the face course. This joint (collar joint) should be completely filled. Level, align, and plumb the course and check the wall thickness.
5. Lay the intersecting wall brick beginning with brick #5 and then brick #6, running each course to the outside wall. Level, align, and plumb each course and check the wall thickness. This completes the first course plan.
6. Lay the second course following the layout shown in **Figure 12-48**. Level, align, and plumb the course. Notice how the intersecting wall is interlocked with the outside wall.

7. Locate the metal Z ties as shown in the second course layout plan in **Figure 12-48**. Embed the metal ties in a mortar bed and lay the third course identical to the first. Metal ties will be added again on the eighth course in the wall.
8. Continue laying courses, alternating the pattern used in the first and second courses until the wall has reached eight courses high. You may want to build end wall leads to serve as guides for the wall.
9. Level and plumb the completed wall.
10. Clean off any mortar splatter.
11. Tool the joints with a concave jointer when the mortar is thumbprint hard. Tool the bed joints first, then the head joints. Use a sled jointer for the bed joints.
12. Remove any mortar tailings (residue) remaining after finishing the joints. Clean the wall by brushing with a bricklayer's brush.

Laying a 12″ Common Bond Solid Wall with Leads

A 12″ solid wall is essentially an 8″ solid wall with a third wythe added. The first course is repeated every sixth or seventh course. It contains three-quarter closures, quarter closures, stretchers, and headers arranged in a particular configuration. See **Figure 12-49**. Headers are visible on the outer face of the wall.

The second course is composed of stretchers, headers, and quarter closures. See **Figure 12-50**. The stretchers are visible on the outer face, and headers show on the inside. The quarter closures are generally placed after the headers and stretchers are laid.

The third course is composed of all stretchers. See **Figure 12-51**. This pattern is continued for the fourth and fifth courses in typical running bond. The sixth course is identical to the first as the process is repeated.

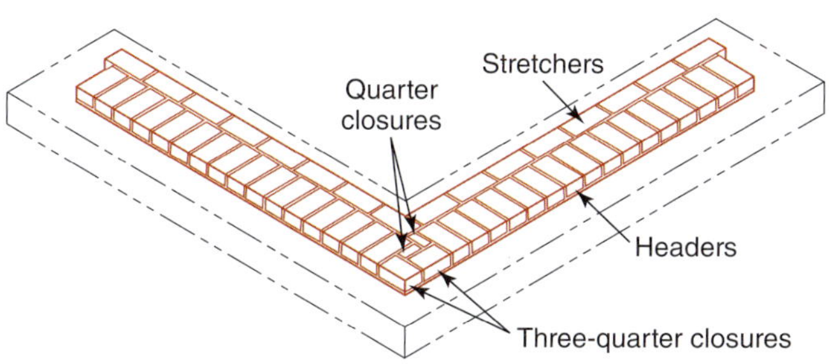

Goodheart-Willcox Publisher

Figure 12-49. The first course of 12″ common bond solid wall.

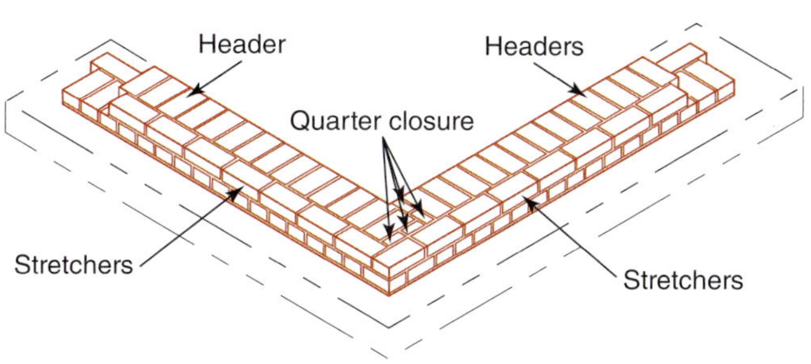

Goodheart-Willcox Publisher

Figure 12-50. The second course of 12″ common bond solid wall.

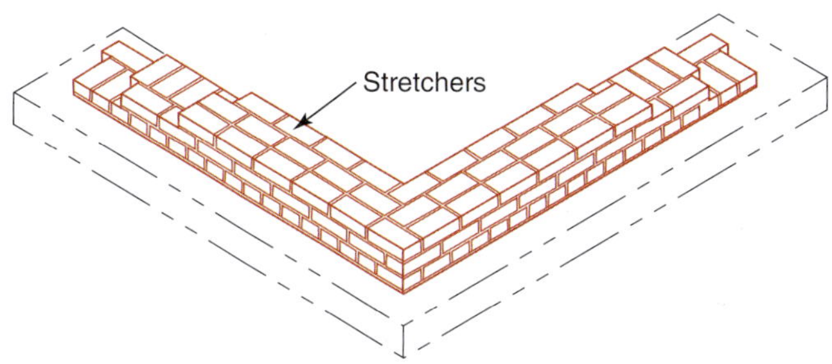

Goodheart-Willcox Publisher

Figure 12-51. The third course of 12″ common bond solid wall.

Twelve-inch brick masonry walls are rare today but may be used as thermal storage units. Solid, dark colored brick should be used for this application, since solid brick provides more mass than hollow or cored brick and dark brick absorbs heat more efficiently than lighter colored units.

Corner Layout in Various Bonds

Laying up a corner is similar for most bonds. After building several of the more common bonds, a mason should have no difficulty raising any corner. In addition to the common bond, the bonds described in the following sections represent some of the most frequently used patterns.

Flemish Bond

The popular Flemish bond is easy to lay and produces a visually interesting and pleasing wall. It is more costly than the common bond and requires careful workmanship. The bond consists of alternate headers and stretchers in each course. The headers are centered on the stretchers between each course. Bonds may be started at the corner with either a 3/4 closure or a 1/4 closure. See **Figure 12-52**.

English Bond

English bond has alternate courses of stretchers and headers. The headers center on the stretchers and on the joints between the stretchers. The stretchers all line

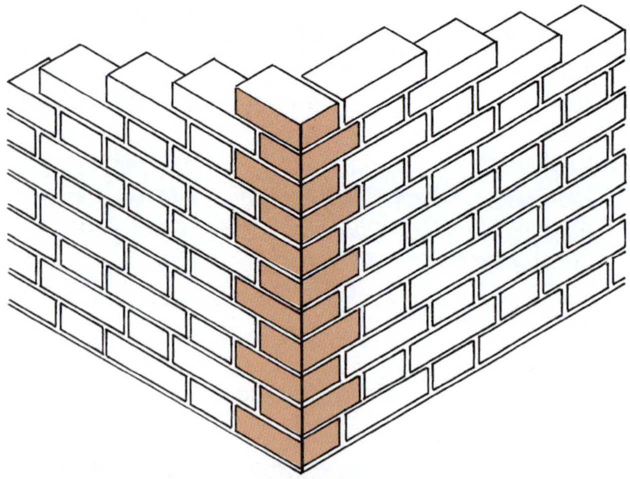

Flemish bond with 3/4 closure at corners

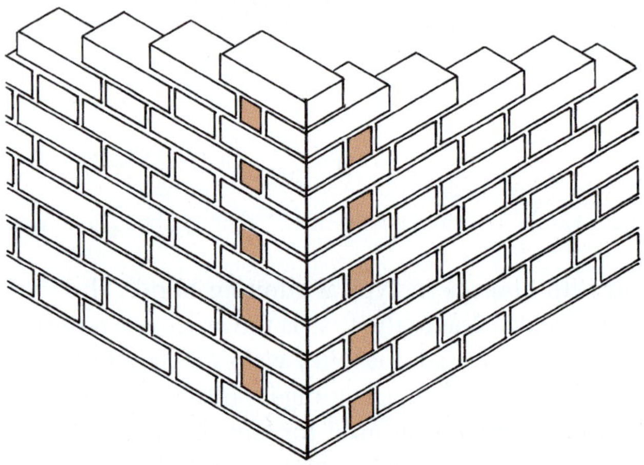

Flemish bond with 1/4 closure

Goodheart-Willcox Publisher

Figure 12-52. Corner layout in Flemish bond with 3/4 and 1/4 closure.

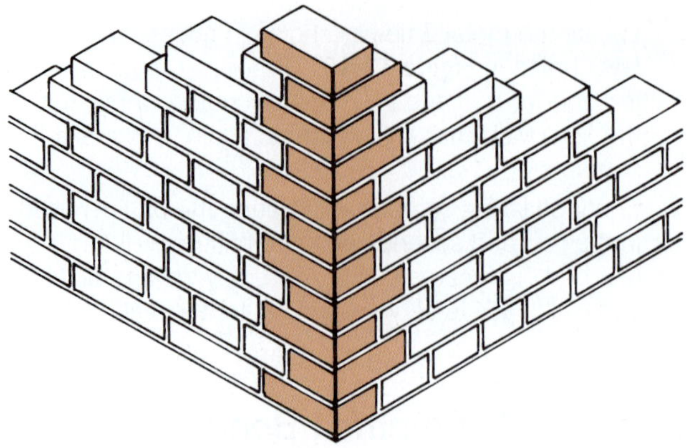

English bond with 3/4 closure at corners

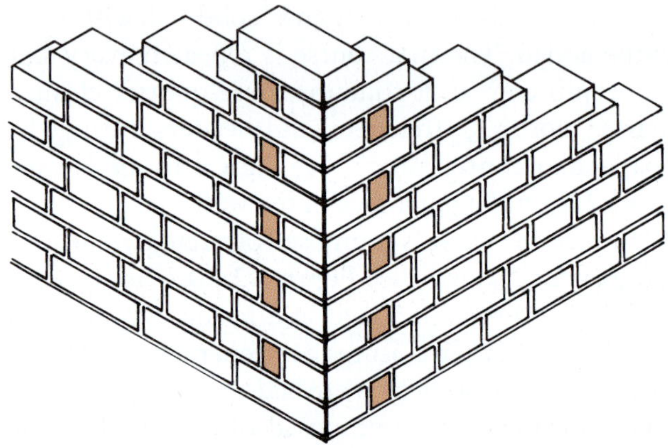

English bond with 1/4 closure

Goodheart-Willcox Publisher

Figure 12-53. Corner layout in English bond with 3/4 and 1/4 closure.

up vertically, one over the other. Two methods can be used to start the English bond at a corner—1/4 closure or a 3/4 closure. See **Figure 12-53**.

Constructing a 10″ Brick Masonry Cavity Wall

Construction of a cavity wall involves basic bricklaying techniques; however, no bridge of solid material capable of carrying water across the minimum 2″ cavity space should be permitted. The construction of two separate wythes, with a clean cavity, is the objective. See **Figure 12-54**.

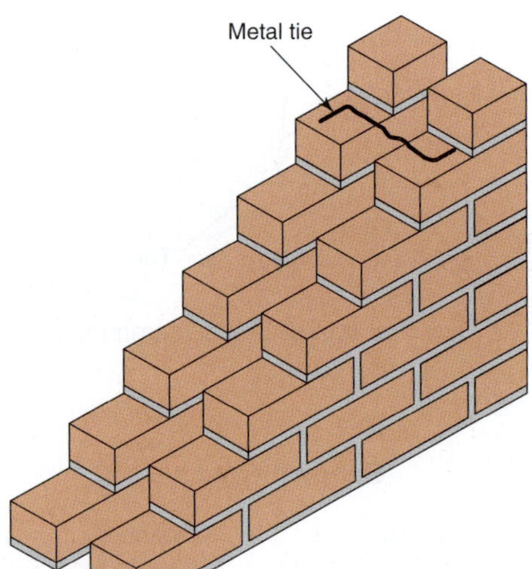

Goodheart-Willcox Publisher

Figure 12-54. Metal ties are used in cavity wall construction.

Workmanship

In order to obtain good masonry performance, all mortar joints that are intended to receive mortar must be completely filled as the units are laid. Leaky walls result from partially filled mortar joints that can contribute to ***spalling*** (the deterioration or flaking off of the brick). Spalling is due to freezing and thawing when excessive moisture is present. Voids in the mortar can be the first location where moisture starts to enter the masonry wall surface.

Trade Tip

Quality joint finishing sets apart a craftsperson from an average mason. Because of the longevity of masonry, the future work requested by customers depends on the quality and appearance of your current work!

Keeping the Cavity Clean

The cavity must be kept clean of mortar droppings and other foreign materials during the construction process. If mortar falls into the cavity, it may form bridges that stop moisture from draining or blocking the weep holes.

One method of keeping the cavity clean is to place a wood or metal strip, slightly narrower than the cavity width, in the cavity or air space. The strip rests on the wall ties as the wall is built. It is pulled out prior to the placement of the next row of ties. In this way, any mortar that has fallen into the cavity can be removed. See **Figure 12-55**.

A considerable amount of mortar can be eliminated from the cavity if these steps are followed:

- After spreading the mortar bed, bevel the cavity edge with the flat side of the trowel. As a result, very little mortar is squeezed out of the bed joints into the cavity when the units are laid.
- Roll the units into place, keeping most of the mortar on the outside.
- After the unit is placed on the bed joint, remove any protruding mortar fins.

A ***mortar fin*** is excess mortar that protrudes from the joint on the inside of the wall cavity. The excess mortar should be flattened over the backs of the unit rather than cut off. This process of flattening the mortar against the back side of the masonry units is referred to as ***plaster back***. This procedure prevents mortar from falling into the cavity and forms a smooth surface that does not interfere with the insulation.

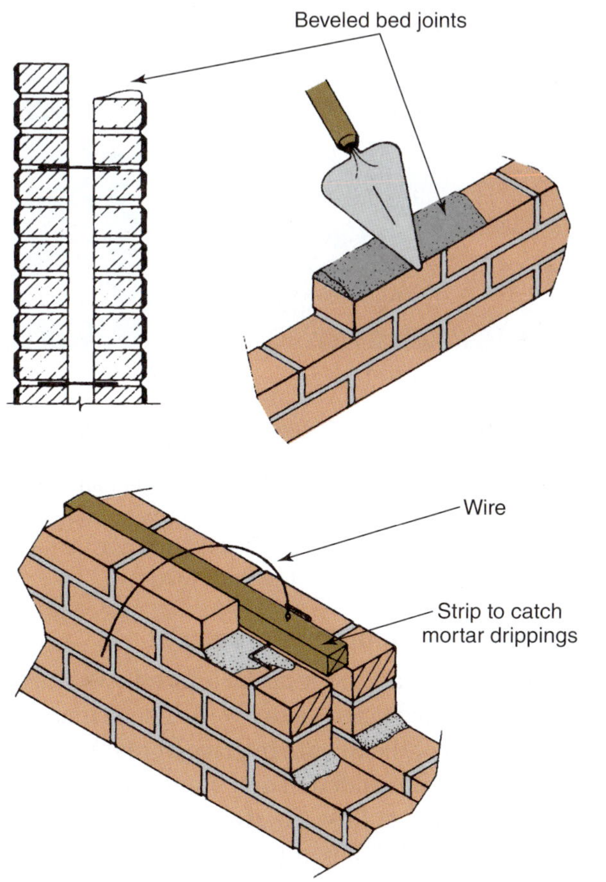

Brick Industry Association

Figure 12-55. Procedures designed to keep the cavity clean in cavity wall construction.

Weep Holes

Weep holes allow moisture to flow from inside the wall cavity to the exterior. Weep holes can be created by leaving out one of the head joints during construction. See **Figure 12-56**.

Weep holes should be placed at the base of the cavity and at all other flashing levels to rid the structure of moisture. Weep holes must be clear of obstructions and placed directly on the flashing to function properly. If excessive moisture accumulates behind the brick, as the moisture migrates through the brick, it will accumulate on the exterior surface creating efflorescence stains. A more damaging occurrence is when moisture inside the brick freezes causing spalling. See **Figure 12-57**. If the process of freezing and thawing continues over a period of time, the brick begins to fall apart.

Barbara Kalbfleisch/Shutterstock.com

Figure 12-56. Shows brick over a door opening. The open weep holes are in the header course directly above doorway. The brick are supported by a metal lintel. A limestone keyway is positioned in the center of the opening.

Itsvan Balogh/Shutterstock.com

Figure 12-57. An example of spalling caused by moisture inside the brick during freeze/thaw cycles.

Weep hole devices come in many different sizes and shapes. **Figure 12-58** shows two examples of weep hole configurations. In the past, weep holes or voids were formed by leaving a head joint out to create a means to drain moisture from the interior of the wall structure. The normal practice is to install weep holes every 16″ O.C.

Code Note **Weep Holes**

The International Residential Code (IRC) **R703.7.2.1 Weep Screeds** specifies that weep holes shall be used to remove trapped moisture or water to the exterior of the masonry wall or building. Also, refer to local building codes or construction specifications to determine the correct procedures and installation practices for the specific project being constructed.

One type of weep hole device is made of plastic or similar material with a stainless-steel screen to allow moisture to escape but stops insects from entering the structure. Another method to remove moisture is to use a specialized rope material usually made of cotton that wicks moisture from behind the wall to the exterior surface.

Tooling

Weather tightness and textural effect are the basic considerations of mortar joint finish selection and execution. Properly striking or tooling the joint helps the mortar and masonry units bond together and seal the wall against moisture. The concave and V joints provide the most weathertight joints. Joints should be tooled when the mortar is thumbprint hard.

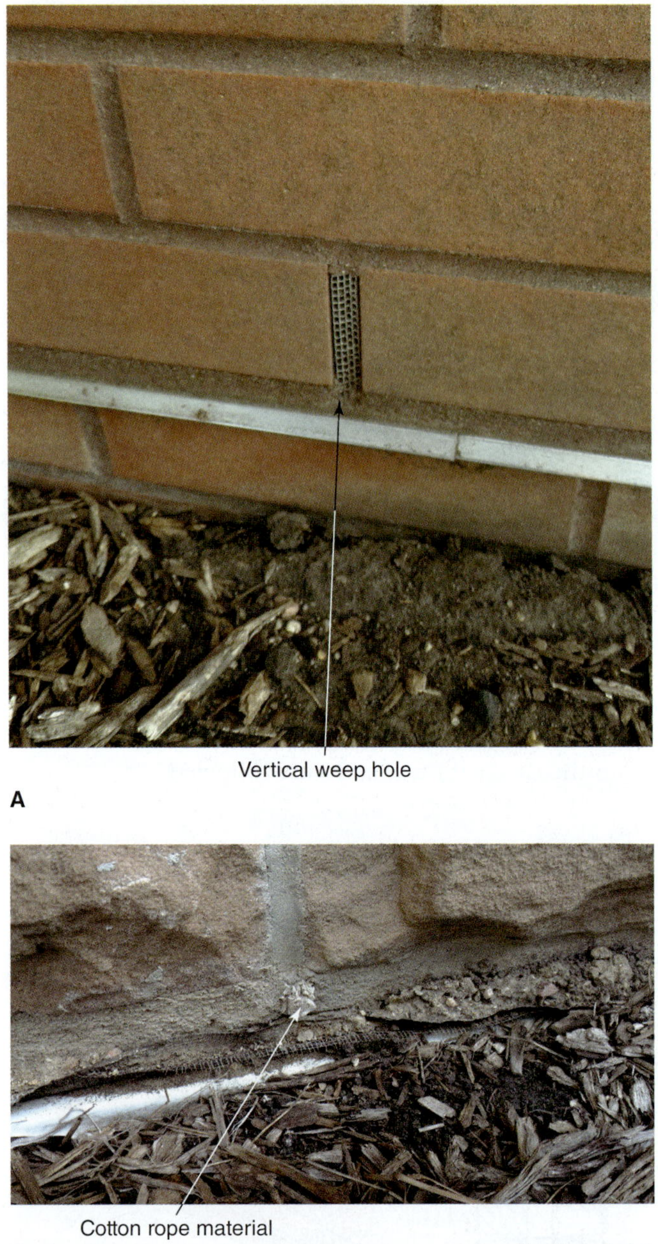

Timothy L. Andera

Figure 12-58. Two different types of weep hole configurations. Both weep holes are directly above the flashing. A—A vertical weep hole device made of plastic that is positioned in the head joint. It has a metal screen to stop insects from entering wall. B—A cotton rope material installed in a plastic tube to collect moisture and transfer it to the exterior.

Tie Placement/Joint Reinforcement

Both wythes of masonry must be properly tied together in a quality constructed cavity wall. The primary concern is that all of the ties are in place, remain operative, and are firmly embedded in and bonded to the mortar. In order to achieve this, the two wythes must be laid with completely filled bed joints and the ties must be in the correct position so later disturbance of the wall assembly is unnecessary. Wall ties do their job if properly embedded.

Workplace Skills
Professional Behavior

No matter what career or skill area you select in the masonry field, there are certain professional behaviors that a new or seasoned employee should possess or strive to develop. A few tips to keep in mind when hired by an employer:

- Show respect for your employer and team members you work with during the day. Everyone will need to work together in a cohesive manner from the laborer to the superintendent to accomplish the task at hand or project being constructed.
- Arrive earlier to the job, at least 15 minutes before the schedule time when work begins. Use the extra time to prepare for the day. For example, grab your masonry tool bag from your pickup and move to the location on the worksite where you will begin the day. Review yesterday's work to make plans for how you will begin working today.
- Continue to work throughout the day without being told to stay busy. If the project you were performing has been completed, contact your supervisor to be assigned a new task.
- When the workday is coming to an end, continue to work and stay busy. The employer is paying for a full day's work. Do not start thinking about what you will be doing after work. There is plenty of time to do this after leaving the jobsite.

These tips will help you advance in your masonry career. They will also help you shine as a valuable employee.

Ties should be 3/16″ in diameter or #9 gauge wire and placed in every 4 1/2 sq ft of wall space when the cavity is no more than 4″. The most common type of wall tie for brick masonry construction is the Z tie. Rectangular and U-shaped ties may also be used when the backup units are hollow masonry units with cells laid vertically. See **Figure 12-59**. The most important factors affecting performance of wall ties are as follows:

- Corrosion resistance.
- Placement of ties at proper spacing. Twice as many ties should be used if they have drips. Crimping the ties reduces the strength.
- Full bedding of the bed joint and placing the wall tie in the mortar 5/8″ from either edge of the masonry unit.

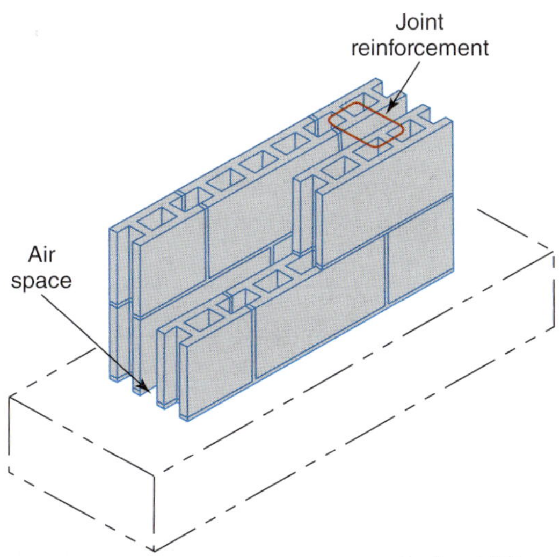

Figure 12-59. Rectangular ties and joint reinforcement used in cavity wall construction.

Prefabricated horizontal joint reinforcement may be used to tie the interior and exterior wythes, but truss type joint reinforcement should not be used to tie the wythes of a brick and block cavity wall together. Rather, ladder type reinforcement, which allows for the in-plane movement between the wythes, is recommended. Horizontal joint reinforcement is generally not required in brick masonry walls because they are not subject to shrinkage stresses.

Storage of Materials

The proper storage of materials at the construction site may have an important influence on their future performance. Masonry units should not be allowed to become wet from rain or snow or to be contaminated by salts or other chemicals that may contribute to efflorescence and staining.

- **Masonry units.** Masonry units should be stored off the ground to avoid contamination by dirt and groundwater that may contain soluble salts. Also, masonry units should be kept dry from rain or snow.
- **Cementitious materials.** Cementitious materials should be protected from the elements and stored off the ground.
- **Sand.** Sand for mortar should be placed on high ground or off the ground to prevent contamination. A protective cover is advisable in cold weather to prevent saturation and freezing.
- **Flashing.** Flashing materials should be stored where they will not be punctured or damaged. Plastic and asphalt-coated flashing materials should not be stored in areas exposed to sunlight. Sunlight causes them to become brittle.

- **Insulation.** Rigid insulation should be stored where it is not exposed to sunlight. Sunlight over time can cause rigid insulation to deteriorate. If rigid insulation is stored outside, it should be off the ground, covered, weighted down, and protected because the lightweight panels are easily damaged.

Protection of Walls

Masonry walls that are exposed to heavy rain during construction can become so saturated with water that they may require weeks or even months to dry out. This increases the likelihood of soluble salts going into solution, which may contribute to efflorescence or contaminate other elements of the structure, such as concrete, plaster, and trim.

All walls should be kept dry by covering the top of the wall with a strong, water-resistant membrane at the end of each working day. The membrane should overhang the wall by at least 24″ on each side and be secured against wind. The cavity wall should remain covered until it is completed and protected by other materials.

It is important to keep masonry from freezing while the mortar is setting or when new material is to be placed on previously laid units. Mortar joints are susceptible to freezing. If freezing occurs, the bonding between the units is affected and damage may occur.

There are many ways to protect newly laid masonry when freezing temperatures are forecasted. Using insulated blankets or erecting shelters are two methods to keep mortar from freezing. To cover masonry, drape blankets over the newly laid units. Place bricks or 2 × 4 lumber as weight on top of the blankets to reduce the chances that wind will move the blankets. Where one blanket ends and another begins, overlap the seams and place the weight, such as a number of bricks or a concrete block, against the blankets to secure them against the wall.

Laying a 10″ Brick Cavity Wall with Metal Ties

A typical 10″ brick cavity wall with a corner is shown in **Figure 12-60**. Use the standard planning and layout procedure in laying out this practice wall. Flashing has purposely been omitted in this exercise. Be especially careful not to drop mortar into the cavity.

PROCEDURE

Laying a 10″ Brick Cavity Wall with Metal Ties

1. Locate the outside wall edges and snap a chalk line to preserve the location.
2. Lay out the brick dry to determine proper spacing and any layout problems. Locate the inside line for the backing wythe such that it is 10″ away from the outside wall edge. This should form a 2″ cavity between the wythes.
3. Lay the inside bed course beginning at brick #1 and continuing to brick #2. Work from both ends toward the middle. Use a full bed joint, checking the bricks to be sure they are plumb and level. Use the chalk line to ensure a straight wall.
4. Continue the outside wythe by laying brick #6 and the closure between brick #6 and brick #1. Level, align, and plumb this part of the wall. Check to be sure the corner is square.
5. Bricks #3, #4, and #5 are three-quarter brick. Cut them.
6. Lay bed bricks #3, #4, and #5 in the position shown in **Figure 12-60**. Level and plumb the brick. Be sure the outside edge of bricks #3 and #4 are 10″ from the inside face. Check the position of bricks #2 and #5 to be sure they are square.

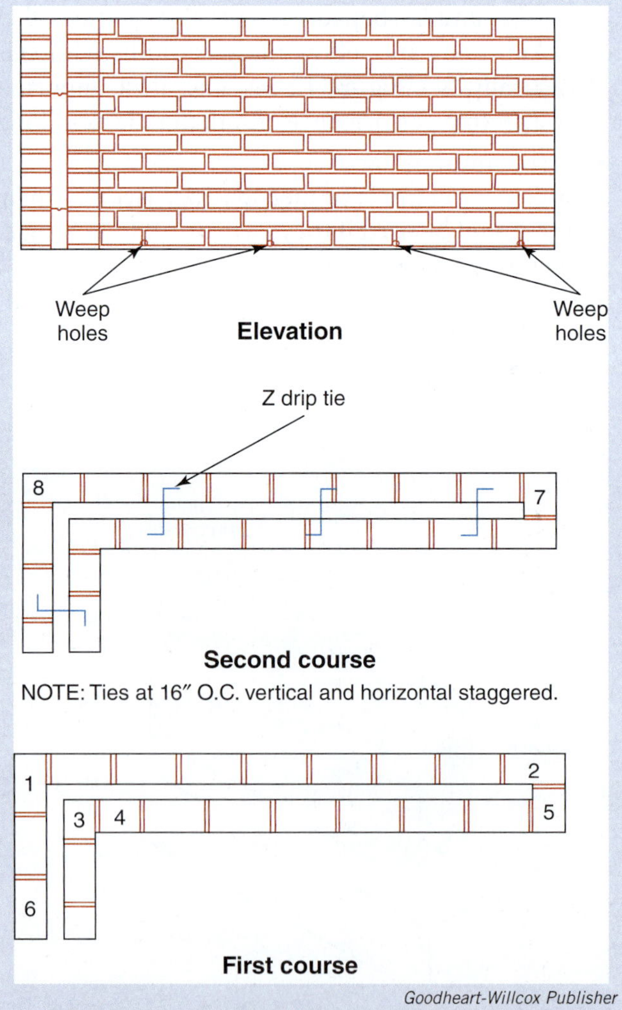

Figure 12-60. Procedure for laying a 10″ cavity wall with metal ties.

7. Lay the brick between bed bricks #4 and #5 and finish the leg from bed brick #3. Level, align, and plumb these sections of the wall. The first course is now complete and should be square at the corners and 10″ wide throughout.
8. If you did not use empty head joints for the four weep holes shown in the elevation, open those joints or insert tubing or wicking at the weep hole locations.
9. Cut three (3) three-quarter bricks for the second course and lay the mortar bed for the second course. Bevel the mortar bed to avoid dropping mortar into the cavity.
10. Lay three-quarter brick #7 at the right end of the wall over bed bricks #2 and #5.
11. Lay second course brick #8 at the other end of the wall and complete the wythe between them.
12. Complete the second course following the procedure established for the first course. Plaster back any mortar that is squeezed out of the inside bed joints. Keep the cavity clean.
13. Locate the four metal Z-shaped ties as shown in the second course plan. Embed the ties in the mortar bed being sure that the drip on the metal ties points downward if your ties are of this design.
14. Build corner leads six courses high on each end of the wall so a mason's line can be used to aid in the construction of the wall.
15. Lay the wall to a height of eight courses and then embed metal Z-shaped ties as previously instructed. The ties should be offset from the ties on the second course.
16. Attach wires at each end of a piece of wood 1″ × 2″ × 4′ long. The board is used to catch any mortar droppings and to flatten any mortar squeezed out inside the cavity.
17. Complete the cavity wall to a height of twelve courses using the wood piece to catch mortar. Check the wall to be sure that it is square and plumb. Remove any mortar splatter on the outside wall and clean with a bricklayer's brush.
18. When the mortar has reached thumbprint hard, finish the joints with a convex jointer. Remove any tailings that are produced from tooling.

Constructing a Single Wythe Brick Bearing Wall

Single wythe concrete masonry construction is a cost-effective and durable form of masonry construction. Single wythe brick bearing walls are sometimes used in one- and two-story home designs and may also be used in commercial construction. In this type of system, the brick masonry serves as both the structural system and the exterior facing. See **Figure 12-61**. A wood, steel, or masonry backing system is not necessary.

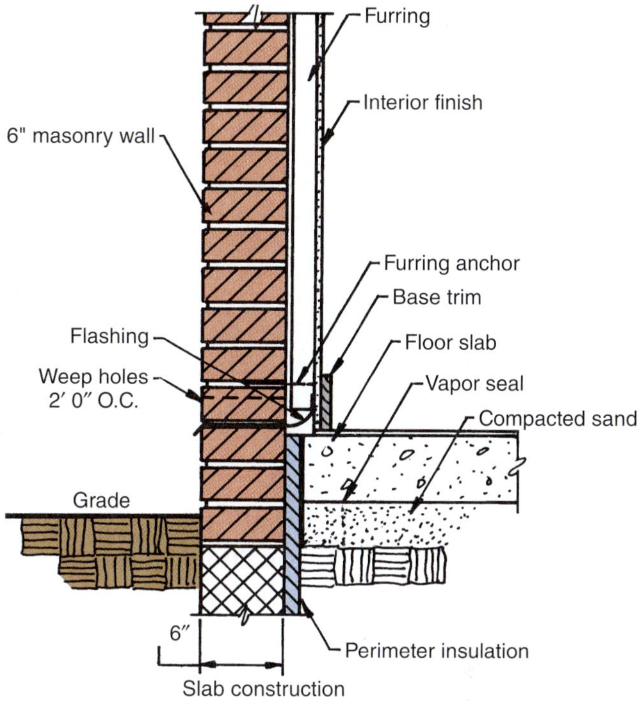

Goodheart-Willcox Publisher

Figure 12-61. Detail of a 6″ exterior loadbearing single wythe wall.

Structural Design

Design considerations for single wythe brick loadbearing walls include minimum loads to be resisted by the structural system, minimum thermal performance requirements, necessary fire resistance of the wall system, and resistance to moisture penetration. These specifications are covered by the model building codes.

Determining the design loads (vertical and horizontal) is generally the first step in the design of a structural system. Vertical loads include loads from the weight of building materials and occupants. Lateral loads include wind, soil, and seismic events. In most regions of the United States, only vertical loads and wind loads are a major concern, but in some areas, seismic loads may control the structural design. Additionally, severe wind loads, as in hurricane prone areas, may dictate the structural design. The size of a building's structural member is determined by model building codes and the structural loads.

Masons are required to familiarize themselves with the International Building Code (IBC) and the International Residential Code (IRC), but other codes may also be used in the design and construction process of a facility. When questions arise related to one of the codes, a local building official can be contacted to clarify construction process rules and regulations.

Code Note **Minimal Nominal Wall Thickness**

The International Residential Code (IRC) **R606.4.1** specifies an 8″ minimum nominal wall thickness for loadbearing masonry structures over one story in height. A nominal 6″ wall is permitted for a single-story home or garage if the wall height does not exceed 9′ and the gable height does not exceed 15′. Minimum wall thickness requirements influence the type and size of brick and block selections. Other model building codes have similar restrictions, so it is important to research the local codes.

Location of Interior Work

Because there is no cavity within the exterior wall to place plumbing, heating/air-conditioning, and electrical systems, the systems must be located between the furring strips, in the floor or ceiling, or in interior frame walls. The type of foundation and floor system used influences the location of interior systems. In structures with basements or crawl spaces, mechanical systems can be located between the floor joists. In slab type structures, it is easier to route mechanical systems through the ceiling or attic space.

Structural Materials

According to BIA Technical Note 26—Single Wythe Brick Bearing Walls, selection of masonry materials for single wythe brick bearing wall systems should consider the structural, energy, and reinforcement performance requirements. The following are some considerations:

- **Brick.** The brick masonry units used in a single wythe bearing wall are solid (ASTM C216) or hollow (ASTM C652). Solid bricks are commonly manufactured in nominal widths of 3″, 4″, and 6″. Hollow bricks are available in nominal widths of 4″, 5″, 6″, and 8″. Nominal 5″ and 6″ wide hollow bricks are most commonly used to build this type of wall for residential purposes. The cells in most hollow bricks are made to accommodate vertical reinforcement and grout.
- **Mortar.** Portland cement-lime mortars with an air content of less than 12% are recommended for their superior bond strength and resistance to water penetration. Masonry cements are prohibited by some codes and all type N mortars are prohibited in seismic performance categories D and E (formerly zones 3 and 4). Type S, M, or N mortar may be generally used in loadbearing brick masonry. Type S is recommended for use in reinforced brick bearing walls.
- **Grout.** Steel reinforcement is bonded to the surrounding brick masonry using grout. Grout may be made using fine or coarse aggregate, but fine aggregate is generally used because it is easier to fill smaller cells. Grout should meet ASTM C476. The water/cement ratio of grout should have a slump of 8″ to 11″. It must be fluid enough to fill voids, but not separate into its constituent parts.
- **Reinforcement.** Lateral loads must be resisted through the use of vertical steel reinforcement or other means. See **Figure 12-62**. The Masonry Standards Joint Committee (MSJC) Code limits the maximum size of reinforcement used in masonry to a No. 11 reinforcing bar. As a rule-of-thumb, the maximum bar size should not exceed the nominal thickness of the wall in inches to ensure proper development of the reinforcement. For example, a maximum reinforcing bar size No. 6 is recommended for nominal 6″ walls.
- **Joint reinforcement.** If used, joint reinforcement should comply with ASTM A82 and be hot-dipped galvanized steel or made from stainless steel to reduce corrosion.

Green Note

Brick and Energy Efficiency

Clay masonry contributes to the energy efficiency of a building or structure. Because of its high thermal mass, clay masonry reduces the heating and cooling energy used in a building. In exterior walls and exposed on interior walls and floors, clay brick reduces temperature swings in the interior.

Because of its porosity, brick absorbs heat in the summer and shields the interior from a rapid rise in temperature. The brick later releases the heat into the cooler night air, reducing peak energy loads. Brick veneer also offers the benefits of thermal mass, and the benefits are even more pronounced when exposed brick is used on the interior of the building. By blocking winter cold and winds, brick retains the warmth generated in the interior. Brick veneer construction, with air space between the brick and the sheathing, attains even better insulating results. By using insulated sheathing (rigid foam insulation), a higher level of energy efficiency can be achieved, which will reduce energy costs to the structure.

In addition to reducing direct energy usage, brick's durability results in minimal exterior maintenance requirements, thus avoiding the energy usage involved in upkeep and replacement. Brick can last over 100 years if installed and maintained appropriately. The long life cycle of brick lessens the manufacturing energy required for replacement products. The local availability, low construction waste, reusability, and recyclability of clay brick also contribute to energy conservation.

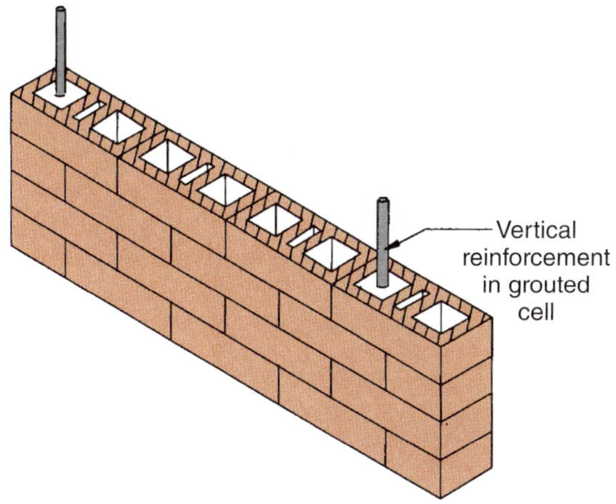

Brick Industry Association

Figure 12-62. Vertical reinforcing used to resist lateral loads in a single wythe wall.

Structural Insulation and Moisture Protection

Model building codes require minimum levels of insulation. For brick bearing wall structures, rigid board insulation is generally placed on the interior face of the brick wythe to produce high insulation values. See **Figure 12-63**. The insulation is installed between furring strips and provides high insulation values. The insulation can also be attached with mechanical fasteners and adhesives or placed in the cells of hollow brick units. Insulation placed in hollow brick is generally not as effective as surface-mounted rigid insulation.

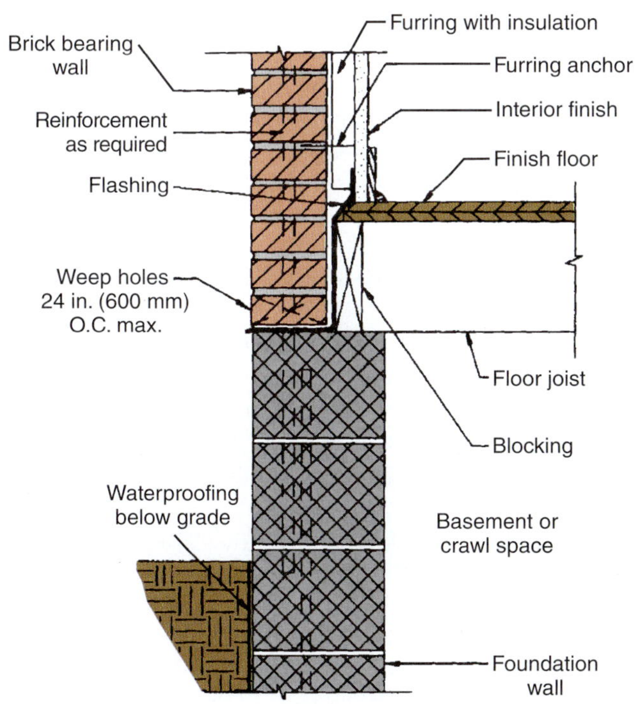

Brick Industry Association

Figure 12-63. A typical method of insulating a single wythe exterior brick wall.

Air leakage is a concern in single wythe exterior brick walls. Some leakage will occur through the weep holes and at the top of the brickwork. Materials such as house wraps are not appropriate for this type of construction. Limiting air leakage must be accomplished with papers or films on the insulation or gypsum board. All joints between panels and around doors and windows must be taped or sealed.

Resistance to water penetration is a major concern in a single wythe wall. Full mortar joints with no voids and a good bond between units are necessary to reduce water penetration. Head joints must be full, not face-shell mortar bedded only. Therefore, a drainage cavity with flashing and weep holes should be provided to prevent water penetration. See **Figure 12-64**. Alternatively, a bituminous, dampproof coating on the inside face of the brick wall can be applied before the insulation and finishes are installed.

12.4 Advanced and Specialized Brickwork

More advanced and specialized forms of brickwork are covered in Chapters 15, 16, and 17. Specific applications include stepped footings, columns, piers, pilasters, solid masonry walls, hollow masonry bonded walls, veneered walls, reinforced masonry walls, retaining walls, wall openings, lintels, sills, arches, window details, door details, steps, paving, fireplaces, chimneys, stone quoins, garden walls, racking and corbels, and caps and copings. Many of these applications involve the integration of multiple materials—concrete block, concrete, tile, stone, and brick.

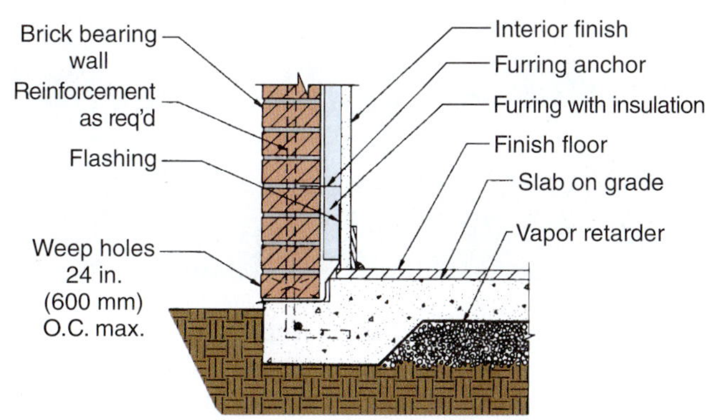

Brick Industry Association

Figure 12-64. Weep holes and flashing are necessary in exterior single wythe masonry walls to reduce moisture penetration.

Corbeling a 12″ Wall

Corbeling is a method used to widen a wall by projecting out masonry units to form a ledge or shelf. When building a corbel, each brick course extends out farther than the one below it. See **Figure 12-65**.

> **Code Note** — **Corbel Projection**
>
> IRC **R606.5.2.1 Corbel Projection** specifies the maximum projection of one unit shall not exceed one-half the height of the unit or one-third the thickness of the unit at right angles to the wall.

Because corbels normally support a load, they must be carefully constructed. Headers are generally used to tie a corbel into the base. Building codes generally require the top course to be a full header course. All joints must be completely filled with mortar.

The following procedure is suggested for corbeling a 12″ wall. See **Figure 12-66**.

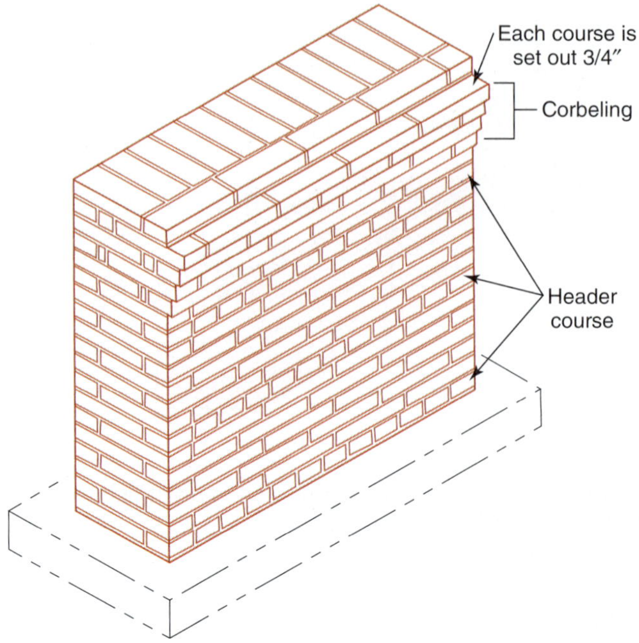

Goodheart-Willcox Publisher

Figure 12-66. Procedure for corbeling a 12″ brick wall.

Author's image taken in Ackley, IA

Figure 12-65. Each course has been corbeled out from the previous one to eventually create a shelf to support the three-sided bay window. The last regular course has been cut to height to accommodate the brick sill. The brick sill is sloped to ensure adequate drainage. The brick sill hangs over the lower course to create a drip edge.

PROCEDURE

Corbeling a 12″ Wall

1. Snap a line on the foundation or base where the front edge of the wall is to be constructed. Also, mark the starting and ending points of the corbel on the foundation or base.
2. Lay out the first course as a dry course to check spacing. Running bond will be used with a header course every seventh course.
3. Lay the first course with headers on the front wythe and stretchers on the back. Use a full mortar bed. Level, plumb, and square the brick. Check the spacing.
4. Lay the second through sixth courses. Check the height of each course as it is laid and be sure bed joints are uniform and level.
5. Lay the seventh course as a header course with full mortar joints.
6. Continue to lay the wall until you reach the height where the corbel course is to begin. Check your progress often to be sure the wall is plumb and straight.
7. Begin the corbel course by projecting headers out 3/4″ beyond the course below. Fill the extra wide head joint with mortar.
8. Lay the second corbel course using stretchers along the front and headers along the back wythe. Fill in the space between with bats. Be sure the course is level and straight.

9. Lay the third corbel course using stretchers on the front and back wythes and three-quarter headers between the two wythes. Each corbel should project out 3/4″ beyond the course below.
10. Lay the next course the same as the first course of the wall—headers on the front wythe and stretchers on the back wythe. Continue the wall to the desired height.
11. Finish all joints when the mortar has set to thumbprint hard.
12. Clean the wall with the trowel and bricklayer's brush.

Hollow Brick Pier

Piers are similar to columns except they are shorter and generally do not support a load. They are commonly used as gateposts at corners or openings or ends of a wall. For example, a garden wall with piers is generally referred to as a *pier and panel wall* because the piers are used to brace the wall panels. Prior to constructing the pier and wall panel, a foundation wall and footing must be constructed at a required depth below the frost line. Check with the local building official to determine the exact depth required for the footing. Footing and foundation construction is discussed in Chapter 15, **Foundation Systems**.

Piers can be constructed of a one-wythe wall 4″ thick. The bond pattern is usually staggered so the wall is tied together from a different side in each course in an interlocking fashion. Piers that are exposed to the weather must be capped to prevent water from entering at the top. Weep holes may be required.

Code Note **Pier Cap**

The International Residential Code (IRC) **R404.1.9.1** specifies that hollow piers should have 4″ (102 mm) of solid masonry or concrete cap, a masonry cap block, or cavities of the top course filled with concrete or grout.

Laying a 16″ by 20″ Hollow Brick Pier

The following procedure is suggested for laying a 16″ by 20″ hollow brick pier. See **Figure 12-67**.

PROCEDURE

Laying a 16″ by 20″ Hollow Brick Pier

1. Prior to constructing the pier, a footing and foundation needs to be built. The depth of the footing is determined by the local or state frost codes. Contact your local building official to determine the correct footing depth.
2. Locate the exact position of the pier on the foundation and snap a chalk line to preserve the location of the outside faces of the brick.
3. Lay the bed course on a generous mortar bed. Check the brick to be sure that they are level and straight and that the corners are square. You may use either 1/2″ or 3/8″ joints. Refer to the first course plan in **Figure 12-67**.
4. Lay the second course as shown in the second course plan. Brick should be positioned so head joints are offset to provide an interlocking connection. Level, plumb, and square the course.
5. Lay successive courses alternating the patterns used in the first and second courses until the pier has reached the desired height. Twelve courses should be adequate for practice. Avoid dropping any mortar inside the pier.
6. When the mortar is thumbprint hard, tool the joints and clean off any fins with the trowel. Clean the surface with a bricklayer's brush.
7. Piers and brick walls need to receive a cap. Pier caps can be made of concrete or stone, depending on the desired effect and application.

12.5 Cleaning New Masonry

The finished appearance of a masonry wall depends not only on the skill used in laying the units but on the cleaning procedure as well. The appearance of a masonry structure can be ruined by improper cleaning. In many instances, the damage caused by faulty cleaning techniques or use of the wrong cleaning agent cannot be repaired. All cleaning should be applied to a sample test area of approximately 20 sq ft.

Take the following general precautions to promote a cleaner wall during construction:

- Protect the base of the wall from rain-splashed mud and mortar splatter.
- At the end of the day, turn scaffold boards that are near the wall on their edges. This prevents possible rainfall from splashing mortar and dirt directly on the completed masonry.
- Cover walls with a waterproof membrane at the end of the workday to prevent mortar joint wash out and entry of water into the completed masonry.
- Protect brick stored on the worksite from dirt and mud. Store brick on pallets or similar materials and place them under protective covering.
- Practice careful workmanship to prevent excessive mortar droppings. Excess mortar should be cut off with a trowel as the brick are laid and returned to the mortar board.

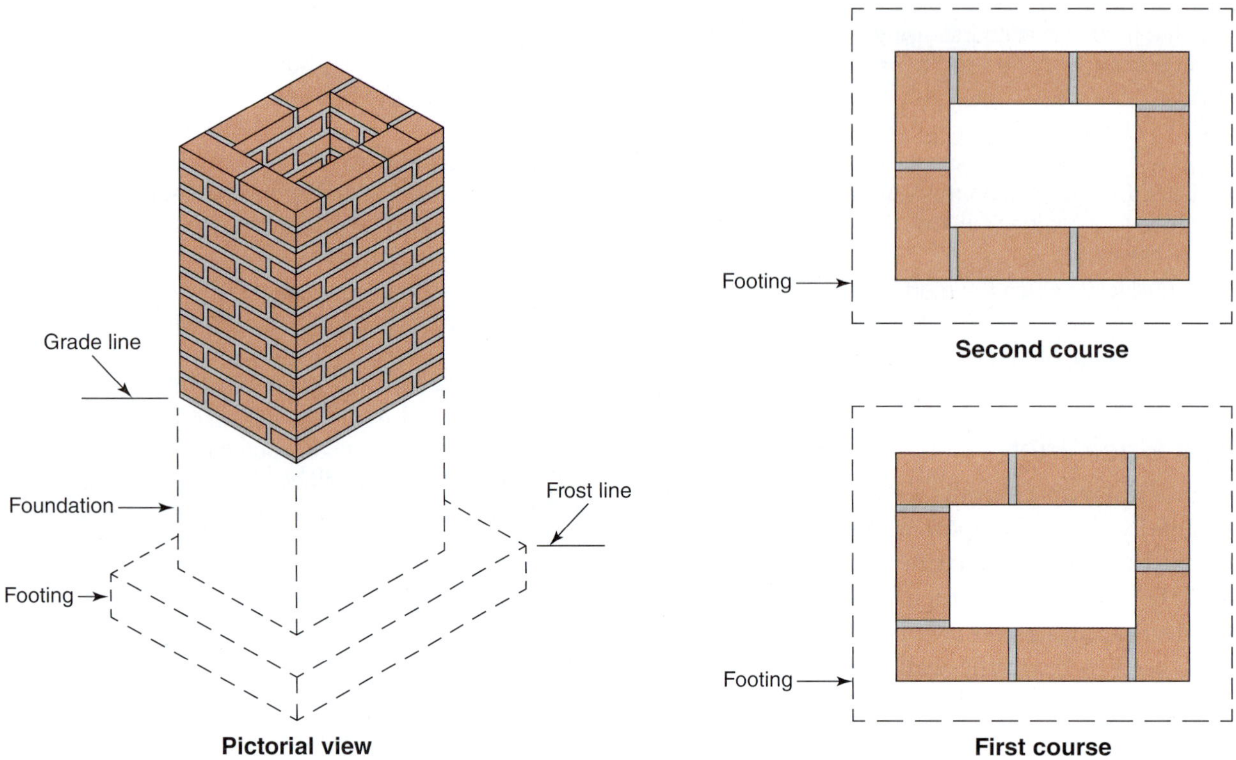

Goodheart-Willcox Publisher

Figure 12-67. Procedure for laying a 16″ by 20″ hollow brick pier. The footing and foundation to support the pier are indicated by dashed lines.

- Tool joints when thumbprint hard. After tooling, brush excess mortar and dust from the surface. A medium soft bristle brush is preferable. Avoid any motion that results in rubbing or pressing mortar particles into the brick faces.

> **Safety Note**
> When cleaning brick with chemicals or acids, refer to the manufacturer's directions for mixing and application instructions. Consult the product's SDS document for specific safety procedures, precautions, and cleanup. Wear personal protective equipment such as eye protection, a face shield, rubber gloves, and boots, as specified in the SDS document and the manufacturer's documentation.

Cleaning Brick

Cleaning new brick masonry involves removing mortar, mortar stains, and other materials such as dirt that are deposited on the masonry during construction. Some mortar stains may result even when the construction work has been done skillfully and carefully. Therefore, most specifications require a final washing down of all masonry work.

Small quantities of various minerals found in some burned clay masonry units react with some cleaning agents to cause staining. These reactions cannot be predicted in advance. Prior to using any cleaning agent on a masonry wall, apply the cleaning agent be applied to a sample section of 10–20 sq ft to judge its effectiveness. Wait a minimum of one week before inspection.

A solution of hydrochloric acid is used extensively as a cleaning agent for new masonry. The following sections describe procedures for cleaning dark and light colored brick.

Cleaning Dark Colored Brick

Dark colored bricks include red, red flash, brown, and black. They are most likely to show light gray, brown, or yellow discoloration from failure to rinse off dissolved mortar or dirt. Dark brick may be cleaned with or without acid.

PROCEDURE

Cleaning Dark Brick without Acid

1. When the mortar is thoroughly set and cured, begin the cleaning operation.
2. Remove large particles of mortar with wooden paddles or scrapers before wetting the wall. A chisel or wire brush might be necessary.
3. Saturate the wall with clean water and flush away all loose mortar and dirt.

4. Scrub down the wall with a solution of 1/2 cup of trisodium phosphate and 1/2 cup of household detergent dissolved in one gallon of clean water. Use a stiff fiber brush.
5. Rinse off all cleaning solution and mortar particles using clean water under pressure.

Acid Cleaning of Dark Brick

1. When the mortar is thoroughly set and cured, begin the cleaning operation.
2. Remove large particles of mortar with wooden paddles or scrapers before wetting the wall. A chisel or wire brush might be necessary.
3. Saturate the wall with clean water and flush away all loose mortar and dirt.
4. Use a clean, stain-free commercial grade of hydrochloric (muriatic) acid. Mix one part of acid to nine parts water in a nonmetallic container. Use a long handled fiber brush to scrub the wall surface.

Safety Note

Before using muriatic acid, read the manufacturer's instructions and the SDS document. Follow manufacturer's specifications for mixing, applying, cleaning, and disposing of leftover solution. Wear appropriate PPE, safety glasses, approved rubber gloves, protective boots, and clothing as specified. When mixing, pour the acid into the water, not the water into the acid!

5. Keep the area not being cleaned flushed free of acid and dissolved mortar. This scum, if allowed to dry, may be impossible to remove later.
6. Scrub the brick, not the mortar joints. Do not use metal tools. Clean only a small area at a time.
7. Rinse the wall thoroughly with plenty of clean water while it is still wet from scrubbing with the acid.

Cleaning Light Colored Brick

Light colored brick includes buff, gray, speck, and pink. Light brick is more likely to be burned by acid than darker brick. Therefore, do not use acid except in extreme cases. The procedure for cleaning light colored brick without acid is the same as the procedure for dark colored brick.

To acid clean light colored brick, follow steps 1–3 of the procedure for acid cleaning of dark brick. Then mix one part acid with 15 parts water and scrub the wall with a fiber brush. Rinse the wall well with clear water. Neutralize the acid wash with a solution of potassium or sodium hydroxide, consisting of 1/2 lb hydroxide to 1 qt of water (2 lb/gal). Allow this solution to remain on the wall for two or three days before washing again with clear water.

Trade Tip

Use the highest grade acid available. It should be free of any yellow or brown discoloration.

Cleaning Failures

Cleaning failures generally fall into one of three categories:

- Failure to thoroughly saturate the brick masonry surface with water before and after application of chemical or detergent cleaning solutions. Saturation of the surface prior to cleaning reduces the absorption rate, permitting the cleaning solution to stay on the surface rather than be absorbed.
- Failure to properly use chemical cleaning solutions. Improperly mixed or overly concentrated acid solutions can etch or wash out cementitious materials from the mortar joints.
- Failure to protect windows, doors, and trim. Many cleaning agents, particularly acid solutions, have a corrosive effect on metal. If permitted to come in contact with metal frames, the solutions may cause pitting of the metal or staining of the masonry surface and trim materials, such as limestone and cast stone. If mortar is on window glass, metal tools or strong chemicals can scratch or etch the glass. Sand in the mortar can also scratch the glass.

Pressurized Water Cleaning

A pressurized power washer is a valuable piece of equipment used to reduce labor and costs. See **Figure 12-68.**

However, because of the extreme pressures that these machines achieve, they can damage masonry and cause injuries to workers who are operating the equipment or who are in close proximity of the spray.

Alan Budman/Shutterstock.com

Figure 12-68. A pressurized power washer can reduce labor and costs when cleaning brick or other surfaces. Wear appropriate PPE when power washing.

Prior to cleaning masonry with a power washer, read the manufacturer's specifications and instruction manuals. Follow procedures for hand cleaning brick as discussed in the Cleaning Brick section of this chapter, such as wetting the surface with water, applying cleaning solution, cleaning the surface, and finally rinsing the area with clean water. From the manufacturer's specifications, determine recommended pressure, size and type of spray nozzle, the distance from wall surface and operation procedures of the equipment. Practice on a section of the wall area that is not as visible. Begin a distance away from the brick, then work closer to the surface as needed to achieve the desired effect. When applying the solution, lower pressure to 30–50 pounds per square foot (psf). Do not allow solution to dry on the wall surface. When the area is clean, rinse the surface at a pressure only between 200 and 300 psf, since higher pressures can adversely affect the brick finish. Pressurized cleaning can reduce time, but care must be taken to ensure that the process does not affect the exterior of the brick surface.

Abrasive Blasting

In abrasive blasting, an abrasive material under air pressure is used to remove unwanted mortar or stains from the surface of brick or other masonry materials.

Code Note **Abrasive Blasting**

When considering abrasive blasting, refer to the Brick Industry Association (BIA) Technical Note 20—Cleaning Brickwork. It refers to cleaning brick and the different methods to remove surface mortar and stains.

Abrasive blasting may also remove certain finishes on brick, such as glazing, sand finishes, or slurry finishes. It may also roughen the exterior of the brick and change the finished appearance. A gentler abrasive than sand should be used. Examples of softer products are crushed nut shells, dry ice, or baking soda. The cost of abrasive blasting is about the same as acid cleaning.

Abrasive blasting or sandblasting is a valuable tool only in certain applications, such as for removing persistent stains or mortar. However, the Brick Industry Association recommends that abrasive blasting *not* be used for brickwork because it may change the appearance of the brick. Particular brick can react with chemicals found in cleaning agents used to wash brick. In certain situations, blasting may be used as a last resort.

If sandblasting is necessary, use a very low pressure (60–100 psi), a 1/4″ sandblast nozzle, and softer materials, such as white urn sand, that will not damage the exterior surface of the brick. According to the Brick Industry Association, the sandblasting machine should be equipped to regulate flow of sand or abrasive material at 300 lb per hour. The secret to successful abrasive blast cleaning is the distance the operator stands from the wall and the manner in which the blast is directed at the brick. Concentrate on hitting the brick rather than the mortar joints. Remember, abrasive blasting is not typically recommended, unless done as a last resort by a trained professional.

Safety Note

Because of the abrasiveness of the process, a competent person must perform abrasive blasting. Manufacturer's specifications and SDS documents should be consulted in order to determine equipment operating procedures and necessary PPE, such as a respirator; eye protection; face mask; and protective clothing, footwear, and gloves.

Chapter Review

Summary

- Mortar on a board should be kept well-tempered with water until it is used.
- Brick can be cut with a brick hammer, brick set chisel, or masonry saw.
- A mason's line is a guide that helps a bricklayer build a straight, level, and plumb wall. When laying units to a mason's line, do not touch the line.
- A building line must be established exactly to prevent a code violation or other complications.
- Batter boards preserve the building lines during excavation and construction.
- Laying the corners is called *laying the leads*.
- Common bond, which consists of headers between every five to seven courses of stretchers, is strong and can be laid up fast.
- Laying up a corner is similar for most bonds.
- When constructing a cavity wall, no bridge of solid material capable of carrying water across the minimum 2″ cavity is permitted.
- When selecting masonry materials for a single wythe brick bearing wall system, consider structural, energy, and reinforcement performance requirements.
- Corbeling is a method used by masons to widen a wall by projecting out masonry units to form a ledge or shelf.
- New brick can be cleaned with acid solutions, pressure washing, or in special circumstances, by abrasive blasting.

Know and Understand

Answer the following questions using the information provided in this chapter.

1. One of the first brick operations to occur is _____.
   A. positioning the mortar mixer
   B. delivery of the brick
   C. organizing the site
   D. spotting brick cubes
2. Mortar is usually furrowed with the _____ of a trowel.
   A. ferrule
   B. heel
   C. toe
   D. shank
3. A _____ is the tool most often used by masons to cut brick.
   A. brick hammer
   B. trowel
   C. brick set chisel
   D. masonry saw
4. *True or False?* Any straight wall longer than 4′ is usually laid to a line.
5. The _____ is generally the face or side of the building being constructed.
   A. mason's line
   B. building line
   C. setback
   D. corner pole
6. Straight batter boards are made with _____.
   A. two boards and three stakes
   B. mason's line and two stakes
   C. two boards and mason's line
   D. one board and two stakes
7. The _____ is used to measure the mortar joint heights of each course of masonry.
   A. corner pole
   B. drop line
   C. gauge stick
   D. plumb bob

8. The overlapping process from one course to another is called _____.
   A. laying the corners
   B. laying the leads
   C. spotting the brick
   D. racking
9. A _____ is used to tool or strike the bed joints of brick masonry.
   A. long jointer
   B. trowel
   C. short jointer
   D. All are correct.
10. *True or False?* The result of too little mortar is called plaster back.
11. _____ allows moisture to flow from inside the wall cavity to the exterior.
   A. An English bond
   B. A weep hole
   C. Spalling
   D. A mortar hole
12. Which of the following is a vertical load?
   A. Seismic events
   B. Occupant weight
   C. Wind
   D. Soil
13. *True or False?* Corbels do *not* typically support a load.
14. *True or False?* A solution of hydrochloric acid is used as a cleaning agent for new masonry.
15. You should use acid to clean _____ brick only in extreme cases.
   A. light
   B. dark
   C. colored
   D. any

Apply and Analyze

1. What is the likely outcome of using a trowel that is too large?
2. List three ways of loading mortar on a trowel.
3. What is the correct way to hold a brick to apply mortar and to lay it on the mortar bed?
4. List four tools used to cut brick.
5. Why is it important to place the same amount of tension on a mason's line each time it is moved?
6. What is the purpose of batter boards?
7. What is the purpose of a corner pole?
8. When starting a wall, why should a dry course of brick be laid out along the wall line?
9. What is the purpose of corbeling?
10. In what two ways do piers differ from columns?
11. When cleaning new masonry by abrasive blasting, what pressure should be used?

Critical Thinking

1. Explain what may occur if weep holes are not used or if the weep holes are blocked by mortar.
2. Explain the process of using corner blocks when laying brick between two corner leads.

Activities

1. (SCIENCE) Locate a safety data sheet for hydrochloric (muriatic) acid. Read about hazards of this material, necessary safety precautions, and PPE that should be worn when handling and using the acid. Prepare a written or oral report summarizing your findings.
2. (ENGINEERING) As a team project, lay out a building that is 10′ × 14′ in size using stakes and batter boards. Measure and adjust until walls are the correct size and diagonals are within 1/4″ or less of each other.

ungvar/Shutterstock.com

CHAPTER 13

Laying Block

LEARNING OUTCOMES

After studying this chapter, you will be able to:

13.1-1 Explain how planning and site preparation impact concrete masonry.
13.2-1 Describe the various types of concrete masonry walls.
13.3-1 Practice basic operations such as spreading mortar, handling concrete blocks, applying head joints, cutting blocks, and using a mason's line.
13.4-1 Construct various types of concrete block walls.
13.5-1 Use appropriate methods to remove mortar from concrete block walls.

TECHNICAL TERMS

cavity wall
collar joint
composite wall
control joints
ears
face shell bedding
hollow masonry walls
metal twig (trig)
nonloadbearing walls
parging
reinforced concrete masonry walls
solid masonry walls

Introduction

Most *concrete masonry units (CMUs)* are used in the construction of basements and foundation walls, exterior above grade walls, interior partitions, or retaining walls. The design and construction of a wall depends on factors such as desired appearance, fire resistance, economy, strength, insulation, and acoustics. The layout of a wall and use of specific masonry units take into account modular planning, internal arrangement of components, provision for expansion and contraction, and weather resistance. All of these factors affect the success of a concrete masonry wall assembly. **Figure 13-1** shows a variety of applications and designs of concrete masonry construction.

13.1 Planning and Site Preparation

Similar to brick laying, block laying requires the planning, preparation, estimating, and ordering of materials. Following these steps ahead of time will help you to be efficient during the construction of a project. Labor costs, material costs, and weather are three elements that can affect the success or failure of a construction company. The processes of estimating quantities of material and developing a schedule are usually performed by the owner, superintendent, project manager, and the estimator of a company. This group is sometimes called the *management team*.

Planning

Planning and implementing are two important activities that contribute to the success of a company. Planning aids in anticipating "what if" scenarios that may occur. For example, what if the weather changes? This may affect the work schedule. What if the material is delayed because of shortages or interruptions in transportation? What if two of your best masons are injured? The management team considers these scenarios and others to plan for circumstances or challenges that could confront a masonry company. Implementation is the activity of taking the planning and putting it to work. Depending on the structure of a company, the management team may give directions and assign tasks to supervisors. These tasks are then carried out by foremen, journeymen, laborers, and apprentices.

Scheduling is developing a timeline that lists work activities and deliveries from the beginning to end of the project. Scheduling includes employees and subcontractors performing work activities for the company. Scheduling also pertains to materials that will be delivered and used during the construction process. For example, the masonry company does not want to stockpile a large inventory of materials. Whatever is needed should be delivered to the jobsite in a timely manner. Specialty block may require extra time for delivery.

Site Preparation

Site preparation is the process of preparing the area where the building will be constructed. The preparation includes site grading to level the construction site. Preparation may include removing trees and excavating the site to install the foundation, such as footings

Besser Company; Trenwyth Industries, Inc.

Figure 13-1. A variety of uses of concrete block construction.

Timothy L. Andera

Figure 13-2. CMUs are stored on pallets and wrapped in plastic to protect them until the construction activities warrant their use.

Safety Note

Back injuries are one of the leading health issues in the construction industry. Improper or overlifting can cause severe injuries to the back and supportive muscles that may affect a person for life. The CDC and OSHA recommend objects 50 lb or greater should be lifted by two workers rather than one employee. A research branch of OSHA is the National Institute for Occupational Safety and Health (NIOSH), whose mission is to "develop new knowledge in the field of occupational safety and health and to transfer that knowledge into practice."

and block walls. Areas need to be cleared for construction equipment and related items. If the site is ready for block to be laid, locations need to be cleared where pallets of block or cubes will be positioned or spotted. See **Figure 13-2**.

All materials and equipment related to mortar must be stored and protected from moisture and the weather. To reduce preparation time, premixed mortar can be purchased in 80 lb bags. For commercial applications, 3000 lb premixed bags can be delivered to the construction site. The contents of the large mortar bags are emptied into gravity silos. See **Figure 13-3**.

Most projects require scaffolding and plank to be brought to the construction site and stationed at appropriate locations for easy access by employees. **Figure 13-4** shows scaffolding wrapping around a masonry building under construction.

Timothy L. Andera

Figure 13-4. Masons are three-stories above the ground receiving materials delivered by a rough terrain forklift. Mortar, masonry units, rigid insulation, scaffold sections, and plank are delivered to the workers using the forklift.

A

B

Timothy L. Andera

Figure 13-3. A—The large, premixed mortar bags are wrapped in plastic for protection against the weather conditions. B—Gravity silos for premixed mortar and mortar mixers are being used to prepare mortar. The mixers are under the silos to receive the appropriate amount of regular and pre-colored dry mixed mortars. Large premixed bags, when needed, will be lifted by a forklift and unloaded into the top of the gravity silos.

A construction office trailer is usually one of the first pieces of equipment brought to a worksite to aid management in preparing the job. Office trailers are usually found on commercial jobs where the trailer will stay for an extended time of six months and longer, depending on the size of the construction project. Some office trailers lay flat on grade. Trailers are transported on a flatbed to the construction site. Other office trailers have wheels similar to a mobile home for making them easy to transport. This type of trailer requires stairs to be fabricated or purchased to access the office. See **Figure 13-5**.

Laying concrete masonry units involves planning, knowledge of the materials, and ability to use the tools

J.D.S/Shutterstock.com

Figure 13-5. Construction office trailer has been brought to the site. The trailer has electricity, heat, and air-conditioning. The trailer is used to plan the project, have safety or other types of meetings, and review the construction blueprints and the project schedule. The office trailer is also a place to warm up when the weather gets cold and to take breaks.

properly. Many procedures for laying concrete blocks are identical to those for laying clay brick, but some differences exist. Procedures that are unique to concrete masonry units are covered in this chapter, but the preceding chapters should be studied prior to this.

13.2 Types of Concrete Masonry Walls

The types of walls that can be built with concrete masonry units are essentially the same as those that are built with clay masonry units. Concrete masonry walls are classified as solid, hollow, cavity, composite, veneered, reinforced, and grouted (nonreinforced) walls.

Solid Masonry Walls

Solid masonry walls are walls built with solid masonry units. All joints are completely filled with mortar or grout. At least 75% of the net area of an ASTM C90 solid unit is solid concrete. Facing units are generally either brick or solid architectural units that are laid with full head and bed joints. See **Figure 13-6**. Backup units are also solid masonry units laid with full head and bed joints.

If flanged end units are used, the end cavity must be filled with grout. The spaces between units are also filled with mortar.

Masonry headers, metal ties, and grout are used between wythes to provide a solid structural bond. Some codes require no less than 4% of the wall area to be headers. Headers should extend at least 3″ or 4″ over the adjacent wythe and should be between 24″ and 36″ apart vertically, depending on the code.

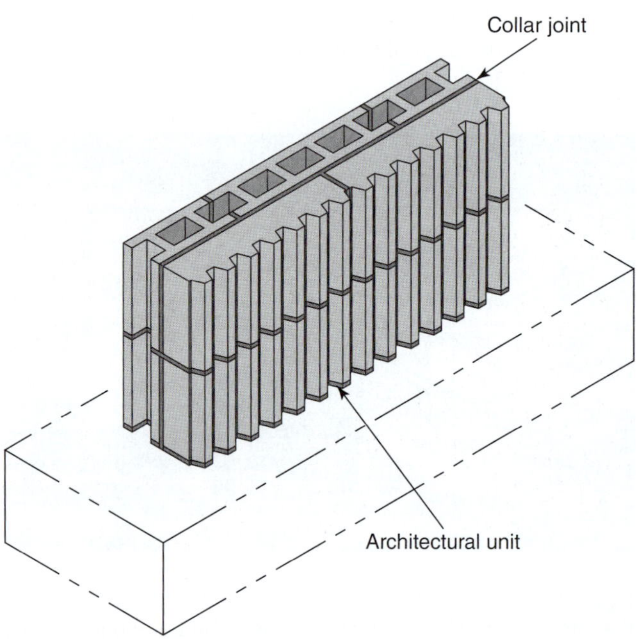

Goodheart-Willcox Publisher

Figure 13-6. An 8″ composite block wall with 4″ × 8″ × 16″ ribbed face lightweight block with 4″ × 8″ × 16″ lightweight concrete block backup. All joints are filled with mortar.

> **Trade Tip**
> The composite wall shown in **Figure 13-6** is not as common as in the past but is still used in certain applications. ASTM considers it obsolete.

Hollow Masonry Walls

Hollow masonry walls are built with hollow masonry units. However, solid units can be combined with hollow units laid in face shell mortar bedding (bedding only on the outside edges of the block). See **Figure 13-7**.

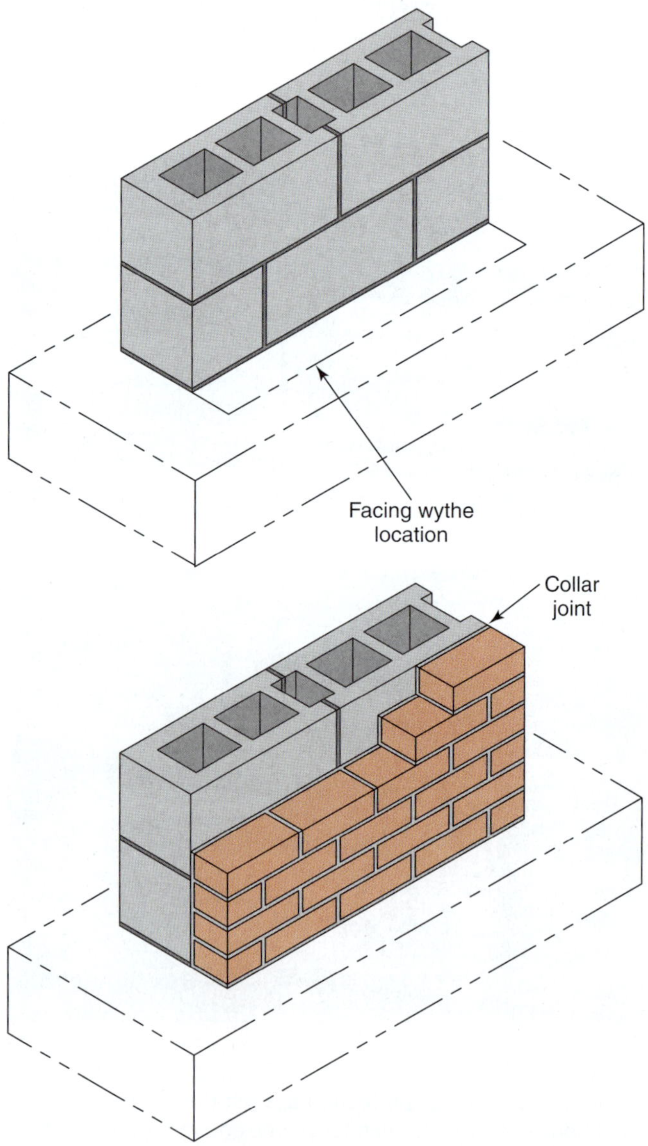

Goodheart-Willcox Publisher

Figure 13-7. Top—An 8″ × 8″ × 16″ lightweight concrete block partition wall. Bottom—A 12″ composite wall with 4″ × 2 2/3″ × 8″ face brick with 8″ × 8″ × 16″ lightweight concrete block backup.

Hollow masonry walls are constructed in any standard thickness (such as 4″, 6″, 8″, 10″, and 12″) using a single or multiple wythe where required. Multiple wythes generally consist of a facing wythe and a backup wythe. Multiple-wythe hollow masonry walls may be classified as composite walls.

Bonding between wythes may consist of masonry headers, metal ties, or grout. Mortar is used to fill all collar joints. A ***collar joint*** is the narrow space between the facing units and the backup units in a wall.

Cavity Walls

A ***cavity wall*** is a two-wythe wall that allows each wythe to react independently to stress. Cavity walls usually consist of two walls separated by a continuous air space that is at least 2″ wide. The 2″ cavity is most common. The walls (wythes) are tied together by rigid metal ties embedded in the mortar joints of both walls. See **Figure 13-8**. The facing wall is generally made from one wythe of solid or hollow masonry units that is 3 1/2″ to 4″ thick. The backing can be a single or multiple wythe wall of solid or hollow masonry units. The backing thickness may be equal to or greater than the facing, depending on the design requirements. The backing usually carries most of the weight, while the facing resists wind loads.

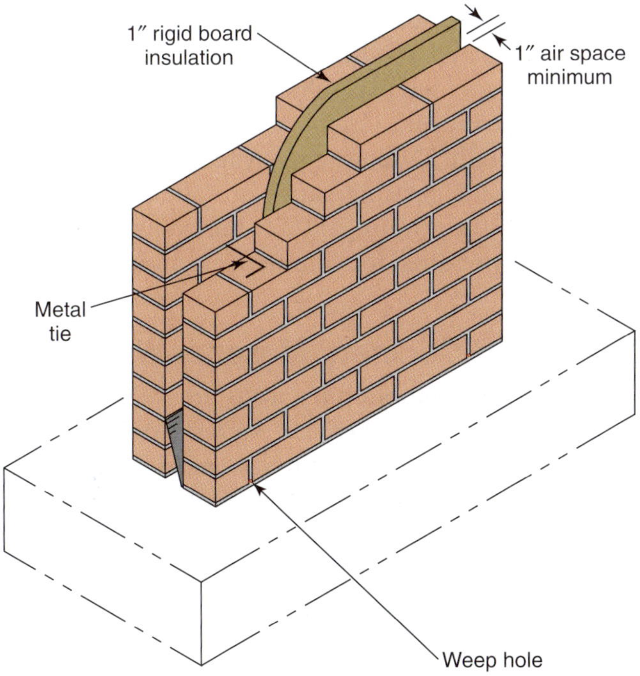

Goodheart-Willcox Publisher

Figure 13-8. A 10″ concrete block cavity wall under construction. There is 1″ of air space between the wythes and 1″ of rigid insulation.

Code Note — Cavity Walls

The International Residential Code (IRC) **R608.1** requires the facing and backing of cavity walls to be at least 3″ (76 mm) nominal in thickness. The cavity wall is required to be less than 4″ nominal in width. The backing must be at least as thick as the facing.

Insulation may be placed within the space between the walls. The insulation must not be water absorbent. Foamed plastic, foamed glass, or glass fiber can be used. Mats or rigid boards are recommended.

Composite Walls

A ***composite wall*** is a multi-wythe wall that is designed to act as a single member when subjected to loads. Stress is shared by the wythes through the collar joint and metal ties or headers. See **Figure 13-9** and **Figure 13-10.**

When ties are used, at least one tie for every 2 2/3 sq ft of wall is required for No. 9 gauge ties. The area expands to 4 1/2 sq ft when 3/16″ diameter ties are used. Regular ties should be spaced no more than 36″ horizontally and 24″ vertically. Adjustable ties should not exceed 1.77 sq ft of wall area or be spaced more than 16″ apart. MSJC Code Requirements for Masonry Structures 2.1.5.2.4 specifies the square footage amount where wall ties need to be placed and the diameter size of wall ties to be used in the masonry wall.

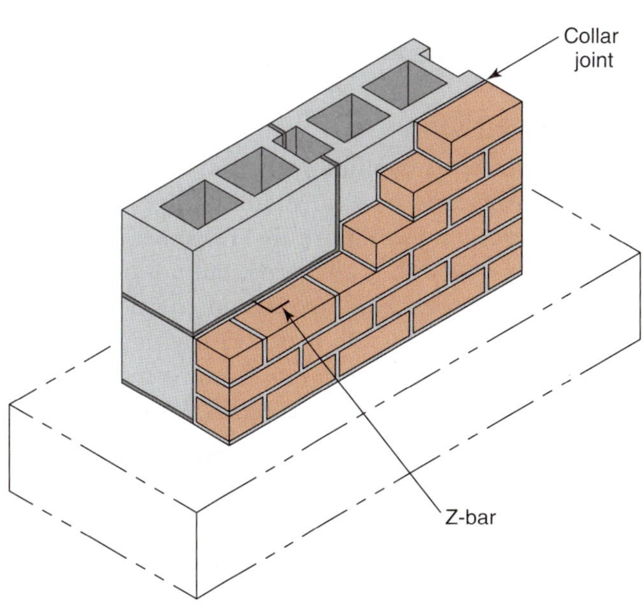

Goodheart-Willcox Publisher

Figure 13-9. This is a composite wall using 8″ × 8″ × 16″ concrete blocks with brick facing. Note the Z-bar and mortar collar joint.

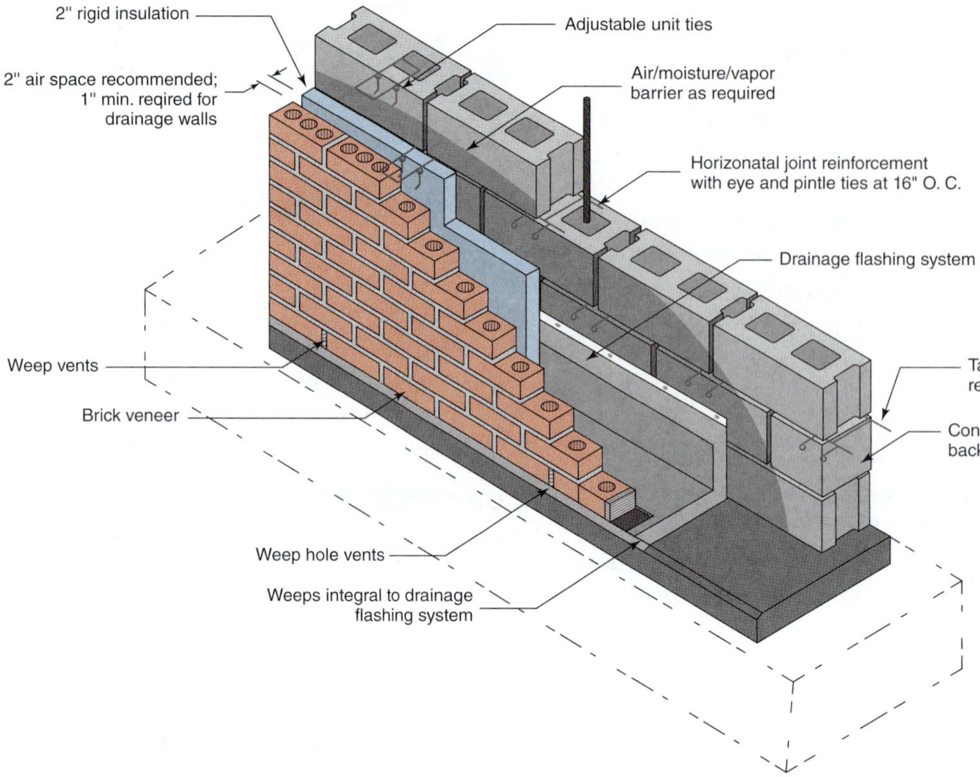

Goodheart-Willcox Publisher

Figure 13-10. A composite wall consisting of concrete blocks, bricks, air space, insulation, and wall waterproofing. Metal ties are used to connect the concrete blocks to the bricks.

Composite walls that use header-bonded wythes of solid masonry should have headers that compose at least 4% of the wall area. Distance between headers should not exceed 24″ in solid masonry and 34″ in hollow masonry. Check with the local codes to determine what is required for flashing and waterproofing masonry walls.

Veneered Walls

Masonry veneer is commonly used as a nonloadbearing facing material in residential and light commercial construction. See **Figure 13-11**. Veneer carries its own weight only and is anchored, but not bonded, to the backing. In some commercial construction, veneer is bonded to the concrete masonry backing through architectural units and joint reinforcement. Thin brick veneer is covered in Chapter 16, **Wall Systems**.

Trade Tip

By definition, if a cavity is less than 2″, the wall is a veneered wall. The space is 2″ or greater in a cavity wall.

The primary objective of veneer is to provide an attractive, durable exterior surface that prevents (or reduces) the entrance of moisture into the building. A system of flashing and weep holes provides for the exit of collected moisture. Other advantages of veneer are its sound barrier qualities and resistance to the effects of wind compared to other wall finishes.

Metal ties are used in residential veneer construction to anchor the veneer to frame or masonry backing. The ties are generally 22 gauge corrugated, galvanized steel strips 7/8″ wide. Spacing varies by code but is generally about 16″ vertically and 32″ horizontally. Adjustable ties are preferred for commercial construction.

Reinforced Concrete Masonry Walls

Reinforced concrete masonry walls are walls designed for applications that experience high stress, high wind loads, or severe earthquakes. Steel is embedded both horizontally and vertically to greatly increase the strength of the wall. A reinforced concrete masonry wall may be either a single or multiple wythe wall. See **Figure 13-12**.

Hollow masonry units can be used for single wythe walls laid with face shell mortar bedding. In this case, the vertical cores are lined up to form continuous vertical spaces that provide space for grouted reinforcement. Two-core blocks are preferred because they provide large spaces in which to locate the reinforcement.

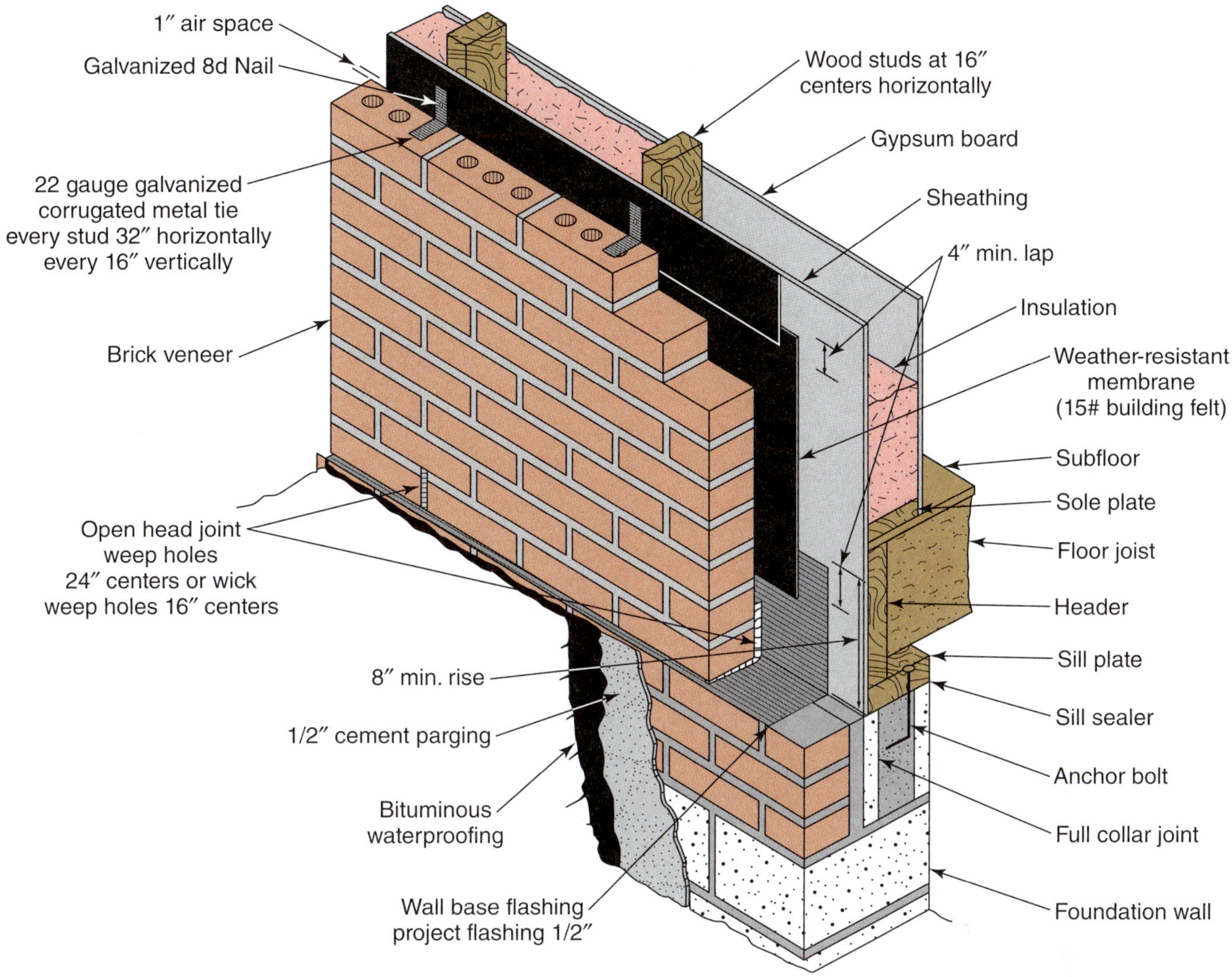

Goodheart-Willcox Publisher

Figure 13-11. A construction detail of a brick veneer wall section.

Two wythes are most common in multi-wythe wall construction. The wythes have a continuous air space from 1″ to 6″ between them. The masonry units are hollow, solid, or a combination of unit types. Reinforcement is placed in the space between the wythes and grouted solid. In commercial construction, more complicated reinforced masonry walls are common and involve three-wythe walls with No. 9 rectangular ties that are 4″ wide and 2″ narrower than the wall thickness.

Higher strength concrete masonry units are sometimes available for use in reinforced concrete masonry walls. These units have compressive strength ranging from 3000 to 5000 psi (net area).

Grouted Masonry Walls

The primary difference between grouted masonry walls and reinforced masonry walls is that grouted masonry walls do not include reinforcement. Grout may be added to loadbearing walls to provide additional strength. It is also used to fill bond beams and the collar joint in two-wythe wall construction.

Code Note **Grouting Requirements**

IRC **R606.3.5.1 Grout Placement** specifies that if work is stopped for one hour or longer, the horizontal construction joints shall be formed by stopping all tiers at the same elevation and with the grout 1″ (25 mm) below the top.

13.3 Basic Operations

A number of basic operations involved in laying concrete block must be mastered in order to build a quality wall. These basic operations include spreading mortar, handling concrete blocks, applying head joints, cutting blocks, and using a mason's line.

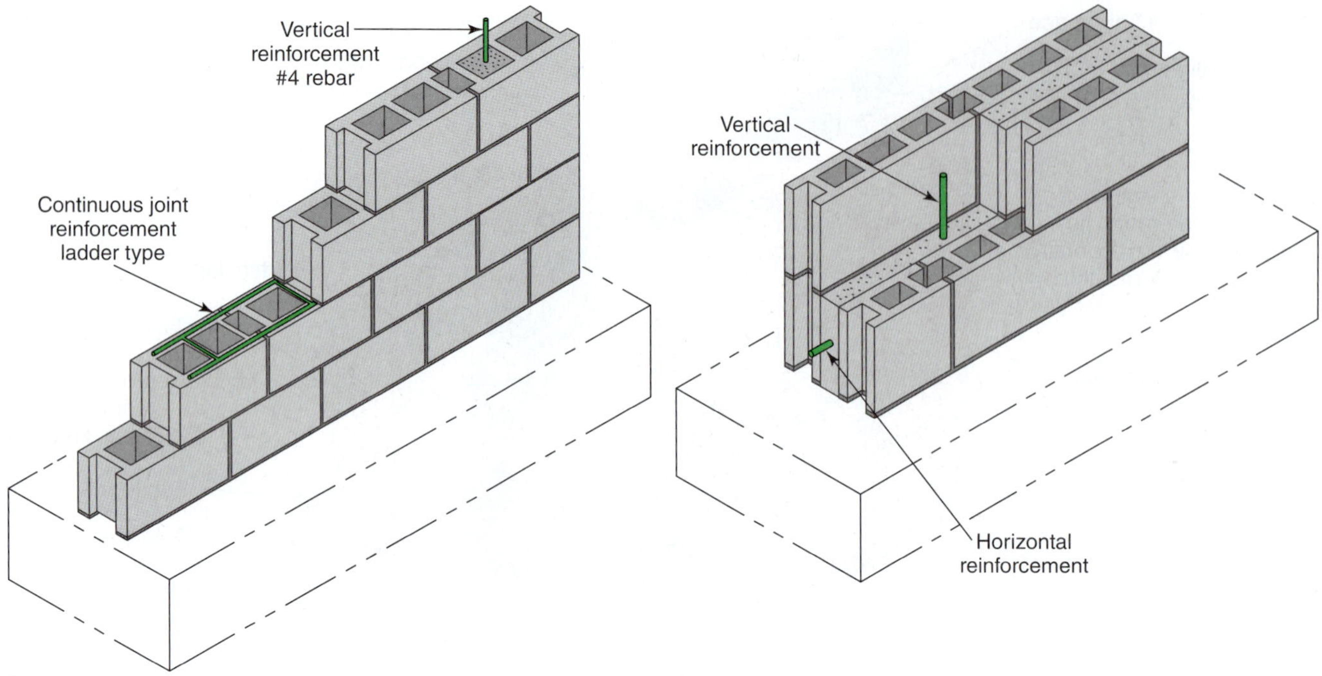

Goodheart-Willcox Publisher

Figure 13-12. A—An 8″ reinforced wall with vertical reinforcing #4 rebar at 48″ O.C. and horizontal continuous reinforcing, ladder type at 16″ O.C. vertically in the wall structure. B—A 10″ reinforced cavity wall with 4″ × 8″ × 12″ concrete masonry face brick on the interior and exterior wythes. The 2″ grouted cavity has vertical reinforcing #4 rebar at 48″ O.C. and horizontal reinforcing at 24″ O.C.

Spreading Mortar

Spreading mortar is a simple process but requires practice to achieve mastery. With practice, you will be able to determine just how much mortar to use for a given joint. Placing the unit is easier if the right amount of mortar has been laid down.

It is important to select a trowel that is best for the particular job and one that fits you. You may want to begin with a standard mason's trowel about 10″ long. Grasp the trowel in your dominant hand. Fingers should be under the handle and the thumb on top of the ferrule as shown in **Figure 12-4**.

Loading the Trowel

Mortar can be loaded on a trowel from the side of the board, from the middle of the board, or from the top of the pile. **Figure 12-5** shows steps that can be used to load the trowel. Develop the method that works best for you and perfect it.

Spread the mortar with a quick turn of the wrist toward the body and a backward movement of the arm. When the trowel is nearly empty, tip the trowel blade even more to help the remaining mortar slide off. The type of mortar and its consistency most often determines the speed of the stroke.

When using solid masonry units or laying the first course of concrete block, spread the mortar and furrow it with the point of the trowel, **Figure 13-13**. This method helps to form a uniform bed on which to lay the masonry unit.

Mortar is generally bedded only on the outside edges of concrete block (***face shell bedding***). A solid bed joint is seldom used. See **Figure 13-14**. Use a swiping

Author's image taken at Job Corps, Denison, IA

Figure 13-13. A full mortar bed is being furrowed with the trowel. It is important that ample mortar be used to ensure a good bond between the footing and the bed course of masonry.

Author's image taken at Job Corps, Denison, IA

Figure 13-14. Face shell of the first course in the corner lead are being bedded with mortar.

Author's image taken at Job Corps, Denison, IA

Figure 13-15. The mason is using a two-hand lift to position the regular concrete block on the mortar bed.

motion to apply the mortar to the shells. Be careful to fill all mortar joints and any holes or voids that appear right after the blocks are laid so all mortar dries at the same time.

Mortar Consistency

Mortar should be the proper consistency. If it is too runny, it will not support the weight of the block. If it is too stiff (dry), it will make setting the block difficult. Blocks should never be moved after they have been laid.

Mortar on the board should be kept well-tempered (sprinkled lightly) with water until it is used. Avoid constant working. This causes the mortar to dry out and become stiff. Stiff mortar does not spread well. Use mortar before it begins to set up. Discard mortar that has not been used within 2 1/2 hours.

Handling Concrete Blocks

Concrete blocks are large units that generally require both hands for placement. For example, a typical 8″ × 8″ × 16″ concrete block made with sand and gravel aggregates weighs about 40 lb. A comparable lightweight block weighs between 22 lb and 28 lb, depending on the specific aggregate used.

Lift a block with both hands, **Figure 13-15**. Grasp the web at each end of the block to lay it on the mortar bed. Use even movements rather than jerking motions so the mortar is not dislodged from the block.

After gaining experience in setting blocks with two hands, try setting them using one hand. This process saves time because the trowel remains in your hand while you are placing the block. Using one hand works well when only one or two blocks are being set. If several blocks have been prepared, lay the trowel aside while placing the blocks on the mortar bed.

Trade Tip

Lifting properly is very important! When lifting block, use your legs, arms, and back muscles to reduce fatigue.

Regular concrete blocks can weigh approximately 40 lb per unit. During a normal day, a mason laying 150 blocks may lift 6000 lb of masonry block. The National Concrete Masonry Association (NCMA) identifies that a mason can lay between 135 and 190 standard concrete masonry units or block (8″ × 8″ × 16″) per day. This amount is based on site conditions and weather. NCMA also determined, as the weight of the block increases the number of block laid by the mason during the day will drop.

By tipping the block slightly forward toward you when you place it on the mortar bed and looking down the face of the block, you can position the block in proper position with respect to the top edge of the course below, **Figure 13-16**. Then roll the block back slightly to correctly align the top of the block with the line. During this movement, the block should be pressed toward the last block to form a good head joint. Mortar should squeeze out slightly.

Position blocks in a wall with the wide flange on the top to provide a wider space for the bed joint. As each block is laid, the excess mortar should be cut off with the trowel held at a slight angle to the block. Return excess mortar to the mortar board.

Applying the Head Joint

When making head joints on concrete blocks, form full head joints on both ***ears*** (end edges) of each block

Author's image taken at Job Corps, Denison, IA

Figure 13-16. The mason is positioning the block in the first course using the mason's line as a guide for placement. They are using a single-hand hold, allowing the right hand to be free to use the trowel to remove excessive mortar and to check flushness with the previous block laid.

to be laid. The ears or end edges extend from the end of the block, **Figure 13-17**. Blocks may have ears on both ends, one end, or no ears at all. The type of block selected depends on where it is to be used in the structure. Stretcher blocks have ears on both ends; corner blocks usually have ears on one end.

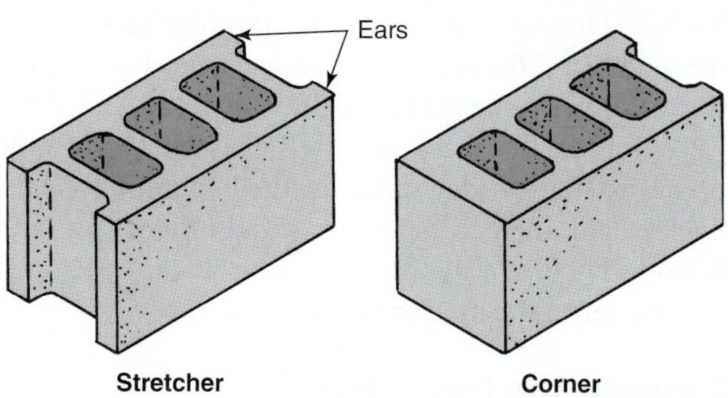

Goodheart-Willcox Publisher

Figure 13-17. These two concrete blocks are the primary units used to construct most concrete masonry walls. Nominal size is 8″ × 8″ × 16″.

First, set the block on end so the ears that are to receive mortar face up. Pick up enough mortar with the trowel to form the head joints on one or more blocks, depending on the situation. The trowel should not be fully loaded. A fully loaded trowel makes it difficult to apply the right amount of mortar to each ear. This process is called *buttering the block* or *buttering the ears.* See **Figure 13-18A**.

Use a downward swiping action to apply mortar to the ears. Next, press the mortar down with the trowel on the inside of each ear of the block to attach it to the block. If the mortar is not sufficiently attached, it will fall off when the block is lifted and turned for placement in the wall. If the mortar falls off during placement, start again with fresh mortar. Apply enough mortar so a full head joint is formed that will be watertight. Too much mortar is wasteful and creates extra work and cost. The correct amount of mortar will be evident after some practice.

Blocks have a narrow surface (bottom) and a wide surface (top). See **Figure 13-18B**. The narrow surface of a block is laid into the fresh mortar. The wide surface is laid in the up positon to receive mortar for the next course. As skills are developed, another method for buttering block is to apply mortar when they are laying horizontally or flat. When using this technique, stage block with the wide surface in the up positon. See **Figure 13-18C**. Stack block close to the mason but not in the way when laying the block in the wall. As a block is buttered, it is laid directly into the wall.

Cutting Block

Even though concrete blocks are available in half-length units as well as full-length units, it is sometimes necessary to cut block to fit. Blocks can be cut with a brick hammer and blocking chisel. See **Figure 13-19**. When using a chisel, hold the beveled edge toward you. The piece of block to be cut off should face away from you. Score the block on all sides to obtain a cleaner break.

If a neater, cleaner cut is desired, use a masonry saw to cut the blocks. See **Figure 13-20**. This method is fast and accurate. The block should be dry when cut with a masonry saw.

Safety Note

Be sure to wear safety glasses and a face shield to protect eyes and the face when cutting blocks. When cutting block dry with a saw, always wear a respirator to protect against the concrete dust that may contain silica.

A

B

C

Author's images taken at Job Corps, Denison, IA

Figure 13-18. A—The ears on the first block have been buttered. Mortar on each of the block ears has been pressed against the block to ensure it adheres. B—The bottom and top surface of two blocks. The left block is the narrow surface or bottom of a block. The right block shows the top surface and is laid facing up. C—Block staged with the wide surface facing up. The mason is buttering the ears of the block to be laid.

Author's image taken at Job Corps, Denison, IA

Figure 13-19. A masonry student is preparing to cut a concrete block using a chisel. They have drawn a guide line to follow when scoring the block on the first side. The backside will be scored next. It is important to score all four surfaces to ensure a clean break.

Author's image taken at Job Corps, Denison, IA

Figure 13-20. A masonry student, wearing appropriate PPE, is preparing to cut a concrete block using a wet masonry saw. The trough is full and its water will be pumped over the blade to eliminate exposure to silica dust when cutting.

Using a Mason's Line

Any straight wall longer than about 4′ is usually laid to a line. Mason's line is a guide that aids in building a straight, level, and plumb wall. Mason's line is a light, strong cord that can be stretched taut with little or no sag. The line should be pulled to the same degree of tautness for each course so all bed joints will be uniform and parallel. See **Figure 13-21**.

Author's image taken at Job Corps, Denison, IA

Figure 13-21. A mason's line is being used to ensure blocks are straight and level.

Trade Tip

When purchasing mason's line, it is best to select braided line. Braided line lasts longer and does not unravel when one of the strands is cut.

Leads are usually built first at each corner to establish the proper height of the block that will be laid between the leads. The leads also provide a place to attach a mason's line when the time is appropriate to fill in between the leads with stretcher blocks. The height of each course in the lead is determined by using a mason's rule or corner pole with height marks on it, **Figure 13-22.**

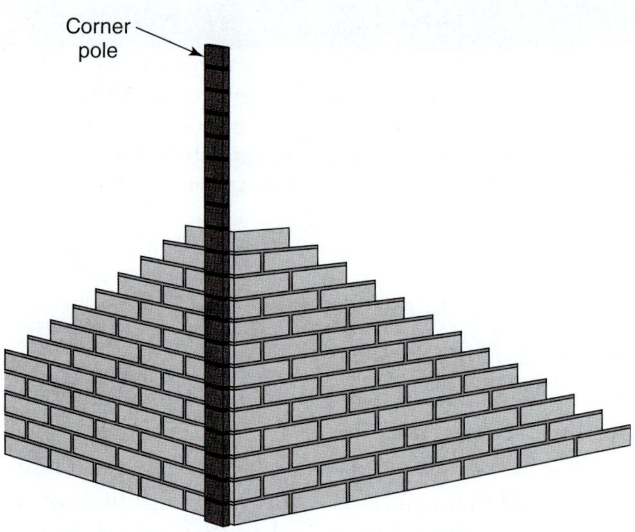

Goodheart-Willcox Publisher

Figure 13-22. A corner pole being used to set height of the first block in each course in the corner lead.

On long walls, it is impossible to eliminate sagging in the line. The course height should be checked with a corner pole to ensure accuracy. A single block can be set approximately midpoint between the two leads to support the line using a ***metal twig (trig)*** in a very long wall. The twig is a thin piece of metal designed to hold the line between its metal fingers. Twigs may also be made of plastic with the line snapped into the twig. The single block is set at the correct height using the proper amount of mortar to ensure that the final top edge of the block is at the same elevation as the corner leads. The height is checked with a mason's rule or corner pole to ensure that the proper height is maintained. The mason's line is then positioned in the twig to hold the line in position. The twig and line are then placed on top of the block and held in position with a smaller piece of block or brick.

Laying block too close to the line (crowding) should be avoided. Stay a line width (1/16″) away from the line so you do not risk moving the line, which would destroy its usefulness as a reference guide.

13.4 Laying Concrete Block Walls

A well-planned concrete block structure mainly involves stretcher and corner blocks. Refer back to **Figure 13-17**. These blocks are nominally 8″ × 8″ × 16″. Actual size is 7 5/8″ × 7 5/8″ × 15 5/8″ to allow for a standard 3/8″ mortar joint. Blocks of other sizes are sometimes used, also with a 3/8″ mortar joint.

Concrete blocks must be protected from excess moisture before use. If blocks are wet when placed, they will shrink when dry and cause cracks. The blocks should be stacked on platforms and covered with plastic or a tarpaulin to protect them from rain.

Mortar for concrete block masonry should be mixed in accordance with the specifications in Chapter 10, **Mortar and Grout**. The consistency of mortar for concrete block masonry is just as critical as it is for brick or stone.

Laying an 8″ Running Bond Concrete Block Wall

To produce a strong wall with a good appearance, adequate reinforcement or staggered joints are required. Advance planning is necessary. The length of the blocks in relation to window and door openings in the wall should be considered. It is a good idea to check building dimensions carefully before beginning work so problems can be identified before the wall is begun.

PROCEDURE

Laying an 8″ Running Bond Concrete Block Wall

1. Establish the outside wall line. A chalk line can be used to provide a straight line for the first course of block. Check the wall line for squareness and proper length before proceeding to the next step.
2. String out the blocks for the first course without mortar to check the layout. See **Figure 13-23**. Allow 3/8″ for each mortar joint. To save time, make a spacer that measures 3/8″ to use as a space block. When you are satisfied with the layout, the blocks are set aside.

Author's image taken at Job Corps, Denison, IA

Figure 13-24. Once the first block in the corner lead is in position, place the second block carefully so as to not move the first.

Trade Tip

Spacer blocks can be cut to different thicknesses to represent mortar joints. The most common mortar joint is 3/8″, but 5/16″ and 7/16″ or larger can be used. The spacers can also be used when laying out brick spacing.

3. Spread and furrow a full mortar bed with a trowel. Refer to **Figure 13-13**. Provide plenty of mortar on which to set the block.
4. Lay the corner block. See **Figure 13-24**. Position it carefully and accurately. Lay concrete blocks with the thicker edge of the face shell up to provide a wider mortar bed.
5. Lay several stretcher blocks along the wall line. To place them, push them downward into the mortar bed and sideways against the previously laid block.

Author's image taken at Job Corps, Denison, IA

Figure 13-23. When the corner leads are located, the first course of blocks are dry laid to check the spacing for the walls. The space between the blocks is equal to the desired mortar joint.

Trade Tip

Several blocks can be buttered on the end of the face shells if they are stood on end. This speeds the operation.

6. After three or four blocks have been placed into position, they may be aligned, leveled, and plumbed with a mason's level. See **Figure 13-25** and **Figure 13-26**. Tap on the block rather than on the level.
7. As the first course is being laid, continue to check for accuracy as shown in **Figure 13-27**. This corner is important since the remainder of the wall depends on its accuracy. The lead corner is usually laid up four or five courses high above the center of the wall. Check each course as it is laid to be sure it is aligned, level, and plumb. See **Figure 13-28** and **Figure 13-29**. The faces of the blocks should be in the same plane. Each block is stepped back half a block. Check the spacing by placing the level diagonally across the corners of the block as shown in **Figure 13-30**. All corners should be lined up on the edge of the level.
8. As the individual courses are laid in the corner lead, check height of the corner block to ensure it is the correct height before laying remaining block in the course or progressing to the next course, **Figure 13-31**.

Continued

PROCEDURE *(continued)*

A B *Author's images taken at Job Corps, Denison, IA*

Figure 13-25. A—A mason is leveling the length but also settling the blocks to the proper elevation. B—After leveling the length, level across the width of the block. Then check the end of the lead for level. Make adjustments accordingly.

A B *Author's images taken at Job Corps, Denison, IA*

Figure 13-26. A—The corner block is plumbed using a 4′ level. B—The end block of the lead is plumbed using a 4′ level. Adjustments are made at both ends of the lead.

9. After the corner leads have been constructed, lay the blocks between the corners. Stretch a mason's line from corner to corner at the proper height for each course. See **Figure 13-32**. Secure the line with line blocks or line pins. Tension should be the same each time to ensure a uniform wall. Work from the corners toward the center of the wall, **Figure 13-33A**. Do not allow a block to touch the line because the line will be

Author's image taken at Job Corps, Denison, IA

Figure 13-27. Alignment is being checked with the level (used as a straightedge) to ensure accuracy before completing the first leg of the corner lead. The mason is lightly tapping on the block to make minor adjustments.

Author's image taken at Job Corps, Denison, IA

Figure 13-28. The last block in the second course has been laid and now both are being leveled.

pushed beyond its normal position, causing a curve in the wall. Keep the block a line width away from the line and even with the bottom of the line. As laying of block progresses toward the closure block, it is important to measure the opening to determine if joint spacing is accurate, **Figure 13-33B**. Measurements should be based on the size of the concrete units and size of joints. The last block in the course should be carefully positioned in the closure space. See **Figure 13-34**.

10. Joints may be tooled when the mortar has become thumbprint hard. To test the mortar, use your thumb to determine hardness of the mortar

Author's image taken at Job Corps, Denison, IA

Figure 13-29. The corner is checked to ensure it is plumb. The mason is making minor adjustments using a trowel.

joint. Press your thumb into the mortar joint. Mortar should resist slightly against the thumb being pressed into the joint. When the mortar does not stick readily to the thumb and leaves thumb impression, the mortar joints in that area of the wall are ready to strike.

11. The most effective joint is one that has been compacted or pressed into place. For this reason, a concave or V-joint is best for exterior work. The tool should be slightly larger than the width of the joint so it will make contact along the edges of the blocks. A 5/8″ diameter bar is usually used for a 3/8″ concave mortar joint. A 1/2″ square bar is used for making a 3/8″ V-joint. When the mortar meets the thumb test, tool the vertical joints first using the correct width striker for the size of the joint. See **Figure 13-35**.
12. After the vertical joints are struck or tooled, the horizontal joints can be finished. Long horizontal joints can be stuck with a sled runner. See **Figure 13-36**. Use a trowel to remove any mortar burrs left from the tooling. See **Figure 13-37**.
13. When the mortar is sufficiently dry, brush the wall with a fiber brush or rub the wall with a burlap bag to remove dried particles. See **Figure 13-38**.

Continued

PROCEDURE *(continued)*

A

B

Author's images taken at Job Corps, Denison, IA

Figure 13-30. A—Spacing of the block is being checked by lining up the corners with a straightedge and making adjustments as needed. B—Checking corner lead joint spacing of block by laying 4′ level over ending points of the blocks in the lead.

Author's image taken at Job Corps, Denison, IA

Figure 13-31. The mason is checking height of corner lead courses.

Author's image taken at Job Corps, Denison, IA

Figure 13-32. The corner leads have been completed. The mason is attaching wooden corner blocks and running the mason's line as a guide to lay blocks between the two corner leads. Take note, both individuals are wearing safety boots, hi-vis vests, and protective gloves.

A B

Author's images taken at Job Corps, Denison, IA

Figure 13-33. A—The wall is being filled in using a mason's line to ensure proper height and a straight wall. The mason's quick movement of the trowel is required to butter the ears of the block. B—The mason is measuring the space for the closure block. Mortar has been placed on all four ears to ensure adequate mortar when placing the closure block.

Author's image taken at Job Corps, Denison, IA

Figure 13-34. The closure block has been installed and the mason is filling in voids to ensure adequate bonding between blocks.

Portland Cement Association

Figure 13-36. Long horizontal joints are being struck last using a sled runner tool.

Author's image taken at Job Corps, Denison, IA

Figure 13-35. The mason is striking the vertical joints first. They are using a barrel jointer with interchangeable barrel heads. The barrel heads range from 1/2″ to 7/8″ and are changed when various projects require different joint sizes.

Author's image taken at Job Corps, Denison, IA

Figure 13-37. The mason is using their trowel to remove any mortar burrs after tooling joints.

Continued

PROCEDURE *(continued)*

Author's image taken at Job Corps, Denison, IA

Figure 13-38. Use a soft fiber brush to clean the wall of dried mortar particles; too stiff of a brush may dig into the soft joints.

Control Joints in a Concrete Block Wall

Control joints are continuous vertical joints built into concrete masonry walls. They are placed where forces might cause cracks. Movement of the wall occurs along the control joint and cracks are not easily seen. Control joints should be the same width as other joints. They should be sealed with caulking compound after the mortar has been raked out to a depth of 3/4″. See **Figure 13-39**.

Other methods of constructing control joints in concrete block walls are as follows:

- Place a Z-bar in the horizontal mortar joint above jamb blocks to provide lateral (sideways) support for the wall above windows and doors. See **Figure 13-40**.
- Use building paper or roofing felt in the end core of a block to break the bond. See **Figure 13-41**. The felt extends the entire height of the control joint.
- Use a special tongue-and-groove block, which allows movement. See **Figure 13-42**.

Wall Intersections

Intersecting walls should not be joined together in a masonry bond as walls are at corners. Instead, one wall should end at the face of the other wall with a control joint. One method that provides for movement and lateral support is to use a metal tiebar.

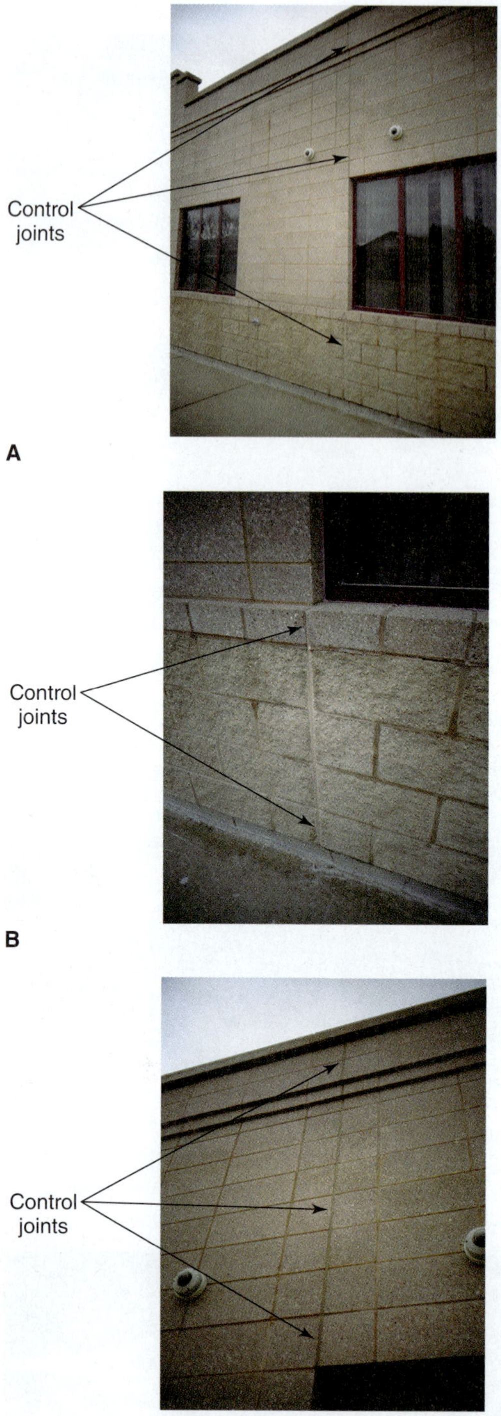

Timothy L. Andera

Figure 13-39. A—A control joint has been installed at the first course of the split face concrete block wall through the edge of the window and continues above the window up to the edge of the roof overhang. B—Special split-face concrete block was manufactured for this job along with half-block control joint CMUs. A control joint is shown at the first course of the split-faced block and continues vertically up through the sill where it forms the edge of the window. C—The control joint has been filled with a matching rubberized type of sealant that expands and contracts with any movement on either size of the joint. Special half-blocks were manufactured to allow control joints at multiple locations on the building to run vertically directly above windows and doorways.

Goodheart-Willcox Publisher

Figure 13-40. Control joint using jamb block and Z-bar.

See **Figure 13-43**. The tiebar is 1/4″ thick, 1 1/4″ wide, and 28″ long, with 2″ right angle bends on each end. These tiebars are usually placed 4′ apart vertically. The bent ends are placed in cores filled with mortar or concrete. Pieces of metal lath are placed under the cores to prevent the mortar from falling through. See **Figure 13-44**.

Nonloadbearing walls (walls that do not carry a structural load) can be tied to other walls using strips of metal lath or hardware cloth. See **Figure 13-45**. The metal strips are put in every other course. Hardware cloth is placed across the joint between the two walls.

Portland Cement Association

Figure 13-41. Building paper is used to break the bond. This method is known as the Michigan control joint.

Goodheart-Willcox Publisher

Figure 13-43. The bent ends of the metal tiebar are embedded in the cores of the block. Metal lath is positioned over the cores to stop mortar or concrete from passing to the lower block. The cores are then filled with mortar or concrete.

A

B

Goodheart-Willcox Publisher

Figure 13-42. A— The mason is installing a control joint half-block. The control joint will be on the right end of the half-block. The half-blocks and full blocks will create a control joint that will run vertically up the complete wall structure. The control joint will be filled later with a specialized sealant. B—This view of the control joint shows that there is a small gap between the tongue-and-groove blocks to allow for slight movement. The sealant will allow movement without breaking the seal to keep exterior moisture from entering the wall cavity.

Goodheart-Willcox Publisher

Figure 13-44. Cores of the next course will be filled with mortar over this metal tiebar.

Goodheart-Willcox Publisher

Figure 13-46. Anchor bolts placed in the cores of the top two courses of block are used to attach the plate to the wall.

Portland Cement Association

Figure 13-45. Metal lath or 1/4″ mesh galvanized hardware cloth can be used to tie nonloadbearing walls together.

Anchorage to Masonry Walls

To attach future framing, a wood plate is often anchored to the top of a concrete block wall. Anchor bolts are generally used for this purpose. When positioning anchor bolts, keep them closer to the inside of the cell of the blocks so they do not interfere with the wood framing that is positioned on top of the blocks. Anchor bolts are usually 1/2″ in diameter, 18″ long, and spaced about 4′ apart. See **Figure 13-46**.

Anchor bolts are placed in the cores of the top two courses of block. The cores should have metal lath under them so they can be filled with mortar or concrete. The threaded ends of the anchor bolts must extend above the top of the wall so they pass through the sill and provide enough thread for the washer and nut. In most situations, the thickness of the sill is 1 1/2″, but a sill sealer is also placed under the sill.

Trade Tip

To determine the appropriate height that anchor bolts will need to be above the top of concrete block wall, you can check with the company that is performing the framing.

Laying a 10″ Concrete Block Cavity Wall

A cavity wall is a wall that consists of two walls (wythes) separated by a continuous air space 2″ to 4 1/2″ wide. The wythes are tied securely together with noncorroding metal ties that are embedded in the mortar joints. See **Figure 13-47**. Unit ties are generally placed at every other block horizontally and in every other horizontal joint. Continuous metal joint reinforcement serves the same purpose.

Cavity walls are popular because they provide good moisture resistance if they are constructed properly. See **Figure 13-48**. Three areas need particular attention—weep holes at the bottom of the wall, flashing to direct water to the weep holes, and cleanliness of the cavity. Weep holes should be located in the bottom course at about every second or third head joint in the outside wythe. The cavity must be kept clear of mortar droppings that could form a bridge from one wythe to the other. Such a condition would transmit moisture across to the interior.

The following procedure is for building a 10″ concrete block cavity wall using two wythes of 4″ × 8″ × 16″ concrete blocks.

Portland Cement Association

Figure 13-47. Rectangular anchors or ties are used in cavity wall construction to bond the wythes together.

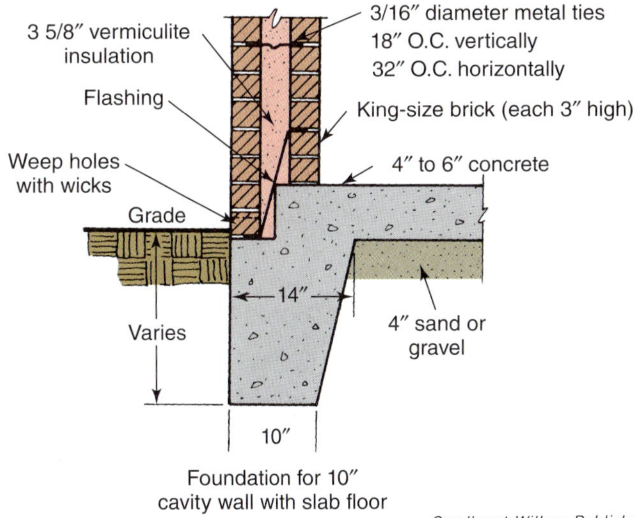

Goodheart-Willcox Publisher

Figure 13-48. A typical cavity wall foundation detail for slab floor construction.

PROCEDURE

Laying a 10″ Concrete Block Cavity Wall

1. Lay down a full bed of mortar for the inside wythe. This bed joint must be watertight.
2. Place the blocks for the first course of the inside wythe. Be sure all head joints are solid and watertight.
3. Position the flashing over the top edge of the first inside course and rest it on the foundation (under the outside wythe). Embed the flashing with mortar between the first and second course of the inside course. See **Figure 13-49**.
4. Lay the first course of the outside wythe, allowing a 2″ continuous air space between the wythes. The flashing will be beneath this course. Maintain weep holes between every second or third block. Insert wick or another acceptable material. Keep the cavity clean as the work proceeds up to the second course.

Trade Tip

Cotton sash cord is the most common material used for creating weep holes. If the cotton cords are removed to leave weep holes, stainless steel wool can be placed in holes to ensure that insects and dirt cannot enter the holes. Plastic tubing can also be used to install weep holes.

5. Lay the second outer course with full bed and head joints. Be sure all joints and holes are filled with mortar. Do not allow any mortar to fall into the cavity. Mortar can be spread about 1/2″ back from the edge of the cavity to reduce the chance of falling into the cavity. Another method is to angle the mortar back to reduce the mortar along the outer edge. If rigid insulation is to be used in the wall, install it now. The insulation should be placed against the inside wythe and should not prevent the escape of moisture from the cavity. Always maintain a minimum of 1″ of air space.

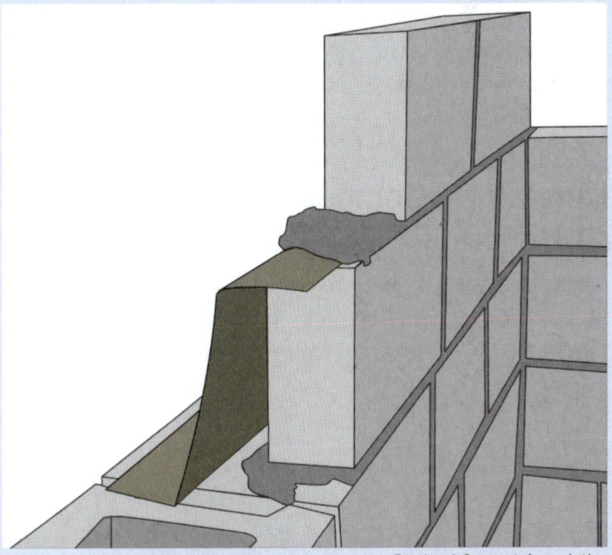

Portland Cement Association

Figure 13-49. Flashing should be used to prevent moisture from entering the inner wall from the cavity. Note how the flashing is embedded in the mortar joint.

6. Position unit ties or continuous joint reinforcement over the second course of the two wythes. Embed the joint reinforcement within the mortar joint under the third course on the inside wythe.
7. Lay the third outer course with the reinforcement embedded in the mortar joint.
8. Place a board on the ties or reinforcement to catch mortar droppings as the next course is laid. It can be

Continued

PROCEDURE *(continued)*

removed when the next reinforcement is placed. See **Figure 13-50**.

9. Repeat the procedure until the wall has reached the proper height.

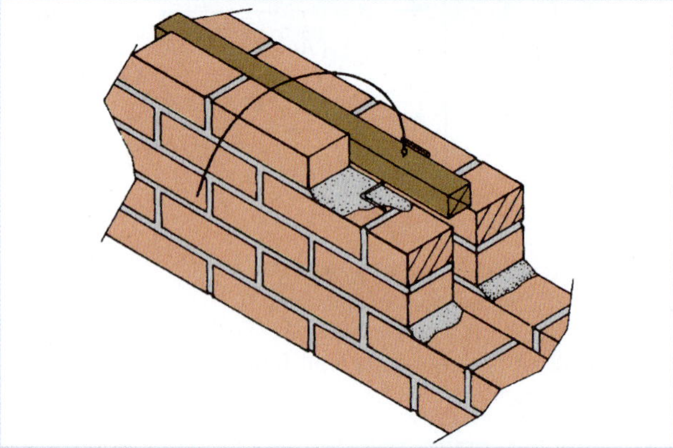

Portland Cement Association

Figure 13-50. A recommended method of keeping a cavity free of mortar droppings is to use a board resting on the joint reinforcement. The board is moved up when the next reinforcement level is reached.

Laying an 8″ Composite Wall with Concrete Block Backup

A composite wall is two wythes bonded together with masonry, metal ties, or joint reinforcement. See **Figure 13-51**. The two wythes are joined together in a continuous mass using a vertical collar joint that prevents the passage of water through the wall. See **Figure 13-52**.

Concrete blocks are often used as backup for bricks to make a composite wall. The basic construction of a composite wall can be similar to a cavity wall, except that the cavity is reduced to the width of the collar joint and filled with mortar. However, a composite wall can be quite different when masonry headers are used to bond the two wythes together, as shown in **Figure 13-53**.

Use the following procedure when building an 8″ composite wall with brick facing and concrete block backup.

Portland Cement Association

Figure 13-52. Bonding a wall across a collar joint by parging the backside of the facing.

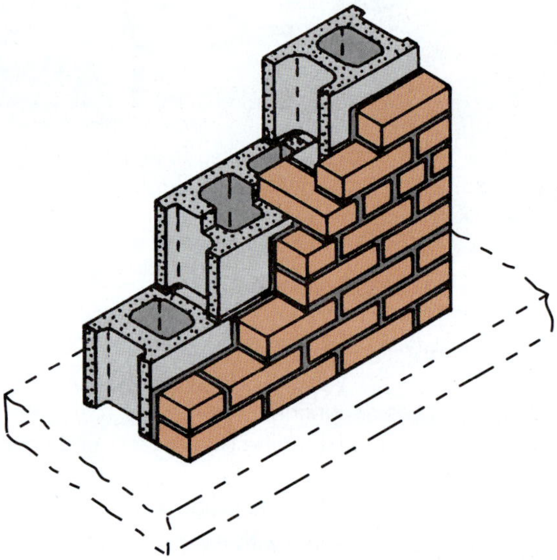

Goodheart-Willcox Publisher

Figure 13-51. Composite walls can be bonded together using a masonry bond to tie the wythes together.

Goodheart-Willcox Publisher

Figure 13-53. A header course being laid in composite wall.

PROCEDURE

Laying an 8″ Composite Wall with Concrete Block Backup

1. Lay up the inner wythe of 4″ concrete block two courses high. See **Figure 13-54**. Parge the side adjacent to the outer wythe, being careful not to upset the bond. Any mortar that has hardened should be cut off with the trowel before parging.

Trade Tip

The term ***parging*** refers to the application of a coating of mortar. The use of a plasterer's trowel is more efficient than a brick trowel for this operation.

2. Next, lay six courses of bricks in running bond to bring the outer wythe to 16″ high. See **Figure 13-55**. Be sure the vertical collar joint is filled with mortar and the bricks are laid with full bed joints.
3. Lay a header course of brick across the two wythes to tie the courses together. See **Figure 13-56**. In header courses, the cross joints should be completely filled with mortar spread over the entire side of each brick before it is placed in the wall. Composite walls with header courses are susceptible to water leakage if not all joints are solid and all holes filled with mortar.
4. Repeat the previous steps until the wall has reached the desired height.

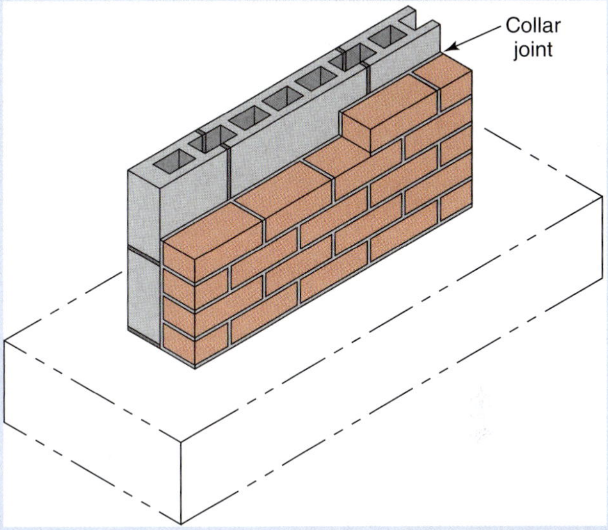

Goodheart-Willcox Publisher

Figure 13-55. Six courses of bricks are laid to bring the outer wythe to 16″ high.

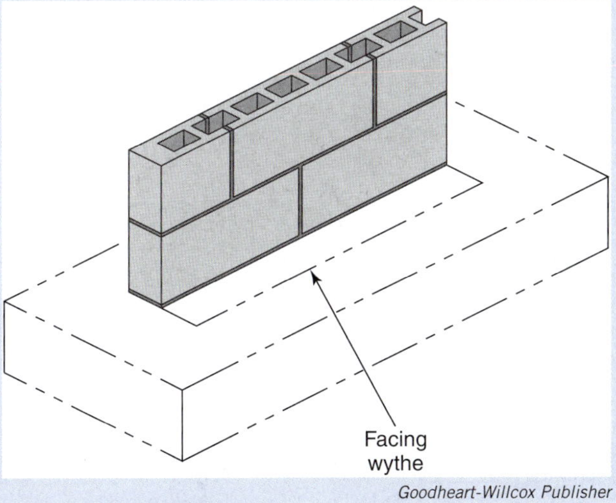

Goodheart-Willcox Publisher

Figure 13-54. Laying up the inner wythe of 4″ concrete block two courses high.

Goodheart-Willcox Publisher

Figure 13-56. Laying a header course to tie the two wythes together.

Workplace Skills

Performance Goals and Project Management

No matter what project or activity an individual intends to accomplish, seeing a project come to fruition, or seeing hard work pay off, is very rewarding! Accomplishments do not happen by accident; they result from hard work, planning, and the ability to stay on task and not to give up when things get tough. The following are a few activities that will help you bring projects to fruition:

- Determine what needs to be done. Identify the tasks or goals to complete the project.
- Develop a list of strategies to accomplish the tasks or goals.
- Set time lines to complete the tasks. In construction, superintendents may use a CPM chart (critical path method) or a Gantt chart. Both list major activities that need to be accomplished and determine the amount of time needed to complete each task.
- Identify tasks as they are completed. If a roadblock occurs with a task, seek help or determine methods that can eventually allow the task to be completed.
- Continue to complete the individual tasks until the complete project has been finished.

Composite walls greater than 10″ in thickness generally use a specially shaped header block to bond the facing headers and backup units with sixth course bonding. The special header block can be laid with the recessed notch up or down and with block backup laid in face shell mortar bedding.

Concrete block is often used as a backup for a stone facing. See **Figure 13-57**. Since the mortar joints in the block and stone do not line up, corrugated metal ties can be used to tie the wythes together.

Piers

Another application for concrete block in construction is the building of piers. See **Figure 13-58**. Piers are commonly used as support members for beams. Piers are also found in areas such as entrances to large buildings, where the piers support trusses or roof structures.

Block piers can be used for smaller structures such as mailboxes or locations to attach light fixtures or signage. In applications such as these, a decorative type of block is selected. In most of these applications, a masonry cap made of concrete is included to shed water and keep moisture from entering the center part of the pier. It is important in the planning stage of constructing piers to contact the local building code official prior to any work being done to determine the depth of the frost line in your area. The footings may also need to be inspected by the building official prior to laying block.

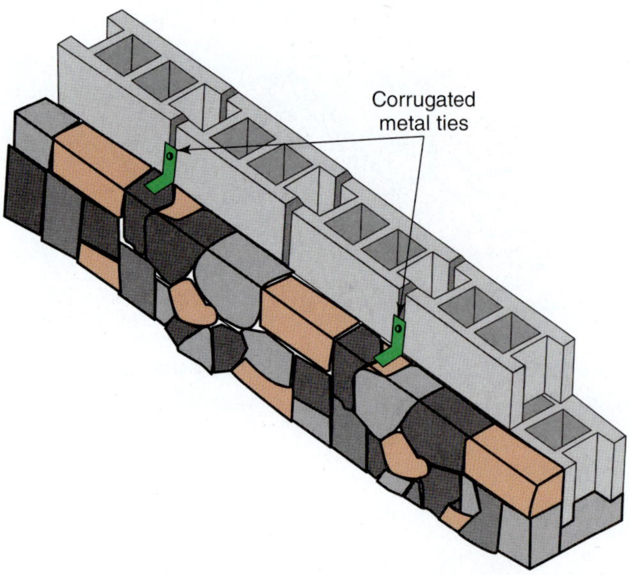

Goodheart-Willcox Publisher

Figure 13-57. A 12″ composite stone wall facing on 8″ × 8″ × 16″ lightweight concrete block backup. Corrugated metal ties are used to add extra strength to tie the stone and block together as a uniform structure.

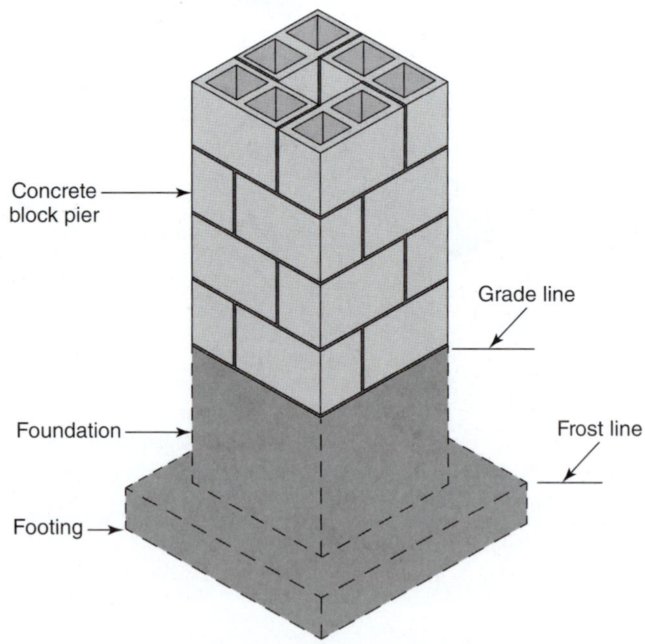

Goodheart-Willcox Publisher

Figure 13-58. Split-faced concrete block pier to be used for a lighting structure. The pier is supported by a concrete footing and foundation. The footing is located below the frost line. A concrete cap is positioned after electrical wiring is installed in the center area of the pier.

13.5 Cleaning Concrete Block Masonry

Concrete block walls are not cleaned with acid to remove mortar smears or droppings. Therefore, take care to keep wall surfaces clean during the construction process.

Any mortar droppings that stick to the wall should be allowed to dry and harden. Large particles of mortar can be removed with a trowel, chisel, or putty knife. If mortar is removed while it is wet, it will most likely smear into the surface of the block.

Rubbing the wall with a small piece of block or a rubbing stone removes practically all of the mortar. In some instances, a commercial cleaning agent such as a detergent can be used. Be sure to follow the manufacturer's directions and try out the product on a small section of the wall to check the results.

Green Note

Recycling Concrete

Concrete rubble can be reused in construction projects. Concrete is put through a crushing machine at a crushing facility or crushed at the construction site using mobile concrete-crushing units. Crushing concrete conserves virgin aggregates and uses less energy than quarry mining. Reusing concrete from demolished structures saves landfill space and reduces the carbon footprint of concrete manufacture.

A large source of concrete for recycling is road-building projects. When an existing concrete roadway is torn up to be replaced by a new surface, the concrete is processed at the site and then stockpiled nearby to be reused as a base under the new pavement. Other uses for recycled concrete include fill, landscape materials, pipe bedding, and soil stabilization.

Chapter Review

Summary

- Planning and site preparation are crucial to the success of a project. Labor costs, material costs, and weather can affect the success or failure of a company.
- Concrete masonry walls include solid masonry walls, hollow masonry walls, cavity walls, composite walls, veneered walls, reinforced concrete masonry walls, and grouted masonry walls.
- Blocks should never be moved after they have been laid.
- Mortar is generally bedded only on the outside edges of concrete block.
- Form full head joints on both ears (end edges) of each block to be laid.
- Blocks can be cut with a brick hammer and blocking chisel or with a masonry saw.
- Control joints are continuous vertical joints built into concrete masonry walls in areas where forces might cause cracks.
- A cavity wall consists of two walls (wythes) separated by a continuous air space 2″ to 4 1/2″ wide. The wythes are tied together with noncorroding metal ties that are embedded in the mortar joints.
- A composite wall is two wythes bonded together with masonry, metal ties, or joint reinforcement.
- Wall surfaces must be kept clean during construction because concrete block walls are not cleaned with acid to remove mortar smears and droppings.

Know and Understand

Answer the following questions using the information provided in this chapter.

1. *True or False?* Weather does *not* affect the potential success of a company.
2. Premixed mortar should be purchased in _____ lb bags for residential sites.
   A. 20
   B. 50
   C. 80
   D. 3000
3. _____ walls are two-wythe walls that allow each wythe to react independently to stress.
   A. Solid masonry
   B. Hollow masonry
   C. Composite
   D. Cavity
4. The primary objective of _____ is to provide an attractive, durable exterior surface that prevents the entrance of moisture.
   A. solid masonry
   B. veneer
   C. composite
   D. cavity
5. *True or False?* A trowel can be loaded from the side of the board.
6. A typical 8″ × 8″ × 16″ concrete block made with sand and gravel aggregates weighs about _____ lb.
   A. 28
   B. 40
   C. 45
   D. 52
7. Any straight wall longer than _____ is laid using a mason's line.
   A. 4′
   B. 1′
   C. 12′
   D. 6′
8. If blocks are wet when placed, they will _____.
   A. remain the proper size and set correctly
   B. create the incorrect consistency
   C. expand when dry
   D. shrink when dry
9. *True or False?* When laying a running bond concrete wall, be sure to allow the blocks to touch the line in order to keep them aligned.
10. _____ should *not* be used to clean concrete block walls.
    A. A trowel
    B. A chisel
    C. Rubbing stone
    D. Acid

Apply and Analyze

1. What four building elements are concrete blocks usually used for?
2. List three materials that are used as insulation in cavity walls.
3. Describe face shell bedding.
4. What problem is caused by mortar that is too runny?
5. What is the proper procedure for lifting a concrete block to lay it on the mortar bed?
6. List two tools used for cutting concrete block.
7. What two devices are used to determine the height of each course of masonry?
8. How is a corner concrete block different from a stretcher?
9. Describe the motion required to place a concrete block in the wall.
10. How many courses is a corner lead laid up above the center of the wall?

Critical Thinking

1. Planning and site preparation are two procedures that can aid in the success of a masonry company. Imagine you have been directed to develop a site preparation document for a new project starting in a couple of weeks. Make a list of the preparations to be used in this document.
2. Imagine you are now overseeing the safety program for on-site projects. A new project will be starting, and it will require masons to lift 55 lb CMUs. You must ensure the safety of the masons who lift the heavy CMUs to avoid possible injuries. What procedures should you implement to ensure your company is in compliance with OSHA and CDC lifting requirements? Be comprehensive when developing your answer. Research OSHA's website if needed for safety literature and suggested procedures for lifting.

Activities

1. (ELA) Research the recycling of concrete. Team up with another student to prepare an oral presentation describing methods of recycling concrete and the uses of recycled concrete.
2. (TECH) The ancient Romans built large-scale structures of concrete. Write a report comparing and contrasting Roman concrete with concrete used today in terms of the composition, properties, and durability.
3. (TECH) Prepare a time line of the history of concrete use in construction from 1800 to the present. Include at least 15 events on the time line.

Vovantarakan/Shutterstock.com

CHAPTER 14
Stonemasonry

LEARNING OUTCOMES

After studying this chapter, you will be able to:

14.1-1 Apply techniques for spreading mortar, handling stone, forming joints, and cutting stone.
14.2-1 Recall the three main classes of stonemasonry construction.
14.3-1 Apply techniques to install limestone panels.
14.4-1 Determine appropriate precautions to promote a clean stone wall during construction.

TECHNICAL TERMS

ashlar
expansion joints
false joint
frost heaving
hydration
nonstaining cement
pointing
quirk joint
rigger
rigging
rubble
tagline
trimmings

Introduction

Stonemasonry is the oldest, most artistic, and most expensive of all types of masonry construction. See **Figure 14-1**. Over the years, stone has declined in importance as a solid wall building material. However, stone is still applied extensively as veneer over other masonry or frame construction. Stone is also often seen as trim on buildings and paving for foyers, patios, drives, and walks.

Setting stone rubble, cut stone, or panels involves planning, knowledge of the materials, and the ability to use stonemasonry tools properly. Stonemasonry is truly an art and requires a great deal of practice and creativity. However, the satisfaction gained from a job well done is worth the effort to learn the proper skills.

Many of the procedures for laying brick and concrete masonry units are applicable to setting stone, but some differences exist. Procedures that are unique to setting stone and placing panels are covered in this chapter, but the preceding chapters should be studied prior to this.

14.1 Basic Operations

A set of basic operations for setting stone must be mastered before quality stonework is possible. These basic operations include spreading mortar, handling stone, forming joints, and cutting stone.

Spreading and Placing Mortar

All masonry construction involves spreading mortar. This is a simple process, but requires practice to gain expertise. With practice, you will be able to determine

Susan Law Cain/Shutterstock.com

Figure 14-1. A—Stonemasonry is used to build garden walls. B—Stone is used as a design accent for places of business. C—Stonemasonry is used to draw attention to one section of this home. D—Stonemasonry is used to make a park pavilion.

just how much mortar to use for a given joint. Placing a unit is easier if the correct amount of mortar has been laid down or applied to the masonry unit.

Select a trowel that is best for the job being performed. The trowel should feel good to you. If the trowel is too small, effort is wasted in needless movements. If it is too large, you will tire easily and be unable to put the mortar where you want it. You may want to select several trowels for the various tasks involved in setting stone. Another tool that is used with stone work is a mortar bag, or grout bag. A mortar bag is used to place mortar around stones or voids in mortar joints. See **Figure 3-4** for examples of trowels and **Figure 3-15** for a mortar bag. For example, a mason's 10″ trowel, 7″ buttering trowel, 5″ margin trowel, and narrow caulking trowel are valuable tools in setting a rubble stone wall.

Grasp the trowel in your dominant hand. Fingers should be under the handle, and the thumb should be on top of the ferrule. See **Figure 12-4**.

Loading a Trowel

Mortar can be loaded on a trowel from the side of the board, the middle of the board, or the top of the pile. **Figure 12-5** shows the steps for one method of loading a trowel. Develop the method that works best for you and perfect it.

Spread the mortar with a quick turn of the wrist toward your body and a backward movement of your arm. As the trowel is nearly empty, tip the trowel blade even more to help the remaining mortar slide off. The type of mortar and its consistency most often determines the speed of the stroke. Mortar used for setting stone may be slightly stiffer (drier) than mortar used for typical masonry units due to the weight of the pieces. Stiff mortar does not smear like a looser mortar, but it is a little harder to spread than thinner mortar.

Mortar Consistency

Mortar on a board should be kept well-tempered (sprinkled lightly) with water until it is used. Avoid

Safety Note

Contact with wet (plastic) concrete, cement, mortar, grout, or cement mixtures can cause skin irritation, severe chemical burns, or serious eye damage. Wear waterproof gloves, a long-sleeved shirt, full-length trousers, and proper eye protection when working with these materials. If you must stand in wet concrete, wear high-top waterproof boots. Wash wet concrete, mortar, grout, cement, and cement mixtures from your skin immediately. Flush your eyes with clear water immediately upon contact with any of these materials. Seek medical attention if you experience a reaction to contact with these materials.

constant working of the mortar, which causes it to dry out and become too stiff. Stiff mortar does not spread well. Use mortar before it begins to set up.

Handling Stone

Stone used for rubble or roughly squared stonework is usually fieldstone (stones in the sizes and shapes that are found in fields and streams). The type of stone in the load varies. A skilled stonemason knows which stones can be split easily and which shapes look best in various locations in the wall. Learn from experience and work with a skilled person to discover the "tricks of the trade." For example, bedding is visible in most sedimentary stones, and an experienced mason knows exactly where to strike the stone to split it.

Trade Tip

Keep the pile of stone close to where you are working but clear of the work area. If the source of stone is too far away, you spend most of your time running back and forth. If it is too close, you do not have enough room to work and may fall over the stones.

Form a particular pattern in your mind before you begin setting stone, taking into consideration the size, shape, and type of stone you have to work with. See **Figure 14-2**. If all of the stones are round and smooth, do not plan a pattern that requires cutting and trimming every stone. If you plan to build a polygonal stone pattern, order broken stone that already has the basic shapes needed.

Trade Tip

It can be helpful to arrange the selected stone on the ground close to the planned project. Try to arrange the stones as you would like to see them in the wall or structure. This process allows time to find the correct size and shape of stone without worrying about the mortar dying out or setting up. Once a number of stones are arranged on the ground, start the process of building the wall or structure.

Use larger stones near the bottom of a wall if some are much larger than average. A very large stone looks out of place near the top. You may decide to break larger stones into smaller pieces to produce a more uniform appearance if not enough large stones are available to develop a theme.

Do not try to lift large stones by yourself. A cubic foot of granite weighs about 170 lb. Use pry bars to move

Goodheart-Willcox Publisher

Figure 14-2. The mix of large and small stones is aesthetically pleasing and in balance.

large stones. Get help to lift large stones into place. Choose the proper tool for each task. For example, use a chisel to chip off an unwanted protuberance on the flat side of a stone. A mason's hammer or brick hammer can be used to *powder-a-point* on a stone, which means to form a stone to a suitable shape or size for its intended application. A trowel should be used to spread mortar, not to cut stone.

Safety Note

Protective clothing and heavy-duty, steel-toed or composite safety work shoes or boots are essential when working with stone. Wear a pair of sturdy leather work gloves to prevent cuts on your hands. Always wear safety glasses or goggles with plastic lenses when splitting or chipping stone. A face shield is suggested when chipping stone to protect the face from fragments. Follow safe work practices. Do not take a chance with a shortcut!

Cut stones are ordered and manufactured according to the construction specification and construction plans. Cut stones are delivered to the jobsite at an appropriate time based on the project schedule and the superintendent's recommendations. See **Figure 14-3**. Cut stone in the form of trimmings, ashlar, or panels must be handled carefully to prevent breakage, staining, or chipping. The process of unloading stone at a jobsite should be coordinated and performed by competent persons who are knowledgeable about procedures for lifting stones safely.

Rigging is a device, such as a sling or chain, attached to stones to lift them from the delivery trailer. A tag on rigging equipment identifies the manufacturer and maximum weight that can be lifted. A ***rigger*** is the

Susan Law Cain/Shutterstock.com

Figure 14-3. A capstone is being lifted by a rough terrain forklift to masons working on scaffolding. Taglines were used until the stone reached the masons on the scaffold. Stones were transported to the construction site using a tractor-trailer.

person who attaches the rigging device to the stone to be lifted. The rigger is required to have training and experience in attaching slings or lifting equipment to the stone in order to perform the lift in a safe and controlled manner.

On large stones or materials, the rigger attaches a tagline to the stone to aid in guiding it from the trailer to the location for storage or being positioned in the project being built. A ***tagline*** is a safety rope or cable that is attached to the material being lifted so the person guiding the stone is a safe distance away from the load to ensure that if it falls or slips from the rigging no one is injured.

The following are helpful suggestions for handling stone:

- Upon delivery, check the stone for damage. Verify that it is the proper stone (color, finish, and grade). Also verify dimensions and any other particular requirements.
- As a general rule, handle stone as carefully as you would handle any material you do not want to chip or scratch. Store stone on sturdy skids, timbers, or pallets intended for stone. Place the stone in a well-drained area, protected against water and mud splatters. In cold climates, cover stone to protect against rain, snow, and ice. When preparing for delivery of stone to the site, select a location where the stone is readily available when needed.
- Lean smooth-finish stones face-to-face and back-to-back. See **Figure 14-4**. Textured finishes should be separated with spacers that will not stain the stones. Spacers can be made of wood, such as white pine or yellow pine that does not show sap or resin. Stones should be covered with a waterproof tarp for added protection.
- When lifting stones, avoid sliding one face surface over another. Make sure lifting holes are easily reached from the stacked position.
- Be careful not to bump stones. Position stones to avoid damage from other stones or equipment.
- Make certain that all lifting devices and all parts involved with hoisting machines are of sufficient capacity for the heaviest stone to be lifted.

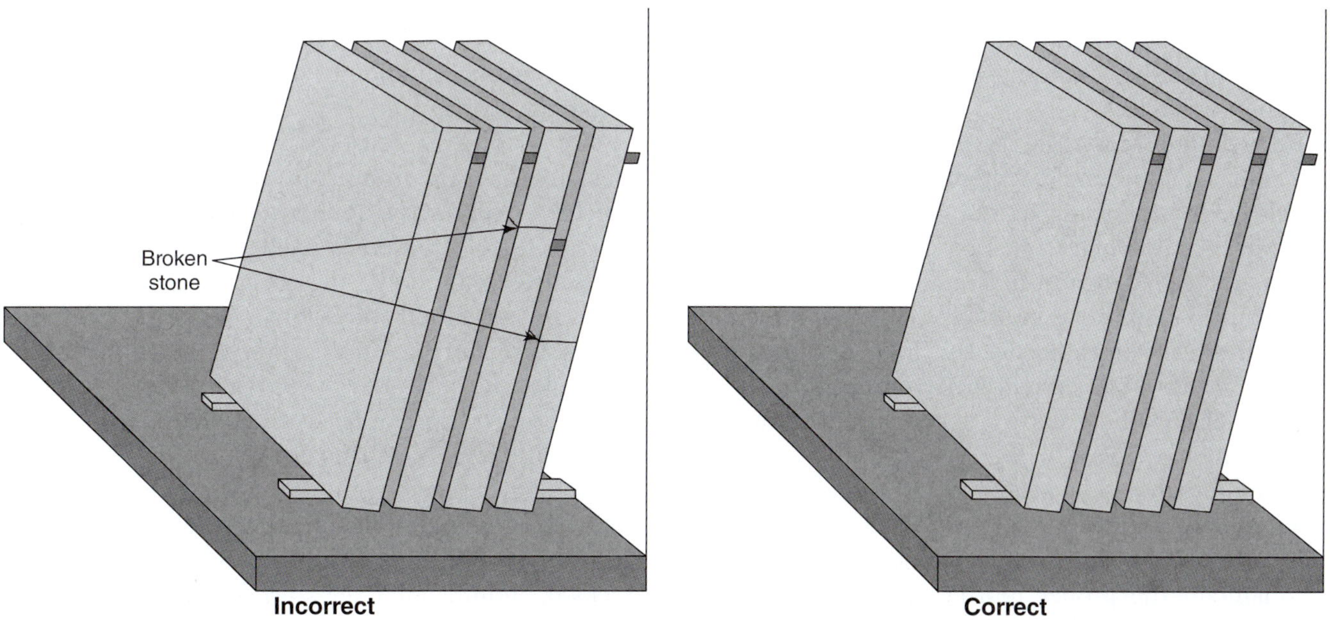

Goodheart-Willcox Publisher

Figure 14-4. Textured stones should be separated using spacers that will not stain the stones. Improper storage causes stones to crack or break. In the image on the left, a spacer is in the incorrect position.

- Be especially cautious in rigging, handling, and lifting large panels. Windy conditions require precautions as well. Use taglines and make certain that tagline operators are instructed in their proper use.
- Always use safety slings under stones as you lift them. Keep safety slings in place until the stones are within a few inches of their final location.
- Protect finished work against construction traffic. Drape the work with a tarp or pile sand on projecting courses to protect them against mortar droppings.

Safety Note

When lifting stones, ensure that all personnel are clear. OSHA specifies that loads or stones are never to be hoisted or suspended over workers or bystanders. If stones or loads must be hoisted over scaffolds or stages, workers must be clear and away from the swing area of the material being lifted and stored on the site. When working from scaffolding without handrails at or above 6′, workers must wear fall protection equipment. It is also important to wear appropriate PPE, such as a hard hat, gloves, and eye protection.

Forming Joints

Joints in stonemasonry are generally set with mortar. Joints in stone panel construction may be mortar or other materials. Whenever mortar is used, it should be a nonstaining type designed for the specific application. A mix of one part nonstaining cement, one part hydrated lime, and six parts clean, sharp, washed sand is recommended. This 1:1:6 mixture provides sufficient compressive strength, good bond strength, and weather resistance.

The width of mortar joints has a profound effect on the finished appearance of stone construction. A frequent error is allowing mortar joints to become too wide, especially when working with rubble or polygonal stones. See **Figure 14-5**. Stones should be cut to fit with mortar joints not exceeding 1″ for rough work and 3/4″ for ashlar. Joints that are 1/2″ for rough work and 3/8″ for ashlar are preferred. A narrow caulking trowel works well for filling narrow spaces and working the mortar into crevices.

Bed joints should be generous. In rubble, polygonal, and roughly squared stonework, stone should be set with wedges or stone chips placed beneath the stone to maintain the proper mortar joint. Heavy stones squeeze out the mortar if they are not supported in some manner. Excess mortar should be removed from

Goodheart-Willcox Publisher

Figure 14-5. Some of the mortar joints in this polygonal stone wall are too wide.

the stone with the trowel. After mortar begins to set, remove the wedges or stone chips. Then fill voids or holes with mortar. The joint can be struck once the initial set of the mortar has occurred. Joints can be raked out to about 1″ deep and pointed later with mortar if preferred. See **Figure 14-6**.

Portland cement used in preparation of cement/lime mortar for limestone should conform to the requirements of ASTM C150. Masonry cement should conform to the requirements of ASTM C91. Either material produces a suitable mortar for stone panel construction.

Cement with a low-alkali content is called ***nonstaining cement***. Nonstaining cement is best for setting limestone. White cements are not necessarily either low-alkali or nonstaining. To avoid staining due to alkali from mortar, wet the joint surfaces of the stone before applying mortar and do not use too much water when mixing the mortar.

Protect stone masonry during the building process to ensure that rain or moisture does not enter the wall structure. This can be done by covering the structure with protective materials such as waterproof tarps or plastic. Weigh down the covering material to ensure wind or other conditions cannot expose walls to moisture during times when work is not being performed.

Protecting walls under construction reduces the chance of stains occurring if moisture were to enter the structure. Once inside the wall, moisture begins to leave the structure through evaporation after a period of time. The evaporation process may continue for weeks or months after the work has been completed. This is why stains sometimes appear on wall surfaces after the project has been completed—moisture is

A

B

Marshalltown Company

Figure 14-6. A—A stonemason is filling joints with a mortar bag. The joints have been previously raked out to a depth of 1″ from the surface of the stones. A colored mortar is being used to fill the raked joints. B—A stonemason is filling the mortar bag with a modified masonry trowel. The front end of the trowel has been rounded to eliminate puncturing the bag.

evaporating and leaving its mark on the exterior surface of the wall.

Splitting, Shaping, and Cutting Stone

Common fieldstone and river rock can be used in their natural state. They may also be split or shaped to fit a particular space in a stone wall, chimney, pavement, or other application. See **Figure 14-7**. Splitting stones takes practice and sharp observation to consistently achieve the desired results.

To split a stone that has a stratified (layered) structure, mark a line along the grain, then chip on the line with the chisel end of a mason's hammer until a crack begins to develop. Widen the crack slowly by driving wedges into the crack at several points. A pry bar may be needed to finish prying apart a very large stone.

Stones such as granite that are not stratified are difficult to split, but not impossible. Begin by drilling holes with the narrow bladed chisel or power drill about 6″ apart along the line where you wish the stone to separate. Then drive thin wedges into the holes. Continue the process until the stone splits.

Split stones can be shaped using a mason's hammer or, if needed, a stonemason's chisel and a maul. A maul is a specialized hammer used to strike chisels

Workplace Skills
Communication

Communication is the process of using words, sounds, signs, or actions to exchange information and express thoughts. We communicate every day in our personal and professional lives. Being an effective communicator, however, takes some work and is essential in the workplace.

At work, you may communicate verbally or nonverbally. An effective communicator recognizes the appropriate method of communication based on the environment. For instance, a noisy jobsite may require earplugs or earmuffs to be worn while working. Therefore, verbal communication with a coworker may not be possible. Some nonverbal alternatives may include writing a note, using hand gestures, or moving to an area of the site with less noise. It is essential that you recognize any barriers that may inhibit effective communication.

Effective communication means using appropriate tone and language to convey your message. It also means using proper body language. Slouching or rolling your eyes, for instance, can communicate an unwillingness to work or be present in a conversation. Being an effective communicator can help make your work, and the work of your team as a whole, more efficient. Honing your communication skills will make you stand out in any workplace environment.

Goodheart-Willcox Publisher

Figure 14-7. These common fieldstones have been split and shaped to fit the contour of surrounding stones.

and is designed for striking hardened steel. For masonry work, the hammer head varies in weight from 2 lb to 4 lb.

A trowel should not be used to cut natural stone. Manufactured stone that is a Portland cement product can be cut with a mason's trowel. To cut a large stone into two pieces, mark a line on the face of the stone where the cut is to be made. Use a chisel or mason's hammer to chip along the line to form the break line, much the same as cutting brick. When a cut has been made along the length of the line, strike the stone with a sharp blow away from the cut. The stone should break along the line. If the break is rough or not exactly where it should have been, chip away the stone with a chisel or hammer until the stone is the proper shape. Be patient—do not try to remove too much material at a time.

Use power tools to cut ashlar veneer, trimmings, and panel stone. Use an ordinary masonry saw to cut smaller pieces. Larger pieces such as panels should be cut to size by the supplier, who has the proper equipment to do this kind of work.

Safety Note

When using a chisel, never strike it with a brick hammer. Brick hammers are not intended for that purpose. Use a heavier hammer, such as a maul.

14.2 Types of Stonemasonry Construction

Natural stone is available in a wide variety of sizes, shapes, colors, and textures. Refer to Chapter 9, **Stone**. These characteristics make many wall patterns possible, **Figure 14-8**. Stonemasonry is time-consuming because the pieces frequently have to be selected individually and trimming is required when building some patterns. Due to the different sizes and uneven surfaces of stone units, they can be difficult to lay and can require several trial placements before the fit is perfect.

The three main classes of stonemasonry construction are rubble, ashlar, and trimmings. Panel construction is not included in this classification.

Rubble Masonry

Rubble is uncut stone or stone that has not been cut to a rectangular shape. Rubble is generally used in walls where a rustic appearance is desired and precise lines are not important. See **Figure 14-9**. Many sizes and shapes of stone are used to create the pattern. A liberal amount of mortar is needed because of irregularities in the stone. Rubble stonework is not as strong as other bonds.

Ashlar Veneer

When stone is precut with enough uniformity to allow some regularity in assembly, the wall is generally called ***ashlar***. See **Figure 14-10**. Ashlar stone is easier to lay than rubble because the individual units are precut to fit. In this respect, it is much like brick or block.

Produced as a standard product, limestone ashlar veneer is broadly available from a network of dealers throughout the United States. Most of these dealers also stock standard trim items that can be cut to length with ordinary jobsite masonry saws or hand tools. Window and door sills, coping, and trim courses are produced in standard heights and thicknesses. They require no additional work other than cutting to proper length on the job. Ashlar veneer and its matching trim are set in mortar like any unit masonry product and may be applied to any type of backup.

When stone is laid in regular courses, it is referred to as coursed ashlar. When it is laid in broken courses without regard to the arrangement of the joints, it is referred to as broken, random, or ranged ashlar. See **Figure 14-11**. Stones for ashlar work are generally produced with thicknesses of 2″ to 8″, heights of 1″ to 48″, and lengths of 1′ to 8′.

Figure 14-12 shows a typical method of anchoring ashlar veneer to concrete block backing and wood frame backing.

A B C D

Eldorado Stone Corporation

Figure 14-8. A—Coursed ashlar. B—Cobbed or polygonal stone. C—Ashlar. D—Rubble.

Goodheart-Willcox Publisher

Figure 14-9. Random rubble stonework with uniform mortar joints.

Nancy Henke-Konopasek

Figure 14-10. Ashlar stonework.

Code Note — Stone and Masonry Veneer

The International Residential Code (IRC) **R703.8** specifies that stone and masonry veneer installed over a wood or cold-formed steel backing be limited to the first story above-grade plane and cannot exceed 5″ (127 mm) in thickness.

Trimmings

Trimmings are stones that are cut on all sides to specific dimensions. They are used for moldings, sills, lintels, and ornamental purposes. See **Figure 14-13**. Trimmings are either set in mortar or anchored using a variety of mechanical fasteners.

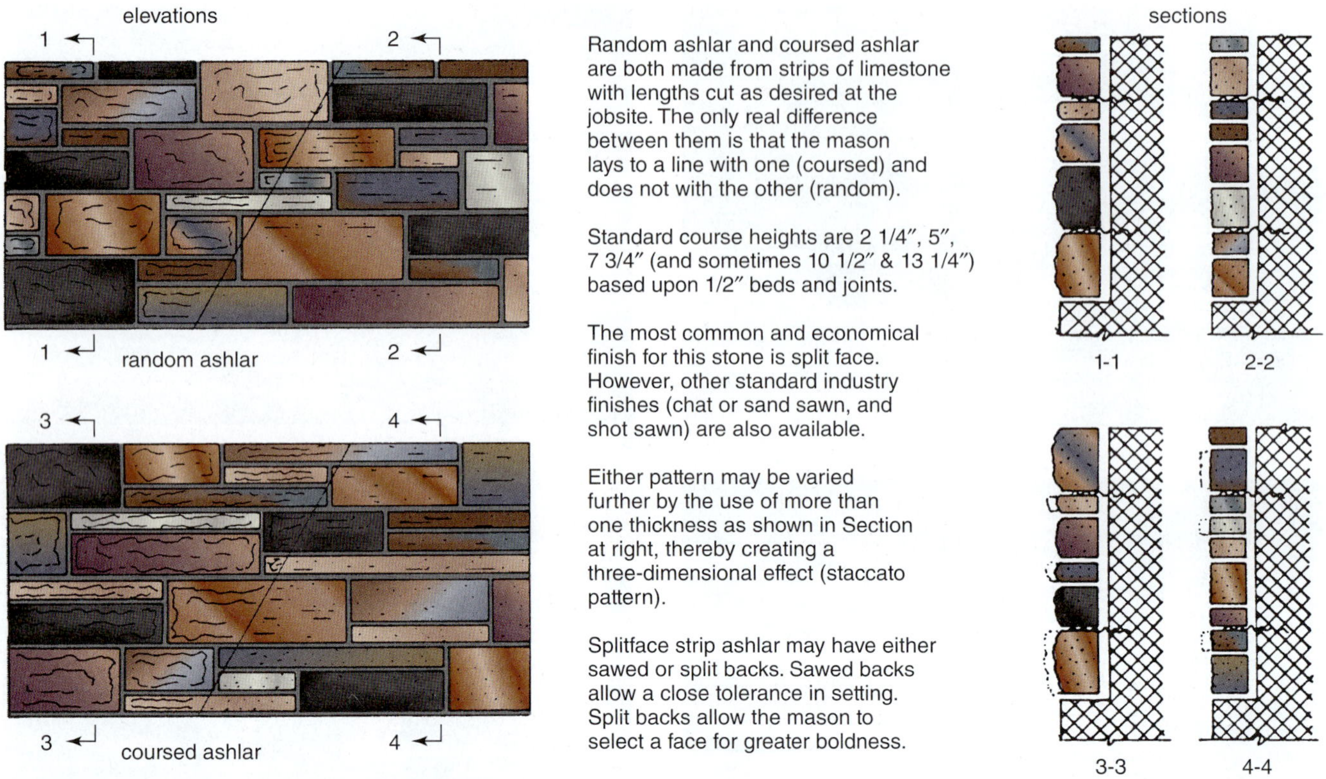

Indiana Limestone Institute

Figure 14-11. Random ashlar and coursed ashlar shown in elevation and section.

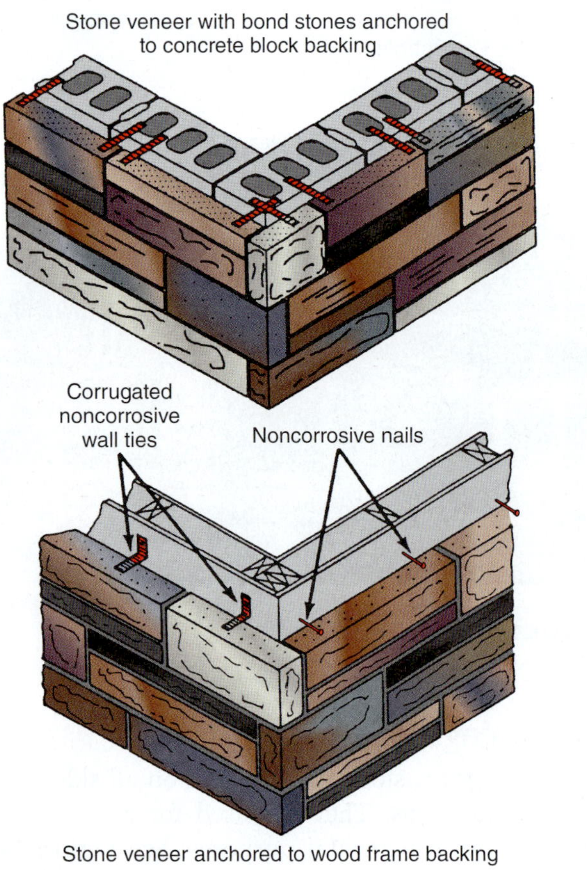

Indiana Limestone Institute

Figure 14-12. Methods of anchoring random ashlar stone veneer.

isabela66/Shutterstock.com

Figure 14-13. A pallet of limestone trimmings and cladding to be used for sills, lintels, and other purposes.

Setting a Rubble Stone Veneer Wall

The following is the procedure for laying a rough stonemasonry veneer wall using rubble.

PROCEDURE

Laying a Rough Stonemasonry Veneer Wall Using Rubble

1. Snap a chalk line where the face of the masonry wall is to be located along adjacent sides.
2. Spread out the stones so each one can be examined for shape, texture, or other characteristics. See **Figure 14-14**.
3. Mix enough nonstaining mortar to last for one hour of work. See **Figure 14-15**. Use the mortar before it begins to set. Mix one part nonstaining cement, one part hydrated lime, and six parts clean, sharp, washed sand. A comparable manufactured mortar cement can also be used.
4. Select several larger stones for the bed course and place them into position dry. With a piece of chalk, mark them for trimming.

moodboard/moodboard/Thinkstock

Figure 14-14. Rubble stone to be used for a stone veneer wall. This stone varies in size from 6″ to 24″. The variance in color enhances the finished wall.

mrkob/Shutterstock.com

Figure 14-15. Mortar should be mixed in small amounts, so it does not begin to set before being used. When mixing, wear gloves and rubber boots for protection.

Trade Tip

Place stones in as natural a position as possible. Have a pattern or special effect in mind as you select and place stones together. Do not use too many varieties and keep textures relatively uniform. Stone veneer is generally 4″ to 8″ thick, depending on the method of bonding and the construction specifications. Large stones must be split so they meet these requirements. A sledgehammer can be used for splitting large stones.

5. Trim each stone as you are ready to place it in the wall. See **Figure 14-16**. Each stone should be thoroughly cleaned on all exposed surfaces by washing with a brush and soap powder, followed by a thorough drenching with clear water.

Safety Note

Be careful of flying chips while trimming stone. When trimming or cutting stone, safety glasses or safety goggles and a face shield should be worn. Also wear ear protection when sawing or cutting stone.

6. Place each trimmed stone in its proper location in the wall and check for proper fit. When you are satisfied with the result, proceed to the next step.
7. Using a trowel, lay a full bed of mortar for the trimmed stones and place them into position. See **Figure 14-17**.

Timothy L. Andera

Figure 14-16. A stonemason's hammer and chisel are used to trim stone to a desired shape. The chisel is a stone tooth 2 1/4″, and a 3 lb maul is used to strike the chisel.

Continued

PROCEDURE *(continued)*

schankz/Shutterstock.com

Figure 14-17. Cut and trimmed stones are placed in a full bed of mortar. Metal bars are used to aid in creating uniform joints.

schankz/Shutterstock.com

Figure 14-19. A trowel is being used to fill spaces between stones. The mortar joints are finished at the appropriate time.

8. Joint spacing should be 3/8″ to 3/4″ for ashlar or cut stone and 1/2″ to 1″ wide for rough work, **Figure 14-18**. Fill in the spaces between the stones with mortar. Use a narrow trowel to fill narrow spaces and work the mortar into crevices. See **Figure 14-19**. Sponge the stone free of mortar along the joints as the work progresses.
9. To perform his duties, a stonemason uses an assortment of tools to build stone walls. See **Figure 14-20**. Remove excess mortar from the stone with a trowel and strike the joint. Joints can be tooled when the initial set has occurred. If desired, they can be raked out 1″ deeper, and ***pointing*** (placing plastic mortar into the raked-out joints) can be done later to add color once the mortar has set. The colored mortar fills the 1″ void, reducing the chances of smearing that could occur if colored mortar was used throughout the process.
10. Trim larger stones for the deeper bed course and fit them into place using the same procedure as was performed when placing smaller stones.
11. Bend corrugated metal ties into place for successive courses of stone. See **Figure 14-21**. All ties must be noncorrosive. Use extra ties at all corners and large stones when possible. Place lead, plastic, wood shims, or pads the thickness of the mortar joint under heavy stones to avoid squeezing mortar out. See **Figure 14-22**. Remove the shims or pads after the mortar has been set and fill the holes. Do not set heavy stones or projecting courses until mortar in the courses below has hardened sufficiently in order to avoid squeezing.

schankz/Shutterstock.com

Figure 14-18. Spacing bars are being used to ensure horizontal and vertical joints are uniform in size. A level is also used to ensure that the stone pier is level and plumb.

Timothy L. Andera

Figure 14-20. Mortar can be added or removed from a stone project using different masonry tools. An assortment of tools are needed to construct a stone wall.

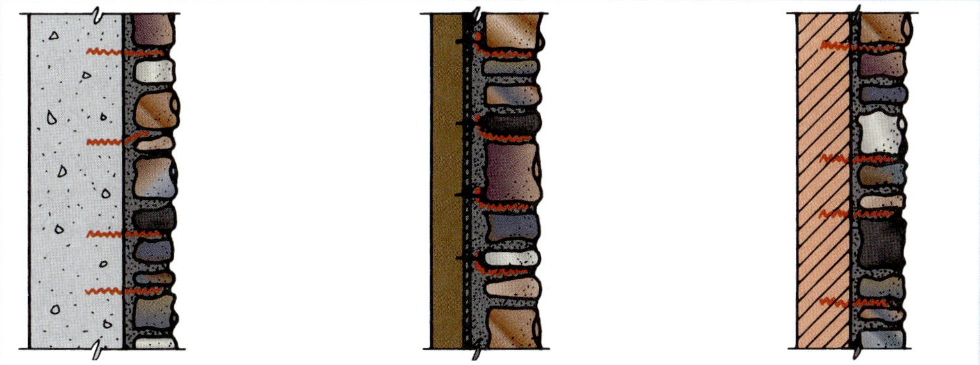

Goodheart-Willcox Publisher

Figure 14-21. Stone veneer sections showing noncorrosive metal ties.

Timothy L. Andera

Figure 14-22. Wooden shims are placed under large heavy stones to avoid squeezing mortar from the joints. Once mortar has set satisfactorily, the shims will be removed and voids will be filled with mortar.

12. After the mortar has set between the stones, continue laying new courses of stones until the wall is the desired height. Work from the corners toward the middle. The masonry should be protected at all times from rain and masonry droppings. Adequate protection must be provided during cold weather construction. Refer to Chapter 10, **Mortar and Grout**.

Trade Tip

A mason's line and level can be used to keep the wall straight and plumb. If the structure is to be inclined or tapered, such as for a chimney, a mason's line should be used to keep the edge straight and moving in the proper direction. See **Figure 14-23**. A mason's line is also used on long walls to keep them from bowing in and out.

Goodheart-Willcox Publisher

Figure 14-23. An example of tapered stonework in a residential structure.

13. When the wall is completed, scrub it with a fiber brush and clear water. The stone should be clean and free of mortar. Avoid strong acid compounds because they burn and discolor certain types of stone, such as limestone.

A stone wall can be waterproofed. Use a nonstaining asphalt emulsion, vinyl lacquer, cement base masonry waterproofing, stearate, or other approved material.

Figure 14-24 shows a close-up of a stone wall with thin mortar joints of uniform width. The stones have been accurately trimmed to fit into a specific place in the wall. Practice and patience are required to develop this talent.

Building a Solid Stone Wall

Walls that use mortar as a bond between stones require a footing below the frost line because frost heaving causes shifting in the wall structure. ***Frost heaving*** results when frost migrates under the footing and causes it to heave because the moisture in the ground has frozen. The resulting shifting in the wall causes cracks between the stones and over time begins to destroy the wall. After the footing is in place, a foundation wall can be built using either concrete block or concrete.

> **Trade Tip**
> When designing a stone wall employing mortar, ensure that the footing that supports the foundation is at the correct depth to protect the structure from frost heaving. Check with the local building official to determine appropriate depth for the footing.

Stone to be dry laid should have a base made of gravel. See **Figure 14-25**. The gravel base makes it easier to position the stones and to achieve a tight fit. Gravel also aids in removing moisture around the base of the stones.

Timothy L. Andera

Figure 14-24. The stone mason has finished the wall with varying sizes of stones. Uniform mortar joints are a mark of good craftsmanship.

Once the first layer of stones is positioned, backfill with gravel as consecutive layers or courses of stones are set in place. Backfilling with gravel assists in removing excess moisture that could affect the stone wall in the future. Drain tile should also be installed at the base of the wall to aid in removing moisture from around the structure. Surplus moisture around the base can have a severe effect on the stability of the wall, causing it to shift. Over time, the wall could be destroyed.

The procedure for building a solid stone wall using rubble, fieldstone, or ashlar is similar to laying a stone veneer wall with a few exceptions:

- A solid stone wall depends on bond stones for strength, stability, and tying the wall together. A bond stone is one that passes through the wall from side to side. Bond stones should be placed as frequently as possible. Every 5 sq ft to 8 sq ft of wall should have a bond stone.
- A solid stone wall is usually thicker than a veneer wall. The backing in a veneer wall provides the structural support and the stone is a protective covering. In a solid stone wall, the stone provides structural support as well as covering. Sometimes a stone wall is capped with a 4″ concrete slab or cut stone covering. The usual wall thickness of a solid stone wall is a minimum of 18″. Larger stones may be used and should be placed in their natural position for best appearance. For longevity of a stone wall, a foundation and footing is placed under the structure. Refer to **Figure 14-25**.

Reinforcement may be used in a solid stone wall to give added strength and support. The material used should be noncorrosive.

14.3 Limestone Panels

Limestone panels are a popular product of choice of architects and builders because of the availability, ease of installation, and final appearance on the structure. Panels can be sold by the square foot or by linear foot, depending on the specific piece being ordered. Limestone panels may be as thin as 1″ thick or as thick as 3 1/2″ or more, depending on the application and location of use. In certain situations, adjustments to thickness can be made by cutting panels with a diamond saw at the jobsite.

Design and Construction Criteria

Designing and building with limestone panels requires attention to numerous details to ensure quality appearance and performance. Some of the criteria are as follows:

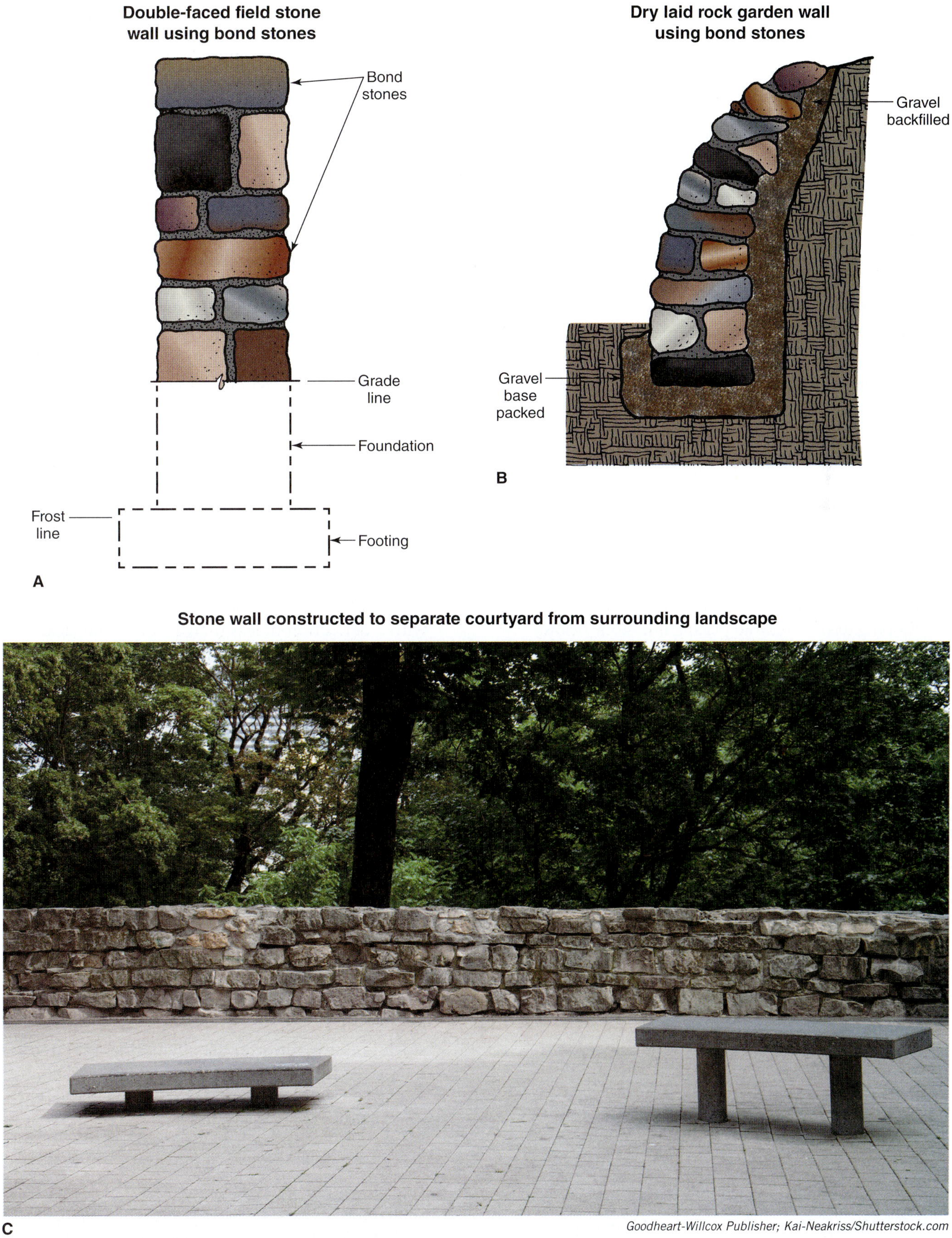

Goodheart-Willcox Publisher; Kai-Neakriss/Shutterstock.com

Figure 14-25. Section views of different types of stone walls. A—A stone wall that has been laid on a concrete foundation and footing. B—A stone retaining wall is built on a gravel base. C—A stone wall that has been built on a concrete foundation with a footing below the frost line.

- Pitch coping stones toward the roof to avoid discoloration on the exposed face and walls, **Figure 14-26**.
- Backs of stones above a roof line do not need to be finished.
- Vent attics over porticos to prevent condensation from entering the wall and filtering through the joints to the face of the stone.
- Provide washes and drips on projecting stones.
- Butt joints are usually less expensive on smooth finishes, while quirk joints are less expensive on textured finishes. In a ***quirk joint***, two surfaces are cut at the same angle to create a uniform corner. In most applications, the stones are cut or beveled at a 45° angle to create a 90° finished outside corner. Each stone is cut at the precise angle for the application and then delivered to the site. See **Figure 14-27**.
- Prevent physical contact between the back face of stone walls and columns, floor slabs, and the end beams. Expansion can push the stone out.
- If exposed stone patios are located over a heated room, insulate the ceiling of the room. Warm stones absorb moisture, and a sudden drop in outside air temperature may freeze and crack the stone.
- Where the design allows, step the foundation in sloping terrain to avoid placing limestone below grade. If such installation is unavoidable, use either the trench or the coating method shown in **Figure 14-28**. The trench method involves excavating a trench next to the foundation wall and panel. After excavation, the area is filled in with gravel to facilitate a means for moisture drainage away from the limestone panel and foundation wall. In the coating method, a sealant or similar material is applied to the limestone panel and continues down below grade to an area past where the foundation begins. The purpose is to protect the panel, seal the joint between the panel and foundation, and direct moisture away from the panel and foundation structures.

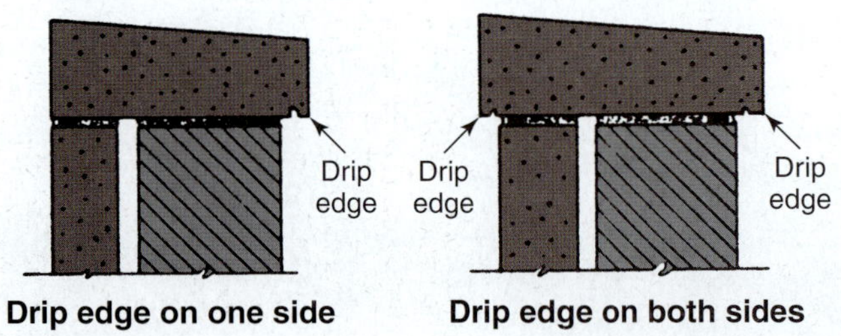

Indiana Limestone Institute

Figure 14-26. Stone coping details. Stones are pitched toward the roof to protect the face of the wall.

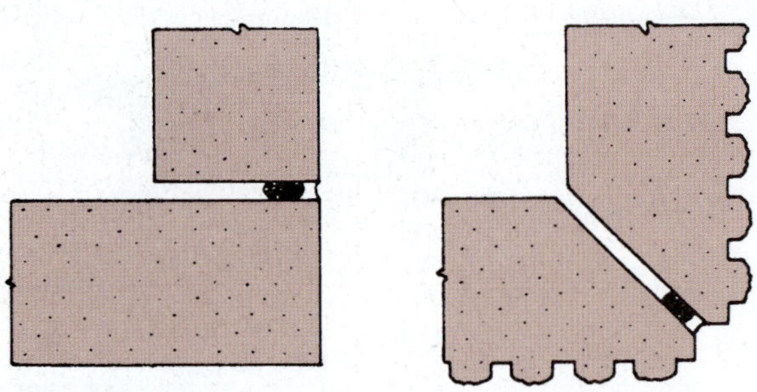

Indiana Limestone Institute

Figure 14-27. Stone panel joints are recommended for smooth and textured finishes.

- Specify mortar joints to be tooled for proper moisture resistance. When using sealants, follow the manufacturer's specifications to obtain the desired joint profile and adherence to inner joint surfaces.
- If false joints are used, they may be pointed to match other joints. ***False joints*** are cut into the stone or panel to create a pattern resembling a true masonry joint.
- Where masonry butts against a stone pilaster or at building corners, specify a sealant joint. Long vertical mortar joints are difficult to compress and may tend to fail.
- Steel lintels carrying stonework should have sufficient stiffness to carry a superimposed load with a deflection of less than 1/360 of the clear span. Deflection refers to the amount that the lintel deflects, or sags, due to the imposed weight being applied across the span of the opening.
- Incorporate a 1″ minimum clearance between stone and all structural members.
- Avoid conditions that place heavy loads on long, narrow stones. Where design requires such conditions, use steel angles to relieve the weight and to avoid cracking the smaller stones.
- Dowel pins must not be embedded solidly when installed through a relieving angle at an expansion joint or in horizontal tie-backs.
- When possible, avoid setting stone with mortar in extreme cold. Stonework set in cold weather can expand and crack the mortar bond in warm temperatures.
- A lean pointing mortar usually performs better than a strong one. While pointing reduces leak potential whether the joint is thoroughly filled or not, a strong mix may cause spalling at joints.

Trade Tip

The difference between a lean and strong mix is the content of cement in the two mixtures. As the cement content increases, so does the strength, but the flexibility of the material decreases.

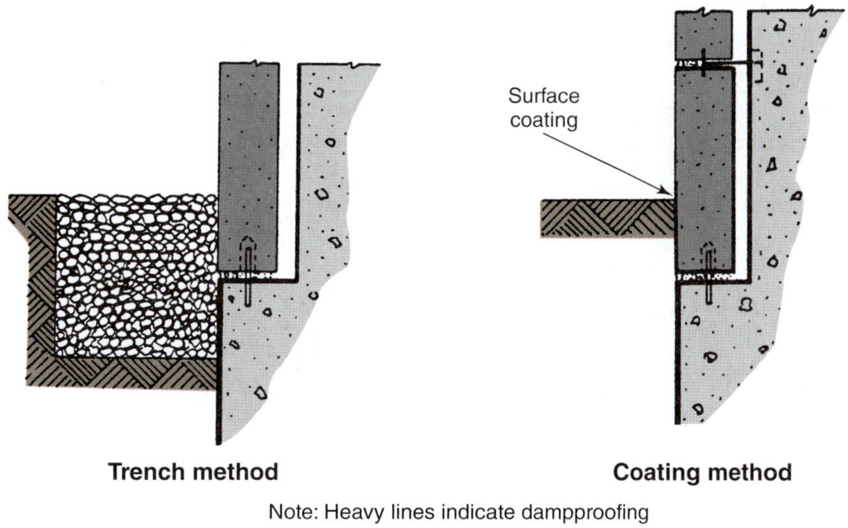

Indiana Limestone Institute

Figure 14-28. Protect limestone panels from contact with soil. Left—Gravel is used to remove excess moisture next to the panel. Right—Dampproofing has been adhered to the panel above grade and continues down below grade. It is attached to the foundation wall to keep soil away from panel.

- Cavity walls should be kept clear of mortar droppings during construction.
- Do not pour the cavity full of grout, **Figure 14-29**.
- In mortar systems, all mortar joints between stones should be thoroughly filled. The front 1/2″ to 1 1/2″ should be filled with pointing mortar after the setting bed has set. See **Figure 14-30**.
- The use of calcium chloride or other salts that act as accelerators or retarders in mortar are not recommended in setting stone.
- Do not pour concrete against unprotected stonework. Alkali from the concrete stains stone.
- Store stone clear off the ground and protect it from the weather and construction traffic.
- Place an adequate number of setting pads in the horizontal joint under heavy stones in order to sustain the weight until the mortar has set. See **Figure 14-31**.
- Be sure that all projecting courses, sills, entrances, columns, column covers, and other stone elements exposed to traffic of the trades and mortar droppings are properly protected during construction. Use galvanized nails in mortar joints or galvanized steel straps for support.
- Do not permit wash from concrete floors or scaffolding to run down onto or behind stone walls.

Trade Tip

Always cover walls and openings adequately at night and during rains to help prevent staining from trapped moisture and efflorescence.

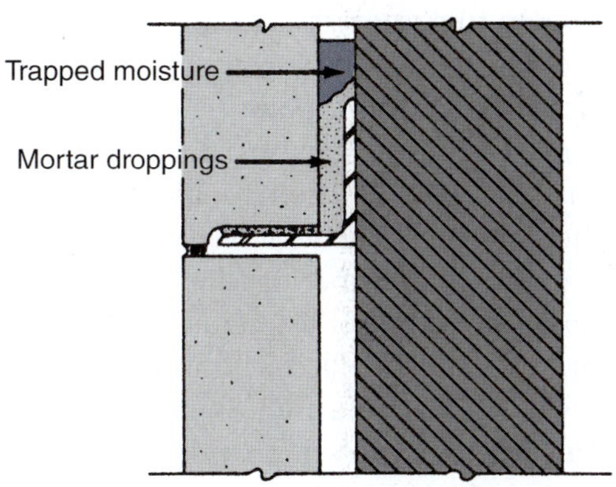

Indiana Limestone Institute

Figure 14-29. To prevent trapped moisture, do not pour the cavity full of grout.

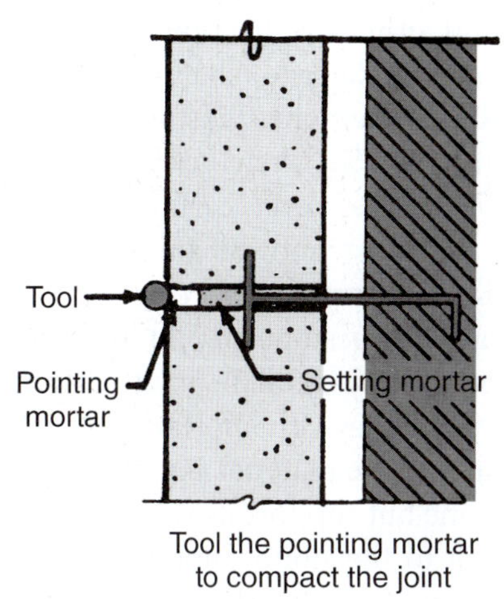

Indiana Limestone Institute

Figure 14-30. Recommended method for tooling joints that have been pointed.

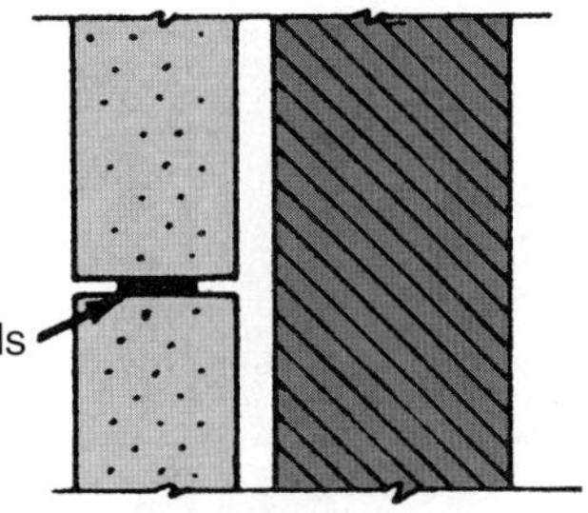

Indiana Limestone Institute

Figure 14-31. Setting pads should be used under heavy stones to support the weight until the mortar sets.

Panel Sizes

The panel sizes commonly produced are the maximum recommended for efficient fabrication and handling. Larger panels are available in either monolithic stones or epoxy assemblies. For information on larger units, consult the Indiana Limestone Institute (ILI) or member companies. Standard panels are available in sizes of 3′ × 5′ × 2″, 4′ × 9′ × 3″, 5′ × 11′ × 4″, 5′ × 14′ × 5″, and 5′ × 18′ × 6″.

> **Trade Tip**
> Two-inch thicknesses in Indiana limestone wall panel work should be considered only in applications where grasping systems specially designed for thin stone can be used. Anchors and supports ordinarily used in 3″ and thicker stones can often be used for thinner materials, but wind loads and other building dynamics become more critical as stone thickness is reduced. Fabrication, handling, and erection procedures associated with conventional erection methods are often more costly when applied to thin stones. For most purposes, conventional support systems should be limited to 3″ and thicker stones.

Anchors and Supports

The term anchor refers generally to the straps, rods, dovetails, and other connections between the stone and the structure. Refer to Chapter 11, **Anchors, Ties, and Joint Reinforcement**. Most anchors are intended to maintain stones in their vertical positions rather than bear weight. However, certain anchors are structural. The use of structural anchors eliminates the need for other support.

All anchors in limestone should be stainless steel or another noncorrosive metal. In practice, anchors are embedded in the stone with mortar, sealant, or another non-expansive, stable material. Supports are not embedded in the stone but support its weight. They may touch or be adjacent to the stone. Supports are typically A36 steel, a standard steel alloy that is a common structural steel in the United States. However, the supports may be made of any metal possessing adequate strength.

> **Trade Tip**
> Supports have a shop coat as well as a job coat of rust protection applied after installation of the supports. Damage to the rust protection coating of supports as a result of installation work, cutting, or drilling must be repaired with a compatible rust protection coating.

Mortars

Portland cement used in preparing cement/lime mortar for setting limestone should conform to the requirements of ASTM C150. If masonry cement is used, it should conform to the requirements of ASTM C91. Either material produces a suitable mortar.

Mortars should be mixed to the proportion requirements of ASTM C270, type N. The mortar's compression resistance is approximately 750 psi when cured. This moderate strength is sufficient for most limestone installations.

Portland cement/lime mortars are mixed with one part cement, one part lime, and six parts sand, all by volume. Lime improves the workability of mortar. A 1:1:6 mixture provides sufficient compressive strength, good bond strength, and good weather resistance.

Low-alkali cement mortar tends to be nonstaining. However, low alkali content in mortar used to set stone is not a guarantee of such performance. The nonstaining cement described in ASTM C91 should be used with limestone products. Stain is best prevented by good masonry practice, especially the protection of stone backs from moisture sources.

Sand should be clean and sharp and washed free of loam, silt, and vegetable matter. Grading should be from fine to coarse, complying with ASTM C144. If the setting mortar is to be the pointing mortar, white sand should be used. Mixing water should be of potable quality.

As setting proceeds, remove mortar tags from joint edges after they have taken their initial set. Avoid smearing pointing mortar—smeared mortar can be difficult to remove. A mortar bag can be used in pointing vertical joints, where pointing mortar cannot be easily scraped off a mortar board. A mortar bag is a small cone-shaped bag used to dispense mortar in joints where mortar is difficult to place using a trowel. For final cleaning, use fiber brushes and detergents or mild soap powder. Do not use acids.

Anchor holes may be filled with mortar or joint sealant. Lead wool may also be used. Shim stock of a noncorrosive material may be used, provided it is not built up beyond a snug thickness. Shim stock is used to support the stone and keep it in place during the time the mortar is hardening or setting. Placing shim stock is necessary when large stones are set in fresh mortar because the weight of the stone may reduce the size of the joint.

Certain types of mortar or packaged grouts expand during the hardening period. Expanding type grout or mortar should never be used in anchor slots and holes because the expansion can cause enough pressure to fracture the stone at the anchor locations.

Pointing

Pointing cut stone after setting, rather than full bed setting and finishing in one operation, reduces a condition that tends to produce spalling and leakage. Full bed setting is the placement of mortar across the complete area where the stone will be positioned in one process. Shrinkage of the mortar bed allows some settling, since the mortar bed hardens from the face in. If set and pointed in one operation, the settling, combined with the hardened mortar at the face, can set up stresses on the edge of the stone. For this reason, it is best to set the stone and rake out the mortar to a depth of 1/2″ to 1 1/2″ for pointing with mortar or sealant application at a later date.

By raking the joints to a depth of about 1/2″ to 1 1/2″, pointing can be done in one, two, or three stages. Each stage seals shrinkage cracks in the preceding stage, and finally the concave tooled joint provides the maximum of protection against leakage.

Cold Weather Setting

The bond strength of mortar is considerably reduced when mortar is frozen prior to hardening. ***Hydration***, the chemical reaction between water and cement that sets and hardens the cement, progresses very slowly below 40°F (12°C). Protection is necessary, including covering masonry work with blankets or performing work in an enclosed structure so the mortar will not freeze.

Admixtures or antifreezes should not be used to lower the freezing point of mortar. The effectiveness of most of these compounds is due to the calcium chloride they contain acting as accelerators. Calcium chloride cannot be used on limestone because salts cause efflorescence and can cause spalling or flaking.

Trade Tip

Never set stone on a snow or ice-covered bed. A bond cannot develop between the mortar bed and frozen supporting surfaces. If setting stone during cold weather, closely follow the cold weather masonry construction recommendations of the Brick Industry Association (BIA) Technical Notes on Brick Construction, Hot and Cold Weather Construction.

Sealant Systems

Sealants serve the same purpose as mortar and pointing in the exclusion of moisture. The most commonly used one-part systems are called *moisture-cure* or *air-cure*. Two-part systems depend on a catalyst or chemical agent to cure. Different sealant products cure at different rates. The temperature at the time of use affects the curing time. Curing is similar to the setting of mortar, but sealants are not intended to bear weight and thus must be handled differently in the construction process. Once sealant is placed, it begins to change physically. When a two-part system is mixed and applied, a chemical change takes place.

The materials of a sealant system consist of a sealant backer or stop, usually a foam rope or rod placed in the joint at a predetermined depth, and the sealant itself, which is gunned in against the stop. Some systems require a primer that must be applied to the joint's inner surfaces in advance of the sealant to ensure its adhesion.

In general, a sealant should not adhere to the backer rod. Joints using a sealant system work best when the sealant is required to adhere to parallel surfaces only. In wider joints especially, omission of a bond-breaking stop may contribute to failure. See **Figure 14-32.**

Trade Tip

Manufacturers of sealant systems are the best source of information on the different types of system and accessories.

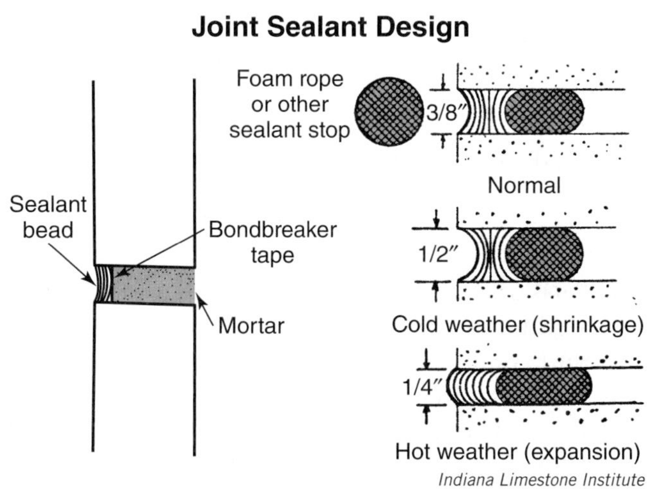

Figure 14-32. Sealant used between stone panels must be elastic and installed properly as described by the manufacturer.

Joint Movement

Limestone products have a low coefficient of expansion, however, a 100′ wall exposed to a temperature variance of 130° will move 7/16″. In the case of a 100′ long wall, seven 1/4″ expansion joints can be incorporated into the wall, allowing 1/16″ movement of each 1/4″ joint. If the sealant's percent of elongation and compression is 25%, the joint design and percent of movement are within the 25% maximum movement recommendations for the joint. The size of the joint should be adjusted to the elongation limits of the sealant. A sealant with a 15% joint movement capability would require a 7/16″ joint size to withstand a 1/16″ joint movement.

These examples assume that movement is equally distributed over all joints. By sizing joints larger than minimum design requirements, a safety factor can be obtained to accommodate atypical movement.

Expansion Joints

In exterior limestone walls, ***expansion joints*** (relief joints) should be provided to reduce the damaging effect of thermal expansion of the building frame. The coefficient of expansion of concrete and steel supporting structures is much greater than that of limestone. Differential movement between the building frame and walls due to thermal changes causes pushing out of header stones at corners, fracturing or spalling of stone at joints, and pulling out of rigid anchors. Stonework erected in cold weather can develop temperature stress in joints in warm weather.

Expansion joints normally consist of a premolded filler and sealant compound. The premolded filler should be compressible to the amount of movement calculated with resiliency to return to its original shape. The sealant compound should be completely elastic and provide lasting adhesion to the surfaces separated.

The best location for an expansion joint is in an offset of a building. An alternate location is at the junctions of the sections comprising a U-shaped, T-shaped, or L-shaped building.

Typical Cornice Detail

The traditional cornice detail in **Figure 14-33** shows the most commonly used method of anchoring a cornice with a projection large enough to be unbalanced in the wall. The bed joint immediately below the heavy cornice is left open back far enough to remove any compressive stress that would have a tendency to break off stone below. The optional rod-and-plate tie-down stabilizes unbalanced, projecting courses.

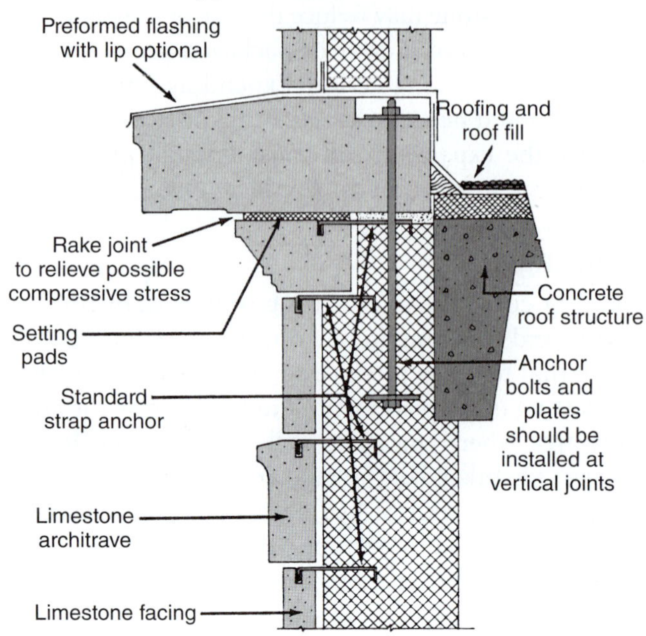

Indiana Limestone Institute

Figure 14-33. A traditional stone cornice detail.

14.4 Keeping Clean and Cleaning New Masonry

The finished appearance of a masonry wall depends not only on the skill used in laying the units but also in taking procedures to keep it clean and using proper techniques to clean the surfaces. The appearance of a masonry structure can be ruined by lack of protection and improper cleaning. In many instances, the damage caused by using faulty cleaning techniques or the wrong cleaning agent cannot be repaired. All cleaning should be applied to a sample test area of approximately 20 sq ft.

Take the following general precautions to promote a cleaner wall prior to and during construction:

- Protect site-stored stone from mud or other contaminants. Store stones off the ground under protective covering.
- Protect the base of the wall from rain-splashed mud, mortar splatter, or concrete that will be placed next to the finished wall surface.
- Cover walls with a waterproof membrane at the end of the workday to prevent mortar joint wash out and entry of water into the completed masonry.
- Turn scaffold boards near the wall on edge at the end of the day to prevent possible rainfall from splashing mortar and dirt on the completed masonry.

- Practice careful workmanship to prevent excessive mortar droppings. Excess mortar should be cut off with the trowel as the stone is placed. Joints should be tooled when thumbprint hard.
- After tooling, brush excess mortar and dust from the surface. Avoid any motion that results in rubbing or pressing mortar particles into the stone faces. A medium soft bristle brush is preferable.

Cleaning failures generally fall into one of the following categories:

- Failure to thoroughly saturate the masonry surface with water before and after application of chemical or detergent cleaning solutions. Saturation of the surface prior to cleaning reduces the absorption rate, permitting the cleaning solution to stay on the surface rather than be absorbed.
- Failure to properly use chemical cleaning solutions. Improperly mixed or overly concentrated acid solutions can etch or wash out cementitious materials from the mortar joints.
- Failure to protect windows, doors, and trim. Many cleaning agents, particularly acid solutions, have a corrosive effect on metal. If permitted to come in contact with metal frames, the solutions can cause pitting of the metal or staining of the masonry surface and trim materials, such as limestone and cast stone.

Keeping Stonework Clean

Planning and preparation prior to laying the first stone can reduce the time and money spent cleaning stonework. Cover or tarp the materials to protect them from weather and jobsite conditions, such as muddy surfaces that can splatter the freshly cut stone. When stones have been laid on walls (either inside or outside), cover the surfaces with plastic sheeting called *visqueen* to protect them from being stained and from construction dust. Visqueen comes in rolls of different lengths and thicknesses.

Cleaning Stonework

Clean stonework with a stiff fiber brush and clean water. If stains are difficult to remove, use soapy water and then rinse with clear water. If the stonework is kept clean by sponging during construction, the final cleaning is easy. Machine cleaning processes should be approved by the supplier before use.

Acids, wire brushes, and sandblasting are usually not permitted on stonework. Strong acid compounds used for cleaning brick burns and discolors many types of stone.

Chapter Review

Summary

- Before beginning to set stone, a stonemason should consider the size, shape, and type of stone to be used.
- Cut stone in the form of trimmings, ashlar, or panels must be handled carefully to prevent breakage, staining, or chipping.
- Cements with low alkali content (nonstaining cements) are best to use when setting limestone.
- Rubble is uncut stone or stone that has not been cut to a rectangular shape.
- Ashlar stone is precut with enough uniformity to allow some regularity in assembly.
- Trimmings are stones that are cut on all sides to specific dimensions.
- Walls that use mortar as a bond between stones require a footing below the frost line because frost heaving causes shifting in the wall structure.
- Designing and building with limestone panels requires attention to many details in order to assure quality performance and appearance.
- The term *anchors* generally refers to the straps, rods, dovetails, and other connections between stone and structure.
- All anchors in limestone products should be stainless steel or another noncorrosive metal.
- An expanding type grout or mortar should not be used. When used in anchor slots and holes, pressure from the expansion can fracture the stone at the anchor locations.
- In exterior limestone walls, relief joints should be provided to reduce the damaging effect of thermal expansion of the building frame.
- The appearance of a masonry structure can be ruined by faulty cleaning techniques or use of the wrong cleaning agent. Stonework should be cleaned with a stiff fiber brush and clean water.

Know and Understand

Answer the following questions using the information provided in this chapter.

1. The best trowel for spreading mortar for stone is _____.
   A. 7″ buttering trowel
   B. mason's 10″ trowel
   C. 5″ margin trowel
   D. the trowel that feels good to you
2. *True or False?* A tagline is a device, such as a sling or chain, attached to stones to lift them from the delivery trailer.
3. A _____ is used for striking hardened steel.
   A. chisel
   B. maul
   C. trowel
   D. mason's hammer
4. _____ is precut stone that is uniform enough to allow regularity in assembly.
   A. Panel construction
   B. Rubble
   C. Ashlar
   D. Trimming
5. _____ is used in walls where a rustic look is desired and precise lines are not important.
   A. Panel construction
   B. Rubble
   C. Ashlar
   D. Trimming
6. *True or False?* When placing rubble, trim each stone after it has been placed.
7. *True or False?* When building with limestone panels, backs of stones above a roof line do *not* need to be finished.
8. _____ cement should be used with limestone products.
   A. Nonstaining
   B. Staining
   C. Pure Portland
   D. Anchor
9. All masonry wall cleaning should be applied to a sample test area of approximately _____ sq ft.
   A. 10
   B. 15
   C. 20
   D. 50

10. To clean stone work, use _____.
    A. an acid solution
    B. wire brushes
    C. sandblasting
    D. stiff fiber brushes

Apply and Analyze

1. Describe the proper way to grasp a trowel.
2. What is the result of constantly working the mortar?
3. What kind of shoes or boots are required when working with stone?
4. What are the three main classes of stonemasonry construction?
5. How can heavy stones be prevented from squeezing mortar out when they are set into place?
6. What simple cleaning procedure is recommended for a completed stone wall?
7. Why should strong acids be avoided when cleaning stone?
8. Why do walls that use mortar as a bond between stones require a footing below the frost line?
9. Who generally provides the anchors used in stone panel construction?
10. What should you do if stains on stonework are difficult to remove?

Critical Thinking

1. Imagine your superintendent explains that an upcoming home project will have stone on the front of the home. A concrete slab will be placed against the wall to create a patio. What procedures will you implement to protect the stone wall against the concrete when placing the slab?
2. Why is pointing cut stone after setting, rather than full bed setting and finishing in one operation, the recommended practice?

Activities

1. (ELA) Research the advantages and disadvantages of natural stone veneer vs. manufactured stone veneers. Prepare a report comparing the two materials.
2. (ENGINEERING) Research the various types of expansion joints used with stonemasonry. Write a report that discusses the different types of materials used in the expansion joints and explain why these materials are used.
3. (MATH) Limestone has a low coefficient of expansion. A wall length of 100′ exposed to a temperature variance of 130°F will move 7/16″. What would the expansion be for a wall 40′ in length if the expansion occurs at the same rate as stated above? Calculate the answer to the nearest 1/16″ of an inch.

SECTION

5 Construction Details

Scot Kelly of Quick Headers, LLC.

Timothy L. Andera

CHAPTER 15

Foundation Systems

LEARNING OUTCOMES

After studying this chapter, you will be able to:

15.1-1 Explain the purpose of foundations and footings.
15.2-1 Describe the different types of spread foundations.
15.3-1 Explain the functions of foundation walls.

TECHNICAL TERMS

column
combined footing
concrete form tubes
continuous footing
dampproofing
footing
foundation
frost line
ice lenses
isolated footing
keyway
matt foundation
pier
pilaster
plaster cove
plain footing
raft foundation
reinforced footing
stepped footing
toothing
waterproofing

Introduction

This chapter covers common construction practices for foundation systems. Footings, floating foundations, foundation walls, columns, and piers are also discussed. The information in this chapter will aid in the selection of appropriate building methods to construct foundations. See **Figure 15-1**.

A main concern in construction is to prevent the settling of buildings as much as possible. Some settling is to be expected and cannot be prevented. The worst enemy of buildings is uneven settling, which causes cracks in finished walls and ceilings. It also tilts floors, causing doors and windows to bind. It damages woodwork. If severe enough, uneven settling can cause total failure of the building.

15.1 Foundations and Footings

The ***foundation*** is the substructure of a building. The foundation resists settling and supports the weight of the building.

The base on which a foundation is built is called the ***footing***. Footings transmit (shift) the weight from the foundation walls to the soil. See **Figure 15-2**. Footings are usually wider than the foundation wall to provide the extra support needed to stop or reduce uneven settling. As a result, the structure remains sound. The two general types of foundations are spread foundations and pile foundations. Spread foundation is built on a wide footing, usually twice the width of the wall being

KalininStudios/Shutterstock.com; swelsh1/Shutterstock.com; Daniel Kourey/iStock/Thinkstock

Figure 15-1. Examples of construction activities involved in building foundation systems. A—Brick pier with rebar being built by a mason. B—Concrete slab with rigid insulation ready for the placement of concrete. C—Foundation wall with footing.

supported. A pile foundation uses slender columns or cylinders to support the above structure. Pilings are made of concrete or steel and can be driven or drilled into the earth to create a foundation that supports different types of structures.

Only spread foundations are covered in this text because pile foundations are not usually used in work performed by masons.

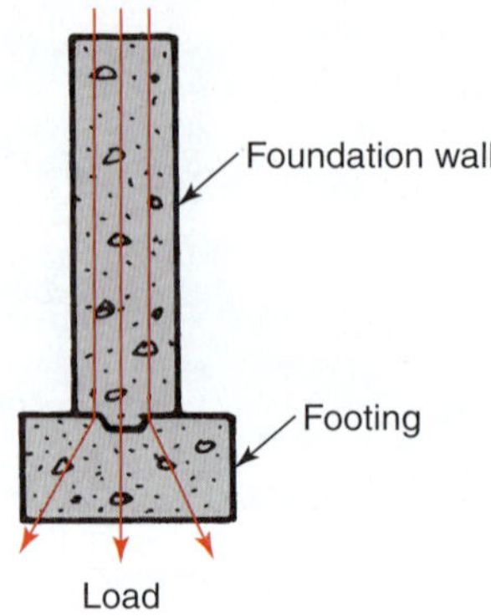

Goodheart-Willcox Publisher

Figure 15-2. Footings spread the load of the structure over a broader area.

15.2 Spread Foundations

Spread foundations consist of elements such as walls, pilasters, columns, or piers that rest on a footing. Spread foundations distribute building loads over a wide area of soil.

Footings

All foundation walls built of masonry units bonded with mortar should rest on footings. These footings are usually made from concrete. They are classified as plain, reinforced, continuous, stepped, and isolated.

Plain footings carry light loads and are not reinforced with steel. They may be continuous, stepped, or isolated.

Steel is embedded in ***reinforced footings*** to strengthen them. See **Figure 15-3**. Reinforcing should be used over weak spots in the soil, such as excavation sites for sewer, gas, or other connections.

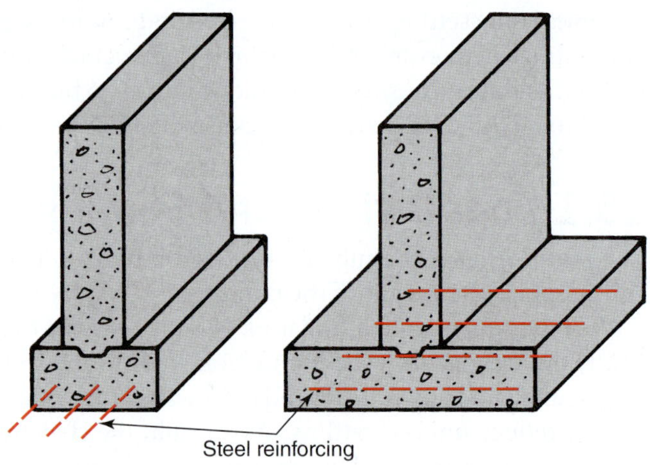

Goodheart-Willcox Publisher

Figure 15-3. Two methods of reinforcing footings to increase strength. Left—Rebar running parallel with the footing. Right—Rebar running perpendicular to the footing.

Continuous footings are constructed to support a foundation wall or several columns in a row. A ***combined footing*** is a single footing supporting more than one column. See **Figure 15-4**.

There are two types of ***stepped footings***. One widens the base to provide added support. Reinforced footings have largely replaced widened footings. The other type occurs when the level of the footing needs to change to accommodate a sloping grade. See **Figure 15-5**. As with any type of footing, including stepped footing, it is important that the footing is below the frost line. The ***frost line*** is a designated point in depth that frost can penetrate. When the temperature is below freezing, frost begins to enter the soil or earth. Freezing causes the soil to expand because there is moisture in the soil, which causes a heaving process. If the footing is not below the frost line, it can cause severe damage to the structure by lifting and shifting the footing and foundation. Southern climates may require the footing to be 1′ below the specified frost depth. In northern climates, the frost may penetrate from 48″ or greater. Check with the local building codes to determine frost depth in a specific area and required depth of the footing.

When a sloping grade exists, it is important to take into consideration the frost line requirement. Also check the maximum step height; certain codes specify a maximum of 8″ per step. To accommodate for sloping grades and frost requirement, the footing may have

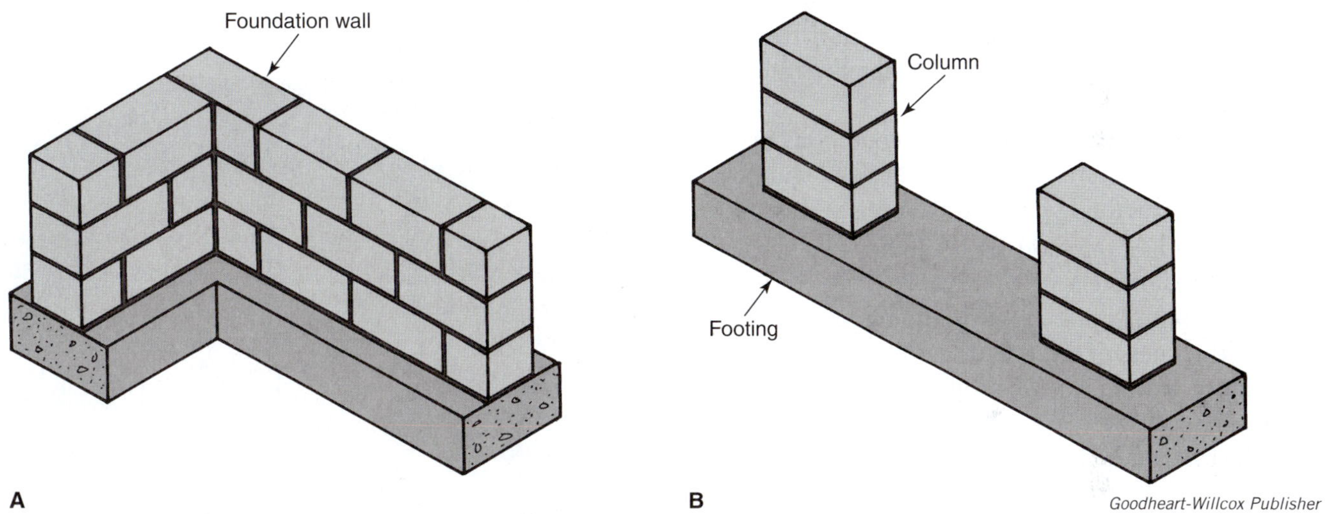

Figure 15-4. Two types of footings. A—Continuous footing. B—Combined footing.

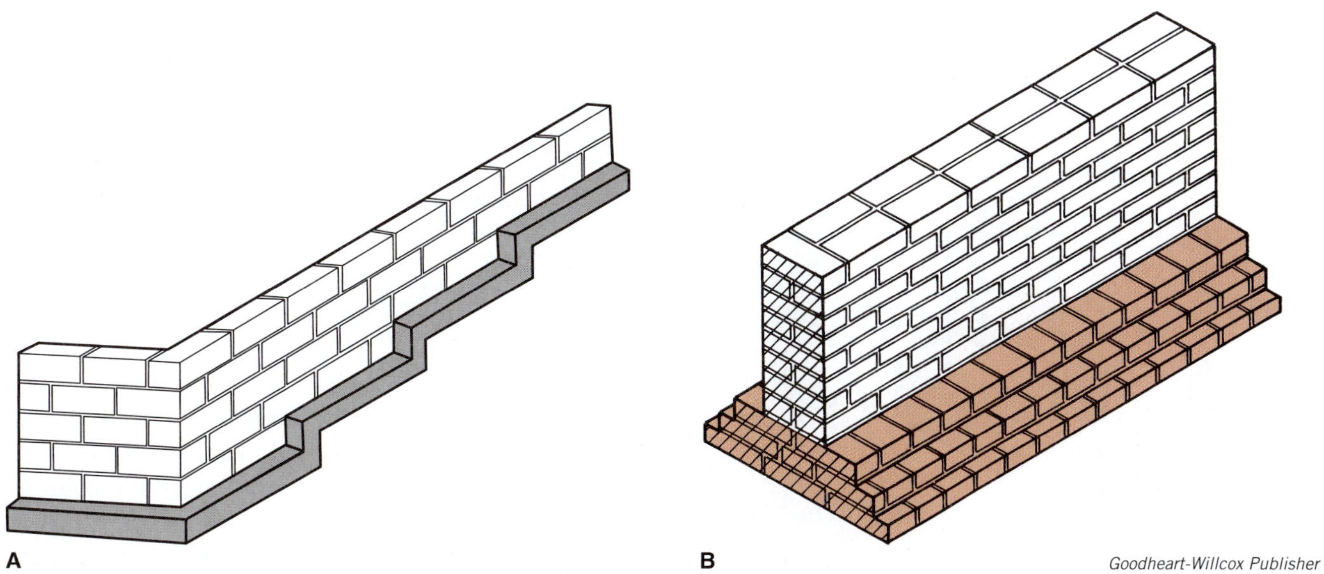

Figure 15-5. Two types of stepped footings. A—This stepped footing changes levels on a sloping grade. B—This stepped footing widens the base for greater support.

multiple steps to ensure the footing stays below the code frost requirement. In situations where concrete blocks are being used for the foundation, common practice uses the height and length of a standard concrete block for the height and length of a step. For example, the footing needs to drop 24″ because of the slope in elevation. To accommodate slope change and the code requirement of 8″, there would be three steps in the footing. The length of steps can vary but because blocks are 16″ length, a step equaling that would work well. Depending on the length of travel for the step footing, using multiples of 16″, such as 32″ or 48″ can be the best practice to reduce the cutting of block and reduce labor time. If the project is using poured concrete, then the length of the step may be based on the width of the concrete forms being used to construct the foundation walls.

Isolated footings are not part of a normal foundation but are separate structures that can support freestanding loads from columns, piers, or other structures. See **Figure 15-6**.

Footing size should relate to soil conditions and the weight of the structure. However, the rule-of-thumb size for residential footings placed on average soil is as follows:

- The width of the footing should be twice the thickness of the wall.
- The depth of the footing should be equal to the thickness of the wall or a minimum of 6″. **Figure 15-7** shows the proportions for a standard footing.
- The base of the footing or column footing should be below the frost line.
- Piers or columns can be created using concrete form tubes. ***Concrete form tubes*** are cardboard or heavy paper tubes for forming a concrete pier or column. They come in different diameters and lengths depending on the requirement of the project. See **Figure 15-8**.

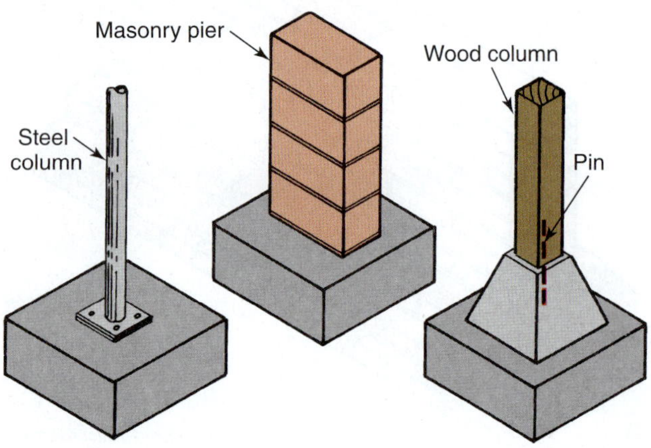

Goodheart-Willcox Publisher

Figure 15-6. Isolated footings with columns and a pier.

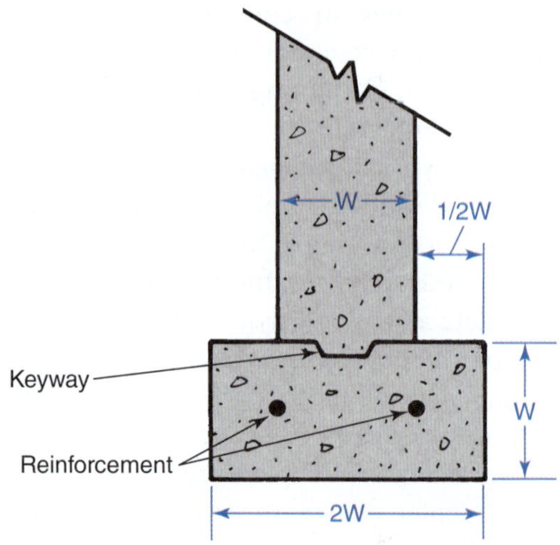

Goodheart-Willcox Publisher

Figure 15-7. Typical rule-of-thumb footing proportions. This footing includes a keyway and reinforcement.

Timothy L. Andera

Figure 15-8. Different diameter sizes of concrete form tubes.

Code Note **Frost Protection**

The International Residential Code **R403.1.4.1** states that a building footing or other types of footings shall be protected against frost. The depth from the bottom of the footing to the grade directly above it shall be based on **Table R301.2(1)**. In the northern Midwest, the depth of frost can vary anywhere from 30″ to 60″ depending on the location.

When designing footings, foundations, and piers, it is important to discuss procedures and submit plans to the local building code officials. States and counties may adopt different codes and versions based on the year the code was adopted by the community or state. It is important to determine the code and version you will be using before building.

To aid in attaching a foundation wall to the footing, a ***keyway*** (recess or groove) is sometimes used to secure the two together. The keyway is usually placed in the middle of the footing prior to the concrete setting up in the footing form. See **Figure 15-9**. The keyway form is made of lumber or preformed metal. After the concrete has set, the wood or metal form is removed, leaving a keyway or depression, in the top of the concrete footing. Once the concrete wall forms are erected on the footings and filled with concrete, the keyway locks the concrete wall to the footing.

Safety Note

When cutting rebar for reinforcement with a cut-off chop saw, make sure to wear safety glasses, face shield, ear protection, and safety footwear.

Best practices are valuable when planning a foundation system, because poor workmanship, carelessness, or lack of knowledge regarding building practices may have a lasting effect on the structure; foundations can last a lifetime. **Figure 15-10** is an example of how workmanship can have an effect on future outcomes of a project. It is a pier that was poured 25 years ago. During this time, the concrete foundation pier has risen out of the ground approximately 3′. Several factors over the years may have contributed to the pier moving vertically:

- The base of the concrete pier was not placed below the frost line.
- Strong winds can get under the deck and put pressure on the structure to lift the deck, which will begin to lift the concrete piers.
- A footing was not created at the base of the pier to aid in stability, support, and to prevent the pier from lifting.
- The concrete was allowed to mushroom at the top of the pier. This permits the ground to push up on the column when the soil freezes under the concrete lip of the pier.

These factors can cause headaches for the contractor many years after the project is completed. Once a foundation, pier, or column begins to lift, it will most likely never return to the original location. So as time passes, the lifting effect continues to worsen.

Frost and freezing can have long-lasting effects on footings, foundations, or piers. **Figure 15-11** discusses the process of freeze cycles and how ice lenses can

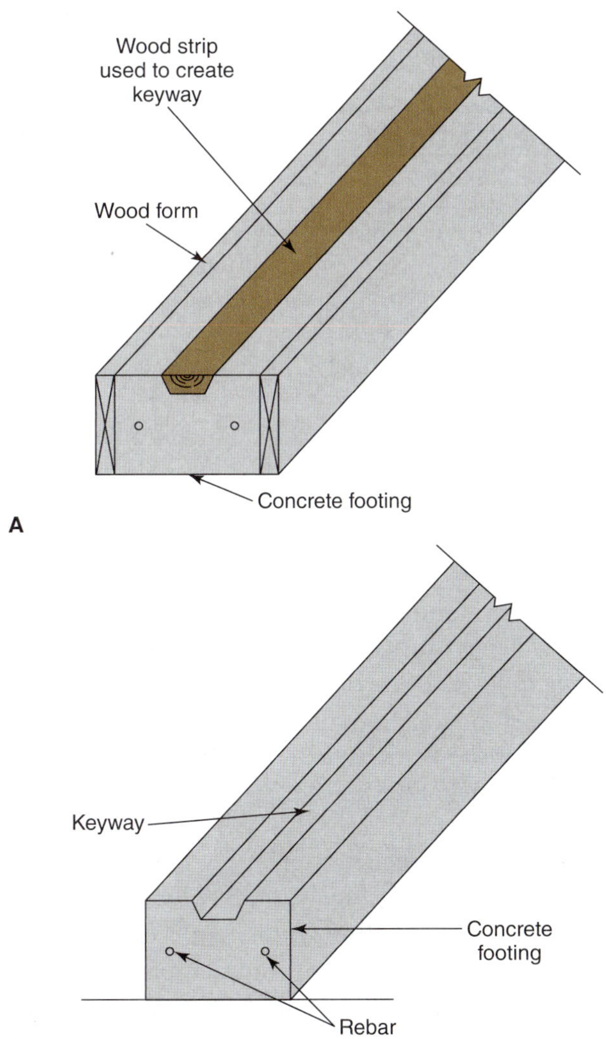

Goodheart-Willcox Publisher

Figure 15-9. A—A wood strip is placed in fresh concrete to form a keyway in the concrete. B—The keyway form has been removed, leaving a depression in the top of the footing.

Timothy L. Andera

Figure 15-10. Concrete was allowed to mushroom over the edge of the hole and harden around the top of the excavation. The pier has lifted approximately 3″ above the surrounding area.

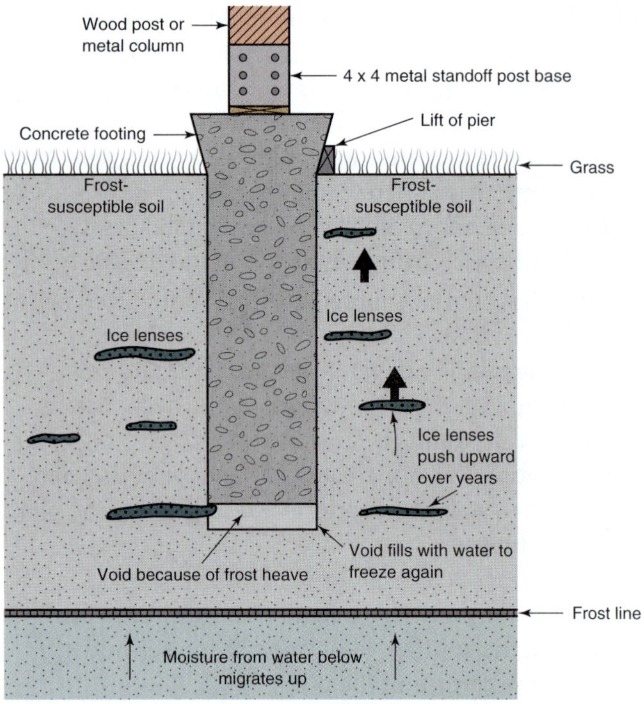

Goodheart-Willcox Publisher

Figure 15-11. Ice lenses, created through the freeze-thaw cycles, contribute to the lifting of the pier. Notice the pier is above the frost line. The resulting void under the pier accentuates the problem.

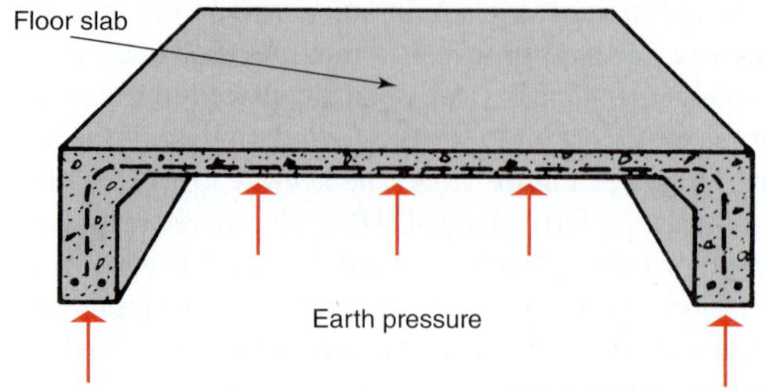

Goodheart-Willcox Publisher

Figure 15-12. A matt foundation ties the entire floor and footings into one unit.

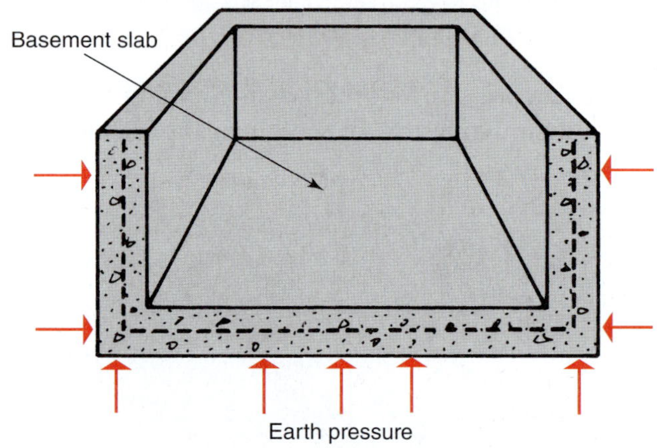

Goodheart-Willcox Publisher

Figure 15-13. A raft foundation ties the walls and floor into one unit.

occur. ***Ice lenses*** are the process of moisture freezing in the ground or soil to create pockets of ice. If a foundation's footing is not below the frost line, then ice lenses can get under the footing. In certain situations, ice lenses attach to rough surfaces of the concrete and contribute to the heaving or lifting process of the footing or foundation. Once the footing or structure begins to lift, it creates a void under the structure. This void becomes a location where moisture or water can accumulate to perpetuate and accelerate the lifting problem.

Raft and Matt Foundations

Raft and matt foundations are classified as spread foundations. They are used over soils with low ability to carry weight. Raft and matt foundations are made of concrete reinforced with steel in such a way that the entire foundation acts as a single unit. They are often referred to as *floating foundations*.

A ***matt foundation*** is a thickened slab that transmits building loads over the entire slab and soil area. See **Figure 15-12**.

A ***raft foundation*** is made with reinforced walls and floor cast into a single unit. See **Figure 15-13**. Soil is excavated that is approximately the weight of the foundation plus the superstructure. (*Superstructure* refers to the part of a building above the foundation.) This amount of excavation minimizes settling of the structure because the building load on the soil is equal to the weight of the excavated soil.

15.3 Foundation Walls

Like footings, foundation walls are load-transferring elements. They must be durable structures that perform the following functions:

- Support the weight of the building above
- Resist pressures from the ground
- Provide anchorage for the superstructure of the building
- Resist moisture penetration
- Must extend above finish grade 4″ for masonry veneer and 6″ for other types

The most common type of foundation wall is the T foundation. The name comes from the shape of the foundation and the footings, which look like an upside-down T. The footing and foundation wall are generally two separate parts, but may be cast as a single unit. **Figure 15-14** shows several styles of foundations.

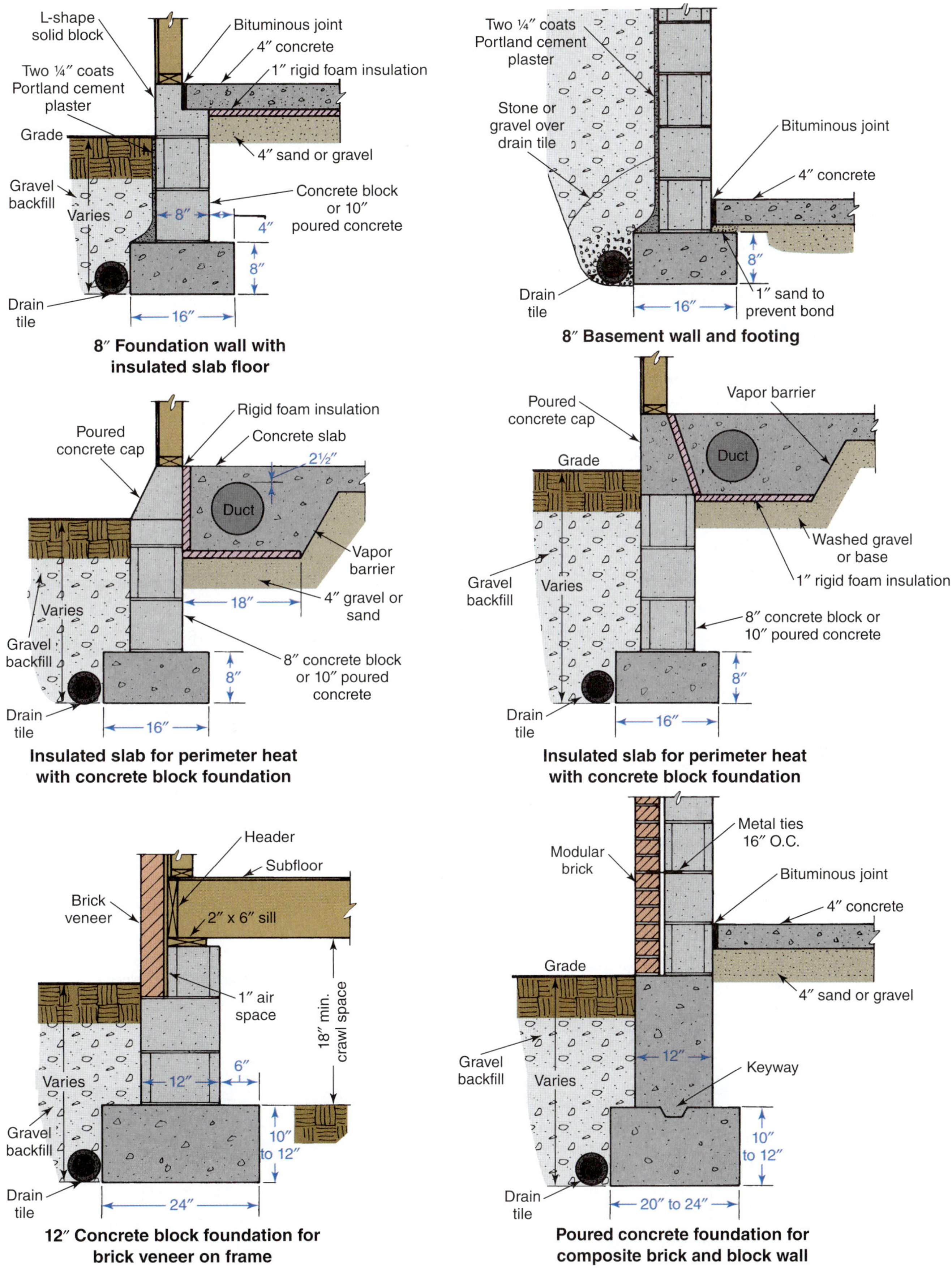

Goodheart-Willcox Publisher

Figure 15-14. Typical foundation details.

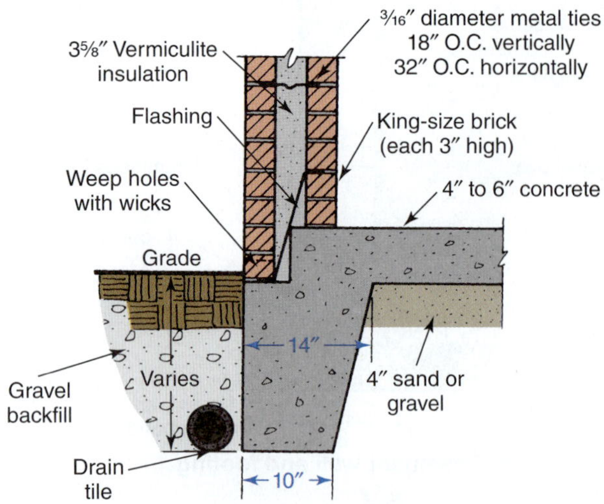

Foundation for 10″ cavity wall with slab floor

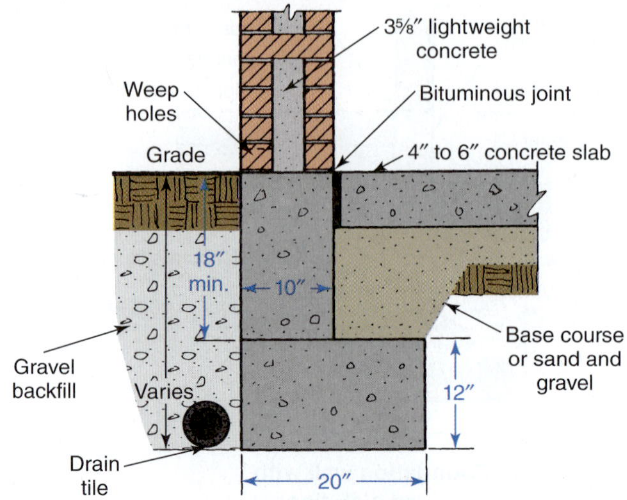

Foundation for 10″ masonry bonded wall with slab floor

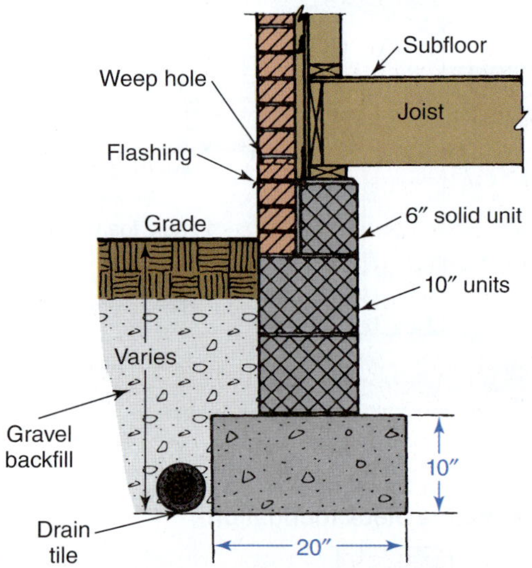

Brick and tile foundation for brick veneer on frame structure

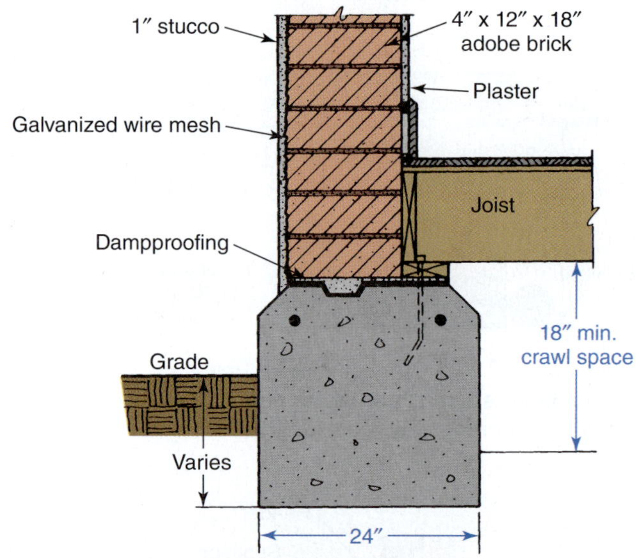

Foundation for 12″ adobe brick with stucco

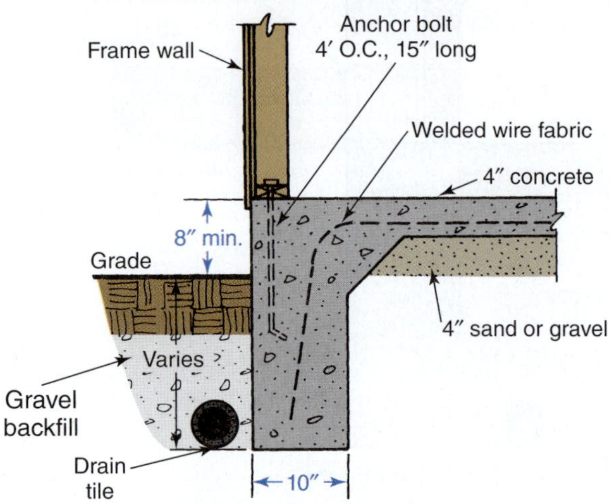

Thickened edge slab foundation for frame wall construction

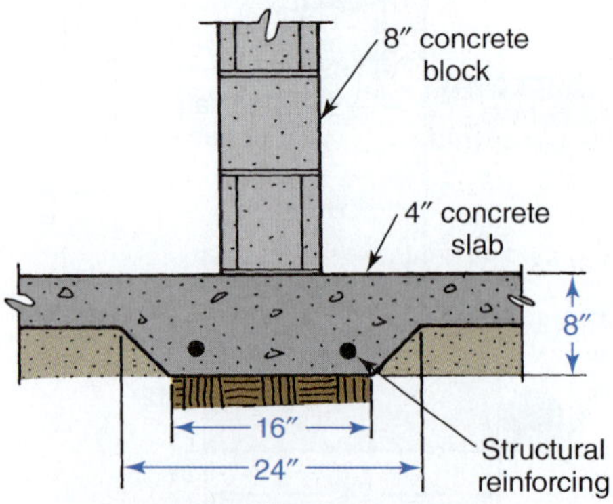

8″ Concrete bearing wall partition on a slab floor

Goodheart-Willcox Publisher

Figure 15-14 (Continued).

When planning foundations, either concrete block or poured concrete should extend above the finish grade of the structure. Masonry needs to extend above the finish grade and the grade needs to slope away from the structure. Both of these processes will aid in keeping the inside of the home or foundation dry when inclement weather occurs.

Figure 15-15 shows a detailed drawing of a composite wall made of brick, poured concrete, and concrete block. A polyethylene (poly) moisture barrier is placed on top of the compacted gravel. The poly reduces the chance of moisture wicking into the concrete from the gravel and in turn lowers the humidity levels inside the living space. By having a dryer environment, it lessens the chance of mold growing. Rigid insulation is added on the inside of the foundation wall and under the concrete slab, on top of the poly vapor barrier. The insulation reduces heat loss and creates a more energy efficient building. Reinforcement for the concrete slab may be added on top of the rigid insulation, if requested by the owner or required by code. If reinforcement is utilized in the concrete floor, jacks or stools are used to elevate the rebar to the proper height before concrete is placed.

In the footing area, reinforcement rebars are placed horizontally to add strength. Reinforcement continues up through the concrete wall into the block wall tying the two structures together. At the top of the block wall, anchor bolts are used to connect the framing structure to the foundation. Drain tiles are placed outside and inside of the footing to remove moisture.

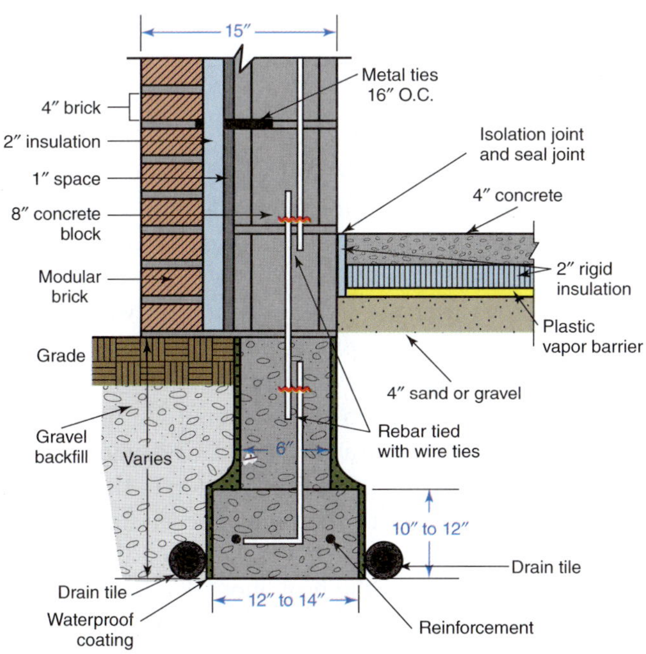

Goodheart-Willcox Publisher

Figure 15-15. A composite wall made up of brick, concrete, and concrete block.

Rigid insulation of 2″ is placed inside the wall cavity and attached to the concrete wall. An air space of 1″ is added to create a void in the wall cavity to allow moisture to move down the insulation and the backside of the brick. Moisture in the cavity is captured and directed out through flashing and weep holes. Weep holes sit directly above the flashing to create a location where moisture can exit the wall cavity.

Code Note **Foundation Extending above Finished Grade**

The International Residential Code (IRC) **R404.1.1.6** states that concrete and masonry foundation walls shall extend above the finished grade adjacent to the foundation at all points a minimum of 4″ (102 mm) where masonry veneer is used and a minimum of 6″ (152 mm) elsewhere.

Foundation walls of hollow concrete block can be capped with a course of solid masonry to help distribute the loads from floor beams and serve as a termite barrier. See **Figure 15-16**.

Solid blocks 4″ thick are available in some areas of the United States. See **Figure 15-17**. If stretcher blocks are used, a strip of metal lath is placed under the cores

Christina Richards/Shutterstock.com

Figure 15-16. A block foundation wall is capped with a course of solid top block.

Christina Richards/Shutterstock.com

Figure 15-17. After the cap blocks are prepped, they are mortared to the 8″ foundation wall. Notice that the foundation wall is insulated with 2″ of rigid insulation. The insulation starts at the top of the footing and runs the height of the wall to be flush with the top of the cap block.

pryzmat/Shutterstock.com

Figure 15-18. Concrete blocks are filled to strengthen the foundation wall. Cores are filled with grout, concrete, or mortar, based on the project specifications.

of the top course except in areas where rebar and anchor bolts are positioned where the cores are filled from the footing to the top of the last block. The remaining cores are then filled with concrete, grout, or mortar and troweled smooth. See **Figure 15-18**.

Dampproofing and Waterproofing Basement Walls

Because concrete is a porous material, moisture migrates over time through the wall and enters the interior of the structure. For this reason, the underground portion of basement walls needs to be dampproofed. ***Dampproofing*** prevents soil moisture surrounding the foundation from entering the structure by the application of products to the exterior surface of the walls.

The outside of masonry basement walls should be parged with a 1/2″ thick coating of Portland cement plaster or mortar. Apply two coats, each 1/4″ thick. The wall surface should be clean and damp, but not soaked, before applying the plaster. Trowel the first coat firmly over the wall. See **Figure 15-19**. When the first coat has partially hardened, roughen it with a notched trowel or a scratcher, **Figure 15-20**.

Keep the first coat moist and allow it to harden for a minimum of 24 hours before applying the second coat. See **Figure 15-21**. Cover the wall from the footing to 6″ above the finished grade line.

Form a cove over the footing to prevent water from collecting at this point. A ***plaster cove*** is plaster placed on top of a footing next to the foundation wall from the outside edge of the footing up the wall about 4″. The plaster is then rounded to create a cove shape. Creating a plaster cove is done by placing plaster on the top of the footing where the wall meets the footing. The plaster is contoured to form a cove (concave) shape from the outside edge of the footing up the wall about 4″. The rounded surface reduces the chance of moisture remaining on the footing and seeping between the foundation wall and footing to enter the structure's interior. The plaster cove will be performed around the complete foundation.

Author's image taken at Job Corps, Denison, IA

Figure 15-19. The first coat of parging is being troweled over the concrete block surface.

Author's image taken at Job Corps, Denison, IA

Figure 15-20. A notched trowel roughens the coat after it has partially hardened.

Author's image taken at Job Corps, Denison, IA

Figure 15-21. The second and final coat of Portland cement plaster is troweled over the roughened surface of the first coat.

The second coat of plaster should be moist cured for at least 48 hours. Moist curing involves applying moisture to the plaster to assist the cement in the mixture to continue the curing process. Mist the surface of the wall periodically so it does not dry out during the curing phase. Inadequate moisture in the plaster would stop curing, reducing the strength of the cement mixture.

Areas of the United States with certain soil conditions require additional dampproofing. Check federal, state, and local building codes to determine the correct procedures for dampproofing prior to construction. A heavy coat of tar, two coats of a cement-based paint, or a covering of thin plastic film may be used in addition to the parge coat. See **Figure 15-22**.

Code Note **Dampproofing of Concrete and Masonry Foundations**

International Residential Code **R406.1** requires dampproofing of foundation walls that enclose interior spaces below grade. The walls must be dampproofed from the top of the footing to the finished grade. If Portland cement parging is used, not less than 3/8″ must be applied to the exterior of the wall. Dampproofing may also be accomplished with a material that is approved for direct application to the wall.

Goodheart-Willcox Publisher

Figure 15-22. Concrete walls and footings have been coated with a liquid dampproofing material. The dampproofing is a minimum of 6″ above the final grade to meet most building codes.

A newer method of dampproofing concrete walls is to spray a sealant similar to tar on the wall and footing surfaces. Spraying the foundation with a product in liquid form reduces the time required to seal concrete surfaces and fills cracks and contours easier than using products applied by hand.

Waterproofing is another method used to improve moisture control when installing basements or other similar structures. Drain tiles are installed around the footing and under the basement floor to remove accumulated water from around and under the foundation. If excessive water is not removed, it moves through voids between the foundation walls and the footings or cracks that exist in the concrete floor.

Drain tiles are commonly made from plastic. In the past, they were made of concrete or clay, and, in certain areas, these types are still used. Drain tiles are generally placed around footings to carry away groundwater, **Figure 15-23**, and are connected to a dry well or storm sewer. Drain tiles are also placed under floors and connected to the perimeter tile to remove excessive groundwater under the concrete floor. Plastic tiles have continuous perforated holes that allow groundwater to enter. In certain situations, a plastic drain tile is slipped into a fabric sleeve to aid in reducing silting that may enter the tile through the holes or slits in the plastic tile.

All three types of drain tile are laid in a bed of gravel. The area is later backfilled with gravel. Backfilling with gravel rather than soil helps to reduce backpressure on the foundation. The gravel allows water to pass through easily to reach the footing area where it can enter the drain tile. Refer back to **Figure 15-14**, which shows a drain tile in place along the footing.

Paul Vasarhelyi/Shutterstock.com

Figure 15-23. Perforated drain tile is installed next to the concrete footing. Note that the footing also has a keyway.

Columns, Piers, and Pilasters

Columns, piers, and pilasters transmit loads to the footing. Columns and piers are freestanding, but pilasters are built into the foundation wall.

A ***column*** is an isolated element with a horizontal dimension measured at a right angle from the thickness dimension and does not exceed three times the thickness dimension. The height of a column is at least three times its thickness.

Brick masonry columns are used to support large axial loads. Axial loads are typically due to the permanent weight of the structure and the transient floor or roof load that is transferred to the column.

According to the MSJC Code and Specifications, columns must be reinforced with a minimum of four reinforcing bars, and the area of reinforcement must be at least 0.0025, but not more than 0.04, times the column's net cross-sectional area. The minimum nominal dimension of a column is 8″. The ratio of height to least lateral dimension must not exceed 25.

Brick or block columns are vertical supports that are frequently used to hold up wood, steel, or masonry beams. See **Figure 15-24**. The bonding for a 12″ × 12″ column is shown in **Figure 15-25**. Columns built with concrete block can be made using one to three blocks in each course. See **Figure 15-26**.

A ***pier*** is an isolated column of masonry or a bearing wall not bonded at the sides to associated masonry. A pier can also be used to support openings in a wall.

Workplace Skills
Working with Community Officials

You will interact with many community officials in a construction business. It is important as a representative of a masonry company to be professional in your dealing with city, state, or federal officials. These individuals are considered agents of the government. As a contractor, or a contractor's representative, you will work with code officials, building inspectors, OSHA Enforcement, OSHA Safety Consultation, and many others. These agents can help you in following existing codes and safety policies and give guidance as it pertains to the building processes. Introduce yourself and invite them to your jobsite or company office when appropriate. It is in your best interest and the interest of the masonry company to have a cordial relationship with these officials.

Code Note **Isolated Masonry Piers**

According to the International Residential Code (IRC) **R404.1.9**, piers should have a nominal thickness of at least 8″ (203 mm). The maximum height of a pier depends on the intended application and design. Footing thickness, width, and length is determined by the intended load. Check local building codes for specific requirements.

Figure 15-27 shows several sections of solid masonry piers. The size of piers, unless governed by a local code, is usually a minimum of 8″ × 12″ for solid units and

Victoria Ditkovsky/Shutterstock.com

Figure 15-24. Brick columns are used to support a steel structure.

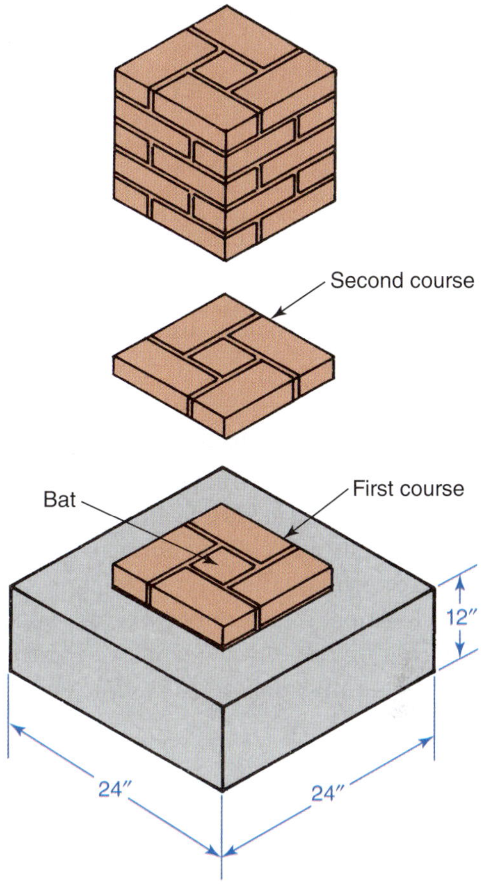

Goodheart-Willcox Publisher

Figure 15-25. Bonding for a 12″ × 12″ brick column. The bat is a piece of brick cut to fill.

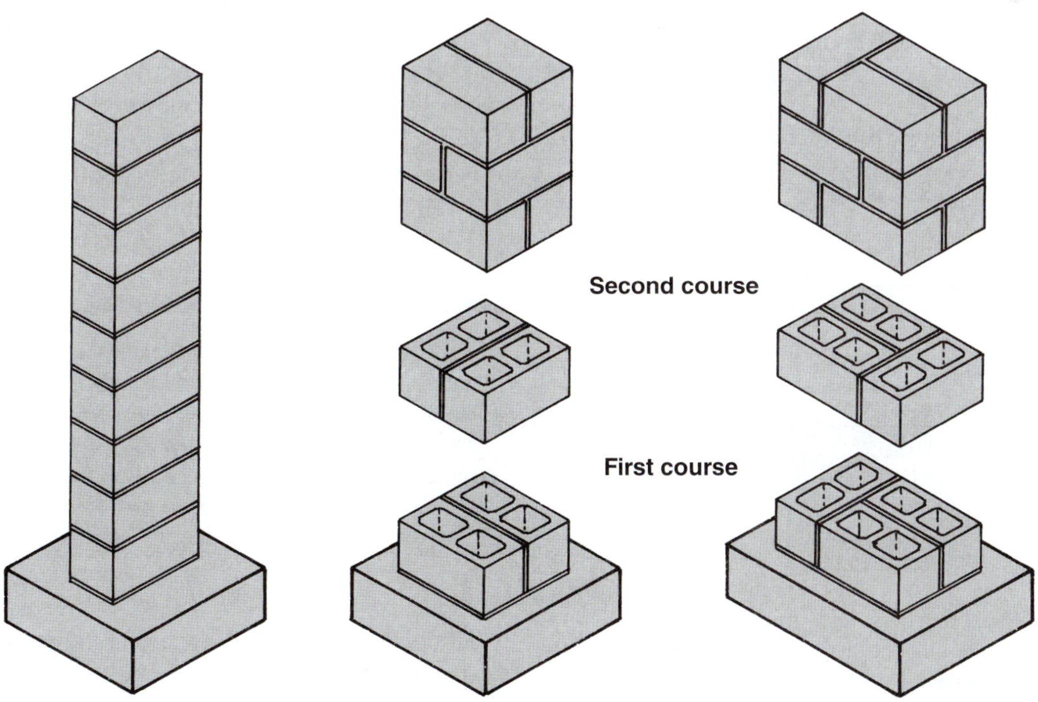

Goodheart-Willcox Publisher

Figure 15-26. Three popular styles of concrete block columns.

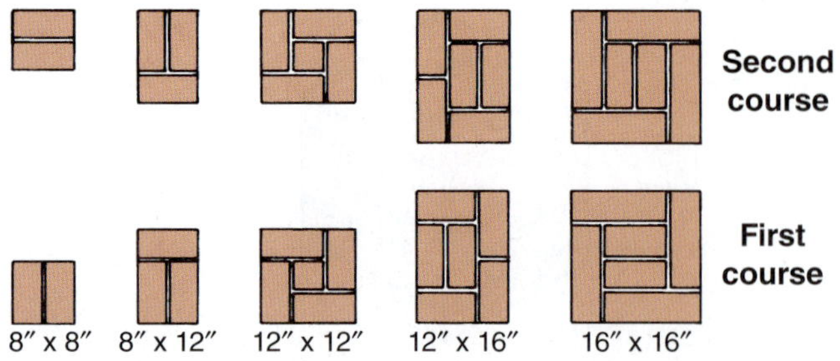

Goodheart-Willcox Publisher

Figure 15-27. Bonding for various sizes of solid brick masonry piers.

8″ × 16″ for hollow units. The height should not exceed 10 times the smallest cross-sectional dimension. Footings for piers should be at least 8″ thick.

A ***pilaster*** is a masonry column bonded to a wall. A pilaster is used to increase the effective thickness of a wall at a specific location. It has a uniform cross-section and serves as a column or vertical beam. See **Figure 15-28.** To work together, the wall and the thickened section must be integrally constructed. MSJC Code Section 5.10 permits three methods of bonding a pilaster to create integral construction:

- Interlocking 50% of the masonry units
- Toothing at 8″ maximum offset and attachment with metal ties
- Providing reinforced bond beams at a maximum spacing of 4′ O.C. vertically

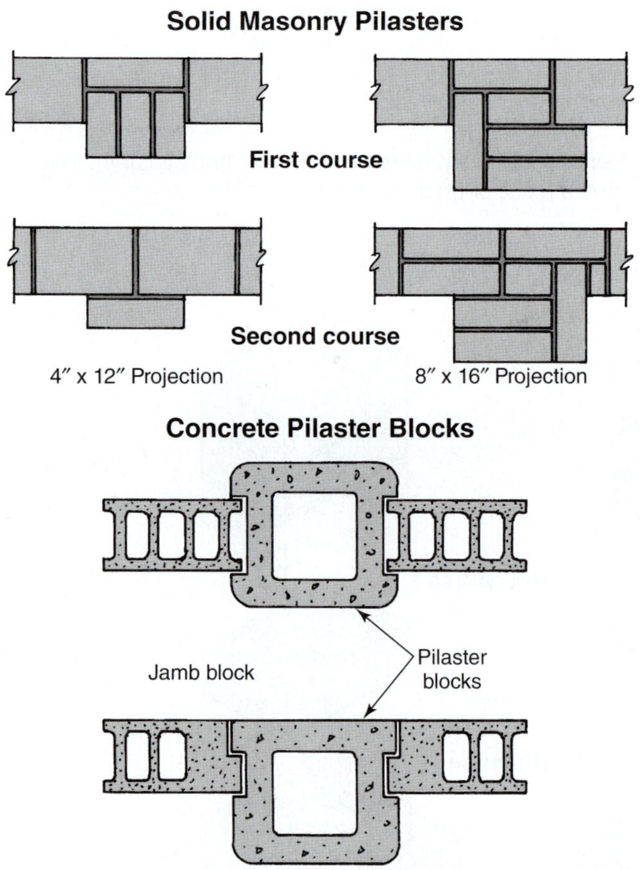

Goodheart-Willcox Publisher

Figure 15-28. Typical pilasters made from solid brick masonry and concrete pilaster block. Their purpose is to provide support for a wall. Pilasters are usually found at intervals in long wall sections.

Trade Tip

Toothing is the creation of temporary wall terminations of masonry units. Alternating course joints are cut out before they have set and then the specific unit is removed. This process creates a tooth-looking segment to the wall. At a later time, a different wall may be started at that location where new units are filled into the tooth block areas to connect the two walls together.

According to Brick Industry Association Technical Notes 3B, the length of a wall or flange that is considered to act integrally with the pilaster from each edge of the pilaster or web is less than six times the thickness of the wall or the actual length of the wall. When pilasters are built with concrete block, the hollow cells are sometimes filled with mortar or grout to increase their strength. See **Figure 15-29.**

Foundation Anchorage

The building superstructure should be anchored to the foundation to resist high winds. Sill plates of wood joist floor systems are generally anchored to masonry walls with 1/2″ bolts extending at least 15″ into the filled cores of masonry units. See **Figure 15-30.** Anchor bolts should be spaced not more than 6′ apart, with one bolt not more than 12″ from each end of the sill plate.

Code Note — Foundation Anchorage

According to International Residential Code (IRC) **R403.1.6**, anchor bolts shall be positioned not more than 1′ from the outside of the masonry wall and spaced a maximum of 6′ apart. The anchor bolts shall be 1/2″ in diameter and imbedded a minimum of 7″ into the concrete wall or grout cavity.

Sill plates should be anchored to poured concrete walls with 1/2″ anchor bolts embedded 7″ with the same maximum spacing as used for masonry walls.

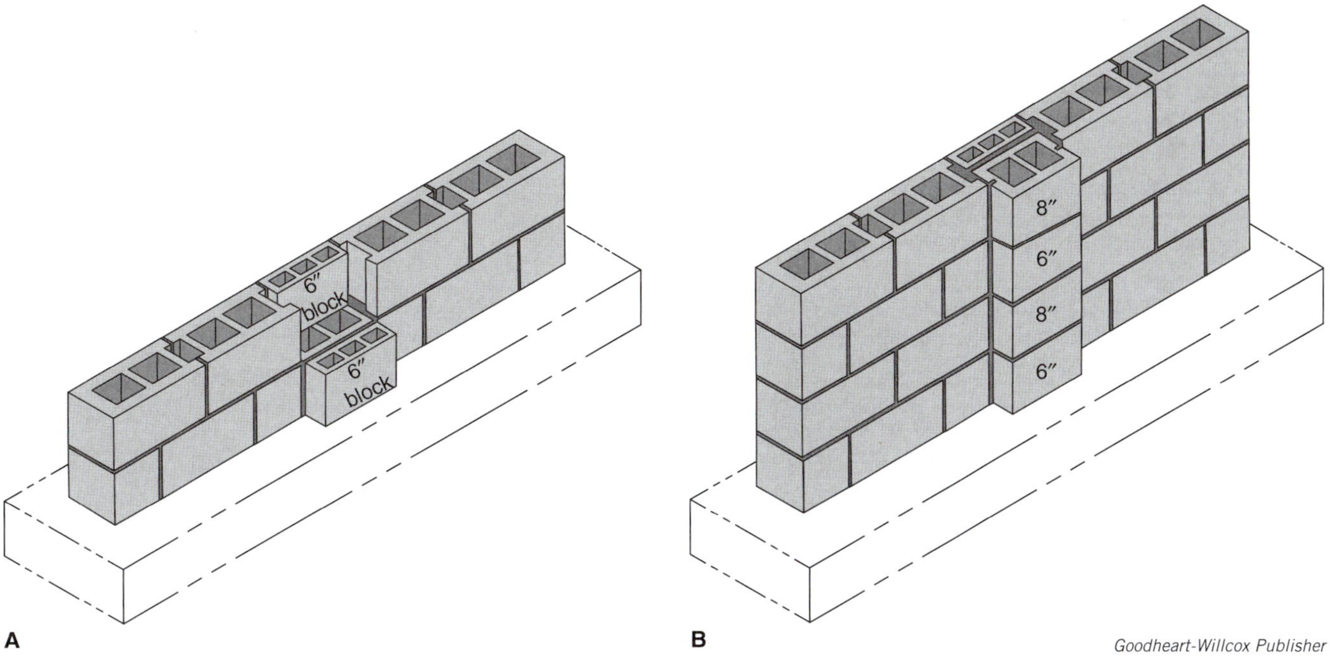

Goodheart-Willcox Publisher

Figure 15-29. A pilaster being constructed from 6″ and 8″ concrete block. The 6″ and 8″ blocks are alternated each course. A—Pilaster with the first course and part of the second course in place. B—Partially completed pilaster. By alternating the positions of the 6″ and 8″ block, the pilaster is tied into the wall.

Christian Delbert/Shutterstock.com

Figure 15-30. A concrete house foundation with anchor bolts ready for wood framing.

Green Note

LEED and Masonry

LEED (Leadership in Energy & Environmental Design) is a green building certification program of the US Green Building Council (USGBC). LEED Certification is a widely recognized standard for buildings that are efficient, cost effective, and better for occupants and the environment.

To become LEED-certified, building projects must earn points in a rating system that pertains to the specific application, such as Building Design and Construction, Building Operation and Maintenance, and Homes. Points can be earned for environmentally friendly building practices in several credit categories, which include Materials and Resources, Energy and Atmosphere, Water Efficiency, Indoor Environmental Quality, Sustainable Sites, and Innovation. Projects can earn points by satisfying credit prerequisites within the various categories. For example, Sustainable Sites credits relate to strategies that minimize a project's impact on ecosystems and water resources. The total number of points earned determines the level of LEED certification.

Masonry construction contributes to LEED points in the following ways:

- Minimum Energy Performance of Building—building to meet energy guidelines
- Clean Construction—heavy equipment to operate efficiently and cleanly
- Construction and Demolition Waste Management—recycling of materials

Chapter Review

Summary

- Footings are usually wider than the foundation wall to provide support that stops or reduces uneven settling.
- Spread foundations consist of elements such as walls, pilasters, columns, or piers that rest on a wider base called a footing.
- Footings are usually made of concrete and are classified as plain, reinforced, continuous, stepped, or isolated footings.
- Footings need to be placed below the frost line. Check with local building officials to determine correct depth.
- Raft and matt foundations, often called floating foundations, are made of concrete reinforced with steel in such a way that the entire foundation acts as a single unit.
- Foundation walls are load-transferring elements, with the most common type being the T foundation.
- The outside of masonry basement walls should be parged with a 1/2″ thick coating of Portland cement plaster or mortar, applied in two coats.
- Drain tiles are generally placed around footings to carry away groundwater. They are connected to a dry well or storm sewer.
- Foundations need to extend above finish exterior grade. Check with local building officials to determine correct amount.
- Columns, piers, and pilasters transmit loads to the footing. Columns and piers are freestanding, while pilasters are built into the foundation wall.
- The building superstructure should be anchored to the foundation to resist high winds.

Know and Understand

Answer the following questions using the information provided in this chapter.

1. *True or False?* The foundation walls are usually wider than footings to help stop uneven settling.
2. A _____ foundation uses cylinders to support the structure.
   A. spread
   B. column
   C. steel
   D. pile
3. *True or False?* Pile foundations are *not* used in work performed by masons.
4. A _____ footing is a single footing supporting multiple columns.
   A. combined
   B. continuous
   C. plain
   D. stepped
5. A _____ footing supports a foundation wall or several columns in a row.
   A. combined
   B. continuous
   C. plain
   D. stepped
6. Footings, including stepped footings, must be _____ the frost line.
   A. below
   B. above
   C. at the same level as
   D. Stepped footings cannot be used with frost lines.
7. _____ footings are not part of the foundation.
   A. Combined
   B. Continuous
   C. Isolated
   D. Reinforced
8. The most common type of foundation wall is the _____ foundation.
   A. floating
   B. matt
   C. raft
   D. T

9. *True or False?* To dampproof basement walls, parge them with a thick coating of Portland cement plaster or mortar.
10. Brick masonry _____ are used to support large axial loads.
    A. columns
    B. piers
    C. pilasters
    D. plaster coves

Apply and Analyze

1. Discuss the two general types of foundations.
2. List the four functions of a foundation wall.
3. Discuss the process of applying the minimum thickness of parging on an exterior block wall. Also, discuss the time required to parge coats to achieve the required minimum thickness.
4. A. What is the recommended diameter of the anchor bolts?
   B. What should be the maximum spacing distance between anchor bolts?

Critical Thinking

1. Explain the process that would be required for a builder to construct a footing for a residential home that has a walkout lower level. The sloping grade for this area of the home is 24″. The footing will be 8″ × 16″ and needs to be kept below the frost line throughout the complete change in slope. The building code frost depth in this community is 4′ and the top of the footing shall be at the frost line depth or greater. Concrete blocks of 8″ × 8″ × 16″ will be used for the foundation walls. The local building code specifies the maximum height change of the footing to be 8″.
   A. Explain the process of what will occur with the footing in this area of the home.
   B. How many steps should there be in the footing to make the elevation change?
   C. What is the height of each step?
   D. What is the length of the steps?

Activities

1. (ELA) Locate the International Residential Code for One- and Two- Family Dwellings on the internet. For each topic, include the corresponding code number. An example is given. Then, choose one of the codes and be prepared to explain its contents in your own words to the class.
   A. Foundation waterproofing and dampproofing—concrete and masonry foundation waterproofing: R406.2
   B. Foundation drainage—concrete or masonry foundations
   C. Foundation and retaining walls—concrete and masonry foundation walls
   D. Footings—general
2. (TECH) The IRC lists "bituminous coating" as an option for dampproofing. Research a bituminous (asphalt) coating product and write a report that details its instructions for use.

Whiteaster/Shutterstock.com

CHAPTER 16

Wall Systems

LEARNING OUTCOMES

After studying this chapter, you will be able to:

16.1-1 Describe different types of masonry wall systems.
16.1-2 Explain the use of ties, flashing, weep holes, lintels, sills, jambs, and expansion joints in masonry wall systems.
16.1-3 Explain the effects of moisture condensation in masonry walls and the methods used to resist condensation.
16.1-4 Describe construction procedures for thin brick veneer, including thin bed set, thick bed set, and precast panels.
16.1-5 Construct a retaining wall.
16.2-1 Explain the design, function, and construction of arches.
16.2-2 Describe the joints used for movement in masonry.
16.2-3 Understand the function of caps and copings.

TECHNICAL TERMS

abutment
arch
bond beam
buck
building expansion (isolation) joint
camber
cap
centering
collar joint
construction joint (cold joint)
control joint
coping
corbel
curtain wall
depth
expansion joint
extrados
flashing
hardscape
insulated concrete forms (ICF)
intrados
jack arch
keystone
lintel
multicentered arch
panel wall
Quick Headers
racking
reinforced masonry walls
removeable masonry form
R-value
rise
segmental arch
segmental retaining walls (SRW)
semicircular arch
skewback
soffit
span
spandrel
spring line
springing
thermal mass
U-value
voussoir

Introduction

This chapter covers some of the common practices used in constructing masonry wall systems. Solid masonry walls, curtain and panel walls, hollow masonry walls, anchor veneered walls, composite walls, reinforced masonry walls, thin brick veneer, openings and arches, and garden walls are discussed. The information in this chapter is intended to aid in choosing appropriate building methods and procedures to construct masonry wall systems.

All buildings must comply with energy code requirements. Energy performance requirements are found in such documents as the US Department of Energy—*Energy Efficiency Building Energy Codes 101: An Introduction* and the RECA

(Responsible Energy Code Alliance) guides to the International Energy Conservation Code.

Most energy codes and standards specify a maximum permissible value for the coefficient of thermal transmittance (U-value) for the building envelope as a function of wall weight. ***U-value*** is the heat flow through a building product based on the difference between external and internal temperatures. The U-value is used in building energy calculations to determine heat loss through the materials of the structure.

Insulating materials used in construction are usually rated based on their ability to resist heat loss. This ability to resist heat loss, or ***R-value***, is the thermal resistance of heat loss through a building product and is the reciprocal of the U-value, which is shown as R = 1/U.

16.1 Masonry Walls and Components

Masonry walls can be single-wythe walls (one unit in thickness) or multiple-wythe walls. The walls may be reinforced or unreinforced. Masonry walls are classified as solid walls, four-inch RBM (reinforced brick masonry) curtain and panel walls, hollow walls, anchored veneered walls, composite walls, and reinforced walls.

Masonry walls are popular because they provide excellent structural performance, are easily maintained, and are attractive. They provide an envelope that is energy efficient due to their high thermal mass. ***Thermal mass*** is the characteristic of heat capacity and surface area capable of affecting building thermal loads by storing heat and releasing it at a later time. Materials with high thermal mass react more slowly to temperature fluctuations and therefore reduce peak energy loads.

Solid Masonry Walls

A solid masonry wall is a wall that is built of masonry units laid close together. All joints between the units are filled with mortar. Solid masonry units, hollow masonry units, or a combination of the two can be used.

Solid masonry walls may be either loadbearing or nonloadbearing. The structural bond of these walls is provided by metal ties, masonry headers, or joint reinforcement. See **Figure 16-1**.

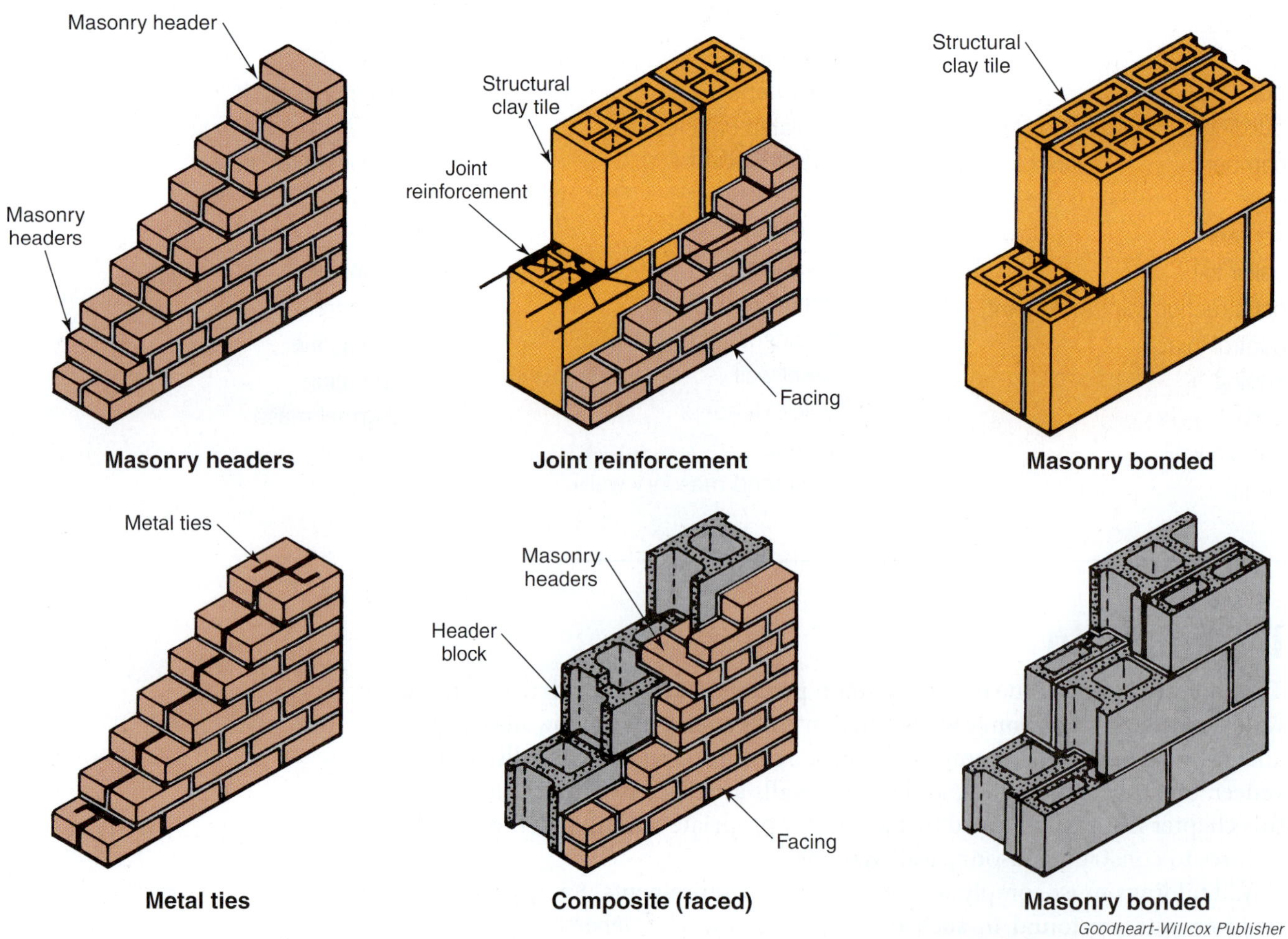

Figure 16-1. Typical solid masonry walls using brick, concrete block, and tile.

Solid masonry walls can be classified by the types of units used in their construction. The following are examples of this classification:

- Solid units of brick or concrete brick and block.
- Hollow units of concrete block or structural clay tile.
- Composite (faced) walls composed of facing and backup units of different materials bonded in such a way that both facing and backup are loadbearing. Header blocks allow masonry headers to overlap block. Refer to the bottom center illustration in **Figure 16-1**.

Nationally recognized building codes, such as ICC and MSJC, permit the use of exterior loadbearing 6″ masonry walls for one-story, single-family dwellings where the wall height does not exceed 9′ to the eaves and 15′ to the gable peak. **Figure 16-2** shows typical sections of 6″ walls. The masonry units most commonly

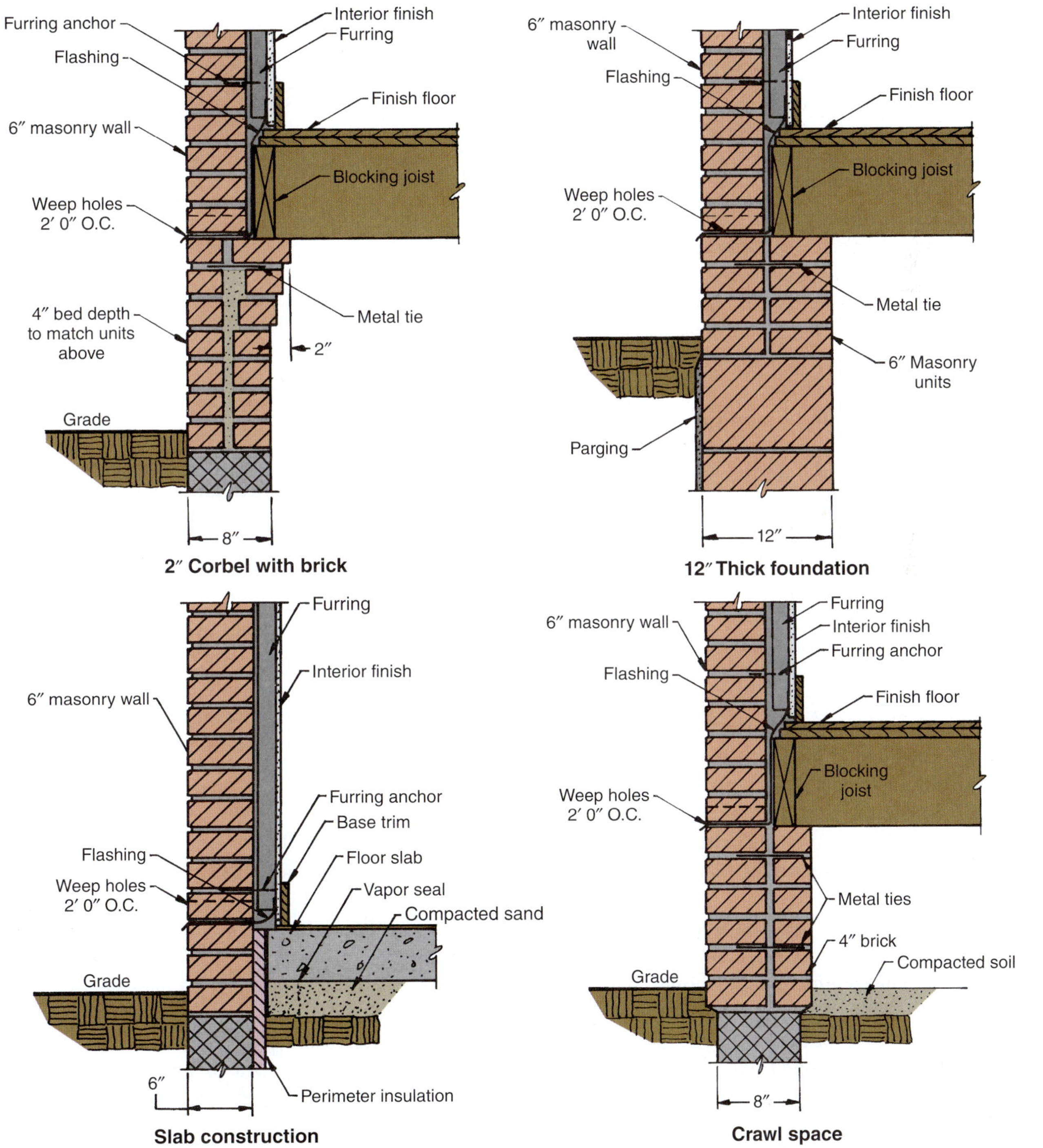

Goodheart-Willcox Publisher

Figure 16-2. Details of 6″ masonry wall construction using SCR brick.

used for this type of construction are SCR (Structural Clay Research) bricks. SCR is a trademark of the Structural Clay Products Institute (now the BIA).

SCR brick has a nominal thickness of 2 2/3″ that produces 16″ in six courses. It is a nominal 6″ wide and 12″ long. These dimensions make it easier to lay in 1/2″ bond. SCR brick is also produced in thicknesses of 3″ (four courses in 12″), 3 1/5″ (five courses in 16″), 4″ (four courses in 16″), and 5 1/3″ (three courses in 16″). Clay tiles are produced in 6″ sizes that can be used with SCR brick.

Four-Inch RBM Curtain and Panel Walls

Exterior nonloadbearing walls may be curtain walls or panel walls. A ***curtain wall*** is not wholly supported at each story. Curtain walls can be anchored to columns, spandrel beams, floors, or bearing walls, but are not necessarily built between structural elements. These walls must support their own weight for the height of the wall. A ***panel wall*** is supported at each story and self-supporting between stories. Both curtain walls and panel walls must be able to resist lateral forces, such as wind pressures. The walls transfer these forces to adjacent structural members.

Several factors need to be considered in the design of 4″ curtain or panel walls. These factors include structural considerations, moisture control, and differential movement.

Four-inch walls are designed as partially reinforced walls. The horizontal joint reinforcement in the wall resists tensile stresses that result from lateral pressures as the wall spans horizontally between structural elements such as columns, pilasters, or cross walls. See **Figure 16-3**.

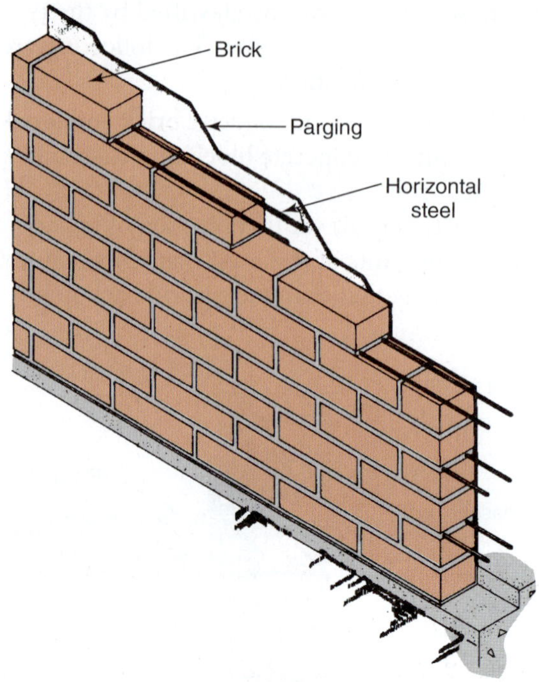

Brick Industry Association

Figure 16-3. Horizontal joint reinforcement is used between every other course in this single-wythe wall.

Lateral loads caused by wind can be either pressure or suction. Therefore, walls must be designed to resist the assumed lateral load acting in either direction. The maximum area of reinforcement should be provided throughout the length of the wall. Lapping steel reinforcement a minimum of 16″ ensures the development of tensile stresses without exceeding the allowable bond stresses.

Ladder- or truss-type joint reinforcement provides two longitudinal bars connected by cross wires. See **Figure 16-4**. The reinforcement should have mortar coverage on both top and bottom and should not be laid directly on top of the brick. For proper vertical placement, the reinforcement wires can be supported on small mortar pads placed prior to full bed-joint mortar placement.

Water may penetrate a 4″ wall due to heavy rains driven by high winds. Water penetration can be managed by providing drainage space on the inside of the wall. This drainage space permits the water to flow to the flashing at the base and to be conducted back to the outside through weep holes. Another method is to apply a water barrier (parging) to the inside of the wall.

Expansion joints may be located at or near corners, offsets, junctures, and openings. Since 4″ walls are considered to span horizontally, any expansion joints must be located at vertical supports. The maximum recommended spacing for expansion joints for 4″ walls is 100′ for a straight wall without openings. However, conditions may require expansion joints as close together as 40′.

Hollow Masonry Walls

Hollow masonry walls are walls built using solid or hollow masonry units, which are separated to form an inner and an outer wall. The two types of hollow masonry walls are cavity walls and masonry bonded type walls.

Cavity Walls

Masonry cavity walls are widely used throughout the United States in all types of low-rise and high-rise buildings. Their popularity is due to their superior rain penetration resistance, excellent thermal capabilities, good sound transmission resistance, and high fire resistance.

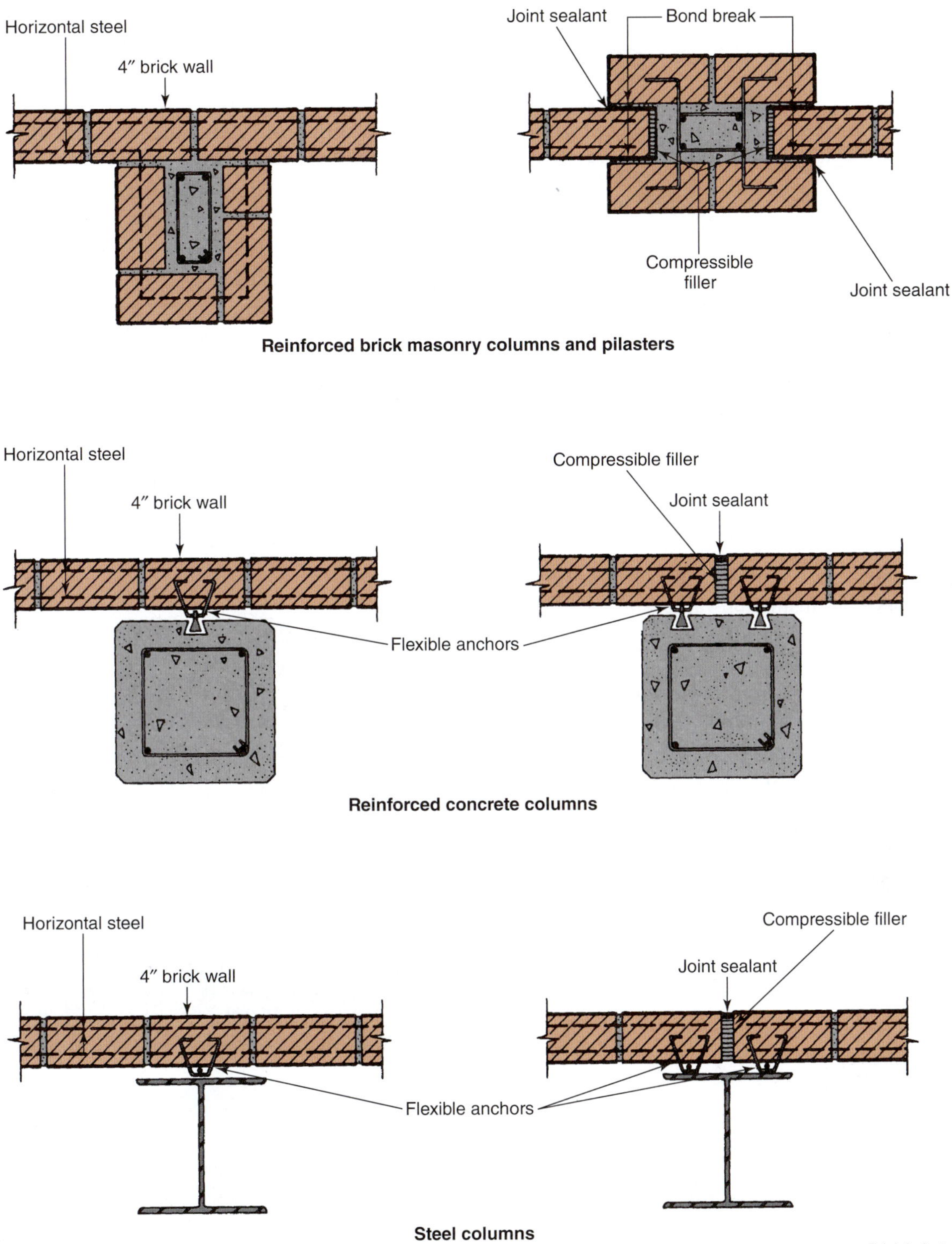

Brick Industry Association

Figure 16-4. Typical 4″ reinforced masonry wall details.

A cavity wall is built of masonry units arranged to provide a continuous air space of 2″ to 3″ wide. See **Figure 16-5**. The facing and backing wythes or tiers are connected with rigid metal ties. The exterior wythe is usually a nominal 4″ thick. The interior wythe may be 4″, 6″, or 8″, depending on the load to be supported and the height and length of wall. The exterior masonry wythe is generally solid or hollow brick, and the interior masonry wythe can be brick, structural clay tile, or hollow or solid concrete masonry units depending on the properties required. When a cavity of nominal 2″ is maintained, the overall thickness will be 10″, 12″, or 14″.

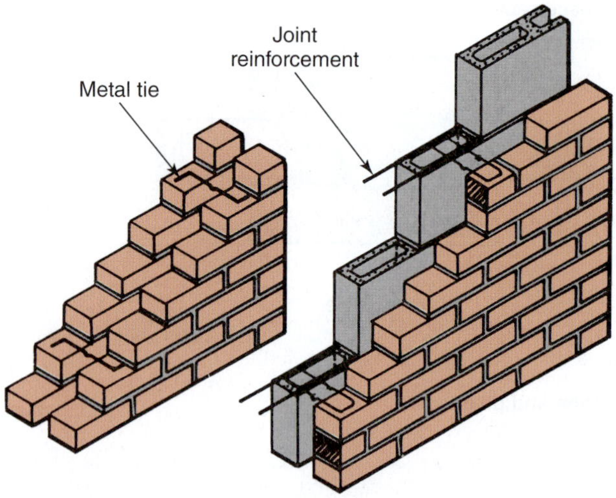

Goodheart-Willcox Publisher

Figure 16-5. A method of cavity wall construction that uses metal ties and joint reinforcement.

The cavity provides two advantages:

- The air space has insulation value, and it can be filled with insulation material for added reduction of heat transfer.
- The cavity acts as a barrier to moisture.

However, to be effective, the cavity must be kept free of mortar droppings during construction. A board may be used to collect the droppings as shown in **Figure 16-6**.

When weep holes are required at the bottom of a cavity wall, flashing should be used. A weep hole is a location in a mortar joint where a device is placed to allow moisture inside the wall cavity to exit to the outside. In the past, a brick's head joint mortar was left out as a location to drain moisture. This creates a void, or opening, in the masonry wall where insects or moisture can enter the structure. Now, devices are used in the head joint to allow moisture to exit. Two different types of weep holes are shown in **Figure 16-7**. The top image is a stainless steel screen weep hole installed in the head joint; the bottom image shows a cotton weep hole added into the mortar joint. The weep hole is usually located immediately above the flashing.

Flashing is a thin, impervious material, usually made of metal, which is placed in mortar joints to prevent water penetration and to direct the drainage of water. The flashing keeps any moisture that might collect in the cavity away from the inner wall. See **Figure 16-8**.

One of the major functions of an exterior wall is to resist rain penetration. A properly designed and built masonry cavity wall is totally resistant to rain

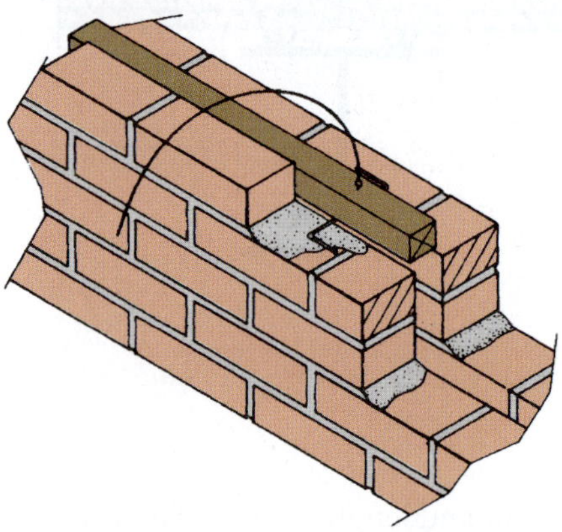

Goodheart-Willcox Publisher

Figure 16-6. A recommended method of keeping a cavity free of mortar droppings is to rest a board on the joint reinforcement (wire ties). The board is moved up when the next reinforcement level is reached.

Timothy L. Andera

Figure 16-7. Weep holes provide an outlet for moisture that has collected in the space between wythes of masonry. Weep holes sit directly above the flashing to aid in moisture removal from the cavity wall.

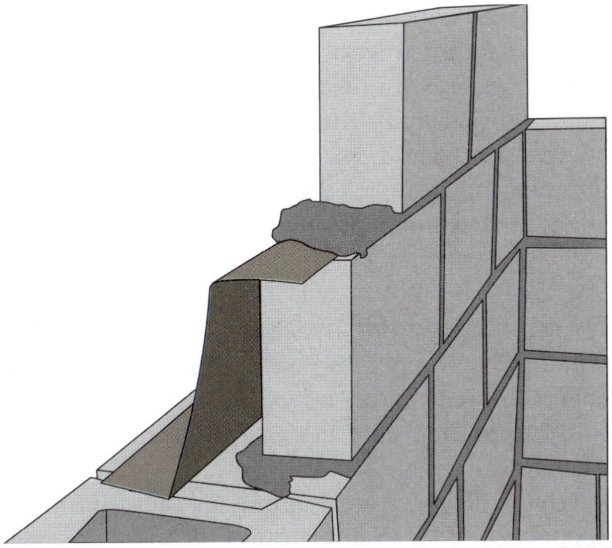

Goodheart-Willcox Publisher

Figure 16-8. Flashing should be installed to prevent moisture from entering the inner wall from the cavity.

penetration through the wall assembly. The outside wythe may not be totally resistant to moisture penetration, but the overall design of the cavity wall assembly takes this into account. In a cavity wall, any moisture that passes through the exterior wythe runs down the cavity face of that wythe. At the bottom of the cavity, the moisture is diverted to the outside by continuous flashing and weep holes. Due to the decreased moisture problem, there is less chance of efflorescence than with solid walls.

Heat losses and heat gains through masonry walls are minimized by the use of cavity wall construction. The complete isolation of the exterior and interior wythes by the air space allows a large amount of heat to be absorbed and dissipated in the outer wythe and cavity before reaching the inner wythe and the building interior. This ability is increased by the use of insulation in the cavity. Thus, considerable energy savings can be realized by proper design, detailing, construction, and use of cavity walls.

Resistance to transmission of sound is accomplished with heavy massive walls or discontinuous construction. A cavity wall employs both of these—the weight of the two masonry wythes plus the partial discontinuity of the cavity.

In cavity wall construction, an air cavity provides a partial isolation between the two wythes. Sound on one side of the wythe strikes it and causes it to vibrate, but because of the separation and cushioning effect of the cavity, plus the massiveness of the wythes, the vibration of the other wythe is greatly reduced.

The results of ASTM E119 fire resistance tests show that masonry cavity walls have excellent fire resistance. Fire resistance ratings of cavity walls range from two hours to four hours, depending on the wall thickness and other factors. Due to their high fire resistance properties, brick walls make excellent fire walls for compartmentation (division into separate sections) in buildings. Compartmentation can halt the spread of fire.

Cavity walls that are properly designed, detailed, and constructed can be used in any building requiring loadbearing or nonloadbearing walls in the same manner as other masonry walls. Separation of the wythes and the use of metal ties results in increased flexibility that permits freedom of differential movement between the wythes. This is critical in construction that makes use of combinations of dissimilar materials. Refer to Chapter 12, **Laying Brick** for cavity wall construction procedures.

Suitable types of insulation materials for cavity walls are rigid boards and granular fills. Each of these, if properly used, produces a thermally efficient wall. See **Figure 16-9**.

Expanded or molded polystyrene, expanded polyurethane, rigid urethane, cellular glass, preformed fiberglass, and perlite board are used as rigid board insulation materials in cavity walls. See **Figure 16-10**. Rigid board insulation is installed horizontally within the air space against the cavity face of the backup wythe. A minimum of 1″ should be left between the cavity face of the external wythe and the insulation board. The 1″ space between the outer wythe and the insulation board facilitates the wall construction and

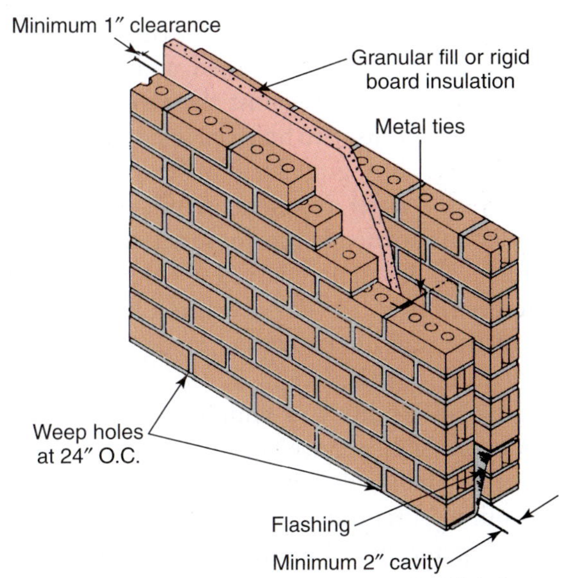

Brick Industry Association

Figure 16-9. Rigid board insulation in a cavity wall.

robcocquyt/Shutterstock.com

Figure 16-10. Rigid insulation in the cavity between masonry walls. The insulation seam between sheets is sealed with special tape to reduce air infiltration. Wall ties protrude through rigid insulation to tie both walls together.

allows for drainage of the cavity. During installation, ensure that all boards are abutted and installed between ties, and fit flush against the inner wythe. An adhesive should be used to hold the insulation in place.

Two types of granular fill insulation are recommended—water repellent vermiculite masonry fill and silicone-treated perlite loose fill insulation.

Granular fill insulation is usually poured directly into the cavity from the bag or from a hopper placed on top of the wall. Pours can be made at any convenient interval, but the height of any pour should be kept below 20′. Rodding or tamping is not necessary, and could possibly reduce the thermal resistance of the material. The insulation in the wall should be protected from weather during construction.

The facing and backing (adjacent wythes) of cavity walls should be tied together with corrosion-resistant 3/16″ diameter steel ties, or metal tie wire of equivalent stiffness, embedded in the horizontal mortar joints. There should be at least one metal tie for each 4 1/2 sq ft of wall area. Individual ties in alternate courses should be staggered—the maximum vertical distance between ties should not exceed 24″, and the maximum horizontal distance should not exceed 36″. The ends of the ties should be bent to 90° angles to provide hooks not less than 2″ long, or ties bent to a rectangular shape should be used. Additional ties should be provided at all openings spaced not more than 3′ apart around the perimeter and within 12″ of the opening. Where the cavity width exceeds 3 1/2″ but is less than 4 1/2″, there should be at least one metal tie for each 3 sq ft of wall area.

Flashing details are important in cavity wall construction. In order to divert moisture out of the cavity through the weep holes, continuous flashing should be installed at the bottom of the cavity, and wherever the cavity is interrupted by elements such as shelf angles or lintels.

Flashing should be placed over all wall openings not protected by projecting hoods and eaves. It should also be properly placed at all window sills, parapet walls, and spandrels. (In a multistory building, the ***spandrel*** is the space between the top of a window of one story to the sill of the window on the second story.)

Since the purpose of the flashing is to collect moisture so it can be diverted to the outside, weep holes must be provided wherever flashing is used. Weep holes are located in the joints of the outer wythe immediately above the flashing. Spacing of weep holes should be approximately 2′ O.C. maximum, except for those using a wick material that should be 16″ O.C. maximum.

The movement of an outer brick wythe due to thermal and moisture expansion can be greater than the movement in solid or composite walls exposed to the same environment. This is due to the greater differences between the mean maximum and mean minimum temperatures of the outer wythe of the cavity wall, as well as the absence of restraint usually provided by dead and live loads, masonry bonders, or filled collar joints in solid walls. For this reason, it is recommended that expansion joints be provided through the outer wythe of the cavity wall on each side of an external corner where the walls are 50′ or more in length. Where possible, the expansion joint can be placed in line with the jambs of the windows nearest the corner.

Hollow masonry bonded walls are walls used as foundation and exterior loadbearing walls. They are popular in some parts of the United States. Though economical, they are not resistant to high moisture. See **Figure 16-11**.

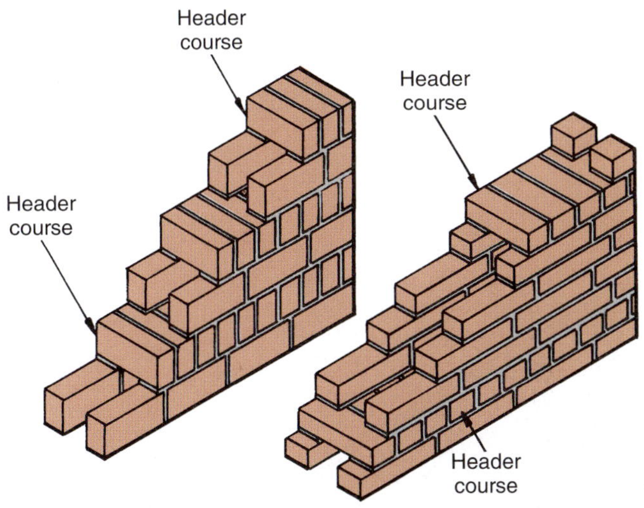

Figure 16-11. Hollow masonry bonded walls.

Anchored Veneered Walls

Brick and stone masonry units are widely used as facing veneer. See **Figure 16-12**. In this application, the loadbearing properties of the materials are not used. The veneer is attached to the backing, but does not act structurally with the rest of the wall.

In anchored brick veneer construction, a nominal 3″ or 4″ thick exterior brick wythe is anchored to a backing system with metal ties. There is a clear air space between the veneer and the backing system. The backing system may be constructed of wood frame, steel frame, concrete, or masonry.

A veneer wall has a facing of masonry units or other weather-resisting, noncombustible materials. The facing is securely attached, but not bonded to, the backing.

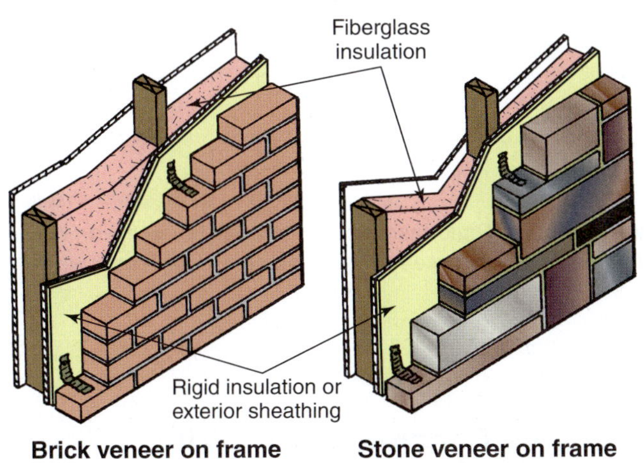

Figure 16-12. Brick and stone veneer attached to frame construction using metal ties. These veneers must rest on an extension of the foundation.

See **Figure 16-13**. Brick veneer is designed to carry loads due to its own weight, but also needs to resist wind and seismic conditions that contribute stresses to the brick veneer. The manufacturing and installation of brick should conform to ASTM C62, ASTM C216, or ASTM C652 for building brick, facing brick, and hollow brick, respectively.

Two sources that discuss brick veneer design, procedures, and practices are MSJC Code Chapter 12 Veneer and BIA Technical Notes on Brick Construction 28B, Brick Veneer/Steel Stud Walls. Both are valuable resources when planning brick veneer.

Trade Tip
Salvaged brick is not recommended for use as brick veneer.

At one time, brick veneer construction was mainly limited to wood frame houses. It is now being used on low-rise commercial and institutional construction and is frequently used for high-rise buildings, especially with concrete masonry or steel stud backing systems.

Veneer wall construction looks much like cavity wall construction in its final form. However, the construction and function are considerably different. For example, the supporting structure in a veneer wall is completed before attaching the veneer, which is not loadbearing. Both wythes in a cavity wall are loadbearing and are built at the same time.

Moisture Resistance

Brick veneer wall assemblies are drainage-type walls. These walls are resistant to rain penetration. Proper drainage requires maintenance of the clear air space between the brick veneer and the backing. Flashing and weep holes work with the air space to resist moisture penetration. Brick veneer with wood or metal frame backing is usually built with a minimum of 1″ of air space.

Brick Veneer Foundations

Brick veneer on a frame backing transfers the weight of the veneer to the foundation. Typical foundation details for brick veneer are shown in **Figure 16-14**. The foundation brick ledge supporting the brick veneer should be at least equal to the total thickness of the brick veneer wall assembly.

Foundations must extend beneath the frost line. Check with your local building official to determine the required depth of the footing as required by local code. Foundation design considerations include differential

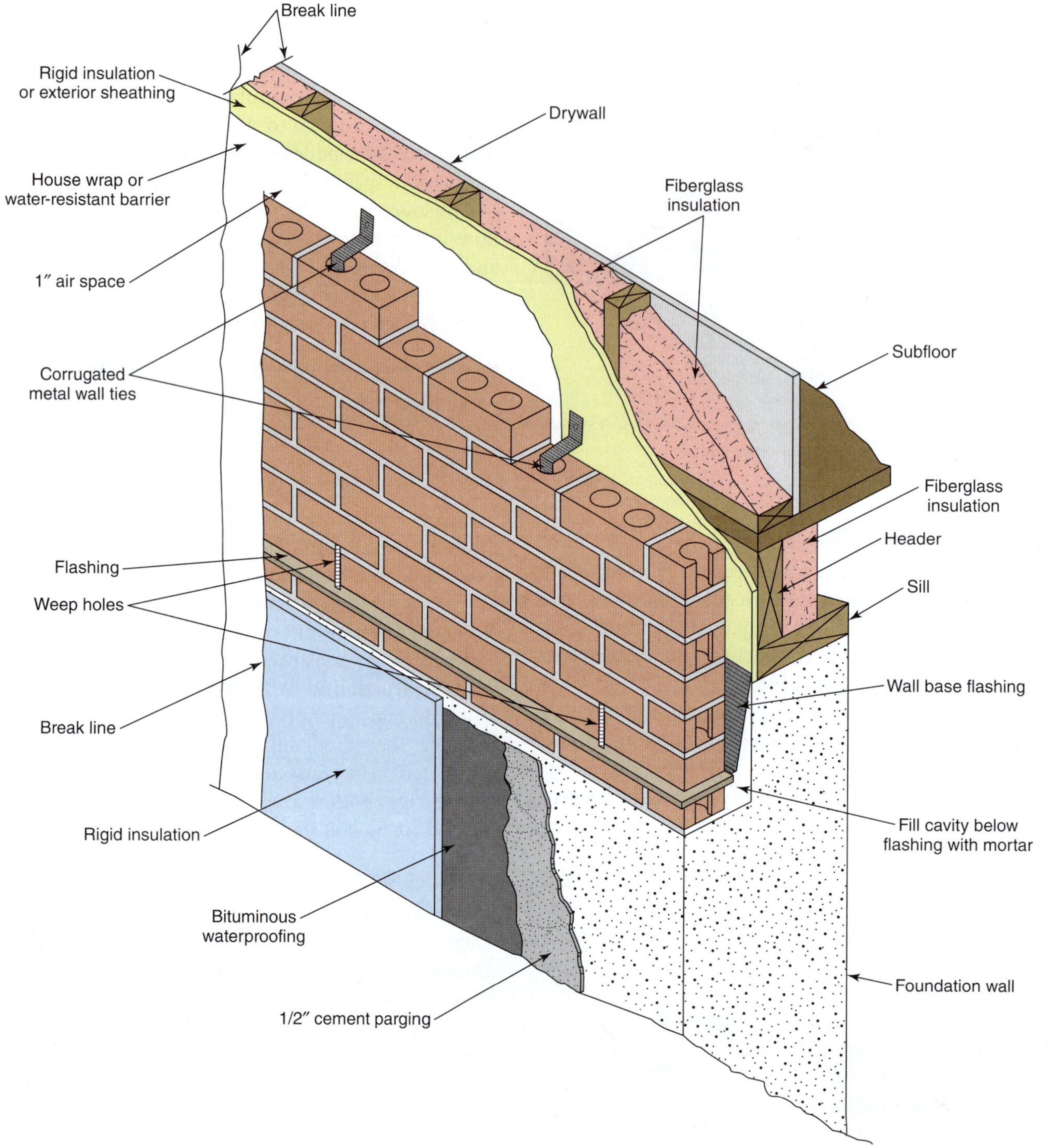

Goodheart-Willcox Publisher

Figure 16-13. Brick veneer attached to wood framing using corrugated wall ties. Rigid insulation over wood framing with house wrap is being used as air infiltration and a water resistant barrier. A 1″ air space between house wrap and brick veneer with flashing directs moisture to weep holes.

settlement and the effect of concentrated loads, such as those from columns or fireplaces. Brick walls that enclose crawl spaces must have openings to provide adequate ventilation. Openings should be located to achieve cross ventilation.

Insulated Concrete Form Walls

New types of energy-efficient construction utilize ***insulated concrete forms (ICF).*** ICFs are made of polystyrene beads that are expanded to create insulated blocks. The interior and exterior insulated panels are held

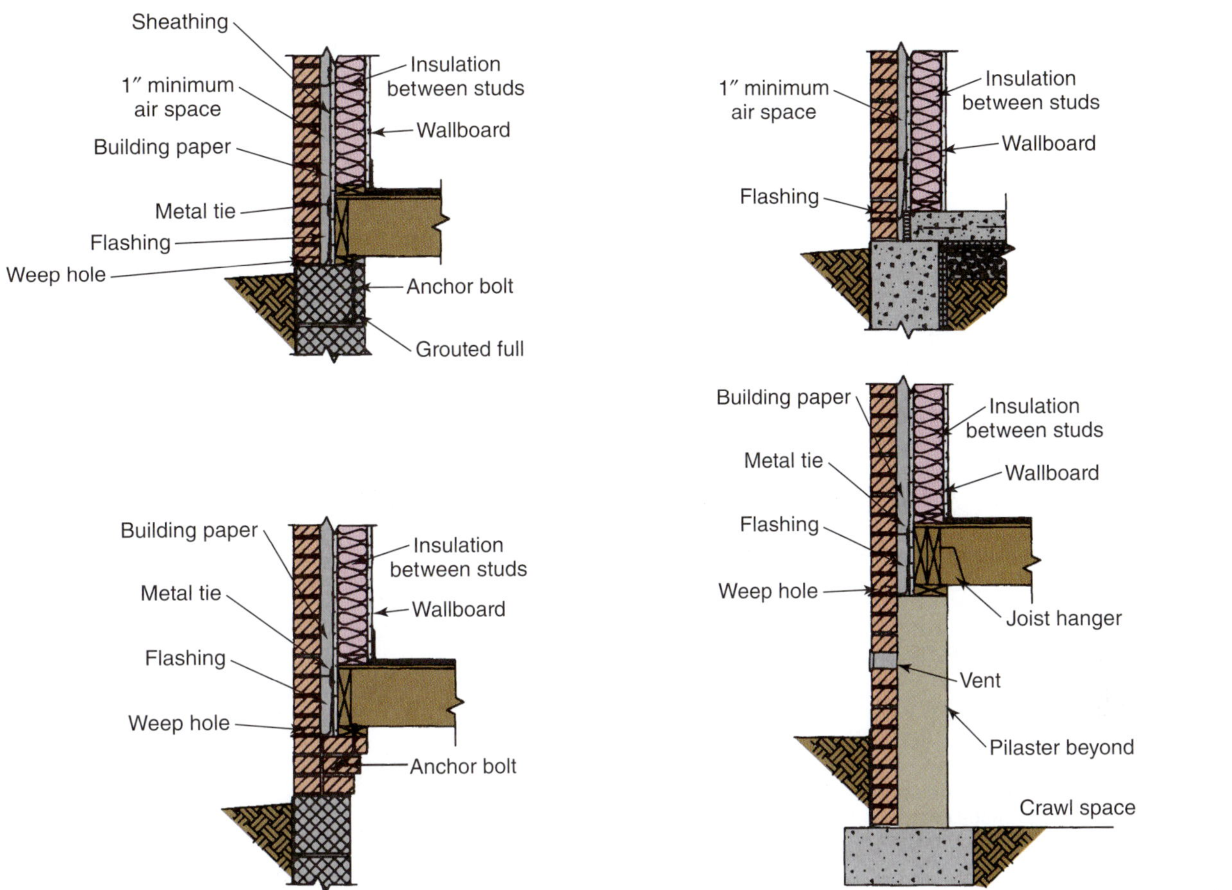

Brick Industry Association

Figure 16-14. Brick veneer details.

together using polypropylene ties. These ties create an internal structure with reinforcing bars placed on top of the ties inside the forms. Horizontal and vertical reinforcement is used to strengthen walls prior to filling the ICF structure with concrete. The exterior sizes of ICF blocks vary from one company to another. As an example, Fox Blocks' standard block is 16″ tall and 48″ long. The internal width between the panels of the ICF blocks varies from 4″ to 12″ and it depends on the type of wall and the strength requirements of the structure.

The ICF building blocks are stacked on top of each other to fabricate a wall structure, similar to toy building blocks. The ICF building blocks should always be stacked according to the manufacturer's directions. **Figure 16-15A** shows a detailed drawing of an ICF wall with face bricks attached to the wall system. A corbel block is positioned at the appropriate location in the ICF wall structure to create a ledge. The interior lower width of the corbel block is 8″ and the upper part is 12″ wide. This forms a 4″ a ledge as a location to lay brick or stone to create a veneered wall structure. **Figure 16-15B** shows the same wall structure but in a 3-D drawing. The drawing displays the external and internal wall structure. On the inside of the building, wood floor framing is attached to the ICF wall structure. The ICF wall is made of foam block, polypropylene ties, rebar, and concrete to support and add strength to the wall system.

Figure 16-16 shows an ICF wall system being constructed from start to completion. The following are the steps of the ICF building process:

1. ICF blocks are placed to create a wall.
2. Internal structure of the ICF with polypropylene ties and rebar is placed in the first and third course.
3. Walls are erected with braces and a walkway to be used to support workers placing concrete in the wall cavity.
4. Concrete is placed in the ICF wall cavity using a concrete pump truck.
5. Once the concrete has set in interior walls, plumbing and electrical runs are placed by cutting out insulation.
6. The structure is now complete.

ICF wall systems will be discussed in greater detail in Chapter 19, **Form Construction**.

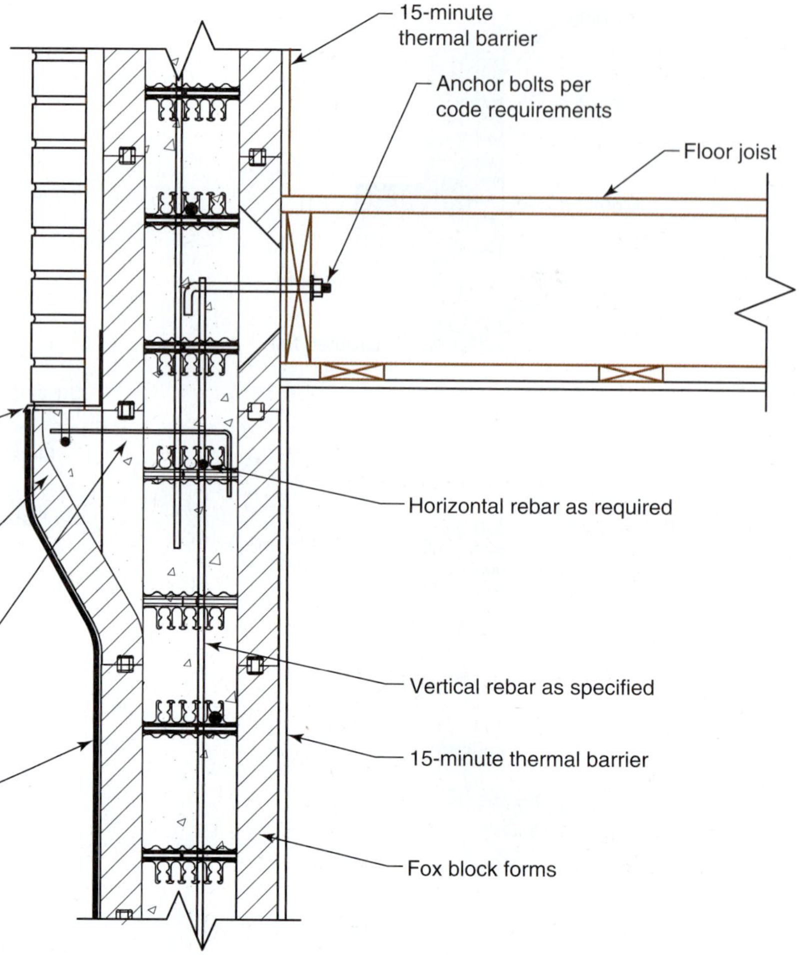

A

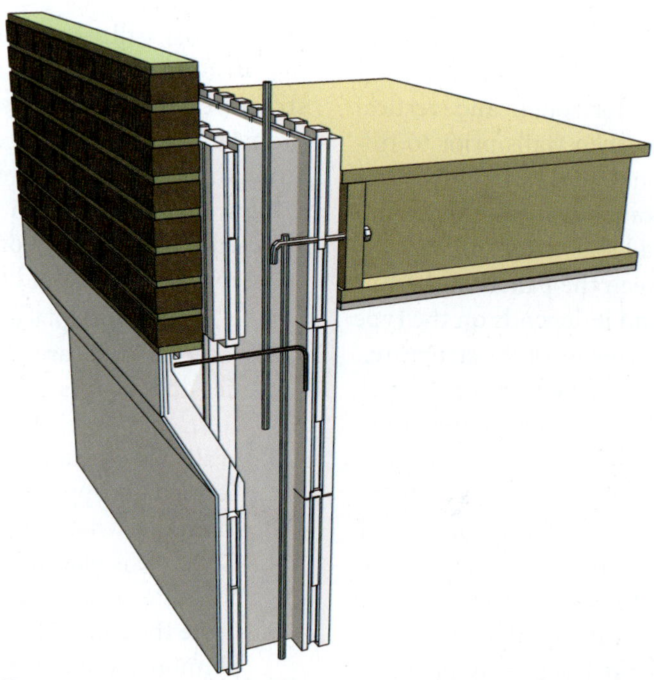

B

Fox Blocks by Airlite Plastic Co.

Figure 16-15. A—ICF detailed drawing showing the external and internal structure of the insulated block wall. B—A 3-D drawing of the same insulated block wall, showing the different components of the wall system.

Fox Blocks by Airlite Plastic Co.

Figure 16-16. ICF building being constructed.

Masonry Wall Applications

Several ties used with wood framing are shown in **Figure 16-17**. Ties, as well as nails, are to be made of a noncorrosive material. There should be one tie for every 2 2/3 sq ft of wall area with a maximum spacing of 24″ O.C. in either direction. The nail attaching a corrugated tie must be located within 5/8″ of the bend in the tie. The best location for the nail is at the bend in the corrugated tie. The bend should be 90°.

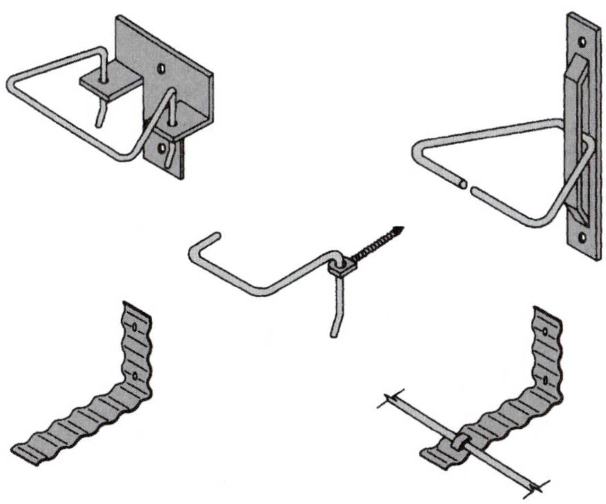

Brick Industry Association

Figure 16-17. Typical ties used with brick veneer on frame construction.

Code Note **Wall Ties**

The International Residential Code (IRC) **R606.3.3** specifies that wall ties are not to be bent after being embedded in grout or mortar.

Wire ties must be embedded at least 5/8″ into the bed joint from the air space. There must be at least a 5/8″ cover of mortar to the exposed face. Corrugated ties must penetrate to at least half the veneer thickness and have at least 5/8″ mortar cover. Ties should be placed so the portion within the bed joint is completely surrounded by the mortar.

Flashing and Weep Holes

Flashing and weep holes must be positioned above grade but as close to the bottom of the wall as possible. In addition, flashing and weep holes should be located as close as possible above openings and below sills. Locate weep holes in head joints directly above the flashing to ensure adequate drainage. Weep holes in which the mortar has been completely removed should be spaced no more than 24″ O.C. Weep holes in which wicking material has been used should be spaced at a maximum of 16″ O.C.

When brick veneer is placed close to the base of the wall and under the flashing, the space between the brick and the backing needs to be filled with grout to the height of the flashing. This practice fills the void below the flashing so moisture does not collect under the flashing in the wall cavity.

When installing flashing, attach it securely to the backing system and then extend through the brick face. The flashing should extend at least 8″ up the wall to ensure adequate moisture protection.

Lintels, Sills, and Jambs

Unless the masonry is self-supporting, brick veneer that is backed by wood or metal frame must be supported by lintels over openings. Loose steel, stone, precast lintels, or lintels formed on the construction site should bear at least 4″ at each jamb. Space at the end of the lintel is required to allow for expansion. The clear span for 1/4″ thick steel angles varies between 5′ and 8′, based on the size of the angle selected and the rating of the steel. Concrete, cast stone, and stone lintels need to be the appropriate size in order to bear the weight of the veneer. Steel lintels with spans greater than 8′ may need lateral bracing for stability. Reinforced brick lintels can also be used.

Expansion Joints

Expansion joints may be needed to allow for horizontal movement. Expansion joints are required in brick veneer when there are long walls, walls with returns, or large openings.

Composite Walls

A composite wall is two wythes bonded together with masonry or wire ties. See **Figure 16-18**. The two wythes are joined together in a continuous mass using a vertical collar joint. A ***collar joint*** is the narrow space between the facing units and the backup units in a wall. See **Figure 16-19**. This joint prevents the passage of water through the wall.

When building an 8″ composite wall, the first course of facing may be either headers, as shown in **Figure 16-20**, or stretchers.

It is important that all facing courses be laid in a full mortar bed with the head joints filled completely.

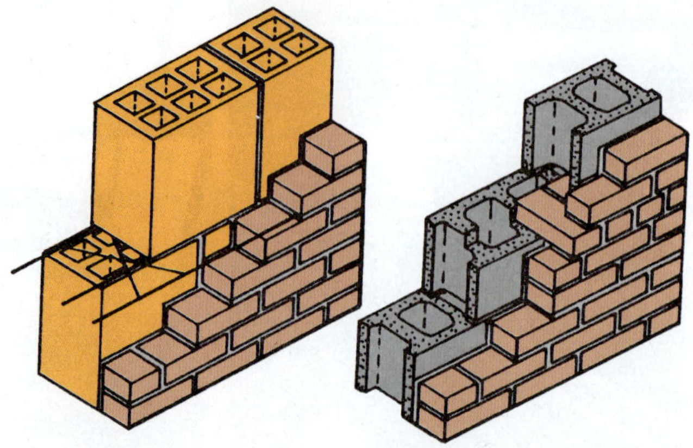

Goodheart-Willcox Publisher

Figure 16-18. Composite walls can be bonded together with masonry or joint reinforcement.

Portland Cement Association

Figure 16-19. The outer and inner wythes of masonry are bonded together by the mortar in the collar joint between them.

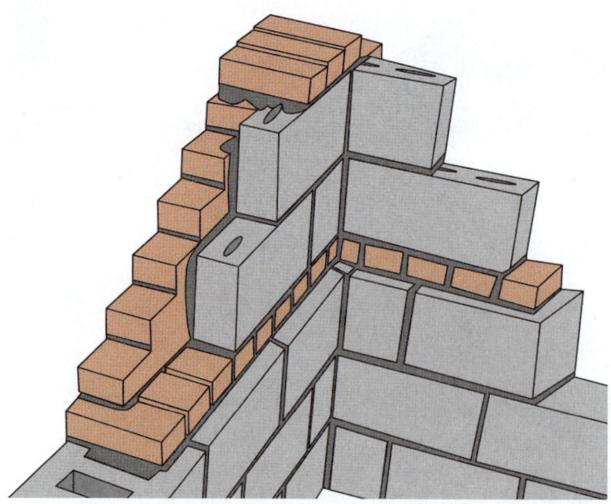

Goodheart-Willcox Publisher

Figure 16-20. Building a composite wall with six courses of brick between each header course.

Mortar that is extruded (squeezed out) on the backside of the facing units should be cut flush with the trowel before it hardens. See **Figure 16-21**.

Parging the backside of the facing is another method of bonding the wythes across the collar joint. See **Figure 16-22**.

Facing headers are laid every seventh course in an 8″ composite wall.

A 12″ composite wall is constructed in a manner similar to the 8″ composite wall. **Figure 16-23** shows the facing header course being laid overlapping the header block. The header block can be laid with the recessed notch up or down, depending on construction requirements.

Figure 16-24 shows the seventh course backup wythe being constructed using stretcher block. Header blocks were used for the sixth course bonding.

Goodheart-Willcox Publisher

Figure 16-23. Header course being laid in composite wall.

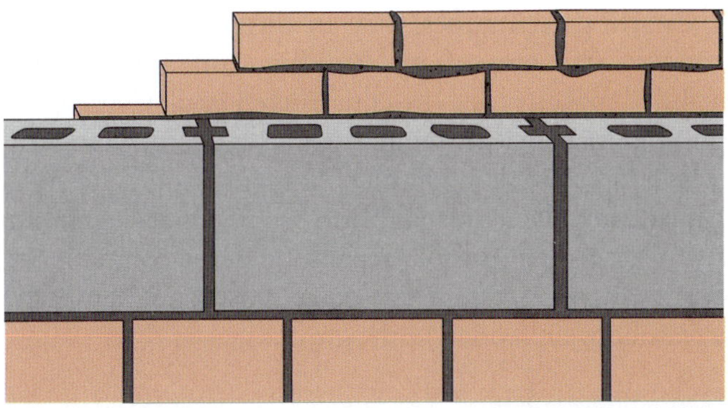

Goodheart-Willcox Publisher

Figure 16-21. These well-placed facing units have extruded mortar on the backside in a composite wall. The mortar must be cut flush with a trowel.

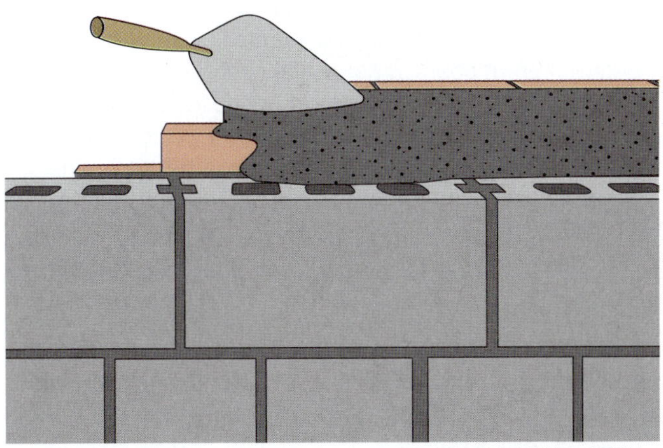

Goodheart-Willcox Publisher

Figure 16-22. Bonding a wall across a collar joint by parging the backside of the facing.

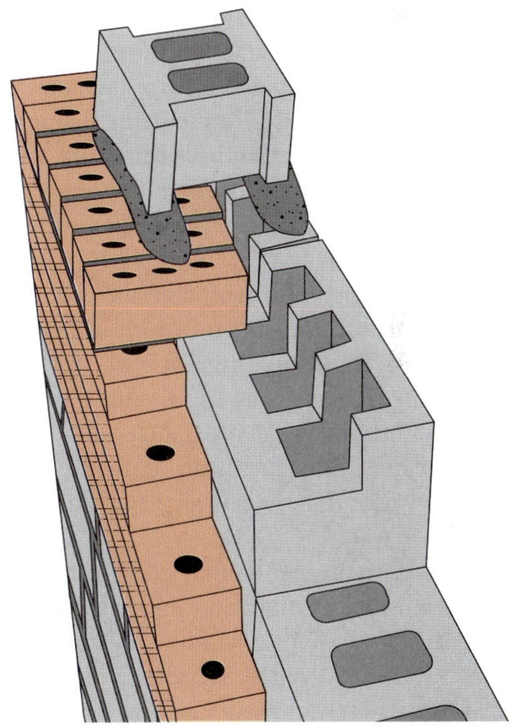

Goodheart-Willcox Publisher

Figure 16-24. Stretcher block being placed as the backup wythe.

Resisting Moisture Condensation

Water vapor condensation can cause moisture problems in masonry walls and should be minimized where possible. When water vapor passes through pores of homogenous walls, which are warm on one side and cold on the other, the vapor may condense into water inside the wall. But if the flow of vapor is blocked by a vapor-resistant material in the wall, the vapor may not reach the point in the wall where the temperature is low enough to cause condensation.

Condensation problems most frequently occur during the heating season in tight, highly insulated buildings. The interior air gains moisture, increasing the interior vapor pressure above the vapor pressure outside. The difference in pressures drives the vapor outward from the building through vapor porous materials in the wall assembly. A vapor barrier or retarder can be used to impede the migration of the water vapor. The vapor pressure differential across the wall section can be decreased through ventilation or dehumidification.

In newer construction, high-moisture locations, such as bathrooms or kitchens, may have mechanical devices that remove high moisture-laden air. Another device used to reduce heat loss in certain buildings is a heat exchanger. A heat exchanger takes moisture-laden air and exchanges it with outside air, which has lower moisture content. As part of the exchanger process, the higher temperature moisture-laden air transfers heat to the colder exterior air being brought into the exchanger. A barrier exists between inside and outside air so they do not mix together, but heat is transferred to the incoming air to increase energy efficiency and air quality.

Water affects many building materials. For example, wood expands with increasing moisture content. Varying humidity conditions in different parts of the cross-section of a single wood framing member may cause warpage. High humidity can also cause wood to decay. Water also promotes the corrosion of metal. The insulating value of most materials is greatly reduced by the presence of free water.

Changes in the volume of fired clay masonry units and concrete masonry units occur due to gains in moisture content. These changes should be considered in the design process. Alternate freezing and thawing of masonry units when saturated may eventually result in cracking and spalling. Moisture caused by condensation may contribute to efflorescence if soluble salts are present in or come in contact with brick.

Reinforced Masonry Walls

Reinforced masonry walls are walls built with steel reinforcement embedded in the masonry units. See **Figure 16-25**. The walls are structurally bonded by grout that is poured into the cavity (collar joint) between the wythes of masonry. The grout core seals the space between the wythes of masonry and bonds the reinforcing steel. Full bed joints are used for reinforced walls.

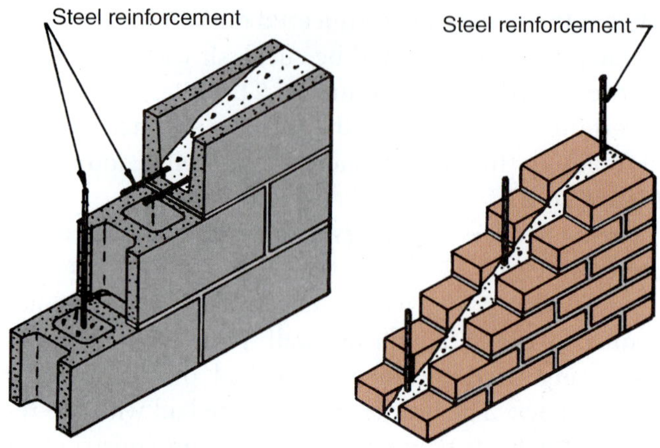

Figure 16-25. Reinforced masonry walls of brick and concrete block.

Grouting techniques vary in different areas of the country. The International Building Code specifies that grouted masonry is to be constructed so all elements of grouted masonry act together as a structural element.

Grout materials and water content should be controlled and mixed thoroughly to provide sufficient fluidity for placement without separation of the components. Reinforcement should be clean, positioned properly, and solidly embedded in the grout.

> **Code Note** — **Grouting**
>
> The International Building Code **2104A.1.3.5** specifies that prior to grouting, all spaces to be filled with grout should not contain mortar projections greater than 1/4″ (6.4 mm), mortar droppings, or foreign material. Spaces to be grouted should not contain voids.
>
> The grouting of a section of wall should be completed in one day with no interruptions greater than one hour. Between grout pours, a horizontal construction joint should be formed by stopping all wythes at the same elevation. The grout should stop at least 1 1/2″ (38 mm) below a mortar joint, except at the top of the wall. Where bond beams occur, the grout pour should be stopped a minimum of 1/2″ (12.7 mm) below the top of the masonry.

Reinforced masonry walls should be reinforced with an area of steel not less than 0.002 times the cross-sectional area of the wall. Not more than two-thirds of this area may be used in either direction. Maximum spacing of principal reinforcement should not exceed 48″. Horizontal reinforcement should be placed in the top of footings, the bottom and top of wall openings, roof and floor levels, and at the top of parapet walls.

The primary use of steel reinforcement is in vertical members, such as columns and walls, lintels, and bond beams. A ***bond beam*** is a device used to strengthen a masonry wall at specific locations. Special blocks, reinforcement, and grout or mortar are used to create a bond beam. See **Figure 16-26**. A bond beam is usually placed at the top of the wall. In some instances, a second bond beam is added in tall walls to strengthen the masonry wall. **Figure 16-27** shows a bond beam block. In this instance, a portion of the block ends have been removed to aid in placing reinforcement in the wall. The bottom of the block is open except for a portion of one cross web that connects the two faces of the block together. Typical methods of reinforcing concrete masonry are shown in **Figure 16-28**.

Timothy L. Andera

Figure 16-26. An example of a bond beam under construction. Two horizontal reinforcement rods are positioned on top of the concrete block webs. A vertical reinforcement rod is held in place using a rebar positioner clip in the cell below. Grout is being placed in the cells to reinforce the wall bond beam. A safety cap is installed on the vertical rebar to protect workers.

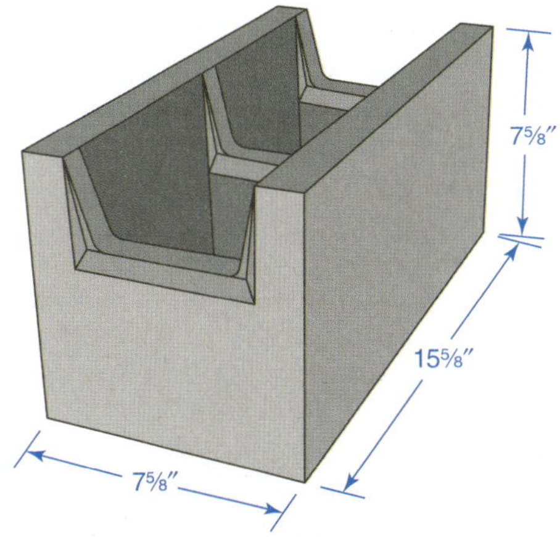

Goodheart-Willcox Publisher

Figure 16-27. A special type of bond beam block where the ends and a portion of the cross web have been completely removed. The block cell is open to allow the mason flexibility to install vertical and horizontal reinforcement.

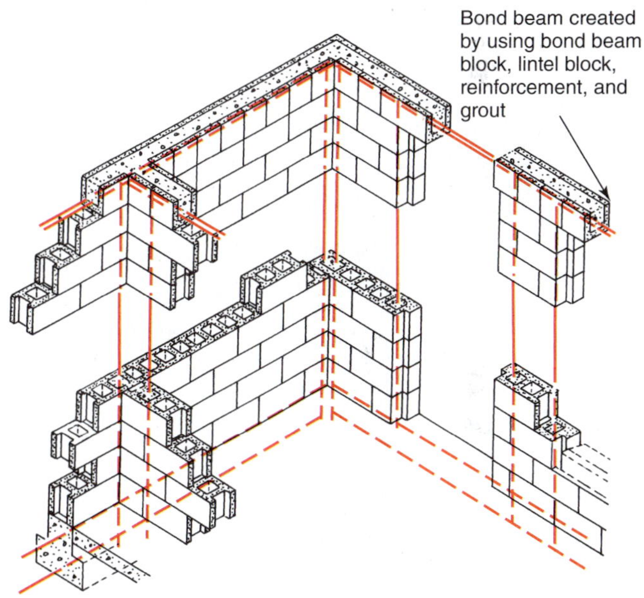

Goodheart-Willcox Publisher

Figure 16-28. Typical locations of reinforcement for concrete block masonry construction.

Retaining Walls

Reinforced masonry is ideal for the construction of retaining walls that must withstand lateral earth pressures. By adding reinforcement, the mass can be greatly reduced and the strength maintained.

Two common types of retaining walls are the gravity and cantilever designs. A gravity-type retaining wall depends primarily on its own weight to hold back the earth pressure. See **Figure 16-29**. A cantilever-type retaining wall uses the weight of the soil together with its own strength to get the same results. See **Figure 16-30**. For walls over 3′ in height, the

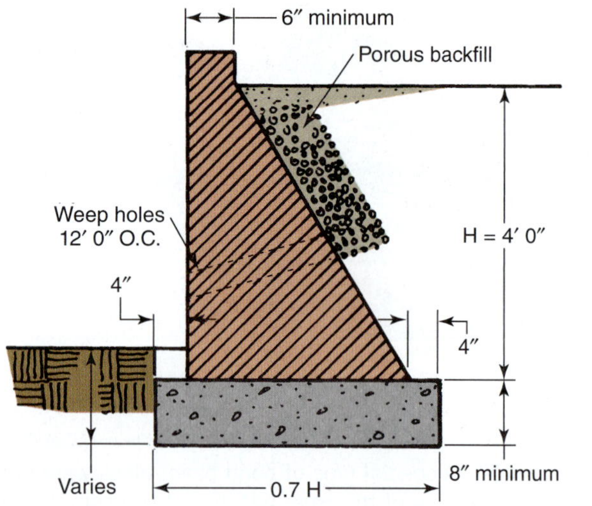

Goodheart-Willcox Publisher

Figure 16-29. Gravity-type retaining wall with specifications. The width of the base must be at least 1/2 to 3/4 of the wall height. This is expressed in the drawings as "0.7H" or "0.7 of the height."

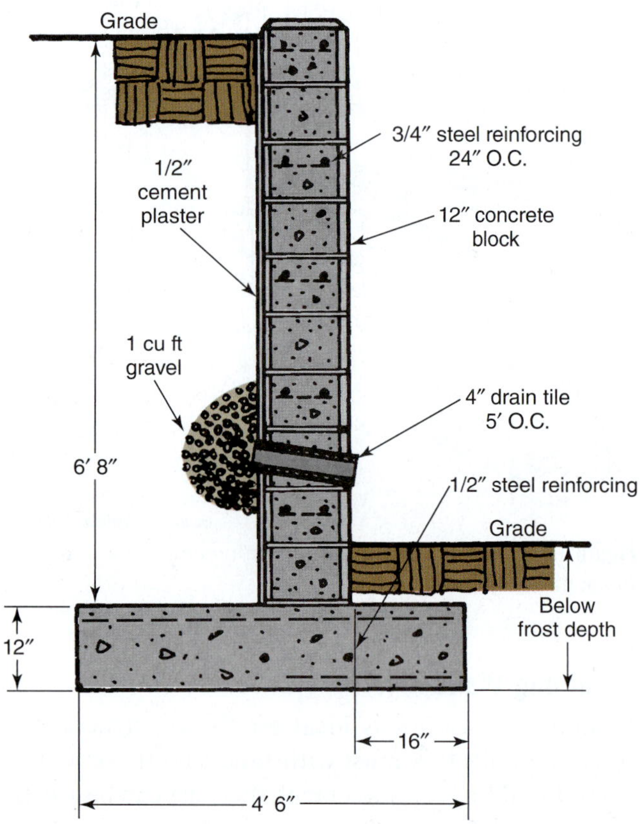

Goodheart-Willcox Publisher

Figure 16-30. Cantilever-type retaining wall constructed of reinforced concrete blocks.

cantilever-type retaining wall provides the best solution. The same principles of good construction should be followed in building a retaining wall as in any other reinforced masonry wall.

Figure 16-31 provides specifications for building reinforced concrete masonry retaining walls up to 9′-4″ in height.

Grouted Masonry Walls

Grouted masonry walls are similar to reinforced masonry walls except that they do not contain reinforcements. Grout is added to the cores in loadbearing masonry walls to provide added strength. Other uses include filling bond beams and the collar joint in a two-wythe wall.

Single-wythe walls are constructed using hollow masonry units laid with face-shell mortar bedding. The vertical cores are aligned to form continuous vertical spaces that can receive grout. Two-core block are generally preferred over three-core block because of the ease of placing grout.

Multi-wythe walls generally have two wythes. The wythes are spaced 1″ to 6″ apart and can be constructed of hollow or solid units depending on the building requirements. The space between the wythes can be grouted solid. Grouting is usually done as the wall is built.

Garden Walls

Garden walls are freestanding structures. There are many types, including straight walls, pier and panel walls, and serpentine walls. Any of the typical types of walls can be perforated. Perforated walls let in light, allow breezes, and do not block all vision, **Figure 16-32**.

Trade Tip

When designing walls that use mortar, ensure that the footing is below the frost line. Contact your local building official to determine the required depth of the footing to meet the building code.

Straight Walls

Straight garden walls depend on texture and color of the masonry for character. The wall must be thick enough to provide lateral stability against wind and impact loads.

It is recommended that for 10 pounds per square foot (psf) wind pressure, the height above grade not exceed 3/4 of the wall thickness squared. For example, an 8″ thick wall could be a maximum height of 48″ ($3/4 \times 8^2 = 48$). This formula does not depend on a bond between the foundation and the wall. Therefore, reinforcing greatly increases the height that can be attained for a given wall thickness.

Specifications for Reinforced Concrete Retaining Walls					
8″ Walls					
Height of Wall	Width of Footing	Thickness of Footing	Distance to Face of Wall	Size and Spacing of Vertical Rods in Wall	Size and Spacing of Horizontal Rods in Footing
3′–4″	2′–4″	9″	8″	3/8″ @ 32″	3/8″ @ 27″
4′–0″	2′–9″	9″	10″	1/2″ @ 32″	3/8″ @ 27″
4′–8″	3′–3″	10″	12″	5/8″ @ 32″	3/8″ @ 27″
5′–4″	3′–8″	10″	14″	1/2″ @ 16″	1/2″ @ 30″
6′–0″	4′–2″	12″	15″	3/4″ @ 24″	1/2″ @25″
12″ Walls					
6′–8″	4′–6″	12″	16″	3/4″ @ 24″	1/2″ @ 22″
7′–4″	4′–10″	12″	18″	7/8″ @ 32″	5/8″ @ 26″
8′–0″	5′–4″	12″	20″	7/8″ @ 24″	5/8″ @ 21″
8′–8″	5′–10″	14″	22″	7/8″ @ 16″	3/4″ @ 26″
9′–4″	6′–4″	14″	24″	1″ @ 8″	3/4″ @ 21″

Goodheart-Willcox Publisher

Figure 16-31. Specifications for reinforced concrete masonry retaining walls up to 9′-4″ high.

Goodheart-Willcox Publisher

Figure 16-32. Perforated brick wall with each unit bonding one-quarter its length on either end. A wide variety of patterns is possible.

Figure 16-33 shows three typical straight walls. Each wall has a different coping.

Pier and Panel Walls

A pier and panel wall is composed of a series of relatively thin panels 4″ thick, which are braced by masonry piers. See **Figure 16-34**. This type of wall is relatively easy to build and is economical because of the reduced panel thickness. It is also ideal for uneven terrain. Foundations are only required for the piers. These foundations should extend below the frost depth.

The panels may be constructed with 2 × 4 forms under the first course. The forms are removed when

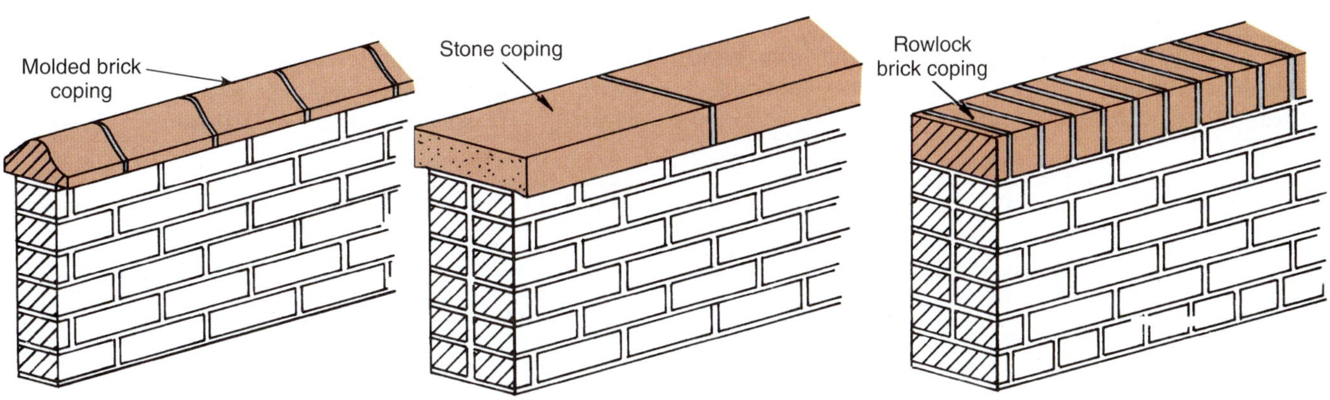

Goodheart-Willcox Publisher

Figure 16-33. Straight garden walls with three commonly used copings.

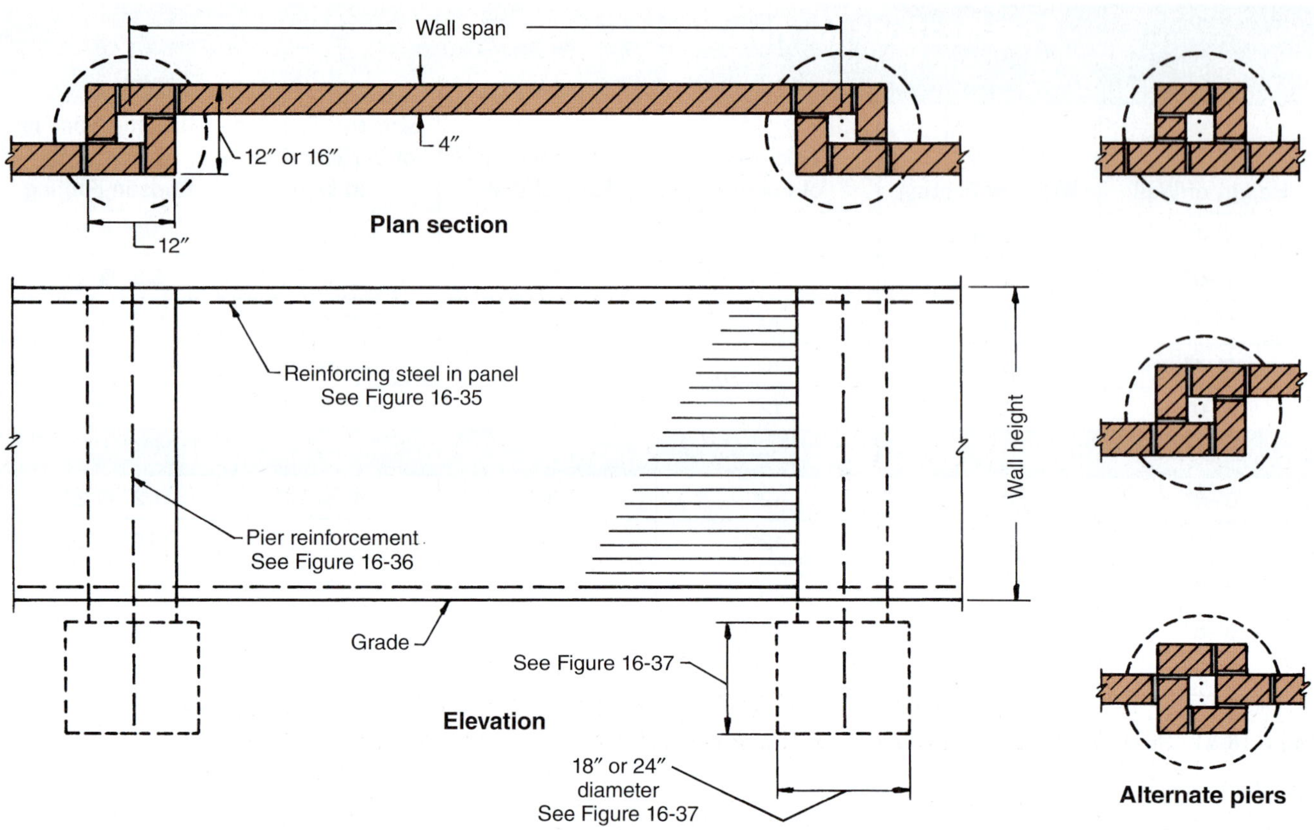

Goodheart-Willcox Publisher

Figure 16-34. Details of a pier and panel garden wall.

the masonry has cured a minimum of seven days. Specifications for building pier and panel walls for varying heights and lengths are found in **Figure 16-35**, **Figure 16-36**, and **Figure 16-37**.

Serpentine Walls

Serpentine walls have been used successfully for hundreds of years. The serpentine shape provides lateral strength so it can normally be built 4″ thick. Since the shape of the wall provides the strength, it is important that the degree of curvature be sufficient. The radius of curvature of a 4″ wall should be no more than twice the height of the wall above the grade. The depth of curvature should be no less than 1/2 of the height. **Figure 16-38** shows the details of a typical serpentine wall.

Panel Wall Reinforcing Steel

Wall Span (ft)	A—Vertical spacing (in), Wind Load, 10 psf	B—Vertical spacing (in), Wind Load, 10 psf	C—Vertical spacing (in), Wind Load, 10 psf	A—Vertical spacing (in), Wind Load, 15 psf	B—Vertical spacing (in), Wind Load, 15 psf	C—Vertical spacing (in), Wind Load, 15 psf	A—Vertical spacing (in), Wind Load, 20 psf	B—Vertical spacing (in), Wind Load, 20 psf	C—Vertical spacing (in), Wind Load, 20 psf
8	45	30	19	30	20	12	23	15	9.5
10	29	19	12	19	13	8	14	10	6
12	20	13	8.5	13	9	5.5	10	7	4
14	15	10	6.5	10	6.5	4	7.5	5	3
16	11	7.5	5	7.5	5	3	63	4	2.5

Note: A = Two No. 2 bars, B = Two 3/16″ diameter wires, C = Two 9 gauge wires

Brick Industry Association

Figure 16-35. Specifications for panel wall reinforcing steel.

Pier Reinforcing Steel									
Wall span (ft)	4—Wind Load, 10 psf, wall height (ft)	6—Wind Load, 10 psf, wall height (ft)	8—Wind Load, 10 psf, wall height (ft)	4—Wind Load, 15 psf, wall height (ft)	6—Wind Load, 15 psf, wall height (ft)	8—Wind Load, 15 psf, wall height (ft)	4—Wind Load, 20 psf, wall height (ft)	6—Wind Load, 20 psf, wall height (ft)	8—Wind Load, 20 psf, wall height (ft)
8	2#3	2#4	2#5	2#3	2#5	2#6	2#4	2#5	2#5
10	2#3	2#4	2#5	2#4	2#5	2#7	2#4	2#6	2#6
12	2#3	2#5	2#6	2#4	2#6	2#6	2#4	2#6	2#7
14	2#3	2#5	2#6	2#4	2#6	2#6	2#5	2#5	2#7
16	2#4	2#5	2#7	2#4	2#6	2#7	2#5	2#6	2#7
Note: Within heavy lines, 12 × 16″ pier required. All other values obtained with 12 × 12″ pier.									

Goodheart-Willcox Publisher

Figure 16-36. Specifications for pier reinforcing steel.

Required Embedment for Pier Foundation									
Wall Span (ft)	4—Wind Load, 10 psf, wall height (ft)	6—Wind Load, 10 psf, wall height (ft)	8—Wind Load, 10 psf, wall height (ft)	4—Wind Load, 15 psf, wall height (ft)	6—Wind Load, 15 psf, wall height (ft)	8—Wind Load, 15 psf, wall height (ft)	4—Wind Load, 20 psf, wall height (ft)	6—Wind Load, 20 psf, wall height (ft)	8—Wind Load, 20 psf, wall height (ft)
8	2′-0″	2′-3″	2′-9″	2′-3″	2′-6″	3′-0″	2′-3″	2′-9″	3′-0″
10	2′-0″	2′-3″	2′-9″	2′-3″		3′-3″	2′-6″	3′-0″	3′-3″
12	2′-3″	2′-3″	3′-0″	2′-3″	3′-0″	3′-3″	2′-6″	3′-3″	3′-6″
14	2′-3″	2′-9″	3′-0″	2′-6″	3′-0″	3′-3″	2′-9″	3′-3″	3′-9″
16	2′-3″	2′-9″	3′-0″	2′-6″	3′-0″	3′-6″	2′-9″	3′-3″	4′-0″
Note: Within heavy lines, 24″ diameter foundation required. All other values obtained with 18″ diameter foundation.									

Goodheart-Willcox Publisher

Figure 16-37. Specifications for embedment for pier foundations.

Thin Brick Veneer

Thin brick veneers are thin brick units formed from kiln-fired shale or clay, or a combination of the two. The units are approximately 1/2″ to 1″ thick. When in place, thin bricks give the appearance of a conventional brick masonry wall. Thin brick is available in various sizes, colors, and textures. See **Figure 16-39**.

The most common face size is standard modular, which has nominal dimensions of 2 2/3″ × 8″. A slight variation in actual face dimensions exists among manufacturers, but 3/8″ to 1/2″ less than the nominal dimensions is typical. The economy modular face size, 4″ × 12″, is commonly used for large buildings because productivity is increased and there are fewer visible mortar joints. Other sizes, such as 3″, nonmodular, and oversize may be available.

Trade Tip

Check with individual manufacturers or distributors regarding thin brick sizes available in your area.

Thin brick veneer is an adhered veneer, which means it relies on a bonding agent between the thin brick units and the substrate. Adhered veneer construction can be classified as thin bed set or thick bed set. A newer technique is to add thin brick to the exterior of precast wall panels prior to casting.

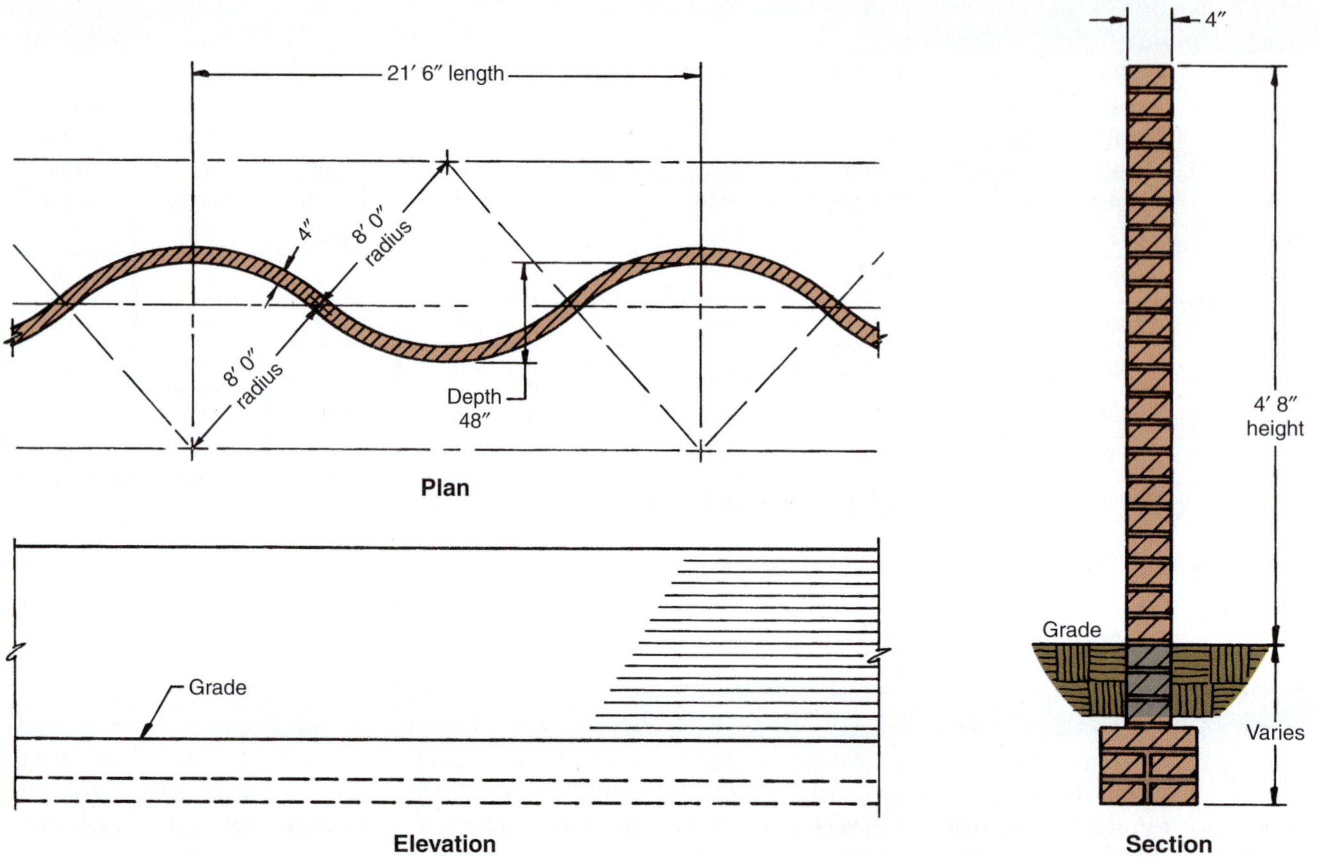

Goodheart-Willcox Publisher

Figure 16-38. Details of a serpentine garden wall showing the plan layout and wall sections.

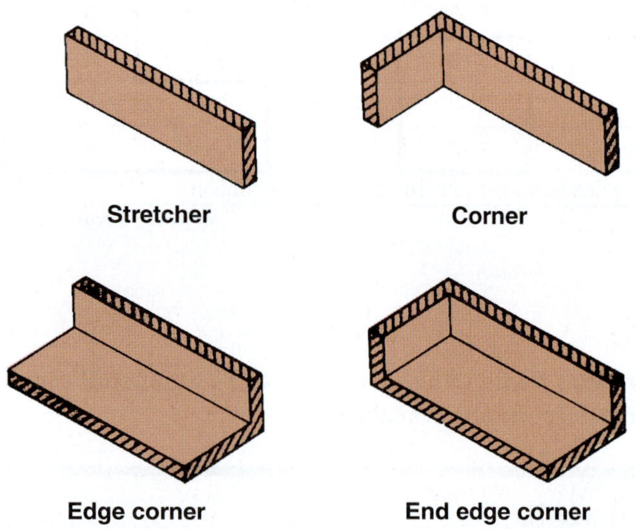

Brick Industry Association

Figure 16-39. Thin brick units.

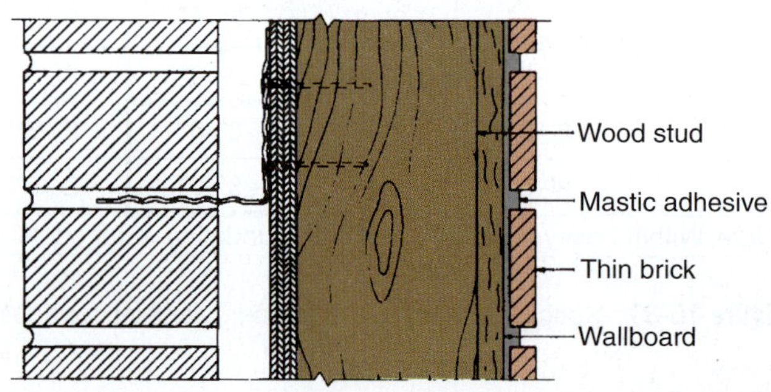

Brick Industry Association

Figure 16-40. A cross-section depicting a wood frame wall on which thin brick veneer is installed using the thin bed set procedure.

Thin Bed Set

In the thin bed set installation procedure, an epoxy or organic adhesive is typically used on interior surfaces only. For dry locations, the backing material (substrate) may be wood, wallboard, or masonry. See **Figure 16-40**. For areas exposed to dampness, only clear and dry masonry surfaces or concrete surfaces should be used for backup.

Thick Bed Set

The thick bed set procedure of installation is used on interior and exterior surfaces. The backing material is either masonry, concrete, steel, or wood stud framing. For applications over steel studs, procedures are similar to those used for concrete or masonry backup. However, wallboard and building felt must be installed

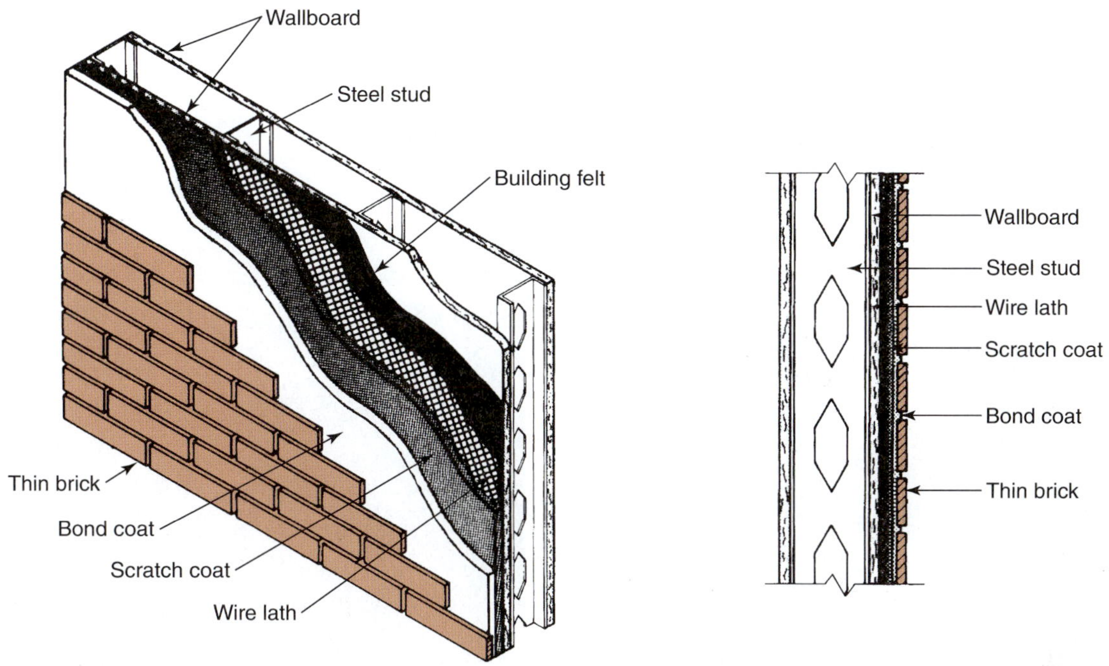

Brick Industry Association

Figure 16-41. Installation of thin brick units over steel stud framing using the thick set method.

over the studs before the lath and mortar bed are placed. See **Figure 16-41**.

Thin Brick Faced Precast Panels

In this process, thin brick are located at the base of the precast form table. See **Figure 16-42**. The bricks are placed in flat rubber mats that hold them in position on the forming table. The form located on the forming table creates the size and shape of the precast wall panel. Thin brick composes the exterior face of the panel and becomes bonded and embedded into the panel when the concrete mixture is poured into the form.

Precast panels are usually constructed indoors in a factory where they cure under ideal conditions. After the panels cure, they are stored at the factory until the construction site is ready for delivery. Once the panels are at the site, they can be lifted into position, **Figure 16-43**. Panels are erected to begin forming the structure of the building, **Figure 16-44**. **Figure 16-45** shows a completed precast building.

Gage Brothers in Sioux Falls, SD

Figure 16-42. Employees are working on a forming table for creating precast panels. Thin bricks are being placed in rubber pattern mats that hold the bricks in place.

Green Note

Precast Concrete—Upcycling and Downcycling

In upcycling, waste materials are used to provide new, high-quality products. Precast concrete can be disassembled with a minimum amount of energy and retains its original qualities after deconstruction. After removal, precast concrete members can be reused in other applications.

Precast concrete can also be downcycled. In downcycling, material is broken down into a material that is useful but of lesser quality than the original. For example, crushed precast concrete can be used as aggregate in new concrete or as base materials for slabs, roads, and sidewalks.

Gage Brothers in Sioux Falls, SD

Figure 16-43. A precast panel being unloaded and positioned using a construction crane. The windows are already built into this panel, which is one of the many panels used to create the exterior of the building. The construction worker has a lead rope to guide the panel into position. Note the safety gear worn by the worker.

Gage Brothers in Sioux Falls, SD

Figure 16-44. Precast panels form the exterior of this building. Window and doorway openings already have sills and jambs as part of the panels.

Gage Brothers in Sioux Falls, SD

Figure 16-45. Finished building using brick-clad precast concrete panels.

Segmental Retaining Walls and Hardscape

Segmental retaining walls (SRW) are constructed of high-strength concrete blocks or units made specifically for dry stacking without the use of mortar. SRW blocks are used to build retaining walls for commercial and residential applications.

Code Note **Segmental Retaining Walls**

The National Concrete Masonry Association (NCMA) has created manuals for segmented retaining walls. For more information on design and installation of SRW's, refer to NCMA and the specific company manufacturing the hardscape products. Also check with the local building officials in the area to determine if there are specific codes for SRW construction.

Retaining walls and other masonry landscaping materials are sometimes called ***hardscape***. See **Figure 16-46**. This term can refer to concrete retaining wall materials, piers, dividing walls, pavers, and caps for walls and piers.

Anchor Wall Systems, Inc.

Figure 16-46. These mortarless masonry units provide a significant amount of flexibility in laying up straight or curved retaining walls.

Hardscape is made similar to concrete block, but is intended for decorative applications and is laid on a gravel or crushed stone and sand base. Hardscape can be purchased in numerous sizes, shapes, and colors.

Retaining walls, tree rings, planters, and edgings can be made from concrete masonry units that have a locking lip on the lower back edge. See **Figure 16-47**. The units are placed so the lip locks behind the course below to provide proper alignment and prevent forward movement exerted by earth pressure. Typical retaining wall products include straight and beveled front units with uniform or tapered widths. See **Figure 16-48**. Retaining walls, sidewalks, courtyards, stairways, and driveways can be made of hardscape units to create a beautiful addition to any home or business. See **Figure 16-49**.

Sizes of units for retaining walls include a 7″ × 18″ × 12″ straight face retaining block and 3″ × 13″ × 12″ reversible cap. Typically, the weight of the straight face blocks is 80 lb and the weight of the caps is 37 lb. Sizes, shapes, and weights vary among manufacturers. It is important to keep safety in mind when working with retaining wall blocks.

Block weights can vary anywhere from 30 lb to 80 lb. Carrying and placing them can be a difficult task. The use of block clamps can reduce injuries and aid in lifting and placing blocks on the wall. Refer to **Figure 3-38**. Pave Tools is a leader in manufacturing hardscape tools and equipment to aid when performing construction activities. **Figure 16-50** shows a larger hardscape clamp rated to support 180 lb. This clamp can be used by hand, or a connection eyelet can be attached to the handle to receive a hook from a chain.

On large projects, such as stairs or heavy retaining wall blocks, a special clamp is needed that can carry heavier loads using an industrial truck, forklift, or mini-crane.

Timothy L. Andera

Figure 16-47. Hardscape retaining blocks are made with a structural lip hangs approximately 3/4″ over the course below. The lip locks the course to the previous course, so it does not move when backfill and soil are filled behind the wall.

Anchor Wall Systems, Inc.

Figure 16-48. The tapered sides of these units make straight or curved walls possible for a variety of designs.

Elmar Langle/Shutterstock.com

Figure 16-49. A hardscape project with retaining walls with caps, stairs, and pavers.

Pave Tool Innovators

Figure 16-50. A BL-180 clamp, a hardscape hand and equipment clamp. It is used to lift heavy retaining wall blocks and concrete steps for hardscape.

Pave Tool Innovators

Figure 16-51. A BL-850 clamp being used to lift a stair block. This clamp is rated to support 850 lb. With special attachments, it can be used to lift large stones or boulders. This clamp is adjustable and attaches to an industrial truck or forklift.

Clamps will have an eyelet that allows a hook on a chain to connect. The clamp is adjustable using clamp pins to open and close the jaws to fit the product being lifted. The use of a forklift attachment called a *fork pocket* can also be implemented where the fork slides into the pocket for quick connection and disconnection. See **Figure 16-51**.

Safety Note

When cutting or working with SRW units wear appropriate PPE, respirator protection, protective gloves, eyewear, and safety footwear.

Segmented Retaining Wall Installation

Before beginning construction, prepare a layout plan that considers the topography, drainage patterns, soil conditions, and local code restrictions. It is important that the subgrade is as level as possible. A laser level is one tool that can be used to level the base grade to ensure the surface is flat and at the same grade elevation. Another instrument that can be used to level the grade is the altimeter level. Two types are discussed in Chapter 3, **Tools and Equipment**. Refer back to review the different types of altimeter levels to determine which one best fits the job being performed.

Code Note — Segmented Retaining Wall Drainage

The NCMA recommends a minimum of 1% slope for adequate drainage. In inches, a 1% slope is 1/8″ or 0.125″ of fall per feet and a 2% slope is 1/4″ or 0.25″ of fall per feet. For example, if the tile run was 50′ with a slope of 1%, the drop elevation would be 1/8″ or 0.125″ × 50′ or 6 1/4″ fall.

PROCEDURE

Installing a Retaining Wall

1. Use a laser level or altimeter level to determine the excavation requirement. Then excavate the trench or area to the appropriate depth. For walls up to 4′ high, one course is generally buried below grade; but the number of courses may depend on the depth of the frost line. The trench should be at least 24″ wide and deep enough to accommodate the amount of wall to be buried (at least one course plus the sand and gravel or crushed stone base). See **Figure 16-52**.
2. It is recommended before beginning a retaining wall project to contact your local building code office to determine the depth of the frost line in the area. Frost can affect retaining walls similar to foundation walls by causing them to change elevation. Over time, the retaining wall may be affected by freezing and thawing cycles of the supporting subgrade.
3. After excavating, prepare the subbase to ensure the surface is uniform and at the same elevation. Roughly grade and level the loose soil and then compact it in the bottom of the trench. Then place 6″ of compactable base aggregate, usually 1/2″ to 3/4″ crushed stone, recycled concrete, or gravel in the bottom of the trench. Grade and level the base and then compact it.
4. If a fabric filter is to be used, install it on top of the base aggregate or stone and continue up the back side of the trench. The fabric helps to keep soil from migrating into the aggregate or crushed stone, which affects the drainage properties of the aggregate or stone.
5. Install the base course. Install the first course of units on the prepared base. Start at the lowest elevation of the trench. Level each unit as it is placed from side to side and front to back.

Trade Tip

If necessary, use a small amount of sand and a rubber mallet to aid in leveling the block. Sprinkle the sand on the gravel base to aid in creating a level base. Then position the block on the sand and place a level on top to check for levelness over the length and width of the unit. If adjustments are needed, tap the high points of the block with a rubber mallet until the unit is level. A mason's line can also be used for elevation of the block to achieve alignment.

6. Install the drain tile behind the first course, gravel, or crushed compacted base. If the tile is to be run underground to a lower elevation away from the retaining wall, it should be installed behind the first course or in the base material where it is easier to hide underground. For proper drainage, ensure the tile slopes away from the retaining wall. Some drain

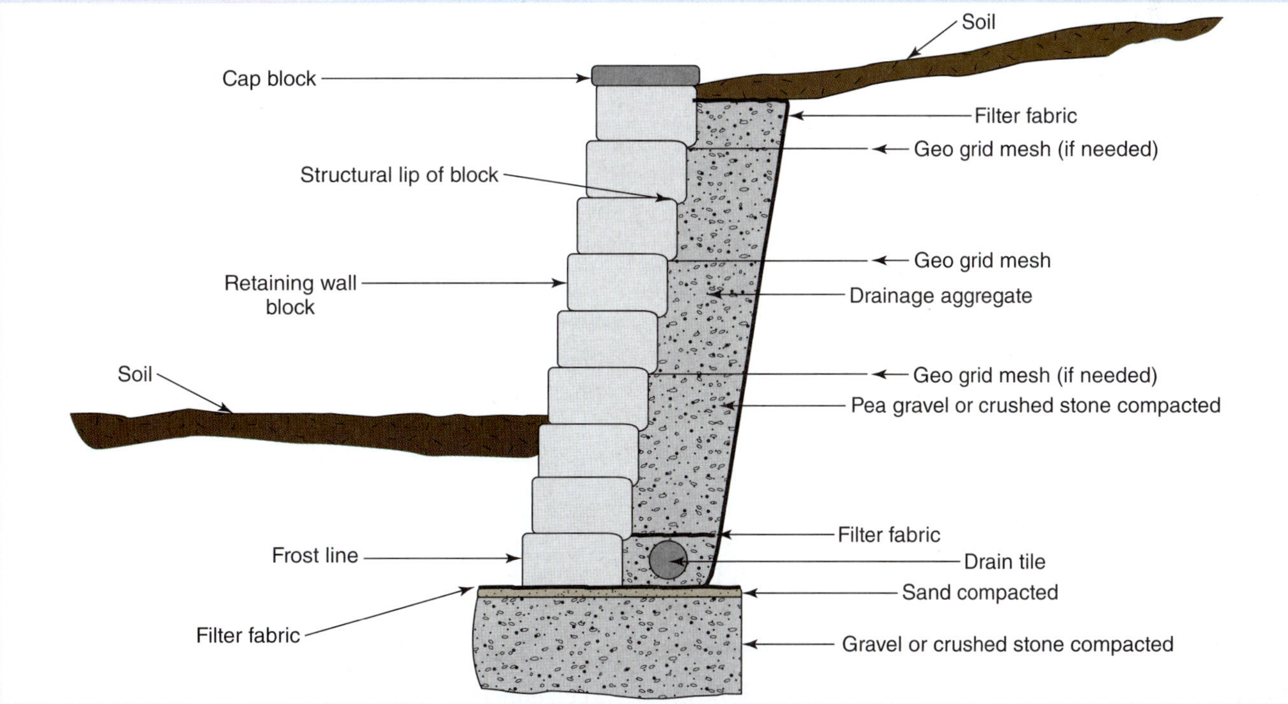

Goodheart-Willcox Publisher

Figure 16-52. A sectional view of a hardscape retaining wall showing block and other materials that compose the wall.

Continued

PROCEDURE *(continued)*

tiles are placed in a protective sock to ensure the holes are not plugged up by sand or silt. Place pea gravel or crushed stone around and on top of the drain tile lying behind the retaining wall or base. Lay a fabric filter over the pea gravel or crushed stone. The fabric filter aids in reducing silt from entering the tile. Backfill with base aggregate to the top edge of the first course and compact. Use 1/2″ or 3/4″ free-drainage aggregate 12″ directly behind the wall in 6″ layers and compact. Fill the void between units with aggregate and compact.

7. If required, apply construction adhesive to the top of the first course and the following courses. Certain hardscape manufacturers recommend a construction adhesive be applied to the top of each course of block prior to placing the next course. This creates a retaining wall where individual units react as one when stresses are placed upon the structure.
8. Install the next course. Stagger the vertical joints. Pull each unit forward until the lip aligns with the back edge of the course below.
9. If a fabric filter is to be used, it should be laid against the back wall of the trench. Bring the remaining fabric forward and lay it over the top of the last regular course. The fabric stops the soil from migrating into the drainage aggregate.
10. Install a geogrid if required. A geogrid is a fabric that is placed level with the course being laid. It ties the retaining wall into the backfill area behind the SRL wall. Determine the appropriate course level and distance to place the geogrid behind the retaining wall. Refer to the manufacturer's installation guide.
11. Install backfill over the fabric filter or aggregate. Fill any remaining voids between the drainage aggregate and the embankment with backfill material (native soil).
12. Install the wall cap on the retaining wall. Certain caps may need to be cut using a masonry or diamond blade to ensure the caps follow the contours of the wall. If a fabric filter was used, it should be placed between the retaining wall block and finishing caps. Use concrete mastic intended for masonry products as an adhesive to glue the caps down to the retaining wall block.
13. Grade and contour the backfill to meet the existing grade and terrain. Firmly compact the backfill soil, fill any voids, and repeat compacting.

Trade Tip

To create partial units, use a masonry saw, splitter, or hammer and chisel to cut each unit to the desired shape and size.

16.2 Wall Elements

Many types of elements are incorporated into walls. These include wall openings and arches. Other wall elements may include stone quoins, caps and copings, and corbels and racking. Window and door details are also considered a part of wall elements.

Wall Openings

Wall openings can extend through the wall as required for a door or window, or they may take the form of a chase or recess. Chases and recesses are horizontal or vertical spaces left in a wall for the purpose of containing plumbing, heating ducts, electrical wiring, or other equipment. See **Figure 16-53**.

Chases and recesses are formed by masons as the wall is built. Chases are generally located on the inside of the wall and vary in size from 4″ to 12″ in width. Recesses reduce the wall thickness and strength and are usually limited to one-third the thickness of the wall.

Code Note — Chases

The International Residential Code (IRC) **R606.8** specifies that chases and recesses in masonry cannot be deeper than one-third the wall thickness.

Lintels

A ***lintel*** is a structural member placed over an opening in a wall used to support the loads above that opening. Steel lintels should be supported on either side of the opening for a distance of at least 4″. Reinforced concrete lintels should have a minimum bearing of at least 8″ at each end. Longer lintels carrying heavier loads should have greater bearing surfaces at the ends. The lintel should be stiff enough to resist bending in excess of 1/360th of the span.

Code Note — Lintels

The International Residential Code (IRC) **R608.8.2** specifies lintels are to be provided over all openings that are equal to or greater than 2′ (610 mm) wide. Masonry over openings must be supported by steel lintels, lintels made of reinforced concrete or masonry, or masonry arches designed to support the imposed load.

Several types of lintels are used in masonry construction, **Figure 16-54**. Specifications for one-piece reinforced concrete lintels are given in **Figure 16-55**. Design data for reinforced concrete split lintels are shown in **Figure 16-56. Figure 16-57** gives the size requirements for concrete reinforced lintels with stirrups. These lintels

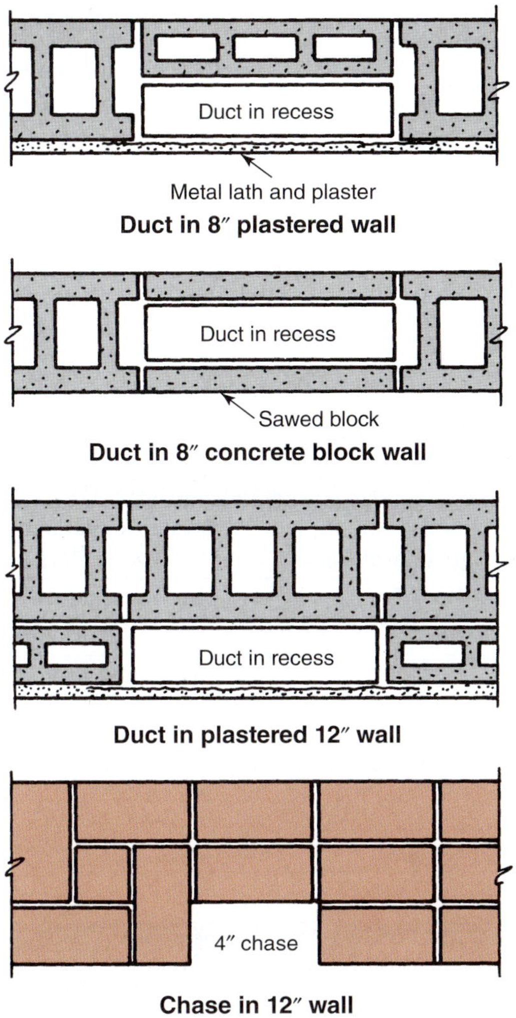

Goodheart-Willcox Publisher

Figure 16-53. Examples of chases and recesses in masonry walls.

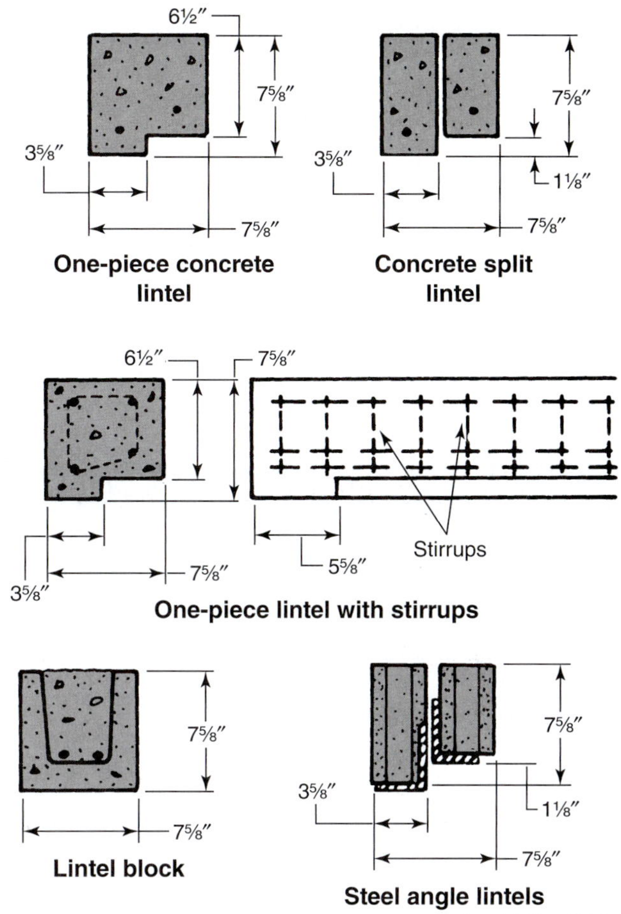

Goodheart-Willcox Publisher

Figure 16-54. Various types of lintels are used in masonry construction.

are made for openings supporting wall and floor loads. Lintels made of angle steel should meet minimum specifications for 4″ masonry veneer. See **Figure 16-58**.

Reinforced lintels can be constructed on the jobsite as the masonry work reaches the stage of door and window openings. An example of reinforced lintels being constructed to support openings is shown in **Figure 16-59**. In the manufacturing of lintel block, cross webs are removed to create an open well in the block. The well area of the block is used for placement of reinforcement after the lintel blocks have been laid. Refer back to **Figure 8-11** to view the two different types of lintel blocks. Depending on the type of lintel being used, some wells are shallow, and others are much deeper, which allows for more reinforcement and grout to be added to strengthen the area above the opening.

Reinforced masonry lintels (brick and tile) are becoming more popular because the steel is completely protected from the elements and the initial cost is less because less steel is required. **Figure 16-60** shows several sizes of reinforced brick masonry lintels. **Figure 16-61** shows a detail of a brick masonry lintel integrated into the wall.

Slight movement often occurs at the location of lintels. For this reason, control joints are often located at the ends of lintels. A noncorroding metal plate is placed under the ends of lintels where control joints occur. See **Figure 16-62**. The metal plate permits the lintel to slip and prevent uncontrolled cracking.

A full bed of mortar should be used over the plate to distribute the lintel load uniformly. When the mortar has hardened sufficiently, it should be raked out to a depth of 3/4″ and filled with caulking.

Removeable Masonry Forms

Carpenters construct temporary framework below the lintel area to support masonry above openings for windows, doorways, or other wall voids. **Figure 16-63** shows wood framing members called a ***buck***. Bucks are left in place under the opening until the masonry can support itself. Constructing bucks can be a time-consuming and expensive job.

Concrete Reinforced One-Piece Lintels with Wall Load Only				
Size of Lintel Height (in)	Size of Lintel Width (in)	Clear Span of Lintel (ft)	Bottom Reinforcement, Size Designation of Bars	Bottom Reinforcement, Size of Bars
5 3/4	7 5/8	Up to 7	No. 2	3/8″ round deformed
5 3/4	7 5/8	7 to 8	No. 2	5/8″ round deformed
7 5/8	7 5/8	Up to 8	No. 2	3/8″ round deformed
7 5/8	7 5/8	8 to 9	No. 2	1/2″ round deformed
7 5/8	7 5/8	9 to 10	No. 2	5/8″ round deformed

Portland Cement Association

Figure 16-55. Specifications for one-piece reinforced concrete lintels.

Concrete Reinforced Split Lintels with Wall Load Only				
Size of Lintel Height (in)	Size of Lintel Width (in)	Clear Span of Lintel (ft)	Bottom Reinforcement, Size Designation of Bars	Bottom Reinforcement, Size of Bars
5 3/4	3 5/8	Up to 7	No. 1	3/8″ round deformed
5 3/4	3 5/8	7 to 8	No. 1	5/8″ round deformed
7 5/8	3 5/8	Up to 8	No. 1	3/8″ round deformed
7 5/8	3 5/8	8 to 9	No. 1	1/2″ round deformed
7 5/8	3 5/8	9 to 10	No. 1	5/8″ round deformed

Portland Cement Association

Figure 16-56. Specifications for reinforced concrete split lintels.

Concrete Reinforced Lintels with Stirrups for Wall and Floor Loads					
Size of Lintel Height (in)	Size of Lintel Width (in)	Clear Span of Lintel (ft)	Top Reinforcement	Bottom Reinforcement	Web Reinforcement No. 6 Gauge Wire Stirrups, Spacings from End of Lintel—Both Ends the Same
7 5/8	7 5/8	3	None	2—1/2″ round	No stirrups required
7 5/8	7 5/8	4	None	2—3/4″ round	3 stirrups, SP. :2, 3, 3 in
7 5/8	7 5/8	5	2—3/8″ round	2—7/8″ round	5 stirrups, SP. :2, 3, 3, 3, 3 in
7 5/8	7 5/8	6	2—1/2″ round	2—7/8″ round	6 stirrups, SP. :2, 3, 3, 3, 3, 3 in
7 5/8	7 5/8	7	2—1″ round	2—1″ round	9 stirrups, SP. :2, 3, 3, 3, 3, 3, 3, 3, 3 in

Goodheart-Willcox Publisher

Figure 16-57. Specifications for reinforced concrete lintels with stirrups.

Steel Angles to Support 4″ Masonry Walls	
Span	Size of Angle
0′–5′	3″ × 3″ × 1/4″
5′–9′	3 1/2″ × 3 1/2″ × 5/16″
9′–10′	4″ × 4″ × 5/16″
10′–11′	4″ × 4″ × 3/8″
11′–15′	6″ × 4″ × 3/8″
15′–16′	6″ × 4″ × 1/2″

Goodheart-Willcox Publisher

Figure 16-58. Specifications for lintels made of angle steel for 4″ masonry veneer.

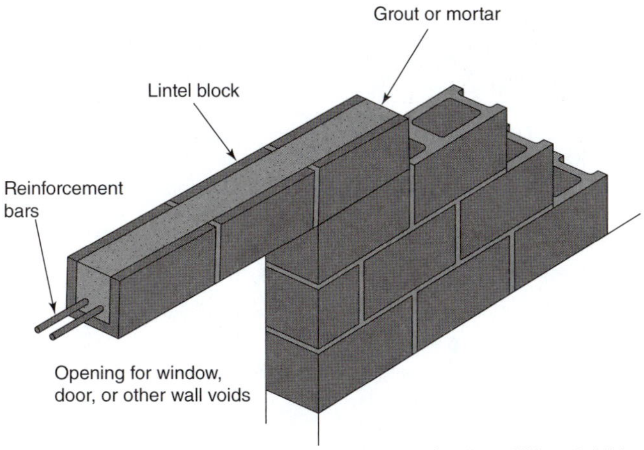

Goodheart-Willcox Publisher

Figure 16-59. Lintel blocks are used to create a reinforced area over the opening in the wall. Reinforcement and grout has been added to aid in supporting the above weight of the wall structure.

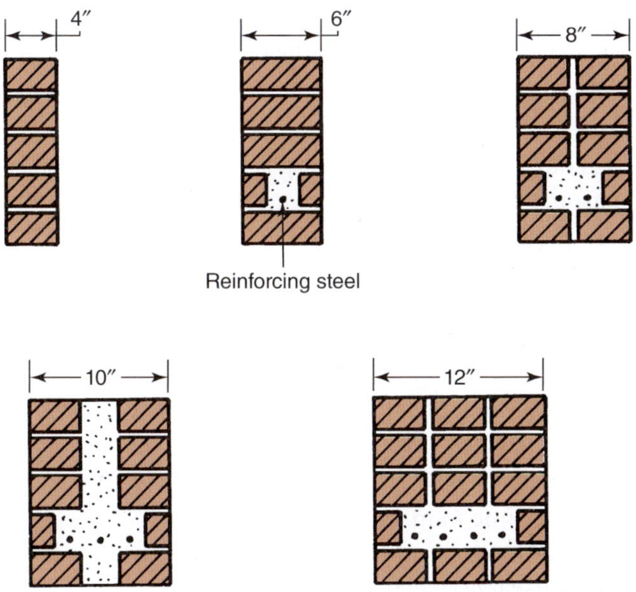

Goodheart-Willcox Publisher

Figure 16-60. Reinforced brick masonry lintels.

Materials such as lumber and plywood will need to be purchased to fabricate the buck. Once the buck is no longer required for support, it will be removed and most likely discarded.

Masonry openings can be supported by ***removable masonry forms***. These forms are made of metal and temporarily support masonry over windows, doors, or other openings. Once the masonry can support itself, the form is removed, and it can be used again at a different location or on another project. One manufacturer of the removeable masonry forms is ***Quick Headers***. See **Figure 16-64**.

The length of Quick Headers is adjustable. This aids in adjusting the device to fit the opening. It also helps in the removal of it from the opening when it is not needed.

Brick Industry Association

Figure 16-61. Detail of a brick masonry lintel with stirrups.

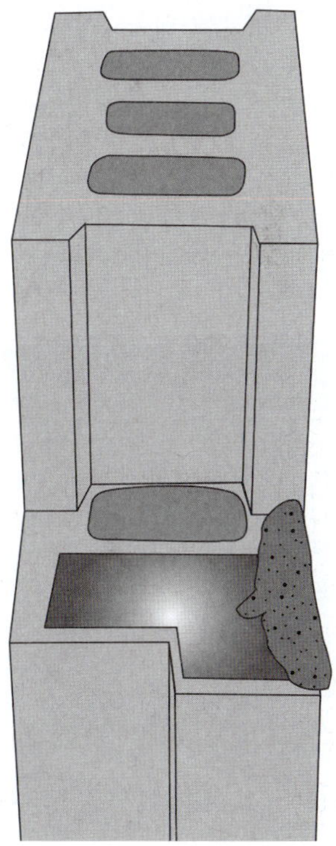

Goodheart-Willcox Publisher

Figure 16-62. A noncorroding metal plate is placed under the ends of lintels where control joints occur. A full bed of mortar will be placed over the plate.

The removeable masonry form tray has a lip that rests on each side of the opening. The tray lip extends out past the edge of the opening by 1″ to rest on the top of the block at each end. The tray lip does not interfere with

sommersby/iStock via Getty Images

Figure 16-63. A wood buck has been constructed to support lintel blocks above the masonry opening.

Scot Kelly of Quick Headers, LLC.

Figure 16-64. A lintel is being constructed over an opening. Reinforcement will be installed later, and grout or concrete will be placed into the lintel. A 3′ Quick Header is resting on the two blocks, which form the opening, and is directly under the lintel blocks.

Scot Kelly of Quick Headers, LLC.

Figure 16-65. A concrete mason is installing a Quick Header in a large opening over 10′. At the upper left, you can see the type of lintel block being used on this jobsite. A mason's line is used to ensure the Quick Header is straight and at the proper height. The mason is turning the adjustable prop to raise the Quick Header in the middle of the opening.

the course of lintel block being laid over the removeable masonry form. If the wall is extremely wide, two forms can be placed next to each other in the opening. Once the form is removed, the small gap created in the joint can be tuckpointed and tooled to close the gap. A metal prop is required at increments of 6′ to support the removeable masonry form. See **Figure 16-65**.

Using Quick Headers or other removeable masonry forms provides several benefits when compared with using wood bucks:

- Less time to install
- Eliminates lumber waste and cost
- Reuseable steel form
- Wide laying surface for blocks

Arches

Arches have been used for centuries to span openings. Some arches span distances of more than 130′. In an ***arch***, masonry units span an opening by transferring vertical loads laterally to adjacent masonry units, and thus to the abutments. An ***abutment*** is the masonry, or combination of masonry and other structural members, which supports one end of the arch at the ***skewback*** (the surface on which the arch joins the supporting abutment).

Several standard types of arches are easily recognized in contemporary architecture. See **Figure 16-66**. These arches include the blind arch, bull's-eye arch, elliptical

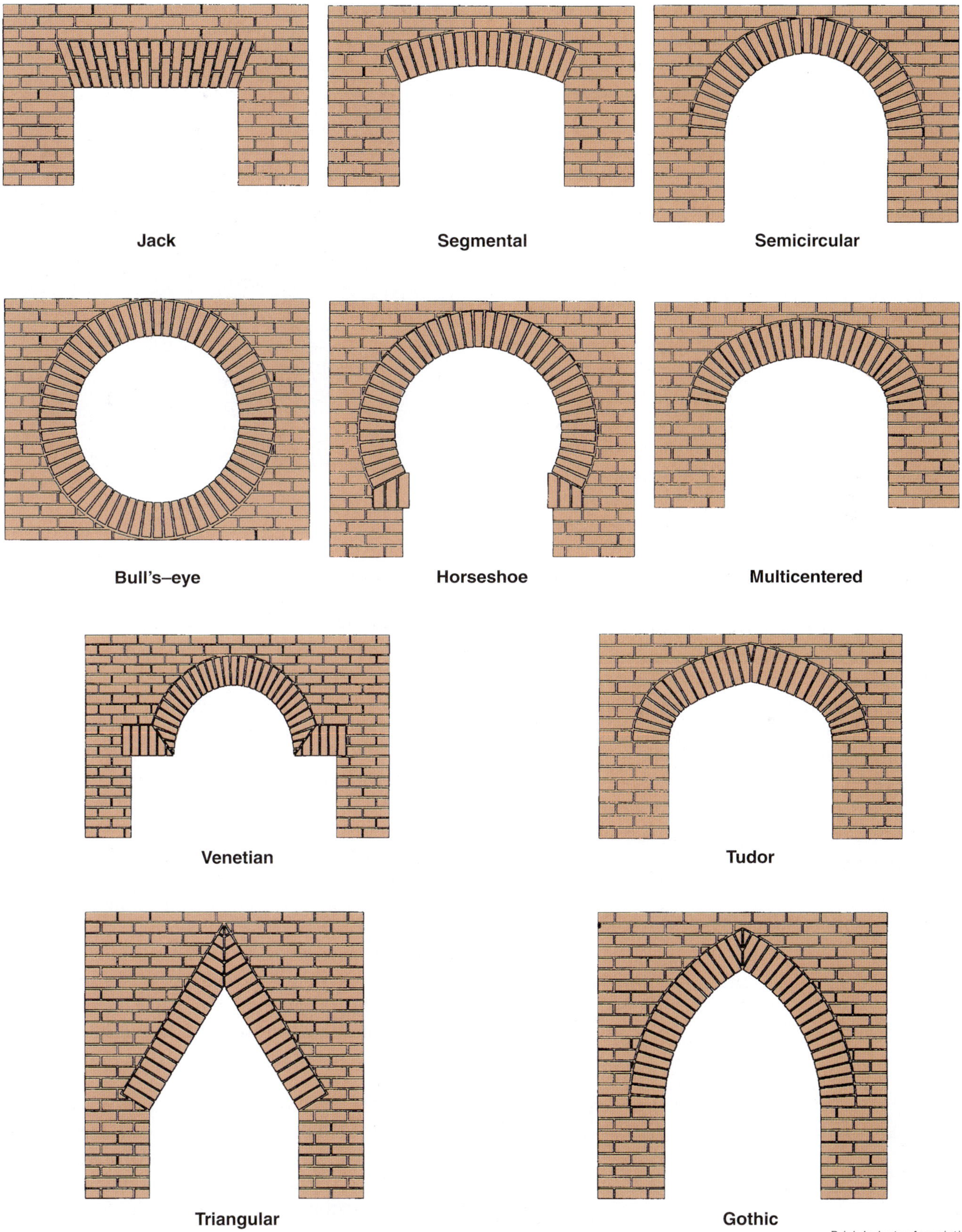

Brick Industry Association

Figure 16-66. Arches used in contemporary construction.

arch, fixed arch, Gothic arch, horseshoe arch, jack arch, multicentered arch, segmental arch, semicircular arch, triangular arch, Tudor arch, and Venetian arch.

Trade Tip

A Gothic arch is also referred to as a *drop arch*, *equilateral arch*, or *lancet arch*, depending on whether the spacings of the centers are respectively less than, equal to, or more than the clear span.

An arch is normally classified by the curve of its ***intrados*** (curve that bounds the lower edge of the arch) and by its function, shape, or architectural style. Arches are also classified as major and minor arches. Minor arches have spans that do not exceed 6′ with maximum rise-to-span ratio of 0.15. The ***rise*** of an arch is the maximum height of the ***soffit***, or lower surface of the arch, above the level of its spring line. The ***spring line*** is the horizontal line that intersects the ***springing*** (point where the skewback intersects the intrados). The ***span*** of an arch is the horizontal clear dimension between abutments. Refer to **Figure 16-67**.

The following types are most commonly used for building arches:

- ***Jack arch***. A flat arch with zero or little rise.
- ***Segmental arch***. An arch with an intrados that is circular but less than a semicircle. See **Figure 16-68**.
- ***Semicircular arch***. An arch with an intrados that is a semicircle (half circle). Because of its structural efficiency, this arch is often used for bridges.
- ***Multicentered arch***. An arch with a curve that consists of several arcs of circles that are normally tangent at their intersections.

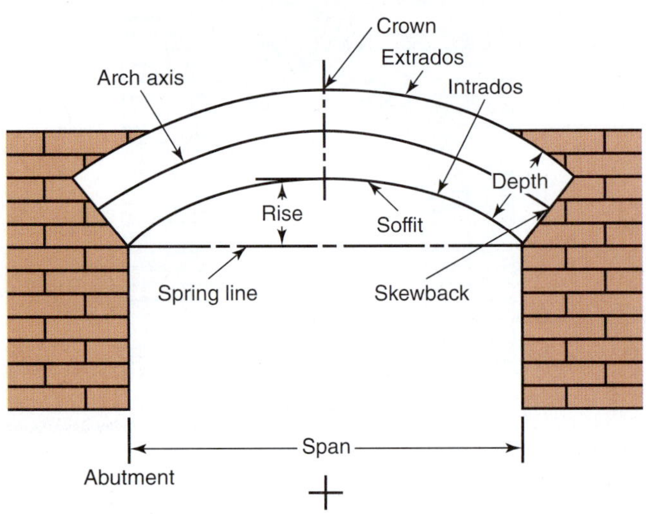

Goodheart-Willcox Publisher

Figure 16-67. Arch terminology.

Allison J. Hahn/Shutterstock.com

Figure 16-68. A typical segmental arch.

Function of Arches

The curvature of an arch provides structural efficiency by transferring vertical loads laterally along the arch to the abutments at each end. Compression is distributed through the arch form and weight is diverted onto the abutments. An arch is designed to keep stresses primarily compressive. The compressive strength of brick masonry allows brick masonry arches to support considerable loads.

Most brick masonry arches are built with unreinforced masonry. However, arches with a small rise and very long span arches may require steel reinforcement to resist tensile stresses. Reinforcement may also be required if abutment size and arch thickness have been reduced for economy.

An arch requires support if it is not structurally adequate. This support is typically provided by a steel angle, which is bent to the curvature of the intrados of the arch. Curved sections of steel angle are welded to horizontal steel angles to form a continuous support. See **Figure 16-69**.

Arch Design

Arches can be built using a variety of depths, brick sizes and shapes, and bond patterns. An odd number of units are used so the ***keystone*** falls at the center of the arch. The keystone is the masonry unit located at the crown of the arch. See **Figure 16-70**. Arch ***voussoirs*** (single masonry units) are usually laid in radial orientation. The voussoirs may be similar in size and color to the surrounding brickwork, or the arch brick can be thinner, wider, or a different color for variation.

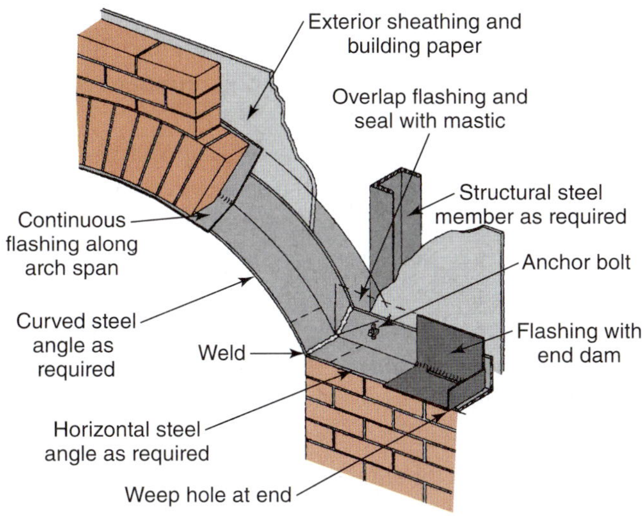

Brick Industry Association

Figure 16-69. A typical method of supporting a brick masonry arch.

Aneese/iStock/Thinkstock

Figure 16-70. The keystone is located in the center of a brick arch to enhance the appearance of the window.

Tapered or rectangular brick may be selected based on the arch type, arch dimensions, and desired appearance. Some arch types require unique shapes and sizes of brick if uniform mortar joint thickness is desired. For example, the bricks in a traditional jack arch or elliptical arch are different sizes and shapes from the abutment to the keystone. Conversely, the voussoirs of a semicircular arch are all the same size and shape.

The arch span is a consideration when selecting the arch brick. For short spans, tapered brick is recommended to avoid excessively wide mortar joints at the ***extrados*** (the curve that bounds the upper edge of the arch). Larger span arches require less taper of the voussoirs, so they can be formed with rectangular brick and tapered mortar joints. The thickness of mortar joints between arch bricks should be a minimum of 1/8″ and a maximum of 3/4″.

Depth is the dimension of an arch at the skewback that is perpendicular to the arch axis, except in jack arches. Arch depth depends on the size and orientation of the brick used to form the arch. The depth is usually a multiple of the brick's width. For structural arches, a minimum arch depth is determined from the structural requirements. If an arch is supported by a lintel, any arch depth can be used.

The depth of an arch is also based on the scale of the arch in relation to the scale of the building and surrounding brickwork. For proper scale and visual balance, arch depth should increase with increasing arch span. As a rule-of-thumb, segmental and semicircular arches should have an arch depth that equals or exceeds 1″ for every foot of arch span or 4″, whichever is greater. For example, the minimum arch depth for an 8′ span should be 8″ for segmental arches.

The depth of a jack arch is the vertical dimension of the arch at the springing. The arch depth of jack arches should equal or exceed 4″ plus 1″ for every foot of arch span or 8″, whichever is greater. For example, the minimum arch depth for an 8′ span should be 12″. Typically, the depth of a jack arch is equal to the height of 3, 4, or 5 courses of the surrounding brickwork, depending upon the course height.

Arch Construction

Two general methods are used in the construction of brick masonry arches. One method entails the use of special shapes that provide for uniform joint thickness. The second method uses regular units of uniform thickness while the joint thickness is varied to obtain the desired curvature. The arch dimensions and desired appearance determine the method to be used.

Because an arch is a structural member, it is important that all mortar joints be completely filled. The crown of the arch is generally laid in a soldier bond or rowlock header bond, and it is frequently difficult to completely fill the joints.

Arches are constructed with the aid of temporary shoring, or ***centering***, which supports the masonry while the arch is being constructed. The centering carries the load of the arch until it has developed sufficient strength to support itself and the imposed load. The centering should not be removed for at least seven days after the arch has been completed.

The centering is usually made from wood. The ribs are 2″ lumber and the lagging is usually 3/4″ × 2″ strips of wood. See **Figure 16-71**. The lagging should be cut 1″ shorter than the width of the masonry wall so it does not interfere with the mason's line. The ribs are cut to the shape of the arch. The centering must be supported adequately for the weight it is to bear.

The jack arch is relatively weak and should be supported by steel if the opening is over 2′ wide. The steel

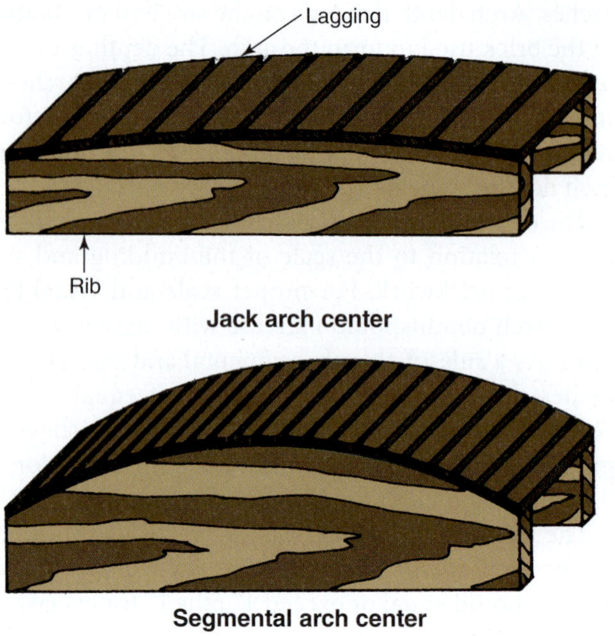

Goodheart-Willcox Publisher

Figure 16-71. Centering for jack and segmental arches.

must be bent to the ***camber*** (small rise of the arch), which is generally 1/8″ per foot of span. See **Figure 16-72**. For best appearance, each joint should be the same width as the entire length of the joint. Special bricks are required or regular bricks must be cut to have a perfect job. The bricks should be shaped so the end joints are horizontal rather than perpendicular to the radius of the arch. See **Figure 16-73**. When constructing any arch, the bricks should be laid out to determine the number required, inclination, and spacing.

The rise and span are the most important features to be considered when constructing a segmental arch. Refer to **Figure 16-74**. The length of the arch must be found in order to determine the number of courses and size of bricks needed.

- The number of courses of bricks in an arch is determined by finding the length of the extrados.
- The size of the brick may be determined by finding the length of the intrados.

The rise of a segmental arch should be 1/6, 1/8, 1/10, or 1/12 of the span. See **Figure 16-75**.

Window and Door Details

Many types of windows and doors are used in structures. Details of the windows and doors for a given situation should be provided so the mason can plan ahead.

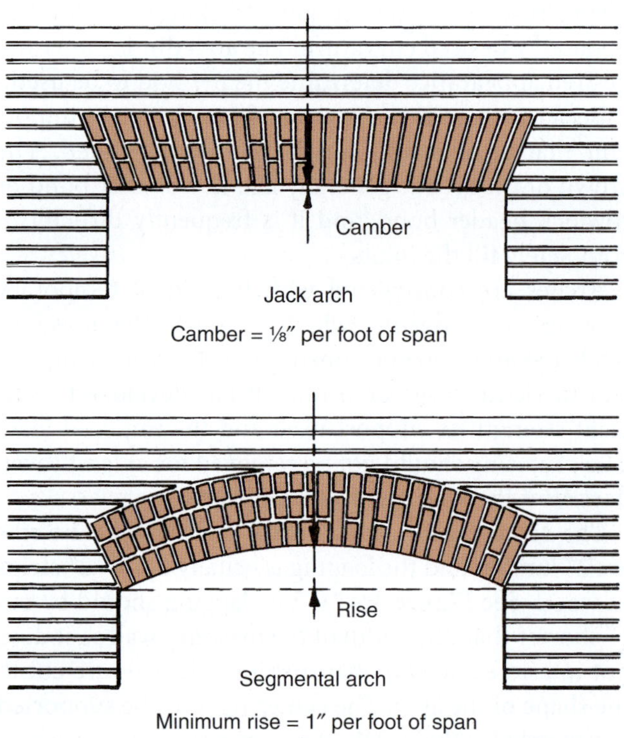

Goodheart-Willcox Publisher

Figure 16-72. The jack and segmental arches are frequently used for short spans.

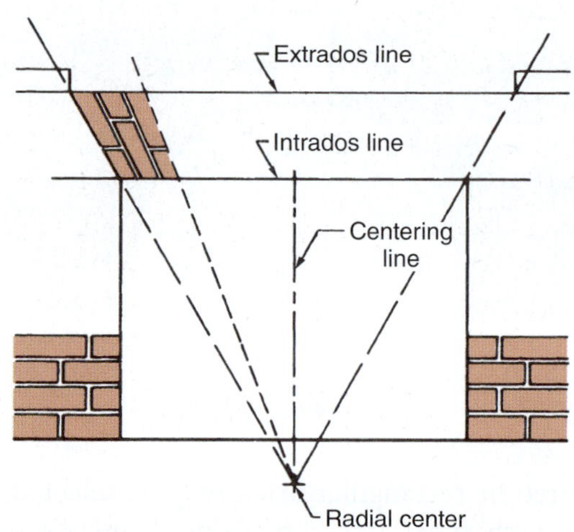

Goodheart-Willcox Publisher

Figure 16-73. Construction layout for a jack arch.

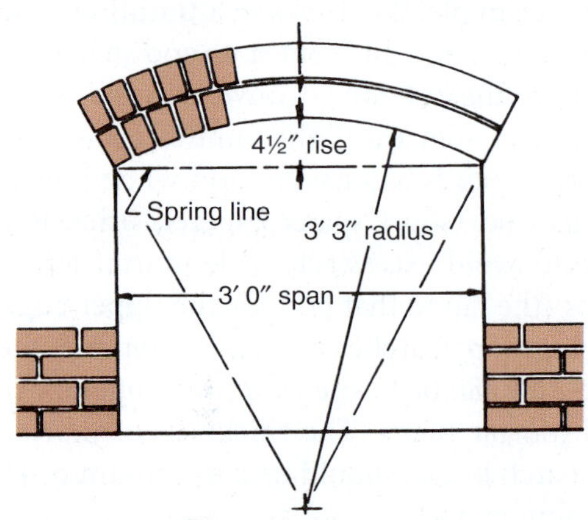

Goodheart-Willcox Publisher

Figure 16-74. Construction layout for segmental arch.

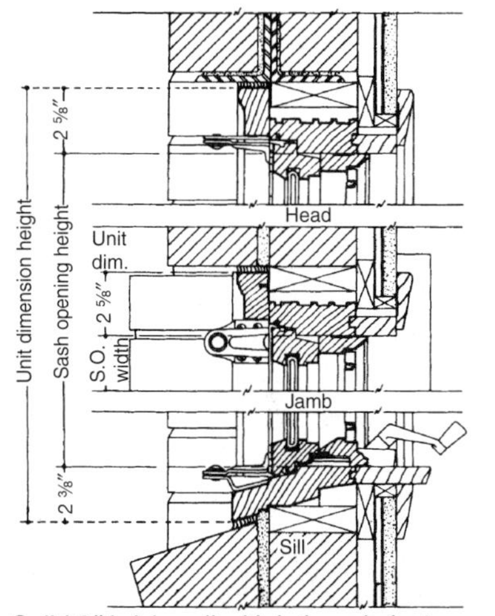

Solid 8″ brick wall with lath and plaster on ¾″ furring strips. Note full 1″ × 3⅝″ wood bucks for securing unit in opening.

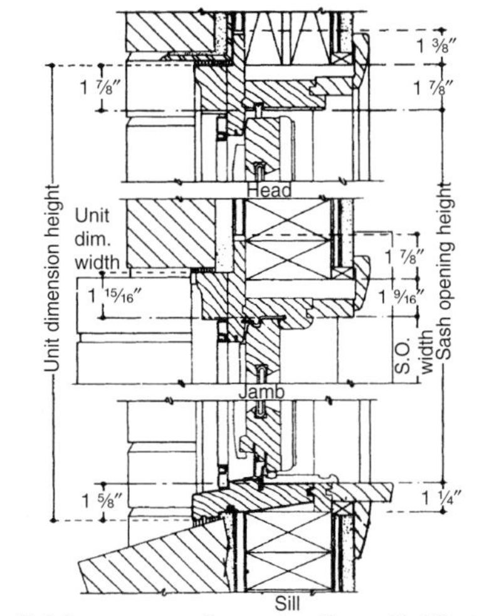

Brick veneer wall construction with blind stop flush with ½″ sheathing. Lath and plaster on interior.

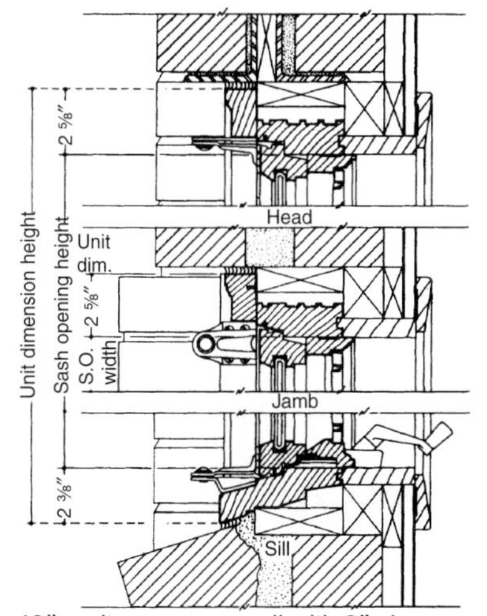

10″ cavity masonry wall with 2″ air space. Drywall interior on ¾″ furring strips.

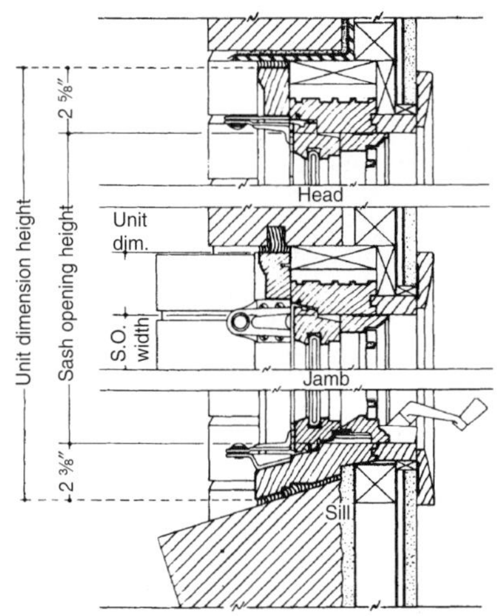

S.C.R. brick wall with lath and plaster on 2″ furring strips. Full 1″ × 3⅝″ wood bucks. Lath and plaster on interior with wood trim.

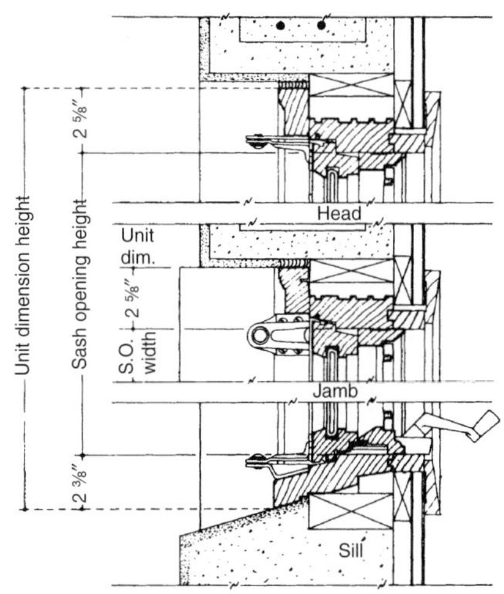

Masonry block wall with drywall applied to furring strips on interior. Stucco applied to masonry on exterior.

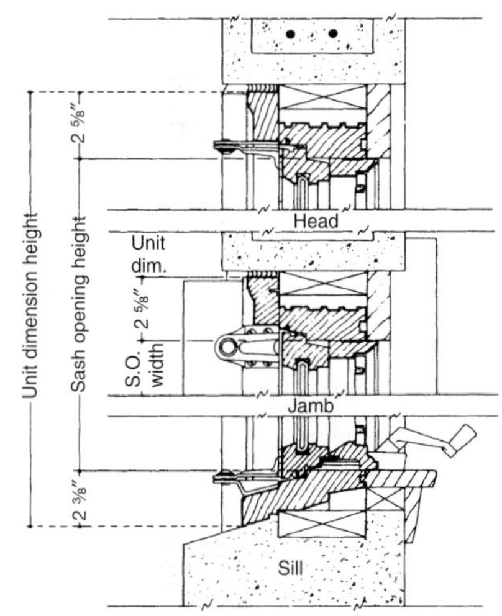

Masonry block wall without interior or exterior finish. Wood bucks for securing unit in wall are full 1″ × 3⅝″.

Anderson Corporation

Figure 16-75. Window details for masonry construction.

Details shown in **Figure 16-75** and **Figure 16-76** illustrate the relationship of the head, jambs, and sill to the masonry wall. The head is the top of a window or door. The jambs are the vertical sides of a door or window. The sill is the bottom piece connecting the jambs.

Sills

A sill channels water away from a building. A sill either consists of a single unit or multiple units. It can be built in place or prefabricated.

Single unit sills are either slip sills or lug sills. See **Figure 16-77** and **Figure 16-78**. Slip sills are the same length as the window opening; lug sills extend into the masonry on either side of the opening. A slip sill can be left out when the opening is being laid and set at a later time. A lug sill, however, must be set when the masonry is up to the bottom of the window or door opening. Sills are usually made of cut stone, brick, concrete, or metal. See **Figure 16-79.**

Joints for Movement in Masonry

Changes in volume of masonry materials result in movement. Joints are needed to allow these movements to occur and to prevent cracks in the construction. Types of movement joints include expansion joints, control joints, building expansion joints, and construction joints. Each joint type is designed to perform a specific task. The various types should not be used interchangeably.

Expansion Joint

An ***expansion joint*** separates individual brick masonry into segments to prevent cracking due to changes in temperature, moisture expansion, elastic deformation

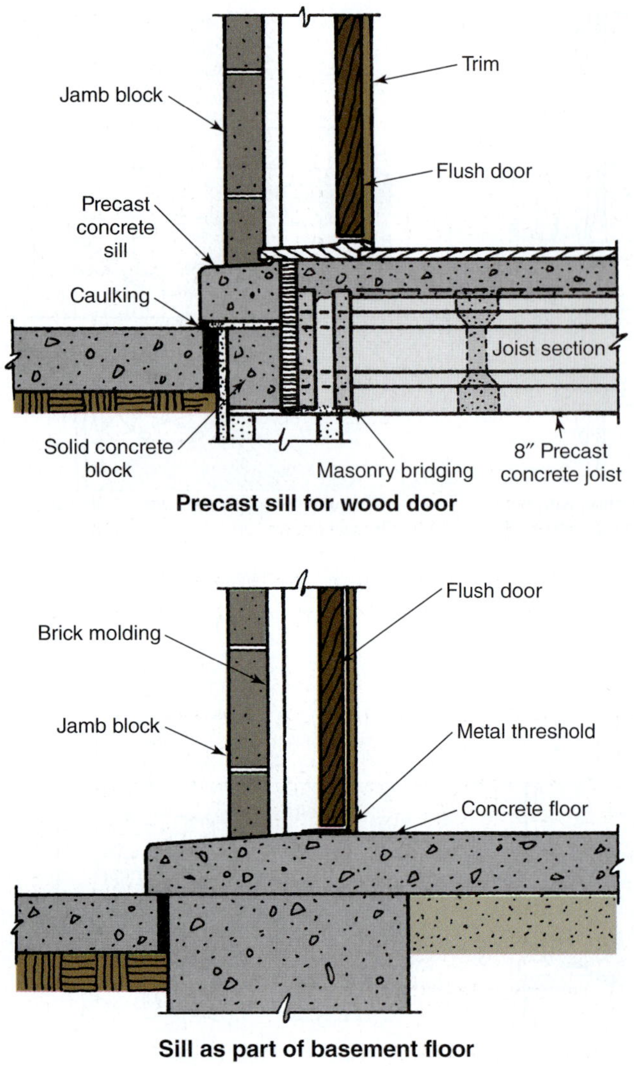

Goodheart-Willcox Publisher

Figure 16-76. Typical door sill construction in masonry walls.

Goodheart-Willcox Publisher

Figure 16-77. Stone slip sill.

due to loads, and creep. Creep is the deformation of materials under load or stress. In brick work, creep usually occurs in the masonry joints.

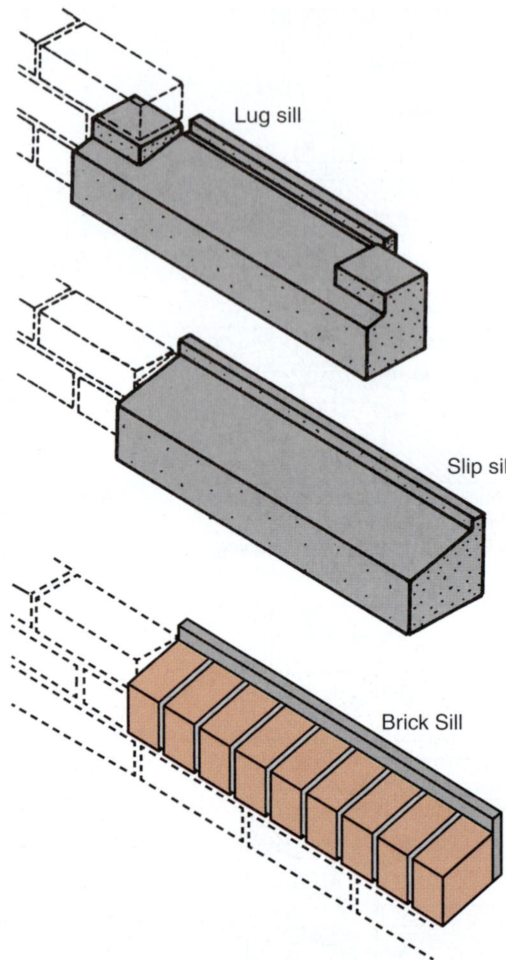

Goodheart-Willcox Publisher

Figure 16-78. How lug and slip sills of stone or cast concrete fit into a brick wall.

Expansion joints are either horizontal or vertical. See **Figure 16-80**. Highly elastic materials are placed in a continuous, unobstructed opening through each wythe, allowing the joints to close as a result of an increase in size of the assemblage. In addition to accommodating movement, expansion joints also resist water penetration and air infiltration. Expansion joints should be formed as the wall is built and must be located so the structural integrity of the masonry is not compromised.

Control Joint

A ***control joint*** creates a plane of weakness in concrete masonry. In conjunction with reinforcement or joint reinforcement, a control joint controls the location of cracks due to volume changes resulting from shrinkage and creep.

A control joint is usually a vertical opening through the concrete masonry wythe and may be formed of inelastic materials. A control joint opens rather than closes. Control joints must be located so the structural integrity of the concrete masonry is not affected.

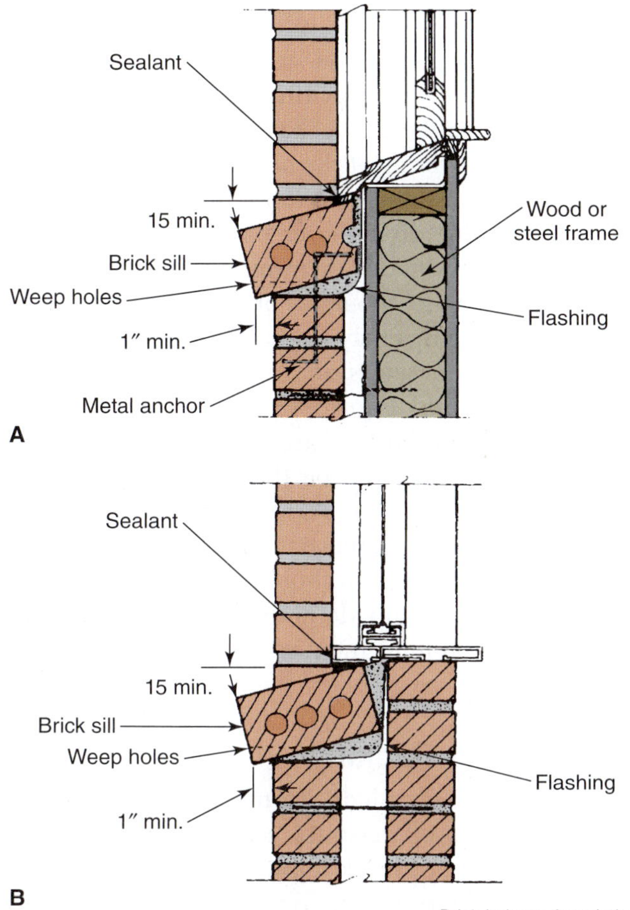

Brick Industry Association

Figure 16-79. Brick sills. A—Brick sill in veneer construction. B—Brick sill in cavity wall construction.

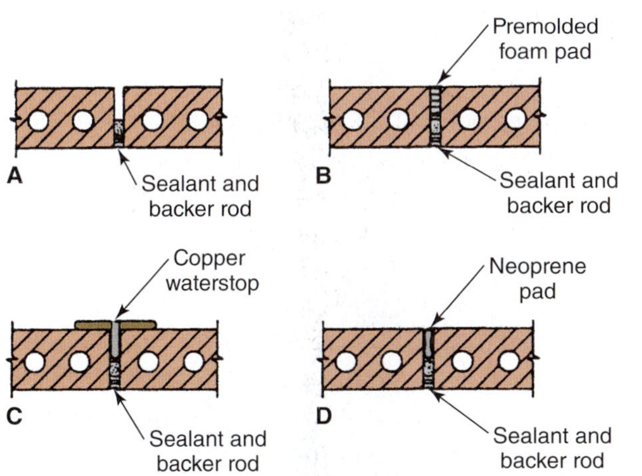

Brick Industry Association

Figure 16-80. Methods of forming vertical expansion joints in a brick wall.

Building Expansion Joint

A ***building expansion (isolation) joint*** divides a building into separate sections so stresses developed in one section do not affect the integrity of the rest of the structure. The isolation joint is a through-the-building joint that allows adjoining parts of a building to move freely from one another.

Construction Joint

A ***construction joint (cold joint)*** is used in concrete construction where masonry work is interrupted. This type of joint occurs at a location suitable to terminate work, such as in the middle of a long wall, because of the limitation of scaffolding or to aid in sheltering smaller sections of work. At a predetermined location, the mason terminates work at a vertical plane and then continues laying brick until that section of the wall is completed. The next section of the wall will then begin at the construction joint to continue until that section is completed. Construction joints are located where they are least likely to impair the strength of the structure or building.

Bond Breaks

Because brick, block, and other masonry products expand and contract differently under stress and temperature, cracks may occur due to differences in the expansion and contraction rates of those materials. Bond breaks are necessary to prevent these cracks. Building paper or flashing is used to separate, or break the bond, between two surfaces of masonry materials, such as concrete block or brick.

Bond breaks are used between floor slabs and foundations, because the concrete slab and foundation are usually under different states of stress due to loading. Bond breaks are also required when dissimilar materials come in contact with each other, such as concrete and clay masonry. Bond breaks allow each element to move somewhat independently while still providing the necessary support. A typical method of breaking the bond between walls and slabs or walls and foundations is shown in **Figure 16-81**.

Brick Masonry Soffits

A brick masonry soffit encloses a building and provides an attractive appearance. Constructing a large soffit in place often requires expensive forming and shoring. However, if only a small soffit area is involved, using a brick masonry soffit may be more efficient.

Brick masonry soffits are usually reinforced and grouted, whether they are prefabricated or built in place. Each soffit presents unique challenges with respect to configuration, support available from the structure, space restrictions, and sequence of construction. A soffit that is constructed in place often requires a complicated system of centering and falsework that must be left in place for a number of days.

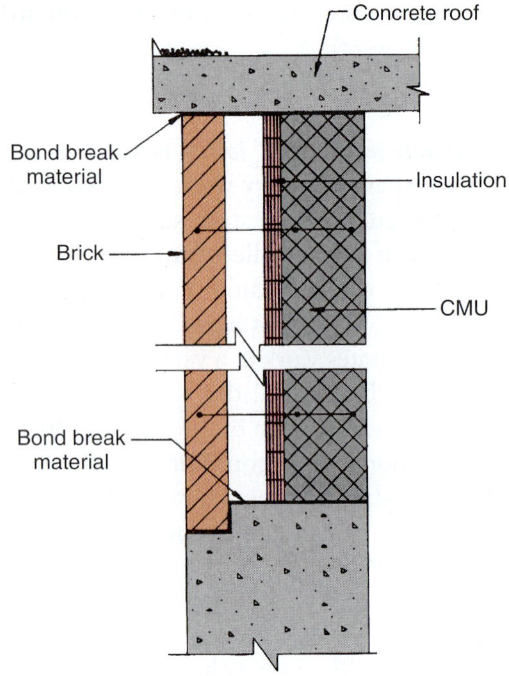

Brick Industry Association

Figure 16-81. Typical method of forming bond breaks in a loadbearing cavity wall.

Stone Quoins

Quoins are large squared stones used at corners and around openings of buildings for ornamental purposes. See **Figure 16-82**. The height of each stone should be three, five, seven, or any odd number of brick courses to permit the brick above and below the stone to have a full lap in bond. The length of each stone should be equal to one or more of the masonry units used in the wall.

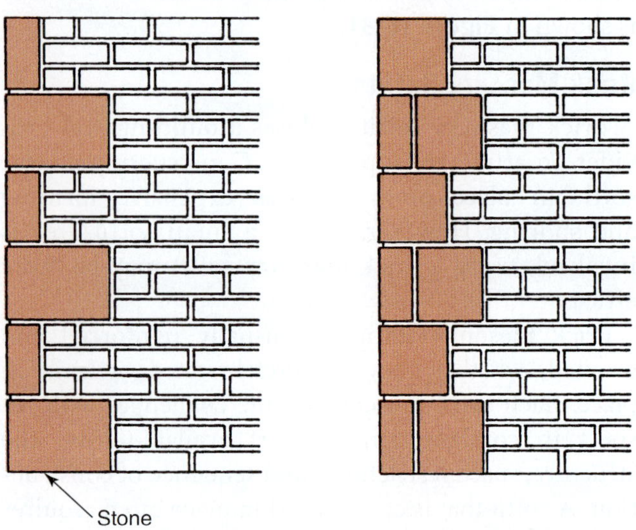

Goodheart-Willcox Publisher

Figure 16-82. Quoins used at corners for ornamental purposes.

Caps and Copings

Caps and copings channel water away from a building. A ***cap*** is a covering within the height of the wall, normally where there is a change in wall thickness. A ***coping*** is the covering at the top of a wall. The cap and coping may be a single unit or multiple units. The tops can slope in one direction or both directions. Where caps are discontinuous, a minimum slope from the ends of 1/8″ per 12″ should be provided.

Caps and copings may be made of brick, precast or cast-in-place concrete, stone, metal, or terra cotta. See **Figure 16-83** and **Figure 16-84**. Because caps and copings are exposed to extreme weather conditions, brick masonry may not be the preferred choice because it requires more joints than other materials. The thermal expansion properties of concrete, stone, and terra cotta are similar to those of brick masonry, and these materials usually do not pose extreme problems with differential movement when applied as caps and copings.

Corbels and Racking

A ***corbel*** is a shelf or ledge formed by projecting successive courses of masonry out from the face of a wall. Corbeling of brick masonry is done to achieve a particular appearance or to provide structural support.

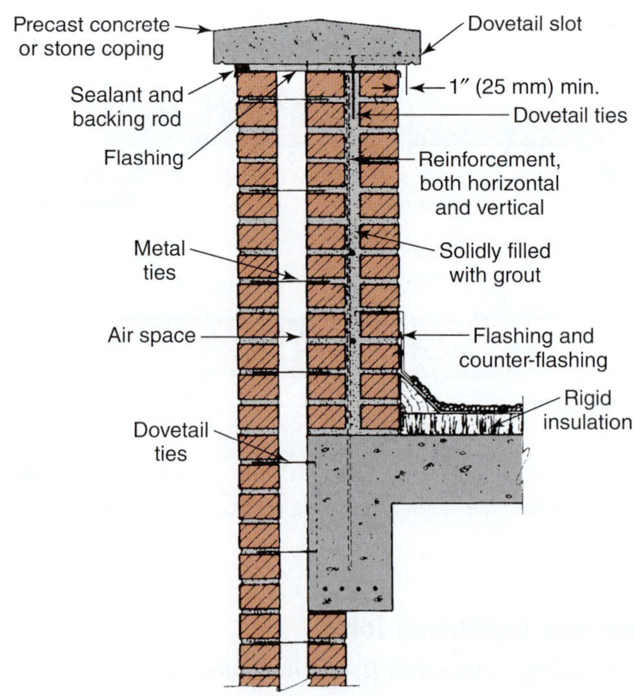

Brick Industry Association

Figure 16-83. Stone or precast concrete coping for parapet wall.

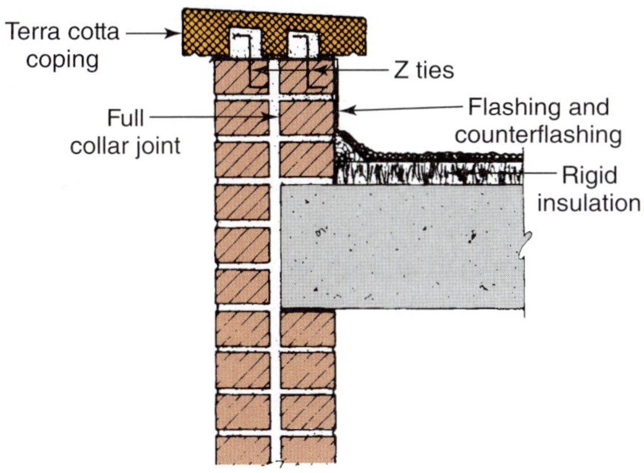

Brick Industry Association

Figure 16-84. Terra cotta coping for a parapet wall. The coping slopes toward the roof to deflect water from the front of the building.

The total horizontal projection should normally not exceed one-half the thickness of a solid wall or one-half the thickness of the veneer of a veneered wall. The projection of a single course should not exceed one-half of the unit height or one-third of the unit bed depth, whichever is less.

Code Note **Corbel Projection Limitations**

The International Residential Code (IRC) **R606.5.2** specifies that the maximum corbeled projection beyond the wall face should not exceed the following:

- One-half of the wall thickness for multiwythe walls bonded by mortar or grout and wall ties or masonry headers.
- One-half the wythe thickness for single wythe walls, masonry-bonded hollow walls, multiwythe walls with open collar joints, and veneer walls.

Racking is masonry in which successive courses are stepped back from the face of the wall. When racking back to achieve the desired dimensions, make sure the cores of the units are not exposed.

Chapter Review

Summary

- Solid masonry walls are built of solid masonry units, hollow masonry units, or a combination of the two. All joints between the units are filled with mortar.
- Exterior nonloadbearing walls may be curtain walls (not wholly supported at each story) or panel walls (supported at each story and self-supporting between stories).
- A cavity wall is built of masonry units arranged to provide a continuous air space of 2″ to 3″ wide.
- Cavity walls exhibit superior rain penetration resistance, excellent thermal capabilities, good sound transmission resistance, and high fire resistance.
- The cavity in a cavity wall must be kept free of mortar droppings during construction.
- In anchored brick veneer construction, nominal 3″ or 4″ thick exterior brick wythe is anchored to a backing system with metal ties. There is a clear air space between the veneer and the backing system.
- In brick veneer wall assemblies, flashing and weep holes work with the air space to resist moisture penetration.
- Moisture problems in masonry walls can be minimized with vapor barriers, dehumidification, ventilation, and heat exchangers.
- Reinforced masonry walls are built with steel reinforcement embedded in the masonry units and are structurally bonded by grout that is poured into the cavity (collar joint) between the wythes of masonry.
- Thin brick units are approximately 1/2″ to 1″ thick. Thin brick veneer is an adhered veneer.
- Adhered veneer construction can be classified as thin bed set or thick bed set. In addition, thin brick can be applied to the exterior of precast wall panels prior to casting.
- Retaining walls and other hardscape units are made from concrete masonry products. Retaining wall units have a lip on the lower back edge. The units are placed so the lip locks behind the lower course, providing proper alignment and preventing forward movement exerted by earth pressure.
- Chases and recesses are horizontal or vertical spaces left in a wall for the purpose of containing plumbing, heating ducts, electrical wiring, or other equipment.
- In an arch, masonry units span an opening by transferring vertical loads laterally to adjacent masonry units, and thus to the abutments.
- Common building arches include the jack arch, segmental arch, semicircular arch, and multicentered arch. An arch is normally classified by the curve of its intrados and by its function, shape, or architectural style.
- For short arch spans, tapered brick is recommended to avoid mortar joints that are too wide at the extrados. Larger span arches can be formed with rectangular brick and tapered mortar joints.
- Window and door details include the head (top of a window or door), the jambs (vertical sides of a door or window), and the sill (bottom piece connecting the jambs).
- Types of movement joints include expansion joints, control joints, building expansion joints, and construction joints.
- Bond breaks use building paper or flashing to separate two surfaces of masonry materials with different expansion and contraction rates to prevent cracks.
- Caps and copings channel water away from a building. A cap is a covering within the height of the wall. A coping is the covering at the top of a wall.

Know and Understand

Answer the following questions using the information provided in this chapter.

1. The characteristic _____ describes the heat capacity and surface area capable of affecting building thermal loads by storing heat and releasing it at a later time.
   A. R-value
   B. T-value
   C. U-value
   D. thermal mass
2. A _____ masonry wall is built of masonry units laid close together.
   A. hollow
   B. solid
   C. reinforced
   D. composite
3. *True or False?* Solid masonry walls can be built of hollow masonry units.
4. *True or False?* Curtain walls are built between structural elements.
5. The two types of hollow masonry walls are cavity walls and _____ walls.
   A. curtain
   B. panel
   C. masonry bonded
   D. composite
6. Which of the following best describes a weep hole?
   A. A barrier to moisture
   B. Space between the top of a window of one story to the sill of the window on a second story
   C. Holes through which metal ties are placed
   D. A void that allows moisture to exit
7. Rigid board materials include all the following *except* _____.
   A. water repellent vermiculite
   B. molded polystyrene
   C. cellular glass
   D. perlite board
8. A _____ is the space between the top of a window of one story to the sill of the window on a second story.
   A. spandrel
   B. weep hole
   C. flashing
   D. granular space
9. A _____ must be positioned above grade but as close to the bottom of the wall as possible.
   A. spandrel
   B. weep hole
   C. head joint
   D. granular space
10. Brick veneer that is backed by wood or metal frame must be supported by _____ over openings.
    A. spandrels
    B. sills
    C. lintels
    D. jambs
11. A _____ wall is two wythes bonded together with masonry or wire ties.
    A. hollow
    B. composite
    C. reinforced
    D. solid
12. *True or False?* Condensation problems most frequently occur during the cooling season.
13. Which of the following methods can be used to resist condensation?
    A. Construct buildings with homogenous walls
    B. Be sure buildings are highly insulated, especially during the heating season
    C. Alternate freezing and thawing of masonry units when saturated
    D. Place mechanical devices in high-moisture locations
14. _____ masonry walls are structurally bonded by grout that is poured into the cavity between the wythes of masonry.
    A. Hollow
    B. Solid
    C. Reinforced
    D. Composite
15. Which type of thin brick veneer installation procedure is commonly used on interior and exterior surfaces?
    A. Modular bed set
    B. Thin bed set
    C. Thick bed set
    D. Precast panels
16. Which of the following construction techniques is *not* adhered veneer construction?
    A. Precast panels
    B. Thick bed set
    C. Thin bed set
    D. None are correct.

17. *True or False?* Retaining walls may be called *hardscape.*
18. When installing a retaining wall, the step that follows excavating is to _____.
    A. contact your local building code office
    B. prepare the subbase
    C. install the base course
    D. compact backfill soil
19. The _____ is the curve that bounds the lower edge of an arch.
    A. soffit
    B. rise
    C. extrados
    D. intrados
20. _____ arches are flat with zero or little rise.
    A. Jack
    B. Segmental
    C. Semicircular
    D. Multicentered
21. The _____ joint divides a building into separate sections so stresses developed in one section do not affect the integrity of the rest of the structure.
    A. control
    B. building expansion
    C. construction
    D. section
22. The _____ joint creates a plane of weakness in concrete masonry.
    A. control
    B. building expansion
    C. construction
    D. section
23. _____ prevent cracks that occur due to differences in the expansion and contraction rates of masonry products.
    A. Expansion joints
    B. Stone quoins
    C. Bond breaks
    D. Soffits
24. A _____ is the covering at the top of a wall.
    A. cap
    B. coping
    C. corbel
    D. racking
25. *True or False?* A cap is a shelf or ledge formed by projecting courses of masonry from the face of a wall.

Apply and Analyze

1. List two insulation materials used in cavity walls.
2. Where should horizontal reinforcement be placed in reinforced masonry walls?
3. Name two types of retaining walls.
4. What is the purpose of an expansion joint?
5. Which type of garden wall provides lateral support through its shape?
6. What is the primary function of caps and copings?
7. What is a corbel?
8. What is the purpose of the locking lip on the lower back edge of masonry units used to construct retaining walls?

Critical Thinking

1. Imagine you are the supervisor of TLA Landscaping Company. The owner has identified you as the manager of a landscaping project being constructed around a new home. The project is located in a northern climate. Prior to beginning work, what three steps of preparation activities should you perform to ensure that the retaining walls to be built are not affected by the severe weather condition in this community?

Activities

1. (ENGINEERING) Research different types of arch bridges. Prepare a report comparing two different types of bridges in terms of their appearance, function, and construction. Include an image of each in your report.
2. (TECH) Research and report on the ways in which precast concrete panel technology contributes to green building practices.

Timothy L. Andera

CHAPTER 17

Paving and Masonry Construction Details

LEARNING OUTCOMES

After studying this chapter, you will be able to:

17.1-1 Describe appropriate materials and procedures used to install floors, pavement, and steps.
17.2-1 Identify the components of fireplaces.
17.2-2 Explain how to construct a chimney according to code specifications.
17.2-3 Describe two types of steel fabricated fireplaces.

TECHNICAL TERMS

combustion chamber
damper
draft
firebox
flue lining
hard-burned brick
Haydite
paver bricks
polymeric sand
riser
smoke chamber
smoke shelf
throat
tread

Introduction

Special applications of brick include the fabrication of floors, walkways, and stairs. Brick is also used to construct fireplaces and related structures. Fireplaces are categorized in this chapter into two major types—masonry construction and steel fabricated units.

17.1 Floors, Pavements, and Steps

Because of their color, pattern, and high resistance to abrasion, brick, tile, and stone are desirable surfaces for roadways, walks, patios, and interior floors. These materials can be used over the ground, over a floor, or incorporated within a structural floor system.

Several factors affect the selection of bricks and jointing materials, including traffic, exposure to moisture and freezing, and desired appearance. Hardened bricks are used for industrial floors and other locations where extreme wear would occur if a softer brick were used. ***Hard-burned bricks*** are resistant to wear because they are fired in a kiln at high temperatures to increase their mechanical strength, moisture resistance, and weather resistance. For residential and nonindustrial floors, Grade SW brick is satisfactory.

For exterior pavements, brick's resistance to freezing and thawing damage is just as important as its resistance to abrasion. Brick that has an average saturation coefficient of 0.78 or less is recommended for severe weathering conditions. The saturation coefficient is a ratio of the weight of water absorbed by brick during immersion in cold water to its weight during immersion in boiling water. This ratio indicates the resistance of the brick to freezing and thawing based on how much moisture the brick will absorb. Refer to ASTM Specifications C216-14, *Standard Specification for Facing Brick (Solid Masonry Units Made From Clay or Shale)* and C62-13a, *Standard Specification for Building Brick (Solid Masonry Units Made From Clay or Shale)*. These standards pertain to heat treatments, strength and moisture resistance of brick, and procedures for selecting appropriate brick.

Finished surface appearance depends largely on the color, the unit sizes, and the pattern in which the units are laid. Bricks laid flat with the largest plane surface horizontal are generally noncored bricks. Cored bricks can be used if they are placed on edge. **Figure 17-1** and **Figure 17-2** show several pattern bonds that can be used for floors and pavements.

Floors

Interior masonry floors are usually laid over concrete slabs. See **Figure 17-3**. The following are the primary elements of an unreinforced masonry floor:

- Base (provides the support)
- Cushion (intermediate layer that aids in leveling)
- Wearing surface
- Joints

Floors or pavements can be laid without mortar joints, but most interior applications have mortar joints. The procedure for laying a masonry floor on a slab begins with choosing a pattern. Next, arrange the materials for easy access. Form a cushion of mortar for each unit as it is laid and leveled. Strike the joints and brush away excess mortar with burlap rags as it begins to harden. The joints can be tooled for maximum hardness when the mortar has sufficiently hardened.

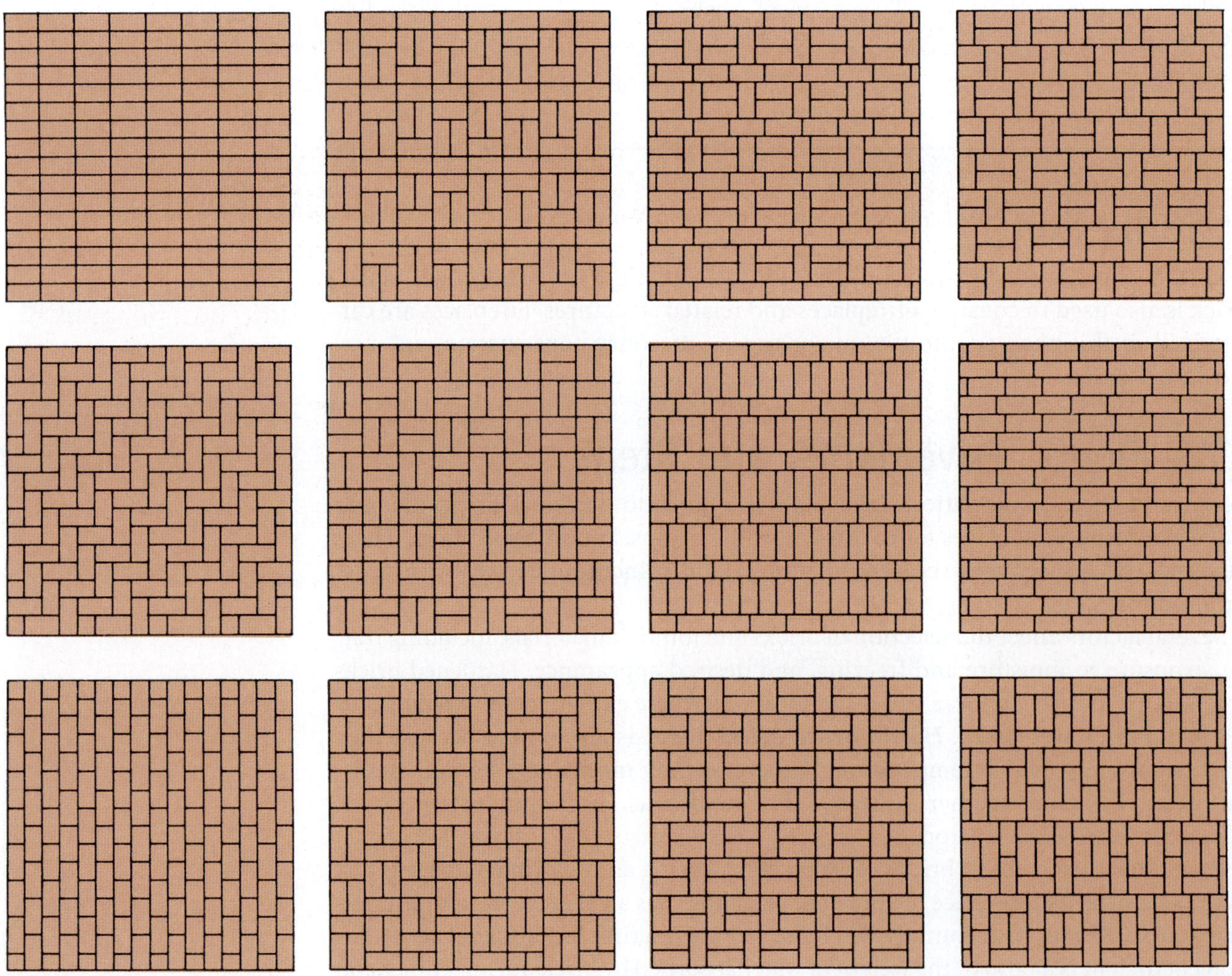

Goodheart-Willcox Publisher

Figure 17-1. Brick pattern bonds commonly used for floors and pavements.

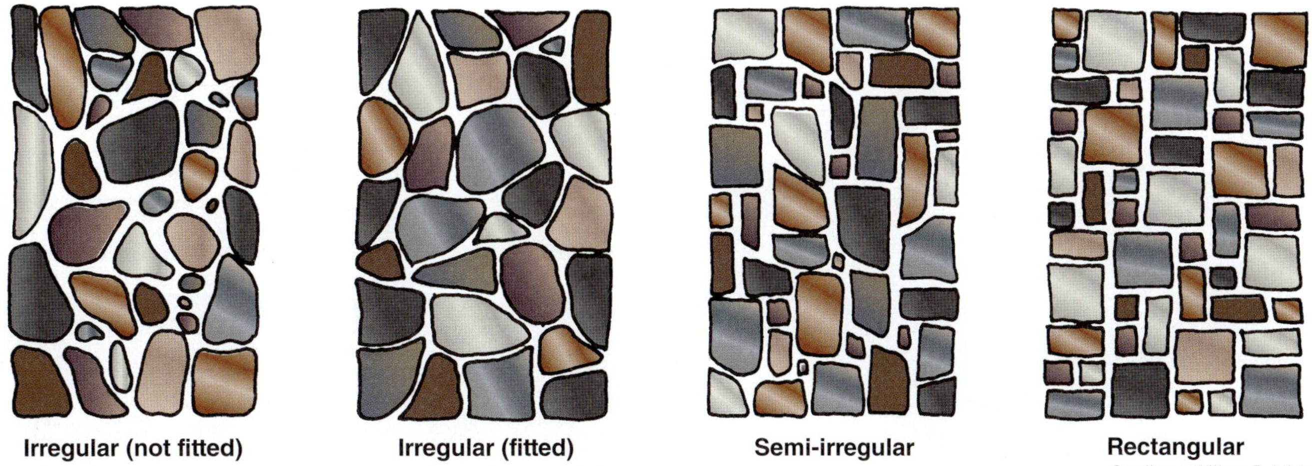

Goodheart-Willcox Publisher

Figure 17-2. Stone pattern bonds frequently used for floors and pavements.

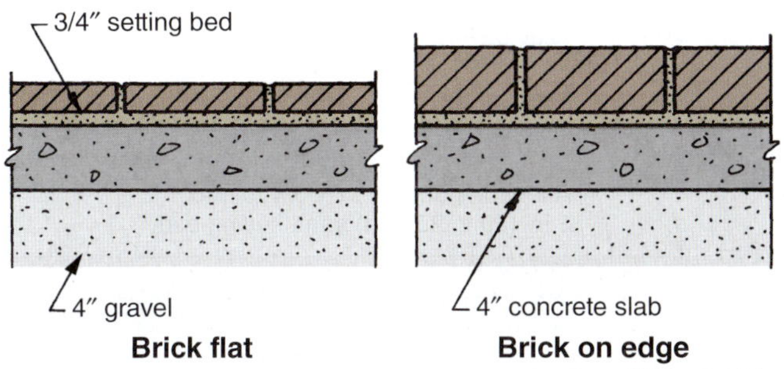

Goodheart-Willcox Publisher

Figure 17-3. Sections of floors with brick laid over a concrete slab.

Sometimes the brick may have a film of grout on the exposed surfaces. A final cleaning can be performed using clean burlap. When cleaning, try to rub diagonally across the joints so the mortar is not removed from the joints.

Pavement

Exterior pavings for walks, patios, and driveways are similar to those used for interior floors. However, care must be taken to select materials that can withstand the weather. It is important when planning any paving project to determine the traffic that will be carried by the pavers and underlying supporting structure of sand and aggregate. **Figure 17-4** displays different procedures based on the intended traffic applied to the brick paving and underlying structural layers. As traffic becomes heavier, the brick type may change depending on manufacturer's specifications. The layers of supporting materials may increase based on codes or engineering requirements. **Figure 17-4** is from the Brick Industry Association (BIA) Technical Notes on Brick Construction 14a, May 2017, "Paving Systems Using Clay Pavers on a Sand Setting Bed." For more detailed information on brick and clay pavers, refer to BIA's Technical Notes.

Employ quality workmanship when building the subgrade, subbase, base, sand bed, and the selected brick. The finished paving job will only be as good as the underlying quality of the supporting substructures of sand, aggregate and soil, and the installation of the paving materials.

Drainage must also be considered. Walks should be sloped to one or both sides to shed surface water. See **Figure 17-5**. Slope patios away from buildings, retaining walls, or other obstructions that hold water on the surface. The slope is generally 1/8″ to 1/4″ per foot for walks and patios.

Below-surface drainage may be required for larger paved areas or locations with a high water table (the level below which the ground is saturated with water). A layer of gravel under the slab, or masonry units if no slab is used, usually provides the necessary drainage and stops the upward capillary flow of moisture. An expansion joint should be installed between the slab and building walls to provide for expansion.

Pavers tend to spread or shift at the edges under normal traffic and weather conditions. Therefore, an edging is desirable. **Figure 17-6** shows a soldier course of bricks used as an edging. The edging is normally placed first and can be used as a guide for elevation and slope.

Paver bricks are normally 1 1/4″ × 3 5/8″ × 7 5/8″ in size. They are approximately 1″ thinner than regular brick and are designed especially for paving. The sizes of pavers in width, length, and thickness will vary from manufacturer to manufacturer and the intended application such as sidewalks, driveways, or streets.

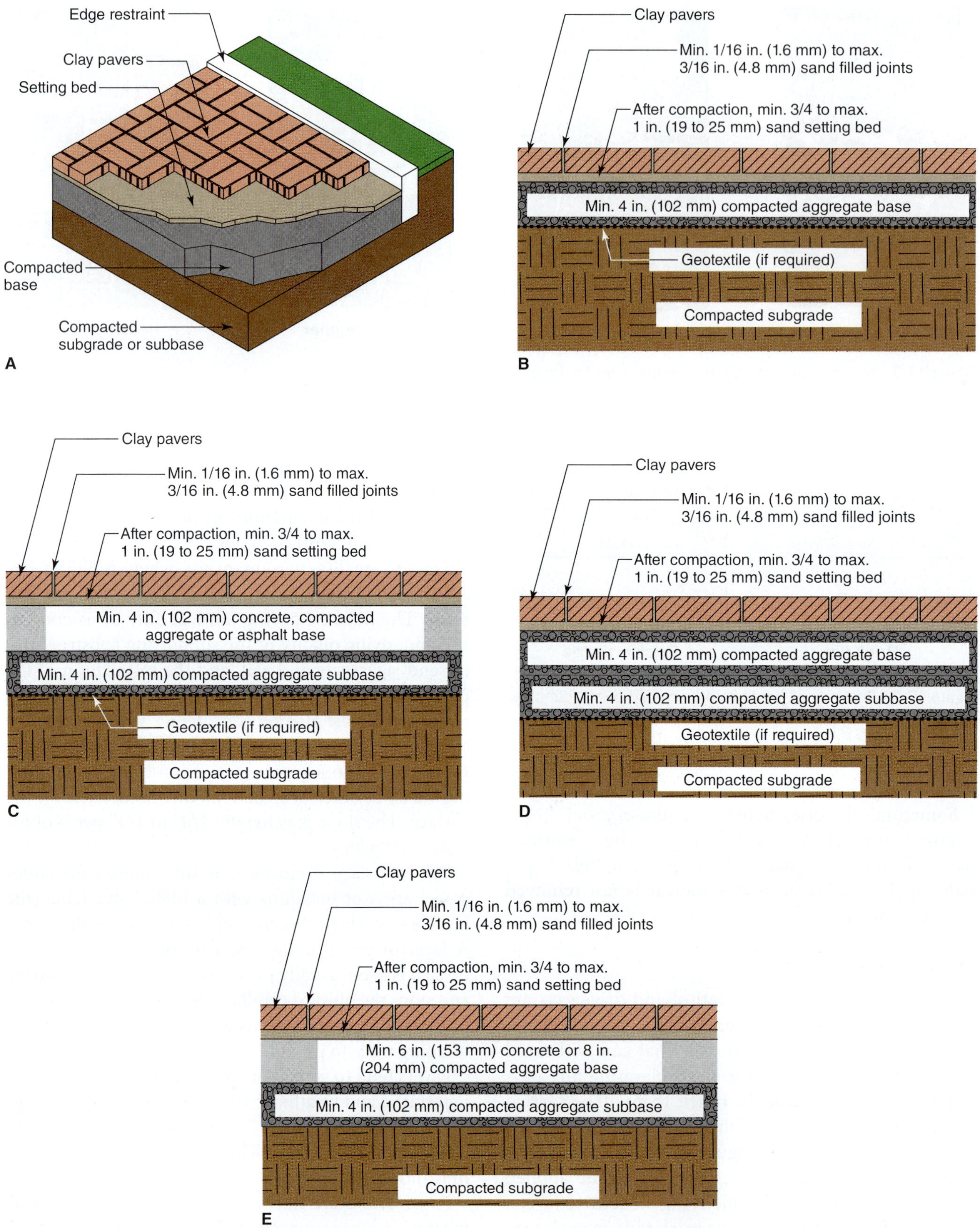

Brick Industry Association

Figure 17-4. A—Typical brick pavement. B—Typical residential patio or walkway. C—Typical residential driveway. D—Typical commercial or pedestrian public sidewalk. E—Typical light-duty vehicular paving.

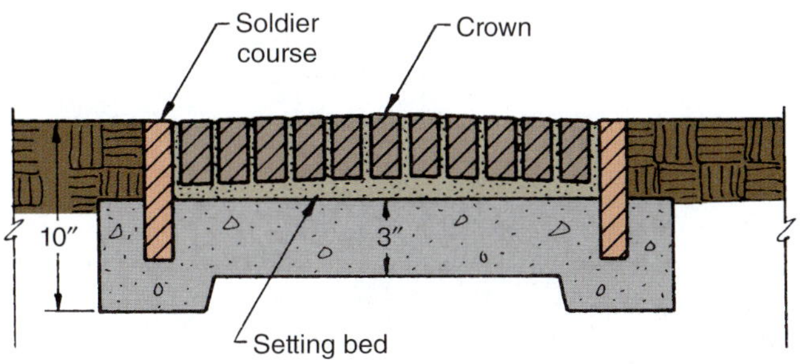

Goodheart-Willcox Publisher

Figure 17-5. Brick sidewalk on a concrete foundation, sloped to both sides.

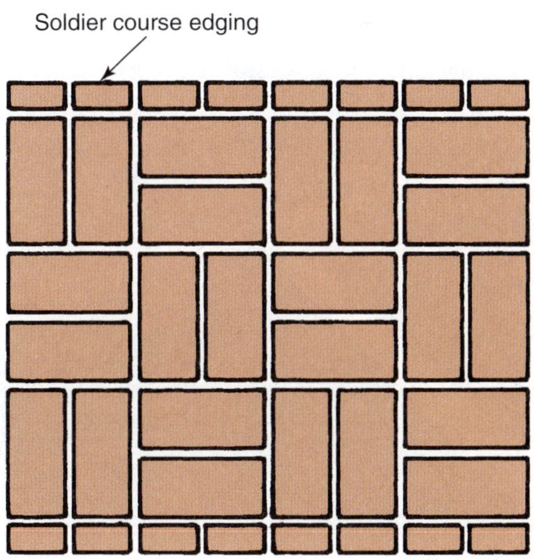

Goodheart-Willcox Publisher

Figure 17-6. This sidewalk uses a soldier course of brick as an edging.

Green Note

Permeable Pavers

Permeable pavers, composed of a layer of concrete or fired clay brick, are separated by joints filled with crushed aggregate. Rainwater seeps through these joints. Below the pavers are layers of different grades of aggregates, which retain the water.

When rainwater falls on impermeable surfaces such as concrete, runoff accumulates toxic pollutants and debris that enter sewer systems. Eventually, the contaminated runoff reaches groundwater sources, streams, and larger bodies of water. In a permeable paver system, rainwater passes around the paver and soaks into the ground, reducing the loss of water through runoff. Permeable paving is used for pedestrian-only areas and for areas with low traffic volume, such as parking lots and residential driveways.

PROCEDURE

Laying Paving Bricks

1. Plan where the paving bricks are to be installed. See **Figure 17-7**.

Timothy L. Andera

Figure 17-7. The yellow mason's line on the right side was used as a guide to spray paint the line on the grass. The painted line identifies the right-side working edge where the excavation and paving project will end. A metal stake on the left identifies the front edge of the pavers. Pavers will butt up against the driveway.

2. Remove existing materials such as sidewalks, landscape rocks, or other materials.
3. Calculate the number of bricks or paver units to be used for the project.
4. Determine the amount of base gravel or crushed stone needed and the amount of sand needed for bedding.
5. Determine the grade slope to ensure adequate paver drainage.

Trade Tip

When determining the slope needed for the final surface of the paving bricks, the excavation base grade should slope 1/8″ to 1/4″ per foot to ensure adequate drainage.

6. Begin to install forms to be used as guides for screeding base material, sand, and positioning pavers. **Figure 17-8**.
7. Complete installation of forms. See **Figure 17-9**.
8. Excavate the area to the appropriate depth. See **Figure 17-10**.
9. Check formwork to ensure the grades and slopes are accurate. Use mason's line and blocks to adjust the forms for sags or bows along the distance of the formwork. See **Figure 17-11** and **Figure 17-12**.

PROCEDURE *(continued)*

Timothy L. Andera

Figure 17-8. Temporary form boards are placed on the garage wall and the end of the project. The blue spray paint line on the grass indicates the end of excavation.

Timothy L. Andera

Figure 17-9. A wood form is attached to the wall between the driveway and walk-in door. This form will aid in screeding the base material and bedding sand.

Timothy L. Andera

Figure 17-10. A backhoe is being used to excavate the project.

Timothy L. Andera

Figure 17-11. A mason's line and wood block are used to straighten the form board. The block lifts the line off the form to ensure accuracy.

Trade Tip

When severe rain or weather is expected, place plastic over the bedding sand to protect it. Lumber can be used to hold down the plastic in an event of strong winds and rains.

10. Place screed in formwork. See **Figure 17-13**. Add base material in 2″ to 3″ lifts inside the forms and screed. Spray the base material with water to aid in compaction. Compact between each lift until the appropriate elevation is reached. See **Figure 17-14**. In areas where the plate compactor cannot operate, use a hand tamper.
11. If necessary, install fabric between base material and bedding sand to aid in keeping sand from migrating into base materials. See **Figure 17-15**.
12. Install bedding sand, screed, then compact. Repeat if needed. See **Figure 17-16**.
13. Determine the pattern in which the brick pavers will be laid.
14. When designing the pattern, paver joints and bricks need to be straight, parallel, and perpendicular with the structure or forms they are laid in. For larger distances, a mason's line can be used as a straight edge or guide for laying pavers. Use a paver at the beginning and end of the layout process to determine the distance from the structure. See **Figure 17-17**. Check the line at multiple locations to determine if the structure has a bow. Readjust the lines on the formwork to compensate for structure variances.

PROCEDURE *(continued)*

Timothy L. Andera

Figure 17-12. A spacer block is used to determine if the form needs to be raised or lowered. The mason's line also identifies if the form needs to be moved in or out to ensure straightness.

Timothy L. Andera

Figure 17-13. The screed allows for easy adjustments when preparing for the next lift of base material. Check base material or screed with a level or transit to determine if the slope is adequate to ensure proper drainage of finished project.

15. Use a square or a mason's line to ensure the pavers are 90° to the structure and form boards. On larger projects the 3-4-5 method discussed in Chapter 6, **Math for Masonry Careers** can be used to determine a 90° line as a guide to follow. See **Figure 17-18**.
16. Adjust pavers to ensure joints are tight, **Figure 17-19A**. When starting a new course, use the mason's line and square to line up the first paver in the course with both guides. Once the bricks are in position, tap them lightly with a mallet or rubber ended hammer to set the pavers into the sand base. See **Figure 17-19B**. Check to ensure that the newly laid paver is flush with previously laid pavers.

Timothy L. Andera

Figure 17-14. A plate compactor is being used to vibrate and pack base material. Note the mason's use of proper PPE.

Timothy L. Andera

Figure 17-15. Fabric is placed over base material prior to adding sand.

Timothy L. Andera

Figure 17-16. A hardscape mason is installing bedding sand. After installation, the sand will be sprayed with water and compacted.

PROCEDURE *(continued)*

A

B

Timothy L. Andera

Figure 17-17. A—A temporary positioning paver is added at the end of the formwork. A mason's line is attached to the formwork to use as a guide. B—A temporary paver placed to position the mason's line. At the initial layout, a pencil line was marked on the wall vertically to identify the location of the form board. A blue chalk line on the wall indicates the elevation of the top side of the pavers.

Timothy L. Andera

Figure 17-18. A large square is used to transfer a 90° line from the wall to the outside form board. The wooden plank is used as a kneeing or walkway, so the bedding sand is not disturbed when installing pavers.

A

B

Timothy L. Andera

Figure 17-19. A—Place pavers carefully making sure to not affect the bedding sand. Notice the mason's line used as a guide. B—Make minor adjustments to ensure gaps are uniform, tops are flush, and pavers are set into the sand base.

17. Continue laying pavers. See **Figure 17-20**. Set pavers flat, as to not move the sand base, then make minor adjustments.
18. Slope is important to factor in for adequate drainage. On this project, pavers were sloped away from house, but also had to follow the slope of the driveway. See **Figure 17-21**. Use a level to determine the slope needed for the project.
19. When pavers need to be cut to fit special locations such doorways, use a masonry saw to perform the task. See **Figure 17-22**.

Safety Note

When cutting pavers with a masonry saw, always wear eye protection, ear protection, face protection, and a respirator. When working with sand, always wear an appropriate respirator because sand may contain silica, which can cause severe health hazards.

PROCEDURE *(continued)*

Timothy L. Andera

Figure 17-20. The last regular size paver has been placed. Red border pavers will be placed around the exterior of the walkway.

Timothy L. Andera

Figure 17-21. An example of the slope on the paving project, this slope is from the doorway to the edge of the lawn.

Timothy L. Andera

Figure 17-22. A wet masonry saw is used to cut a paver which will be placed next to a walk-in doorway.

20. Install edge restraints. They can be made of plastic, stainless steel, or brick. If bricks are used, they are laid vertically. Edge restraints' purpose are to keep pavers from shifting when walked on or driven on. When designing the layout, add extra space for the edge restraints to be inside the boundary lines of the form boards. Doing this allows the edge restraints to be installed prior to the removal of the form boards for less of a chance of disturbing the pavers. See **Figure 17-23**.

Timothy L. Andera

Figure 17-23. Restraint edging made of plastic with 10″ plastic spikes. The form boards were installed about 5″ outside of the pavers in the initial setup.

21. After all the paver restraints have been installed, remove the form boards. Fill the areas where forms were with soil and compact against restraints. This will reinforce the edge restraints, so they do not move when pavers are compacted.
22. Use a vibrating plate compactor or hand tamper to set the pavers into the sand base. See **Figure 17-24**. A plate compactor or hand tamper vibrates and settles pavers uniformly to create a smooth surface.

Trade Tip

When preparing to use a plate compactor for setting pavers, it is important that the plate has a protective cover. The metal plate can cause scratches on the exposed area of the pavers.

When using a hand tamper, it is important to protect the surface when the tamper strikes the surface of the pavers. A sheet of plywood or oriented strand board (OSB) will protect the surface and spread impact force over a larger surface area.

23. Place polymeric sand over the pavers and work it into joints with a broom. See **Figure 17-25**. ***Polymeric sand*** is a mixture of fine sand and water activated polymers that are intended to fill, bind, and seal joints. Always wear proper PPE, as the sand contains silica dust.

PROCEDURE *(continued)*

Timothy L. Andera

Figure 17-24. A hand tamper is being used to set pavers into the sand base. A 20″ × 20″ × 3/4″ sheet of plywood is used as a protection plate. The plywood ensures that multiple pavers are leveled at the same time to create a smooth surface.

Timothy L. Andera

Figure 17-25. Broom polymeric sand in multiple directions to adequately spread the sand into the paver joints.

24. Retamp or vibrate the pavers to work the polymeric sand into the joints. Resand joints to fill voids and broom again.
25. Clean the paver surface to remove any excess polymeric dust. See **Figure 17-26**. The dust can stick to the surface of the paver after it becomes wet and can discolor the face.
26. Lightly spray water on polymeric sand joints to activate the binding agents in the sand. After a minute or two, respray the surface to ensure water has soaked all the way into the sand joint. See **Figure 17-27**. Too much water or high-water pressure can wash polymeric sand from joints.
27. Prepare for spraying the sealer by taping off walls and covering other surfaces to protect them from the sealer. Apply the sealer in light coats so the product does not pool on the surface of the pavers. See **Figure 17-28**. Wear the proper PPE identified in the sealer instructions or the related SDS forms. **Figure 17-29** shows the completed project.

Timothy L. Andera

Figure 17-26. Use a vacuum or hand blower to lightly blow dust off pavers. Do not use excessive air as it can remove polymeric sand from joints.

Timothy L. Andera

Figure 17-27. A gentle spray of water will activate polymeric sand in joints.

PROCEDURE *(continued)*

Timothy L. Andera

Figure 17-28. Set the nozzle of the sprayer to create a mist when spraying the pavers and joints. Multiple coats of sealer will aid in protecting pavers and joints.

Timothy L. Andera

Figure 17-29. The completed paving project. The sheen on the surface of the pavers is from the sealer protection.

Steps

Steps made of brick, block, or stone offer a practical solution to a sloping terrain. They also present a particular appearance in an interior setting. A good approach to building steps or stairs with masonry units is to pour a concrete foundation for the structure and cap the ***treads*** (horizontal portions of the steps) and ***risers*** (vertical members between treads) with the material selected. **Figure 17-30** shows a typical set of steps capped with brick.

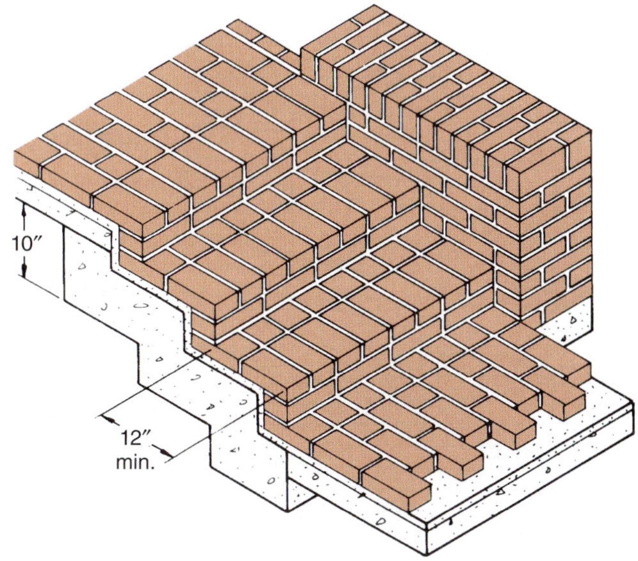

Goodheart-Willcox Publisher

Figure 17-30. Brick steps laid over a concrete foundation.

When designing steps, ensure that each tread is the same size and each riser is identical. Any change in height of riser or depth of tread invites an accident. Treads for all outside steps are usually 12″. A riser height equal to two regular bricks plus mortar joints is an appropriate height. Treads should be sloped about 1/4″ to the front edge to aid in water drainage.

17.2 Fireplace and Chimney Construction

The design and construction of fireplaces and chimneys has a long history. Much data has been collected regarding the most efficient designs for these structures. Follow recommended practices and procedures to ensure the finished product performs satisfactorily. The International Residential Code (IRC) Chapter 10, *Chimneys and Fireplaces*, is devoted to their design and construction. The codes and procedures listed in the chapter are valuable resources for individuals interested in building fireplaces and chimneys.

Fireplace Design and Construction

Several basic fireplace designs are currently in use. They include the single face, the two face-adjacent, the two face-opposite, and the three face. See **Figure 17-31**.

Other fireplaces with exotic shapes and designs are also possible, but they require special engineering to function properly. Because of this, these fireplaces are not covered in this text. All fireplaces, regardless of their design, have basically the same component parts. See **Figure 17-32**.

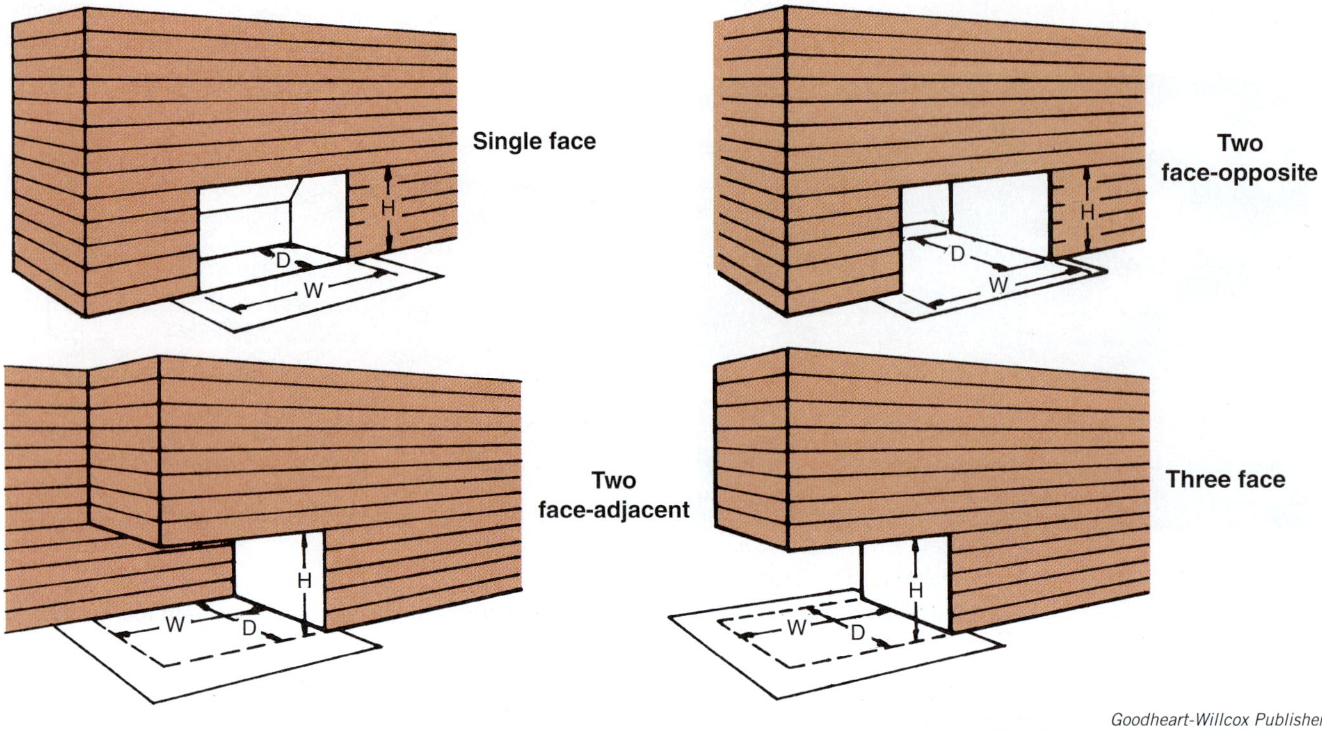

Figure 17-31. Basic fireplace designs.

Sizes of Fireplaces

The size of a fireplace is important from an aesthetic (how it looks) as well as functional (how it works) point of view. A fireplace that is too small may work well but not provide enough heat. A fireplace that is too large would be too hot for the room. A too-large fireplace would also require a large chimney and an abnormal amount of air to supply its needs for combustion (burning). A room with 300 sq ft of floor area can be adequately served by a fireplace with an opening 30″ to 36″ wide.

Fireplace openings are seldom greater than 36″. Fireplace opening sizes, hearth dimensions, and flue sizes are shown in **Figure 17-33**. The specified dimensions can be changed slightly to meet brick courses and joints if necessary but should not be changed significantly.

Combustion Chamber and Firebox

The ***combustion chamber*** is the space or area where combustion occurs to create heat in the fireplace. The ***firebox*** is the internal assembly fabricated or built in the fireplace where combustion occurs. The firebox consists of the fireplace opening, combustion chamber, hearth, throat, and smoke shelf. See **Figure 17-34**.

The shape and depth of the combustion chamber affects both the ***draft*** (movement of air for combustion and heated gases through the chimney) and the amount of heat radiated into the room. The size of the firebox should be balanced with the size of the room where the fireplace is to be built. The draft that occurs in the fireplace is affected by the size of the opening, the size of the combustion chamber, and the size of the flue. If a fireplace opening is too large for the room, the space will be overheated. The slope of the back of the firebox forces the flame forward and forces the gases with increasing velocity into the throat. The firebox should be lined with firebricks that are laid with thin joints of fire clay mortar.

Code Note **Firebox Dimensions**

The International Residential Code (IRC) **R1001.6** requires a depth of not less than 20″ (508 mm) for the firebox of a masonry fireplace.

The back and end walls of an average size fireplace are generally 8″ thick. Large sizes have thicker walls to support the load above.

Throat and Damper

The ***throat*** of a fireplace is a narrow passage that connects the firebox with the smoke chamber. The throat should not be less than 8″ above the highest point of the fireplace opening. The sloping back is the same height and supports the back of the damper.

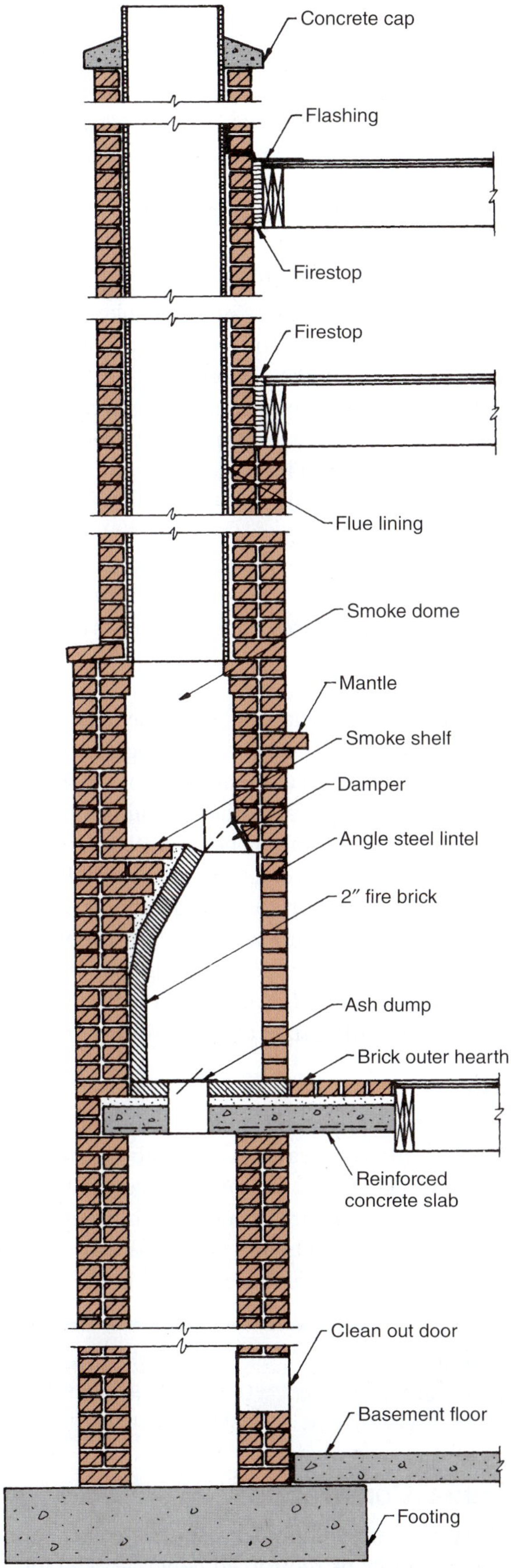

Goodheart-Willcox Publisher

Figure 17-32. Fireplace and chimney components.

Fireplace Dimensions			
Fireplace Type	**Opening Height (inches)**	**Hearth Size (W × D, inches)**	**Modular Flue Size (inches)**
Single face	29	30 × 16	12 × 12
	29	36 × 16	12 × 12
	29	40 × 16	12 × 16
	32	48 × 18	16 × 16
Two face-adjacent	26	32 × 16	12 × 16
	29	40 × 16	16 × 16
	29	48 × 20	16 × 16
Two face-opposite	29	32 × 28	16 × 16
	29	36 × 28	16 × 20
	29	40 × 28	16 × 20
Three face	27	36 × 32	20 × 20
	27	36 × 36	20 × 20
	27	44 × 40	20 × 20

Goodheart-Willcox Publisher

Figure 17-33. Specifications for fireplace opening height, hearth size, and flue size.

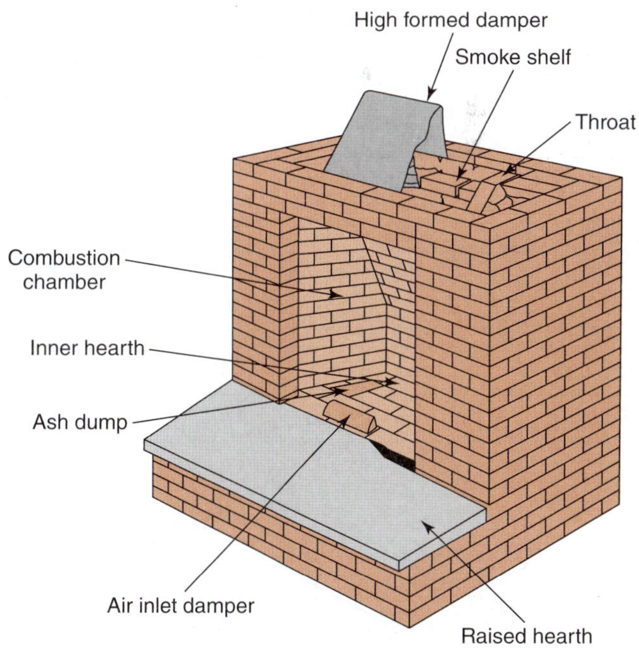

Goodheart-Willcox Publisher

Figure 17-34. Firebox assembly of components.

A ***damper*** is a metal plate just above the throat that controls the flow of air and gaseous products through the fireplace and up the chimney. It closes when the chimney is not in use to reduce air-infiltration down the chimney. The damper extends the full width of the fireplace opening. The best designs open upward toward the back and form a barrier to stop downdrafts. Typical damper sizes are shown in **Figure 17-35**.

Steel Dampers (inches)					
Width of Fireplace (inches)	A	B	C	D	E
24 to 26	28 1/4	26 3/4	13	24	9 1/2
27 to 30	32 1/4	30 3/4	13	28	9 1/2
31 to 34	36 1/4	34 3/4	13	32	9 1/2
35 to 38	40 1/4	38 3/4	13	36	9 1/2
39 to 42	44 1/4	42 3/4	13	40	9 1/2
43 to 46	48 1/4	46 3/4	13	44	9 1/2
47 to 50	52 1/4	50 3/4	13	48	9 1/2
51 to 54	56 1/4	54 3/4	13	52	9 1/2
57 to 60	62 1/2	60 3/4	13	58	9 1/2
Cast Iron Dampers (inches)					
Width of Fireplace (inches)	A	B	C	D	E
24 to 26	28	21	13 1/2	24	10
27 to 30	34	26 3/4	13 1/2	30	10
31 to 34	37	29 3/4	13 1/2	33	10
35 to 38	40	32 3/4	13 1/2	36	10
39 to 42	46	38 3/4	13 1/2	48	10
43 to 46	52	44 3/4	13 1/2	48	10
47 to 50	57 1/2	50 1/2	13 1/2	54	10
51 to 54	64	56 1/2	14 1/2	60	11 1/2
57 to 60	76	58	14 1/2	72	11 1/2

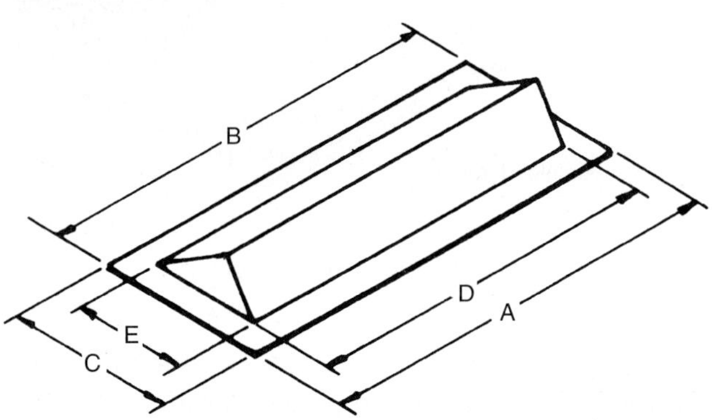

Donley Brothers Company

Figure 17-35. Steel and cast iron damper sizes for fireplace construction.

Code Note — Dampers

The International Residential Code **R1001.7.1** specifies that ferrous metal dampers should be installed in masonry fireplaces. The damper should be located at least 8″ above the top of the fireplace opening.

Smoke Shelf and Smoke Chamber

The ***smoke shelf*** is located in the throat of the fireplace structure and should be directly under the bottom of the flue and the lowest part of the smoke chamber. It extends the full width of the throat and is constructed horizontally. The ***smoke chamber*** is located directly above the firebox and compresses gases created during the combustion process of burning of wood or other flammable materials. The back and front walls of the chamber are vertical with the side walls sloping to the bottom of the flue lining.

Metal lining plates can be purchased for the smoke chamber. The plates give the smoke chamber the proper form with smooth surfaces, which simplifies bricklaying.

Fireplace Flue

Each fireplace should have its own clay flue (the duct that carries the exhaust gases out through the chimney) that is free of openings and connections. A ***flue lining*** is usually a fired clay pipe, either round or square, that lines the flue of a chimney. The first section of the flue lining must start at the center line of the fireplace opening. If the lining is offset to one side, the draft will not be uniform over the full width of the fireplace. The lining should be supported on at least three sides by a ledge of projecting bricks that is flush with the inside of the lining. Flue lining sizes are given in **Figure 17-33.**

Hearth

Hearths are usually constructed by building a reinforced brick masonry cantilevered slab or a reinforced concrete slab. The simplest method of constructing this slab is to cover the entire area of the chimney base the width of the hearth. Openings for the ash chute and basement flue (if required) are formed in the hearth.

Code Note — Hearth Thickness

The International Residential Code (IRC) **R1001.9.1** requires that hearths be at least 4″ (102 mm) thick.

Chimney Construction

Masonry chimneys should be supported on foundations of masonry or reinforced concrete. The foundation wall should be a minimum of 8″ thick. Reinforcement in the footing is recommended.

Code Note — Footings and Foundations for Fireplaces and Chimneys

The International Building Code (IBC) **R1001.2** specifies that footings for masonry chimneys are to be constructed of concrete or solid masonry at least 12″ (305 mm) thick. The footings should extend at least 6″ (153 mm) beyond the fireplace face or foundation wall on all sides. Footings should be at least 12″ (305 mm) below finished grade in areas not subjected to freezing.

A chimney should not change in size or shape as it passes through the roof. Changes must not be made within 6″ above or below the roof joists or rafters.

Chimney walls must be at least 4″ thick for residential structures. The walls must be lined with approved fire clay flue liners not less than 5/8″ thick. Flue liners should be installed ahead of construction of the chimney as it is laid up. The joints must be close fitting and smooth on the inside. In masonry chimneys with walls less than 8″ thick, liners are separate from the chimney wall. The space between the liner and masonry is not filled. Liners should extend from the throat of the fireplace to the top of the chimney as straight up as possible.

When a chimney has two flues not separated by masonry, the joints of adjacent flue linings should be staggered at least 7″. If more than two flues are located in the same chimney, masonry wythes at least 4″ wide should be built between them so that no more than two flues are grouped together. See **Figure 17-36**.

The tops of chimneys above the roof must meet certain specifications. The flue lining should project 4″ above the chimney cap. The cap should be finished with a straight or concave slope to direct air current upward and drain water from the top of the chimney. The top of the flue liner should extend a minimum of 3″ above the highest point where the chimney passes through the roof.

Code Note — Chimneys

The International Residential Code (IRC) **R1003.18.** specifies procedures for building chimneys and their distance from combustible materials. For example, the chimney is required to be at least 2″ higher than any portion of a building within 10″. See **Figure 17-37**.

At the intersection of the chimney and roof, the connection should be weathertight. Flashing and counterflashing made of copper or another rust-resisting material should be used. See **Figure 17-38**.

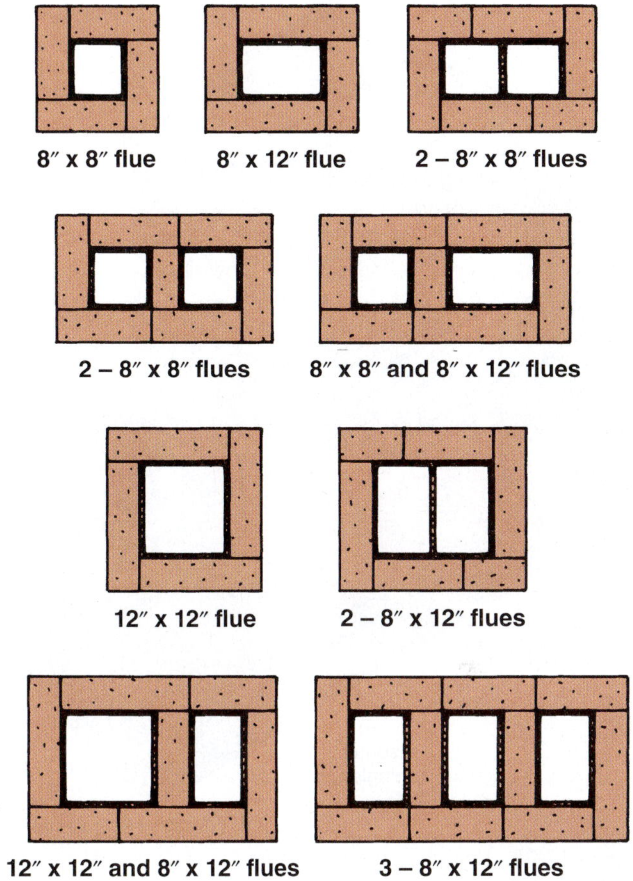

Goodheart-Willcox Publisher

Figure 17-36. Masonry chimney section details.

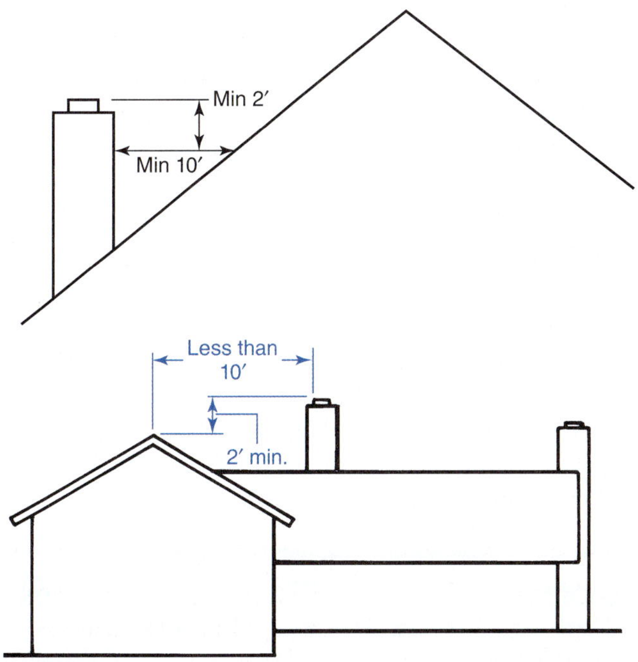

Goodheart-Willcox Publisher

Figure 17-37. Minimum recommended chimney heights above the roof.

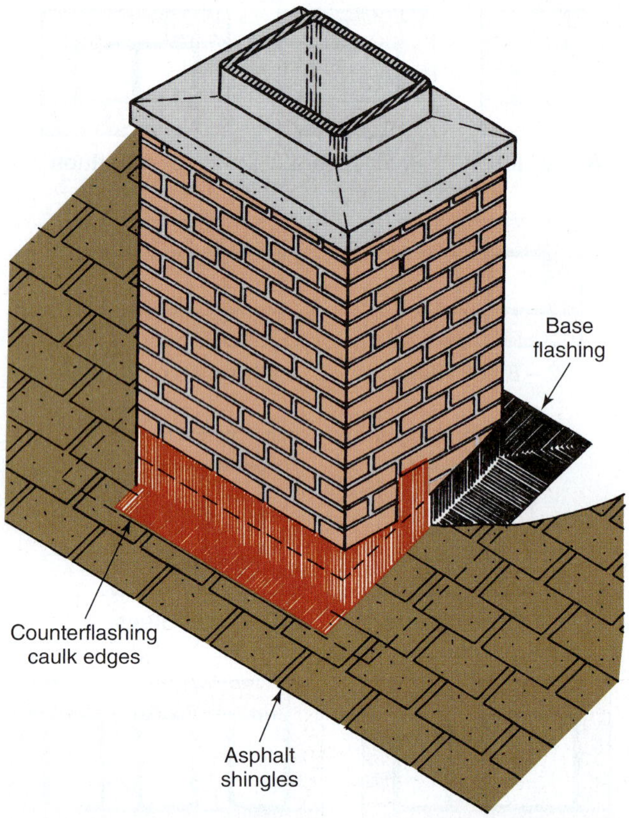

Goodheart-Willcox Publisher

Figure 17-38. Recommended flashing for chimneys at the roof intersection.

Prefabricated Steel Heat Circulating Fireplaces

Prefabricated steel heat circulating fireplaces are popular in residential construction. They include not only the firebox and heating chamber, but also the throat, damper, smoke shelf, and smoke chamber. See **Figure 17-39**. Prefabricated units are efficient because the sides and back are double-walled where air is heated and returned through vents. Certain fireplaces are designed to burn wood with the doors closed to increase energy efficiency and to improve the output of the fireplace. These units will also bring outside air into the combustion chamber for burning through a piping system to improve combustion and to eliminate the use of inside air, which reduces efficiency. See **Figure 17-40**. Always follow manufacturers' instructions when installing prefabricated fireplaces because one company may have different procedures than another.

> **Code Note** — **Fireplace Clearances**
>
> International Residential Code (IRC) **R1001.11** discusses fireplace clearances in relationship to wood beams, joists, studs or other combustible materials. It specifies that at least a 2″ clearance is needed to prevent a fire hazard.

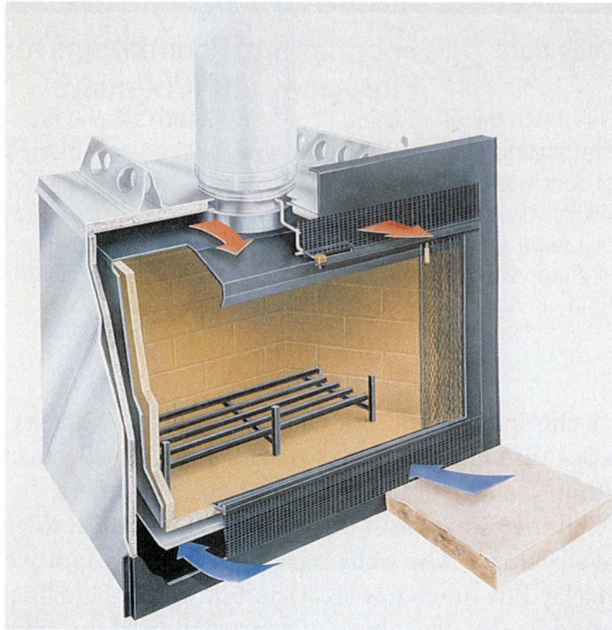

Heatilator, Inc.

Figure 17-39. Prefabricated steel heat circulating fireplace with complete firebox, damper, and smoke shelf.

Author's image taken at Fireplace Professionals, Inc., in Sioux Falls, SD

Figure 17-40. A prefabricated wood burning fireplace ready to be delivered to the construction site. The firebrick shown in the firebox will be correctly positioned once the unit is installed.

Design specifications are shown in **Figure 17-41**. Fireplace inserts are popular because some of them can burn natural gas, propane, corn, rye, wheat, and wood pellets. See **Figure 17-42**. Venting can be through a typical chimney or through the wall based on manufacturer's requirements.

Prefabricated Steel Heat Circulating Fireplace Sizes (in inches)					
Overall Dimensions Width	Overall Dimensions Height	Overall Dimensions Depth	Finished Opening Width	Finished Opening Height	Flue Size (Modular)
34 3/4	44	18	28	22	8 × 12
38 1/2	48	19	32	24	8 × 12
42 1/2	51	20	36	25	12 × 16
47 3/4	55	21	40	27	12 × 16
55	60	22	46	29	12 × 16
63	65	23	54	31	16 × 16

The Majestic Co., Inc.

Figure 17-41. Design specifications for a prefabricated fireplace.

tab62/Shutterstock.com

Figure 17-42. Example of a clean burning natural gas fireplace insert.

Workplace Skills
Teamwork

An important part of being professional in any workplace is being a valued team member. Often, many masonry projects are too big for one person to undertake on their own, and may require various talents and skills. Being part of a team not only means working alongside other professionals to reach a common goal, but also understanding each team member's responsibilities toward reaching that goal. Having this shared awareness creates an environment where team members can take pride in their individual work, and the work of their team.

Businesses are constantly looking for effective team members because a team is only as strong as its members. Effective team members contribute ideas, share common goals, and encourage other team members to succeed in their work. Being an effective team member also means being able to collaborate with the team. Good collaborators share ideas and work with others to achieve a common goal, which often includes finding a solution to a problem. These are all important skills needed to be a valued team member, and if you are constantly working on these skills, then you will also be a valued worker.

Masonry Enclosed Steel Firebox Fireplaces

One of the most energy efficient fireplace designs incorporates a masonry surround and support structure with a steel firebox assembly. This type of fireplace joins an almost airtight set of doors on the 1/4″ plate steel firebox. See **Figure 17-43**. Combustion air is brought into the firebox through an exterior vent arrangement.

Wilkening Fireplace Company

Figure 17-43. Exterior brick face of an energy-efficient masonry enclosed steel firebox fireplace. The intake grill is located in the raised hearth and the upper exhaust grill is located above the mantle.

The fireplace structure is a double wall system with the exterior made of brick or stone and an inner layer of concrete block, similar to regular fireplace structures. At this point, the similarities between fireplaces end because of the unique nature of the heat chamber and airtight firebox. The inside wall chamber is usually made of Haydite masonry block. ***Haydite*** blocks are made from expanded shale aggregate, which produces air pockets in the aggregate to create a higher R-value. Between the outside walls and inside Haydite block walls is 4″ of insulation. The cavity between the two Haydite block walls creates the heat chamber where the fireplace box rests. The ceiling of the heat chamber is created by forming and placing 6″ of reinforced concrete slab. Above the slab is 6″ to 10″ of insulation to reduce heat loss from the chamber. See **Figure 17-44** and **Figure 17-45**.

When a masonry enclosed steel firebox fireplace is in use, it is recommended that the doors be kept closed to improve energy efficiency. Once the heat chamber begins to increase in temperature, cool air from the living area is brought into the chamber through convection from the lower grill. The heated air in the chamber is then expelled out the upper grill into the living area. To increase airflow performance, an optional fan blower can be purchased and installed in the hearth of the fireplace.

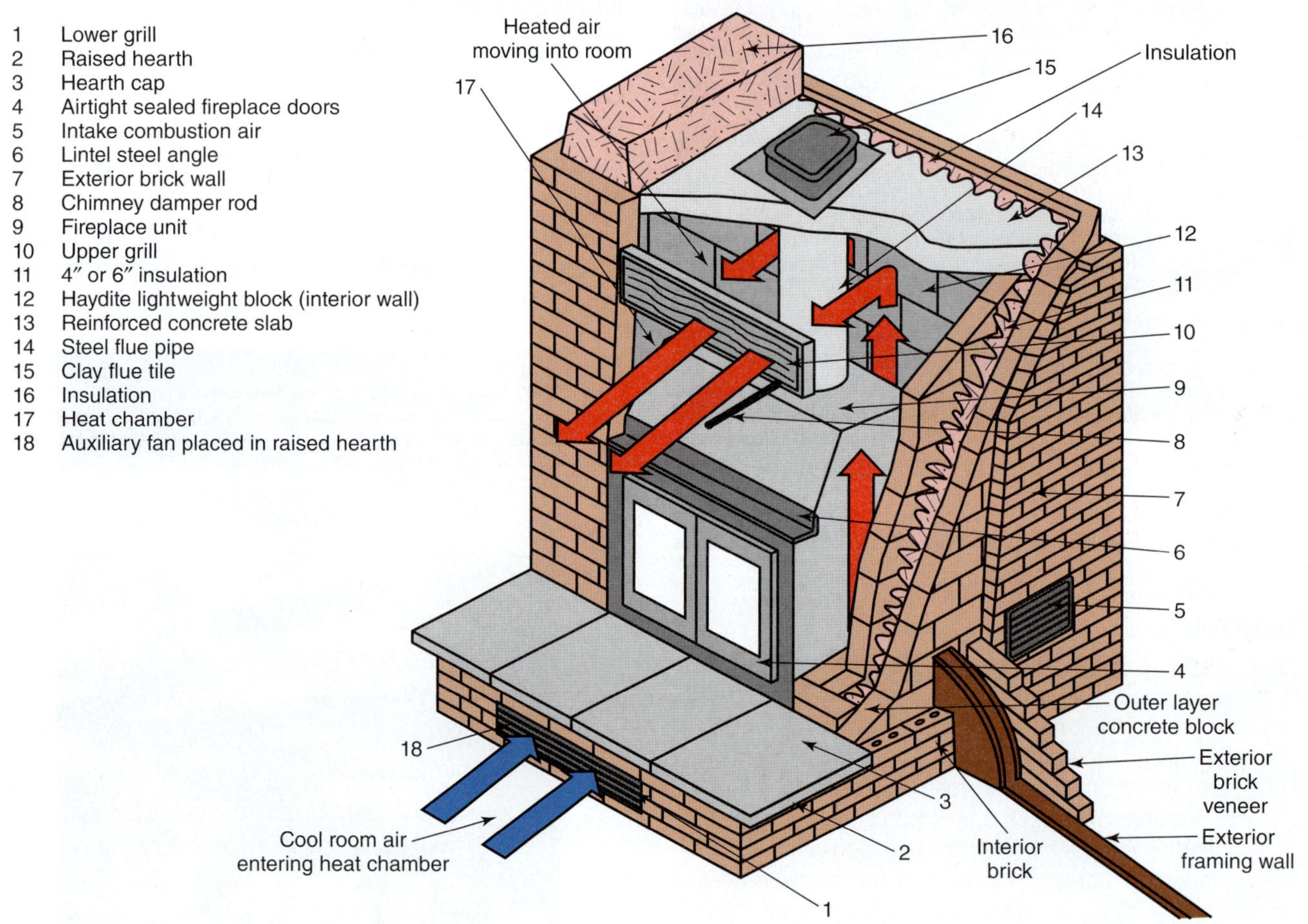

Wilkening Fireplace Company

Figure 17-44. The external and internal construction of the GranView Masonry Fireplace™.

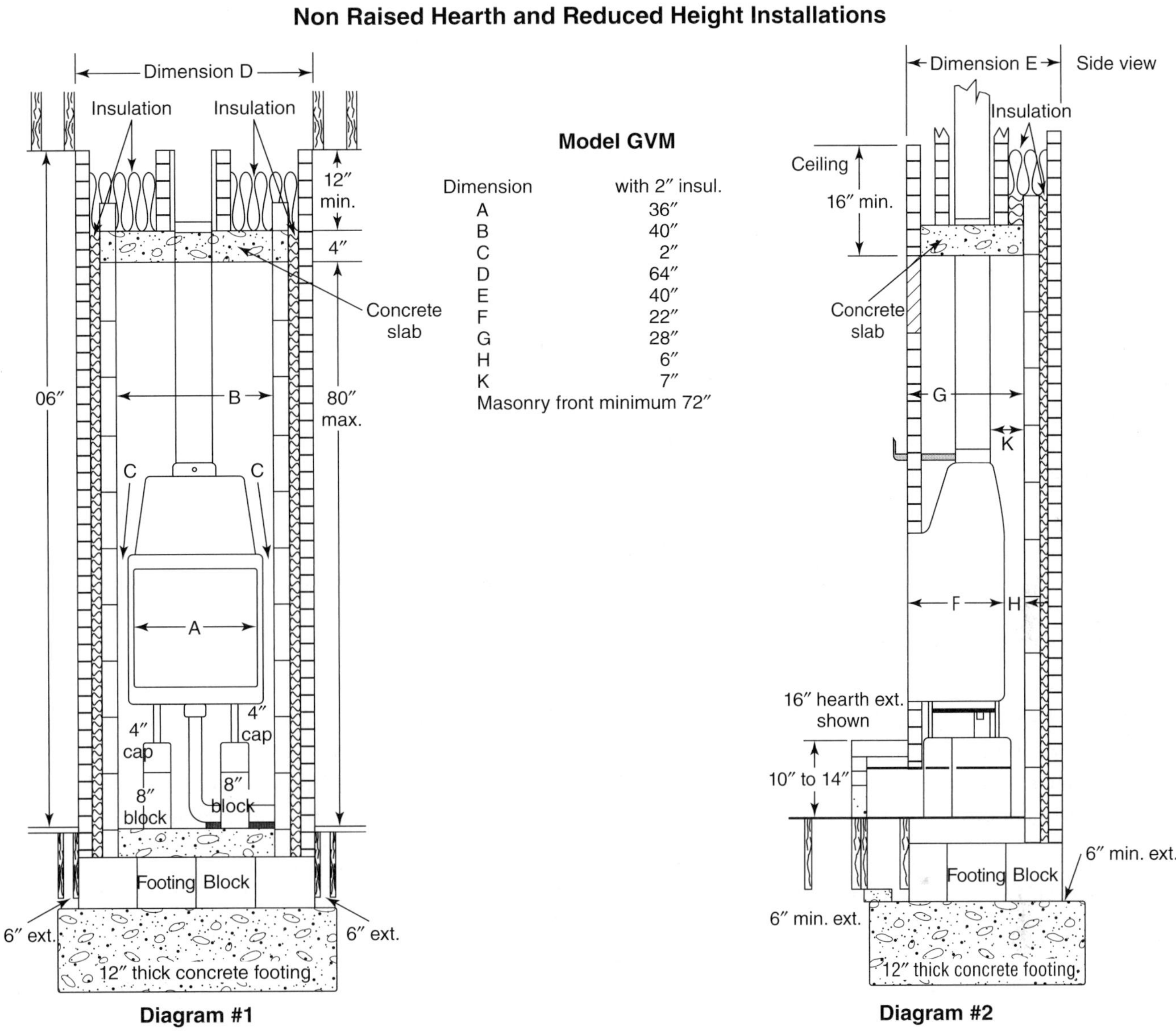

Wilkening Fireplace Company

Figure 17-45. Installation drawings of the GranView Masonry Fireplace™.

Chapter Review

Summary

- Factors affecting the selection of bricks and jointing materials for roadways, walks, patios, and interior floors include traffic, exposure to moisture and freezing, and desired appearance.
- The quality of a paving job will depend on the brick selection, the workmanship of installing the supporting sand layer, and the base and subgrade structures.
- A slope of 1/8″ to 1/4″ per foot is generally used for walks and patios in order to shed surface water. Below-surface drainage may be required for large paved areas or locations with a high water table.
- Basic fireplace designs include the single face, the two face-adjacent, the two face-opposite, and the three face.
- A firebox consists of the fireplace opening, combustion chamber, hearth, throat, and smoke shelf.
- The shape and depth of the firebox affect both the draft and the amount of heat radiated into the room.
- The fireplace throat should not be less than 6″ and preferably 8″ above the highest point of the fireplace opening.
- Each fireplace should have its own clay flue (the duct that carries the exhaust gases out through the chimney) that is free of openings and connections. The first section of flue lining must start at the center line of the fireplace opening.
- A chimney should not change in size or shape as it passes through the roof.
- Flashing and counterflashing should be used to make the intersection of the chimney and roof weathertight.
- Prefabricated steel heat circulating fireplaces include the firebox and heating chamber, throat, damper, smoke shelf, and smoke chamber.
- Masonry enclosed steel firebox fireplaces incorporate a masonry surround and support structure with a steel firebox assembly. Combustion air is brought into the firebox through an exterior vent arrangement.

Know and Understand

Answer the following questions using the information provided in this chapter.

1. The intermediate layer of an unreinforced masonry floor that aids in leveling is called the _____.
   A. joint
   B. wearing surface
   C. base
   D. cushion
2. When using brick in exterior pavements, each of the following must be considered *except* _____.
   A. weather
   B. layers of supporting materials
   C. drainage
   D. All are correct.
3. Paver joints and brick need to be _____.
   A. straight
   B. parallel
   C. perpendicular
   D. All are correct.
4. When cutting pavers with a masonry saw, you should always wear _____ protection.
   A. eye
   B. ear
   C. face
   D. All are correct.
5. *True or False?* Polymeric sand is a mixture of fine sand and water activated polymers that are intended to fill, bind, and seal joints.
6. *True or False?* Regardless of design, all fireplaces basically have the same component parts.
7. The narrow passage that connects the firebox with the smoke chamber is called the _____.
   A. surround
   B. damper
   C. firebox
   D. throat
8. The International Residential Code specifies that a _____ should be installed in masonry fireplaces and located at least 8″ above the top of the fireplace opening.
   A. combustion chamber
   B. surround
   C. damper
   D. throat
9. Where is the smoke chamber located in a fireplace?
   A. Above the smoke shelf
   B. In the throat
   C. Above the throat
   D. Above the firebox

10. *True or False?* A chimney must change in size and shape to be uniform as it passes through the roof.
11. When a chimney has two flues not separated by masonry, the joints of adjacent flue linings should be staggered at least _____.
    A. 6″
    B. 7″
    C. 10″
    D. 1″
12. *True or False?* A prefabricated fireplace is inefficient because it is designed to burn wood with the doors open.
13. The inside wall of a masonry enclosed steel firebox fireplace is made of _____.
    A. Haydite
    B. steel
    C. fire clay
    D. reinforced concrete

Apply and Analyze

1. Why are hard-burned bricks used for industrial floors?
2. When laying paving brick, why do masons tap the pavers with a rubber mallet?
3. What is the purpose of a smoke shelf?

Critical Thinking

1. You are the supervisor of a brick paving company. The company has been hired to install a residential brick driveway in front of a new home. Discuss the procedures the company will implement to follow the Brick Industry Association's guidelines for constructing this type of residential driveway. Then sketch a section view detailing the different layers of brick and supporting materials that make up the driveway.

Activities

1. (ENGINEERING) Various materials can be used for chimney flashing. Research four flashing materials and explain the advantages and disadvantages of each.
2. (SCIENCE) Research "cold hearth syndrome." What does the term mean? What are the possible causes? What can be done to remedy the problem?
3. (MATH) Calculate the number of paving bricks needed to lay an L-shaped patio with a brick size of 4″ × 8″ and the bricks are laid flat. Round up if the answer is a portion of a brick.

SECTION 6 Concrete

Timothy L. Andera

sonsam/iStock via Getty Images

CHAPTER 18

Concrete Materials and Applications

LEARNING OUTCOMES:

After studying this chapter, you will be able to:

18.1-1 Identify the basic components of concrete.
18.2-1 Describe the effect of the water-cement ratio on the strength, durability, and watertightness of concrete.
18.3-1 Use proper techniques for measuring materials for concrete.
18.4-1 Use proper procedures for mixing and remixing concrete.
18.5-1 Identify different types of reinforcement for concrete.
18.6-1 Explain the differences between concrete and lightweight concrete aggregate.

TECHNICAL TERMS

accelerating admixture
air-entraining admixtures
high-slump concrete
low-slump concrete
pozzolan
reinforced concrete
retarding admixture
slump
tensile strength
water-cement ratio
water-reducing admixtures

Introduction

Concrete is one of the most important building materials. It is used in almost every type and size of architectural structure. See **Figure 18-1**. Concrete is a major material in footings and foundations, exterior and interior walls, floors and roof systems, walks, and driveways. The plastic quality of concrete lends itself to almost unlimited design possibilities in form, pattern, and texture. Hardened concrete is durable, needs little maintenance, and can be used in many ways.

18.1 Basic Materials of Concrete

Concrete is a mixture of four basic materials in proper amounts (proportions). The basic materials are Portland cement, fine aggregate (sand or finely crushed stone), coarse aggregate (gravel or crushed stone), and water.

Lightweight manufactured materials are used to produce lightweight concrete. Sometimes other ingredients are added to the basic concrete mix in order to produce certain properties in the finished product. These ingredients are called *admixtures*.

Portland Cement

Portland cement is hydraulic because it sets and hardens by reacting with water. This chemical reaction, hydration, forms the cement into a stone-like mass.

Portland cement is manufactured in a specialized plant where raw materials are mixed. Portland cement contains varying amounts of lime, silica, alumina, and iron components. The raw materials are pulverized and combined into a mixture that has the desired chemical makeup. The mixture is placed in a kiln and heated

Tilt-Up Concrete Association

Figure 18-1. Concrete is used in a variety of ways in construction. A—The primary building material for this contemporary Frank Lloyd Wright designed structure is cast concrete. B—Tilt-up wall construction (concrete panels) is the system used to construct this building. C—Scored concrete paving. D—An elegant building made with cast concrete.

to approximately 2700°F (1482°C). The mixture exits the kiln as clinkers (lumps or nodules) about the size of marbles. Once cooled, the clinkers are ground into a fine powder, producing Portland cement.

Manufacturers of Portland cement use their own brand names. However, all Portland cements are manufactured to meet ASTM standard specifications. ASTM designations are as follows:

- C150 Portland Cement, Types I through V
- C175 Air-Entraining Portland Cement
- C205 Portland Blast Furnace Slag Cement
- C340 Portland-Pozzolan Cement

Types of Portland Cement

Portland cement is produced to meet various physical and chemical requirements for specific purposes. The strength of concrete made with different cements is shown in **Figure 18-2**. Type I or normal Portland cement concrete is used as the basis for comparison. There are several types of Portland cement:

- **Type I—Normal Portland cement.** General-purpose cement that is suitable for all applications where the special properties of the other cement types are not required. This cement is used where it is not subject to attack by sulfate from the soil or water or to an objectionable temperature rise from the heat generated by hydration. It can be used for pavements, sidewalks, reinforced concrete structures, bridges, tanks, reservoirs, culverts, water pipe, and masonry units.
- **Type II—Modified Portland cement.** Cement that produces less heat due to hydration and can resist sulfate attack. This cement can be used in structures such as piers, heavy abutments, and heavy retaining walls.
- **Type III—High early strength Portland cement.** Cement that produces high strength quickly, usually in a week or less. It is used when forms must be removed as soon as possible or when the structure is to be used quickly. Its controlled curing period may also be reduced in cold weather.

Compressive Strength, Percent of Strength of Type I or Normal Portland Cement Concrete					
ASTM Designation	Canadian Standards Assoc.	1 Day	7 Days	28 Days	3 Months
I	Normal	100	100	100	100
II	—	75	85	90	100
III	High early strength	190	120	110	100
IV	—	55	55	75	100
V	Sulfate resisting	65	75	85	100

Portland Cement Association

Figure 18-2. Relative strengths of concretes made with different cements. These concretes were moist-cured until tested.

- **Type IV—Low-heat Portland cement.** Cement used when the heat generated by hydration must be kept low and when a slower developing strength is not objectionable. This cement is designed for use in massive concrete structures such as dams.
- **Type V—Sulfate-resistant Portland cement.** Cement that gains strength more slowly than normal cement. This cement is used only in construction exposed to severe sulfate action, such as in some western states with water and soil of high alkali content.
- **Air-entraining Portland cements.** These cements improve resistance to freeze-thaw action and scaling caused by chemicals used in snow and ice removal. Designated as IA, IIA, and IIIA, these types correspond in composition to Types I, II, and III except that small quantities of air-entraining materials are added to the cement during production. Concrete made with these cements contains billions of tiny, well-distributed and completely separated air bubbles.
- **White Portland cement.** Cement recommended wherever white or colored concrete or mortar is desired. The manufacturing process is controlled so the finished product will be white instead of gray. This cement is manufactured to conform to the specifications of ASTM C150 and C175. It is used mainly for architectural purposes, such as precast wall and facing panels, terrazzo, stucco, cement paint, tile grout, and decorative concrete.
- **Portland blast-furnace slag cements.** Cements that can be used in general concrete construction when the specific properties of other types are not required. In these cements, granulated (sand-like) blast-furnace slag is ground and mixed with Portland cement.
- **Portland-pozzolan cements.** Cements in which pozzolan is blended with ground Portland cement. These cements are used principally for large hydraulic structures such as bridge piers and dams.

Aggregates

Sand, gravel, crushed stone, and air-cooled blast furnace slag are the most commonly used aggregates in concrete. See **Figure 18-3**. The aggregates usually make up 60% to 80% of the volume of concrete and can be considered filler material.

Aggregates must meet certain requirements. They must be clean, hard, strong, durable particles free of chemicals and coatings of clay, which could affect the bond. Since aggregates are such a large proportion of the volume of concrete, they greatly affect the quality of the finished product. Requirements for concrete aggregates are specified in ASTM C33. These specifications place limits on allowable amounts of damaging substances. They also cover requirements pertaining to grading, strength, and soundness of particles.

Fine Aggregate

Fine aggregate contains sand or other acceptable fine materials. A desirable concrete sand contains particles varying uniformly in size from very fine up to 1/4″ in diameter. Fine particles are necessary for good workability and smooth surfaces. These fine particles help fill the spaces between the larger particles. However, an excess of fine particles requires more cement-water paste to attain the right strength and increases the cost.

Coarse Aggregate

Coarse aggregate consists of gravel, crushed stone, or other acceptable materials larger than 1/4″ in diameter. Coarse aggregate suitable for use in concrete should be sound, hard, and durable. It should not be soft, flaky, or wear away rapidly. Coarse aggregate should also be clean and free of loam, clay, or vegetable matter. These foreign materials prevent the cement paste from properly binding the aggregate particles together and will reduce the strength of the concrete.

The maximum size of coarse aggregate is usually 1 1/2″ in diameter. This large size aggregate can be

Concrete Weight with Various Aggregates		
Concrete Weight	Aggregate Used	Unit Weight, pcf
Normal	Sand and gravel concrete	130–145
Normal	Crushed stone and sand concrete	120–140
Normal	Air-cooled slag concrete	100–125
Lightweight	Coal cinder concrete	80–105
Lightweight	Expanded slag concrete	80–105
Lightweight	Pelletized-fly ash concrete	75–125
Lightweight	Scoria concrete	75–100
Lightweight	Expanded clay, shale, slate, and sintered fly ash concretes	75–90
Lightweight	Pumice concrete	60–85
Lightweight	Cellular concrete	25–44

Portland Cement Association

Figure 18-3. The weight of a cubic foot of concrete varies from 25 lb/cu ft to 145 lb/cu ft, depending on the aggregate used in the mix.

used in a thick foundation wall or footing. As a rule of thumb, the largest piece should never be greater than one-fifth the thickness of a finished wall section, one-third the thickness of a slab, or three-fourths the width of the narrowest space through which the concrete will be required to pass during placing.

Well-graded coarse aggregate ranges uniformly from 1/4″ up to the largest size that is satisfactory for the kind of work to be performed. Using the maximum allowable particle size usually results in less drying shrinkage and lower cost.

Mixing Water for Concrete

The primary purpose of adding water to cement is to cause hydration (a chemical reaction) of the cement. Water usually makes up from 14% to 21% of the total volume of concrete. The amount of water used definitely affects the quality of the concrete.

Concrete mixing water should be clean and free of oil, alkali, and acid. Generally, water that is potable (good for drinking) is suitable for mixing concrete. However, water with too many sulfates should not be used even though it may be potable.

Admixtures

Admixtures are any materials added to the concrete batch before or during mixing other than Portland cement, water, and aggregates. Admixtures are classified as air-entraining, water-reducing, retarding, accelerating, pozzolans, workability agents, and miscellaneous.

The desirable qualities of concrete can often be attained conveniently and economically through proper design of the mix and selection of suitable materials without using admixtures (except air-entraining admixtures).

Air-Entraining Admixtures

Air-entraining admixtures are materials used to improve the durability of concrete exposed to moisture during cycles of freezing and thawing. Concrete resists surface scaling caused by ice-removal agents far better with air-entraining admixtures. The workability of fresh concrete is also improved significantly. Segregation (separating of materials) and bleeding are reduced or stopped. Entrained air can be produced in concrete using an air-entraining cement or through an air-entraining admixture.

To determine levels of air content in the concrete, an air meter testing device is used to measure the levels of air-entrainment in the concrete. The test determines the percentage of air that is contained in the mixture. The standard amount commonly requested by the contractor is 3% to 5% air-entrainment in the mixture. A mix containing 2% air-entrainment or less will not take advantage of the capabilities inherent in the admixture. A mixture of over 7% air-entrainment can affect the strength of the finished concrete. The air-entrainment test can be performed either at the concrete ready-mix facility or at the construction site.

Prior to performing the test, a technician taps the sides of the container unit with a rubber mallet to remove air pockets and settle the concrete. A tamping rod is used to eliminate air pockets in the concrete. The concrete is leveled off to the top of the container with a strike-off plate (1/2″ plastic plate). Water is then added into the petcocks in the lid to purge excess air in the container. An electronic scale is used to determine the weight of the concrete when placed in the air-entrainment unit. See **Figure 18-4** and **Figure 18-5**.

After filling the container with concrete, rodding, and screeding off the excess amount of concrete, the technician wipes the lip of the container to remove any concrete residue prior to placing the lid on the unit. See **Figure 18-6**. The test is then run on the air-entrainment unit to check for air content. See **Figure 18-7**.

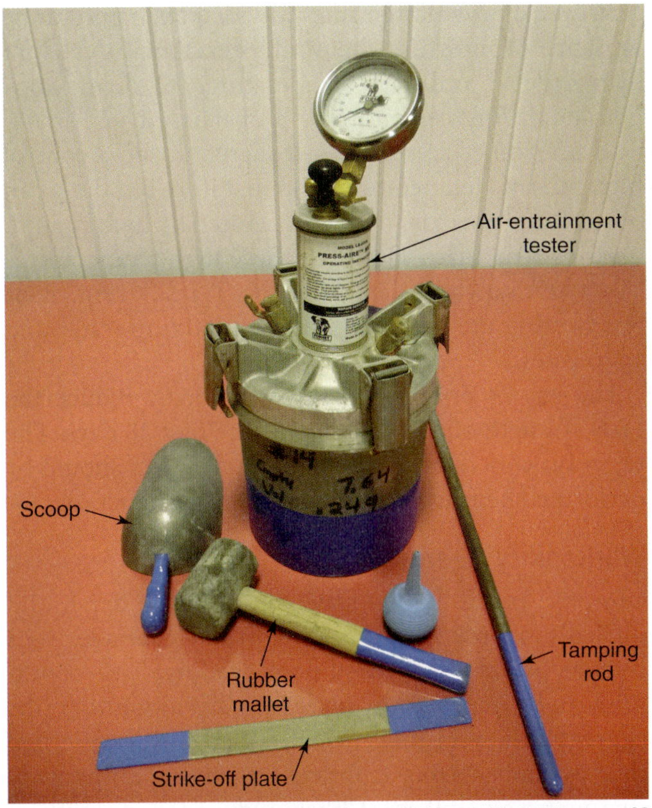

Author's image taken at GCC Ready Mix in Watertown, SD

Figure 18-4. An air-entrainment tester used to determine air content in concrete.

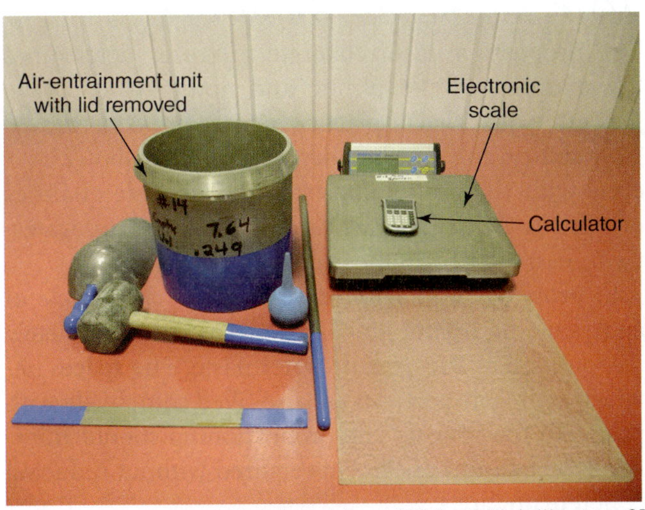

Author's image taken at GCC Ready Mix in Watertown, SD

Figure 18-5. Other tools and items for air-entrainment testing. The weight of the empty container and its volume are written on the side of the container.

Code Note **Air Entrainment**

The International Residential Code (IRC) **R402.2** specifies that concrete exposed to a freezing and thawing environment should be air entrained with a minimum total air content of 5%.

Author's image taken at GCC Ready Mix in Watertown, SD

Figure 18-6. The technician is wiping the lip of the container to remove residue prior to placing the lid on the unit.

Author's image taken at GCC Ready Mix in Watertown, SD

Figure 18-7. The air-entrainment tester is ready to run the test to check for air content.

Water-Reducing Admixtures

Water-reducing admixtures are materials that reduce the quantity of mixing water required to produce concrete of a given consistency. These materials increase the slump of concrete with less water content. ***Slump*** is the relative consistency or stiffness of plastic concrete.

Some water-reducing admixtures retard the setting time of concrete. Others are modified to result in varying degrees of retardation, and some do not have much effect on the setting time. Still others can entrain air in the concrete.

An increase in strength is usually gained with water-reducing admixtures if the water content is reduced for a given mix and the cement content and slump are kept the same. Even though water content is reduced with these admixtures, increases in drying shrinkage are probable.

Retarding Admixtures

A ***retarding admixture*** is a material used to retard (slow down) the setting time of concrete. High temperatures of plastic concrete often cause it to harden too fast. This increased hardening rate makes placing and finishing of concrete difficult. One of the most practical ways to combat this difficulty is to reduce the temperature of the concrete by cooling the mixing water and/or aggregates. The retarders do not lower the initial temperature of the concrete.

Retarders can be used when it is desirable to offset rapid setting up (hardening) of concrete due to hot weather. It is also used to delay the initial set of concrete when difficult conditions of placement are encountered.

Retarders usually reduce the strength of concrete for the first one to three days. The effect on other properties, such as shrinkage, may not be predictable. It is advisable to test the retarders under actual job conditions to be sure of results.

Accelerating Admixtures

An ***accelerating admixture*** is a material used to speed up the setting and the strength development of concrete. However, most of the commonly used accelerators cause an increase in the drying shrinkage of concrete. The development of concrete strength can also be increased by the following methods:

- Using Type III Portland cement
- Lowering the water-cement ratio
- Increasing the cement content or curing at higher temperatures

Calcium chloride is the most commonly used accelerating admixture. The amount added should not be more than 2% of the cement weight. Calcium chloride should not be used in prestressed concrete, in concrete containing embedded aluminum, where galvanized steel remains in contact with the concrete, or in concrete subjected to alkali or sulfates.

Pozzolans

A ***pozzolan*** is a siliceous or siliceous and aluminous material that possesses little or no cementitious value. However, a pozzolan in finely divided form and in the presence of moisture chemically reacts with calcium hydroxide at ordinary temperatures to form compounds having cementitious properties. These materials are sometimes used in concrete to help control high internal temperatures. Temperature buildups can occur from slow loss of heat generated by hydration in massive structures. However, these high temperatures can often be controlled by using Type II cement or by lowering the temperature of the mixing water and aggregates.

The use of pozzolans can substantially reduce the early strength of concrete during the first 28 days. The effect on concrete varies considerably. Tests should be performed before using pozzolans in construction.

Workability Agents

Concrete in the plastic state can often be harsh and difficult to work with due to aggregate characteristics such as particle shape and poor grading. This condition can be improved by increasing the amount of fine aggregate or entraining air. Entrained air is considered the best and acts like a lubricant. Excessive amounts of fine materials will most likely reduce the strength of the concrete and increase shrinkage.

18.2 Concrete Mixtures

The proportioning of concrete mixtures determines the qualities of the plastic and hardened concrete. Desirable qualities of plastic concrete are consistency, workability, and finishability. **Figure 18-8** shows concrete with the proper consistency. Desirable qualities of hardened concrete are strength, durability, watertightness, wear resistance, and economy.

In properly proportioned concrete, each particle of aggregate, no matter how large or small, is completely surrounded by the cement-water paste. The spaces between aggregate particles are also filled completely with paste. To lower costs, the proportioning should minimize the amount of cement required without reducing concrete quality.

Before a concrete mixture can be proportioned, the size and shape of concrete members, the concrete strength required, and the conditions of exposure must

SuphonPhoto/Shutterstock.com

Figure 18-8. This concrete has the proper consistency.

be known. For example, if concrete is being placed in forms where reinforcement is being used, the spacing between rebar members and forms must be considered so concrete flows appropriately and is not segregated because of obstructions. When designing the mix, the size or the aggregate may need to be reduced to accommodate spacing. Most of the properties of hardened concrete depend on the quality of the cement paste. The first step in determining the mix proportions is to select the right water-cement ratio.

Water-Cement Ratio

The ***water-cement ratio*** is the proportion of water to cement. This ratio is stated in gallons of water per bag (94 lb) of cement. The binding properties of cement paste are due to chemical reactions between the cement and water.

The chemical reactions require time and favorable conditions of temperature and moisture. They proceed rapidly at first and then much more slowly for a long period of time if conditions are satisfactory. Only a small amount of water (about 3 1/2 gallons per bag of cement) is necessary to complete the chemical reactions. However, more water (4 gallons to 9 gallons per bag of cement) is used to increase workability and aid placement. More water increases the amount of aggregate that can be used and makes the mix less expensive but will affect the strength of the concrete mix.

Good cement paste requires a proper proportion of water to cement. If too much water is used, the paste is thin and weak when it hardens. Too little water in the mix makes placing and finishing difficult.

Cement paste made with the proper amount of water is strong, watertight, and durable. The concrete will be strong and durable if the aggregates and cement paste are strong and durable. The quality of the concrete is therefore largely determined by the water-cement ratio in the cement paste.

Strength of concrete is directly related to the water-cement ratio. **Figure 18-9** shows the results of many tests performed on cured concrete. The results were similar for other types of cements. More water resulted in less strength and less water resulted in greater strength.

After concrete is in place, the strength continues to increase as long as moisture and satisfactory temperatures are present. The highest curing temperature is 73°F (23°C). When concrete is permitted to dry, the chemical reactions cease and lower strength results. Concrete should therefore be kept constantly moist as long as possible after placing.

In certain situations, concrete needs to be tested for strength. Engineers and architects may require samples of concrete to be selected and kept for strength testing using a concrete compression testing machine. The machine operates as a press that registers the point at which the concrete fractures. See **Figure 18-10**.

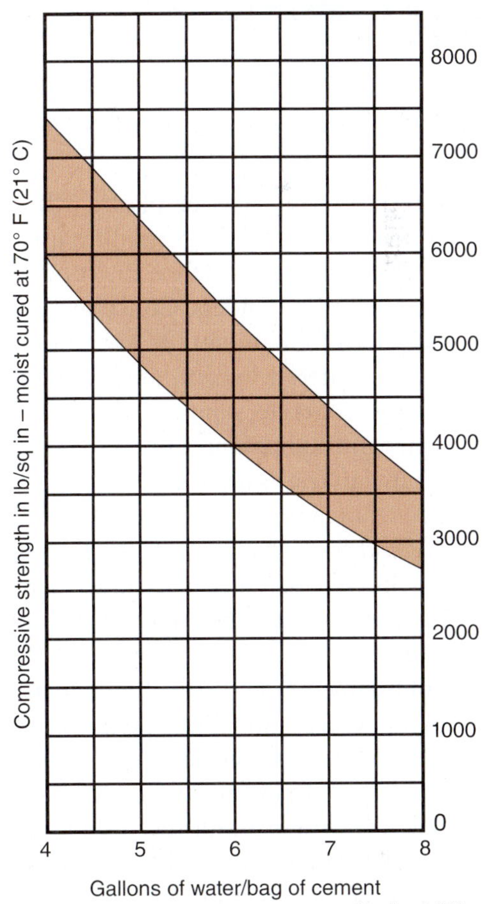

Goodheart-Willcox Publisher

Figure 18-9. The area within the curves represents the age-compressive strength relationship for Type I Portland cement after 28 days moist curing at 70°F (21°C).

Author's image taken at GCC Ready Mix in Watertown, SD

Figure 18-10. A concrete compression testing machine is used to determine the compression strength of sample test cylinders.

Author's image taken at GCC Ready Mix in Watertown, SD

Figure 18-11. The technician is placing a protective wrap around a concrete cylinder prior to testing.

In the testing process, steel or plastic cylinders are filled and rodded similar to the procedure used to perform an air-entrainment test. Once the concrete is set, it is removed from the cylinders. Concrete test samples are then kept moist to maintain the hydration process, which allows the concrete to continue gaining strength.

Testing of the concrete samples may occur at the ready-mix plant if equipment is available. Usually, a concrete technician tests samples after a 28-day period following procedures prescribed in ASTM C39, *Standard Test Method for Compressive Strength of Cylindrical Concrete Specimens*. Once a cylinder sample has been selected, a wrap may be attached to protect the technician during the test. See **Figure 18-11**. The wrap helps prevent fragments of crushed concrete from passing through the screen in the press door.

The cylinder is then placed into steel retainer rings to protect the top and bottom surfaces of the cylinder to ensure that it fractures uniformly. See **Figure 18-12**. Inside the steel retainer rings, 1/2″ thick pad caps are placed to add cushion when the force of the compression machine is applied to the concrete cylinder. The pad caps are made from neoprene and natural rubber and are reusable for multiple tests.

Author's image taken at GCC Ready Mix in Watertown, SD

Figure 18-12. The technician is placing steel retainer rings on the top and bottom of a concrete cylinder prior to testing. Black rubber pad caps are installed in the steel retainer rings.

The concrete sample is placed on the bottom plate of the concrete compression machine, and then the top compression piston is lowered over the concrete cylinder. Once the top piston is in the correct position over the cylinder, the protection door is closed and the technician prepares to test the sample. See **Figure 18-13.** When ready, the technician activates the compression piston to apply concentrated pressure on the test cylinder. The compression piston continues to apply pressure to the cylinder until it fractures. The pressure required to fracture the cylinder is registered on the meter. Samples of test cylinders after a compression test was performed are shown in **Figure 18-14**.

There is a direct relationship between the durability of concrete and the relative quantities of water and cement in the mixture. Freezing and thawing are the most destructive natural forces of weathering on wet or moist concrete. During this process, water expands as it changes to ice. If the water-cement ratio was high during mixing, the concrete is more porous and is more subject to damage by freezing and thawing.

Watertightness also depends on the water-cement ratio in the mix. The more water used in the mix, the less likely the concrete is to be watertight.

Author's image taken at GCC Ready Mix in Watertown, SD

Figure 18-13. The technician has placed the cylinder to be tested inside the concrete compression machine and has prepared it for testing.

Author's image taken at GCC Ready Mix in Watertown, SD

Figure 18-14. Concrete cylinder samples after being tested in the concrete compression machine. The compression machine recorded the strength and the technician submits this information to the architect or engineer.

Selecting Concrete Mixtures

The right concrete mixture for any job is the one that is most economical and workable while giving the desired properties. **Figure 18-15** shows concrete mixtures for various construction applications. Rule-of-thumb approximation of volumes such as 1:2:3 of cement to sand to gravel are not recommended for quality concrete mixtures. Most fine aggregates contain some water. Therefore, allowance should be made for this factor in determining the amount of water to be added to the mix. Adjustments to the water-cement ratio are needed for damp, wet, or very wet fine aggregate. See **Figure 18-16.**

Mix Consistency

Very fluid (wet) mixes are called ***high-slump concrete***. Stiff (dry) mixes are called ***low-slump concrete***. Slump is related primarily to water-cement ratio. A high water content causes a high slump. Generally, a low-slump concrete produces a better concrete product.

Slump Test

A slump test can be performed in the field with a metal slump cone. See **Figure 18-17**. The cone is 8″ in diameter at the bottom, 4″ in diameter at the top, and 12″ high. Specific tools are required to conduct the slump test. See **Figure 18-18**.

The cone is placed on a clean dry surface and filled one-third full with the concrete being tested. The concrete is rodded 25 times with a metal rod that is 24″ long and 5/8″ in diameter. Concrete is then added until the cone is two-thirds full and is again rodded 25 times with the rod penetrating the lower layer. The cone is then overfilled and the mixture rodded 25 more times. Finally, the top of the cone is screeded off. See **Figure 18-19**.

Construction Application	Water/cement ratio (gal per bag)	Consistency (amount of slump)	Maximum size of aggregate	Appropriate cement content (bags per yd)	Probable 28th day strength (psi)
Footings	7	4″ to 6″	1 1/2″	5.0	2800
8″ basement wall moderate ground water	7	4″ to 6″	1 1/2″	5.0	2800
8″ basement wall severe ground water	6	3″ to 5″	1 1/2″	5.8	3500
10″ basement wall moderate ground water	7	4″ to 6″	2″	4.7	2800
10″ basement wall severe ground water	6	3″ to 5″	2″	5.5	3500
Basement floor 4″ thickness	6	2″ to 4″	1″	6.2	3500
Floor slab on grade	6	2″ to 4″	1″	6.2	3500
Stairs and steps	6	1″ to 4″	1″	6.2	3500
Topping over concrete floor	5	1″ to 2″	3/8″	8.0	4350
Sidewalks, patios, driveways, porches	6	2″ to 4″	1″	6.2	3500

Goodheart-Willcox Publisher

Figure 18-15. Recommended concrete mixtures for various construction applications.

Amount of water to use based on wetness of sand			
Water-cement ratio of mix	Damp	Wet	Very wet
5 gal/bag	4 3/4 gal	4 1/2 gal	4 1/4 gal
6 gal/bag	5 1/2 gal	5 gal	4 1/2 gal
7 gal/bag	6 1/4 gal	5 1/2 gal	4 3/4 gal

Goodheart-Willcox Publisher

Figure 18-16. Recommended adjustment to the water-cement ratio when the fine aggregate is damp, wet, or very wet. The amount of water specified in the first column plus the water in the fine aggregate equals the amount of water required for the mix. If the sand is dry, then use the full amount indicated in Figure 18-15.

The cone is then removed by lifting it straight up and the slump is measured immediately. Slump is measured in inches. The rod is placed across the top of the cone extending over the concrete sample. The distance from the bottom edge of the rod to the concrete is measured with a rule. See **Figure 18-20**. This distance is the slump rating for the batch.

A slump test is a rough measure of the consistency of concrete, but should not be considered a measure of workability. The test should not be used to compare mixes of entirely different proportions or mixes containing different aggregates. However, the slump test does help to signal changes in grading or proportions of the aggregate or water content in the mix.

18.3 Measuring Materials

The ingredients of each batch of concrete must be measured accurately in order to produce uniform batches of proper proportions and consistency. The problem of varying amounts of moisture in the aggregate has already been discussed. This must be taken into account if accurate control is to be obtained.

Measuring Cement

If bagged cement is used, the batches should be sized so only full bags are used. If this is not possible, then the proper amount should be weighed out each time. Volume is not an accurate method of measuring cement. Bulk cement (not bagged) should always be weighed for each batch.

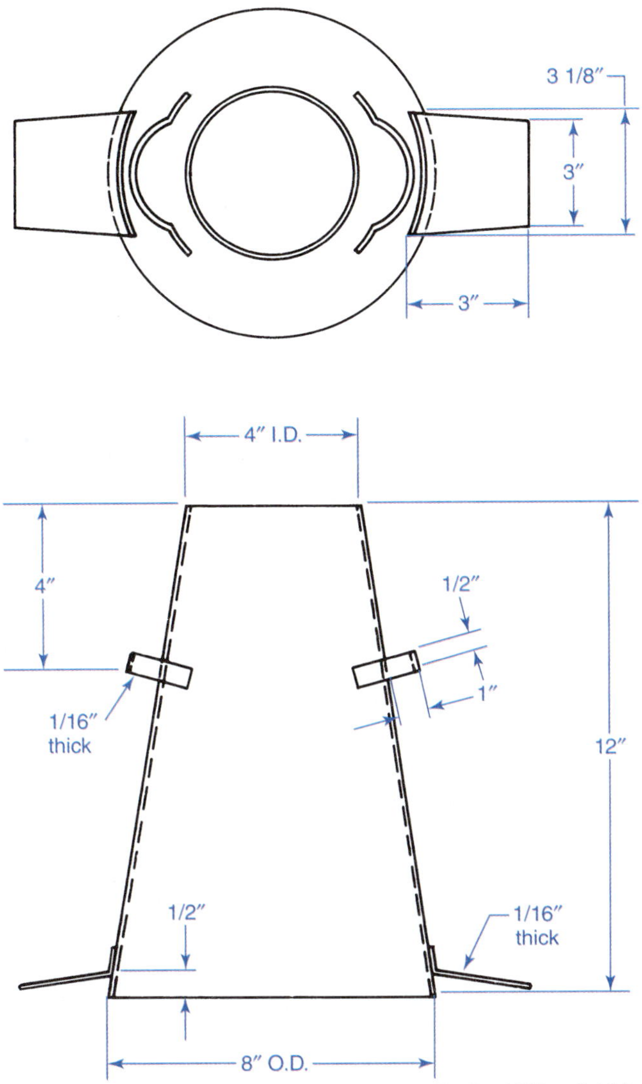

Goodheart-Willcox Publisher

Figure 18-17. A sheet metal slump cone showing dimensions.

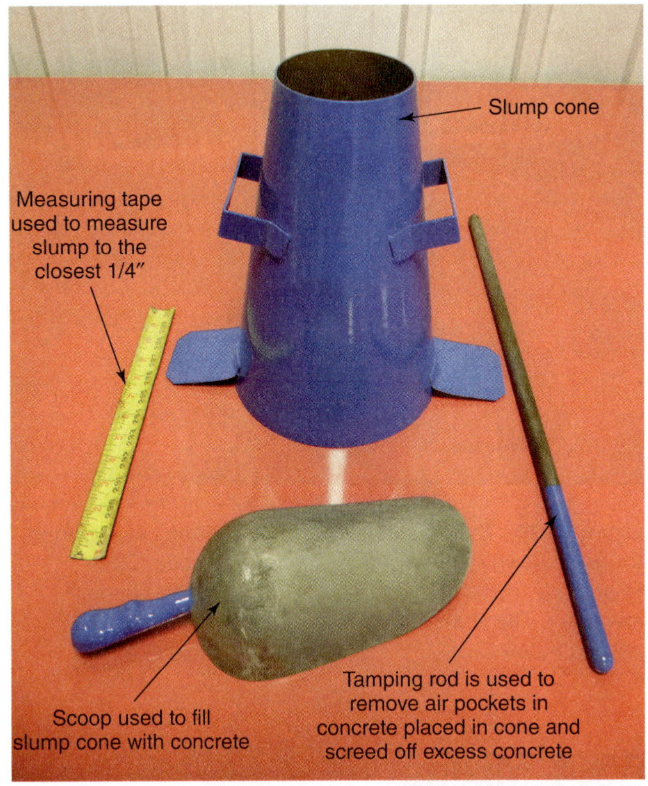

Author's image taken at GCC Ready Mix in Watertown, SD

Figure 18-18. Slump cone with tools to perform test.

Author's image taken at GCC Ready Mix in Watertown, SD

Figure 18-19. A technician is holding the slump cone in place with his feet while rolling a rod across the top of the cone to remove excess concrete.

Measuring Water

The effects of the water-cement ratio on the qualities of concrete make it necessary to accurately measure the water used. Water is measured in gallons, and any method that ensures accuracy is acceptable.

Measuring Aggregates

Measurement of aggregates by weight is the recommended practice and should be required on all jobs that require a high degree of consistency in each batch. Measurement of fine aggregate by volume is not accurate because a small amount of moisture is nearly always present. This moisture causes the fine aggregate to bulk or fluff up. The degree of bulking depends on the amount of moisture present and the fineness of the sand. A fine sand with a 5% moisture content increases in volume about 40% above its dry volume.

Another problem with aggregates, especially coarse aggregates, is separation of the various sizes. This problem can be reduced by weighing the coarse aggregate in two or more sizes.

Author's image taken at GCC Ready Mix in Watertown, SD

Figure 18-20. The technician has removed the cone, turned it upside down, and placed the rod on the cone. Once in position, measurements for the slump of the concrete are taken. The slump for this sample of concrete is approximately 2 1/2″.

Author's image taken at GCC Ready Mix in Watertown, SD

Figure 18-21. A ready-mix plant with a loading ramp that moves sand, gravel, and rock into holding bins in the building. Technicians measure specific materials by weight to build batches specified by customers.

18.4 Mixing

All concrete should be mixed thoroughly until it is uniform in appearance and all of the ingredients are uniformly distributed. Mixing time depends on the following factors:

- Speed of the machine
- Size of the batch
- Condition of the mixer

Generally, the mixing time should be at least one minute for mixtures up to 1 cu yd with an increase of 15 seconds for each 1/2 cu yd or fraction thereof. Mixing time should be measured from the time all materials are in the mixer.

Mixers should not be loaded above their capacity and should be run at the speed for which they were designed. Mixers coated with hardened concrete or having badly worn blades do not perform as efficiently as they should. These conditions should be corrected.

Generally, about 10% of the mixing water is placed in the mixer before the aggregate and cement are added. Water should then be added uniformly along with the dry materials. The last 10% of the water is added after all the dry materials are in the mixer.

Ready-Mixed Concrete

Ready-mixed concrete is purchased directly from a central plant. See **Figure 18-21** and **Figure 18-22**. It is convenient and usually of high quality. Concrete delivered to the jobsite arrives in specially designed trucks. A rear discharge concrete truck is shown in **Figure 18-23**.

Author's image taken at GCC Ready Mix in Watertown, SD

Figure 18-22. Area of the ready-mix plant where trucks are driven under the loading chute. Plant technicians load the ready-mix truck with the concrete batch specifically designed for the customer.

Anton Foltin/Shutterstock.com

Figure 18-23. A ready-mix truck has been positioned along the side of a foundation wall. The concrete in the mixer is being unloaded into forms for a garage under construction. This type of ready-mix truck unloads from the rear of the truck. Other specialized trucks can unload from the front.

A ready-mix truck batch size varies from 5 to 11 cu yd, depending on the specific truck design. The amount a truck can carry may also be affected by weather conditions. Also, certain municipalities may place restrictions on the weight that a concrete truck can carry. **Figure 18-24** shows a front discharge ready-mix truck. The advantage of a front discharge truck is the chute and mixer drum can be controlled through a joystick inside of the ready-mix truck. The operation of the chute can be controlled from the outside of the truck and by a remote control. The controller device can direct the chute left and right and up and down. The operator can also control the speed of the mixer and discharge of concrete. The operator can climb the side of the truck using an attached ladder, then enter the elevated platform where there is a better view to move the discharge chute to the proper location. A device called a *concrete conveyor* can be added to the front discharge truck to extend the reach of the concrete truck. The conveyor increases the distance concrete can be discharged to over 36′ from the front of the ready-mix truck. Front discharge concrete trucks can be purchased with an all-wheel drive feature that increases the mobility of the truck when the construction site is muddy, or the terrain is uneven.

Depending on the job size, only one truck may be required to deliver the required yardage or multiple trucks may be needed to fulfill the order. When scheduling a ready-mix project that requires more than one truck, it is important to ensure that adequate workers are available. Concrete deliveries should be scheduled so ready-mix trucks are not being delayed too long to unload their pre-mixed concrete.

Kimble Manufacturing Company

Figure 18-24. A front discharge ready-mix truck. This model can carry 11 yards of concrete and has all-wheel drive to aid in transporting concrete to difficult locations on the construction site.

In some ready-mixed operations, the materials are dry-batched at the central plant and then mixed en route to the site in truck mixers. Another method involves mixing the concrete in a stationary mixer at the central plant just enough to intermingle the ingredients. The mixing is then completed in a truck mixer en route to the jobsite. ASTM C94, *Standard Specification for Ready-Mixed Concrete* requires that concrete must be delivered and discharged from a truck mixer within 1 1/2 hours after the water has been added to the mix.

Remixing Concrete

Initial set of concrete usually takes place two or three hours after the cement is mixed with water. During this time, the concrete may dry out and stiffen if left standing. It can be remixed if it becomes sufficiently plastic and workable. Water cannot be added to make the mix more workable (retempering). Adding water would reduce the quality of the concrete just as a larger amount of water in the original mix would.

Green Note

Cement Manufacture

Cement manufacturers are implementing ways to reduce energy consumption. An enormous amount of energy is necessary to transfer raw products, such as limestone, shale, clay, silica, and sand into cement. These ingredients must be heated to approximately 2700°F. Cement companies are investing in new types of furnaces that burn more efficiently to reduce the amount of energy used in creating Portland cement.

In addition, cement manufacturers are looking to reduce their consumption of fossil fuels like coal and coke by burning industrial products and consumer wastes as alternative fuels. According to the Portland Cement Association, using tires for combustion in a cement kiln produces 25% more energy than coal and can also result in lower emissions. Using alternative fuels conserves fossil fuels while diverting waste from landfills.

18.5 Reinforced Concrete

Steel reinforcing is frequently used in footings, foundations, slabs, and other concrete work to add strength and control stresses. ***Reinforced concrete*** is a composite material in which the concrete resists compression forces and some other material, usually steel bars or wires, resists the tension forces.

Concrete has great compressive strength, meaning that it has the ability to support great loads placed directly on it. However, concrete has little ***tensile strength***, or ability to resist stresses or forces that tend to bend or pull it apart. The compressive strength of concrete is about ten times its tensile strength. The use of steel reinforcing may be used to greatly increase concrete's tensile strength.

Types of Reinforcement

Steel is regarded as the best reinforcing material for concrete. It has almost the same contraction and expansion rate due to temperature changes as concrete.

Steel can also be purchased in many sizes and forms. The two main types of steel reinforcement used in concrete are reinforcing bars (rebars) and welded wire reinforcement. See **Figure 18-25**. Bars are either smooth or deformed. Smooth bars are generally smaller in diameter. Deformed bars have ridges along the sides. These ridges increase the bond between the concrete and steel. See **Figure 18-26**.

Wire Reinforcement Institute

Figure 18-25. Reinforcing bars and welded wire reinforcement are being used to strengthen the slab joist in this skip pan joist system. The floor also has welded wire reinforcement.

Timothy L. Andera

Figure 18-26. An example of number 4 reinforcing bars. These bars are 1/2″ in diameter and are used to strengthen concrete. Notice the ribs on the reinforcing bars.

Bars are produced in standard sizes. They are designated by a number. See **Figure 18-27**. Larger bars are made for extremely heavy construction. Bars usually come from the mill in 60′ lengths, but most local suppliers stock 20′ and 40′ lengths.

A type of rebar, fiberglass reinforced polymer (FRP), is made from fiberglass strands, resins, and polymers. See **Figure 18-28**. One manufacturer of FRP rebar is GatorBar® in Michigan. They manufacture rebar in two sizes: #3 (3/8″ diameter) and #4 (1/2″ diameter).

Some advantages of FRP are as follows:

- Stronger in tensile strength and it may be able to replace #4 steel rebar with #3 FRP rebar
- More lightweight than steel with less of a transportation cost
- Corrosive resistance and rustproof, resulting in a longer life expectancy
- Nonconductive and nonmagnetic

The bar size, steel or FRP, is based on the requirements for the job. It depends on the amount of tensile strength needed for the project. Discuss the type of

Bar Number	Bar Diameter (in)	Bar Area (sq in)	Approximate Weight of 100 Feet
2	1/4	0.05	17
3	3/8	0.11	38
4	1/2	0.20	67
5	5/8	0.31	104
6	3/4	0.44	150
7	7/8	0.60	204
8	1	0.79	267

Wire Reinforcement Institute

Figure 18-27. Size, area, and weight of reinforcing bars from 1/4″ to 1″ in diameter.

Photo Courtesy of Neuvokas Corporation – Manufacturers of GatorBar®

Figure 18-28. FRP reinforcement is lighter than steel reinforcement and is stronger in tensile strength.

project with the supplier or manufacturer to determine the size of reinforcement needed. On large jobs or commercial projects, reinforcement sizes are determined by construction engineers who are trained in the areas of concrete and reinforcement requirements.

Welded wire reinforcement is a prefabricated reinforcement consisting of parallel series of high-strength, cold-drawn wires welded together in square or rectangular grids. Smooth wires, deformed wires, or a combination of both can be used in welded wire reinforcement, which is manufactured in many types and sizes.

Welded wire reinforcement provides more uniform stress distribution and more effective crack control in slabs and walls than larger diameter bars more widely spaced. The range of wire sizes and spacings available is broad. Thus, it is possible to furnish almost exactly the cross-sectional steel area required for a specific job.

Cross-sectional area is the basic element used in specifying wire size. Smooth wire sizes are specified by the letter *W*, and deformed wires are specified by the letter *D*. This letter is followed by a number indicating the cross-sectional area of the wire in hundredths of a square inch. For example, W10 is a smooth wire with a cross-sectional area of 0.10 sq in. See **Figure 18-29**.

Spacings and sizes of wires in welded wire reinforcement are identified by style. A typical style designation, 6 × 12—W16 × W8, specifies a welded wire

Metric Wire Area, Diameter & Mass with Equivalent US Customary Units***								
Size**** (MW = Plain) (mm^2)	Area* (mm^2)	Diameter* (mm)	Mass* (kg/m)	Size**** (MW = Plain) (in^2 × 100)	Area** (in^2)	Diameter** (in)	Weight** (lb/ft)	Gage Guide
MW200	200	16.0	1.570	W31	0.310	0.628	1.054	—
MW130	130	12.9	1.020	W20.2	0.202	0.507	0.687	—
MW120	120	12.4	0.941	W18.6	0.186	0.487	0.632	7/0
MW100	100	11.3	0.784	W15.5	0.155	0.444	0.527	6/0
MW90	90	10.7	0.706	W14.0	0.140	0.422	0.476	5/0
MW80	80	10.1	0.627	W12.4	0.124	0.397	0.422	—
MW70	70	9.4	0.549	W10.9	0.109	0.373	0.371	4/0
MW65	65	9.1	0.510	W10.1	0.101	0.359	0.343	3/0
MW60	60	8.7	0.470	W9.3	0.093	0.344	0.316	—
MW55	55	8.4	0.431	W8.5	0.085	0.329	0.289	2/0
MW50	50	8.0	0.392	W7.8	0.078	0.314	0.263	—
MW45	45	7.6	0.353	W7.0	0.070	0.298	0.238	1/0
MW40	40	7.1	0.314	W6.2	0.062	0.283	0.214	1
MW35	35	6.7	0.274	W5.4	0.054	0.262	0.184	2
MW30	30	6.2	0.235	W4.7	0.047	0.245	0.160	3
MW26	26	5.7	0.204	W4.0	0.040	0.226	0.136	4
MW25	25	5.6	0.196	W3.9	0.039	0.223	0.133	—
MW20	20	5.0	0.157	W3.1	0.031	0.199	0.105	—
MW19	19	4.9	0.149	W2.9	0.029	0.192	0.098	6
MW15	15	4.4	0.118	W2.3	0.023	0.171	0.078	—
MW13	13	4.1	0.102	W2.0	0.020	0.160	0.068	8
MW10	10	3.6	0.078	W1.6	0.016	0.143	0.054	—
MW9	9	3.4	0.071	W1.4	0.014	0.135	0.048	10

* Metric wire sizes can be specified in 1 mm^2 increments.

** US customary sizes can be specified in 0.001 in^2 increments.

*** For other available wire sizes, consult other WRI Publications or discuss with WWF manufacturers.

**** Wires may be deformed, use prefix MD or D, except where only MW or W is required by building codes (usually less than MW26 or W4).

Wire Reinforcement Institute

Figure 18-29. This chart shows metric wire area, diameter, and mass with equivalent US customary units.

reinforcement that has a 6″ spacing for horizontal wires, a 12″ spacing for transverse wires, longitudinal wires (W16) of 0.16 sq in, and transverse wires (W8) of 0.08 sq in.

A welded deformed wire reinforcement style would be specified the same way, except that D-number wire sizes would be substituted for the W-number wire sizes. The terms *longitudinal* and *transverse* are related to the manufacturing process. They bear no relationship to the position of the wires in a concrete structure. Some of the stock sizes of welded wire reinforcement used in light construction are shown in **Figure 18-30**.

Spacing of wires in welded wire reinforcement allows a wide variety of combinations. Some of the many spacing combinations possible are as follows:

- 2″ × 4″
- 4″ × 8″
- 4″ × 16″
- 6″ × 12″
- 16″ × 16″

Common Styles of Metric Welded Wire Reinforcement (WWR) with Equivalent US Customary Units[3]

Group	A_s (mm²/m)	Metric Styles (MW = Plain wire)[2]	Wt. (kg/m²)	Equivalent Inch-Pound Styles (W = Plain wire)[2]	A_s (in²/ft)	Wt. (lb/CSF)
Group A[1 & 4]	88.9	102 × 102–MW9 × MW9	1.51	4 × 4–W1.4 × W1.4	0.042	31
	127.0	102 × 102–MW13 × MW13	2.15	4 × 4–W2.0 × W2.0	0.060	44
	184.2	102 × 102–MW19 × MW19	3.03	4 × 4–W2.9 × W2.9	0.087	62
	254.0	102 × 102–MW26 × MW26	4.30	4 × 4–W4.0 × W4.0	0.120	88
	59.3	152 × 152–MW9 × MW9	1.03	6 × 6–W1.4 × W1.4	0.028	21
	84.7	152 × 152–MW13 × MW13	1.46	6 × 6–W2.0 × W2.0	0.040	30
	122.8	152 × 152–MW19 × MW19	2.05	6 × 6–W2.9 × W2.9	0.058	42
	169.4	152 × 152–MW26 × MW26	2.83	6 × 6–W4.0 × W4.0	0.080	58
Group B[1]	196.9	102 × 102–MW20 × MW20	3.17	4 × 4–W3.1 × W3.1	0.093	65
	199.0	152 × 152–MW30 × MW30	3.32	6 × 6–W4.7 × W4.7	0.094	68
	199.0	305 × 305–MW61 × MW61	3.47	12 × 12–W9.4 × W9.4	0.094	71
	362.0	305 × 305–MW110 × MW110	6.25	12 × 12–W17.1 × W17.1	0.171	128
Group C[1]	342.9	152 × 152–MW52 × MW52	5.66	6 × 6–W8.1 × W8.1	0.162	116
	351.4	152 × 152–MW54 × MW54	5.81	6 × 6–W8.3 × W8.3	0.166	119
	192.6	305 × 305–MW59 × MW59	8.25	12 × 12–W9.1 × W9.1	0.091	69
	351.4	305 × 305–MW107 × MW107	9.72	12 × 12–W16.6 × W16.6	0.166	125
Group D[1]	186.3	152 × 152–MW28 × MW28	3.22	6 × 6–W4.4 × W4.4	0.088	63
	388.7	152 × 152–MW52 × MW52	5.61	6 × 6–W8 × W8	0.160	115
	186.3	305 × 305–MW57 × MW57	3.22	12 × 12–W8.8 × W8.8	0.088	66
	338.7	305 × 305–MW103 × MW103	5.61	12 × 12–W16 × W16	0.160	120
Group E[1]	177.8	152 × 152–MW27 × MW27	3.08	6 × 6–W4.2 × W4.2	0.084	60
	317.5	152 × 152–MW48 × MW48	5.52	6 × 6–W7.5 × W7.5	0.150	108
	175.7	305 × 305–MW54 × MW54	3.08	12 × 12–W8.3 × W8.3	0.083	63
	317.5	305 × 305–MW97 × MW97	5.52	12 × 12–W15 × W15	0.150	113

[1] Group A–Compares areas of WWR at f_y = 60,000 psi with other reinforcing at f_y = 60,000 psi
Group B–Compares areas of WWR at f_y = 70,000 psi with other reinforcing at f_y = 60,000 psi
Group C–Compares areas of WWR at f_y = 72,500 psi with other reinforcing at f_y = 60,000 psi
Group D–Compares areas of WWR at f_y = 75,000 psi with other reinforcing at f_y = 60,000 psi
Group E–Compares areas of WWR at f_y = 80,000 psi with other reinforcing at f_y = 60,000 psi

[2] Wires may also be deformed, use prefix MD or D, except where only MW or W is required by building codes (usually less than a MW26 or W4). Also wire sizes can be specified in 1 mm² (metric) or 0.001 in² (inch-pound) increments.

[3] For other available styles or wire sizes, consult other WRI publications or discuss with WWR manufacturers.

[4] Styles may be obtained in roll form. Note: It is recommended that rolls be straightened and cut to size before placement.

Wire Reinforcement Institute

Figure 18-30. Common styles of metric welded wire reinforcement with equivalent US customary units.

Certain styles of welded wire reinforcement are recommended for specific applications. See **Figure 18-31**. It is very important that welded wire reinforcement is placed appropriately in the concrete being installed. If the reinforcement is allowed to settle to the bottom of the concrete, then the benefit of it is lost. Make sure the location of the reinforcement is where it is intended and not at bottom or close to the top of the concrete.

Placing Steel Reinforcement

Placing steel reinforcement is performed at the jobsite. See **Figure 18-32**. Reinforcement should be placed so it is protected by an adequate coverage of concrete. See **Figure 18-33**.

Because reinforcing steel carries the tensile load in concrete, it must be lapped at a splice. The overlap distance for deformed bars should be at least 24 bar diameters. For example, if the job requires 5/8″ diameter rebar, 5/8″ × 24 bar diameters equal 15″. In any case, the distance of overlap should never be less than 12″. Smooth bars should be lapped a greater distance than deformed bars. Welded wire reinforcement should be lapped at least one full stay spacing plus 2″. An 8″ × 8″ material would require a minimum of 10″ overlap. See **Figure 18-34**.

18.6 Lightweight Concrete

The weight of concrete depends on the type of aggregate used. Aggregates used in normal weight concrete include sand and gravel, crushed stone, and air-cooled blast-furnace slag. Lightweight concrete aggregates include expanded shale, clay, and slate; expanded blast furnace slag; sintered fly ash; coal cinders; and natural lightweight materials, such as pumice and scoria. Local availability generally determines the use of a particular aggregate.

A cubic foot of concrete is considered lightweight if it has a density or weight between 105 and 125 pcf (pounds per cubic foot). Normal weight concrete weighs more than 125 pcf. The weight of concrete is based on the density or oven-dry weight per cubic foot of concrete.

Using lightweight aggregates reduces the weight of concrete about 20% to 45% compared with the weight

Recommended Styles of Welded Wire Reinforcement		
Type of Construction	**Recommended Style**	**Remarks**
Basement floors	6 × 6–W1.4 × W1.4 or 6 × 6–W2.1 × W2.1 or 6 × 6–W2.9 × W2.9	For small areas (15′ maximum side dimension) use 6 × 6–W1.4 × W1.4. As a rule of thumb, the larger the area or the poorer the subsoil, the heavier the gauge.
Driveways	6 × 6–W2.9	Continuous reinforcement between 25′ to 30′ contraction joints.
Foundation slabs (Residential only)	6 × 6–W1.4 × W1.4	Use heavier gauge over poorly drained subsoil, or when maximum dimension is greater than 15′.
Garage floors	6 × 6–W2.9 × W2.9	Position at midpoint of 5″ or 6″ thick slab.
Patios and terraces	6 × 6–W1.4 × W1.4	Use 6 × 6–W2.1 × W2.1 if subsoil is poorly drained.
Porch floor A. 6″ thick slab up to 6′ span B. 6″ thick slab up to 8′ span	6 × 6–W2.9 × W2.9 or 4 × 4–W4 × W4	Position 1″ from bottom form to resist tensile stresses.
Sidewalks	6 × 6–W1.4 × W1.4 or 6 × 6–W2.1 × W2.1	Use heavier gauge over poorly drained subsoil. Construct 25′ to 30′ slabs as for driveways.
Steps (Free span)	6 × 6–W2.9 × W2.9	Use heavier style if more than five risers. Position reinforcement 1″ from bottom of form.
Steps (On ground)	6 × 6–W2.1 × W2.1	Use 6 × 6–W2.9 × W2.9 for unstable subsoil.

Wire Reinforcement Institute

Figure 18-31. Recommended styles of welded wire reinforcement for specific concrete applications.

Wire Reinforcement Institute

Figure 18-32. Flat sheets of welded wire reinforcement strengthen the floor of this large structure. The welded wire reinforcing is 6 × 6–W4 × W4.

Recommended Protection for Reinforcement	
Application	**Minimum Concrete Protection**
Footings	3″
Concrete surface exposed to weather	1 1/2″ for No. 5 bars and smaller
Slabs and walls	3/4″
Beams and girders	1 1/2″
Joists	3/4″
Columns	Not less than 1 1/2″ or 1 1/2 times the maximum size aggregate
Corrosive atmosphere or severe exposures	Protection shall be suitably increased

From: "Building code requirements for reinforced concrete," ACI 318

Goodheart-Willcox Publisher

Figure 18-33. The American Concrete Institute recommends minimum concrete protection for reinforcement. Rods used for reinforcing must be covered by the amount of concrete shown in this chart.

Wire Reinforcement Institute

Figure 18-34. Sheets of welded wire reinforcing are overlapped at least one full stay spacing plus 2″.

of concrete made with normal weight aggregates, without compromising the structural properties. Study comparisons have also shown that worker efficiency increases with lightweight masonry units when compared with normal weight units.

Safety Note

Contact with wet (plastic) concrete, cement, mortar, grout, or cement mixtures can cause skin irritation, severe chemical burns, or serious eye damage. Wear waterproof gloves, a long-sleeved shirt, full-length trousers, and eye protection when working with these materials. If you must stand in wet concrete, wear high-top waterproof boots. Wash wet concrete, mortar, grout, cement, or cement mixtures from your skin immediately. Flush your eyes with clear water immediately upon contact. Seek medical attention if you experience a reaction to contact with these materials.

Chapter Review

Summary

- The basic components of concrete are Portland cement, fine aggregate (sand or finely crushed stone), coarse aggregate (gravel or crushed stone), and water.
- Portland cement is produced in various types to meet physical and chemical requirements for specific purposes.
- Fine aggregate contains sand or other acceptable fine materials. Coarse aggregate consists of gravel, crushed stone, or other acceptable materials larger than 1/4″ in diameter.
- Concrete mixing water should be clean and free of oil, alkali, and acid.
- Types of admixtures include air-entraining admixtures, water-reducing admixtures, retarding admixtures, accelerating admixtures, pozzolans, and workability agents.
- The proportioning of concrete mixtures determines the qualities of the plastic and hardened concrete.
- Cement paste made with a proper proportion of water to cement (water-cement ratio) is strong, watertight, and durable. Strength and durability of concrete is directly related to the water-cement ratio.
- Adjustments to the water-cement ratio are needed for damp, wet, or very wet fine aggregate.
- A high water content causes a high slump. Generally, a low-slump concrete produces a better concrete product.
- Measurement of aggregates by weight is recommended to achieve a high degree of consistency in each batch.
- Concrete mixing time depends on size of the batch, speed of the machine, and condition of the mixer.
- In reinforced concrete, the concrete resists compression forces, and steel, FRP bars, or wires resist the tension forces.
- In welded wire reinforcement, smooth wire sizes are specified by the letter *W*, and deformed wires are specified by the letter *D*. The letter is followed by a number that indicates the cross-sectional area of the wire in hundredths of a square inch.
- Steel reinforcement should be placed so it is protected by an adequate coverage of concrete.
- Smooth bars should be lapped a greater distance than deformed bars.
- Waterproof gloves and boots, a long-sleeved shirt, full-length trousers, and eye protection should be worn when working with plastic concrete, cement, mortar, grout, or cement mixtures.

Know and Understand

Answer the following questions using the information provided in this chapter.

1. Each of the following is a basic material in concrete *except* _____.
   A. Portland cement
   B. aggregate
   C. clay
   D. water
2. A general cement suitable for all uses when special properties are *not* required is _____.
   A. Type I
   B. Type II
   C. Type III
   D. Type IV
3. Aggregates usually make up _____% of the volume of concrete.
   A. 0
   B. 15–25
   C. 30–50
   D. 60–80
4. *True or False?* The use of pozzolans substantially increases the strength of concrete during the first 28 days.
5. In properly proportioned concrete, each particle of _____ is surrounded by cement-water paste.
   A. pozzolan
   B. admixture
   C. Portland cement
   D. aggregate
6. *True or False?* To complete the chemical reactions between water and cement, only a small amount of water is necessary.
7. *True or False?* A slump test is considered a measure of workability.

8. Which of the following is *not* true when measuring cement?
   A. Use only full bags of cement.
   B. Weigh out the proper amount of cement.
   C. Use volume for an accurate measurement of cement.
   D. Weigh bulk cement for each batch.
9. Measurement of aggregates by _____ is the recommended practice.
   A. size of bulk bag
   B. weight
   C. volume
   D. pressure
10. Concrete mixing time does *not* depend on _____.
   A. type of cement used
   B. speed of the machine
   C. size of the batch
   D. condition of the mixer
11. *True or False?* Ready-mixed concrete is convenient and usually of high quality.
12. *True or False?* Concrete has little tensile strength.
13. The two main types of steel reinforcement used in concrete are reinforcing bars and _____.
   A. ridged bars
   B. tensile steel
   C. spacing steel
   D. welded wire reinforcement
14. *True or False?* The weight of concrete depends on the type of cement used.
15. *True or False?* A cubic foot of concrete is lightweight if it has a weight of between 125 pcf and 150 pcf.

Apply and Analyze

1. What are the four basic materials used to make concrete?
2. What type of cement produces high strength quickly, usually in a week or less?
3. How do air-entraining admixtures affect the workability of fresh concrete?
4. What is the result of using the maximum allowable particle size of coarse aggregate?
5. Identify four types of admixtures used in concrete.
6. What is the purpose of adding air-entraining admixtures to concrete?
7. List three desirable qualities of plastic concrete.
8. List five desirable qualities of hardened concrete.
9. What is *slump*?
10. Why is measurement of fine aggregate by volume inaccurate?
11. Generally, how long should concrete be mixed?
12. Identify five types of aggregates used in lightweight concrete.

Critical Thinking

1. Imagine you are the owner of a masonry construction building. Your company is building a residential foundation for a home that has a 4″ basement concrete floor. List the following characteristics of the concrete mixture for this project based on the recommendations listed in this chapter.
   A. Recommended slump
   B. Recommended strength after 28 days, in psi (pounds per square inch)
   C. Water/cement ratio
   D. Maximum size of aggregate
   E. Appropriate cement content (bags per yd)

Activities

1. (SCIENCE) What is external sulfate attack and how does it affect concrete? Where do sulfates occur? How can concrete be protected from sulfate attack? Research these questions and be prepared to provide answers in class.

photovs/iStock via Getty Images

CHAPTER 19

Form Construction

LEARNING OUTCOMES

After studying this chapter, you will be able to:

19.1-1 Identify the materials used for concrete formwork and form facing.
19.2-1 Explain considerations relating to form design.
19.3-1 Construct forms for specific applications.
19.4-1 Practice maintenance on forms.
19.5-1 Describe the advantages and considerations when using insulated concrete wall forms.

TECHNICAL TERMS

adjustable post base
braces
break back
builder's level
builder's transit
bulkhead
concrete column footing form
Plyform
sheathing
sill
spreaders
stiff back
strongback
studs
ties
wales

Introduction

Concrete structures require forms to provide the desired shape and surface texture. See **Figure 19-1**. Forms are made of wood, steel, fiberglass, hardboard, plastics, or other materials. The forms must be strong enough to resist the forces developed by the plastic (liquid) concrete. Regular concrete weighs about 150 lb/cu ft. Poured in a form 8′ high, the concrete would create a pressure of about 1200 lb/sq ft along the bottom side of the form.

Safety Note
Safe construction procedures must be followed when constructing forms for concrete. Two excellent sources of information are the American Concrete Institute (ACI) and the Occupational Safety and Health Administration (OSHA).

19.1 Form Materials

Wood is the most popular form material. Both lumber and plywood are used in form construction. Construction lumber is used for frames, bracing, and shoring, while plywood is used for the form surface. Boards may be used instead of plywood if certain surface patterns are desired. However, plywood is manufactured in more than 40 surface textures, ranging from glass-smooth to board-and-batten panels. Its large, smooth surface, resistance to change in shape when wet, and

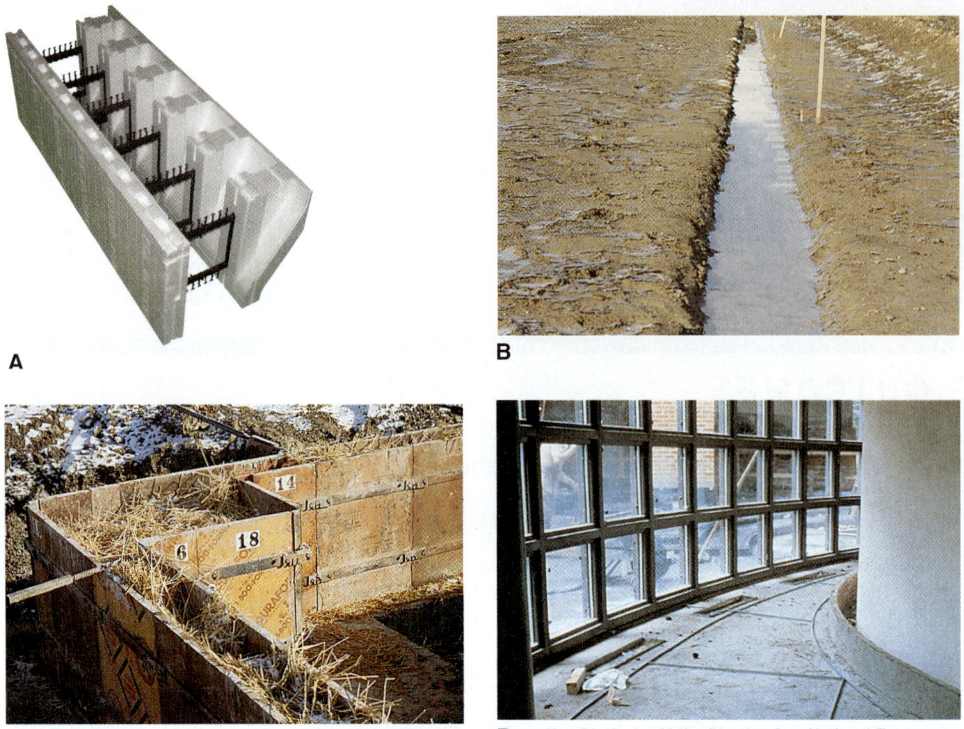

A B C D Fox Blocks by Airlite Plastics Co.; National Terrazzo and Mosaic Association, Inc.

Figure 19-1. A—Insulated form for concrete wall. B—Soil as a form material for footing. C—1 1/8″ thick plywood forming system designed especially for residential and low industrial concrete walls. D—Bronze strips used to form terrazzo sections.

ability to resist splitting has made plywood the most popular material for form construction.

Virtually any exterior-type plywood can be used for concrete formwork because it is made with waterproof glue. However, the plywood industry markets a product called ***Plyform***, which is specially designed for concrete forming. Plyform is an exterior-type plywood made from special wood species and veneer grades to ensure high performance.

Plyform is available as Plyform Class I and Plyform Class II. Class I is stronger and stiffer than Class II. Either type can be purchased with a high-density overlaid surface on each side. This surface is smooth, grainless, and resistant to abrasion and moisture penetration.

Structural 1 Plyform is also available in some areas of the United States. It is especially designed for engineered applications. It is stronger and stiffer than either Plyform Class I or II. This plywood is also made with high-density overlay faces. All Plyform panels are made in thicknesses of 19/32″, 23/32″, and 3/4″. Panel size is 4′ × 8′.

Plywood is also valued as a form material because of its ability to bend. Curved surfaces can be formed easily if the thickness of the sheet and the direction of face grain are considered. **Figure 19-2** shows the minimum bending radii for plywood panels. The rate of bending can be increased for a given thickness sheet by sawing kerfs across the inner face at right angles to the curve. See **Figure 19-3**. Curves can also be formed by using two or more thinner sheets bent one at a time and then fastened together.

Plywood Thickness	Minimum Bend Radius—Perpendicular to Face Grain	Minimum Bend Radius—Parallel to Face Grain
1/4″	16″	48″
5/16″	24″	60″
3/8″	36″	72″
1/2″	72″	96″
5/8″	96″	120″
3/4″	120″	144″

Goodheart-Willcox Publisher

Figure 19-2. Minimum bending radii for plywood panels used in form construction.

Hardboard that is tempered, specially treated, and 1/4″ thick can be used as a form facing material. The hardboard is coated with plastic to prevent water penetration. This material has a smooth surface, but it is essentially a form liner and must be supported by

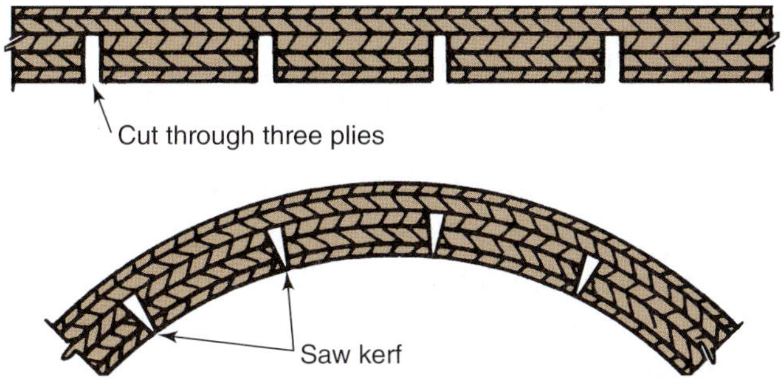

Figure 19-3. Saw kerfs through three plies help bend short radius curves.

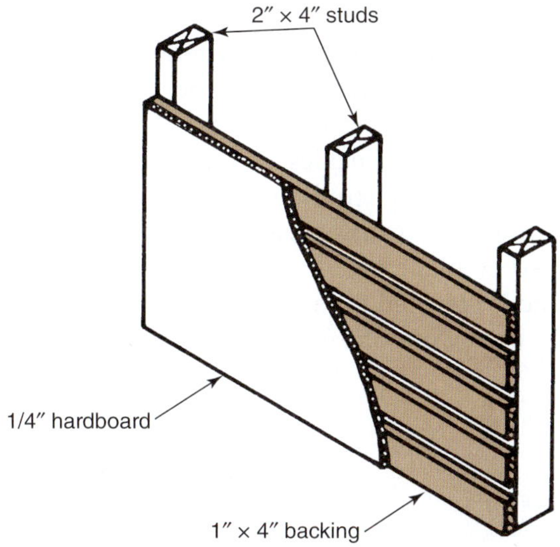

Figure 19-4. Recommended backing for hardboard lined forms.

a backing of lumber. **Figure 19-4** shows the recommended method of backing for hardboard liners.

Steel and aluminum are frequently used for form construction, for both frames and form facing. See **Figure 19-5**. Steel and aluminum frames provide greater strength and support heavier loads than wood frames. Metal frame forms can be faced with either plywood or sheet metal. Metal forms have a longer life than wood forms, which somewhat offsets their greater cost.

Fiberglass form facing is increasing in use. This material is usually prefabricated in the size and shape desired. It is being used extensively in precast concrete plants. Pan forms for ribbed concrete floors are frequently made of fiberglass as well.

Insulating board and rigid foam can be used as form liners, but they have little or no strength. Most often, they are left attached to the concrete when the form is removed. Clips can be used for this purpose.

Figure 19-5. Steel-framed, plywood-faced panels with standard walkway.

19.2 Form Designs

Forms should be designed for practicality and economy. They must be in the correct shape, width, and height and be strong enough to resist the pressure of the plastic concrete. They need to retain their shape during the pouring and curing phases. Wet concrete should not leak from joints and cause fins and ridges.

Forms should be as simple as possible to build and use in sizes that can be easily handled and stored. They must be designed to be removed without damaging the concrete. Forms must also be safe for those who work around them.

The primary consideration in form design is usually its strength. A form must support its own weight, and the weight of the liquid concrete and any other loads that may be placed on it, such as wind, workers, and equipment. For most general form requirements, concrete that is made with natural sand and gravel aggregates weighs about 150 lb/cu ft. Designs presented in this chapter are based on that weight of concrete and average conditions.

Form Elements

Forms generally have five elements. See **Figure 19-6**. These elements are sheathing, studs, wales, braces or supports, and ties or spreaders.

Sheathing gives the concrete surface its shape and texture. It keeps the concrete in place until it hardens. Sheathing is usually a special type of plywood intended to be used for concrete formwork. Thicknesses vary from 5/8″ to 1 1/8″ and lengths can be purchased in 8′ or 10′ sheets.

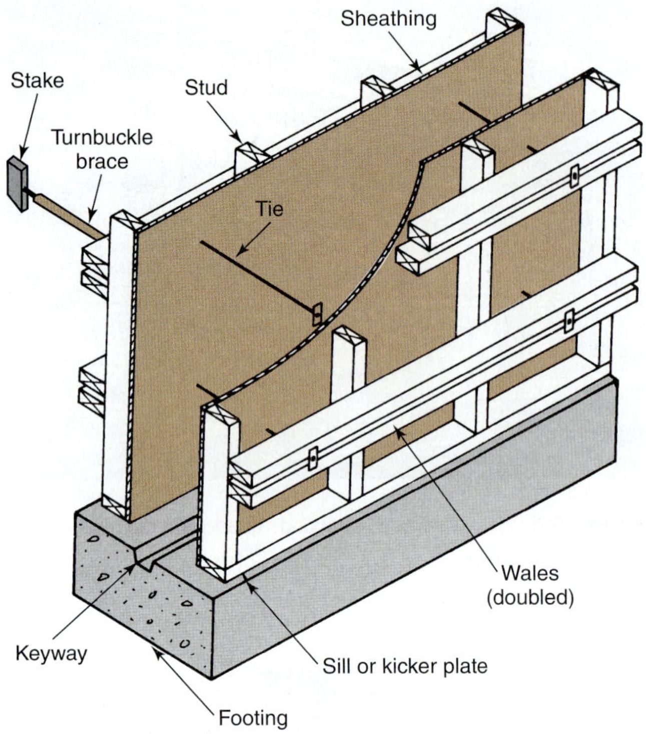

Goodheart-Willcox Publisher

Figure 19-6. Parts of the concrete form.

The ***sill*** is the base of the wall form that is temporarily attached to the footing. The sill is usually made of treated 2 × 4 lumber that is nailed to the bottom of the studs to form the wall.

Studs support the sheathing and prevent bowing. Some types of forms do not require studs, but most wooden forms do. Studs are usually made of 2 × 4 lumber that is treated to withstand exposure to weather.

Wales are used to align the forms, secure the ties, and support the studs. Wales are usually made of treated 2 × 4s that run perpendicular to the studs to prevent studs from bowing when concrete is placed in the forms. Wales can be doubled by laying the second 2 × 4 on top of the first to increase the strength of the concrete form system.

Braces and supports provide lateral support against wind and other forces. One type of brace that is easy to use is the turnbuckle. See **Figure 19-7** and **Figure 19-8**. Turnbuckles can be used to align the form and are only required on one side of the form. These braces are usually spaced 8′ to 10′ apart. Braces can also be made of 2 × 4 lumber. To secure braces, one end is attached to formwork and the other end to a stake driven into the earth.

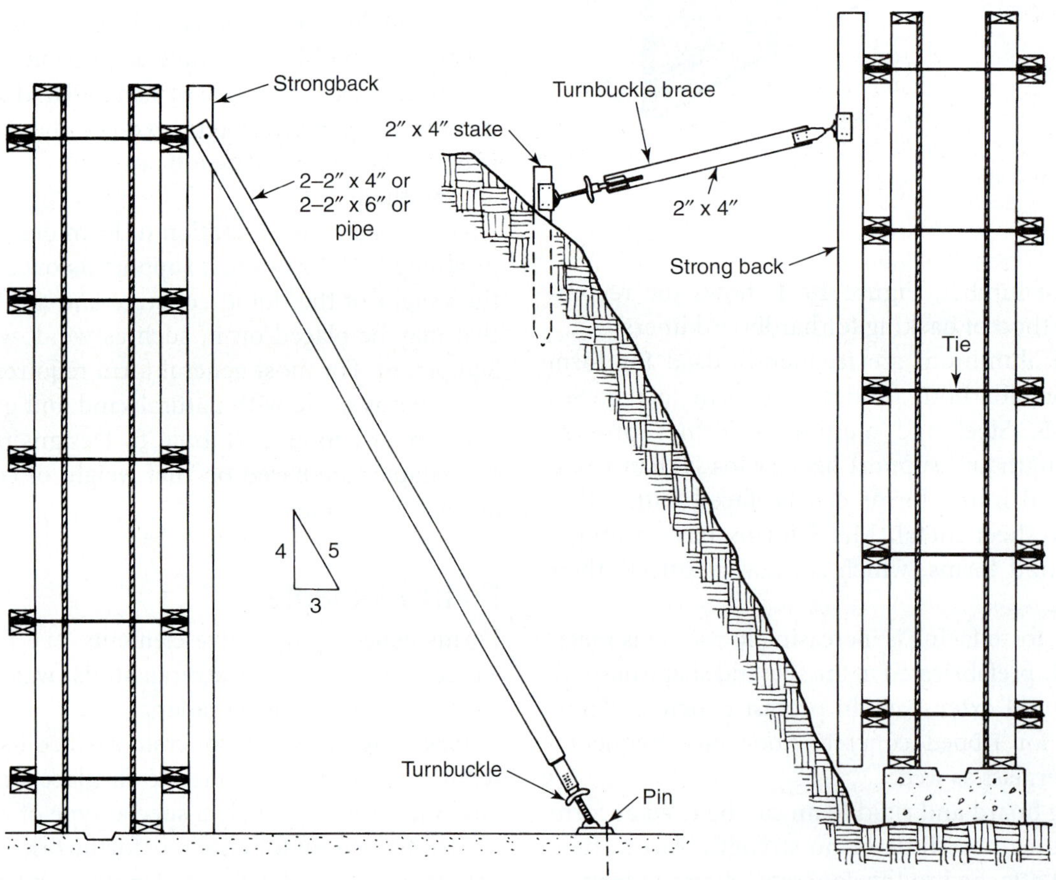

Goodheart-Willcox Publisher

Figure 19-7. The turnbuckle brace can be used to align the form as well as support it.

Jack Klasey

Figure 19-8. A turnbuckle brace is being used to support one side of the form. Straw has been added to the top of the wall form cavity to insulate the freshly poured concrete.

Ties and ***spreaders*** prevent the sides of a form from spreading or moving together. See **Figure 19-9**. Wooden spreaders keep the walls of the forms separated until concrete is in place and then are removed from the inside of the form as the concrete is poured to their level. The metal or fiberglass ties remain in the concrete to become a permanent part of the structure. Fiberglass ties are becoming popular because they are not as conductive as metal ties, making them more energy efficient.

When the concrete has cured and the forms are removed, the ties project out from the surface of the concrete. They may be broken off or disconnected internally, depending on the type. See **Figure 19-10** and **Figure 19-11**.

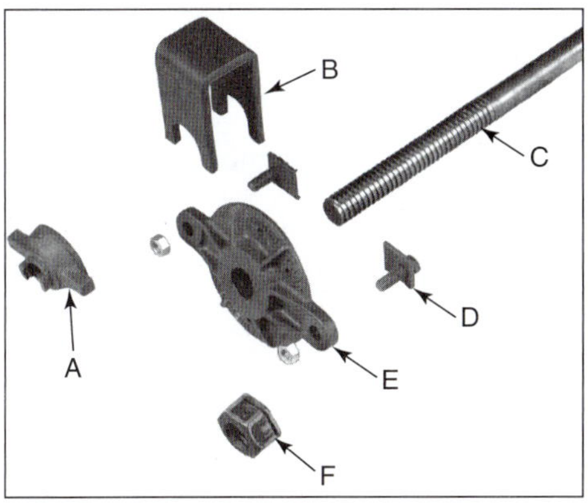

Tapered tie showing parts. A—Cast knuckle nut. B—Spreader clip. C—Tapered tie rod. D—Friction clamp. E—Cast bearing washer. F—Hex knuckle nut.

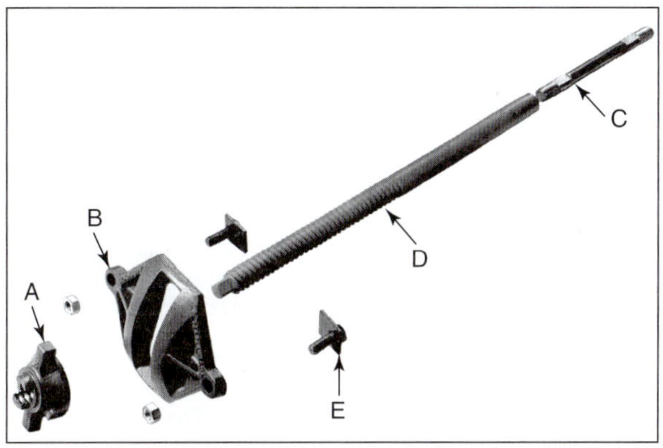

She-bolt tie with parts. A—Cast knuckle nut. B—Cast bearing washer. C—Inner unit. D—15″ or 20″ she-bolt knuckle thread. E—Friction clamp.

Tie bracket for securing the tie to the walls or vertical stud.

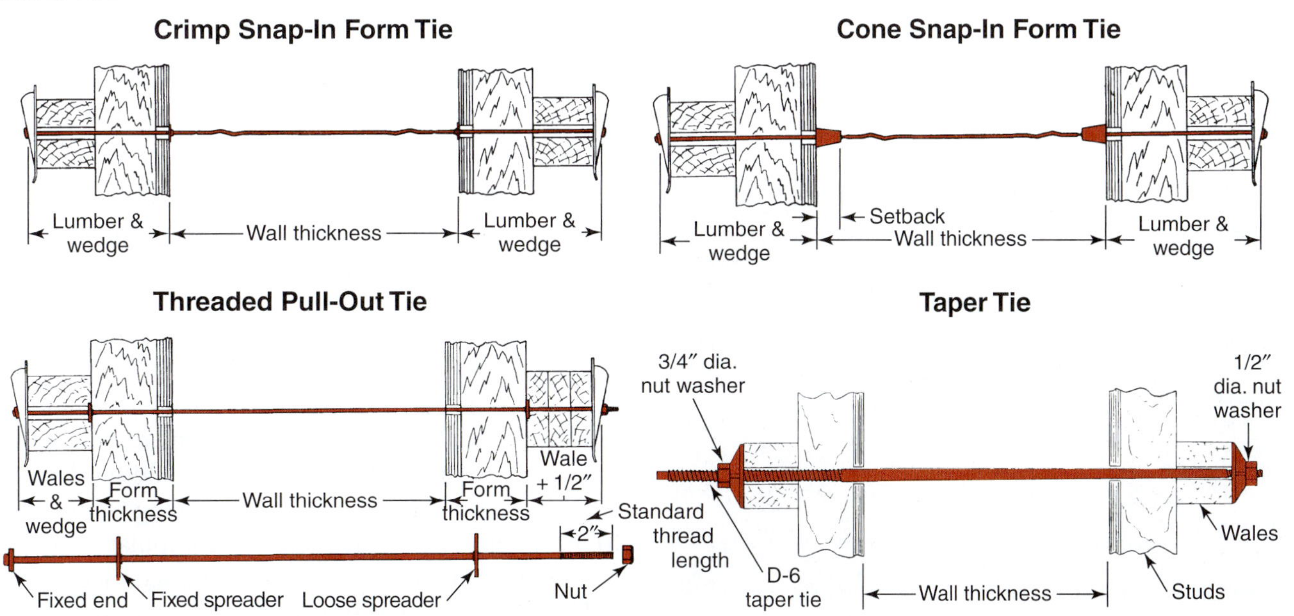

Dayton Superior

Figure 19-9. Ties, tie accessories, and tie assemblies. Top—Panel accessories. Bottom—Drawings show cross-sectional view of tie assemblies attached to the form.

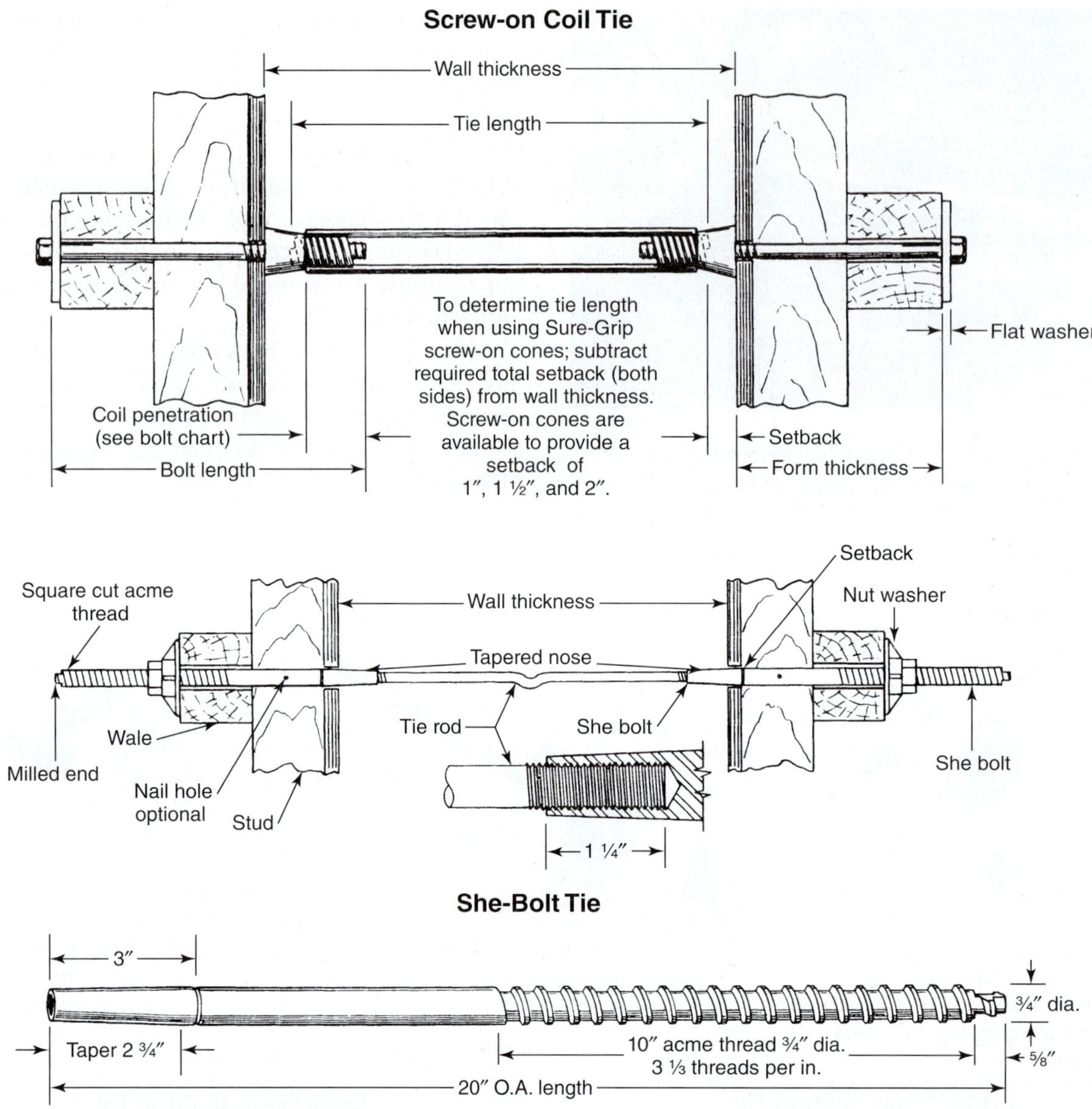

Dayton Superior

Figure 19-10. Threaded ties such as these can be disconnected. Protruding portions can be removed from the wall.

Goodheart-Willcox Publisher

Figure 19-11. Concrete form tie.

When specifying ties, it is necessary to know the type required; the wall thickness; the actual size of sheathing, studs, and wales; and the required break back. The ***break back*** is the depth the tie is broken off in the concrete. Some ties have spreader washers that take the place of wood spreaders.

A variety of patented ties are available to secure and hold form walls apart. A typical patented tie has a rod that passes through the wall with a holding device at each end. They can support loads ranging from 1000 lb to 50,000 lb. Two broad classes are popular—continuous single-member ties and internal disconnecting ties.

The most common type of continuous single-member tie is the snap tie. Refer to **Figure 19-9**. Snap ties can be used with either single or double waler systems and are made for wall thicknesses from 6″ to 26″. Wedges or wedge brackets are used to secure these ties, which are nothing more than a metal rod that extends from the outside surfaces of the opposing wales. Small plastic cones or metal washers along the rod act as spreaders to hold the walls at the proper spacing. The protruding sections of the ties are snapped off after the forms are removed.

Another type of continuous single-member tie is the loop and tie. This type of tie is secured with a tapered steel wedge driven through the loop at the end of the tie and against a plate that is outside the waler. These ties are frequently used with prefabricated metal forms.

Internal disconnecting ties are designed for heavier construction work where heavier loads are expected. These ties have external sections that screw into an internal threaded section. Internal disconnecting ties are used for wall thicknesses from 8″ to 36″.

Dayton Superior

Figure 19-13. Strippable inside corner assemblies provide a means of releasing forms from inside captive conditions such as counterfort walls, box conduits, and vertical, square, and rectangular shafts.

Prefabricated Forms

Prefabricated forms (panels) are prebuilt, usually in modular sizes. Prefab forms can be fastened together on the job to produce the desired wall or column size. These forms can be purchased, rented, or constructed by a contractor.

The most common type of prefab form has a metal frame with a plywood facing. See **Figure 19-12** and **Figure 19-13**. Special connectors attach the modular units together. Some connectors have hinged corners that aid in removing the forms.

The most common size of module is 2′ × 8′. The modules can be combined to make gang forms. Gang forming is used on large jobs where repetitive forming is required. Gang forms are generally lifted with a crane since they are too heavy to be lifted by hand.

Slip Forms

Slip forms are used for casting very large structures of great height. They are raised slowly by jacks as the concrete is poured. An average speed is about 15″ per hour. See **Figure 19-14**. Interior forms along with the exterior forms move up the walls as previous placed concrete has begun to cure and new concrete is added to the top of the movable slip form. Concrete is usually placed nonstop 24 hours a day to ensure a uniform

Dayton Superior

Figure 19-12. Modular form components are produced in several sizes. Plywood facing (3/4″) is attached to a steel frame.

tracy tucker/iStock/Thinkstock

Figure 19-14. A grain elevator under construction using the slip-form process.

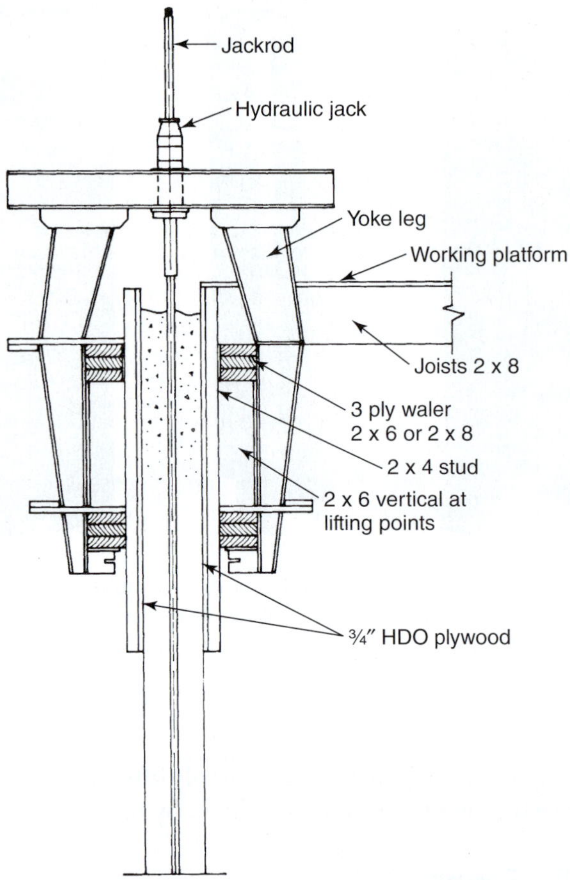

American Plywood Association

Figure 19-15. Section through a typical slip form.

bond throughout the wall structure. Slip forms are also used for casting long, low walls such as median walls between lanes of traffic.

Figure 19-15 shows a section through a slip form. The form facing is 3/4″ high-density overlaid plywood. The wales are lumber and the yokes are steel. The form is pulled upward by the jacks that move up jackrods embedded in the concrete. Jackrods can be recovered after the pouring is finished by pulling them from a recovery pipe.

19.3 Form Applications

Building construction forms can be grouped into those used for footings, walls, slabs, steps, beams and columns, and masonry support. This text focuses on forms that can be built on-site.

Footing Forms

Footings are not usually visible, and therefore appearance is of little concern. However, footings must be located accurately and built to the specified dimensions. If the soil is firm and not too porous, it may serve as the form for footings. However, inferior results are possible if the soil absorbs too much water from the concrete or if soil falls into the concrete. See **Figure 19-16** and **Figure 19-17**.

22August/Shutterstock.com

Figure 19-16. A footing form has been created in clay soil by using a large mechanical excavator. Rebar has been formed for footing reinforcement. Rebar for the pier extends above the footing rebar. After the footing is poured and hardens, the pier form will be built around the pier rebar.

Concrete footing forms are usually constructed from 2″ construction lumber the same width as the footing thickness. The boards are placed on edge and held in place by stakes and cross-spreaders. See **Figure 19-18**. Soil may be used to fill cracks around the forms and provide some support.

Goodheart-Willcox Publisher

Figure 19-17. Two types of earth footing forms.

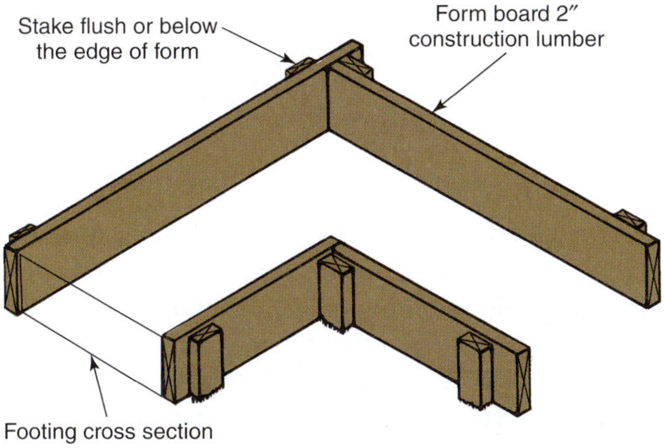

Goodheart-Willcox Publisher

Figure 19-18. Footing forms made of 2″ construction lumber.

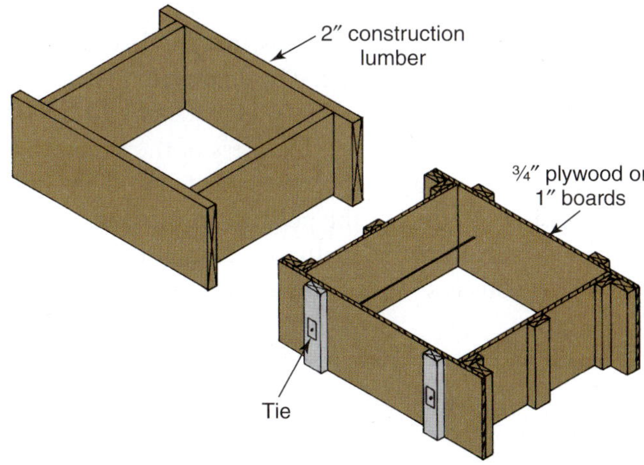

Goodheart-Willcox Publisher

Figure 19-20. Isolated footing forms made of 2″ construction lumber and 3/4″ plywood.

Trade Tip

Remember that footings must be poured on undisturbed or well-compacted earth. Do not fill in low spots under the footing with loose earth.

Occasionally, a site requires stepped footings. **Figure 19-19** illustrates one method of forming stepped footings. The height and length of the steps should coincide with the length and course height of one or more masonry units being used. Always study the plans carefully.

Isolated footings for piers and columns are formed with bottomless boxes, which are usually made with construction lumber. See **Figure 19-20**. The width of the boards should be the same as the thickness of the footing.

Tapered footing forms are built like a hopper. Two sides must have the exact dimensions of the footing while the other two are slightly larger so cleats may be attached along their edges. See **Figure 19-21**. Side and end sections are held together with these 2 × 4 cleats. The form must be securely anchored because it tries to rise as the concrete is poured.

Wall Forms

Wall forms can be built in place or prefabricated. The type selected depends on the complexity of the walls, need to reuse, or preference.

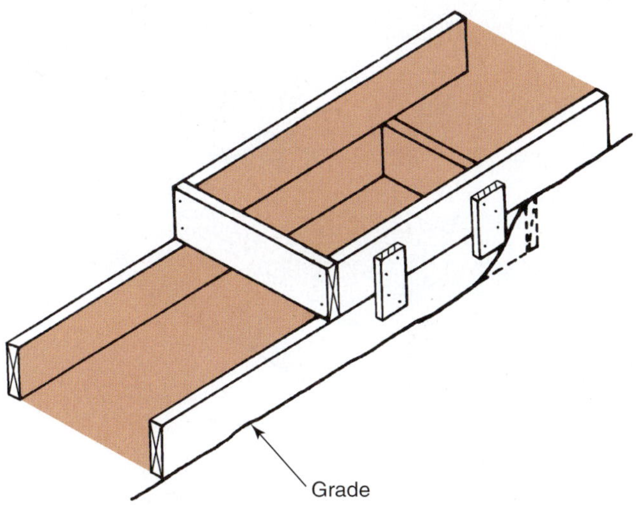

Goodheart-Willcox Publisher

Figure 19-19. Form for a stepped footing.

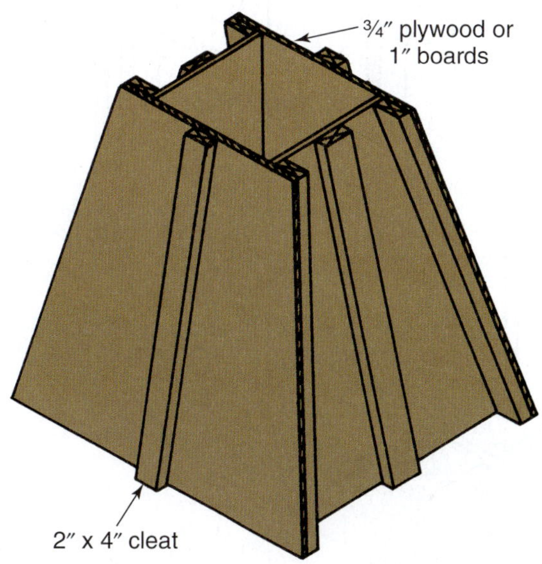

Goodheart-Willcox Publisher

Figure 19-21. Tapered footing form of 3/4″ plywood or 1″ boards. This form will be held in place by stakes driven into the ground.

To build in-place forms, first attach a sole plate to the footing using concrete nails, power driven nails, or stakes. See **Figure 19-22**. The sole plate provides a place to attach the studs. The sole plate should be set back from the wall line the thickness of the sheathing. Be sure the sole plate is properly located and straight because this determines the position of the wall.

Cut studs to desired length and toenail them to the top of the plate as shown in **Figure 19-23**. In a form for a typical basement wall, spacing of the studs is generally 24″ with 3/4″ plywood or 1″ boards used for sheathing. Stud spacing should be decreased as the height of the wall increases.

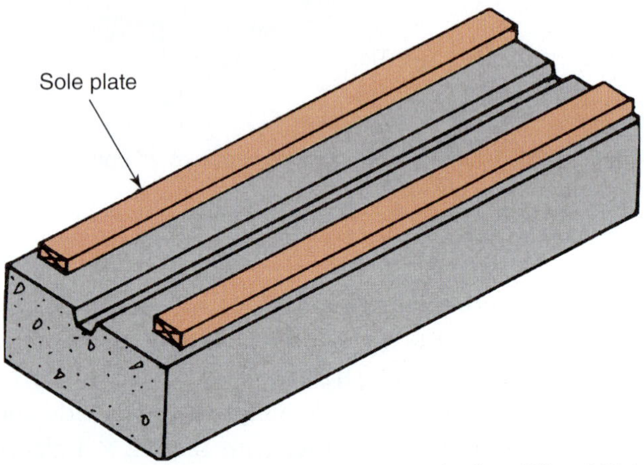

Goodheart-Willcox Publisher

Figure 19-22. The first step in form building is to attach the sole plate to the footing with power-driven nails, concrete nails, or concrete screws.

Sheathing is nailed to the studs when they are in place. The first sheet should be leveled so remaining sheets will be level. Holes for ties can be drilled as the sheathing is being attached. Placement of the ties determines the location of the wales. See **Figure 19-24**.

Wales are attached to the outside of the studs using nails, clips, or wires. Either single or double wales can be used. One end of the braces is attached to the wales. The other end is anchored in the ground or some other solid support.

If the wall is too narrow or deep for a person to work inside, the inside wall can be built in a horizontal position and tilted into place. Also, any reinforcing steel should be placed in position before the inside form is placed.

Use of prefab forms is becoming more common. These forms can be assembled on the ground and then positioned. Bracing is required for them just as it is for built-in-place forms.

Corners of forms are weak and must be given special attention. The forms must be tied together firmly to prevent concrete leakage. Corner brackets or tie rods can be used to give the needed strength. See **Figure 19-25**.

Goodheart-Willcox Publisher

Figure 19-23. The second step in building wall forms is to toe-nail the wall studs to the sole plate.

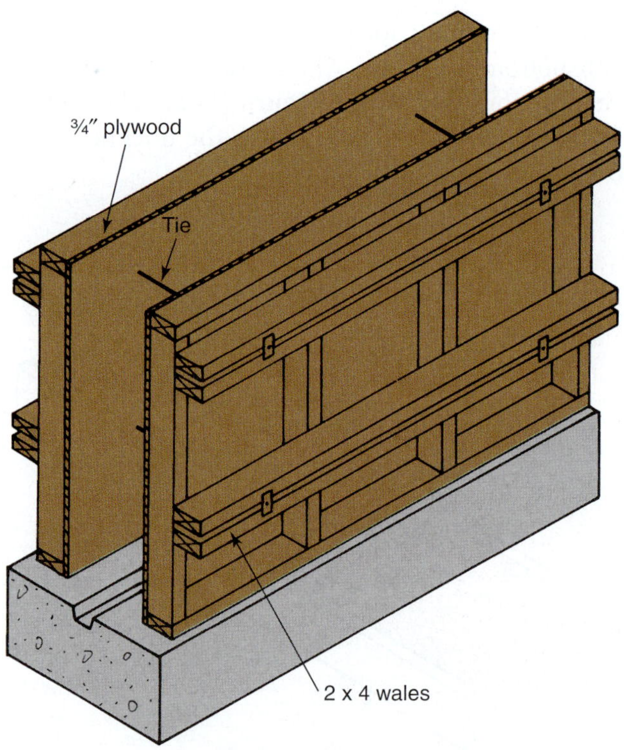

Goodheart-Willcox Publisher

Figure 19-24. A completed section of a wall form with sheathing, ties, and wales in place.

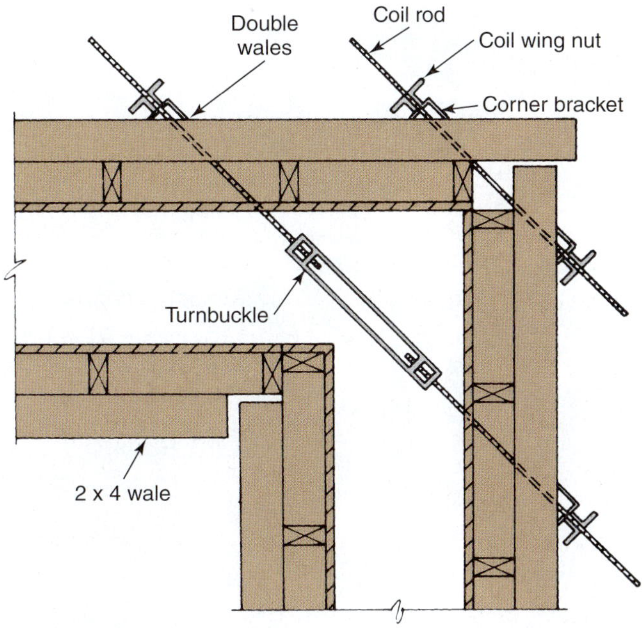

Goodheart-Willcox Publisher

Figure 19-25. Form units held together at the corner with corner brackets and tie rods.

Some forms have a series of metal eyes fastened to the sheathing panels that allow a metal rod to secure them together. Another method of tightening the corners is to lap the wales at the corner and secure them. Wooden wedges can be driven behind the wales to close any openings.

Wall Openings

Any wall openings, such as for windows, can be formed using bucks. Bucks are wood or steel frames set in the form between the inner and outer form to make an opening in the wall. See **Figure 19-26**.

A typical buck design consists of an outside frame reinforced by 2 × 4s placed on edge with horizontal cross braces. For construction of bucks, 3/8″ plywood, 1″ boards, or 2″ construction lumber is used. Thinner material requires more bracing. If a recess is required to receive the frame, a recess strip is nailed to the outside of the buck's frame. If a wood nailing strip is required in the concrete, a wedge-shaped strip of wood is attached to the frame.

Bucks for arches are generally made from two pieces of plywood that are fastened at the top of a rectangular buck. The arched form is enclosed with 1/4″ plywood or heavier material as needed.

Some metal window and door frames can be set in the forms and serve as bucks. In this instance, they must be carefully plumbed, aligned, and well-braced to maintain proper placement.

Small openings for vents, ducts, and large diameter pipes may be required. These openings can be formed using small wooden box frames that have been built to the proper size of the openings. Rigid foam plastic blocks and fiber or sheet metal sleeves can also be used.

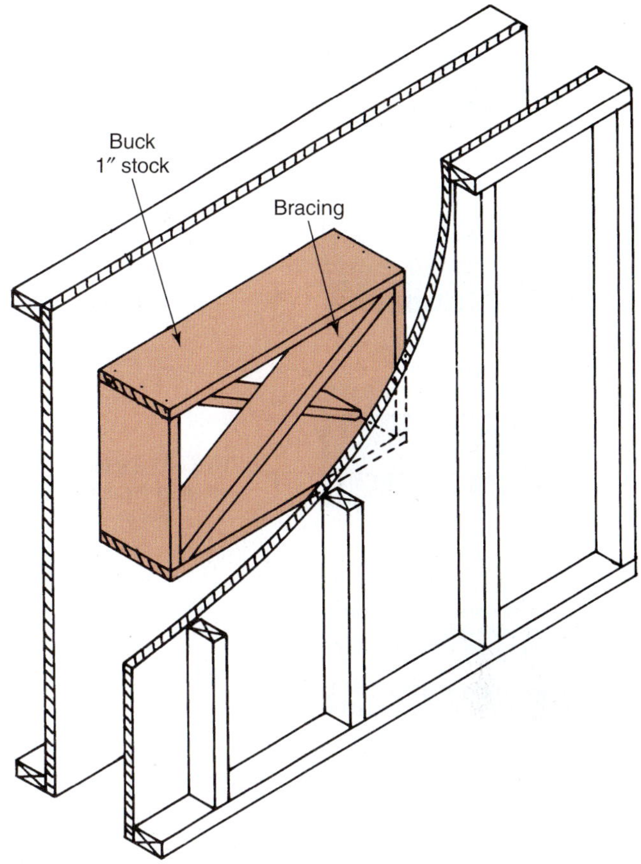

Goodheart-Willcox Publisher

Figure 19-26. A well-braced buck in place in the form casts a void for a window or other opening through the wall. Bucks are braced to prevent distortion.

Slab Forms

Forms for drives, walks, patios, or other flatwork are either wood or metal. Two-inch lumber is commonly used with wood or steel stakes for bracing.

For slabs 4″ thick, 2 × 4 lumber is recommended. One-inch lumber can be used, but stakes must be much closer together to keep the form straight. A 5″ slab can be formed with 2 × 4s, but 2 × 6 materials are preferred. Slabs that are 6″ thick require 2 × 6 lumber. See **Figure 19-27**.

Form material for slabs should be smooth, straight, and free of knotholes and other surface imperfections. Remember that dressed (machined smooth) 1″ lumber is actually 3/4″ thick and 2″ lumber is 1 1/2″ thick. Widths are usually 1/2″ less than the nominal size. It is important to know the actual size of materials when building forms.

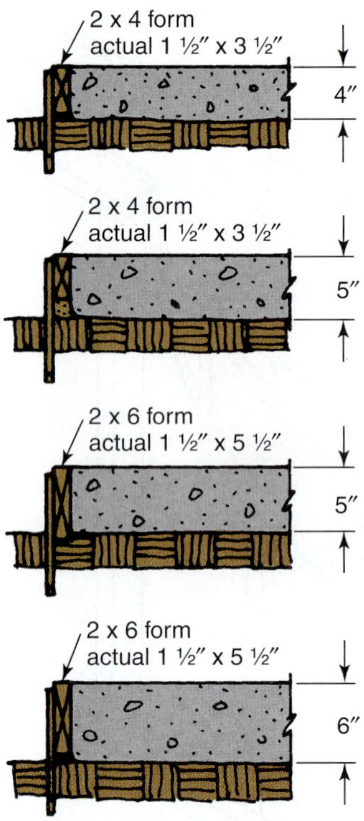

Goodheart-Willcox Publisher

Figure 19-27. Recommended form lumber size for typical slab thicknesses.

Stacey Ellen/Shutterstock.com

Figure 19-28. 2 × 4 form boards are being used to create a concrete driveway. A mason's line is used to ensure the forms are straight and at the appropriate height.

On large projects, metal forms are generally used to form slabs. They save time, withstand rough handling, and can be used many more times than wood.

Wood stakes are made from 1 × 2, 1 × 4, 2 × 2, or 2 × 4 material. See **Figure 19-28**. They are spaced about 4′ apart for 2″ thick formwork. A spacing of about 2′ is recommended for 1″ thick formwork. Steel stakes are available for use with wood or metal forms. They are easier to drive and much stronger. Even though they cost more than wood stakes, steel stakes can be cheaper over a period of time due to their long life.

Stakes should be driven slightly below the top of the forms to aid in screeding and finishing the concrete slab. Drive the stakes straight and plumb so the form is true. Use double-headed (form) nails to attach the stake to the form. Drive the nail through the stake into the form.

Forming Curves

Horizontal curves are easy to form if plywood, 1″ boards, hardboard, or sheet metal is used. Curves with a short radius can be formed with 1/4″ or 1/2″ plywood with the face grain vertical. If heavier lumber is desired or a sharper bend is necessary, saw kerfing is helpful. Wet lumber bends more easily than dry. **Figure 19-29** shows some details for forming horizontal curves. Long gentle curves can be formed using 2″ thick material. Staking must be secure because it must support a heavier load. It is a good idea to use extra stakes for all types of curves.

You can also use multiple layers of 1/4″ Masonite or similar materials to make tight curves. See **Figure 19-30**. Masonite comes in 4′ × 8′ sheets and can be cut to varying widths to accommodate different thicknesses of concrete slabs. Hardboard siding is another option that can be used when forming tight curves. This product can be purchased in varying widths from 6″ to 12″ and can be cut down to the correct size or into multiple pieces. The thickness of hardboard siding varies by manufacturer, but 7/16″ is a common thickness with a length of 16′.

The proper grade can be maintained by using a mason's line or builder's level. A ***builder's level*** is an accurate spirit level combined with a telescope on a circular base. The builder's level sits on a tripod. Readings are taken in a horizontal plane. See **Figure 19-31**. The grade is determined at a given point, and a reading is performed by looking through the level and taking a reading on a rod or rule. The rod or rule is then moved along the form line and measurements are made on the stakes, indicating proper grade. When enough stakes

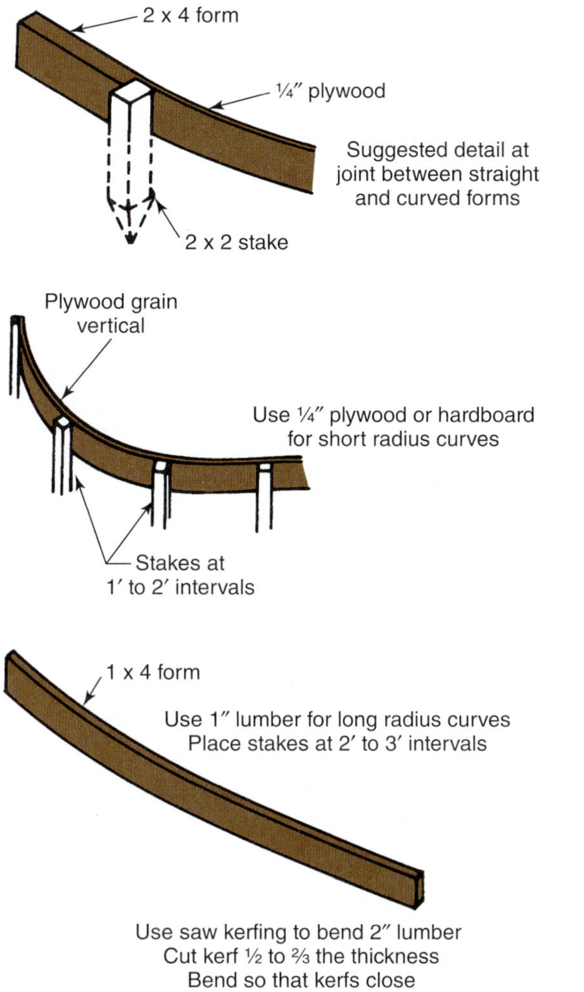

Goodheart-Willcox Publisher

Figure 19-29. Forming horizontal curves using plywood.

Author Image, Cotton Ranch Volga, SD

Figure 19-30. Three layers 1/4″ Masonite are used to form the curves on a concrete patio. 1 × 4 stakes are placed every 2′ to ensure the forms do not move when concrete is placed inside the forms. The tops of the stakes are cut flush with the forms to aid when screeding the concrete.

Photodiem/Shutterstock.com

Figure 19-31. A contractor is performing the setup procedures for the builder's level by adjusting the leveling screws. After completing the setup, they will take a grade reading of the different locations on the construction site.

have been sighted, the forms can be attached at the height indicated.

An instrument used to take elevation readings is a ***builder's transit***. See **Figure 19-32**. A builder's transit is similar to a builder's level, but it can also move in the vertical plane. For example, the transit can shoot multiple points in a straight line that goes up a hillside where a sidewalk will be installed. Another instrument used to measure elevation is an altimeter level. See **Figure 19-33**. Altimeter levels are precise and measure the differences in elevation between the base unit and the readout device. The advantage of an altimeter level is it allows an individual by themselves to take reading for planning and developing the construction layout. For more information, refer to Chapter 3, **Tools and Equipment**.

Tom Wang/Shutterstock.com

Figure 19-32. A contractor is using a builder's transit to determine grade heights. In the background, their partner is holding a grade rod so the contractor can take height measurements.

Workplace Skills
Digital Citizenship

Being a good citizen means behaving ethically, honestly, and responsibly. The same applies for our behavior online. Digital citizenship is the responsible use of technology, especially within a digital society.

Online activities are never entirely deleted from the internet. Anyone online is said to leave behind a *digital footprint*, or a record of their online activities. You are responsible for anything you post on social media sites. Be aware of your actions online, as they may impact your future career opportunities.

Communication is also an important aspect to be aware of when you are online. Maintaining proper etiquette in emails, on social media, and in any other contact with people is an essential part of being a good digital citizen. Proper online etiquette includes using proper grammar and spelling, using appropriate language, and being respectful of others. Etiquette is especially important when communicating with coworkers, your boss, and clients.

Digital citizenship is an important role to uphold not only in your personal life but also in your professional life. Behave online as you would in a professional working environment. Additionally, if your company provides technology, use it respectfully and for business purposes only. Respect any additional policies set in place by your company.

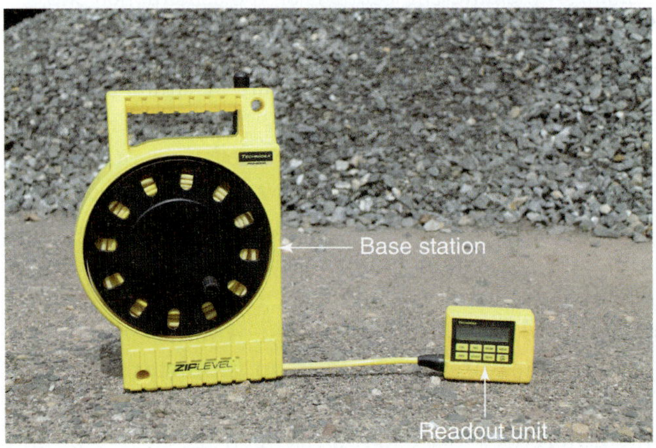

Pave Tool Innovators

Figure 19-33. ZipLevel is an altimeter level that allows a mason to work alone and record elevations accurately when designing a masonry project.

Permanent Forms

Divider strips in a patio are an example of permanent forms. They are left in place for decorative purposes or to serve as control joints. These forms should be made from 1 × 4 or 2 × 4 cypress, redwood, or cedar that has been treated with wood preservative.

The strips should be anchored to the concrete with 16d galvanized nails. Drive them through the board from alternate sides at 16″ intervals. Outside forms would be nailed from only one side.

Masking tape can be used on the top edge of divider strips to protect the wood from abrasion and stain while placing the concrete. The tape is easily removed when the slab has set.

Another type of permanent form is a hollow tube, which is used to cast a void in concrete slabs, beams, or other applications. Refer back to **Figure 15-8**. The tubes are generally made of fiber material or rigid cardboard and are produced in sizes from 2 1/4″ to 48″ in diameter. Lengths vary to meet specifications.

Construction Joints

A large slab or wall is often poured in sections that require a construction joint. Construction joints are formed by placing a bulkhead in the form. A ***bulkhead*** is a piece of material that prevents the concrete from moving past a certain point in the form. See **Figure 19-34**.

Forms for Steps

Two basic types of concrete steps can be constructed. One type is built directly on the slope of the ground. The other type is supported at the top and bottom with an open space under the steps.

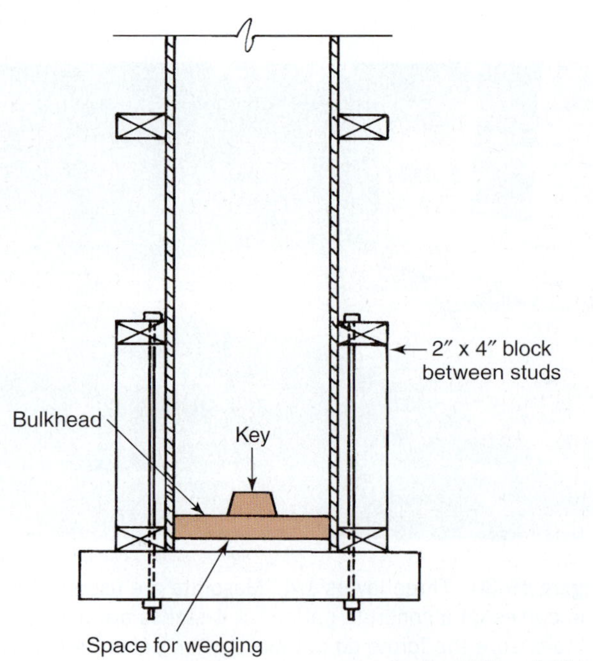

Goodheart-Willcox Publisher

Figure 19-34. Bulkhead in a concrete wall form.

Figure 19-35 shows a method of forming steps built directly on sloping ground. The sides of the form are usually made from 2 × 6 or 2 × 8 lumber. The riser forms are 1″ boards. The beveled bottom edge allows troweling of the step surface right up to the riser.

Figure 19-36 shows one forming method for self-supporting steps. These steps must be reinforced with steel and must rest on a foundation. This form needs to be securely braced to hold the weight. The basic procedures for building concrete steps are subgrade preparation, layout, formwork, placement of concrete, striking off, wait time, floating and edging, tread and riser finishing, form removal and curing.

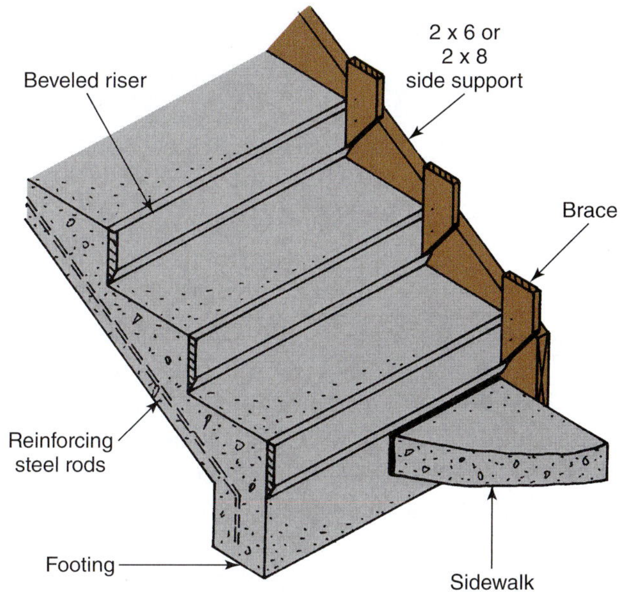

Goodheart-Willcox Publisher

Figure 19-35. Forming technique for steps cast on sloping terrain.

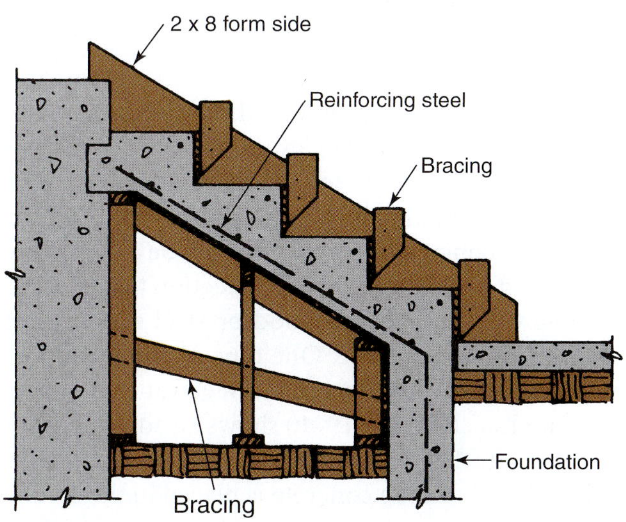

Goodheart-Willcox Publisher

Figure 19-36. Form for a freestanding set of concrete steps.

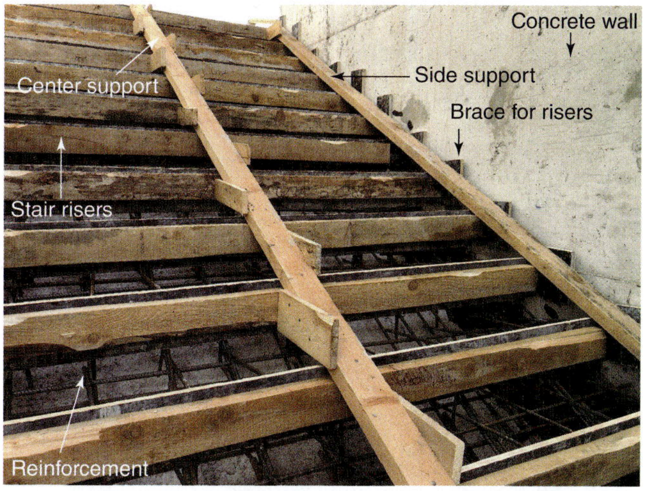

phil_berry/Shutterstock.com

Figure 19-37. Concrete steps being formed.

Figure 19-37 shows a set of concrete steps being formed between two foundation walls. Individual step braces are attached by concrete screws to the concrete walls. The braces position the riser horizontally and vertically to create the angle of the stairs. The risers are 2 × 6 framing material. To create the proper riser height, plywood has been cut to the correct width and added to the front side of the 2 × 6 frame risers. The sides and middle supports are made of 2 × 6 framing material. The supports aid in securing and supporting the risers to ensure they do not move when concrete is placed in the form. The middle support also ensures the risers do not bow out under the pressure of the concrete when it is placed in the form and later when it is vibrated to consolidate the concrete.

To add strength to the concrete stairs, reinforcement has been installed prior to the stair framing materials being fabricated. Reinforcement is installed horizontally and sloped at the angle of the stairs along with the concrete to create a monolithic stair assembly. This means once the stairs have been finished and cured, the stair assembly acts as one structure. Building concrete steps and forming concrete will be discussed further later in the chapter and in Chapter 20, **Concrete Flatwork and Formed Shapes**.

Column Forms

Concrete columns reinforced with steel are commonly cast in square, rectangular, and round shapes. Forms for the square and rectangular columns are easily constructed from wood. See **Figure 19-38**. Round column forms are usually metal, fiberboard, or paper. See **Figure 19-39**. Round prefabricated forms are produced for various lengths and diameters.

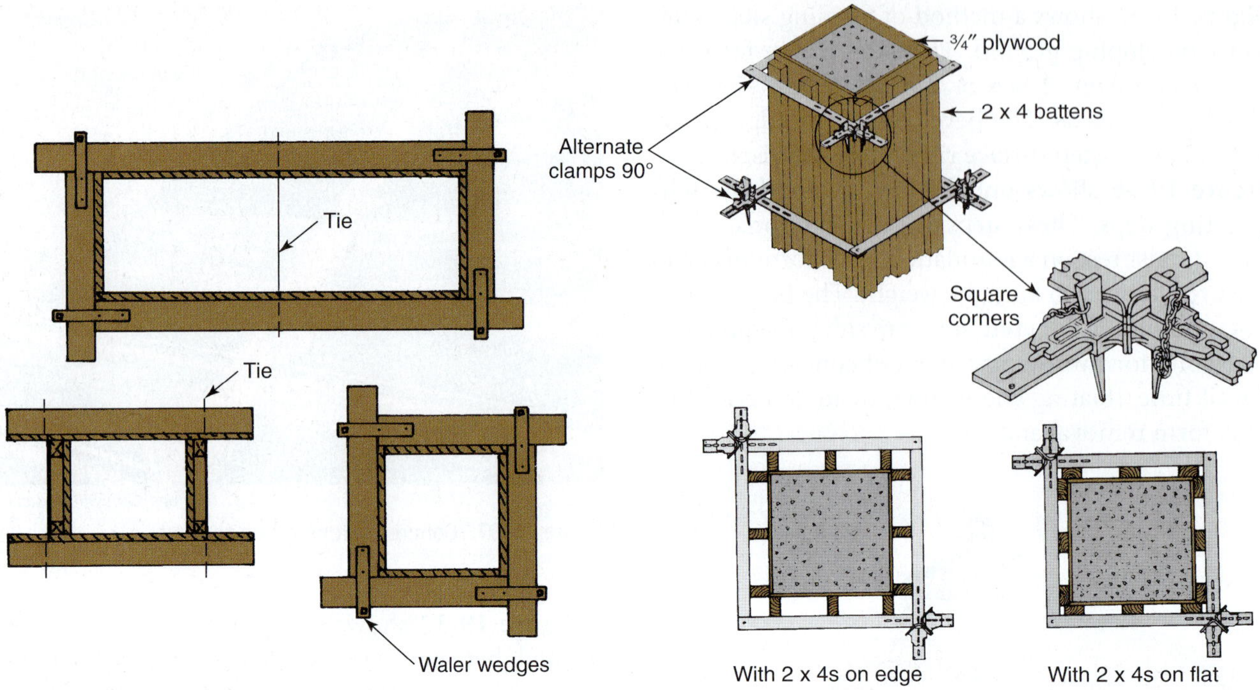

Goodheart-Willcox Publisher

Figure 19-38. Square and rectangular column forms.

Goodheart-Willcox Publisher

Figure 19-39. Cardboard or fiber paper can be used for round column forms. Top—Column form made of fiber paper. The form is only used once because it is destroyed by the removal process. Bottom—Concrete column after the fiber paper form has been removed. This form leaves groves on the surface of the column. If this is objectionable, they now make forms that are smooth inside.

Some forms are reusable, but forms made of fiberboard and paper are not reusable. Fiberboard and paper forms have some advantages. They are economical and lightweight. Paper surfaces produce a smooth concrete surface with no seams. Fiberboard forms often leave a visible seam. The form can be removed from cured concrete by sawing almost through the form material and then making the final cut with a sharp knife blade. Some newer form tubes have a stripping cord that runs the full length of the forming tube to aid in removing the form from concrete. Once the concrete is cured, the stripping cord is pulled to open a seam where the form can be removed from the concrete pier or column.

If the column is to support a post, it is recommended that the concrete form tube be three times the diameter of the post width. For example, if a 4″ × 4″ post is to be positioned on top of the concrete column, then the diameter of the form should be 12″. If using 6″ × 6″, the concrete form tube should be 18″. It is recommended to have a positive connection that locks the concrete column and the wood or steel support post together as one structure. One type of connection is a metal ***adjustable post base*** made of galvanized or powder coated steel. **Figure 19-40** shows an adjustable post base that attaches to the concrete column through an anchor bolt. After the concrete is placed in the column form, an anchor bolt is worked down into the fresh concrete so the threads extend out of the concrete.

Timothy L. Andera

Figure 19-40. An adjustable post base made of galvanized steel. The hole in the base is where the anchor bolt is to be placed. A galvanized washer is slipped over the bolt threads, then the galvanized nut is threaded down to secure the base to the concrete pier. Holes in the vertical plates are used to nail or screw the unit to a wood or steel column post.

Once the concrete is cured, an adjustable post base is positioned over the threads of the anchor bolt and tightened down using the corresponding washer and nut. Not only does it provide a positive connection, but it also reduces the change of a wood post rotting or a steel post rusting.

Concrete form tubes may require a footing, check with local and state building codes. To save time and labor costs, instead of building footing forms by hand, construction companies purchase manufactured concrete forms. ***Concrete column footing forms*** are made from plastic or similar materials to create a concrete footing form. See **Figure 19-41**. The plastic footing form is attached to the concrete column form tube, usually with screws to hold the two together. The footing form has vent holes on the angled sizes that allow trapped air to be vented when they are being filled with concrete. There are a number of companies that manufacture plastic footing forms.

Timothy L. Andera

Figure 19-41. This plastic footing form is 22″ × 22″ × 16 1/2″ tall. Depending on column size, the top of the plastic form will be cut to the appropriate diameter of the concrete tube. An 8″ × 4′ tube is next to the concrete footing form.

Certain codes and municipalities require footing to be included when columns or piers are constructed. Forces that are supported by the column are transferred down through the column into the footing. Not having a column footing reduces the support footprint of the column and over time the column may be pushed down into the subgrade. If the column supports a deck, strong winds get under the deck and over time it can cause the column to move vertically. When the column has a footing, it is very difficult for the uplift to occur.

PROCEDURE

Installing a Concrete Column or Pier with a Concrete Footing

1. Layout the location and finished height for the column or pier using stakes and a mason's line. See **Figure 19-42**.
2. Determine the frost level for the site, make sure the footing is below the frost line.
3. Determine the center of the concrete column and the location on the mason's line attached to the stakes.
4. Determine the depth of the excavation; this should include 3″ to 4″ for base material.
5. Excavate the area 1 1/2′ to 2′ larger than diameter of plastic footing form.
6. Add gravel or crushed stone as the base material and compact in layers.
7. Transfer the center of the footing down from the mason's line using a plumb bob and line to identify

Continued

PROCEDURE *(continued)*

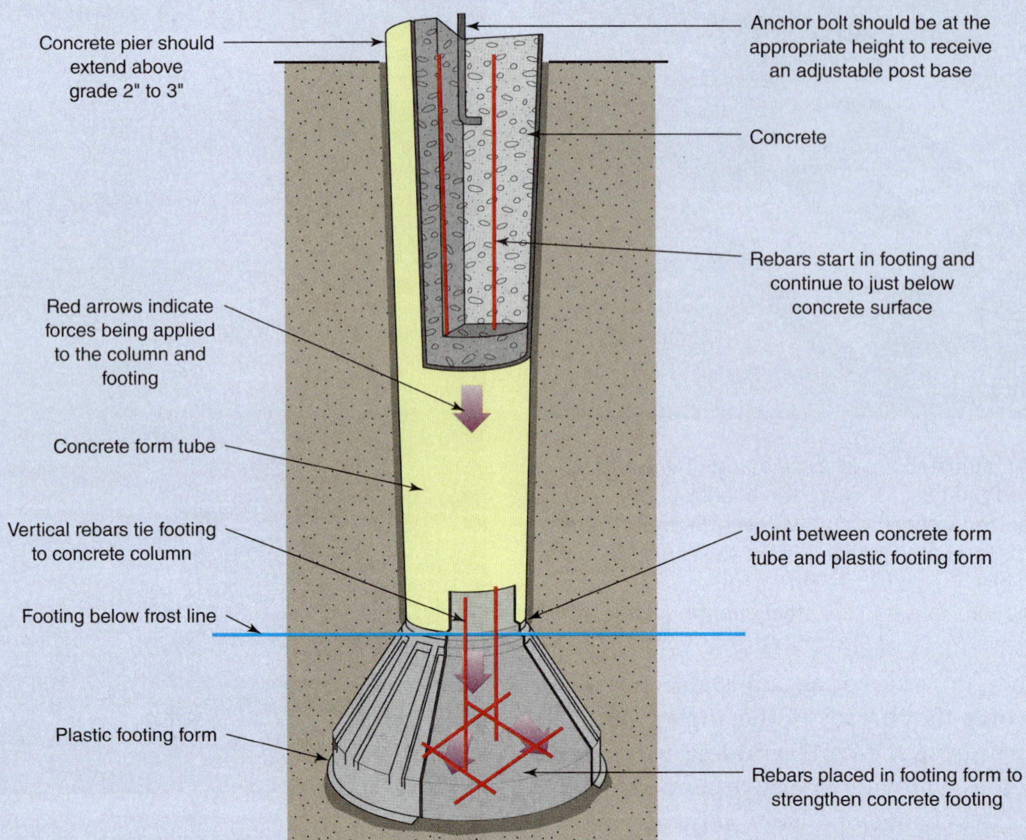

Goodheart-Willcox Publisher

Figure 19-42. The different parts of a concrete column or pier with an attached plastic concrete footing form. Red arrows indicate forces being applied to column and footing.

the center of the footing form. Mark the location by driving a stake in the base material.

8. If necessary, place rebar on top of the base material (check local and state code requirements).
9. Install chairs under rebar to elevate to the appropriate height.
10. Place plastic footing form over rebar and center over stakes in the base material. Check the top of the plastic form for levelness. Then rotate level 90° and check again. Adjust accordingly.
11. Stake down the plastic footing form to secure the position.
12. Measure down from the mason's line to the inside edge of the footing form to determine the length of the form tube. Cut the form tube to the appropriate length.
13. Position the concrete tube over the footing form and slide into place.
14. Use screws to secure the concrete form tube to the plastic footing form.
15. Fabricate formwork at ground level. Use 2 × 4 lumber to hold the top of the concrete form tube from moving when concrete is placed into the form. Temporarily attach the form tube to 2 × 4 formwork.
16. Plumb one side of the form while keeping the level parallel and against the tube. Reattach to the form. Move the level 90° around the tube and plumb the other side of the form. Keep adjusting until the form tube is plumb. Secure the tube to the form.
17. Determine the length of the vertical reinforcement (if needed) and cut to length, usually at least two rebars per column. Install the rebar, the top of the rebars should be 1″ to 2″ below the finished surface of the concrete column.
18. Determine volume of concrete needed for the column and footing, order and schedule the delivery.
19. Fill the footing in lifts of concrete. Vibrate the concrete and lightly tap the sides of the footing form to consolidate and remove air pockets around reinforcement and the plastic form.
20. Fill pier form in lifts, vibrate and tap sides of the form.
21. Bring concrete to the top of the form tube. Screed and float concrete at the top of the form tube.
22. Install the anchor bolt in the center of the form or previously determined location.
23. Once concrete cures, remove the paper form tube from concrete column.
24. Attach the adjustable post base to the anchor bolts and secure.

Supporting Steel Reinforcing in Forms

The amount, type, and spacing of steel reinforcing required in a slab, wall, or other concrete structure depends on many factors. Frequently, residential construction requires no reinforcing at all if not in a seismic (earthquake) risk zone. The concern in form construction, however, is how to maintain the proper spacing between the forms and the rebars. Common methods include spacer blocks, plastic snap-on devices, and wood strips. See **Figure 19-43**. Spacer blocks and snap-on devices remain in the concrete, but wood strips are removed as the concrete is placed.

One approach to supporting horizontal steel reinforcing bars in a concrete wall form is to place the rebar on the tie rods. Vertical rebar, if required, can be attached to the horizontal bars. This method maintains proper placement of the rebar and involves no special devices.

Form Removal

Concrete form release agents are manufactured exclusively for wood and steel forms to aid in removal of forms from the concrete structure. There are two types of release agents—barrier release agents and reactive release agents. Barrier release agents develop a film between the forming surface and the concrete to prevent adhesion. Reactive release agents contain ingredients that produce a chemical reaction between the release agent and the free lime in fresh concrete.

It is important to read and follow the manufacturer's directions before using these products. In general, when prepping new plywood or wood, a liberal amount of the product should be applied to the surfaces a few days prior to use to allow it to soak into the material. Reapplication should continue until the wood is saturated. Just prior to installation, a thin film should be applied to all surfaces. For metal, wipe a thin film just before using the forms.

Goodheart-Willcox Publisher

Figure 19-43. A spacer block supports the horizontal rebar at the proper height in the form.

Green Note

Form Release Agents

In the past, form release agents were created using ingredients such as diesel fuel, home heating oil, and mineral oil to prevent wood from sticking to poured concrete. Those "homemade" release agents have been replaced with commercial products.

Barrier release agents may be oil-based or water-based. Form release agents with a petroleum or solvent base can cause health and environmental problems. Emission of volatile organic chemicals (VOCs) into the air can cause odor or health problems among workers, especially when used indoors for manufacturing of precast concrete products. When using commercial form release agents, ensure they comply with current federal, state, and local maximum allowable VOC requirements.

Water-based release agents can prevent wood forms from sticking to concrete without releasing VOCs into the air. Commercial water-based products are made from natural oils.

19.4 Form Maintenance

Form maintenance is important because it reduces the cost by extending the life of forms. Many forms are damaged during stripping (removal from cured concrete). Do not use metal pry bars because they will damage the forms—especially wood forms. Use wood wedges and tap lightly if necessary.

Forms should be inspected, cleaned, repaired, and lightly oiled after they are removed. Use a stiff fiber brush on wood forms. A wire brush raises the grain and roughens the surface. Mill oil that is 100 or higher viscosity and pale in color can be used for oiling forms.

Forms should be solid stacked or stacked in small packages with faces together for future use. Panels should be protected from the sun and rain.

19.5 Insulated Concrete Forms

An alternative to building a poured concrete wall that uses conventional forming materials and techniques is to use stay-in-place insulated forms. These products are known as insulated concrete forms (ICFs). Refer back to **Figure 16-16**.

The use of ICFs eliminates the need for traditional metal or wood forms. Constructing a foundation using ICFs is similar to stacking building blocks. The insulated blocks are designed in such a way that the end of each block fits into the previous one. Once the first course has been positioned, the second course is placed on the first. The second course fits into grooves or slots of the first course, and so on. See **Figure 19-44**.

Code Note **ICF Construction**

Insulated concrete forms (ICF) for construction are addressed in the International Building Code (IBC) **Section R611 Insulating Concrete Form Wall Construction**. This code is a valuable resource when considering the use of insulated foam construction.

As with regular construction, ICF wall construction requires reinforcement. Rebars are installed horizontally and vertically in the footing during the time of construction. The rebar extends up into the wall cavity of the ICF to tie the wall to the footing, as shown in **Figure 19-45**. As the ICF wall is constructed, new rebar is added vertically and horizontally to existing bars to tie reinforcement together for the purpose of strengthening the wall. This process continues up through the wall structure as it is fabricated.

Fox Blocks by Airlite Plastics Co.

Figure 19-44. The cavities of the ICF walls of this residential foundation are filled with concrete. Like regular concrete foundations, ICF foundations must be sealed against moisture entering the wall. The exterior of this ICF foundation was sealed with Resisto Waterproofing Membrane prior to backfilling of the walls.

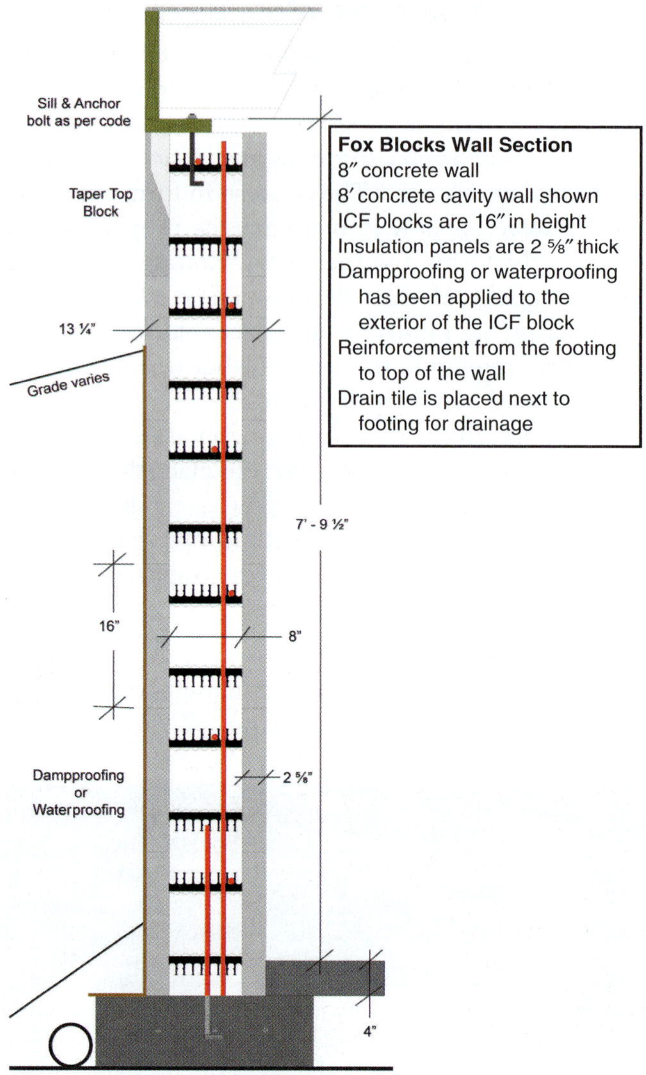

Fox Blocks by Airlite Plastics Co.

Figure 19-45. Fox Blocks wall section showing the layout structure of an insulated concrete form (ICF).

Safety Note

Installation of a mid-rail and handrail made of 2 × 4s to scaffolding reduces the chances of a worker falling off the planking to the concrete below. Add a toe-board to ensure tools do not slide off the scaffold.

Building an Insulated Concrete Form Structure

The first step in building an ICF structure is to research companies that manufacture the ICF products in your area. Select the company that best fits your needs. Each company has specific procedures and guidelines to follow when building. Discuss with company representatives the support materials available, such as ICF manuals, instructional videos, and in-house trainings. Trainings can consist of aiding in designing, estimating, and building an ICF structure. Usually, an ICF structure is built on a standard concrete footing. After the concrete footing has cured it is time to begin the ICF build.

PROCEDURE

Building an ICF Structure

1. Determine the locations of the outside corners and temporarily mark. Measure diagonally to ensure the layout is square.
2. Adjust the layout until the dimensions of the walls are accurate and diagonal measurements are the same.
3. Snap chalk lines on the concrete footing for wall locations. Use red or another permanent color chalk. Snap lines marking the exterior side of the ICF panel. See **Figure 19-46A**.
4. Beginning at the corners, install the ICF blocks. Install the first two courses of ICF blocks.
5. Certain areas along the bottom course of ICF blocks may need to be shimmed or trimmed to ensure the top surface of the second course is straight. Use a mason's line and spacers on top of the second course to check if the wall is straight. These adjustments are done to compensate for variations in the footing, such as having depressions or humps.
6. Once the ICF walls are straight across the top and walls are on the snapped chalk lines, spot glue the forms to the concrete footing at locations approximately 16″ apart. See **Figure 19-46B**.
7. A small amount of foam is needed, later in construction the complete seam between the ICF block and the footing on both sides of the wall will be sealed with low expansive foam.
8. Continue installing the ICF blocks. Install horizontal and vertical reinforcement as block walls are stacked. Make sure to follow building codes and manufacturer's recommendations regarding reinforcement.
9. Straight, plumb, and brace the ICF formwork. Vertical support braces, such as ***strongbacks*** or ***stiff backs***, are attached to the ICF wall structure. See **Figure 19-47**. The spacing is based upon thickness, height, and winds that can affect the structure. Strongbacks or stiff backs are usually placed every 4′ to 6′ to tie the wall brace into the ICF internal vertical studs.
10. Attach an anchor foot to the concrete floor or if the floor is unfished, stake it to the ground. Attach a diagonal brace, sometimes called an *outrigger*, to the anchor foot and other side of the strongback. See **Figure 19-48**.
11. To straighten the ICF, you must start at the corners. Drive a 3″ wood screw into the vertical internal polypropylene stud with the head sticking out 1/2″ from the face of the ICF foam. The screw location should be close to the top of the last course of ICF block.
12. Attach a mason's line to the screw head and run along the length of the wall. Place a second screw at the other end of the ICF wall. Attach the other end of the mason's line to the second screw with a slip knot and tighten the line around the second screw to create a straight and taut line to be used to straighten the ICF wall.
13. Place a small block of wood 3/4″ × 1 1/2″ or similar gauging material under the mason's line next to the screws at each corner. By starting at one end of the

A

B

Courtesy of BuildBlock Insulated Concrete Forms

Figures 19-46. A—The outside of the ICF wall with a red chalk line as a guide to position the wall. Low expansion foam-in caulk was used to spot glue the forms to the concrete. B—Caulk spaced about 16″ apart gluing the ICF block to the footing.

Continued

PROCEDURE *(continued)*

Fox Blocks by Airlite Plastics Co.

Figure 19-47. The anchor foot of the brace system is attached to the concrete floor. A diagonal brace, sometimes called the *outrigger*, is attached to the strongback. An attached walkway has been created to allow the concrete mason a safe and secure surface to walk on when straightening and filling the ICF wall with concrete.

Courtesy of BuildBlock Insulated Concrete Forms

Figure 19-48. A metal brace system is being installed. Strongbacks are screwed to the ICF wall every 4′. The outrigger has a threaded mechanism or turnbuckle that allows the device to be turned clockwise or countercloskwise to move the wall in or out, assisting in plumbing the ICF wall.

wall, use the same size gauge to move down the ICF wall to the first strongback brace. Slide the gauge next to the line, it should slide under. If there is a gap or the gauge will not fit between the line and wall, move the turnbuckle to adjust the wall so the gauge is almost touching the line. Perform this operation at each strongback on the ICF wall. Do this on all ICF walls to make them plumb and straight.

14. Place the concrete in the ICF wall. See **Figure 19-49**. At the start of the pour, the concrete is allowed to flow and spread inside the form. Work around the wall to

A

B

Courtesy of BuildBlock Insulated Concrete Forms

Figure 19-49. A—A mason placing concrete into the ICF form between the insulated panels. B—Strongbacks are on the outside of the ICF structure and are attached to vertical polypropylene studs with exterior construction screws. Masons are now placing concrete into the ICF form between the insulated panels.

continue to fill the ICF to 1/3 of the wall height. Depending on the size of the project, this may take a couple of hours.

Trade Tip

A special mixture of concrete is used to allow it to flow through the pumping tube and reinforcement in the forms without consolidating the aggregate. The mixture should be discussed with the ICF block manufacturer and concrete ready-mix supplier to ensure the proper mix and slump.

15. As the forms are being filled, have another mason vibrate the concrete to make sure it is consolidated and has filled all voids around reinforcement and openings. See **Figure 19-50**.
16. Once the first lift is completed, move back to the starting point and pour the concrete up another third, filling the form to 2/3 full. The purpose of the lifts is not to overwhelm the forms and allow the concrete to begin to set before the next lift is placed into the forms.

Courtesy of BuildBlock Insulated Concrete Forms

Figure 19-50. Concrete masons are working together to place and consolidate concrete in the ICF project. One mason is operating the concrete pump hose while the other mason is operating a concrete vibrator to consolidate the concrete.

17. Using a mason's line, recheck the walls for plumb and straightness. Make minor adjustments if needed.
18. In the top last course of the ICF blocks, insert anchor bolts when the cavity is filled with concrete. See **Figure 19-51**. The anchor bolts tie the wood framing to the ICF masonry structure.

Courtesy of BuildBlock Insulated Concrete Forms

Figure 19-51. Anchor bolts have been installed in the fresh concrete ICF wall. A concrete mason will smooth out the surface of the concrete to prepare for a wood sill to be installed later.

Trade Tip

The number of lifts required to fill the ICF cavity when concrete is placed in the walls depends on the ICF manufacturer's guidelines and local and state building codes. Check these regulations prior to installation of the forms since the scheduling of the project may be affected.

Advantages of Insulated Concrete Forms

In most situations, the use of ICFs reduces the time needed to construct concrete walls with the benefit of creating a structure that has insulation on both sides. Another benefit of using ICFs is the exterior wall forms are actually the insulation that stays in place. Because of this, the time required to dismantle formwork is considerably less than that of a conventional poured foundation.

Unlike conventional form ties that are made of metal, on most ICF walls the form ties are made of polypropylene. These materials have less thermal transferability; therefore, the energy efficiency of an ICF wall is higher compared to standard construction. See **Figure 19-52.** The use of ICFs allows the construction season to be extended in cold climates. The insulation permits the concrete to be cast in cold weather.

The R-value of an ICF wall is considerably higher than that of a regular concrete block or concrete poured wall. The enhanced thermal performance is due to the insulation on the exterior and interior surfaces of the poured concrete structure.

Another advantage of ICF construction when placed above grade is the sound resistance of the structure. Sound transmission is reduced because of the foam insulation layers and the concrete mass between the two layers of foam.

Manufacturers of ICFs have designed many different block patterns to accommodate unique wall layouts and architectural designs. See **Figure 19-53.**

Examples of ICF block manufacturers are Fox Blocks, BuildBlock, and Lite-Form®. As builders become more familiar with these and other similar products, the traditional removable type of forms may become obsolete in residential and some light commercial construction.

Considerations When Using Insulated Concrete Forms

Construction with ICFs is more expensive than conventional concrete or wood construction. Foundations built with foam blocks cost more than conventional construction. However, other factors should be considered, such as having insulation already installed as part of the foundation.

When using ICFs, construction of a concrete wall requires less time than wood frame construction. The light weight of the forms permits factory or on-site

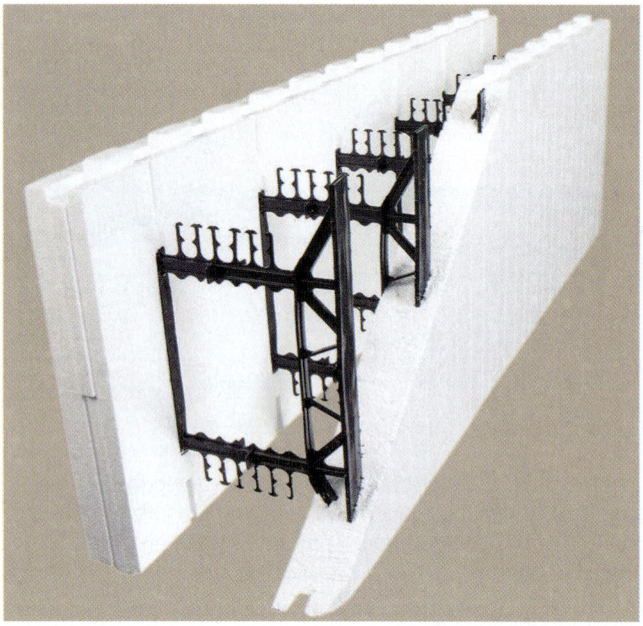

Fox Blocks by Airlite Plastics Co.

Figure 19-52. Form cutaway showing the internal parts of the ICF. The insulated block is 48″ in length with an 8″ cavity. Cavity sizes come in 4″, 6″, 8″, 10″ and 12″. The insulation thickness is 2 5/8″ on each side. Tie flanges connect the two insulating panels together and receive rebar or other reinforcement. Flanges have dividers molded into them to receive rebar and are designed tall enough to have rebar stacked on top of each other to keep the interior space as open as possible to improve concrete flow when filling the wall.

assembly performed by trained installers. Installation of ICF forms requires contractors or installers to follow manufacturer's standards and specifications. See **Figure 19-54**. Each manufacturer has specific procedures for installation because products vary from one manufacturer to another. Installation training is usually available through the manufacturer or local distributor of the product.

ICF buildings create a strong reinforced concrete structure. As ICF blocks are laid, reinforcement is installed in the courses to strengthen the concrete and to tie the exterior walls together as one complete structure. As the courses are laid and reinforcement is installed in the courses, the locations of windows and doors need to be positioned accordingly. The installer creates cutouts, called *blockouts*, by leaving out blocks to create voids in the wall.

Blockouts formed from lumber, ICF foam, or other materials need to be positioned at certain locations to stop the flow of concrete out of the forms. The contractor or installer determines in advance where blockouts will be positioned for windows, doors, and other special locations in the structure. Blockouts may be left in the opening to create a nailing surface for doors, windows, or other special framing units that will be installed later in the construction process, or they may be removed after the concrete has cured. Whether the blockouts are left in or removed depends on the design or specific application.

A pump truck with a hydraulic boom is usually used to fill the forms properly with concrete. See **Figure 19-55**. The concrete must not be placed into the ICF structure too rapidly or a blowout may occur. A blowout results

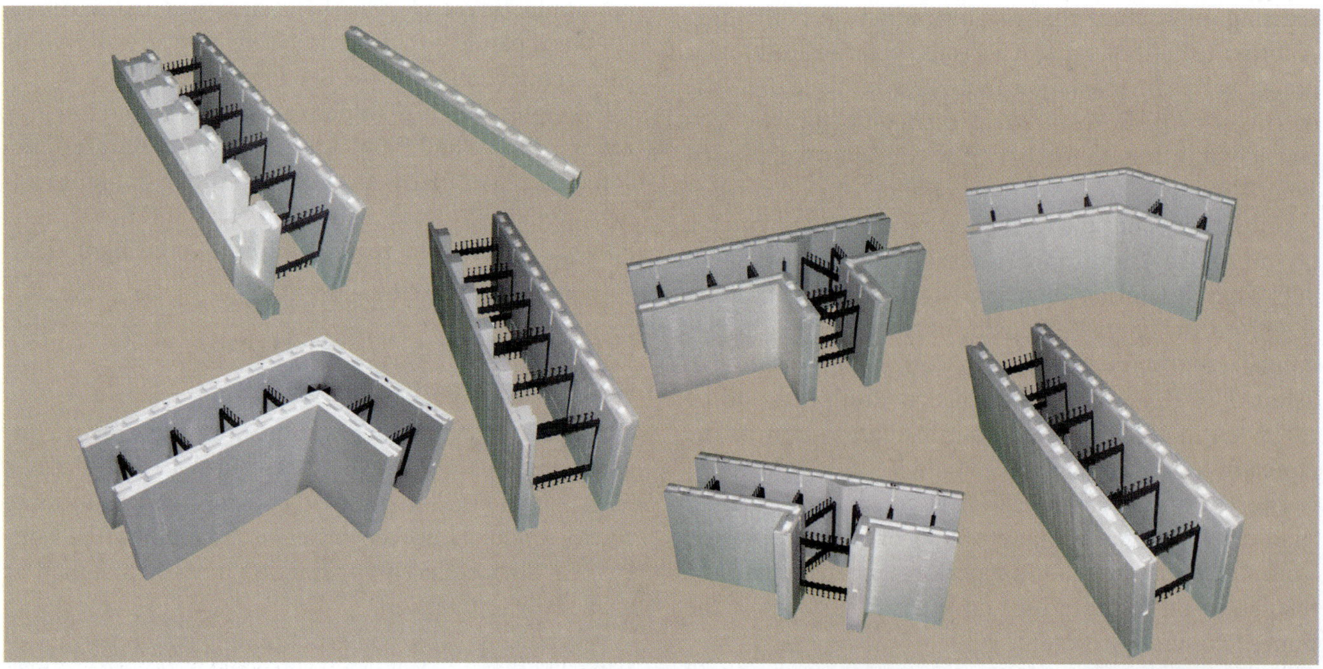

Fox Blocks by Airlite Plastics Co.

Figure 19-53. Examples of ICFs used to construct foundations and multistory buildings.

A B

Fox Blocks by Airlite Plastics Co.

Figure 19-54. A multistory commercial building under construction. A—As the insulated blocks are laid, reinforcement is installed horizontally and vertically. Vertical rebar extends above the current height of the wall for a continuous reinforcement as new forms are placed and then concrete is added to the center core of the insulated block. B—The completed building.

in a delay in the placement of concrete and a waste of labor, time, concrete, and money.

Trade Tip

The term *blowout* refers to fresh concrete escaping from the formwork when too much pressure is placed on the form, causing a fracture or small portion of a form to fail during a concrete pour.

Costly stripping of forms is not required since the concrete forms remain in place. In most applications, furring strips are not required because drywall or interior wall finishing can be attached directly to the polypropylene ties or studs that are an internal part of the ICF.

The interior surface of an ICF must be covered with a thermal barrier to protect the insulation from fire. This is usually performed by placing 1/2″ gypsum drywall or equivalent material on the foam surface. Check local and state building codes to determine specific requirements for adequately covering exposed insulation against fire when it is located on the inside of a structure.

Fox Blocks by Airlite Plastics Co.

Figure 19-55. This large complex is being constructed with Fox Blocks. Pump trucks stationed at different locations around the site pump concrete into the cavities of the insulated block forms.

On the exterior of an ICF structure, exposed foam that comes in contact with backfilling materials must be dampproofed similar to a conventional foundation. Check with local and state building codes to determine specific requirements for ICF structures. The contractor should also contact the ICF manufacturer to determine appropriate materials and procedures required for dampproofing insulation when it is exposed to backfilling and grading materials. Where insulation is exposed above grade, the area between the grade and exterior wall covering must be protected from the weather and ultraviolet sunlight. Refer to the manufacturer's specifications and recommendations to determine options that can be used to cover exposed ICF surfaces.

Finishing ICF wall surfaces is similar to finishing wall framing construction. The exterior of ICF walls can accommodate finishes such as metal siding, wood siding, or other exterior sheathing materials. See **Figure 19-56**. Brick and stone can also be incorporated in most ICF structures with special design considerations. A ledge is built with ICF special block units on which the brick or stone is placed as the finish wall surface. Refer back to **Figure 16-15**. Check with the specific ICF manufacturer to determine if that company makes the specialized ledger block.

Fox Blocks

Fox Blocks are reversible, preassembled foam blocks. The strong interlock of the blocks minimizes form movement during placement.

Fox Blocks are designed during the manufacturing process to achieve a close tolerance when the structure is fabricated. An assortment of block units is available for a variety of applications, such as the previously mentioned ledger block. Since Fox Blocks are made of

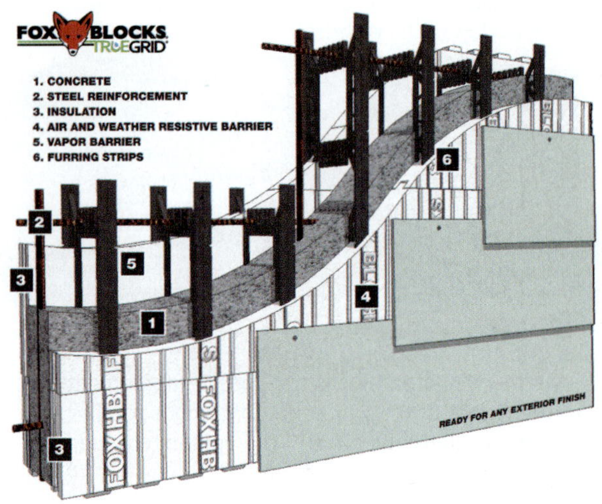

Fox Blocks by Airlite Plastics Co.

Figure 19-56. The different parts of an ICF wall structure and a wall covering on the exterior of the ICF structure.

expanded polystyrene, which is not affected by moisture, leaving the blocks out in the weather during construction is not a concern.

The layout process using Fox Blocks is started at the corners with specialized corner blocks. The block cavity for straight runs comes in standard widths of 4″, 6″, 8″, 10″ and 12″ thick concrete walls. Both sides of the insulated cavity forms are 2 5/8″ in thickness. The blocks are made of expanded polystyrene (EPS) with an R-value of 4.11 per inch of foam. Standard blocks are 48″ in length and 16″ in height. Blocks of different sizes can be used together when fabricating layouts that create unique designs or configurations. Specialized blocks are manufactured to create a masonry ledge where brick or other masonry can be laid.

Ties that hold the insulated panels together are 1 1/2″ in height. Ties give the block strength and provide rebar positioning in the wall layout. The ties are made of polypropylene with designed elevated notches to accommodate stacking of rebar.

Reversible interlocks are used to hold the insulated panels together. The interlock is 1/2″ in height. Reinforcement saddles provide a location for rebar to be positioned in the interior cavity wall layout. Furring strips of 1 1/2″ wide by 15″ in height are imbedded in each of the ICF faces. The furring strips give strength to the forms during concrete placement and become the vertical studs in the ICF to provide attachment points on the inside and outside of the ICF structure.

As the courses of block are assembled, bracing needs to be installed according to manufacturer's specifications. The contractor or installer will need to ensure that adequate bracing and bracing systems are installed to straighten and secure walls during the pouring of concrete in the forms.

BuildBlock

BuildBlock has an assortment of training materials and videos available online with training modules on different topics related to ICF construction. BuildBlock also provides hands-on training at their regional locations. BuildBlock has multiple manufacturing locations around the country.

BuildBlocks are 48″ in length and 16″ in height. Each side panel is 2 1/2″ thick, creating 5″ of total insulation to the block. When working with corner blocks, one leg of the block is 31″ long and the other leg is 22″. The height of the corner blocks are the same as a standard block. BuildBlocks are completely reversible blocks. The blocks will fit any way they are stacked. When installing corners, they are the same for left or right corners. The blocks are tight interlocking units that stack easily and lock securely to the block below. Blocks can be purchased in different core sizes such as, 4″, 6″, 8″, 10″ or 12″. Corbel blocks can be purchased for brick or stone ledges. Corbels can also be used for ledges to support floors, sidewalks, driveways, and other purposes in residential or commercial construction.

BuildBlock forms are designed on a 6″ web spacing adding strength to the ICF form. Blocks have 1″ vertical cut lines imbedded in the forms to aid when cutting block to certain lengths. The block also has horizontal cut lines every 2″ to aid when cutting the block height for windows or door openings. See **Figure 19-57**.

Deep internal web bars in the cavity of the block allow for stacking rebar on top of each other when installing reinforcement. BuildBlocks can be used to fabricate gable ends on ICF projects to create unique structures. See **Figure 19-58**.

Lite-Form

Lite-Form is an insulated wall form system that uses panels of plastic foam insulation as formwork. See **Figure 19-59** and **Figure 19-60**. The panels are separated with plastic ties so the space between the panels can be filled with concrete. The insulation panels remain in place and become part of the wall. This results in a concrete wall that is insulated on both sides.

Form sections are built from 8″ high by 8′ long strips of polystyrene. Special form ties and corner ties are required. See **Figure 19-61**. The wall form is built course by course until the desired height is reached. A typical footing is generally used. The completed form is braced and reinforced before filling with concrete.

Exterior siding or interior paneling can be screwed into the plastic ties. Any polystyrene that is above the grade and exposed should be covered with a trowel-applied protective coating. Electrical and plumbing lines can be installed in the 2″ thick insulation. Conduit may be required by some codes.

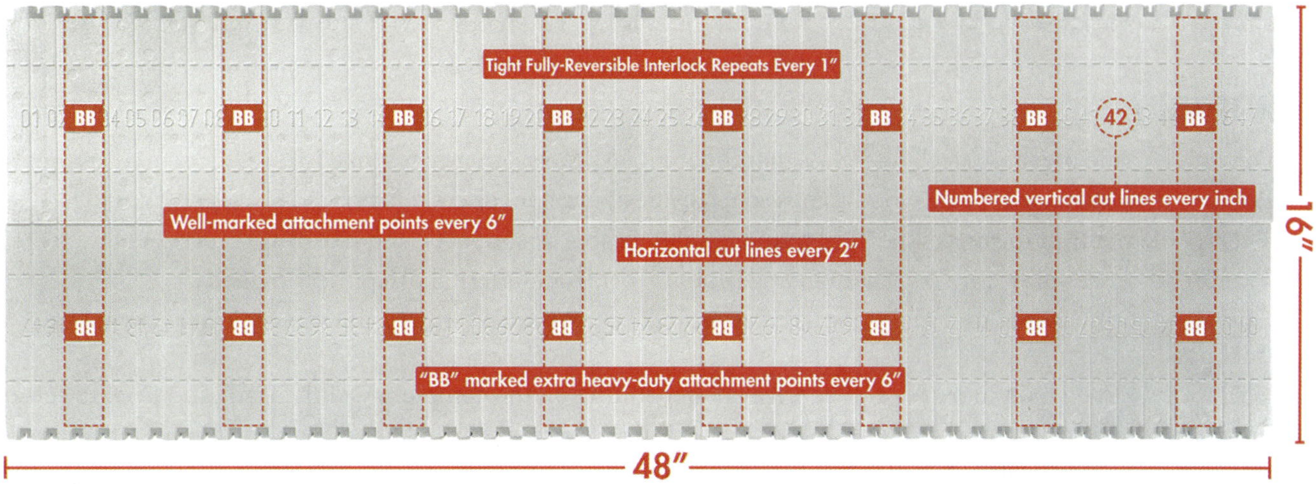

Courtesy of BuildBlock Insulated Concrete Forms

Figure 19-57. BuildBlocks have identification marks to aid in cutting and determining locations of internal studs and heavy-duty attachment points marks as BB.

Courtesy of BuildBlock Insulated Concrete Forms

Figure 19-58. A two-story gable end is being braced using the BuildBrace ICF bracing system. Two walkways have been created to allow masons access to building the structure and placing concrete in the wall cavities.

LiteForm Technologies

Figure 19-60. This building technique allows the insulation panels to remain in place and become a part of the wall, which results in an insulated wall.

LiteForm Technologies

Figure 19-59. The forms to be used to cast the foundation of this structure are made from rigid foam plastic modular units.

LiteForm Technologies

Figure 19-61. The patented form ties that are the trademark of this forming system are clearly visible as the wall forms are being put into place.

Chapter Review

Summary

- Forms are made of wood, steel, fiberglass, hardboard, plastics, or other materials.
- Plyform is an exterior-type plywood that is specially designed for concrete forming.
- The five elements of forms are sheathing, studs, wales, braces or supports, and ties or spreaders.
- Two broad classes of ties are continuous single-member ties and internal disconnecting ties.
- Concrete footing forms are usually constructed from 2″ construction lumber the same width as the footing thickness.
- Wall forms can be built in place or prefabricated, depending on the complexity of the walls and need to reuse.
- Forms for the square and rectangular columns are easily constructed from wood. Round column forms are usually metal, fiberboard, or paper.
- Forms should be inspected, cleaned, repaired, and lightly oiled after they are removed.
- Each course of insulated blocks fits into grooves or slots of the previous course. As the courses are stacked on each other, the walls are braced to add strength and to keep the wall straight and plumb.
- The number of lifts required to fill an ICF cavity when concrete is placed in the walls depends on the ICF manufacturer's guidelines and local and state building codes.
- Advantages of ICF construction include reduced construction time, extension of the construction season in cold climates, and higher energy efficiency of the wall.

Know and Understand

Answer the following questions using the information provided in this chapter.

1. The most popular form material is _____.
   A. wood
   B. steel
   C. fiberglass
   D. plastic
2. A type of form that is usually prefabricated in the size and shape desired and then used in precast concrete plants is _____.
   A. wood
   B. steel
   C. fiberglass
   D. plastic
3. The form elements that support the sheathing and prevent bowing are _____.
   A. sills
   B. studs
   C. wales
   D. braces
4. *True or False?* A turnbuckle is the depth a tie is broken off in concrete.
5. *True or False?* The appearance of footings is of little concern.
6. Forms for drives, walks, and patios are typically _____.
   A. wood
   B. hardboard
   C. fiberglass
   D. plastic
7. *True or False?* Wood forms should be removed with metal pry bars.
8. Mill oil that is _____ or higher viscosity can be used for oiling forms.
   A. 5
   B. 25
   C. 50
   D. 100
9. Which of the following is *not* true of ICFs?
   A. Traditional metal or wood forms are not needed.
   B. Walls must be braced as courses are stacked.
   C. Anchor bolts are inserted in the top of the last course when the cavity is filled with concrete.
   D. ICF wall construction requires no reinforcement.
10. When first pouring concrete into an ICF wall, fill it to _____ of the wall height.
    A. 1/4
    B. 1/3
    C. 1/2
    D. 2/3

11. *True or False?* To straighten an ICF, you start at the corner.
12. *True or False?* Construction of a concrete wall using ICFs requires more time than wood frame construction.

Apply and Analyze

1. Name three materials commonly used for concrete form construction.
2. What is the function of wales?
3. What type of soil can be used as a footing form material?
4. Give two examples of permanent forms (forms that are left in place).
5. List three advantages of insulated concrete wall forms compared to conventional construction.
6. Why do walls need to be braced as courses of ICF blocks are stacked on each other?
7. What determines the number of lifts required to fill an ICF cavity when concrete is placed in walls?

Critical Thinking

1. List and then identify the purpose of the main elements of traditional concrete forms.

Activities

1. (ELA) Using the US Green Building Council website, prepare a presentation that explains the LEED rating systems, credit categories, and levels of certification. Explain three ways in which concrete masonry can contribute to achieving LEED credits. Include images and charts in your presentation.
2. (ENGINEERING) Explain the ways in which homes constructed with insulated concrete forms are energy efficient.
3. (SCIENCE) Research three different types of form release agents. Compare their ingredients, the way they work, application instructions, and precautions for use. Be prepared to answer questions about the products in class.

Photo Courtesy of Neuvokas Corporation – Manufacturers of GatorBar®

CHAPTER 20

Concrete Flatwork and Formed Shapes

LEARNING OUTCOMES

After studying this chapter, you will be able to:

20.1-1 Identify how to place and finish concrete forms.
20.2-1 Explain the ideal conditions to cure concrete.
20.3-1 Describe three basic types of joints used in concrete construction.
20.4-1 Explain how to build a variety of patterns, textures, and decorative finishes into concrete during construction.
20.5-1 Describe how cast-in-place concrete is used in walls.
20.6-1 Explain how cast-in-place concrete is used in roof and floor systems.
20.7-1 Describe the construction of precast/prestressed concrete systems.

TECHNICAL TERMS

dowels
edging
ponding
screeding
shotcrete
tilt-up construction

Introduction

Concrete can be cast in practically any shape as long as a form can be built to contain it while it is in a plastic state. See **Figure 20-1**. This is one reason concrete is valuable as a building material. However, most concrete is formed into slabs on grade, cast in place as walls or floors, or used in the manufacture of precast units, such as beams and panels. This chapter covers the practical aspects of using concrete in these applications.

When planning concrete flatwork or creating unique shapes, check national, state, and local codes to ensure that concrete projects meet specific building codes. Consider these items when planning concrete projects:

- Intended use or uses
- Type of finish on the surface of the concrete
- Color to be added to the concrete
- Weight that will be supported by the project

Code Note — **Concrete Flatwork Thickness**

The International Building Code (IBC) **R506.1** states when concrete slabs are supported by the ground, it shall not be less than 3 1/2″ in thickness to ensure the slab has strength to adequately support intended loads.

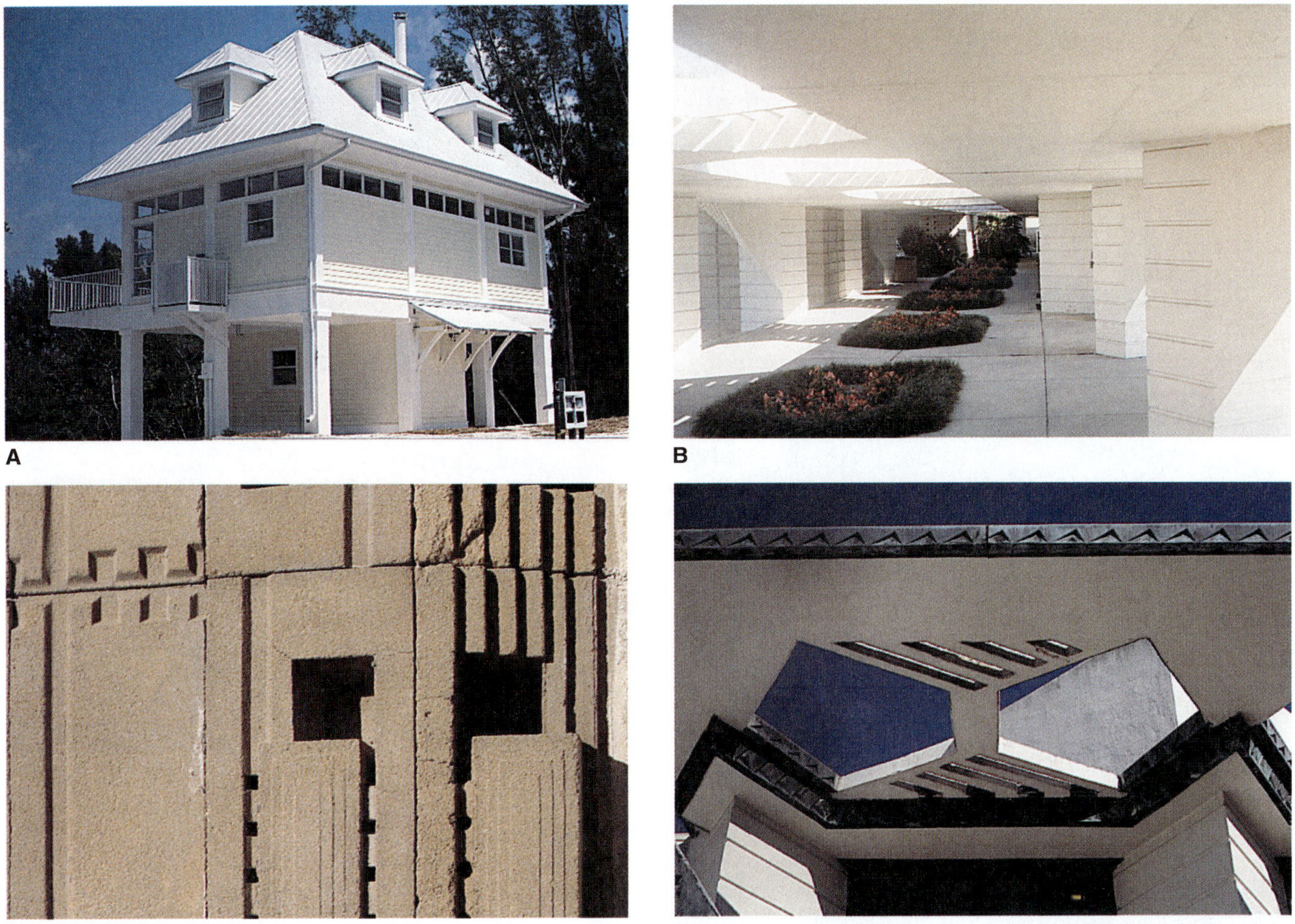

A B C D

Tilt-Up Concrete Association

Figure 20-1. A—This beach cottage uses cast concrete piers and a cast concrete floor system. B—Concrete is the building material in this contemporary structure. C—Colored concrete is cast in complex shapes to simulate stone. D—Geometric shapes sculpted in concrete form this roof overhang.

20.1 Placing and Finishing Concrete

Concrete is moved from a delivery truck to its final location by a variety of methods. Methods include chutes, push buggies, and buckets. The concrete can be handled by cranes or pumping through pipes. The method used should not restrict the consistency of the concrete. Consistency is governed by the placing conditions and the application. In other words, if conditions permit use of a stiff mix, the equipment should be designed to handle it.

Preparation

Subgrade should be properly prepared before the concrete arrives. Subgrade is the base on which the concrete is placed. Although a gravel subgrade is not required, it cushions the concrete slab when heavy loads are applied.

A subgrade of gravel also helps remove moisture that has accumulated in the subgrade under the slab. Soil with excessive moisture expands when it freezes. The pressures created by freezing of the substrate raise sections of the slab, resulting in cracking.

The subgrade must be compacted uniformly to reduce settling. Settling can cause cracking of the concrete slab when heavy loads are applied. If gravel is used as a base under the slab, the soil should be excavated to a uniform depth equal to the inches of gravel to be used as the base. After adding a layer of gravel, tamp it to compact the material. See **Figure 20-2**. On large jobs, mechanical plate compacters can be used. Refer to **Figure 3-46**. This procedure is the same as that performed when hand tamping. It is important to layer the subgrade and compact each layer before new gravel or sand is added to bring the surface to the appropriate level.

Author's image taken at Cotton Angus Ranch, Volga, SD

Figure 20-2. A concrete mason is hand tamping a gravel base. Four layers of 1″ gravel will be compacted to create a 4″ subgrade. The subgrade will be sprinkled with water to settle the base, if needed fresh gravel will be added and compacted to keep the subgrade at the appropriate level. Rebars have been added to strengthen the concrete.

Portland Cement Association

Figure 20-3. This worker is checking the subgrade for smoothness and proper height.

Before receiving concrete, forms should be prepared, leveled, and adequately braced. See **Figure 20-3**. Subgrades should be smooth, compacted, and moist. Moistening the subgrade prevents rapid loss of water from the concrete when pavements, floors, and similar flatwork are being placed. This is especially important in hot weather.

Any previously attached cement should be removed from forms. The forms should be tight in the seams and clean of dirt or other materials. Poorly constructed forms will sag, leak, and possibly fail. Forms should be constructed from materials that will furnish the desired texture to the finished concrete.

Treating forms with oil or a release agent facilitates their removal after the concrete has hardened. Wooden forms exposed to the sun for long periods of time should be thoroughly dampened with water to tighten the joints. Prior to placing concrete in forms, the forms should be brushed or sprayed with a release agent or oil.

If you use steel or FRP reinforcement, it should be clean and free of loose dirt, rust, and scale. Reinforcement can strengthen concrete and reduce cracks from forces applied to the surface or from below. Reinforcement aids in transferring loads from one part of the concrete to another. For example, if a heavy truck was to drive on a concrete slab, reinforcement can aid in distributing the forces throughout the area of the floor rather than being concentrated in a small region where the truck is setting or operating. Fiber reinforced polymer (FRP) rebars can be used as reinforcement. See **Figure 20-4**. Large forklift trucks will be operated inside this warehouse building to move heavy pallets and materials. Also, tractor and trailer trucks will be loaded inside the warehouse so the strength of the floor is paramount.

Photo Courtesy of Neuvokas Corporation – Manufacturers of GatorBar®

Figure 20-4. When the FRP rebars were in a bundle, pink spray was used to mark the rebars at 2′ increments. Marking the FRP prior to laying makes it easier to form the 2′ × 2′ grid on the poly. Yellow poly was used to reduce the moisture transfer into the concrete. Rigid insulation was used under the poly to reduce heat loss.

Code Note **Tie Wires or Other Bar Supporting Devices**

The International Residential Code (IRC) **R403.1.3.5.3** states that reinforcement in concrete forms shall be secured with tie wires or other bar supporting devices to prevent displacement of reinforcement.

When installing reinforcement in floors and walls where rebars cross, the bars can be tied together at determined locations using tie wires. The tie wire is intended to connect reinforcement together so displacement of the rebars does not occur when concrete is placed in the form. To hand tie rebars together takes a considerable amount of time and labor in order to do the job correctly. Tool manufacturers have developed battery powered tie wire tools to reduce labor time and decrease chances of employee's receiving soft tissue damage or wrist injuries because of repetitive tying. See **Figure 20-5**. Contact your local building inspector with questions about tie wire spacing when connecting reinforcement.

Photo Courtesy of Neuvokas Corporation – Manufacturers of GatorBar®

Figure 20-5. A rebar tying tool is capable of tying four #4 diameter bars or two #5 bars together. It can roughly tie 120 joints with one spool of wire. The tool can tie one cross joint of multiple bars together in 1 to 2 seconds of time.

Determine if dowels or other devices will be used in construction joints between concrete slabs. ***Dowels*** are steel bars, square plates, or similar mechanisms installed in construction joints between concrete slabs. See **Figure 20-6**.

NorGal iStock/Getty Images Plus via Getty Images

Figure 20-6. This concrete highway shows dowels installed to transfer load from one pour of concrete to the next. Dowels were placed at the end of this pour and along the right side to tie this concrete slab to the next sections to be installed.

Dowels are used to transfer load from one concrete slab to another, across the construction joint in concrete driveways, sidewalks, or other concrete flatwork. Dowels allow horizontal movement of slabs but restrict vertical movement, **Figure 20-7**. One of the main reasons to use dowels is to minimize the vertical movement of one concrete slab without the same action occurring to the corresponding slab. Frost and heavy loads can affect individual slabs, but connecting slabs using dowels will aid in transferring loads and minimizing vertical movement from slab to slab, **Figure 20-8**. A type of dowel being used to transfer loads between two concrete slabs is called a *diamond* or *square shaped tapered dowel*. Plastic sleeves are temporarily attached to concrete forms where construction joints are to occur. Once the forms are removed, the sleeves will stay in the concrete slab. Prior to placing concrete in the adjoining slab, tapered metal plates (tapered dowels) are slid into the sleeves. These sleeves will allow the slabs and metal plates to move horizontally but restrict vertical movement.

Allen Engineering Corporation

Figure 20-7. Dowel bars have been positioned in holes drilled into the concrete floor to connect the shop floor to the new concrete slab being placed. Plastic sleeves (red devices) have been temporarily attached to the concrete form. After the forms are removed, the plastic sleeves remain in the slab. Metal plates are installed in the sleeves prior to placing the second slab.

Timothy L. Andera

Figure 20-8. Horizontal lift has occurred in the construction joint between the two adjacent concrete slabs identifying the need for doweling or other construction procedures to have happened at the time of construction.

Moisture Barrier and Insulation under Concrete Slabs

During the planning stages of placing concrete, determine if a moisture barrier or rigid insulation will be used on top of the subgrade. Moisture barriers are used to reduce migration of moisture from the subgrade aggregates to the concrete slab. Concrete is somewhat porous and will absorb moisture from the subgrade. Then, moisture transfers from the underside of the concrete slab to the top surface. Certain basements feel damp because the moisture coming through the concrete turns into vapor, increasing the humidity inside the structure. To reduce the effect of moisture transfer, a moisture barrier is used between the subgrade and the concrete slab. See **Figure 20-9**. While installing a moisture barrier, it is important to seal joints in the film with an appropriate tape to reduce transfer from the subgrade to the concrete slab. Also, it is essential to seal the film to the foundation walls with a suitable sealant that will adhere the moisture barrier to the foundation wall, in order to eliminate as much moisture transfer as possible.

serato/Shutterstock.com

Figure 20-9. A moisture barrier is being installed in a commercial project over the subgrade prior to placing the concrete.

As the building industry has become more conscious of energy use and of reducing energy costs for the consumer, the application of rigid insulation under the concrete slab has become commonplace. To conserve energy, a number of homes and commercial buildings use the application of rigid insulation on top of the moisture barrier. In most instances, 1″–2″ of rigid insulation is installed to reduce the transfer of ambient subgrade temperature into the concrete slab. **Figure 20-10** shows panels of rigid insulation laid over a moisture barrier to conserve energy and to make the living space more comfortable.

Placing Concrete

When concrete is being placed, it should be as near as possible to its intended location. Concrete should not be placed in large quantities in one area, be allowed to run, or be worked over a long distance in the form. Segregation (separation) of ingredients can occur in areas where the grade slopes. The concrete can pool or bunch up from the force of concrete running down the truck chute.

Generally, concrete should be placed in horizontal layers having uniform thickness. If the form is deep, the concrete should be compacted or vibrated after each layer is placed. See **Figure 20-11**. Layers or lifts are usually 6″ to 12″ thick in reinforced concrete and up to 18″ thick for nonreinforced applications.

Concrete should not be allowed to drop freely more than 3′ or 4′. Drop chutes of rubber or metal can be used when placing concrete in thin vertical sections.

sonsart/Shutterstock.com

Figure 20-10. Rigid insulation has been combined with a moisture barrier in a commercial application. Reinforcing bar panels are being used to strengthen the concrete slab. Notice rebar chairs are used to lift the panels off the insulation to allow concrete to flow freely under the panels and to ensure they are located properly in the concrete slab.

Dmitry Kalinovsky/Shutterstock.com

Figure 20-11. A concrete worker is using a vibrating machine to compact freshly placed concrete. Note that the worker is wearing appropriate PPE—protective clothing and gloves, safety shoes, safety glasses, and a hard hat.

On construction projects, concrete can be moved by wheelbarrow or through mechanical means such as a wheeled or track buggy. Refer to **Figure 20-7**. Buggies can be used for numerous tasks or activities around a construction site from carrying tools, aggregate, or concrete.

On large projects, concrete can be dumped from a ready-mix truck into a pump truck that pushes the concrete through a hose attached to a boom. The use of a pump truck aids in moving concrete to isolated areas where other means can be almost impossible. See **Figure 20-12**. A hose can be worked down into the wall forms to reduce the distance the concrete drops or can be moved around with a boom to deliver concrete to the forms around the structure. Use of a boom truck allows much faster placing of concrete compared to using wheelbarrows or other devices.

Slab Construction

Steps involved in placing and finishing concrete for slab construction include the following:

1. Determine size and location of the slab.
2. Excavate area to the appropriate depth and slope; include the area of the forms.
3. Construct forms for the concrete slab.
4. Add aggregate and compact in layers as required.
5. Install moisture barrier and insulation if specified.
6. Place reinforcement if required or needed.

Photo Courtesy of Neuvokas Corporation – Manufacturers of GatorBar®

Figure 20-12. An aerial view of a large concrete placement project. A pump truck is delivering concrete in a hose on the left side of the floor area. A ready-mix truck is backed up to the pump truck to unload approximately 10 cubic yards of concrete into the hopper. Many more trucks will be needed to complete the pour. Poly and FRP rebars are used in this concrete project.

7. Add doweling or other construction joint devices if required or needed.
8. Determine the amount of concrete and type; order for delivery.
9. Place concrete inside forms.
10. Level concrete in the forms (screed).
11. Wait; if bleed water appears, then wait until it evaporates.
12. Edge the concrete slab.
13. Joint the concrete slab.
14. Float the concrete slab.
15. Trowel the surface of the concrete.
16. Texture the surface if required or needed.
17. Cure concrete as soon as the finishing operation is completed.
18. If needed, cut or saw control joints in concrete slab. Check with local building officials on code requirements.

Trade Tip

When working with concrete, never sprinkle or spray water on the surface of the concrete. This may cause the concrete surface to scale or small pieces of concrete to flake off in the future.

Placement of concrete in slab construction should be started at the most distant point of the work so each batch is dumped against the previously placed concrete, not away from it. See **Figure 20-13.** Care should be taken to prevent stone pockets (areas of excessive

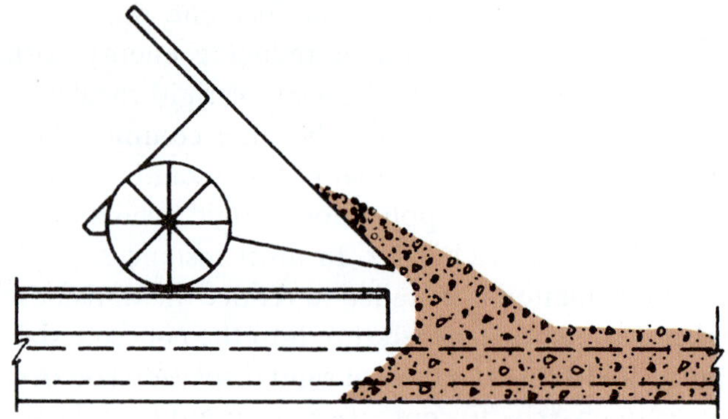

Proper placement of concrete slab

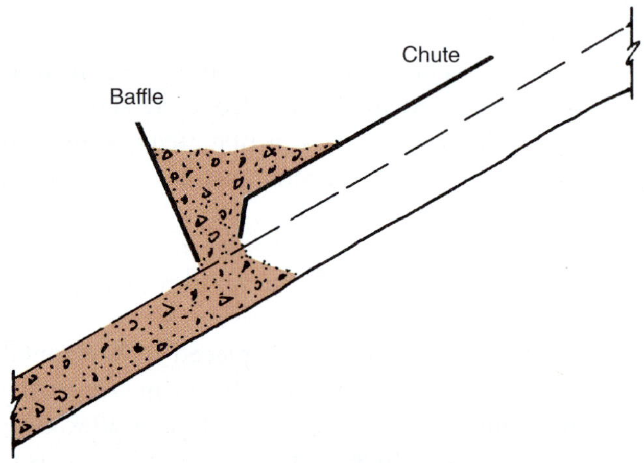

Proper placement of concrete on a sloping surface

Goodheart-Willcox Publisher

Figure 20-13. The proper method of dumping concrete when pouring a slab.

large aggregate) from occurring. If this happens, some of the aggregate can be moved to areas where there is more cement paste to surround them.

Concrete placement in walls should begin at either end and progress toward the center. The same order should be used for each successive layer.

Concrete should be placed around the perimeter (outer edges) first in large flat open areas. Whatever method of placement is used, do not allow water to collect at the ends and corners of forms.

Compacting the concrete is always necessary. Work the mix with a spade, shovel, or rod to be sure all spaces are filled and air pockets are worked out. This is called *puddling*, *spading*, or *rodding*. This process should help eliminate stone and air pockets and consolidate each layer with the previously placed layer. Compacting also brings fine material to the faces and top for proper finishing.

Mechanical concrete vibrators can be used either in the concrete or on the forms. Mechanical vibration does not make the concrete stronger, but it does permit the use of a stiffer mix that is stronger than a wet mix. Too much vibration causes segregation of particles. Indicators of sufficient vibration are the appearance of a line of mortar along the forms and sinking of the coarse aggregate into the mortar.

Placing on Hardened Concrete

When placing fresh concrete on hardened concrete, it is important to produce a good bond and a watertight joint. To ensure this result, the hardened concrete should be reasonably level, rough, clean, and moist. Leaving some coarse aggregate exposed aids bonding. All loose or soft mortar should be removed from the top surface of the hardened concrete before placing the fresh batch.

For floors that require two courses of concrete, the top of the first course (lower level) can be broomed with a steel bristle broom or stiff bristle broom just as it sets. The surface should be heavily scored and cleaned before the grout coat and top course are placed. The grout coat is a mixture of Portland cement and water. This coat usually has the consistency of thick paint and is scrubbed into the surface of the slab just before the top course is placed.

If old concrete is to receive a new topping, it must be thoroughly roughened and cleaned of dust and loose particles, grease, oil, and other materials. The surface can be chipped with pneumatic tools or sandblasted to expose sound concrete. Hardened concrete must be moistened thoroughly before new concrete is placed on it, but no pools of water should be left standing on the surface of the existing concrete.

Where concrete is to be placed on hardened concrete or rock, a layer of mortar 1/2″ to 1″ thick is placed on the hard surface. This mortar provides a cushion for the new concrete, prevents stone pockets, and aids in securing a tight joint. The mortar is generally made of the same materials as the concrete, but without the coarse aggregate. The mortar should have a slump of less than 6″.

Pneumatic Application of Concrete

Pneumatically applied concrete, frequently called ***shotcrete***, is a mixture of Portland cement, aggregate, and water shot into place by means of compressed air. Aggregate up to 3/4″ in size can be used with some equipment. Shotcrete can be applied by either the wet mix or dry mix process.

In the dry mix process, the cement and aggregates are mixed in a relatively dry condition. This mix is pumped through a hose to a nozzle where water is added. At least 45 psi air pressure and a nozzle velocity of about 400 feet per second force the dry materials through the hose. Water pressure at the nozzle should be at least 15 psi higher than the air pressure. In the wet mix process, the concrete is premixed before it is applied pneumatically.

Water content can be kept to a minimum using the pneumatic process. As a result, high-strength, durable concrete may be obtained. The key element in this process is the person controlling the nozzle. In both processes, the worker directs the nozzle and controls the thickness of the concrete layer and the angle of application. Pneumatic application can be used for new construction and repair work in difficult locations. It also works well where relatively thin sections and large areas are involved.

Finishing Concrete Slabs

Concrete slabs can be finished several ways, depending on the effect desired and the use of the product. Some surfaces are left rough, while others are broomed, floated, or troweled. Surfaces can be textured, colored, or have exposed aggregate.

Screeding

Screeding is usually the first finishing operation after the concrete is placed in the forms. It is performed with a tool called a *screed*.

Screeding is the striking off of excess concrete to bring the top surface to the proper grade or elevation. See **Figure 20-14**. The edge of the screed is either straight or curved, depending on the surface requirements. The screed rests on the top of the form and is moved across the concrete with a sawing motion. The screed is advanced forward slightly with each movement.

Timothy L. Andera

Figure 20-14. Concrete masons are screeding a 12′ sidewalk. They are using a 16′ aluminum screed. The sidewalk is 6″ thick to accommodate light motored vehicles as well as people.

An excess of concrete should be carried along in front of the screed to fill low places as the tool is moved forward. However, if too much concrete is allowed to build up in front of the screed, hollows may be left behind it. As the screeding process continues, remove the excess concrete in front of the screed and place it in areas that need to be filled. Try to keep a 1/2′ to 1′ of excess concrete in front of the screed as it is pulled forward.

Wooden screeds, either 2″ × 4″ or 2″ × 6″, are commonly used. Prior to use, ensure the screed is straight without twists or warps to produce a quality job. Another screed material is magnesium. The advantage of magnesium is moisture does not cause warpage or twisting. Magnesium screeds can be purchased in an assortment of widths and lengths. When using either a wood or metal screed, be sure the edges are not damaged, since this can affect the quality of the job.

When placing concrete against walls in an interior space such as a basement, the concrete needs to be hand troweled or floated in order to position it to the correct height and to level the surface. See **Figure 20-15**. Hand floating this area will also aid when it is time to screed and bull float the concrete.

Prior to the placement of concrete, snap lines are usually placed on walls as a guide to indicate the final elevation or finished concrete floor height. A laser level can also be used to determine elevations. See **Figure 20-16**. As the laser rotates in the horizontal plane, the light beam revolves in a 360° circle so the worker can mark locations on the walls as reference points for determining heights.

Pukhov Konstantin/Shutterstock.com

Figure 20-16. A rotary laser level creates a light beam (laser beam) either in the horizontal or vertical plane. Some lasers automatically level themselves and can also shoot a vertical beam out of the top of the device. The color of the laser beam, red or green, depends on the type of laser.

In larger concrete jobs, screeding can be accomplished with a power screed. See **Figure 20-17**. Power screeds have a gas or electric motor that operates a mechanism that causes the base to vibrate. As the screed is pulled over the concrete surface, the vibration drives the surface stones down into the mixture and brings up the "cream" to the top of the concrete. On very large concrete projects, wheeled laser screed machines can be used to reduce time and labor on the screeding process. Usually, the machines have four-wheel drive capability. Two types are a boom operated concrete screed and a drive-in concrete screed.

Marshalltown Company

Figure 20-15. A worker is using a hand float to level the concrete next to a foundation wall prior to screeding and bull floating. The pattern found in the foundation wall is being used as a guide line for the top edge of the concrete floor. The worker is smoothing the area next to the wall because the screed and bull float cannot get that close to the wall.

Marshalltown Company

Figure 20-17. Concrete workers are leveling the concrete with a power screed. The power screed is riding on screed pipes that are embedded just under the surface of the freshly placed concrete. The screed pipes are resting on plastic or metal jacks or wood stakes to keep the pipes elevated to the correct height. When the workers complete the section of floor, the pipes are removed, and the voids are filled in with concrete.

Boom Operated Concrete Screed

Boom operated screeds use a boom that reaches over the freshly placed concrete and the screed head is lowered to where the laser eye's pick up the rotating laser beam signal. On each side of the screed head is a control pole and laser eye that receives data from the laser to adjust the head to the appropriate height. The machine operator lowers the screed head to the finished elevation of the concrete slab. Once both sides of the screed head are level, the boom is drawn back toward the machine, creating a smooth screeded surface. The laser eyes continue to adjust automatically on each side of the screed according to the laser level beam.

Drive-In Concrete Screed

A drive-in screed is similar to the boom operated concrete screed, but it is driven into the freshly placed concrete. See **Figure 20-18**. The screed operates the same by automatically adjusting the screed head to the laser beam signal. Once the machine is positioned correctly in the concrete, the screed is lowered to pick up the laser beam signal. The screed head is adjusted according to the laser signal and the operator backs the machine up, leveling and screeding the fresh concrete.

Edging and Jointing

Edging, if necessary, could be the next operation. See **Figure 20-19**. Edging provides a rounded edge or radius to prevent chipping or damage to the edge. An edger is run back and forth until the desired finish is obtained. Care should be taken to cover all coarse aggregate. Do not leave too deep a depression on the top of the slab, because this indentation could be difficult to remove during subsequent finishing operations.

Stanley Goldblatt

Figure 20-19. Edging produces a radius on the edge of the slab that prevents chipping. A trowel may be inserted between form and concrete to provide a track for the edging operation shown here.

As soon as edging has been completed, the slab is jointed or grooved. See **Figure 20-20**. The bit (cutting edge) of the jointing tool cuts a groove in the slab that is called a *control joint* or *contraction joint*. Any cracking due to shrinkage caused by drying or temperature change is intended to occur at the joint. The joint

Photo Courtesy of Neuvokas Corporation – Manufacturers of GatorBar®

Figure 20-18. A four-wheel, drive-in laser level screed is being used to screed a large concrete project. On the left side of the project, a concrete mason is using a bull float to smooth the surface after being screeded by the machine. Note the two laser poles and laser eyes at the top of each pole on the screed machine.

ungvar/Shutterstock.com

Figure 20-20. A concrete mason just finished cutting a joint in the sidewalk. The mason weakened the slab to ensure cracks occur at the bottom of the joint. Jointing a slab also helps control any cracking due to shrinkage caused by drying out or temperature change.

weakens the slab and induces cracking at the bottom of the joint rather than some other place. These cracks are not noticeable when controlled. It is important that the groove continues across the entire slab to ensure a controlled crack in the joint.

In sidewalk and driveway construction, the tooled or cut joints are generally spaced at intervals equal to the width of the slab, but not more than 10′ apart. The joints should be perpendicular (at right angles) to the edge of the slab. Concrete has a tendency to crack in square shapes. For example, if a sidewalk was 6′ wide, it would be smart to place or cut the joint equal to 6′. If the slab is 4″ thick, the groove or cut should be 1/4 the thickness or a minimum cut of a 1″ bit. It is very important the joints are placed according to the correct size of the project.

Use a wood or magnesium straightedge as a guide when making grooves. A 1 × 8 or 1 × 10 board is ideal. Be sure the board is straight and not cupped or twisted.

Large concrete surfaces can be jointed by cutting grooves with a power saw with an abrasive or diamond blade. When grooves are cut rather than jointed, the operation should be performed 4–12 hours after the slab has been finished. The cutting must be done before random shrinkage cracks develop but after the concrete is hard enough not to be torn or damaged by the blade. Cutting concrete joints will be discussed in greater detail later in the chapter.

Safety Note

When dry sawing, wear an appropriate respirator. Concrete may contain silica, which is a severe health hazard. When sawing either wet or dry, wear safety glasses and a face shield to protect from flying particles. Wear ear protection to protect your hearing from the noise created by sawing.

Floating

There are three reasons why concrete is floated:

- To embed aggregate particles just beneath the surface
- To remove slight imperfections, waves, and voids
- To compact the concrete at the surface in preparation for other finishing operations

After concrete has been edged and jointed, it should be allowed to harden enough to support a person and leave only a slight foot imprint. In normal concrete, the dry materials used are heavier than water. They begin to sink or settle to the bottom of a plastic concrete mixture shortly after placement. This settling action causes bleeding, which is excess water that rises to the surface. See **Figure 20-21**. Bleeding can also occur if excessive water is added during concrete mixing. This usually occurs at the jobsite in an attempt to make the mix easier to place. Adding water on the jobsite also reduces the strength of the mix. Bleeding does not usually occur with air-entrained concrete.

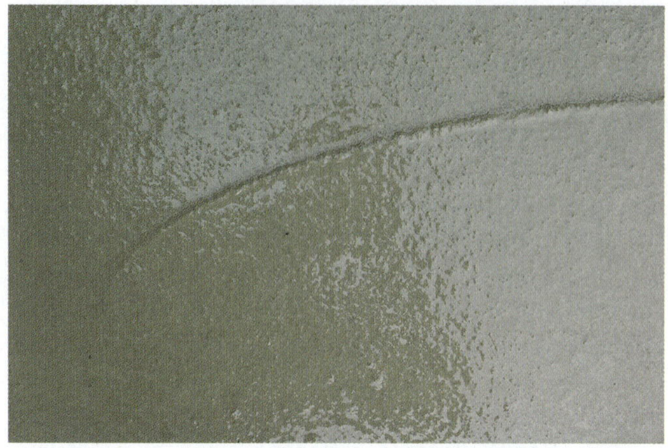

Craig Jewell/Hemera/Thinkstock

Figure 20-21. This concrete has been screeded and bull floated. Bleed water has appeared on the surface.

Trade Tip

The first operations of placing, screeding, and darbying must be performed before any bleed water appears on the surface of the concrete. If a finishing operation is performed on the surface while the bleed water is present, serious scaling, dusting, or crazing can result.

Some surface finishes may not need any further finishing, but most require operations such as edging and jointing, floating, troweling, or brooming.

Metal floats reduce the amount of work required by the finisher. Drag is reduced because the float slides more readily over the surface. A wood float tends to stick to the surface and produces a tearing action. A light metal float also forms a smoother surface texture than a wood float.

Trade Tip

Aluminum or magnesium floats work better than a wood float, especially on air-entrained concrete. Air-entrained concrete can be placed with less water in the mix. Because of this, the concrete mix may be more tacky or sticky and have a tendency to adhere to wood.

A bull float or darby is used to eliminate high and low spots and to smooth the surface after it is leveled with a power screed. A darby float is smaller than a bull float and is used to float areas the bull float cannot access.

When bull floating, make sure the float does not dig into the concrete. It should float on the top to smooth the surface and to drive exposed stones down into the cement. See **Figure 20-22** and **Figure 20-23**. On certain slabs, it may be easier to access the perimeter of the slab of the concrete first. See **Figure 20-24**.

Be sure not to overwork the concrete. Overworking brings excess water and fine aggregate material to the surface, resulting in surface defects.

Floating can be done to provide a coarse texture as the final finish. If this is done, then a second floating may be necessary after the concrete has partially hardened.

Marshalltown Company

Figure 20-22. This worker is pushing the bull float. Notice the angle at which the float is held—it is riding on the back side when being pushed.

Marshalltown Company

Figure 20-23. This worker is pulling the bull float back toward him. The back side of the trowel is lifted up to allow the float to smooth the cream of the concrete rather than digging into it and creating a dip.

Vicki Reid/iStock/Thinkstock

Figure 20-24. This worker is bull floating around the perimeter of a driveway. In some situations, it is easier to float the perimeter than to bull float between the edges.

Troweling

When a smooth, dense, hard surface is desired, steel troweling is performed after floating. Cement masons frequently float and trowel an area before moving their kneeboards. See **Figure 20-25**.

Troweling can be done by hand or with a troweling machine. During the first troweling, whether by hand or machine, the trowel blade must be kept as flat against the surface as possible. If the blade is held at too great an angle, a washboard effect results. A new trowel is not recommended for the first troweling. An old trowel that is broken in can be worked quite flat without the edges digging into the surface.

As the concrete progressively hardens, it can be troweled several times to obtain a very smooth and hard surface. Usually, smaller size trowels are used for successive applications so sufficient pressure can be exerted for proper finishing. If necessary, tooled joints and edges can be rerun after troweling to maintain uniformity.

A Alice Scully/iStock/Thinkstock

B John Sarlin/iStock/Thinkstock

C Ari N/Shutterstock.com

Figure 20-25. Finish troweling produces a smooth, hard surface on the concrete. A—Workers are finishing the job using hand trowels. B—This worker is finishing troweling the concrete surface. They are using two trowels to be able to put extra weight on the finishing trowel and for added balance. The rounded edges of the finishing trowel reduce the chances of the trowel catching the fresh concrete surface, which would leave marks. C—Final troweling is done with a power trowel to create a smooth and hard finish on the concrete floor.

Brooming

Steel troweling produces a very smooth surface that is often slippery when wet. Sidewalks, driveways, and other outside flatwork frequently are broomed or brushed to produce a slightly roughened surface. The broomed surface is made by drawing a soft-bristled push broom over the surface of the slab after steel troweling. See **Figure 20-26**. The concrete must be hard enough so brooming does not damage the edges of tooled joints. Coarser textures for steep slopes can be produced by using a stiffer bristled broom. Brooming is usually perpendicular to the traffic to provide the most resistance.

TFoxFoto/Shutterstock.com

Figure 20-26. A concrete mason is using a special broom to put the final finish on a concrete sidewalk. Brooming produces a slightly roughened surface that reduces the danger of slipping.

Form Removal

It is generally best to leave forms in place as long as possible for better curing. However, it is sometimes desirable to remove the forms as soon as possible. In either case, leave them in place until the concrete is strong enough to support its own weight and any other loads that may immediately be placed on it. The concrete should be hard enough to resist damage from form removal.

Usually, the side forms of relatively thick sections can be removed 12–24 hours after placing. Testing the concrete to determine hardness is better than relying on an arbitrary age for form removal. The age-strength relationship should be determined from representative samples of concrete used in the structure and cured under job conditions. **Figure 20-27** shows the times required to attain certain strength under average conditions for air-entrained concrete. Keep in mind that these are averages, and strength is affected by materials, temperature, and other conditions.

If forms are tight and require wedging, only wooden wedges should be used. Do not place a pinch bar or other metal tools against the concrete to wedge forms loose. Start removing forms some distance from a projection. This method relieves pressure against projecting corners and reduces the likelihood of breaking off the edges.

Finishing Air-Entrained Concrete

Air-entrained concrete has a slightly different consistency from standard concrete and, therefore, requires a small change in finishing operations. Since air-entrained concrete contains many tiny air bubbles that hold all the materials in concrete in suspension, it requires less mixing water than concrete that is not air-entrained. It also bleeds less.

There is no waiting for the evaporation of free water from the surface before floating and troweling. If floating is done by hand, an aluminum or magnesium float is essential. A wooden float drags and increases the work necessary to finish the surface. If floating is done with a power finishing machine, there is practically no difference in the finishing procedure for air-entrained and standard concrete, except that finishing can begin sooner with air-entrained concrete.

Almost all horizontal surface defects and failures in standard concrete are due to finishing operations performed while bleed water or excess surface moisture is present. Better results are usually obtained with air-entrained concrete.

20.2 Curing Concrete

Curing a concrete slab is one of the most important operations in producing quality work, but also one of the most often neglected operations. Even if the concrete is mixed, placed, and finished properly, poor quality work results if proper curing operations are not followed.

Little or no moisture loss should be allowed during the early stages of hardening. If necessary, the concrete should be protected during this critical period. Newly placed concrete should be protected from the sun and not allowed to dry out too fast. This can be accomplished with damp burlap, canvas, or polyethylene film coverings. See **Figure 20-28**. Coverings can be applied as soon as the surface is hard enough so it will not be marred. Coverings should be kept moist for at least three days.

Another method of curing is called ponding. ***Ponding*** involves keeping approximately an inch of water on the concrete surface, usually by earth dikes around the edges of the slab. Basement concrete floors lend themselves to ponding since walls surround the perimeter of the slab. Once the concrete basement floor is finished, floor drains, sump pump areas, and any other openings in the floor need to be sealed or diked. To increase the hardness of the floor surface, keep fresh concrete covered with water for 3 to 5 days or longer if possible.

Age-Strength Relationship of Air-Entrained Concrete

Strength (psi)	Age—Type I (Normal) Cement	Age—Type III (High Early-Strength) Cement
500	24 hours	12 hours
750	1 1/2 days	18 hours
1500	3 1/2 days	1 1/2 days
2000	5 1/2 days	2 1/2 days

Goodheart-Willcox Publisher

Figure 20-27. This chart shows the age-strength relationship of air-entrained concrete that must be considered when removing forms.

Portland Cement Association

Figure 20-28. Polyethylene film can be used to prevent newly placed concrete from drying out too fast.

Membrane curing compounds are frequently sprayed on concrete surfaces. See **Figure 20-29**. Uniform coverage is necessary. Two coats are often required to provide adequate protection.

Curing Temperatures

The rate of chemical reaction between cement and water is affected by temperature. Therefore, temperature affects the rate of hardening, strength, and other qualities of concrete.

Cold Weather Construction

According to the American Concrete Institute (ACI), cold weather is considered air temperature that averages less than 40°F (5°C) and is below 50°F (10°C) more than half of each day for three weeks in a row. In cold weather, concrete placement may require heated materials, a covering for the fresh concrete, or a heated enclosure. See **Figure 20-30**.

Under cold weather conditions, concrete sets up more slowly, takes longer to finish, and gains strength more slowly. The hydration of the cement generates some heat, but this may not be enough. If the concrete freezes before it hardens, the damage done by freezing can reduce the final compressive strength by as much as 50%. Ice starts to form in plastic concrete when the concrete temperature approaches 27°F (–3°C). The freezing point can be as low as 20°F (–7°C) if there are admixtures in the mix. Ice requires more space than water, and this expansion in wet concrete weakens the product by creating void spaces that disrupt the bond between the cement paste and the aggregate.

The temperature of concrete at the time of placing should generally be 50°F to 70°F (10°C to 21°C). The materials should never be heated to the point that the fresh concrete temperature is above 70°F (21°C), or strength will be reduced.

Vladimir Zlotnik/iStock/Getty Images Plus via Getty Images

Figure 20-29. A membrane curing compound has been sprayed on the new section of highway. The curing compound is applied immediately after the concrete has had its final finishing operation.

Jack Klasey

Figure 20-30. Straw is used to protect newly placed concrete in cold weather.

Concrete should never be placed on a frozen subgrade. When subgrades thaw, uneven settling and cracking of the slab usually results. Snow and ice in the form take up space intended for the concrete. Thawed subgrade should be recompacted before placing concrete to avoid uneven settlement and cracked concrete. Forms, reinforcing steel, and embedded fixtures should be free of ice when the concrete is placed. A thin layer of warm concrete should be placed on cold, hardened concrete when an upper layer is to be poured. The thick upper layer shrinks as it cools, and the lower layer expands as it warms. Failure of the bond results if care is not taken.

In cold weather, moisture for curing is still very important. Keep the concrete moist, especially near heating units. Maintain the temperature of normal concrete at 70°F (21°C) for three days or at 50°F (10°C) for five days. Keep the temperature of high early-strength concrete at 70°F (21°C) for two days or at 50°F (10°C) for three days. Do not allow the concrete to freeze for the next four days.

It is estimated that for every 10°F drop in concrete temperature, set time increases by approximately one-third. Hydration stops completely at 14°F (–10°C) but resumes when the temperature rises.

Set time can be shortened and the early strength of the concrete increased by ordering the concrete with extra cement, Type III cement, chemical accelerators, or a combination of the three. Type III cement is similar to ordinary Type I cement, but it is almost 50% stronger after 24 hours because the particles are finer. Type III cement may not be readily available from smaller companies.

Chemical accelerators provide the same benefits as adding an extra 100 lb of cement per cubic yard or using Type III cement. Accelerators (Type E and Type C) are available in chloride-based or nonchloride-based formulations. Calcium chloride can increase the potential for corrosion of steel reinforcing if water is present.

Air-entrained concrete is not as susceptible to damage caused by freezing and thawing as standard concrete because ice crystals form in the tiny air spaces in the mix. For this reason, air-entrained concrete is generally a good choice for exterior concreting that will be exposed to freezing and thawing.

Cold Weather Protection

Fresh concrete should be protected from freezing, and curing conditions should be maintained to ensure adequate strength of the concrete. Two methods are generally used to protect concrete and maintain curing conditions—insulating the concrete and providing a protective cover with heat.

The most common way to protect concrete against low temperatures is by insulating it. Unless the temperature is too low, the heat of hydration can provide enough heat to protect the concrete. However, 6″ of straw held in place with polyethylene sheeting or tarps provides extra protection. An alternative method is to use insulating or curing blankets. Insulating blankets are easy to install and can be secured with weights or ties. Most blankets are made from fiberglass or closed-cell polypropylene insulation that is laminated to canvas or some other material. To be effective, insulating blankets must lie flat on the concrete surface with edges secured to prevent air movement under the blanket. See **Figure 20-31**. Blankets or tarps vary in thickness, width, and length depending on the application. Thickness sizes vary from 3/8″, 1/2″ to 1 1/2″. Overall dimensions vary from 6′ × 25′ to 12′ × 25′.

In very cold weather, another option is heated tarps or blankets made to protect concrete from freezing. A normal size electric blanket is 4 1/2′ × 15′, but larger sizes are available. Most blankets do not require a thermostat and are set to maintain a temperature of 60°F.

Another option is to tent the concrete. If severe cold weather is to occur, heating the enclosure can be used to protect the concrete. Be sure to use vented heaters or electric heaters. Heaters that produce carbon dioxide can cause a soft, chalky layer to form on the surface of the concrete.

Safety Note
Carbon monoxide is a toxic gas formed when combustion is incomplete. Carbon monoxide poisoning can result from unvented heaters.

Timothy L. Andera

Figure 20-31. Insulated blankets have been added to a freshly laid sidewalk. It is important to secure blankets down with lumber on the concrete surface and secure the edges of the blankets with lumber or other materials to protect from strong winds. It is important to carefully lay lumber or other materials on fresh concrete so as not to damage it.

Hot Weather Construction

There is a good reason not to place concrete in hot, dry, windy weather. The concrete sets much faster and may result in cracked or poorly finished slabs. The American Concrete Institute (ACI) states the following:

"If the initial setting time for a concrete mix is 2 1/2 hours at 60°F (16°C), that time is likely to be reduced to about an hour or less at 95°F (35°C)."

Concrete loses its slump faster in hot weather and can become unworkable before a large load can be placed. In addition, moisture evaporates rapidly from the slab surface, which reduces finishing time. Plastic shrinkage cracking is a potential problem for any placement done in hot, dry, windy weather. These cracks can appear after the slab has been placed, screeded, and bull floated. The only thing that can be done in this situation is to continue sealing the cracks by troweling.

To prevent rapid drying in extremely hot weather, avoid high temperatures in fresh concrete. The temperature of the mixing materials can be cooled by using chilled water or ice. The ice should be melted by the time the concrete leaves the mixer. When concreting in hot weather, get an early start while it is cooler, break big placements into smaller sections if possible, and shade the operation if practical.

Some contractors use additives like set retarders and superplasticizers to reduce the problems of rapid set or slump loss. However, these additives are not complete solutions. Although superplasticizers can make a stiff mix flow more easily, the effect wears off suddenly—sometimes before the finishing process is completed. The use of set retarders requires experienced cement finishers due to the possibility of unexpected results with these chemicals.

Subgrades should be saturated some time in advance and sprinkled just ahead of placing the concrete. Wood forms should be treated or wetted thoroughly. Placing should not be delayed, and the concrete should be screeded and darbied immediately after placing. Covers, such as burlap, which are kept constantly wet, should be placed over the concrete as soon as it is darbied. When the surface is ready for final finishing, a small section should be uncovered immediately ahead of the finishers and recovered as soon as possible.

The purpose of curing is to maintain conditions in the setting concrete that encourage complete hydration. The water in the concrete must be kept from evaporating. Moisture must be present for at least seven days. Longer than seven days is desirable because the curing process can continue for 28 days. The longer the curing process, the stronger, harder, and more durable the finished product will be.

Several techniques are used to trap moisture in fresh concrete. The most common method is to spray on a liquid curing compound that forms a thin film over the slab. Some contractors use ponding; sprinkling or fogging; covering the concrete with sheets of plastic; or applying wet sand, hay, or burlap to the surface after the finishing is completed. All of these methods work, but each has its advantages and disadvantages depending upon the specific situation.

Air-entrained concrete requires special expertise, especially during hot weather. It develops a rubberlike surface in hot weather if finishing is delayed. This concrete is then very hard, if not impossible, to surface smoothly.

20.3 Joints in Concrete

Three basic types of joints are frequently used in concrete construction. These joints are as follows:

- **Isolation joints.** Isolation joints, sometimes called *expansion joints*, separate different parts of a structure to permit both vertical and horizontal movement. This type of joint is used around the perimeter of a floating slab floor and around columns and machine foundations. See **Figure 20-32**.
- **Control joints.** Control joints provide for movement in the same plane as the slab or wall is positioned. They compensate for contraction caused by drying shrinkage. Control joints should be constructed in such a way that they permit the transfer of loads perpendicular to the plane of the slab or wall. If control joints are not used in slabs or lightly reinforced walls, random cracks occur due to drying shrinkage. Control joints are sometimes called *contraction joints* or *dummy joints*. See **Figure 20-33**.

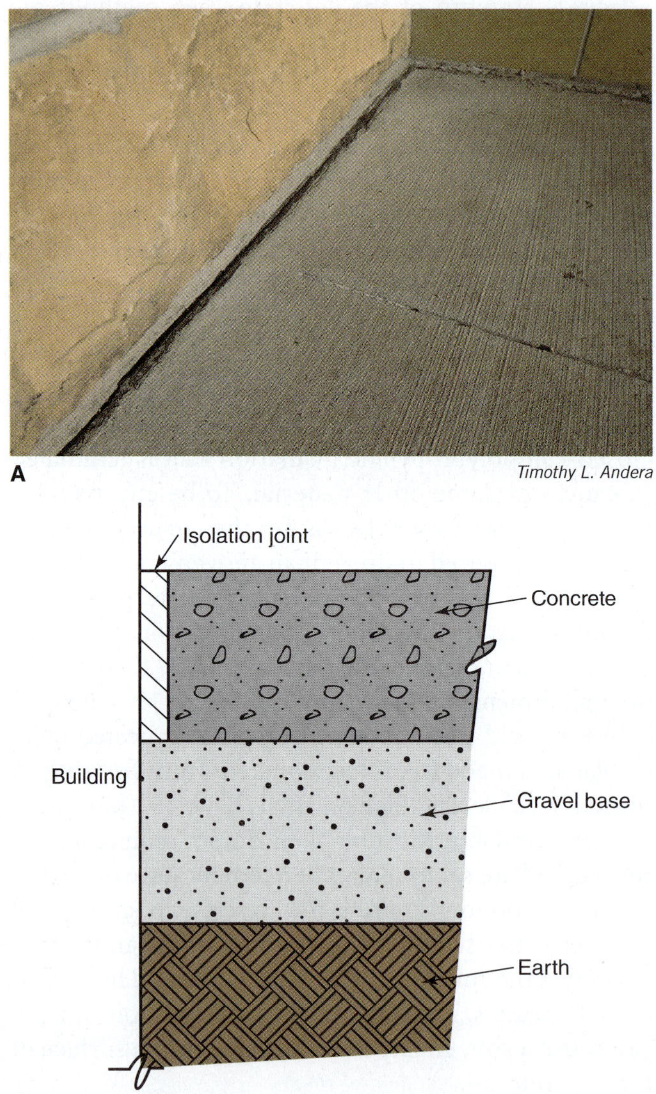

A Timothy L. Andera

B Goodheart-Willcox Publisher

Figure 20-32. Isolation joints separate different parts of a structure. A—Isolation joint material was placed against a building wall to separate the wall from the floating slab. The floating slab was then poured against the joint material. B—Detail of an isolation joint. The joint material can be flush in areas where no safety hazard from tripping exists.

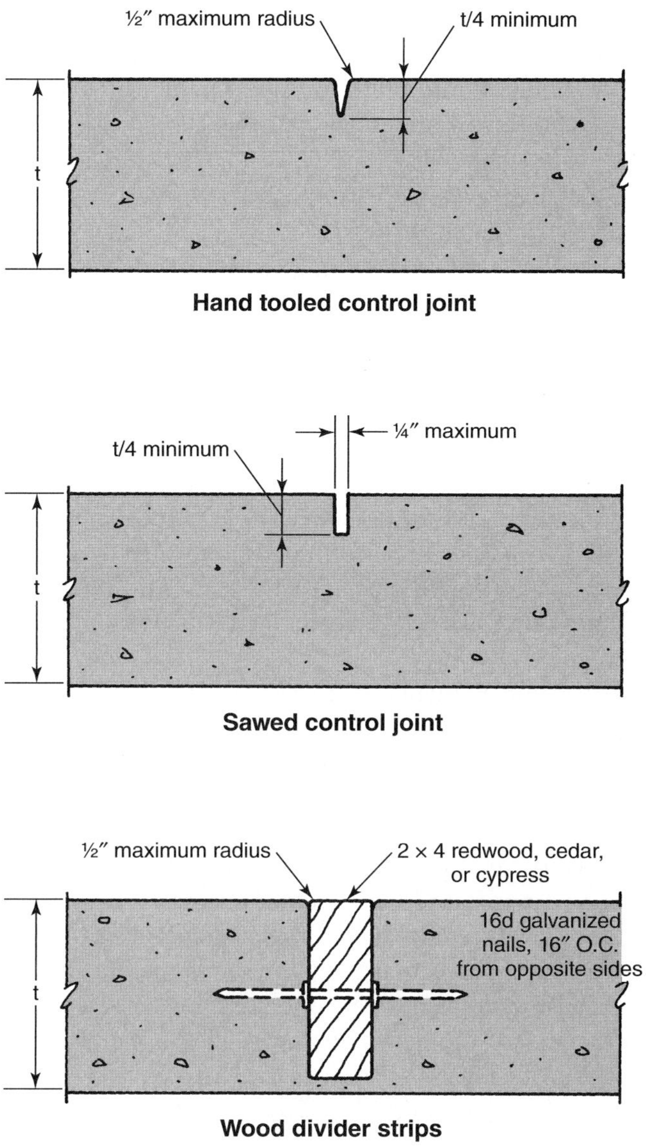

Goodheart-Willcox Publisher

Figure 20-33. These types of control joints are used in sidewalks, drives, and patios.

- **Construction joints.** Construction joints provide for no movement across the joint. They are only stopping places in the process of casting. In a construction joint, the two surfaces are not allowed to bond together. This can be accomplished by applying oil, paint, or curing compound to one surface of the joint to allow the two surfaces to move independently. Construction joints, however, may be made to perform as control joints. See **Figure 20-34**.

Sawing Control Joints in Concrete

Sawing concrete can happen between 6 to 18 hours after it is placed. If the weather is particularly hot, then sawing can begin as early as 4 hours. A practice cut should be performed to determine if the concrete slab is ready to saw. The saw cut should look crisp and clean. If cut too early, the cut will be raveled or muddy. If cut too late, shrinkage cracks may appear in the surface of the concrete. The timing of the cuts will depend on the outside temperature and wind conditions. Hot temperatures and strong winds can begin to dry the surface of the concrete and create shrinkage cracks.

When determining the layout for joints or cut patterns on concrete, it is important to look at the slab or sidewalk to determine the weakest or most likely location the concrete will crack. See **Figure 20-35**. Concrete tends to crack at the inside corners. Concrete cracking can be caused by numerous things; but a common cause is stress placed upon the slab because of the freeze thaw cycles and the movement of the earth under and surrounding the concrete. Another cracking cause can be attributed to poor base compaction under the concrete slab. Cracks do not look pretty but cut joints can be attractive if done correctly. See **Figure 20-36**. After the process of cutting has occurred, joints should be filled with a sealant manufactured to caulk or seal the concrete joints. Sealing joints serve two purposes, to keep

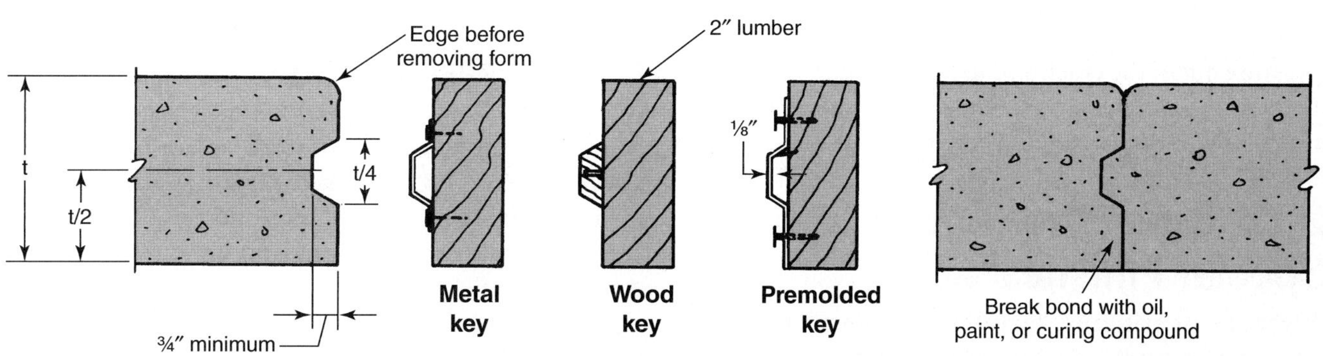

Goodheart-Willcox Publisher

Figure 20-34. Details of a construction joint. The wood key can be made from a beveled 1″ × 2″ strip and is used for slabs 4″ to 6″ thick. The premolded key can be left in the slab permanently by tacking it lightly to the bulkhead.

Timothy L. Andera

Figure 20-35. A sidewalk crack that appeared shortly after the concrete was placed. The sidewalk should have been cut diagonally to release stresses that caused the crack. When this crack occurred, the concrete was less than a year old and it cracked during the first winter season.

Timothy L. Andera

Figure 20-36. An example of control joints being sawed in a concrete sidewalk. The longitudinal joint has been sawed first. Only one diagonal joint has been sawed, but the remaining joints will be cut shortly. This sidewalk is 16′ wide and will be cut into sections of 8′ × 8′ squares to reduce random cracks.

moisture out of the crack and to fill the joints so foreign materials do not collect and increase the deterioration of the joint and surrounding concrete.

20.4 Decorative and Special Finishes

A variety of patterns, textures, and decorative finishes can be built into concrete during construction. Color can be added to the concrete. Aggregates can be exposed. Textured forms can be used. Concrete can be ground to produce a polished appearance. Geometric patterns can be scored or stamped into the concrete to resemble stone, brick, or tile. Divider strips can be used to form interesting patterns. The possibilities are endless.

Colored Concrete

Concrete can be colored using any of three methods—one-course, two-course, and dry-shake.

The first two methods are similar. In both, the concrete mix is colored by adding a mineral oxide pigment prepared especially for use in concrete. White Portland cement produces brighter colors or light pastel shades when used with light-colored sand. Normal gray cement can be used for black or dark gray colors.

All materials must be accurately controlled by weight to attain a uniform color in each batch. The amount of pigment should never exceed 10% of the weight of the cement.

Mix color pigments with the cement while dry. Dry mixing of cement and color pigments ensures that the color mixes uniformly in the cement prior to other ingredients being added to the mix. Use clean tools and a separate mixer to prevent streaking. When using the one-course method, uniform moistening of the subgrade is important for good color results.

The only difference between the one-course and two-course methods is that the two-course method uses a base coat of conventional concrete. The surface of the base coat is left rough to produce a good bond. The top coat may be placed as soon as the surface water disappears. The top coat of colored concrete is generally 1/2″ to 1″ thick.

In the dry-shake method, a commercially prepared dry color material is applied over the concrete surface after floating, edging, and grooving. Apply two coats of the dry-shake to the surface of the slab. Perform the finishing operations after each application of dry-shake. The color must be thoroughly worked into the concrete.

Exposed Aggregate

A popular decorative concrete finish is the exposed aggregate finish, which provides an unlimited color selection and a broad range of textures. Not only are exposed aggregate finishes attractive, but they can be rugged, they are slip resistant, and they are immune to weather. See **Figure 20-37**.

Exposed aggregate finishes can be produced in several ways. One of the most common is called the *seeding method*. The procedure is to place, screed, and bull float or darby the concrete in the usual manner,

Green Note

Concrete Sealers

Concrete sealers protect concrete from surface damage, staining, and corrosion. Some of these sealers are considered "green" but many are not. Many types of sealers release volatile organic compounds (VOCs) as vapors into the atmosphere. Toxic chemicals in sealers can produce adverse health effects in humans and affect wildlife when used outdoors. Water-based sealers are considered environmentally safe because they do not contain VOCs.

Local, state, and federal governments are increasing regulation of VOC emissions. To check whether a product meets VOC guidelines, consult your regional EPA office or local government environmental office to learn the VOC limits in your area.

Manufacturers are reformulating their products to conform to new limits. A common change is a reduction in xylene, which produces VOCs, and an increase in acetone, a non-VOC producer. The performance of sealers and coatings is not affected by this change, but other differences may be noted—difficulty in rolling out, faster drying times, surface blistering, and an increase in cost. Low-VOC coatings often require a change in application techniques.

keeping the level of the surface about 3/8″ to 1/2″ lower than the forms. This space will be filled with the extra aggregate.

When these finishing operations have been completed, spread the aggregate uniformly with a shovel or by hand so the entire surface is completely covered with a layer of stone. Next, embed the aggregate by tapping with a wood hand float or straightedge. Work the surface with the hand float until the surface is similar to that of a normal slab after floating.

Starting the next operation requires accurate timing. When the slab can bear the weight of a worker on kneeboards with no indentations, brush the slab with a stiff nylon bristle broom to remove the extra mortar over the stones. Next, apply a fine spray of water along with brushing. If the aggregate becomes dislodged, stop the operation for a while. Continue washing and brushing until the water is clear and there is no noticeable cement film left on the aggregate. A surface retarder can be used for better control of the exposing operations, but is not necessary.

Another method of producing an exposed aggregate finish is to expose the stone in conventional concrete. No extra stone is added, but a high proportion of coarse to fine aggregate is necessary. The coarse aggregate should be uniform in size, bright in color, closely

A Kathie Nichols/Shutterstock.com

B Wilm Ihlenfeld/Shutterstock.com

Figure 20-37. Exposed aggregate finishes are attractive and slip-resistant. A—Exposed aggregate finishes provide a wide range of colors and textures. B—An attractive walkway of concrete with exposed aggregate has been created for a garden. To accentuate the exposed aggregate, a course of pavers was placed around the outer edge of the wandering walkway slab.

packed, and uniformly distributed. The slump of this concrete must be between 1″ and 3″.

A third method is to place a thin topping course of concrete containing special aggregates over a base of regular concrete. This technique is used for terrazzo construction. Terrazzo toppings on outdoor slabs are generally 1/2″ thick and contain marble, quartz, or granite chips. Random cracking is eliminated by using brass or plastic divider strips set in a bed of mortar. This type of terrazzo is called *rustic terrazzo* or *washed terrazzo*.

Textured Finishes

A wide variety of interesting decorative textures can be produced on vertical concrete surfaces by using textured form materials. See **Figure 20-38**. Special procedures may be required for these finishes.

Goodheart-Willcox Publisher

Figure 20-38. Cast concrete textured to imitate stone.

Bomanite Corporation

Figure 20-39. These colored concrete slabs have been scored to give the appearance of terra cotta tile.

Textured surfaces on slabs can be achieved by brooming and the use of mortar, dash coat, or rock salt, or other materials. Brooming can be executed in a fine or coarse straight line pattern or a wavy texture. The procedure for brooming a surface was discussed earlier in this chapter.

A travertine finish, sometimes called a *keystone finish,* is produced by applying a dash coat of mortar over freshly leveled concrete. The dash coat is mixed to the consistency of thick paint and usually contains a yellow pigment. It is applied in a splotchy manner with a dash brush. Numerous ridges and depressions are formed by this procedure.

After the coating has hardened a little, the surface is troweled slightly to flatten the ridges and spread the mortar. The resulting finish is smooth on the high spots and coarse grained in the low areas. This effect resembles travertine marble, from which it gets its name.

Another texture can be produced by scattering rock salt over the surface after hand floating or troweling. The salt is pressed into the surface so the top of each grain is exposed. When the concrete has hardened, the surface is washed and brushed. The salt pellets dissolve, leaving holes in the surface.

Trade Tip

Neither the rock salt nor travertine finish is recommended for areas that experience freezing weather. Water frozen in the recesses ruins the surface.

Geometric Patterns

A variety of geometric patterns can be stamped, sawed, or scored into a concrete surface to enhance the beauty of walks, drives, or patios. See **Figure 20-39**. Random flagstone or ashlar patterns are popular. They may be produced using a piece of 1/2″ or 3/4″ copper pipe bent into a flat S-shape to score the surface. Scoring must be done before the concrete becomes too hard to push the coarse aggregate aside. The best time is just after darbying or bull floating. After hand floating, a second scoring is required to smooth the joints.

Patterns such as stone, brick, or tile can be cut into partially set concrete with special stamping tools. Color can also be added to create varying effects. See **Figure 20-40**.

Other patterns can be created using divider strips of wood, plastic, metal, or masonry units. These divider strips help to create interest. They also aid in the placement of concrete, provide for combinations of various surface finishes, and greatly reduce random cracking.

Keith Muratori/Shutterstock.com

Figure 20-40. A stamped concrete pool patio.

Wood divider strips should be made from pressure-treated lumber, redwood, or cypress.

Stamping of concrete usually begins after screeding and finishing have been completed. The following procedure provides the general steps for stamping concrete. Consult the stamping equipment manufacturer's specifications for detailed instructions.

PROCEDURE

Stamping Concrete

1. Apply color hardener on the concrete surface, following the manufacturer's procedures for coloring.
2. Apply the release agent on the concrete surface, following the manufacturer's specifications.
3. Check the surface to determine if the concrete is ready for stamping.
4. Begin stamping in a corner, then work around the perimeter of the slab. The starting point is usually where the concrete was first placed in the forms.
5. Once the perimeter is completed, go back to the corner to begin the first row of stamping.
6. Continue to stamp the rows of concrete until the slab is completed.
7. Perform detailing work, completing joints by hand where the stamping forms cannot access or when the coursing begins to meet the perimeter stamping.
8. Remove residual concrete, extra release agent or curing compound, and any other foreign materials from the slab.
9. Install control joints. Cutting with a power saw usually works best. Cut to a depth of 1/4 the thickness of the slab.
10. Seal the surface of the concrete according to the manufacturer's specifications.

Nonslip and Sparkling Finishes

Nonslip finishes can be applied to surfaces that are frequently wet or are especially dangerous if slippery. Abrasive grains can be dry-shaken on the surface and lightly troweled. The two most widely used abrasive grains are silicon carbide and aluminum oxide. The silicon carbide grains are sparkling black in color and are also used to make "sparkling concrete." Aluminum oxide is usually gray, brown, or white and does not sparkle. The grains should be spread uniformly over the surface. Use from 1/4 lb to 1/2 lb of grains per sq ft of slab surface.

Combination Finishes

Concrete is a versatile material. It can be applied in many ways to create a beautiful walk, drive, or patio to complement the style of any architectural design. See **Figure 20-41**.

Imagenet/Shutterstock.com

Figure 20-41. The concrete driveway has been stamped and colored to complement the architectural design of this home.

Striking effects can be attained by combining several colors, textures, and patterns in concrete. Alternate areas of exposed aggregate and plain or colored concrete provide an exciting combination. Ribbons and borders of masonry or wood create a dramatic effect with plain concrete or exposed aggregate surfaces.

20.5 Cast-in-Place Concrete Walls

Cast-in-place (CIP) concrete walls are made with ready-mix concrete that is transported to the jobsite and placed into forms that are erected on the site. Cast-in-place concrete permits structures of all shapes, sizes, and heights. Exterior wall surfaces can be rough or smooth, natural or colored. Popular types of cast-in-place (CIP) concrete walls include rustic, grid-patterned, colored aggregate, window walls, and sculptured.

Rustic Concrete Walls

Rustic concrete walls can be produced by using rough form boards, bush hammering the surface, or casting vertical fins in the concrete surface.

Trade Tip

Bush hammering exposes the aggregate that lies just beneath the surface of the concrete. The process is usually performed with a special hammering tool that removes the outer surface of the concrete to expose the aggregate, leaving a rough exterior texture to the concrete. Bush hammering is usually performed three to five weeks after the concrete is poured to ensure that the surface is adequately hardened.

Forms made from rough-sawn boards have been used for years to produce a textured surface in concrete walls. See **Figure 20-42**.

Concrete faithfully reproduces the wood texture, and the rough board appearance tends to hide tie rod holes and other imperfections. Use one of the following methods to ensure uniformity of surface texture:

- Use lumber of the same type throughout.
- Use a form coating to seal the surfaces of the boards.
- Bush hammering produces a coarse concrete texture. This is a method of mechanically breaking away the wall surface of hardened concrete to expose coarse aggregate. Interesting color variations and surface textures are produced by bush hammering.

Vertical fins or ribs can produce shadow effects on a concrete wall. See **Figure 20-43**. The fins can be smooth, sandblasted, or hammered. Inserts of wood, metal, or plastic can be used to create the ribs.

Timothy L. Andera

Figure 20-43. Fins or ribs on concrete add interest to the design of a structure.

Grid-Patterned Concrete Walls

Well-planned joint patterns provide a low-cost architectural treatment for cast-in-place walls or other structural elements. Construction joints can be inconspicuous or hidden by rustication strips. These strips produce grooves in the concrete surface, which add something to the overall architectural effect and serve as control joints.

LePetitJournal26/Shutterstock.com

Figure 20-42. This concrete wall texture was the result of using rough sawn form boards.

Removable architectural ties can be used to provide minimum-size, easily patched holes, or the tie holes can be accentuated to reduce cost and enhance the appearance. When form joints and tie holes are left unfinished, they are placed at predetermined locations and the pattern is repeated throughout the structure to create an architectural effect. See **Figure 20-44**.

Colored Aggregate Surfaces

One of the best methods of obtaining color in a cast-in-place concrete wall is by using exposed aggregate. A large percentage of coarse aggregate is used in the mix. The surface is sandblasted, bush hammered, or chisel-textured to expose the colored aggregate.

Another method of producing exposed aggregate cast-in-place walls is to preplace dry aggregate in the form and then grout under pressure with a cement-sand-water slurry. A hole near the bottom of the form is used for pumping in the grout and an external vibrator

KPG_Payless/Shutterstock.com

Figure 20-44. Tie holes form a grid pattern in the structure and are left exposed as part of the surface texture.

is used. The aggregate is exposed by sandblasting two to seven days after grouting.

An alternate, patented method is to wrap wire mesh around the reinforcement and preplace aggregate between the mesh and the outside form. A special concrete mix is cast into the core of the mesh and dispensed to the outside.

Cast-in-Place Window Walls

Repetition of window openings makes an attractive design in large structures such as the one shown in **Figure 20-45**. White Portland cement concrete is well suited for window walls because the color is permanent. Any coarse aggregate can be used, but it should be reasonably uniform in color. White aggregate is preferred if maximum whiteness is desired.

Reinforced plastic forms provide a concrete surface that is smooth and hard with few air voids and defects. The plastic forms are reinforced with fiberglass and are usually from 3/16″ to 5/8″ thick. Normal

S.Tatiana/Shutterstock.com

Figure 20-45. Unique and futuristic designs can be achieved using concrete as a building material. White Portland cement concrete and white aggregates were used to create this stylish structure.

thickness of wall forms is 3/8″. Column forms are usually 1/2″ thick.

Sculptured Concrete Walls

Sculptured concrete provides aesthetic qualities. See **Figure 20-46**. A wide range of materials can be used to form decorative patterns in concrete walls. A few such materials are stone, wood, corrugated metal sheets, plastic form liners, fiberglass sheets, formed plastic, and tempered fiberboard. Plaster waste molds can be used for fine sculpturing.

20.6 Cast-in-Place Concrete Roof and Floor Systems

There are four basic cast-in-place concrete roof and floor structural systems. They are pan joists, waffles, flat plates, and flat slabs.

Alisa_Ch/Shutterstock.com

Figure 20-46. Sculptured concrete walls are used as part of a waterfall in this stunning memorial.

Pan Joist Roof and Floor System

Pan joist construction is a one-way structural system using a ribbed slab formed with pans. See **Figure 20-47**. This system is economical because the standard forming pans may be reused. Standard pan forms produce inside dimensions of 20″ to 30″ and depths from 6″ to 20″.

Waffle Roof and Floor System

Waffle pans or forming domes are normally used to form the ceiling between levels of one floor and another. The pans or domes are placed between structural members to form the ceiling in the lower level and the floor directly above. Concrete is placed on top of the pans or domes. After the concrete cures, the pans or domes are removed. Waffle and forming domes are available in standard sizes but may be custom-made for a particular job. Like pan joist construction, the forms can be reused. **Figure 20-48** shows a structure built with the waffle technique.

Standard 30″ × 30″ square domes have a depth of 8″, 10″, 12″, 14″, 16″, or 20″. They have 3″ flanges that provide for 6″ wide joist ribs on 36″ centers. Standard domes 19″ × 19″ square have a depth of 6″, 8″, 10″, 12″, or 14″ and form 5″ wide joist ribs on 24″ centers using 2 1/2″ flanges. Spans from 25′ to 60′ are possible using waffles.

Goodheart-Willcox Publisher

Figure 20-48. Waffle plate construction used to form the roof slab. Spans of up to 60′ are possible.

Flat Plate Roof and Floor System

The main features of a flat plate system are minimum depth and architectural simplicity. See **Figure 20-49**. A flat plate is a two-way reinforced concrete framing system that uses the simplest structural shape—a slab of uniform thickness.

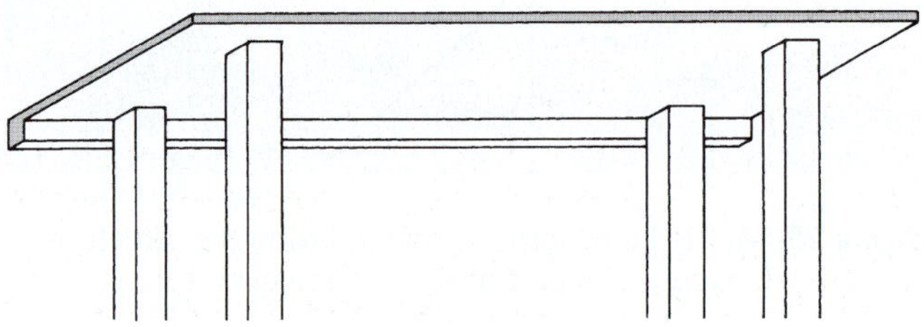

Goodheart-Willcox Publisher

Figure 20-49. A flat plate roof is a two-way reinforced concrete framing system having a slab of uniform thickness.

The flat ceiling is economical to form and may be used for the finished ceiling without any additional treatment. Cantilevers (beams supported at only one end that carry a load at the other end or along their length) are easily produced, as well as other architectural projections. Slabs generally range from 5″ to 14″ thick and provide spans up to 35′. Flat plates provide a continuous solid ceiling with complete flexibility for locating partitions and mechanical equipment. Supporting columns need not be in a straight line. This adds flexibility. Electrical ducts and conduits may be embedded in the slab. The flat plate system is well suited for heavy loads such as roof parking.

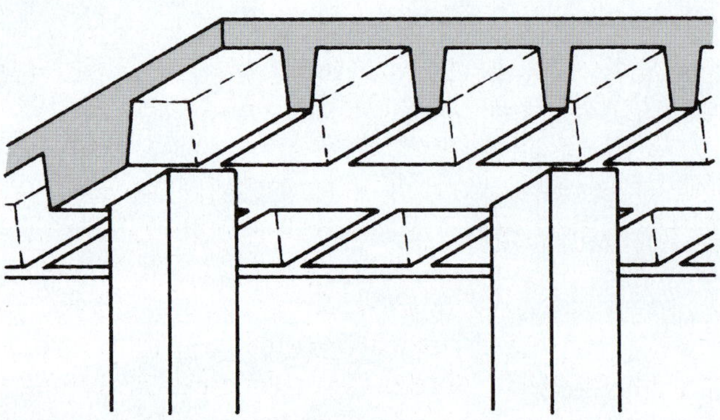

Goodheart-Willcox Publisher

Figure 20-47. Pan joist roof construction is a one-way structural system using a ribbed slab formed with pans. Spans of up to 50′ are common in this type of construction.

Flat Slab Roof and Floor System

The flat slab system is designed for heavy roof loads with large open bays below. The difference between the flat plate and the flat slab systems is the flat slab has a supporting panel in the area of each column for added support. See **Figure 20-50**.

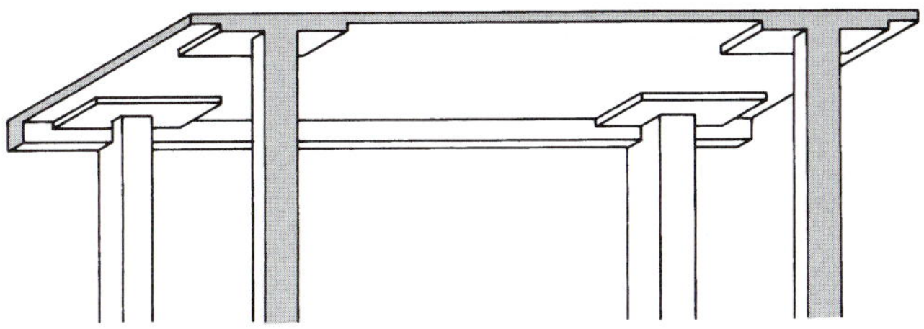

Goodheart-Willcox Publisher

Figure 20-50. A flat slab roof is a two-way reinforced structural system that includes either drop panels or column capitals to carry heavier loads.

Flat slab thickness is usually 2.5% or 3% of the span and a minimum of 4″ thick. The size of the supporting panel is about 33% of the span and 25% to 50% of the slab thickness. Flat slabs are designed to span up to 40′ with columns an equal distance apart. This system has all the advantages of the flat plate system and is stronger.

20.7 Precast/Prestressed Concrete Systems

Expansion of industrial processes in building construction has resulted in greater use of modular layout, planning, and cost-conscious construction. Precast units for walls, floors, ceilings, and roofs can be mass-produced at the factory or job site. Precast concrete units can be cast as tilt-up panels, standard-shaped concrete panels, or concrete window walls. A variety of prestressed panels and shapes are produced for use in wall, roof, and floor systems.

Only trained and competent individuals are involved in the unloading, storing, placing, erecting, and bracing of concrete panels. OSHA safety standards pertaining to lifting of precast and tilt-up panels include the following:

- The lifting of panels must be designed and planned by a registered engineer who has knowledge and experience in lifting of slabs. Designated panel lifting procedures are supplied by the manufacturer prior to their delivery to the construction site.
- Slab weight must be known to ensure adequate lifting equipment is used.
- A crane or other lifting equipment must be rated to lift 2 1/2 times the load to be lifted.
- Only a designated competent person is allowed to rig the slab for lifting.
- Only essential individuals are allowed in the lifting area during lifting and placing of the slab.
- No one is allowed under the load during lifting or placing of the slab.
- The height of wall plus 4′ must be cordoned off as clear space around the perimeter in the event of a wall tipping over.
- Bracing of walls must be retained until permanent supporting members of the structure are in place.

Tilt-Up Construction

Tilt-up construction is the process of casting concrete walls in a horizontal position on-site or in another location and lifting them from the casting position or truck to their final location in the building. Tilt-up construction is a popular construction method in the United States due to the reasonable cost, low maintenance, durability, and speed of construction.

Tilt-up construction is especially suited for buildings greater than 10,000 sq ft with 20′ or higher side walls that incorporate repetition in panel size and appearance. See **Figure 20-51.**

Panel connections to the footings, floors, roof, and between panels must be determined before construction. See **Figure 20-52**. Lifting loads are a major consideration because they generally have the most influence on the panels. Reinforcement must be designed to accommodate these loads.

A drawing of each panel should be made showing the front and back as well as inserts and embeds. See **Figure 20-53**. Inserts provide the attachment points for lifting hardware and braces. Embeds are prefabricated steel plates with lugs that are cast into the panel to attach it to the footing, other panels, or roof system.

Spread footings are used for most tilt-up buildings, but pier footings can be used. See **Figure 20-54**. The footing must be straight and smooth if the panels are to fit properly. Footing size depends on the soil bearing capacity, weight to be supported, and other factors specific to the site and building.

Tilt-up panels can be cast as individual panels or as larger slabs that are sawed into smaller panels after the concrete has been placed. Tilt-up panels have a grid of reinforcing steel embedded within the panel to provide the necessary strength. Standard Grade 40 or 60 bars are used. Plastic chairs are used to support rebars instead of steel chairs to avoid rust on the panel face. Concrete must meet all mix specifications for the application.

Many panels are cast with a textured surface or pattern. The surface treatment must be planned carefully, but the results are impressive. Other finishes such as exposed aggregate, sandblasting, and brick or stone facings are commonly used with tilt-up panels to "dress up" the building.

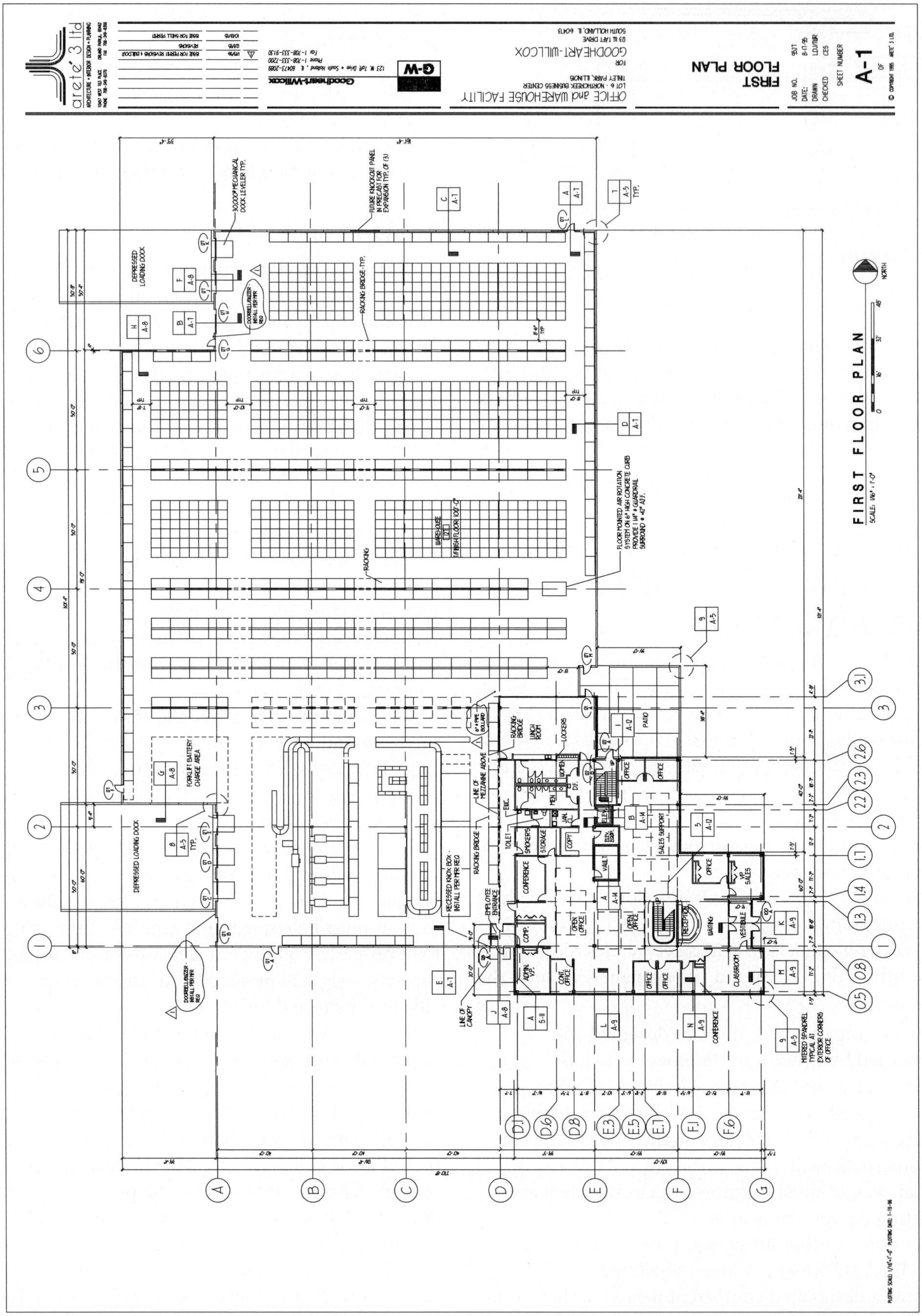

Areté 3 Ltd.

Figure 20-51. First floor plan of a warehouse and office building constructed using tilt-up panels.

Goodheart-Willcox Publisher

Figure 20-52. This metal connection will be used to attach a panel to the footing.

Goodheart-Willcox Publisher

Figure 20-54. Typical footing and foundation wall used in tilt-up construction.

The erection process should be well-planned for efficiency and safety. Braces should be attached before the panels are lifted and not removed until the roof and decking are installed. See **Figure 20-55**. Installing braces once the panels are in the vertical position is difficult and dangerous. The result of a well-planned erection process is a spectacular finished structure.

Tilt-up panels are set in place similar to erecting building blocks that were played with as children except on a larger scale. As panels are set, depending on the

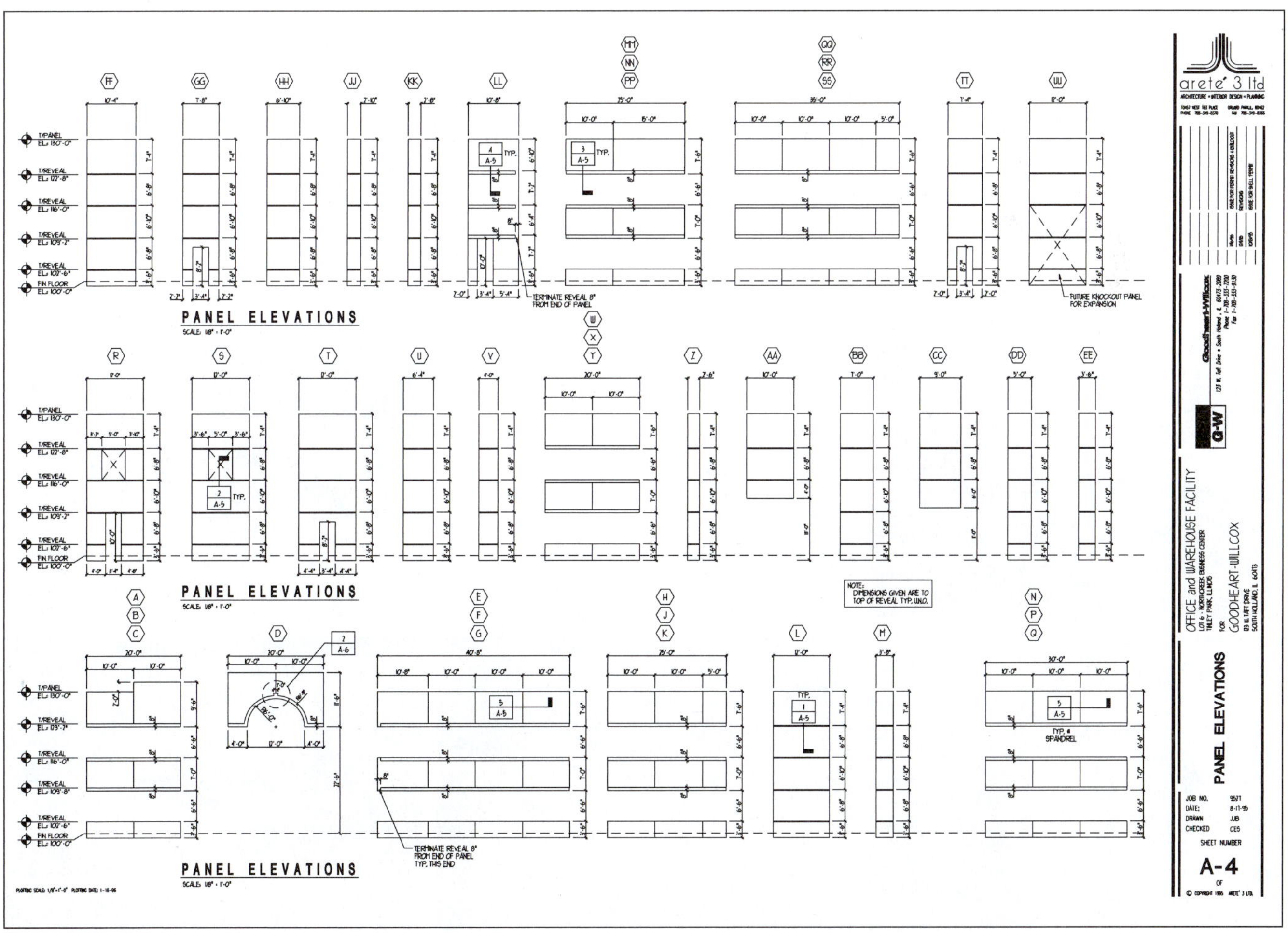

Areté 3 Ltd.

Figure 20-53. An elevation drawing of each panel type used in the construction of the building shown in Figure 20-56.

Goodheart-Willcox Publisher

Figure 20-55. Proper bracing of tilt-up panels is important for both safety and efficient erection of the structure. The diagonal braces attached to the wall panels are not removed until the roof decking is installed to tie the structure together.

type, a certain amount of space is left between the panels for insulating and sealing. As panels are erected, they need to be braced similar to concrete block walls under construction. See **Figure 20-56**. Once wall panels are in place and braced, construction can begin on the roof.

Prestressed Panels

Standard-shaped concrete panels that are prestressed have brought a new concept to architectural precast concrete. These units are crack-free and highly resistant to deterioration. Common designs include

Goodheart-Willcox Publisher

Figure 20-56. Crane lifting panels in place. As more panels are erected, the final building structure becomes more identifiable.

double-tee units, single-tee units, and hollow-core panels. **Figure 20-57** shows a structure with double-tee roof units.

Double-tees are the most widely used prestressed, precast units. Widths are generally 4′, 5′, 6′, 8′, and 10′. The 4′ and 6′ widths are the most popular sizes. Depths range from 6″ to 36″. Spans up to 60′ are possible with the 4′ wide unit. Greater spans are possible with wider units since they usually have a deeper web.

Single-tee units are generally used where long spans, beginning at about 30′, are needed. They can be placed flange to flange, as shown in **Figure 20-58**, or spread apart with concrete planks or cast-in-place concrete completing the enclosure. Flange widths range from 4′ to 10′ with depths from 12″ to over 36″ in 4″ increments. Spans of 30′ to 100′ are common with single-tee units.

Gage Brothers in Sioux Falls, SD

Figure 20-57. Double-tee concrete units are the most widely used prestressed concrete product for medium range spans. Four-foot wide double-tees are commonly used for spans up to 60′.

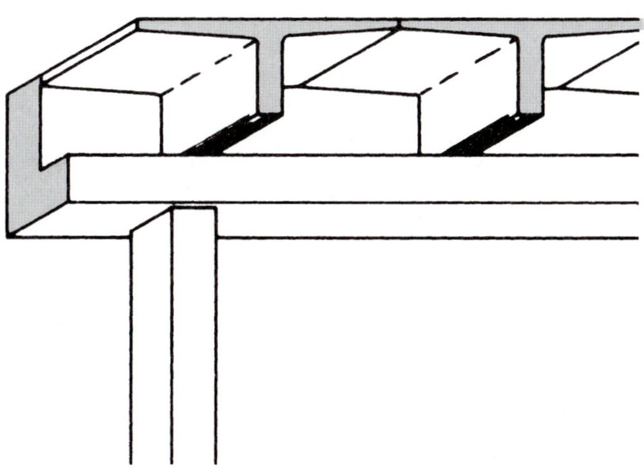

Goodheart-Willcox Publisher

Figure 20-58. Single-tee precast, prestressed concrete units are generally used for very long spans. Typical spans are from 30′ to 100′.

Cored slabs provide a flush ceiling with minimum depth required for the roof or floor system. See **Figure 20-59**. Slab thicknesses range from 4″ to 12″ in increments of 2″. A variety of widths are available, including 1′-4″, 1′-8″, 2′-0″, 3′-4″, and 4′-0″. Not all sizes are available in a single core style. Lengths generally range from 12′ to 40′, but the length of a particular slab depends on the thickness, width, and reinforcing. Generally, the thicker the slab, the longer the length available.

Precast concrete window walls may be cast as curtain walls or loadbearing walls. Forms or molds used to produce complicated designs are made from plastic, wood, or steel. These forms are custom units and must be designed by architects or engineers.

Typically, fiberglass molds are used for smooth concrete with sculptured mullions (wall members supporting a window). Precast window walls or wall openings can be one-story or multistory. See **Figure 20-60**. Precast window walls can even be preglazed (the glass installed) at the plant.

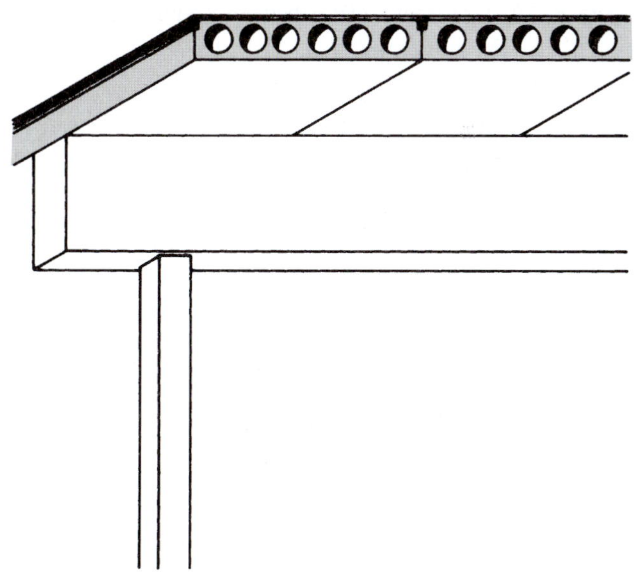

Goodheart-Willcox Publisher

Figure 20-59. Hollow-core slabs are low in cost and widely produced. Depths range from 4″ to 12″. Widths are available from 16″ to 48″. Spans possible with cored slabs are up to 50′, but generally under 35′.

Kawneer Company, Inc.

Figure 20-60. A precast panel is being raised by a crane to be positioned in its final location in the structure.

Chapter Review

Summary

- To receive concrete, forms should be prepared, leveled, and adequately braced. Subgrades should be smooth, compacted, and moist.
- If a form is deep, the concrete should be compacted or vibrated after each layer is placed.
- After concrete is placed, it is screeded. Excess concrete is struck off to bring the top surface to the correct grade or elevation.
- Floating of concrete should follow immediately after screeding to prevent bleeding.
- A jointing tool is used to cut a control joint or contraction joint in the concrete. Cracking due to shrinkage is intended to occur at the joint.
- Floating embeds aggregate particles just beneath the surface of the concrete, removes imperfections, and compacts the concrete at the surface in preparation for other finishing operations.
- When a smooth, dense, hard surface is desired, steel troweling is performed after floating. Brooming produces a slightly roughened surface.
- The temperature of concrete at the time of placing should generally be 50°F to 70°F (10°C to 21°C).
- Air-entrained concrete is not as susceptible to damage caused by freezing and thawing as standard concrete.
- Three basic joints used in concrete construction are isolation joints, control joints, and construction joints.
- Decorative and special concrete finishes include colored concrete, exposed aggregate, textured finishes, geometric patterns, nonslip and sparkling finishes, and combination finishes.
- Exposed aggregate finishes can be created by using the seeding method (keeping the level of the concrete surface about 3/8″ to 1/2″ lower than the forms, then filling the space with the extra aggregate), exposing the stone in conventional concrete, or placing a thin topping course of concrete containing special aggregates over a base of regular concrete.
- Popular types of cast-in-place (CIP) concrete walls include rustic, grid-patterned, colored aggregate, window walls, and sculptured.
- Four basic cast-in-place concrete roof and floor structural systems are pan joists, waffles, flat plates, and flat slabs.
- In tilt-up construction, concrete walls are cast in a horizontal position on-site or in another location and lifted from the casting position or truck to their final location in the building.

Know and Understand

Answer the following questions using the information provided in this chapter.

1. Which of the following does *not* describe the ideal preparation of a subgrade before receiving concrete?
   A. Dry
   B. Compacted
   C. Smooth
   D. Moist
2. *True or False?* Dowels are used to help prevent moisture from migrating from the subgrade to the concrete.
3. _____ is the striking off of excess concrete to bring the top surface to the proper grade or elevation.
   A. Floating
   B. Brooming
   C. Edging
   D. Screeding
4. _____ is a method of curing that involves keeping an inch of water on the concrete surface.
   A. Brooming
   B. Screeding
   C. Ponding
   D. Subgrading

5. *True or False?* The most ideal weather to place concrete is hot, dry, windy weather.
6. _____ joints are also called expansion joints.
   A. Isolation
   B. Control
   C. Contraction
   D. Construction
7. *True or False?* Control joints provide for no movement across the joint.
8. Textured surfaces on slabs can be achieved by _____.
   A. brooming
   B. screeding
   C. ponding
   D. subgrading
9. *True or False?* Stamping of concrete usually begins after screeding and finishing have been completed.
10. Which of the following CIP concrete walls reproduce the wood texture?
    A. Grid-patterned
    B. Rustic
    C. Window
    D. Sculptured
11. *True or False?* Aggregates are the best methods for creating CIP window walls.
12. An economical construction method using a ribbed slab is the _____ system.
    A. flat slab
    B. waffle pan
    C. pan joist
    D. flat plate
13. *True or False?* The main features of a waffle pan system are minimum depth and architectural simplicity.
14. _____ is especially suited for buildings greater than 10,000 sq ft with 20′ or higher side walls that incorporate repetition in panel size and appearance.
    A. Concrete window wall construction
    B. Standard-shaped concrete panel construction
    C. Prestressed panel construction
    D. Tilt-up construction
15. Double-tees are the most widely used _____ units.
    A. concrete window wall
    B. standard-shaped concrete panel
    C. prestressed panel
    D. tilt-up

Apply and Analyze

1. Why should the subgrade be moistened before placing concrete?
2. Why are forms treated with oil or other preparation?
3. What problems may result from performing finishing operations on a concrete slab while the bleed water is present?
4. List three reasons why concrete is floated.
5. When may forms be removed?
6. How is air-entrained concrete finished differently than normal concrete?
7. What is the most important consideration in curing concrete properly?
8. Why should concrete *not* be placed during hot, dry, windy weather?
9. What is the purpose of isolation (expansion) joints?
10. Identify three decorative and special finishes that are used on concrete.
11. In addition to providing an attractive finish, what are three other attributes of exposed aggregate surfaces?
12. List the four basic cast-in-place concrete roof and floor structural systems.
13. Identify the three most common designs of prestressed panels.

Critical Thinking

1. You are the foreman of a concrete crew that will be installing a parking lot during late summer. Identify and explain three things that can be done to reduce the effect of hot weather on placing and curing concrete.

Activities

1. (TECH) Investigate trends in decorative concrete. Examples include glow-in-the-dark concrete, acid-stained concrete, and metallic concrete. Prepare a visual presentation that illustrates applications for several types of decorative concrete.
2. (MATH) Determine the amount of concrete needed to place a driveway that is 5″ thick, 24′ wide, and 36′ long. Round up the answer to the closest 1/4 cu yd of concrete. One cu yd is 3′ × 3′ × 3′.

Metric Measures in Construction

Spacing of wall studs	300, 400, and 600 mm (400 mm = 15 ¾″)
Kitchen countertops	600 mm wide (600 mm = 1′-11 5/8″) 900 mm high (900 mm = 2′-11 7/16″)
Size of 2 × 4 stud	38 × 89 mm
Standard metric brick	67 × 100 × 200 mm (100 mm = 3 15/16″)
Standard mortar joint	10 mm
Bag of cement	40 kg (40 kg = 88.2 lb)
Temperature for pouring concrete	Not less than 10°C nor more than 27°C
Insulation value	R20 = RSI 3.5

Goodheart-Willcox Publisher

Grade-Use Guide for Concrete Forms

Use These Terms to Specify Plywood	Description	Typical Grade Trademarks	Veneer Faces	Veneer Inner Plies
B-B Plyform Class I & II*	Specifically manufactured for concrete forms. Many reuses. Smooth, solid surfaces. Edge-sealed. Mill-oiled unless otherwise specified.	B-B PLYFORM CLASS I EXTERIOR DFPA	B	C
High density overlaid Plyform Class I & II*	Hard, semiopaque resin-fiber overlay, heat-fused to panel faces. Smooth surface resists abrasion. Up to 200 reuses. Edge-sealed. Light oiling recommended between pours.	HDO - PLYFORM- I - EXT - DFPA - PS 1 -66.	B	C plugged
Structural I Plyform*	Especially designed for engineered applications. All Group 1 species. Stronger and stiffer than Plyform Class I and II. Recommended for high pressures where face grain is paralled to supports. Also available with High Density Overlay faces.	STRUCTURAL I B-B PLYFORM CLASS I EXTERIOR DFPA	B	C or C plugged
Special overlays, proprietary panels and medium density overlaid plywood specifically designed for concrete forming*	Produce a smooth uniform concrete surface. Generally mill treated with form release agent. Check with manufacturer for design specifications, proper use, and surface treatment recommendations for greatest number of reuses.	—	—	—

Note: Commonly available in 5/8″ and 3/4″ panel thicknesses (4′ × 8′ size).
* Check dealer for availability in your area.

Goodheart-Willcox Publisher

Resistivity to Heat Loss of Common Building Materials

Size	Material	Resistivity	Size	Material	Resistivity
4″	Concrete or stone	0.32	1/2″	Plywood	0.65
6″	Concrete or stone	0.48	5/8″	Plywood	0.80
8″	Concrete or stone	0.64	3/4″	Plywood	0.95
12″	Concrete or stone	0.96	3/4″	Softwood sheathing or siding	0.85
4″	Concrete block	0.70	—	Composition floor covering	0.08
8″	Concrete block	1.10	1″	Mineral batt insulation	3.50
12″	Concrete block	1.25	2″	Mineral batt insulation	7.00
4″	Common brick	0.82	4″	Mineral batt insulation	14.00
4″	Face brick	0.45	2″	Glass fiber insulation	7.00
4″	Structural clay tile	1.10	4″	Glass fiber insulation	14.00
8″	Structural clay tile	1.90	1″	Loose fill insulation	3.00
12″	Structural clay tile	3.00	1/2″	Gypsum wallboard	0.45
1″	Stucco	0.20	1″	Expanded polystyrene, extruded	4.00
15 lb	Building paper	0.06	1″	Expanded polystyrene, molded beads	3.85
3/8″	Sheet rock or plasterboard	0.33	—	Single thickness glass	0.88
1/2″	Sand plaster	0.15	—	Glassweld insulation glass	1.89
1/2″	Insulation plaster	0.75	—	Single glass with storm window	1.66
1/2″	Fiberboard ceiling tile	1.20	—	Metal edge insulation glass	1.85
1/2″	Fiberboard sheathing	1.45	4″	Glass block	2.13
3/4″	Fiberboard sheathing	2.18	1 3/8″	Wood door	1.92
—	Roll roofing	0.15	—	Same with storm door	3.12
—	Asphalt shingles	0.16	1 3/4″	Wood door	1.82
—	Wood shingles	0.86	—	Same with storm door	2.94
—	Tile or slate	0.08	—	—	—

Goodheart-Willcox Publisher

Load Table for Standard Steel Pipe Columns

Maximum Allowable Concentric Loads											
Nominal size in inches	Weight per ft in pounds	Unbrace length in feet									
		6	7	8	9	10	11	12	14	16	18
3	7.58	38	36	34	31	28	25	22	16	12	10
3 ½	9.11	48	46	44	41	38	35	32	25	19	15
4	10.79	59	57	54	52	49	46	43	36	29	23
5	14.62	83	81	78	76	73	71	68	61	55	47
6	18.97	110	108	106	103	101	98	95	89	82	75

Loads are in kips. 1 kip = 1,000 pounds

American Institute for Steel Construction

Minimum Thickness and Maximum Span with Concrete Masonry Screen Walls

Maximum Distance between Lateral Supports (height or length, not both)					
Construction	Minimum Nominal Thickness (t)	Nominal Thickness of Wall, inches			Other Nominal Thicknesses (t)
Nonloadbearing Reinforced:*					
Exterior	4″	10′-0″	15′-0″	20′-0″	30 t
Interior	4″	16′-0″	24′-0″	32′-0″	48 t
Nonreinforced:					
Exterior	4″	6′-8″	10′-0″	13′-4″	20 t
Interior	4″	12′-0″	18′-0″	24′-0″	36 t
Loadbearing:					
Reinforced*	6″	Not recommended	12′-6″	16′-8″	25 t
Nonreinforced	6″		9′-0″	12′-0″	18 t

* Total steel area, including joint reinforcement, not less than 0.002 times the gross cross-sectional area of the wall, not more than two-thirds of which may be used on either vertical or horizontal direction.

Goodheart-Willcox Publisher

Construction Details of Screen Walls

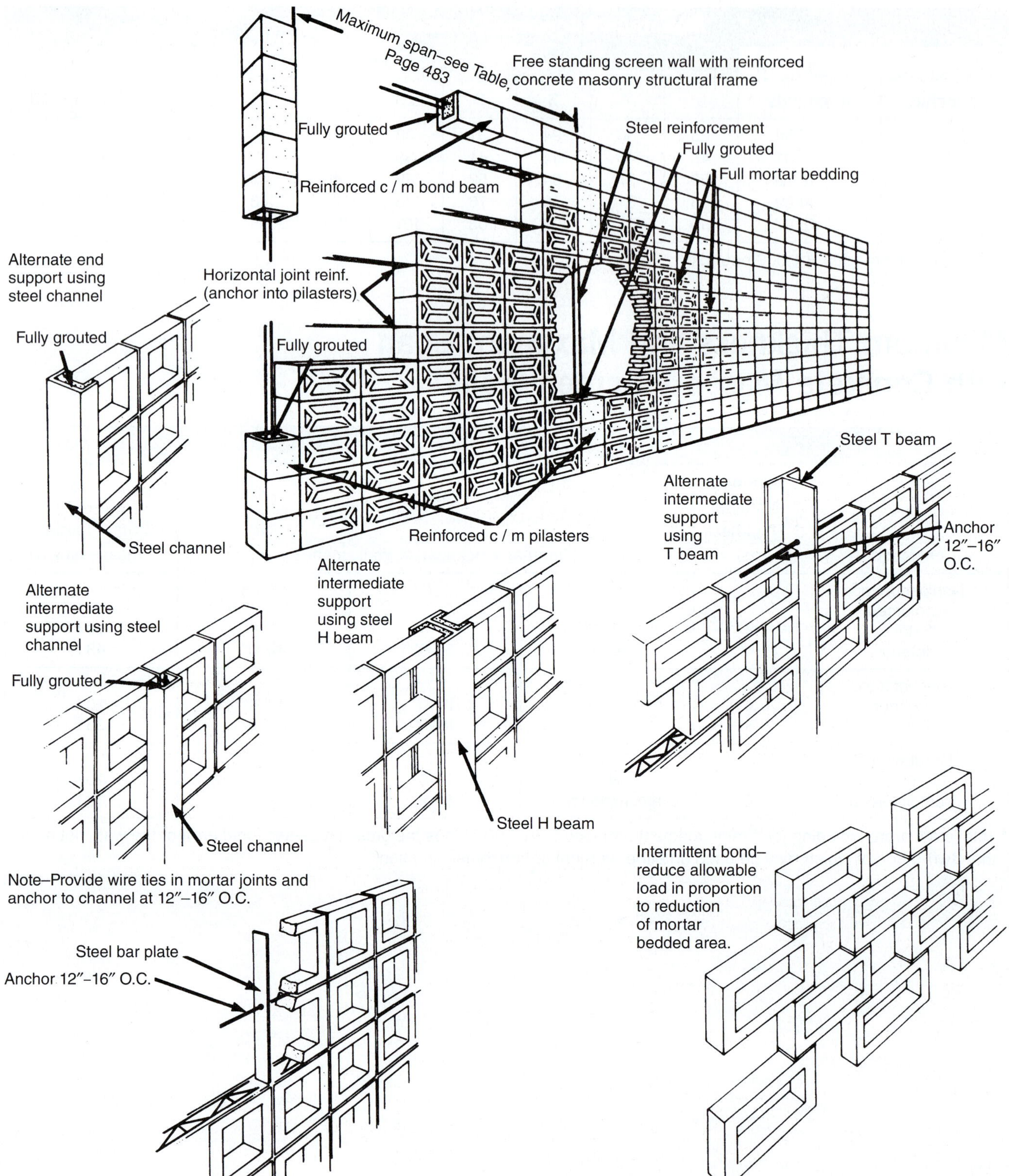

National Concrete Masonry Association

Goodheart-Willcox Publisher

Vertical Coursing Table for Nonmodular Brick[1]

	Nominal Height (h) of Unit[2]				
No. of Courses	2″	2 2/3″	3 1/5″	4″	5 1/3″
1	0′ - 2″	0′ - 2 11/16″	0′ - 3 3/16″	0′ - 4″	0′ - 5 5/16″
2	0′ - 4″	0′ - 5 5/16″	0′ - 6 3/8″	0′ - 8″	0′ - 10 11/16″
3	0′ - 6″	0′ - 8″	0′ - 9 5/8″	1′ - 0″	1′ - 4″
4	0′ - 8″	0′ - 10 11/16″	1′ - 0 3/16″	1′ - 4″	1′ - 9 5/16″
5	0′ - 10″	1′ - 1 5/16″	1′ - 4″	1′ - 8″	2′ - 2 11/16″
6	1′ - 0″	1′ - 4″	1′ - 7 3/16″	2′ - 0″	2′ - 8″
7	1′ - 2″	1′ - 6 11/16″	1′ - 10 3/8″	2′ - 4″	3′ - 1 5/16″
8	1′ - 4″	1′ - 9 5/16″	2′ - 1 5/8″	2′ - 8″	3′ - 6 11/16″
9	1′ - 6″	2′ - 0″	2′ - 4 13/16″	3′ - 0″	4′ - 0″
10	1′ - 8″	2′ - 2 11/16″	2′ - 8″	3′ - 4″	4′ - 5 5/16″
11	1′ - 10″	2′ - 5 5/16″	2′ - 11 3/16″	3′ - 8″	4′ - 10 11/16″
12	2′ - 0″	2′ - 8″	3′ - 2 3/8″	4′ - 0″	5′ - 4″
13	2′ - 2″	2′ - 10 11/16″	3′ - 5 5/8″	4′ - 4″	5′ - 9 5/16″
14	2′ - 4″	3′ - 1 5/16″	3′ - 8 13/16″	4′ - 8″	6′ - 2 11/16″
15	2′ - 6″	3′ - 4″	4′ - 0″	5′ - 0″	6′ - 8″
16	2′ - 8″	3′ - 6 11/16″	4′ - 3 3/16″	5′ - 4″	7′ - 1 5/16″
17	2′ - 10″	3′ - 9 5/16″	4′ - 6 3/8″	5′ - 8″	7′ - 6 11/16″
18	3′ - 0″	4′ - 0″	4 - 9 5/8″	6′ - 0″	8′ - 0″
19	3′ - 2″	4′ - 2 11/16″	5′ - 0 13/16″	6′ - 4″	8′ - 5 5/16″
20	3′ - 4″	4′ - 5 5/16″	5′ - 4″	6′ - 8″	8′ - 10 11/16″
21	3′ - 6″	4′ - 8″	5′ - 7 3/16″	7′ - 0″	9′ - 4″
22	3′ - 8″	4′ - 10 11/16″	5′ - 10 3/8″	7′ - 4″	9′ - 9 5/16″
23	3′ - 10″	5′ - 1 5/16″	6′ - 1 5/8″	7′ - 8″	10′ - 2 11/16″
24	4′ - 0″	5′ - 4″	6′ - 4 13/16″	8′ - 0″	10′ - 8″
25	4′ - 2″	5′ - 6 11/16″	6′ - 8″	8′ - 4″	11′ - 1 5/16″
26	4′ - 4″	5′ - 9 5/16″	6′ - 11 13/16″	8′ - 8″	11′ - 6 11/16″
27	4′ - 6″	6′ - 0″	7′ - 2 3/8″	9′ - 0″	12′ - 0″
28	4′ - 8″	6′ - 2 11/16″	7′ - 5 5/8″	9′ - 4″	12′ - 5 5/16″
29	4′ - 10″	6′ - 5 5/16″	7′ - 8 13/16″	9′ - 8″	12′ - 10 11/16″
30	5′ - 0″	6′ - 8″	8′ - 0″	10′ - 0″	13′ - 4″
31	5′ - 2″	6′ - 10 11/16″	8′ - 3 3/16″	10′ - 4″	13′ - 9 5/16″
32	5′ - 4″	7′ - 1 5/16″	8′ - 6 3/8″	10′ - 8″	14′ - 2 11/16″
33	5′ - 6″	7′ - 4″	8′ - 9 5/8″	11′ - 0″	14′ - 8″
34	5′ - 8″	7′ - 6 11/16″	9′ - 0 13/16″	11′ - 4″	15′ - 1 5/16″
35	5′ - 10″	7′ - 9 5/16″	9′ - 4″	11′ - 8″	15′ - 6 11/16″
36	6′ - 0″	8′ - 0″	9′ - 7 3/16″	12′ - 0″	16′ - 0″
37	6′ - 2″	8′ - 2 11/16″	9′ - 10 3/8″	12′ - 4″	16′ - 5 5/16″
38	6′ - 4″	8′ - 5 5/16″	10′ - 1 5/8″	12′ - 8″	16′ - 10 11/16″
39	6′ - 6″	8′ - 8″	10′ - 4 13/16″	13′ - 0″	17′ - 4″
40	6′ - 8″	8′ - 10 11/16″	10′ - 8″	13′ - 4″	17′ - 9 5/16″
41	6′ - 10″	9′ - 1 5/16″	10′ - 11 13/16″	13′ - 8″	18′ - 2 11/16″
42	7′ - 0″	9′ - 4″	11′ - 2 3/8″	14′ - 0″	18′ - 8″
43	7′ - 2″	9′ - 6 11/16″	11′ - 5 5/8″	14′ - 4″	19′ - 1 5/16″
44	7′ - 4″	9′ - 9 5/16″	11′ - 8 13/16″	14′ - 8″	19′ - 6 11/16″
45	7′ - 6″	10′ - 0″	12′ - 0″	15′ - 0″	20′ - 0″
46	7′ - 8″	10′ - 2 11/16″	12′ - 3 3/16″	15′ - 4″	20′ - 5 5/16″
47	7′ - 10″	10′ - 5 5/16″	12′ - 6 3/8″	15′ - 8″	20′ - 10 11/16″
48	8′ - 0″	10′ - 8″	12′ - 9 5/8″	16′ - 0″	21′ - 4″
49	8′ - 2″	10′ - 10 11/16″	13′ - 0 13/16″	16′ - 4″	21′ - 9 5/16″
50	8′ - 4″	11′ - 1 5/16″	13′ - 4″	16′ - 8″	22′ - 2 11/16″
100	16′ - 8″	22′ - 2 11/16″	26′ - 8″	33′ - 4″	44′ - 5 5/16″

[1] Brick positioned in wall as stretchers.
[2] For convenience in using table, nominal 1/3″, 2/3″, and 1/5″ heights of units have been changed to nearest 1/16″. Vertical dimensions are from bottom of mortar joint to bottom of mortar joint.

Goodheart-Willcox Publisher

Vertical Coursing Table for Nonmodular Brick[1] (continued)

No. of Courses	2 1/4 in High Units—3/8″ Joint	2 1/4 in High Units—1/2″ Joint	2 5/8 in High Units—3/8″ Joint	2 5/8 in High Units—1/2″ Joint	2 3/4 in High Units—3/8″ Joint	2 3/4 in High Units—1/2″ Joint
1	0′ - 2⅝″	0′ - 2¾″	0′ - 3″	0′ - 3⅛″	0′ - 3⅛″	0′ - 3¼″
2	0′ - 5¼″	0′ - 5½″	0′ - 6″	0′ - 6¼″	0′ - 6¼″	0′ - 6½″
3	0′ - 7⅝″	0′ - 8¼″	0′ - 9″	0′ - 9⅜″	0′ - 9⅜″	0′ - 9¾″
4	0′ - 10½″	0′ - 11″	1′ - 0″	1′ - 0½″	1′ - 0½″	1′ - 1″
5	1′ - 1⅛″	1′ - 1¼″	1′ - 3″	1′ - 3⅝″	1′ - 3⅝″	1′ - 4¼″
6	1′ - 3¾″	1′ - 4½″	1′ - 6″	1′ - 6¾″	1′ - 6¾″	1′ - 7½″
7	1′ - 6⅜″	1′ - 7¼″	1′ - 9″	1′ - 9⅞″	1′ - 9⅞″	1′ - 10¾″
8	1′ - 9″	1′ - 10″	2′ - 0″	2′ - 1″	2′ - 1″	2′ - 2″
9	1′ - 11⅝″	2′ - 0¾″	2′ - 3″	2′ - 4⅛″	2′ - 4⅛″	2′ - 5¼″
10	2′ - 2¼″	2′ - 3½″	2′ - 6″	2′ - 7¼″	2′ - 7¼″	2′ - 8½″
11	2′ - 4⅞″	2′ - 6¼″	2′ - 9″	2′ - 10⅜″	2′ - 10⅜″	2′ - 11¾″
12	2′ - 7½″	2′ - 9″	3′ - 0″	3′ - 1½″	3′ - 1½″	3′ - 3″
13	2′ - 10⅛″	2′ - 11¾″	3′ - 3″	3′ - 4⅝″	3′ - 4⅝″	3′ - 6¼″
14	3′ - 0¾″	3′ - 2½″	3′ - 6″	3′ - 7¾″	3′ - 7¾″	3′ - 9½″
15	3′ - 3⅜″	3′ - 5¼″	3′ - 9″	3′ - 10⅞″	3′ - 10⅞″	4′ - 0¾″
16	3′ - 6″	3′ - 8″	4′ - 0″	4′ - 2″	4′ - 2″	4′ - 4″
17	3′ - 8⅝″	3′ - 10¾″	4′ - 3″	4′ - 5⅛″	4′ - 5⅛″	4′ - 7¼″
18	3′ - 11¼″	4′ - 1½″	4′ - 6″	4′ - 8¼″	4′ - 8¼″	4′ - 10½″
19	4′ - 1⅞″	4′ - 4¼″	4′ - 9″	4′ - 11⅜″	4′ - 11⅜″	5′ - 1¾″
20	4′ - 4½″	4′ - 7″	5′ - 0″	5′ - 2½″	5′ - 2½″	5′ - 5″
21	4′ - 7⅛″	4′ - 9¾″	5′ - 3″	5′ - 5⅝″	5′ - 5⅝″	5′ - 8¼″
22	4′ - 9¾″	5′ - 0½″	5′ - 6″	5′ - 8¾″	5′ - 8¾″	5′ - 11½″
23	5′ - 0⅜″	5′ - 3¼″	5′ - 9″	5′ - 11⅞″	5′ - 11⅞″	6′ - 2¾″
24	5′ - 3″	5′ - 6″	6′ - 0″	6′ - 3″	6′ - 3″	6′ - 6″
25	5′ - 5⅝″	5′ - 8¾″	6′ - 3″	6′ - 6⅛″	6′ - 6⅛″	6′ - 9¼″
26	5′ - 8¼″	5′ - 11½″	6′ - 6″	6′ - 9¼″	6′ - 9¼″	7′ - 0½″
27	5′ - 10⅞″	6′ - 2¼″	6′ - 9″	7′ - 0⅜″	7′ - 0⅜″	7′ - 3¾″
28	6′ - 1½″	6′ - 5″	7′ - 0″	7′ - 3½″	7′ - 3½″	7′ - 7″
29	6′ - 4⅛″	6′ - 7¾″	7′ - 3″	7′ - 6⅝″	7′ - 6⅝″	7′ - 10¼″
30	6′ - 6¾″	6′ - 1½″	7′ - 6″	7′ - 9¾″	7′ - 9¾″	8′ - 1½″
31	6′ - 9⅜″	7′ - 1¼″	7′ - 9″	8′ - 0⅞″	8′ - 0⅞″	8′ - 4¾″
32	7′ - 0″	7′ - 4″	8′ - 0″	8′ - 4″	8′ - 4″	8′ - 8″
33	7′ - 2⅝″	7′ - 6¾″	8′ - 3″	8′ - 7⅛″	8′ - 7⅛″	8′ - 11¼″
34	7′ - 5¼″	7′ - 9½″	8′ - 6″	8′ - 10¼″	8′ - 10¼″	9′ - 2½″
35	7′ - 7⅞″	8′ - 0¼″	8′ - 9″	9′ - 1⅜″	9′ - 1⅜″	9′ - 5¾″
36	7′ - 10½″	8′ - 3″	9′ - 0″	9′ - 4½″	9′ - 4½″	9′ - 9″
37	8′ - 1⅛″	8′ - 5¾″	9′ - 3″	9′ - 7⅝″	9′ - 7⅝″	10′ - 0¼″
38	8′ - 3¾″	8′ - 8½″	9′ - 6″	9′ - 10¾″	9′ - 10¾″	10′ - 3½″
39	8′ - 6⅜″	8′ - 11¼″	9′ - 9″	10′ - 1⅞″	10′ - 1⅞″	10′ - 6¾″
40	8′ - 9″	9′ - 2″	10′ - 0″	10′ - 5″	10′ - 5″	10′ - 10″
41	8′ - 11⅝″	9′ - 4¾″	10′ - 3″	10′ - 8⅛″	10′ - 8⅛″	11′ - 1¼″
42	9′ - 2¼″	9′ - 7½″	10′ - 6″	10′ - 11¼″	10′ - 11¼″	11′ - 4½″
43	9′ - 4⅞″	9′ - 10¼″	10′ - 9″	11′ - 2⅜″	11′ - 2⅜″	11′ - 7¾″
44	9′ - 7½″	10′ - 1″	11′ - 0″	11′ - 5½″	11′ - 5½″	11′ - 11″
45	9′ - 10⅛″	10′ - 3¾″	11′ - 3″	11′ - 8⅝″	11′ - 8⅝″	12′ - 2¼″
46	10′ - 0¾″	10′ - 6½″	11′ - 6″	11′ - 11¾″	11′ - 11¾″	12′ - 5½″
47	10′ - 3⅜″	10′ - 9¼″	11′ - 9″	12′ - 2⅞″	12′ - 2⅞″	12′ - 8¾″
48	10′ - 6″	11′ - 0″	12′ - 0″	12′ - 6″	12′ - 6″	13′ - 0″
49	10′ - 8⅝″	11′ - 2¾″	12′ - 3″	12′ - 9⅛″	12′ - 9⅛″	13′ - 3¼″
50	10′ - 11¼″	11′ - 5½″	12′ - 6″	13′ - 0¼″	13′ - 0¼″	13′ - 6½″
100	21′ - 10½″	22′ - 11″	25′ - 0″	26′ - 0½″	26′ - 0½″	27′ - 1″

[1] Brick positioned in wall as stretchers. Vertical dimensions are from bottom of mortar joint to bottom of mortar joint.

Goodheart-Willcox Publisher

Weights and Measures

Water									Cement						Water-Cement Ratio		
Weight lb	US gallons	Imp. gallons	Weight lb	US gallons	Imp. gallons	Weight lb	US gallons	Imp. gallons	US bags	Can. bags	Weight lb	US bags	Can. bags	Weight lb	W/C by weight	Can. w/c, gal/bag	US w/c, gal/bag
10	1.2	1.0	185	22.2	18.5	360	43.2	36.0	1.0	1.1	94	4.5	4.8	423	.30	2.6	3.4
15	1.8	1.5	190	22.8	19.0	365	43.8	36.5	1.1	1.2	103	4.6	4.9	432	.31	2.7	3.5
20	2.4	2.0	195	23.4	19.5	370	44.4	37.0	1.2	1.3	113	4.7	5.0	442	.32	2.8	3.6
25	3.0	2.5	200	24.0	20.0	375	45.0	37.5	1.3	1.4	122	4.8	5.2	451	.33	2.9	3.7
30	3.6	3.0	205	24.6	20.5	380	45.6	38.0	1.4	1.5	132	4.9	5.3	461	.34	3.0	3.8
35	4.2	3.5	210	25.2	21.0	385	46.2	38.5	1.5	1.6	141	5.0	5.4	470	.35	3.1	3.9
40	4.8	4.0	215	25.8	21.5	390	46.8	39.0	1.6	1.7	150	5.1	5.5	479	.36	3.2	4.1
45	5.4	4.5	220	26.4	22.0	395	47.4	39.5	1.7	1.8	160	5.2	5.6	489	.37	3.2	4.2
50	6.0	5.0	225	27.0	22.5	400	48.0	40.0	1.8	1.9	169	5.3	5.7	498	.38	3.3	4.3
55	6.6	5.5	230	27.6	23.0	405	48.6	40.5	1.9	2.0	179	5.4	5.8	508	.39	3.4	4.4
60	7.2	6.0	235	28.2	23.5	410	49.2	41.0	2.0	2.1	188	5.5	5.9	517	.40	3.5	4.5
65	7.8	6.5	240	28.8	24.0	415	49.8	41.5	2.1	2.3	197	5.6	6.0	526	.41	3.6	4.6
70	8.4	7.0	245	29.4	24.5	420	50.4	42.0	2.2	2.4	207	5.7	6.1	536	.42	3.7	4.7
75	9.0	7.5	250	30.0	25.0	425	51.0	42.5	2.3	2.5	216	5.8	6.2	545	.43	3.8	4.8
80	9.6	8.0	255	30.6	25.5	430	51.6	43.0	2.4	2.6	226	5.9	6.3	555	.44	3.9	5.0
85	10.2	8.5	260	31.2	26.0	435	52.2	43.5	2.5	2.7	235	6.0	6.4	564	.45	3.9	5.1
90	10.8	9.0	265	31.8	26.5	440	52.8	44.0	2.6	2.8	244	6.1	6.6	573	.46	4.0	5.2
95	11.4	9.5	270	32.4	27.0	445	53.4	44.5	2.7	2.9	254	6.2	6.7	583	.47	4.1	5.3
100	12.0	10.0	275	33.0	27.5	450	54.0	45.0	2.8	3.0	263	6.3	6.8	592	.48	4.2	5.4
105	12.6	10.5	280	33.6	28.0	455	54.6	45.5	2.9	3.1	273	6.4	6.9	602	.49	4.3	5.5
110	13.2	11.0	285	34.2	28.5	460	55.2	46.0	3.0	3.2	282	6.5	7.0	611	.50	4.4	5.6
115	13.8	11.5	290	34.8	29.0	465	55.8	46.5	3.1	3.3	291	6.6	7.1	620	.51	4.5	5.7
120	14.4	12.0	295	35.4	29.5	470	56.4	47.0	3.2	3.4	301	6.7	7.2	630	.52	4.6	5.9
125	15.0	12.5	300	36.0	30.0	475	57.0	47.5	3.3	3.5	310	6.8	7.3	639	.53	4.6	6.0
130	15.6	13.0	305	36.6	30.5	480	57.6	48.0	3.4	3.7	320	6.9	7.4	649	.54	4.7	6.1
135	16.2	13.5	310	37.2	31.0	485	58.2	48.5	3.5	3.8	329	7.0	7.5	658	.55	4.8	6.2
140	16.8	14.0	315	37.8	31.5	490	58.8	49.0	3.6	3.9	338	7.1	7.6	667	.56	4.9	6.3
145	17.4	14.5	320	38.4	32.0	495	59.4	49.5	3.7	4.0	348	7.2	7.7	677	.57	5.0	6.4
150	18.0	15.0	325	39.0	32.5	500	60.0	50.0	3.8	4.1	357	7.3	7.8	686	.58	5.1	6.5
155	18.6	15.5	330	39.6	33.0	505	60.6	50.5	3.9	4.2	367	7.4	7.9	696	.59	5.2	6.7
160	19.2	16.0	335	40.2	33.5	510	61.2	51.0	4.0	4.3	376	7.5	8.1	705	.60	5.3	6.8
165	19.8	16.5	340	40.8	34.0	515	61.8	51.5	4.1	4.4	385	7.6	8.2	714	.61	5.3	6.9
170	20.4	17.0	345	41.4	34.5	520	62.4	52.0	4.2	4.5	395	7.7	8.3	724	.62	5.4	7.0
175	21.0	17.5	350	42.0	35.0	525	63.0	52.5	4.3	4.6	404	7.8	8.4	733	.63	5.5	7.1
180	21.6	18.0	355	42.6	35.5	530	63.6	53.0	4.4	4.7	414	7.9	8.5	743	.64	5.6	7.2

Note: Larger quantities can be secured by addition of above figures.

Equivalent weights and measures		
1 gal of water, US	=	8.33 lb
1 gal of water, Imp.	=	10.00 lb
1 cu ft of water	=	62.40 lb
1 bag of cement, US	=	94.00 lb
1 bag of cement, Can.	=	87.50 lb

Metric conversions					
To convert from:	US gal	To:	cu meters	Multiply by:	0.003785
	Imp. gal		cu meters		0.004546
	US bags		kilograms		42.637
	Can. bags		kilograms		39.689
	pounds		kilograms		0.4536
	cu ft		cu meters		0.02832
	cu yd		cu meters		0.7646

Goodheart-Willcox Publisher

Patterns for Concrete Masonry

1. Running bond, 8 × 16 in

2. Offset bond, 8 × 16 in

3. Offset bond, 8 × 16 in

4. Running bond, 4 × 16 in

5. Offset bond, 4 × 16 in

6. Coursed ashlar, 4 × 16, 8 × 16 in

7. Coursed ashlar, 4 × 16, 8 × 16 in

8. Coursed ashlar, 4 × 16, 8 × 16 in

9. Coursed ashlar, 4 × 16, 8 × 16 in

10. Coursed ashlar, 8 × 16 in, brick size

11. Coursed ashlar, 8 × 16 in, brick size

12. Coursed ashlar, 4 × 16, 8 × 18 in

(Continued)

Goodheart-Willcox Publisher

Patterns for Concrete Masonry (continued)

13. Coursed ashlar, 4 × 16, 8 × 16 in

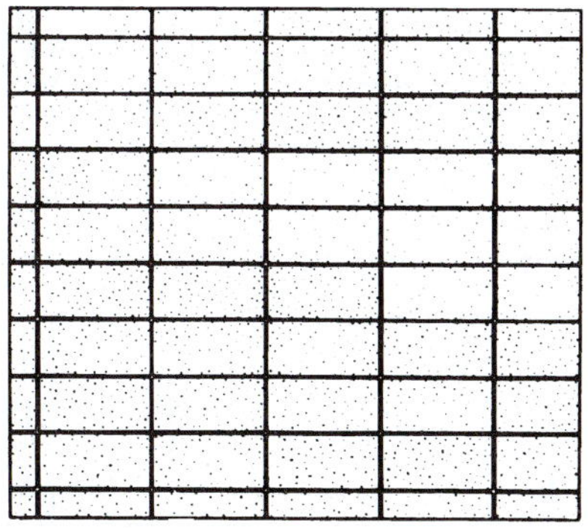
14. Horizontal stacking, 8 × 16 in

15. Horizontal stacking, 4 × 16 in

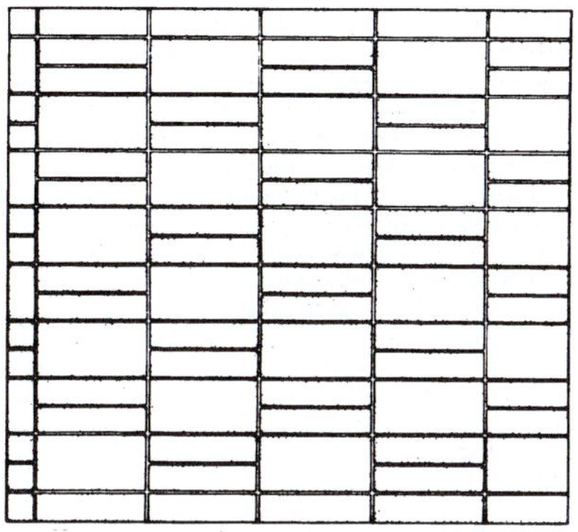
16. Coursed ashlar, 8 × 16 in, brick size

17. Coursed ashlar, 8 × 16 in, brick size

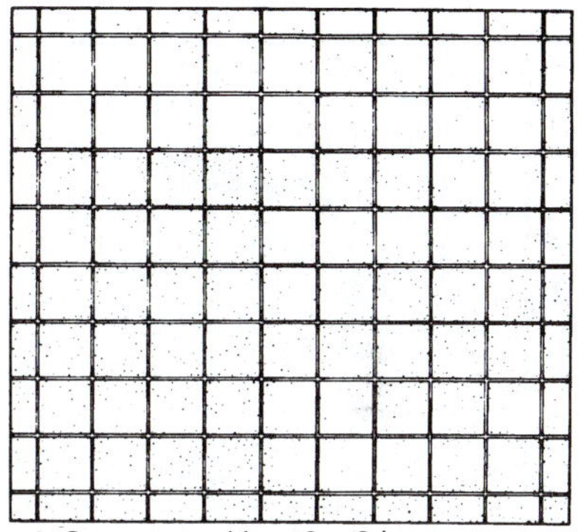
18. Square stacking, 8 × 8 in

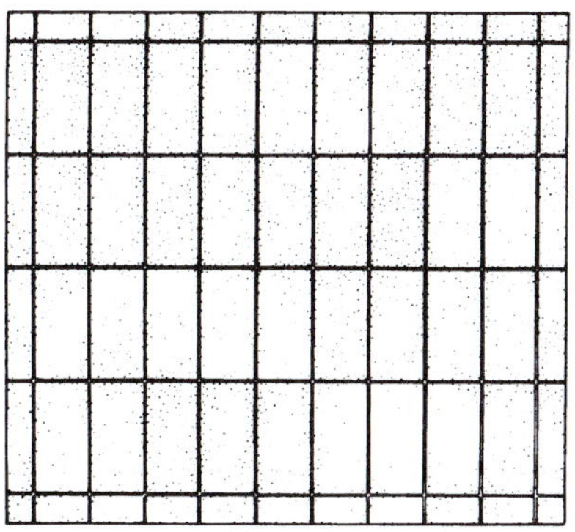
19. Vertical stacking, 8 × 16 in

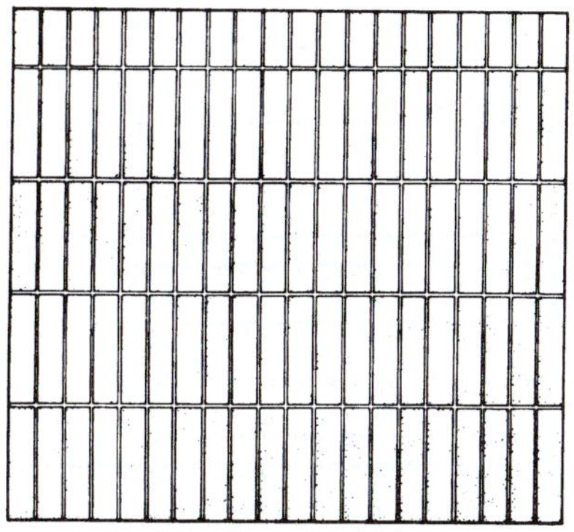
20. Vertical stacking, 4 × 16 in

21. Diagonal stacking, 8 × 8 in

22. Diagonal bond, 8 × 16 in

23. Diagonal basket weave, 8 × 16 in

24. Diagonal basket weave, 8 × 16 in

(Continued)

Goodheart-Willcox Publisher

Patterns for Concrete Masonry (continued)

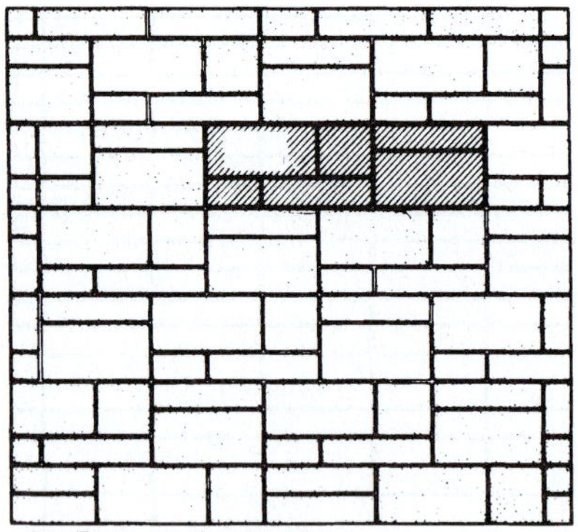

25. Patterned ashlar, 4 × 8, 4 × 16, 8 × 8, 8 × 16 in

26. Patterned ashlar, 4 × 8, 4 × 12, 4 × 16, 8 × 12, 8 × 16 in

27. Patterned ashlar, 4 × 8, 4 × 16, 8 × 8, 8 × 12, 8 × 16 in

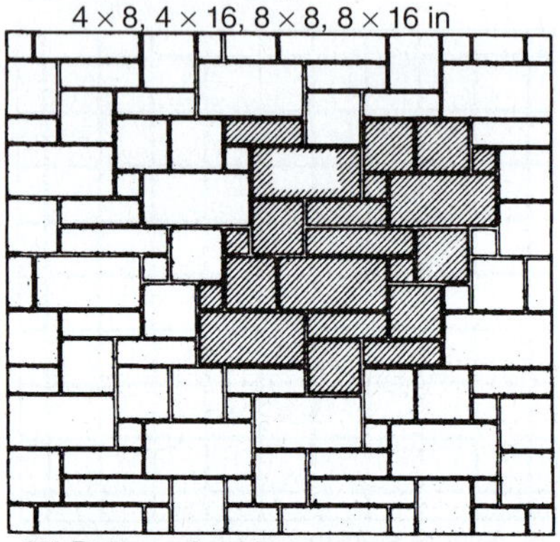

28. Patterned ashlar, 4 × 4, 4 × 12, 4 × 16, 8 × 8, 8 × 16 in

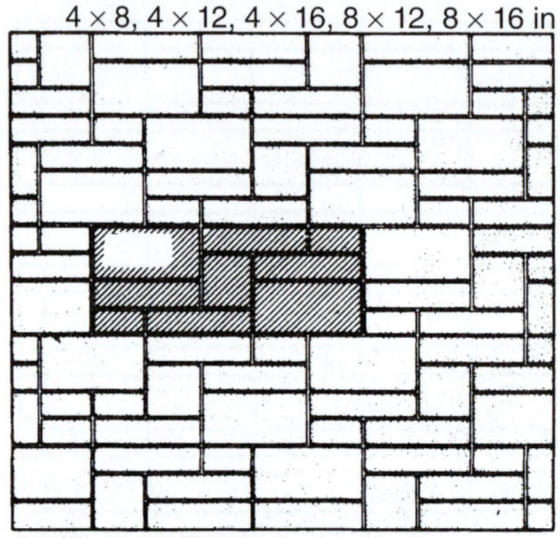

29. Patterned ashlar, 4 × 8, 4 × 16, 8 × 8, 8 × 16 in

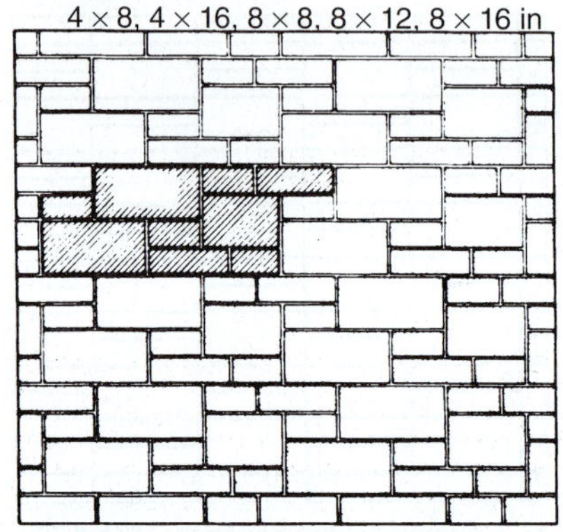

30. Patterned ashlar, 4 × 8, 4 × 12, 8 × 12, 8 × 16 in

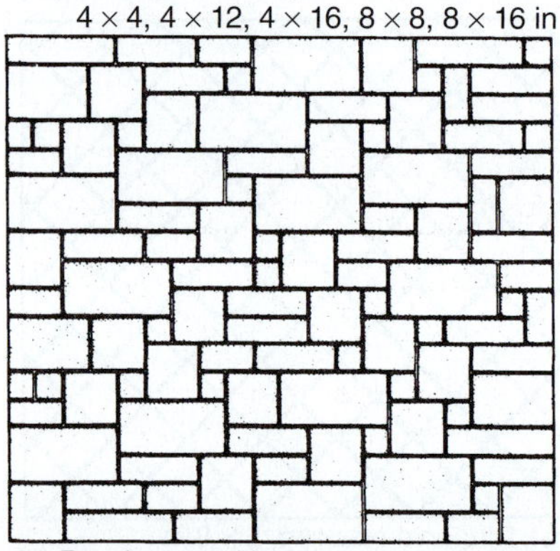

31. Random stacking, 4 × 4, 4 × 8, 8 × 8, 8 × 16 in

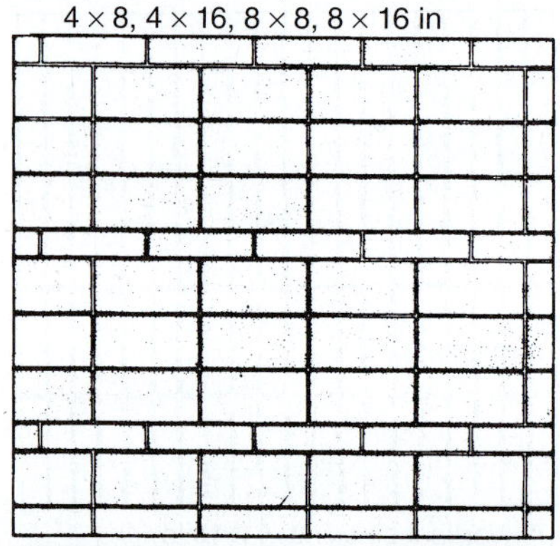

32. Patterned stacking, 4 × 16, 8 × 16 in

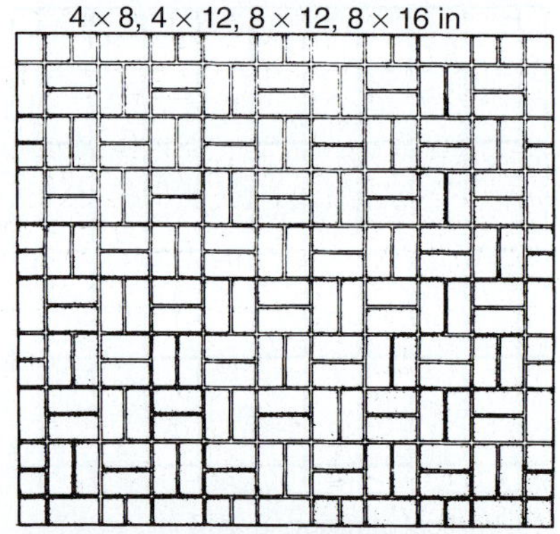

33. Basket weave, 4 × 8 in

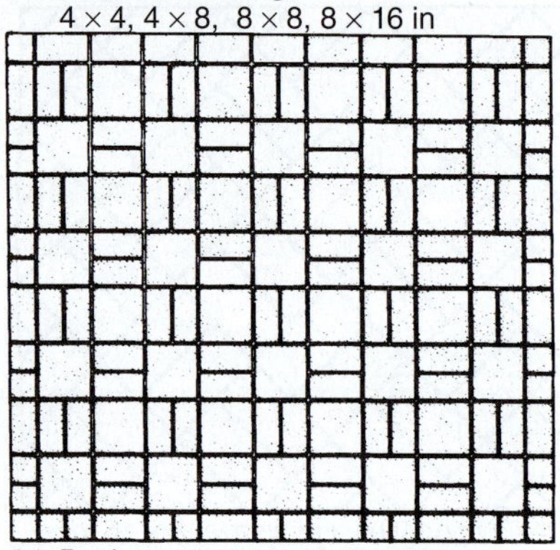

34. Basket weave, 4 x 8, 8 x 8 in

35. Basket weave, 8 × 8 in, brick size

36. Basket weave, brick size

(Continued)

Goodheart-Willcox Publisher

Patterns for Concrete Masonry (continued)

37. Basket weave, 4 × 8, 4 × 16 in

38. Basket weave, 4 × 8, 4 × 16 in

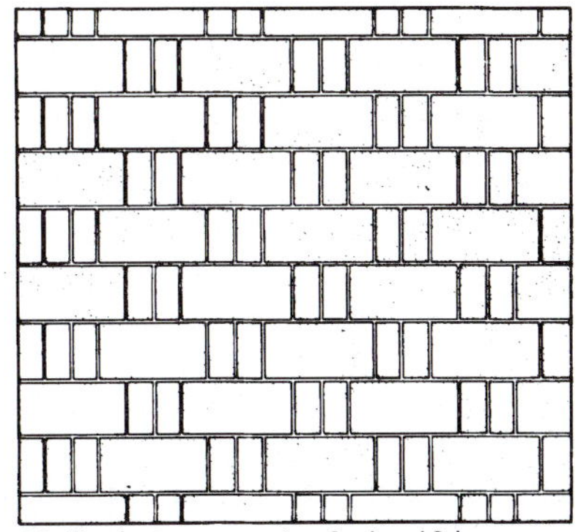
39. Basket weave, 4 × 8, 4 × 16 in

40. Basket weave, 4 × 16 in

41. Basket weave, 8 × 8, 8 × 16 in

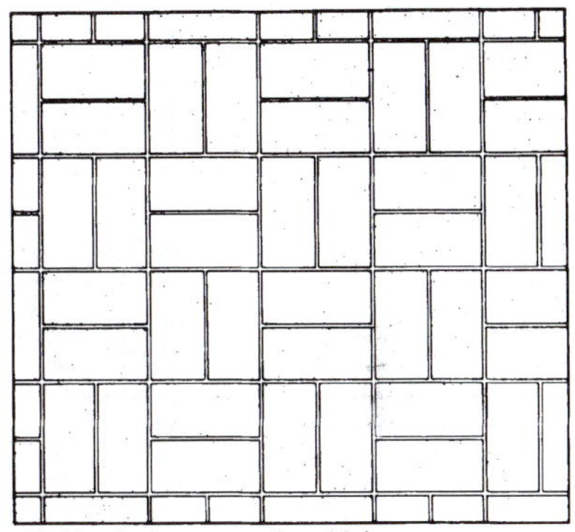
42. Basket weave, 8 × 16 in

43. Basket weave, 4 × 16, 8 × 16 in

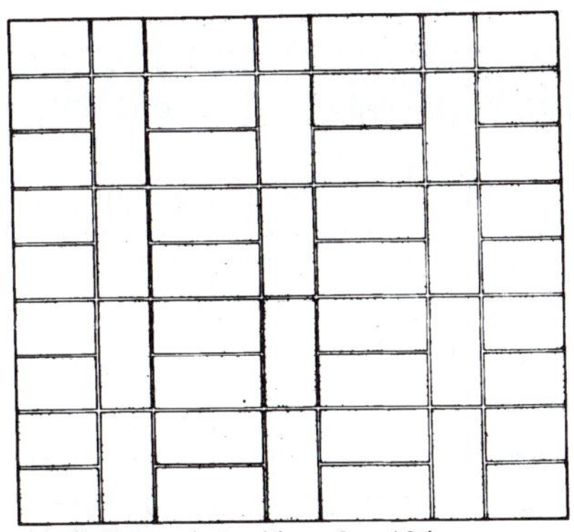
44. Patterned stacking, 8 × 16 in

45. Basket weave, 8 × 16 in

46. Basket weave, 4 × 8, 4 × 16, 8 × 16 in

47. Basket weave, 4 × 4, 8 × 16 in

48. Basket weave, 8 × 8, 8 × 16 in

(Continued)

Goodheart-Willcox Publisher

Specifications for Single-Face Fireplaces

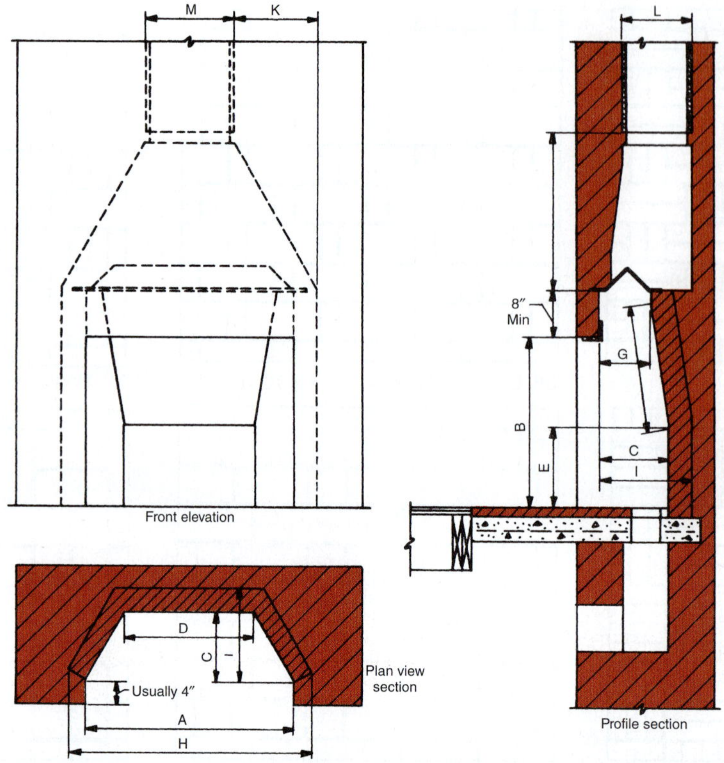

Goodheart-Willcox Publisher

Design Data for Single-Face Fireplaces														
Width	Hgt.	Depth	Back	Vert. Back	Slope Back	Throat	Width	Depth	Smoke Chamb.	Flue Lining Sizes Rect.		Flue Lining Sizes Rnd.	Flue Lining Sizes Modular	
A	B	C	D	E	F	G	H	I	J	K	L × M		K	L × M
24	24	16	11	14	15	8 ¾	32	20	19	11 ¾	8 ½ × 8 ½	8	10	8 × 12
26	24	16	13	14	15	8 ¾	34	20	21	12 ¾	8 ½ × 8 ½	8	11	8 × 12
28	24	16	15	14	15	8 ¾	36	20	21	11 ½	8 ½ × 13	10	12	8 × 12
30	29	16	17	14	18	8 ¾	38	20	24	12 ½	8 ½ × 13	10	13	12 × 12
32	29	16	19	14	21	8 ¾	40	20	24	13 ½	8 ½ × 13	10	14	12 × 12
36	29	16	23	14	21	8 ¾	44	20	27	15 ½	13 × 13	12	16	12 × 12
40	29	16	27	14	21	8 ¾	48	20	29	17 ½	13 × 13	12	16	12 × 12
42	32	16	29	14	23	8 ¾	50	20	32	18 ½	13 × 13	12	17	16 × 16
48	32	18	33	14	23	8 ¾	56	22	37	21 ½	13 × 13	15	20	16 × 16
54	37	20	37	16	27	13	68	24	45	25	13 × 18	15	26	16 × 20
60	37	22	42	16	27	13	72	27	45	27	13 × 18	15	26	16 × 20
60	40	22	42	16	29	13	72	27	45	27	18 × 18	18	26	16 × 20
72	40	22	54	16	29	13	84	27	56	33	18 × 18	18	32	20 × 20
84	40	24	64	20	26	13	96	29	67	36	20 × 20	20	36	20 × 20
96	40	24	76	20	26	13	108	29	75	42	24 × 24	22	42	20 × 20

Goodheart-Willcox Publisher

Abbreviations

aggr	aggregate
alum.	aluminum
L	angle
arch.	architectural
asb.	asbestos
asph	asphalt
AT.	asphalt tile
avg	average
bbl	barrel, barrels
basmt	basement
B	bathroom
BR	bedroom
BM	benchmark
bev	beveled
BP	blueprint
BRKT	bracket
br	brass
BRK	brick
Btu	British thermal unit
bro	bronze
Bldg	building
Buz	buzzer
CAB	cabinet
CP	candlepower
CI	cast iron
clk	caulk
clg	ceiling
cem	cement
c or CL	centerline
cm	centimeter, centimeters
cer	ceramic
chfr	chamfer
chkd	checked
cin bl	cinder block
cir	circular
CIRC	circumference
c – o	cleanout
cir	clean
CL	closet
col	column
com	common
con	concrete
conc b	concrete block
conc clg	concrete ceiling
conc fl	concrete floor
cnd	conduit
const	construction
cop	copper
cu	cubic
cu ft	cubic foot, feet
cu in	cubic inch, inches
cu yd	cubic yard, yards
DP	dampproofing
det	detail
diag	diagram
dia	diameter
dim.	dimensions
DR	dining room
d – h	double hung
dn	down
ds	downspout
dr	drain
dwg	drawing
drwn	drawn
dw	dry well
E	east
elec	electric
el	elevation
elev	elevator
ent	entrance
equip.	equipment
est	estimate
exc	excavate
exist.	existing
exp bt	expansion bolt
exp jt	expansion joint
EXT	extension
ext	exterior
xh	extra heavy
fab	fabricate
F	fahrenheit
fig.	figure
fil	fillet
fin.	finish
FAO	finish all over
frbk	firebrick
fdr.	fire door
fp	fireplace
fprf	fireproof
flge	flange
flr	floor
flg	flooring
fluor	fluorescent
fl	flush
′ or ft	foot, feet
ft lb	foot pound
ftg	footing
fdn	foundation
fr	frame
gal.	gallon, gallons
galv	galvanized
GA	gage
gl	glass
gr	grade
g	granite
grtg	grating
gyp	gypsum
hdw	hardware
hdwd	hardwood
hgt	height
hor	horizontal
HB	hose bibb
hw	hot water
C	hundred
I	I beam
″ or in	inch, inches
incl	include
ID	inside diameter
insul	insulation
int	interior
jt	joint
j	junction
jb	junction box
kg	kilogram
km	kilometer
kw	kilowatt, kilowatts
kwh	kilowatt-hour
k	kip (1000 lb)
K	kitchen
ldg	landing
lth	lath
Lau	laundry
LT	laundry tray
lav	lavatory
LDR	leader
ld	left-hand
LH	left-hand
lgth	length
lev	level
lt	light
lwc	lightweight concrete
Ls	limestone
lin ft	linear feet
L CL	linen closet
lino	linoleum
LI	live load
LR	living room
l	lumen
mip	malleable iron pipe
mfr	manufacture
mr	marble
MO	masonry opening
matl	material
max	maximum
mm	millimeter, millimeters
min	minimum
(′)	minute (angular measure)
moldg	molding
ni	nickle
nom	nominal
N	north
No.	number
O.C.	off center
opg	opening
oz	ounce, ounces
out.	outlet
OD	out diameter
oa	overall
ovfl	overflow
ovhd	overhead
ptn	partition
d	penny
perp	perpendicular
pc	piece
plas	plaster
pl	plate
PL GL	plate glass
lb	pound, pounds
psi	pounds per square inch
P/C	poured concrete
prect	precast
prefab	prefabricated
PC	pull chain
PB	push button
r or rad	radius
recp	receptacle
rect	rectangle
ref	refrigerator
reg	register
reinf	reinforce
ret	return
RH	right-hand
R	riser
rm	room
rgh	rough
rnd	round
rub.	rubber
sad.	saddle
sch	schedule
smls	seamless
sec	second
″	second (angular measure)
shthg	sheathing
ship	ship lap
sh	shower
sdg	siding
sc	sill cock
S	sink
sl	slate
sp	soil pipe
S	south
SE	southeast
SW	southwest
sq	square
sq ft	square foot, feet
sst	stainless steel
Stp	standpipe
stl	steel
stiff.	stiffener
stn	stone
SP	sump pit
susp	suspending
clg	ceiling
sw	switch
T	tee
tel	telephone
TC	terra cotta
ter	terrazzo
thermo	thermostat
thk	thick, thickness
thd	thread
T & G	tongue and groove
tr	tread
trnbkl	turnbuckle
unfin	unfinished
V	vent, ventilator
vent.	ventilation
vert.	vertical
vol	volume
WV	wall vent
wp	waterproofing
w	watt, watts
whr	watt-hour
WS	weather stripping
wp	weatherproof
whp	weep hole
W	west
WF	wide flange
wth	width
wdw	window
w/	with
wf	wood frame
wrt	wrought
WI	wrought iron
yd	yard, yards
z	zinc

Goodheart-Willcox Publisher

Quality Assurance Levels and Requirements

Minimum Tests and Submittals	Minimum Inspection
Level 1 Quality Assurance	
Certificates for materials used in masonry construction indicating compliance with the contract documents	Verify compliance with the approved submittals
Level 2 Quality Assurance	
Certificates for materials used in masonry construction indicating compliance with the contract documents Verification of f'_m prior to construction, except where specifically exempted by this Code	As masonry construction begins, verify the following are in compliance: • proportions of site-prepared mortar • construction of mortar joints • location of reinforcement and connectors Prior to grouting, verify the following are in compliance: • grout space • placement of reinforcement and connectors • proportions of site-prepared grout • construction of mortar joints Verify grout placement is in compliance Observe preparation of grout specimens, mortar specimens, and/or prisms Verify compliance with the required inspection provisions of the contract documents and the approved submittals
Level 3 Quality Assurance	
Certificates for materials used in masonry construction indicating compliance with the contract documents Verification of f'_m • prior to construction • every 5000 sq ft (450 m^2) during construction Verification of proportions of materials in mortar and grout as delivered to the site	From the beginning of masonry construction and continuously during construction of masonry, verify the following are in compliance: • proportions of site-mixed mortar and grout • placing of masonry units and construction of mortar joints • placement of reinforcement and connectors • grout space prior to grouting • placement of grout Observe preparation of grout specimens, mortar specimens, and/or prisms Verify compliance with the required inspection provisions of the contract documents and the approved submittals

Note: Adapted from the 1999 MSJC Code

Goodheart-Willcox Publisher

Hazard Communication Standard Pictograms and Hazards

HCS Pictograms and Hazards		
Health Hazard  Carcinogen Mutagenicity Reproductive toxicity Respiratory sensitizer Target organ toxicity Aspiration toxicity	Flame  Flammables Pyrophorics Self-heating Emits flammable gas Self-reactives Organic peroxides	Exclamation Mark  Irritant (skin and eye) Skin sensitizer Acute toxicity Narcotic effects Respiratory tract irritant Hazardous to ozone layer (nonmandatory)
Gas Cylinder  Gases under pressure	Corrosion  Skin corrosion/burns Eye damage Corrosive to metals	Exploding Bomb  Explosives Self-reactives Organic peroxides
Flame Over Circle  Oxidizers	Environment (nonmandatory)  Aquatic Toxicity	Skull and Crossbones  Acute toxicity (fatal or toxic)

Occupational Safety and Health Administration

Average Depth of Frost Penetration in Inches

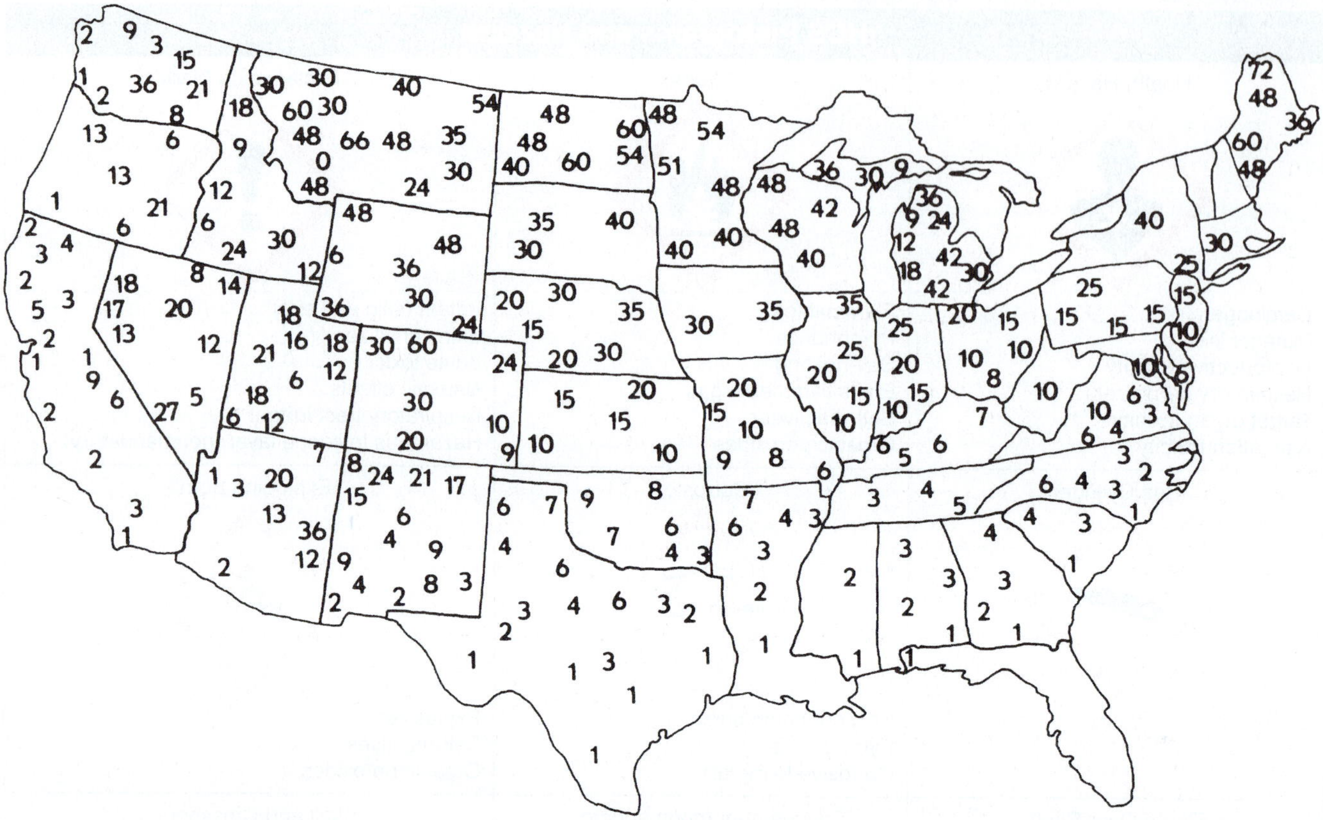

US Dept. of Commerce Weather Bureau

ASTM Standards Used in the Masonry Industry

The following is a partial list of ASTM standards used in the masonry industry. The standards in the list are referenced in the various chapters of the text.

- C5 Specification for Quicklime for Structural Purposes
- C32 Specification for Sewer and Manhole Brick (Made from Clay or Shale)
- C55 Specification for Concrete Building Brick
- C62 Specification for Building Brick (Solid Masonry Units Made from Clay or Shale)
- C73 Specification for Calcium Silicate Brick (Sand-Lime Brick)
- C90 Specification for Loadbearing Concrete Masonry Units
- C91 Specification for Masonry Cement
- C94 Specification for Ready-Mixed Concrete
- C126 Specification for Ceramic Glazed Structural Clay Facing Tile, Facing Brick, and Solid Masonry Units
- C129 Specification for Nonloadbearing Concrete Masonry Units
- C139 Specification for Concrete Masonry Units for Construction of Catch Basins and Manholes
- C143 Test Method for Slump of Hydraulic-Cement Concrete
- C144 Specification for Aggregate for Masonry Mortar
- C150 Specification for Portland Cement
- C207 Specification for Hydrated Lime for Masonry Purposes
- C216 Specification for Facing Brick (Solid Masonry Units Made from Clay or Shale)
- C270 Specification for Mortar for Unit Masonry
- C279 Specification for Chemical-Resistant Masonry Units
- C404 Specification for Aggregates for Masonry Grout
- C410 Specification for Industrial Floor Brick
- C476 Specification for Grout for Masonry
- C652 Specification for Hollow Brick (Hollow Masonry Units Made from Clay or Shale)
- C744 Specification for Prefaced Concrete and Calcium Silicate Masonry Units
- C902 Specification for Pedestrian and Light Traffic Paving Brick
- C1019 Test Method for Sampling and Testing Grout for Masonry
- C1088 Specification for Thin Veneer Brick Units Made from Clay or Shale
- C1261 Specification for Firebox Brick for Residential Fireplaces
- C1272 Specification for Heavy Vehicular Paving Brick
- C1405 Specification for Glazed Brick (Single Fired, Brick Units)
- E119 Test Methods for Fire Tests of Building Construction and Materials

Building Code Requirements for Masonry Structures

The Masonry Standard Joint Committee (MSJC) is a joint committee sponsored by The Masonry Society (TMS), the American Concrete Institute (ACI), and the Structural Engineering Institute of the American Society of Civil Engineers (SEI/ASCE). The adoption of the Building Code Requirements for Masonry Structures, TMS 402-22/ACI 530/ASCE 6, represents the consolidation of several masonry codes into one consistent set of provisions. The code is maintained by the MSJC. The following is an outline of the code document.

PART 1—GENERAL

Chapter 1—General Requirements

1.1 Scope
1.2 Contract Documents and Calculations
1.3 Alternative Design or Method of Construction
1.4 Standards Cited in this Code

Chapter 2—Notation and Definitions

2.1 Notation
2.2 Definitions

Chapter 3—Quality and Construction

3.1 Quality Assurance Program

PART 2—DESIGN REQUIREMENTS

Chapter 4—General Analysis and Design Considerations

4.1 Loading
4.2 Material Properties
4.3 Specific Compressive Strength
4.4 Section Properties
4.5 Connection to Structural Frames
4.6 Definitions of Beams Supporting Unreinforced Masonry
4.7 Masonry Not Laid in Running Bond
4.8 Embedded Conduits, Pipes, and Sleeves

Chapter 5—Structural Elements

5.1 General
5.2 Walls
5.3 Beams
5.4 Columns
5.4 Pilasters
5.6 Corbels

Chapter 6—Reinforcement, Metal, Accessories, and Anchor Bolts

6.1 Reinforcement
6.2 Metal Accessories
6.3 Anchor Bolts

Chapter 7—Seismic Design Requirements

7.1 Scope
7.2 General Analysis
7.3 Element Classification
7.4 Seismic Design Category Requirements

PART 3—ENGINEERING DESIGN METHODS

Chapter 8—Allowable Stress Design of Masonry

8.1 General
8.2 Unreinforced Masonry
8.3 Reinforced Masonry

Chapter 9—Strength Design of Masonry

9.1 General
9.2 Unreinforced Masonry
9.3 Reinforced Masonry

Chapter 10—Prestressed Masonry

10.1 General
10.2 Design Methods
10.3 Permissible Stress in Prestressing Tendons
10.4 Axial Compression and Flexure
10.5 Design in Walls
10.6 Design of Beams and Lintels
10.7 Axial Tension
10.8 Shear
10.9 Deflections
10.10 Prestressed Tendon Anchorages, Couplers, and End Blocks
10.11 Protection of Prestressing Tendons and Accessories
10.12 Development of Bonded Tendons

Chapter 11—Strength Designs of Autoclaved Aerated Concrete

11.1 General
11.2 Unreinforced AAC Masonry
11.3 Reinforced AAC Masonry

Chapter 12—Design Masonry Infill

12.1 General
12.2 Non-Participating Infills
12.3 Participating Infills

PART 4—PRESCRIPTIVE DESIGN METHODS

Chapter 13—Veneer

13.1 General
13.2 Anchored Veneer
13.3 Adhered Veneer

Chapter 14—Glass Unit Masonry

14.1 General
14.2 Panel Size
14.3 Support
14.4 Expansion Joints
14.5 Base Surface Treatment
14.6 Mortar
14.7 Reinforcement

Chapter 15—Masonry Partition Walls

15.1 General
15.2 Prescriptive Design of Partition Walls
15.3 Lateral Support
15.4 Anchorage
15.5 Miscellaneous Requirements

PART 5—APPENDICES

Appendix C—Limit Design Methods

C – General
C.1 Yield Mechanism
C.2 Mechanism Strength
C.3 Mechanism Deformation

Appendix D—Limit

D.1 General
D.2 Material Property
D.3 Reinforcement
D.4 Flexural Members
D.5 Creep Rupture

Equation Conversions
Conversions of Inch-Pound Units to SI Units
Prefixes
References for the Code Commentary

Specification for Masonry Structures

The Specification for Masonry Structures was developed by the MSJC. The following is an outline of this document.

PART 1—GENERAL

1.1 Summary
1.2 Definitions
1.3 Reference Standards
1.4 System Description
1.5 Submittals
1.6 Quality Assurance
1.7 Delivery, Storage, and Handling
1.8 Project Conditions

PART 2—PRODUCTS

2.1 Mortar Materials
2.2 Grout Materials
2.3 Masonry Unit Materials
2.4 Reinforcement, Prestressing Tendons, and Metal Accessories
2.5 Accessories
2.6 Mixing
2.7 Fabrication

PART 3—EXECUTION

3.1 Inspection
3.2 Preparation
3.3 Masonry Erection
3.4 Reinforcement, Tie, and Anchor Installation
3.5 Grout Placement
3.6 Prestressing Tendon Installation and Stressing Procedures
3.7 Field Quality Control
3.8 Cleaning

FORWARD TO SPECIFICATION CHECKLISTS

Mandatory Requirements Checklist
Optional Requirements Checklist

REFERENCES FOR THE SPECIFICATION COMMENTARY

Construction requirements and quality assurance provisions in the code—Building Code Requirements for Masonry Structures—are included by reference to Specification for Masonry Structures, TMS 602-22/ACI 530/ASCE 6.

GLOSSARY

3-4-5 method. Method of layout that uses the Pythagorean Theorem ($a^2 + b^2 = c^2$). (6)

A

abutment. The masonry or combination of masonry and other structural members that support one end of the arch at the skewback. (16)

accelerating admixture. Material used to speed up the setting and the strength development of concrete. (18)

acidic. Term describing igneous stone that is composed primarily of light minerals (quartz and potash feldspar). (9)

acute angle. Angle less than 90°. (6)

addition. Process of combining two or more quantities (numbers) to find a total. (6)

adhesion ceramic veneer. Veneer usually produced in thin sections. It is held in place without metal anchors by a mortar backing. (7)

adjustable post base. A metal connection made of galvanized or powder-coated steel that locks the concrete column and the wood or steel support post together. (19)

adjustable ties. Devices used to attach a masonry facing to a backup system of masonry or typical stud construction. (11)

admixtures. Coloring agents, air-entraining materials, accelerators, retarders, and water repellents added to concrete. (8)

aggregate. Inert granular particles that are mixed with Portland cement and water to form concrete. (8)

air-entraining admixtures. Materials used to improve the durability of concrete exposed to moisture during cycles of freezing and thawing. (18)

Alphabet of Lines. The different types of lines used on construction drawings. (5)

altimeter level. Precision instruments that measure differences in elevation heights. (3)

American bond. Variation of the running bond that has a course of full-length headers at regular intervals to provide structural bonding as well as pattern. (7)

anchor. Device, usually made of steel, that connects masonry to a structural support or frame. Anchors usually connect two dissimilar materials. (11)

anchor bolt. Metal bolt placed in fresh concrete to be used later in construction to attach sill plates to the masonry work. (11)

anchorage point. A point of attachment for lanyards or other devices when a person needs to connect to a secure location during the time of wearing fall protection equipment. (4)

anchored ceramic veneer. Veneer that is usually produced in thick sections held in place by grout and wire anchors connected to its backing. (7)

angle. Geometric figure formed by two intersecting lines. (6)

apprentice. A person at least 16 years old who signs an agreement to work and learn the trade. (1)

arc. Any portion of the circumference of a circle. (6)

arch. A form of construction in which masonry units span an opening by transferring vertical loads laterally to adjacent voussoirs and, thus, to the abutments. (16)

architectural concrete masonry units (ACMU). Concrete masonry units that have been designed for aesthetic appeal. (8)

architectural rendering. A drawing that is natural in appearance, similar to a photograph. The drawing illustrates the appearance of a building at the time of completion. (5)

ashlar. A wall pattern using stones that are precut with enough uniformity to allow some regularity in assembly. (9, 14)

B

basic. Term describing igneous stone that is composed primarily of ferro-magnesian minerals. (9)

bat. A half brick. (7)

batter boards. Boards that preserve the building lines during excavation and construction. (12)

bearing surfaces. The tops and bottoms of the bricks or blocks. (7)

Note: The number in parentheses following each definition indicates the chapter in which the term can be found.

bleeding. The separation of water from the mortar during the time it is in the plastic state prior to setting. (10)

block bond. Pattern bond in which all vertical joints are aligned and there is no overlapping of units. Also called *stack bond.* (7)

block clamp. A tool used to carry blocks and hardscape blocks. (3)

bond. The property of hardened mortar that holds masonry units together. (10)

bond beam. A device used to strengthen a masonry wall at specific locations. (16)

border lines. Very heavy lines that are drawn to form a boundary for the drawing. (5)

braces. Devices, made of 2 × 4 lumber or metal, that add strength and a means to straighten walls. One end is attached to the wall and the other end is attached to a stake driven into the earth. (19)

break back. The depth that a tie is broken off in the concrete. (19)

brick hammer. Hammer designed to both drive nails and break or chip stone or masonry units. (3)

brick panel. A structural panel combining thin brick, reinforcement, and concrete. Insulation may also be added if desired. (7)

brick set. Type of chisel used to cut brick to specified length. (3)

brick tongs. Adjustable tools designed to carry brick without chipping it. (3)

buck. Wood or steel frames set in the form between the inner and outer form to make an opening in the wall. (16)

builder's level. A device that includes an accurate spirit level combined with a telescope on a circular base to shoot grades. The builder's level is attached to a tripod to take readings in the horizontal plane. (19)

builder's transit. A device, similar to a builder's level, that can also move in the vertical plane to shoot multiple points. (19)

building expansion (isolation) joint. Through-the-building joint used to separate a building into discrete sections so stresses developed in one section do not affect the integrity of the entire structure. (16)

bulkhead. Piece of material that prevents concrete from moving past a certain point in the form. (19)

bull float. Large, flat, rectangular piece of wood, aluminum, or magnesium that does essentially the same job as the darby but enables the mason to float a much larger area. (3)

bull float extension handle. Device used to attach to a bull float. It can be made of different materials; aluminum and fiberglass are two popular products. (3)

buttering. The application of mortar on a masonry unit with a trowel. (12)

C

camber. The relatively small rise of a jack arch. (16)

cap. Covering within the height of the wall, normally where there is a change in wall thickness. (16)

cardiopulmonary resuscitation (CPR). An emergency technique used on individuals when the heart has stopped beating. (4)

cavity wall. Two-wythe wall consisting of two walls separated by a continuous air space that is 2″ or more wide. Each wythe reacts independently to stress. (13)

cells. Openings in concrete blocks. Also called *cores* and *voids.* (8)

center lines. Thin lines that indicate the center of symmetrical objects. (5)

centering. Temporary shoring used to support an arch until the arch becomes self-supporting. (16)

ceramic glazed bricks. Units with a ceramic glaze fused to the body used as facing brick. (7)

chalk box and chalk. A device used to mark lines on concrete or concrete footings to identify where walls or other structures are to be located. Chalk is used to fill the box container to color the line inside the box. (3)

chemical-resistant masonry units. Brick or tile composed of clay, shale, or a suitable combination, which is made for locations where chemicals or acids are present and may be at a significant level. (7)

circle. A continuous curve maintaining an equal distance from its center at all points. (6)

circumference. Distance around a circle. (6)

climate control plan. A plan that shows the plumbing, heating, and cooling systems. (5)

collar joint. The narrow space between the facing units and the backup units in a wall. (13, 16)

column. An isolated element with a horizontal dimension measured at a right angle from the thickness dimension that does not exceed three times the thickness dimension. The height is at least three times the thickness. (15)

combined footing. Single footing supporting more than one column. (15)

combustion chamber. The space where combustion occurs to create heat in the fireplace. (17)

common bond. Variation of the running bond that has a course of full-length headers at regular intervals to provide structural bonding as well as pattern. (7)

compaction. Squeezing action that makes a material denser. (8)

competent person. A person who is capable of identifying existing and predictable hazards that are unsanitary, hazardous, or dangerous to employees and who is authorized to take prompt corrective measures to eliminate unsafe conditions. (4)

composite wall. A multi-wythe wall that is designed to act as a single member when subjected to loads. Two wythes are bonded together with masonry, metal ties, or joint reinforcement. (13)

compressive strength. The ability of a masonry unit to stand up under heavy weight without crumbling. (7)

concave joint. A joint created by first striking off the excess mortar on the joint, then running a steel jointing tool that has a concave surface to compact the mortar to create a concave appearance to the joint. (7)

concrete broom. Broom used to place a textured finish on the surface of the still pliable concrete. (3)

concrete column footing form. A form made of plastic or similar materials used to create a concrete footing form. (19)

concrete form tubes. Cardboard or heavy paper tubes used for forming a concrete pier or column. (15)

concrete masonry unit (CMU). A hollow or solid unit made of Portland cement and suitable aggregates. (8)

concrete spacers. A tool used to hold rebars in the appropriate location in concrete framework. (11)

concrete vibrator. Tools that masonry contractors use to aid in removing air pockets or voids in concrete that have been placed in walls, floors, footings, or other types of concrete forms. (3)

construction drawings. Drawings that describe the size, shape, location, and specifications of the elements of a structure. (5)

construction joint (cold joint). Joint used primarily in concrete construction where construction work is interrupted. Construction joints are located where they will least impair the strength of the structure. (16)

continuous footing. Footings constructed to support a foundation wall or several columns in a row. (15)

contractor. The person who organizes people and work, prepares bids for jobs, inspects work, and runs the business. (1)

control joint. Joint used in concrete or concrete masonry to create a plane of weakness in places where forces might cause cracks. (13, 16)

coping. Covering at the top of a wall. (16)

corbel. Shelf or ledge formed by projecting successive courses of masonry out from the face of the wall. (16)

corbeling. A method used to widen a wall by projecting out masonry units to form a ledge or shelf. (12)

cores. Openings in concrete blocks. Also called *cells* and *voids.* (8)

corner pole. A narrow board, about one story in height, that is used to measure the mortar joint heights of each course of masonry. (12)

course. A continuous level row of bricks. (7)

coursed stone. Stone laid in regular levels or courses to form a specific pattern or style. (9)

critical thinking skills. Higher-level skills that enable you to interpret information and make judgments. (2)

cross joint. Another name for the brick head joint in courses being laid. Mortar is placed on the leading end of the brick, and then set on top of the bed of mortar. After being positioned, the brick is pushed against the previously laid brick to create the cross joint. (12)

cross web. Wall between the cells or cores of a concrete block. (8)

crosshatch lines. Very thin lines used to show that a feature has been sectioned. These lines may represent a specific material or may be a general symbol. Also called *section lines.* (5)

crystalline silica. A hard, glassy mineral that can be found in a number of forms, including quartz, opal, and sand. (4)

curing. The process of controlling the moisture loss of cementitious material over time to attain designed standards of strength and bond adherence. (10)

curtain wall. Exterior nonloadbearing wall not wholly supported at each story. (16)

curved claw hammer. A framing hammer that works well for pulling out and pounding nails. The hammer head can vary from 16 ounces to 28 ounces. (3)

cutting plane lines. Heavy lines used to indicate where the object has been sectioned to show internal features. (5)

D

damper. A metal plate just above the throat that controls the flow of air and gaseous products through the fireplace and up the chimney. (17)

dampproofing. Applying products to the exterior surface of basement walls to prevent soil moisture surrounding the foundation from entering the structure. (15)

darby. Long, flat, rectangular piece of wood, aluminum, or magnesium used to remove any high or low spots left by the screed. (3)

decimal mixed number. Number with numerals on both sides of the decimal. (6)

decimals. Types of fractions that only have numerators. The denominators are implied by the place of the last digit to the right of the decimal point. (6)

denominator. Bottom number of a fraction. (6)

depth. The dimension of an arch at the skewback that is perpendicular to the arch axis. The depth of a jack arch is the vertical dimension of the arch at the springing. (16)

detail drawings. Drawings that display the internal and external view of walls, windows, floors, and other structures in a larger scale than what is found on a plan view. (5)

diameter. Distance across a circle through its center. (6)

difference. Portion that remains after subtraction. (6)

dimension lines. Thin lines used to show size or location of a feature of the structure. (5)

dimensions. Indications on a drawing of the size and location of building elements. (5)

dividend. Number that is to be divided by another number (divisor). (6)

division. Process of finding how many times one number is contained within another number. (6)

divisor. Number by which another number (dividend) is divided. (6)

double stretcher garden wall bond. Pattern bond that has two stretchers between headers. (7)

dowels. Devices used to transfer load from one concrete slab to another across the construction joint in concrete driveways, sidewalks, or other concrete flatwork. Dowels allow horizontal movement of slabs but restrict vertical movement. (20)

draft. The movement of combustion air and gases through the fireplace and chimney. (17)

drawing notes. Information on a drawing that is needed to build a structure but cannot be represented by conventional dimensions or symbols. (5)

dressed. Shaped to the desired dimension and size with the face of the stone being smooth and ready for installation. (9)

drop line. A line or mason's line attached to a plumb bob to transfer line intersections of batter board lines down into the excavation. (12)

dryout. The process of mortar or grout losing moisture to masonry units or bricks when they are extremely hot and dry. (10)

Dutch bond. Variation of English bond in which vertical joints between the stretchers in alternate courses do not align vertically. Also called *English cross*. (7)

E

ears. Extended ends that protrude from a concrete block. (8, 13)

edgers. Tools used to produce a radius on the edge of a slab. (3)

edging. Using an edger to provide a rounded edge or radius to prevent chipping or damage to the edge of the concrete. (20)

efflorescence. A white powder or salt-like deposit on masonry walls, which is caused by water-soluble salts. (7)

electrical plan. A plan that locates switches, convenience outlets, ceiling outlet fixtures, TV jacks, service entrance location, and panel box, as well as giving general information concerning circuits and special installations. (5)

elevation drawings. Drawings made for each side of a structure. The drawings show outside features as well as exterior materials and important vertical dimensions. (5)

end construction. Structural clay tile that is laid with the cells in a vertical plane. (7)

English bond. Pattern bond made up of alternate courses of headers and stretchers. The headers are centered on the stretchers, and joints between stretchers in all courses are aligned vertically. (7)

English cross. Variation of English bond in which vertical joints between the stretchers in alternate courses do not align vertically. Also called *Dutch bond*. (7)

entrepreneur. Someone who organizes, manages, and assumes the risks of a business. (2)

expansion joint. Relief joint in exterior walls used to separate brick masonry into segments to prevent cracking due to changes in temperature, thermal and moisture expansion, elastic deformation due to loads, and creep. (14, 16)

extension lines. Thin lines used to show where a dimension line ends. (5)

extrados. Curve that bounds the upper edge of the arch. (16)

extruded joint. Joint made by squeezing out the mortar as the unit is laid. The mortar is not trimmed off but is left to harden. (7)

F

face shell. Side wall of a hollow concrete masonry unit. (8)
face shell bedding. Bedding on the outside edges of blocks. (13)
face shield. Face protection to protect from flying particles because of sawing, grinding, or other tasks. (4)
facing bricks. Bricks used on exposed surfaces where appearance is an important consideration. (7)
fall protection harness. A device that is worn by an individual which contains straps that fasten around the user to distribute the fall arrest forces over the upper thighs, pelvis, chest, and shoulders in an event of a fall. (4)
false joint. A joint cut into the stone or panel to create a pattern resembling a true masonry joint. The joint may be left as is or partially filled to resemble a regular masonry joint. (14)
feldspar. A crystalline mineral made up of aluminum silicates with either potassium, sodium, calcium, or barium. (7)
ferrule. Metal band around the shank end of a trowel. (3)
fiber reinforced polymer (FRP). A type of reinforcing bar that is lighter weight, higher tensile strength, and more durable than steel reinforcement. (11)
firebox. The internal assembly created in the fireplace where combustion occurs. (17)
firebox brick. Brick manufactured from clay, fireclay, shale, or other similar materials, for use in the lining of fireboxes of residential fireplaces. (7)
fireclay. A type of clay that resists the effects heat can have on regular mortar. It can withstand temperatures up to 3000°F. (10)
first aid. The assistance provided to someone who has been injured or become ill. (4)
flashing. A thin, impervious material, usually made of metal, that is placed in mortar joints to prevent water penetration and to direct the drainage of water. (16)
Flemish bond. Pattern bond that alternates courses of stretchers and headers. The headers in alternate courses are centered over the stretchers in the course between them. (7)
floor plan. A plan that shows all exterior and interior walls, doors, windows, patios, walks, decks, fireplaces, built-in cabinets, and appliances. (5)
flue lining. A fired clay pipe, either round or square, that lines the flue of a chimney to protect the outer masonry from the extreme heat. (17)
flush joint. Joint that is made by holding the trowel flat against the brick and cutting in any direction. The joint is not compacted. (7)
footing. The base on which a foundation is built. (15)
foreman. A journeyman who has the responsibility of supervising a group of workers. (1)
foundation. The substructure of a building. (15)
foundation plan. A plan that shows the foundation walls and their sizes and materials. It may also give information about the excavation, reinforcing materials, special wall construction procedures, waterproofing, and supporting structures. (5)
fraction. One or more parts of a whole. (6)
framing plan. A plan designed to show in detail the framing required for a roof, floor, or other framed area. (5)
frost heaving. Frost migrates under a footing and causes it to heave because the moisture in the ground has frozen. (14)
frost line. The depth that frost can penetrate the soil or earth in a specific area of the country. When the temperature is below freezing, frost begins to enter the soil or earth. Freezing causes the soil to expand because there is moisture in the soil, which causes a heaving process. (15)

G

garden wall bond. Pattern bond that has three stretchers alternating with a header. (7)
gauge stick. A device similar to a corner pole, but shorter (approximately 4′ long and 3/4″ square). It is used during the building process to space masonry joints and brick so courses are uniform in size. (12)
general conditions. Specifications that deal with items such as terms of the agreement, insurance, responsibility for permits, supervision, payment for work, and temporary utilities. (5)
glazed brick. Brick units that have been fired once with a glaze body and are used as facing brick. (7)
glazed clay facing tile. Facing tile produced from high-grade, light-burning fire clay suitable for the application of ceramic glaze. (7)
gneiss. Coarse-textured stone with minerals in parallel streaks or bands. It is relatively rich in feldspar and usually contains mica. (9)
Grade MW. Sand-lime bricks that can be exposed to freezing conditions, but not to water saturation. (8)
Grade MW brick. Brick used where there may be exposure to temperatures below freezing, but where the brick is not likely to be permeated with water. (7)

Grade N. Concrete brick used for architectural veneer and facing units in exterior walls. It is also used where high strength and resistance to moisture penetration and severe frost action are desired. (8)

Grade NW. Sand-lime bricks that can be exposed to light frost action with very little moisture present. (8)

Grade NW brick. Brick designed to be used as a backup on interior masonry. (7)

Grade S. Concrete brick used where moderate strength and resistance to frost action and moisture penetration are required. (8)

Grade SW. Sand-lime bricks that can be exposed to freezing temperatures and water saturation. (8)

Grade SW brick. Brick that is highly resistant to frost action and is used where bricks are exposed to freezing weather. (7)

granite. Light-colored igneous stone that is formed mainly of potash feldspar, quartz, and mica. (9)

green bricks. Bricks that have not been fired. (7)

groovers. Tools used to cut a joint partly through fresh concrete to control the location of any possible cracks. (3)

ground face units. Oversized and specialty units that may be used to create walls, columns, and arches of monumental proportions. (8)

ground-fault circuit interrupter (GFCI). Electrical device that protects workers from receiving a severe electrical shock or being electrocuted by activating and tripping to interrupt the electrical current. (4)

grout. A liquid cement mixture used to fill voids. A mixture of cementitious materials, aggregates and water, used to fill cells of hollow masonry units or spaces between wythes of solid masonry. (7, 10)

H

hand floats. Tools used to prepare concrete surfaces for troweling. (3)

hand tampers, aggregate or soil. A tool used to compact aggregate, sand, soil, or other materials during the time of backfilling around a foundation or structure. (3)

hand tampers, concrete. Tools used on flatwork construction concrete to move stones and rocks under surface of the concrete to aid in producing a smooth finished surface. (3)

hard conversion. Conversion from US customary to metric that requires a physical change in the size of the product. (8)

hard metric conversion. The resizing of products to new dimensions based on a specific metric measurement. (6)

hard-burned brick. Brick that is fired in a kiln at high temperatures to increase the mechanical strength, moisture resistance, and weather resistance. (17)

hardscape. Concrete masonry materials used to create structures that are part of an outside landscape and include freestanding walls, retaining walls, piers, dividing walls, pavers, caps on finish walls, stairs, walkways, driveways, and patio areas. (8, 16)

hardware cloth. Metal fabric that is placed on masonry units to add extra bonding strength and to reinforce masonry joints. The cloth is normally used where one block wall butts up against another wall. Also called *mesh wall tie (MWT)*. (11)

Haydite. Type of block made from expanded shale aggregate, which produces air pockets in the aggregate to create a higher R-value block. (17)

header high. The distance of masonry up to the top of the brick course directly under the header course. (12)

hidden lines. Medium-weight lines that represent an edge or intersection of two surfaces that are not visible in a given view. (5)

high-lift grouting. The process of grouting a structure in one pour, usually after the walls are constructed. (10)

high-slump concrete. Very fluid (wet) concrete mixes. (18)

high-visibility vest. Special clothing worn on construction sites that allows the wearer to be seen more readily. (4)

hollow bricks. Bricks that are identical to facing bricks but have a larger core area. (7)

hollow masonry unit. A brick with less than 75% solid material in its net cross-sectional area in any plane. (7)

hollow masonry walls. Walls built with hollow masonry units. Solid units can be combined with hollow units laid in face shell mortar bedding. (13)

hydration. Chemical reaction between water and cement that sets and hardens the cement. (14)

hypotenuse. Longest side of a right triangle. (6)

I

ice lenses. The process of moisture freezing in the ground or soil to create pockets of ice. (15)

Indiana limestone. Well-known, even-textured building stone that is quarried in south-central Indiana. (9)

industrial floor brick. Brick manufactured to withstand extremely heavy traffic and commercial use that occurs on the surface of industrial floors. The bricks are made of clay, shale, or a combination of those mixtures. (7)

insulated concrete forms (ICF). Forms that are expanded to create insulated blocks that are stacked on top of each other to fabricate a wall structure. (16)
interest. Cost of borrowing money. (6)
International System of Units (SI). System of measurement that uses a base of 10 to form the various units of measure. (6)
intrados. Curve that bounds the lower edge of the arch. (16)
isolated footings. Footings that are not part of the foundation. They receive the loads of freestanding columns or piers. (15)

J

jack arch. A flat arch with zero or little rise. (16)
job hazard analysis (JHA). A technique in which site walk-throughs are conducted to identify, document, and discuss hazards found that currently exist or may appear on the job, then to determine procedures to stop them from reoccurring. (4)
joint raker. Tool used to remove a portion of the mortar from a joint just before the mortar hardens. (3)
jointer. Tools used to compact and finish the surface of mortar joints. (3)
journeyman. An experienced craftsperson who has successfully completed an apprenticeship in the trade. (1)

K

keystone. The masonry unit located at the crown of the arch. Also called *key*. (16)
keyway. A recess or groove placed in the middle of a footing prior to the concrete setting up in the footing form to aid in attaching a foundation wall to the footing. (15)
kiln. An oven or furnace that heats brick to a predetermined temperature. (7)
king closure. The brick remaining when one corner is cut off. (7)

L

landscaping plan. A plan that locates and identifies trees, plants, hardscape, and other elements included in landscaping. (5)
lead. Part of a wall that is built first at each corner to establish proper height and provide a place to attach a mason's line. (12)
leadership. The ability to guide and motivate others to complete tasks or achieve goals. (2)
level case. A device used to protect delicate levels when not in use or being transported from one jobsite to another. Level cases come in different styles and sizes; the most common are 2′ and 4′. Most level cases are made of a plastic composite. (3)
lightweight aggregate. Aggregate made up of granular, puffy materials such as cinders, pumice, or expanded shale. (8)
limestone. Sedimentary stone that consists mainly of calcite and is usually marine in origin. (9)
line. Geometric object that has only length and is the distance between two or more points. (6)
line block. A tool used in combination with mason's line and line pins to aid the mason in laying bricks or concrete blocks at the proper height. (3)
line pin. Devices made of hardened steel used to secure a mason's line from one outside masonry corner to the other corner of a building. (3)
lintel. Structural member placed over an opening in a wall used to support the loads above that opening. (16)
lockout. The process of locking a machine or locking the electrical power out so it will not operate. (4)
long break lines. Thin lines used to show that all of the part is not shown. (5)
low-lift grouting. The process of grouting in multiple lifts or segments as the wall is constructed. (10)
low-slump concrete. Stiff (dry) concrete mixes. (18)

M

magma. Molten mixture of minerals. (9)
manufactured stone. Simulated stone veneer made from lightweight concrete. (9)
marble. Recrystallized limestone that can be white, yellow, brown, green, or black. (9)
mason tender. A person at least 16 years of age or older who supports the mason or journeyman. A tender mixes mortar, keeps materials on hand, and maintains the job area to keep it clean and safe. (1)
mason's chisel. Tool designed to cut brick, block, and stone. (3)
mason's level. Level used to check a wall to ensure it is built vertical and level. (3)
mason's line. Strong nylon cord used to keep each course level and the wall true and out-of-wind. (3)
mason's rule. A standard tool used by masons for measurement. The front is in standard US customary measurements and the back is a scale for laying bricks and blocks. (3)
masonry saw. Device used to cut masonry units using a diamond, carbide, or other cutting type blades. (3)

matt foundation. A thickened slab that transmits building loads over the entire slab and soil area. (15)
maul. A special hammer used to strike chisels. (3)
mechanical plan. A plan that shows the plumbing, heating, and cooling systems. (5)
mesh wall tie (MWT). Metal fabric placed on masonry units to add extra bonding strength and to reinforce masonry joints. MWT is normally used where one block wall butts up against another wall. Also called *hardware cloth.* (11)
metal twig (trig). A thin piece of metal designed to hold a mason's line between fingers. (13)
mixed number. A whole number followed by a fraction. (6)
modular unit. Brick based on a measurement of 4″; however, the dimension is not exactly 4″ since allowance is made for the thickness of the mortar joints. (7)
mortar. Bonding agent that ties masonry units into a strong, well-knit, weathertight structure. (10)
mortar bag. A small cone-shaped bag used to dispense mortar in joints, reducing the chances of smearing mortar on surfaces. Also used to place mortar into joints where old mortar has been removed because it was damaged or aged by weather conditions. The bag is filled with mortar and then is squeezed to push mortar to the pointed front of the bag. The bag may be filled with grout to be used to fill hollow cells of masonry units during the masonry construction process. (3)
mortar board. The purpose of a mortar board is to hold mortar during the time a mason is laying bricks or concrete blocks. A mortar board can be made of plywood, plastic, metal, or a composite of materials. (3)
mortar bond. The adhesion of mortar to the masonry units or reinforcing. (7)
mortar box. Device used to hold mortar. Mortar can be mixed and stored in the box until ready to be used by the mason. (3)
mortar fin. Mortar that protrudes from the back side of the mortar joint on the inside a wall cavity after the masonry unit has been laid in the bed joint. (12)
mortar hoe. A tool used to mix different mortar components in a mortar box or wheelbarrow. The blade of a hoe may have holes in it to allow the user to pull and push the mortar to reduce time in mixing the different components. The blade may also be rounded to aid in mixing mortar in a wheelbarrow. (3)
mortar mixer. A machine used to mix dry materials and water to produce mortar or grout. (3)
mortar stand. A device used to support a mortar board and mortar placed on top of the board when laying bricks or concrete blocks. (3)
multicentered arch. Arch with a curve that consists of several arcs of circles that are normally tangent at their intersections. (16)
multiplication. Process of repeated addition using the same numbers. (6)

N

net wall area. The gross area minus all wall openings. (6)
networking. The exchange of information or services among individuals or groups. (1)
nonloadbearing walls. Walls that do not carry a structural load. (13)
nonstaining cement. Cement with a low-alkali content. (14)
normal weight aggregate. Class of dense aggregates that includes sand, gravel, and crushed limestone. (8)
numerator. Top number of a fraction. (6)

O

object lines. Heavy lines that show the outline of the visible features of an object. (5)
obtuse angle. Angle greater than 90°. (6)
Occupational Safety and Health Administration (OSHA). Part of the US Department of Labor responsible for developing and enforcing safety and health standards through providing education, training, outreach, assistance and enforcement. (4)
ornamental terra cotta. Terra cotta that is produced in sculptured patterns or freestanding sculpture. (7)

P

panel wall. Exterior nonloadbearing wall supported at each story. (16)
parging. The application of a coating of mortar. (13)
pattern bond. The pattern formed by the masonry units and the mortar joints on the face of a wall. (7)
paver bricks. Bricks that are designed especially for paving. They are 1 1/4″ × 3 5/8″ × 7 5/8″ in size and approximately 1″ thinner than regular bricks. (17)
percentage. A number used to denote parts of one hundred. (6)
perpendicular. Relationship between two lines that form a 90° angle, or right angle, with respect to each other. (6)

personal protective equipment (PPE). The protective clothing, footwear, eye protection, head protection, and other safety equipment that an employee or worker will need to wear during the time spent at the jobsite or shop to protect against possible injury. (4)

perspective drawing. A drawing in which parallel lines move closer together as they move away from the foreground. (5)

pier. An isolated column of masonry or a bearing wall not bonded at the sides to associated masonry. (15)

pigment. Coloring agent that can be added to mortar or concrete to change the color or to tint the mortar color to match an existing material. (10)

pilaster. A masonry column bonded to a wall. It has a uniform cross-section and serves as a column or vertical beam. (15)

placer tool. A tool used to position concrete either by pulling or pushing the mix once it has been placed in the form. Also called *concrete placement tools.* (3)

plain footing. Footing that carries light loads and is not reinforced with steel. (15)

plaster back. The process of flattening mortar fins or excess mortar against the back sides of the masonry units rather than cutting it off, preventing mortar from falling into the cavity. (12)

plaster cove. Plaster placed on top of a footing next to a foundation wall from the outside edge of the footing up the wall about 4″. The plaster is then rounded to create a coved shape. (15)

plasticity. The ease of working with and applying mortar to units. (10)

plumb bob. A tool used in conjunction with mason's line to transfer building lines from batter board lines to footing and foundations. (3)

plumb rule. Level used to check a wall to ensure it is built vertical and level. (3)

Plyform. An exterior-type plywood made from special wood species and veneer grades to ensure high performance in concrete forming applications. (19)

pointed up. A wood or metal striking tool is used to work the joint area between the stones. (9)

pointing. Placing plastic mortar into a joint. (14)

points. Basic elements of lines that have no width, length, or height, but merely indicate a location. (6)

polymeric sand. A mixture of fine sand and water activated polymers used to fill, bind, and seal joints. (17)

ponding. Concrete curing method that involves keeping approximately an inch of water on the concrete surface, usually by earth dikes around the edges of the slab. (20)

power concrete joint saw. A power saw used to cut joints in concrete slabs. It can be either gasoline or electric. (3)

power screed. A tool used to drive surface rock into the concrete mixture and brings a layer of liquid concrete to the surface. (3)

power trowel. A trowel machine used to give a concrete slab a smooth, dense finish. It can be either gasoline or electric. (3)

pozzolan. Siliceous or siliceous and aluminous material that, in finely divided form and in the presence of moisture, chemically reacts with calcium hydroxide at ordinary temperatures to form compounds having cementitious properties. (18)

prefaced concrete masonry units. Concrete masonry units that have a specially prepared surface material applied to one or more surfaces to provide color, pattern, or texture. (8)

preliminary drawings. Drawings on which corrections and revisions are recorded during the promotional stage of a building's design. (5)

presentation drawing. A drawing, commonly a two-point perspective, that shows how the finished structure will appear. (5)

principal. Amount of money borrowed. (6)

product. Answer to a multiplication problem. (6)

project manager. A professional who is responsible for planning and executing the construction project in the field and oversees the work being performed on the jobsite. (1)

pull crete tool. A tool used to keep the concrete mix together when moving it in forms. Also called *placer tool.* (3)

Pythagorean Theorem. A mathematical formula ($a^2 + b^2 = c^2$) that aids masons in laying out a building. The square of the hypotenuse of a right triangle is equal to the sum of the squares of the other two sides. (6)

Q

quarter closure. One-fourth of a brick. (7)

quartzite. Sandstone that has been recrystallized. (9)

queen closure. A brick cut lengthwise across the end. (7)

Quick Headers. A reuseable metal form that temporarily supports masonry over windows, doors, and other openings. Also called *removeable masonry forms.* (16)

quirk joint. Stones or masonry panels that have been beveled or cut at the same angle to create a uniform corner. In most applications, the stones are cut at a 45° angle to create a 90° finished outside corner. (14)

quoins. Large squared stones set at the corners of buildings. (9)
quotient. Answer of a division problem. (6)

R

racking. 1. Masonry in which successive courses are stepped back from the face of the wall. 2. The process of stepping back brick at the end of the brick lead. This creates a stepping effect at the tail end of each of the corner leads to aid in laying brick between the two corners. Also called *racking the lead*. (12, 16)
radius. Distance from the center to any point on the circumference or one-half the diameter. (6)
raft foundation. A foundation made with reinforced walls and floor cast into a single unit. (15)
raked joint. Joint made by removing some of the mortar while it is still soft. (7)
range. Squared stones laid in horizontal courses of even height. (9)
rate of interest. Percentage charged per year for borrowing money. (6)
rattle stick. A device attached between a bull float and the handle, used to vibrate the bull float to aid in creating a smooth surface on fresh concrete during the floating process. (3)
rebar bender. A tool used to bend reinforcement bars to fit concrete forms, flatwork, and vertical walls. (3)
rebar cutter. A tool used to cut reinforcement to the correct lengths when forming concrete work. (3)
rebar positioners. Devices used to hold reinforcement in a specific location in the core or cell of the block. (11)
reinforced concrete. Composite material that uses the concrete in resisting compression forces, and some other material, usually steel bars or wires, to resist the tension forces. (18)
reinforced concrete masonry walls. Walls in which steel is embedded both horizontally and vertically to increase strength, designed for applications that experience high stress, high wind loads, or severe earthquakes. (13)
reinforced footing. Footing embedded with steel to increase its strength. (15)
reinforced masonry walls. Walls built with steel reinforcement embedded in the masonry units. (16)
reinforcing bars (rebars). Reinforcement used to increase the tensile strength of concrete. (11)
reinforcing chairs or stools. Reinforcement used to ensure rebars do not sink below the recommended or required height above the base material. (11)
reinforcing rods. Reinforcing steel used to increase the tensile strength of concrete. Also called *re-rods*, *bars*, or *rebars*. (11)
remainder. Amount left over if a number is not contained in another number an equal number of times. (6)
removeable masonry form. A reuseable metal form that temporarily supports masonry over windows, doors, and other openings. Also called *Quick Headers*. (16)
repointing. The complete process of removing or cutting out damaged mortar joints between masonry units and replacing them with new mortar. (10)
retaining wall block. Blocks that allow changes to be made in grade elevation when the terrain has a large slope or hill. They can hold back earth, gravel, and soil. (8)
retarding admixture. Material used to slow down the setting time of concrete. (18)
retempering. Adding water to mortar and mixing briefly to replace water lost by evaporation. (10)
retractable steel tape. A measurement tool used to quickly measure long distances. (3)
rigger. A worker who has been trained as a competent person capable of safely attaching rigging slings or lifting equipment to lift stone or other materials. (14)
rigging. Attaching slings, ropes, or other devices to lift stone or other material into the air to unload, load, or position the materials. (14)
right angle. An angle with exactly 90°. (6)
right triangle. Triangle that has one 90° angle. (6)
rise. Maximum height of the arch soffit above the level of its spring line. (16)
riser. A vertical member between two stair treads. (17)
roller tamper. A tool used in large surface areas to compact concrete into a dense mass. (3)
roof plan. A plan showing ridges, valleys, skylights, and other features of the roof. The plan also displays the sheathing and roofing material. (5)
rotary laser level. Devices used in site preparation for grading, construction of batter board and stake layout, and concrete leveling. (3)
rough cut joint. Joint that is made by holding the trowel flat against the brick and cutting in any direction. The joint is not compacted. (7)
rubble. Stone that consists of broken fragments of irregular shapes and textures. (9, 14)
running bond. Simplest basic pattern bond that consists of all stretchers and is used largely in cavity wall and veneer wall construction. (7)
R-value. The thermal resistance of heat loss through a building product and the reciprocal of the U-value. (16)

S

safety data sheet (SDS). A written document that outlines product information, inherited hazards, and safety procedures to be used when working with or handling a product or chemical. (4)

safety hard hat. A type of PPE worn to protect a worker's head from a direct impact from above. (4)

safety helmet. Head safety protection created to protect the worker's head from above impacts and side impacts. (4)

safety glasses. Eye protection that can fit over prescription eyewear. (4)

safety goggles. A type of eye protection that protects from dust and particles. Some goggles are made to fit over prescription eyewear and can be clear or tinted. (4)

safety meetings. Scheduled meetings, usually attended by superintendents, subcontractors, foremen, supervisors, to schedule activities and address safety concerns, previous injuries, and hazard identification, and corrections. (4)

safety programs. Ongoing efforts between a contractor, crew leaders, and workers on company sites. Emergency procedures are discussed and practiced in the event of an accident or emergency on the jobsite. (4)

sandstone. Sedimentary stone composed mainly of grains of quartz cemented by silica, lime, or iron oxide. (9)

scaffold. A temporary structure that holds workers and materials while they are erecting, repairing, or doing other work on a building. (4)

schist. Coarse-grained stone that contains large amounts of mica. (9)

scoria. Scrap left after the melting of metals. (8)

SCR acoustile. A type of facing tile that is designed to absorb sound. (7)

screed. Straightedge or strike-off rod made of any straight piece of wood or metal that has sufficient rigidity (stiffness). It is the first finishing tool used after the concrete is placed. (3)

screeding. The striking off of excess concrete to bring the top surface to the proper grade or elevation. (20)

section lines. Very thin lines used to show that a feature has been sectioned. These lines may represent a specific material or may be a general symbol. Also called *crosshatch lines.* (5)

sector. Portion of a circle between two radii (more than one radius). (6)

segment. Portion of a circle contained by a straight line and the circumference that it cuts off. (6)

segmental arch. Arch with an intrados that is circular but less than a semicircle. (16)

segmental retaining walls (SRW). High-strength concrete blocks or units made specifically for dry stacking without the use of mortar. (16)

semicircle. Half the circumference of a circle. (6)

semicircular arch. Arch with an intrados that is a semicircle. (16)

sheathing. Materials, such as plywood or metal, used in concrete forms to keep concrete in place until it hardens. (19)

shop drawings. Drawings that provide information about such things as reinforcing steel, complex cabinetwork, and electronic systems. (5)

short break lines. Heavy lines used when part of the object is shown broken away to reveal a hidden feature. (5)

shotcrete. A mixture of Portland cement, aggregate, and water shot into place by means of compressed air. (20)

side construction. Structural clay tile that is laid with the cells in a horizontal plane. (7)

sill. A wall form member that is temporarily attached to the footing and attached to the studs until the concrete hardens. (19)

site plan. A plan that shows the location of the building on the site. It may also show utilities, topographical features, site dimensions, other buildings on the property, landscaping, streets, walks, drives, retaining walls, and property lines. (5)

skewback. Surface on which the arch joins the supporting abutment. (16)

slaked. Formed into putty through combination with water. (10)

slate. Metamorphic stone that is frequently a blue-gray color, but may be green, red, or brown. (9)

sledgehammer. A large hammer used for striking wooden or steel stakes, chisels, stones, and concrete. The weight can vary from 6 lb to 20 lb. (3)

slump. The relative consistency or stiffness of plastic concrete. (18)

Smart Level. A type of altimeter level used for 3D terrain mapping and image mapping. (3)

smoke chamber. An area located directly above the firebox that compresses gases created during the combustion process of burning wood or other flammable materials. (17)

smoke shelf. A ledge in the smoke chamber that aids in eliminating downdrafts that can occur in the flue and fireplace. Downdrafts can cause smoke to enter the room where the fireplace is located. (17)

snap headers. Half bricks used when headers are not needed for structural bonding. (7)
soffit. Lower surface of an arch or vault. (16)
soft conversion. Metric conversion in which inch-pound dimensions are stated in metric units, but the material sizes currently produced are not required to change. (8)
soft metric conversion. The process of relabeling products from the US customary to the metric system without physically changing the product dimension. (6)
solid masonry walls. Walls built with solid masonry units. (13)
spalling. The deterioration or flaking off of the brick face. This is caused by excessive moisture in the brick during freezing and thawing cycles. Over time, this can cause the brick to fall apart. (12)
span. Horizontal clear dimension between abutments. (16)
spandrel. The space between the top of a window of one story to the sill of the window on the second story. (16)
specifications. Written descriptions of the construction project that are prepared by architects and engineers. (5)
split. A brick cut lengthwise parallel to the bed. (7)
split face masonry units. Concrete units that look like rough quarried stone. The units are produced by splitting solid concrete blocks lengthwise. (8)
spotting the brick. A technique where an intermediate lead or single brick is positioned or laid with the appropriate masonry joint between the corner leads to reduce line sagging. (12)
spreaders. Blocks, usually made of wood, that separate the inside walls of forms until concrete is placed in the form. (19)
spring line. Horizontal line that intersects the springing. (16)
springing. The point where the skewback intersects the intrados. (16)
stack bond. Pattern bond in which all vertical joints are aligned and there is no overlapping of units. Also called *block bond.* (7)
steel square. A device used to aid in the layout process of building or other structures. The two arms (the body and the tongue) form a 90° angle. The body is 2″ wide and 24″ long. The tongue is 1 1/2″ wide and 16″ long. (3)
stepped footing. A footing that changes in elevation to correspond to the change in grade elevation or a footing that is increased at the base for greater support and then is stepped back in increments to reduce the overall width at the top of the footing. (15)
stiff back. A vertical support base that can be attached to an ICF wall structure. Also called *strongbacks.* (19)
straight angle. An angle of 180°. (6)
straight claw hammer. A type of framing hammer that can be used as a pry bar. (3)
stretchers. Blocks with ears on both ends. (8)
strong back. A vertical support base that can be attached to an ICF wall structure. Also called *stiff backs.* (19)
struck joint. Joint that is struck with the trowel, resulting in some compacting. (7)
structural bond. The method by which individual units are interlocked or tied together to form a wall. (7)
structural clay products. Products made from clay or shale, such as brick, tile, and terra cotta. (7)
studs. Devices, usually made of treated 2 × 4 lumber, that are placed vertically in forms that support the sheathing to prevent sheathing from bowing. (19)
subtraction. Taking one number away from another number. (6)
sum. The total in addition. (6)
superintendent. The person who is in charge of all the activities and work being performed in the field for the construction company or corporation. (1)

T

tagline. Rope or cable attached to a large stone or other item being lifted to guide the item being unloaded or positioned. The line allows the rigger to be a clear distance away in case the stone slips from the rigging. (14)
tagout. The process of placing a tag within a tagout lock to identify the reason why a machine is out of service. (4)
tamping rammer compactor. A machine used to compact aggregate or other materials. (3)
tangent line. Line that touches the circle, but does not cut it and is at right angles to a straight line from the center. (6)
technical sections. Specifications that deal with the actual building of the structure and the tradework involved in the building. (5)
tensile strength. A material's ability to resist stresses or forces that tend to bend or pull it apart. (18)
terra cotta. A product made of fired clay, usually brownish-red in color. Terra cotta is used for architectural and structural units. The products or units can be made by machine, by hand, or

by a combination of the two depending on the design. Terra cotta can be unglazed or glazed, depending on the application and desired finished appearance. (7)

thermal mass. The characteristic of heat capacity and surface area capable of affecting building thermal loads by storing heat and releasing it at a later time. (16)

thin brick veneer. Fired clay units with normal face dimensions, but a reduced thickness. (7)

three-quarter closure. Three-fourths of a brick. (7)

throat. The narrow passage that connects the firebox with the smoke chamber. (17)

tie wire twister. A tool used to twist or tie form wire around reinforcement. (3)

ties. Metal or fiberglass devices that connect the two exterior wall forms together to hold them at the appropriate width when placing concrete and remain in the concrete to become a permanent part of the structure. (19)

tilt-up construction. The process of casting concrete walls in a horizontal position on-site or in another location and lifting them from the casting position or truck to their final location in the building. (20)

tool bag. A bag used to hold hand tools and other items needed on the job. They can be made from canvas, leather, or other synthetic material. (3)

toolbox talks. Short safety talks usually conducted by a foreman or supervisor and held at the jobsite or work area at the beginning of the day. (4)

tooled joint. Joint made with a special tool to compress and shape the mortar in the joint. (7)

toothing. The creation of temporary wall terminations of masonry units where alternating course joints are cut out before setting and then the specific unit is removed. (15)

track-driven power buggy. Device used to carry materials around a construction site. The machine is powered by a gasoline engine that can carry two to three times the amount of a wheelbarrow. (3)

traprock. Hard and durable igneous stone. A term used for diabase, basalt, or gabbro. (9)

tread. Horizontal portion of a step. (17)

triangle. Geometric shape with three angles that total 180°. (6)

trimmings. Stones that are cut on all sides to specific dimensions. (14)

trowel. The most commonly used tool by a mason. They can be used for a variety of purposes such as applying mortar to block and laying brick or block. (3)

troweled joint. Joint in which the excess mortar is struck (cut off) with a trowel and finished with the same tool. (7)

tuckpointing. Filling in cut out or defective mortar joints in masonry with fresh mortar. (10)

Type K mortar. Very low-strength mortar that is best used in interior nonloadbearing partition walls. (10)

Type M mortar. Mortar with high compressive strength and somewhat greater durability than other mortar types. (10)

Type N mortar. Medium-strength mortar suitable for general use in exposed masonry above grade. (10)

Type O mortar. Low-strength mortar suitable for general interior use in nonloadbearing masonry. (10)

Type S mortar. Medium-high-strength mortar intended for use where Type M is recommended but where bond and lateral strength are more important than high compressive strength. (10)

twig. A tool used to hold the mason's line at the appropriate location and height when laying masonry units between the corners. (3)

U

uncoursed stone. Stone laid in an irregular pattern where stones are placed randomly in the wall. (9)

unglazed facing tile. Facing tile made from either light-burning fire clay or shale and other darker-burning clays. These tiles are not glazed and have either a smooth or a rough-textured finish. (7)

utility knife. A type of knife with a blade that can be opened to reveal the blade and closed to conceal it into the handle. (3)

U-value. The heat flow through a building product based on the difference between external and internal temperatures. (16)

V

vibratory plate compactor. A gas-powered machine that uses a vibrating base plate to compact aggregate, soil, or other materials. The vibration created and weight of the machine aids in the compaction process. The primary use of this compactor is to prep subbases for footings, driveways, sidewalks, and basement floors. (3)

voids. Openings in concrete blocks. Also called *cells* and *cores*. (8)

voussoir. One masonry unit of an arch. (16)

V-shaped joint. Mortar joint formed by striking the joint to remove excess mortar, then running a "V" shaped steel joint tool to compact the mortar and form a recessed joint. (7)

W

wales. Treated 2 × 4s that run perpendicular to the studs to align the forms, secure the ties, and support the studs. (19)

wall ties. Strips of metal or metal wire used to tie masonry wythes together or to tie masonry veneer to concrete frames, wood frame walls, or structures. (11)

water-cement ratio. The proportion of water to cement. (18)

waterproofing. The installation of drain tile to remove accumulated water from around a foundation and under a concrete floor. (15)

water-reducing admixtures. Materials that reduce the quantity of mixing water required to produce concrete of a given consistency. (18)

weathered joint. A compacted joint that readily sheds water. (7)

weep hole. A small opening in a masonry wall that allows moisture to escape from behind the surface of the masonry to the exterior of the structure. The opening is usually formed by leaving out a head joint. Newer construction will install a weep hole device in the head joint, usually made of plastic, rope, or similar materials. (12)

wheelbarrow. Device used to move or carry mortar, sand, gravel, soil, concrete, or a number of other products around the jobsite. (3)

whole number. Any one of the natural numbers, such as 1, 2, and 5. (6)

work ethic. A belief in working diligently. (2)

working drawings. A group of drawings used to construct a building. (5)

wythe. Each continuous vertical section of masonry one unit in thickness. (6)

Z

ZipLevel. An altimeter level that measures change in elevation between the base unit and readout unit. (3)

INDEX

C

D

E

F

M

S

T

U

V

W

Z